Karl Heinz Höhne Ron Kikinis (Eds.)

Visualization in Biomedical Computing

4th International Conference, VBC '96
Hamburg, Germany, September 22-25, 1996
Proceedings

With 18 Color Pages
and Numerous Halftones

Springer

Series Editors

Gerhard Goos, Karlsruhe University, Germany

Juris Hartmanis, Cornell University, NY, USA

Jan van Leeuwen, Utrecht University, The Netherlands

Volume Editors

Karl Heinz Höhne
University of Hamburg, University Hospital Eppendorf
Martinistraße 52, D-20246 Hamburg, Germany
E-mail:hoehne@uke.uni-hamburg.de

Ron Kikinis
Harvard Medical School and Brigham and Women's Hospital
Department of Radiology
75 Francis Street, Boston, MA 02115, USA
E-mail: kikinis@bwh.harvard.edu

Cataloging-in-Publication data applied for

Die Deutsche Bibliothek - CIP-Einheitsaufnahme

Visualization in biomedical computing : 4th international conference ; proceedings / VBC '96, Hamburg, Germany, September 22 - 25, 1996. Karl H. Höhne ; Ron Kikinis (ed.). - Berlin ; Heidelberg ; New York ; Barcelona ; Budapest ; Hong Kong ; London ; Milan ; Paris ; Santa Clara ; Singapore ; Tokyo : Springer, 1996
(Lecture notes in computer science ; Vol. 1131)
ISBN 3-540-61649-7
NE: Höhne, Karl H. [Hrsg.]; VBC <4, 1996, Hamburg>; GT

CR Subject Classification (1991): J.3, I.4, I.5, I.2.10

ISSN 0302-9743
ISBN 3-540-61649-7 Springer-Verlag Berlin Heidelberg New York

Printed in Germany

Typesetting: Camera-ready by author
SPIN 10513550 06/3142 – 5 4 3 2 1 0 Printed on acid-free paper

Preface

This book represents the proceedings of the Fourth International Conference on *Visualization in Biomedical Computing*. Following the Georgia Institute of Technology, the University of North Carolina and the Mayo Institutions, it took place at the University of Hamburg, Germany. As in previous meetings its goal was to promote the science of computer based visualization of biomedical information, especially of the human body. An interdisciplinary international group of scientists, engineers, and clinicians gathered in order to present and discuss the state of the art concerning algorithms, the rapidly increasing number of applications, and practical problems of system design.

The increasing interest in the field of biomedical visualization was manifested by 232 paper submissions from which only 73 could be selected in a peer review process as oral or poster presentations. While this fact was disappointing for many applicants with good papers, the high degree of selection certainly led to a high quality book. It reflects the recent achievements and also the current problems in the field of computer based visualization in medicine and biology. Substantial improvements in rendering techniques and multi-modality imaging have occurred. The critical field of automated image segmentation has seen significant progress as well. Concerning applications we observe a shift from pure visualization to therapeutic interventions using visualization tools. Not surprisingly, the brain is the most popular organ for development of tools and applications, since it is certainly the most complex organ and relatively easy to image.

A book like this one is, of course, not possible without the help of many people. First of all we have to thank the members of the program committee for their hard work. With three reviewers per paper most of them had to review more than 25 papers! We are especially grateful to Professor Siegfried Stiehl who helped us in solving all kinds of unexpected problems. The financial risk of the meeting was minimized by generous support of the German Reseach Council (DFG). The Institute of Electrical and Electronics Engineers (IEEE), the German Society of Computer Science (GI), and the German Society of Medical Informatics, Biometrics, and Epidemiology (GMDS) helped us in advertising the conference.

Last but not least we want to thank the conference coordinators Bernhard Pflesser, Thomas Schiemann, and Rainer Schubert, and the staff of the Department of Computer Science in Medicine for their excellent work in organizing the entire conference.

Hamburg and Boston, July 1996

Karl Heinz Höhne Ron Kikinis

Conference Committee

Chairman
Karl Heinz Höhne, University of Hamburg

Program Committee
Ron Kikinis (Chair), Harvard Medical School

Nicholas Ayache, INRIA, Sophia-Antipolis
Jean Louis Coatrieux, University of Rennes
Takeyoshi Dohi, University of Tokyo
James Duncan, Yale University
Alan C. Evans, Montreal Neurological Institute
Norberto Ezquerra, Georgia Tech, Atlanta
Henry Fuchs, UNC, Chapel Hill
Guido Gerig, ETH Zürich
Eric Grimson, MIT, Cambridge
David Hawkes, Guy's Hospital London
Arie Kaufman, State University of New York
Andres Kriete, University of Giessen
Hal Kundel, University of Pennsylvania
Heinz Lemke, Technical University of Berlin
David Levin, University of Chicago
Heinrich Müller, University of Dortmund
Chuck Pelizzari, University of Chicago
Stephen Pizer, UNC, Chapel Hill
Richard Robb, Mayo Foundation / Clinic
H. Siegfried Stiehl, University of Hamburg
Paul Suetens, University of Leuven
Andrew Todd-Pokropek, University College London
Jun-Ichiro Toriwaki, Nagoya University
Michael Vannier, Washington University, St. Louis
Max Viergever, Utrecht University

Coordinators
Bernhard Pflesser, Thomas Schiemann, Rainer Schubert, IMDM

Invited Lecturers

Peter E. Peters, Westfälische Wilhelms-Universität, Münster
Alan C. Evans, McGill University, Montreal
Russell H. Taylor, Johns Hopkins University, Baltimore

Table of Contents

Visualization

GI Tract Unraveling in Volumetric CT

Ge Wang, Michael W. Vannier
Elizabeth G. McFarland, Jay P. Heiken and James A. Brink

Mallinckrodt Institute of Radiology
Washington University School of Medicine
St. Louis, Missouri 63110, USA

Abstract. Gastrointestinal (GI) tract examination with computed tomography (CT) is currently performed by slice-based visual inspection despite the volumetric nature of the problem. The entire abdomen may now be continuously scanned within a single breath-hold with spiral CT. The GI tract can be computationally extracted from spiral CT images, the lumen explicitly unraveled onto a plane, and colonoscopy virtually simulated via fly-through display. A semi-automatic approach was developed for fly-through and unraveling *in vivo*. Automation of this approach was investigated in numerical simulation. The techniques are promising for colon cancer screening.

1 Introduction

Colon carcinoma ranks second as a cause of cancer death in the United States. Approximately, 60,000 deaths and 150,000 new cases occur annually in this country. The total cost of colorectal cancer is about $10 billion per year. This disease usually involves a progressive change of normal mucosa into adenomatous polyps and then cancer. Currently, screening for colon cancer is performed with fecal occult blood test and sigmoidoscopy. The whole colon can be examined by colonoscopy and barium enema. Colonoscopy is more sensitive than barium enema for detecting polyps less than 1 cm in diameter and offers therapeutic options. However, invasive and expensive colonoscopy often gives negative results, sometimes fails to complete examination, and occasionally results in perforation. Therefore, it is desirable to have a noninvasive, inexpensive, reliable and accurate imaging method for selecting only those patients who do need colonoscopy.

Spiral computed tomography (CT) is a major advance [1, 2], and has become the standard in diagnostic X-ray CT. Recently, it was established that with overlapping reconstruction spiral CT allows substantially better longitudinal resolution than conventional CT [3–5]. Both improved longitudinal resolution due to overlapping reconstruction and improved temporal resolution due to fast scanning make spiral CT the method of choice for volumetric imaging.

Tomographic examination of the gastrointestinal (GI) tract is currently performed by slice-based visual inspection despite the volumetric nature of the anatomical components, tumors/lesions, and imaging modalities. The entire abdomen may now be continuously scanned within a single breath-hold with spiral

CT. Image processing algorithms can be developed to unravel the highly convoluted tubular structure into a planar volume or "fly-through" the colon in a colonoscopic fashion. Both techniques have potential for detection and evaluation of colonic polyps. Recently, screening colorectal carcinoma using spiral CT attracts increasing attention [6–10]. We are developing a spiral CT image unraveling system for global comprehension and quantitative analysis, as illustrated in Figure 1.

In this paper, we report our results on fly-through and unraveling of the GI tract. In the second section, a semi-automatic approach is demonstrated to determine the central axis of the GI tract *in vivo*, and then perform fly-through and unraveling. In the third section, key algorithms needed in semi-automatic unraveling are described. In the fourth section, fully automatic tracking and unraveling of the GI tract is investigated in simulation with a mathematical GI tract phantom. The last section concludes the paper.

2 Semi-Automatic Unraveling *In Vivo*

It is well known that segmentation of the GI tract can be difficult. In practice, complete segmentation may even be impossible. Therefore, a divide-and-conquer strategy was adopted for GI tract unraveling. This task is decomposed into *central axis determination* and *orthogonal section formation*. We developed a semi-automatic approach to determine the central axis of the colon in spiral CT images. Our method fits a GI tract central axis through a sequence of user selected intraluminal points using cubic spline interpolation. Colon tracking follows the fitted axis. Cross-sections perpendicular to this axis are produced and stacked to straighten the colon computationally. The colon is then mapped into an unraveled image volume for further analyses.

In vivo fly-through and unraveling of four patients with suspected colonic polyps by barium enema were performed. A typical study on a patient is presented in Figure 2. An abdominal spiral CT scan was acquired after insufflation of air via a rectal catheter. The imaging parameters were 5 mm collimation, 5 mm table feed and 1 mm reconstruction interval. The image was compressed into a 256 by 256 by 297 volume of 1.4 mm isotropic voxels and 256 gray-levels. Figure 2(a) is a surface rendered view of the anterior abdominal wall and three orthogonal sections. These orthogonal sections can be interactively changed with the screen cursor in the Analyze^{TM} system [11]. Due to presence of the air contrast, the GI tract lumen was obvious. Using a point-and-click interface, representative points on an intraluminal GI tract central axis can be easily and quickly selected with a mouse. These coordinates were recorded in order as listed in Figure 2(b). Since these selected central axis points were relatively sparse, cubic spline interpolation was performed to generate a smooth central axis through the points. Figures 2(c) and 2(d) show a surface rendering of a synthetic hollow rope along the fitted central axis and three coronal views, respectively. This rope can be turned on and off at users' choice. Perpendicular to the fitted axis, cross-sections of the GI tract were generated incrementally along the axis at a

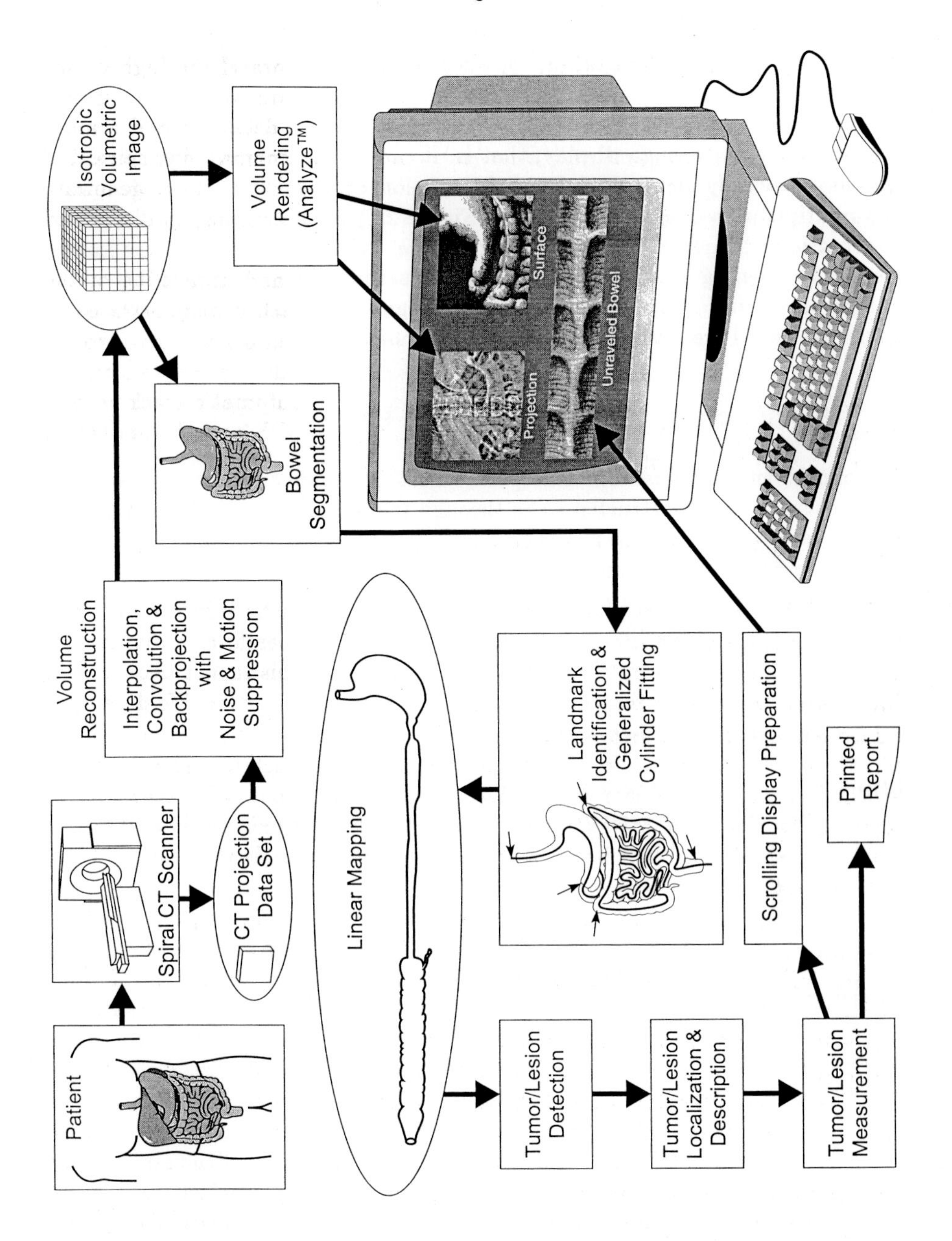

Fig. 1. Abdominal spiral CT image unraveling system. Visualization and analyses of GI tract components and tumors/lesions are automatically performed in 3D via a user-friendly interface. The GI tract is transformed onto a strip volume and surface/volume-rendered in a rectangular rolling window with its conventional views. Features of normal and abnormal bowel (especially focal mucosal lesions such as adenocarcinoma of the colon) are on-line quantified.

step equal to the voxel length, and stacked to straighten the tubular structure. We found that orthogonal sections provided better visualization for polyp examination, as shown in Figure 2(e). This straightened GI tract image volume is suitable for fly-through display; that is, it can be inspected in a controllable cine loop (*directly* viewing sections of the colon orthogonal to its central axis in a multiplanar sequence) as shown in Figure 2(f), thus avoiding limitations of volumetric and surface rendering. Furthermore, "virtual pathology" was implemented by opening up the straight tubular structure and extending it into a flat image volume, as illustrated in Figure 2(g). Figure 2(h) includes two sections of an unraveled GI tract segment with a polyp indicated by arrows.

3 Algorithms in Semi-Automatic Unraveling

To connect smoothly central axis points that are manually selected, a cubic spline curve can be constructed to go through the points with continuous second derivative everywhere. Given function and argument values $f[i]$ and $p[i]$, $i = 1, \cdots, n$, programs *spline* and *splint* in C can be utilized to determine a cubic spline interpolation formula [12]. The *spline* takes $f[i]$ and $p[i]$, $i = 1, \cdots, n$, $f'[1]$ and $f'[n]$ at $p[1]$ and $p[n]$ respectively, returns $f''[i]$, $i = 1, \cdots, n$. The *splint* computes $f(p)$, given $f[i]$, $p[i]$ and $f''[i]$, $i = 1, \cdots, n$.

3.1 Central Axis Generation

Input: Selected central axis points $\mathbf{P}[i] = (x[i], y[i], z[i])^t$, $i = 1, \cdots, n$, and a pre-specified arc length increment Δs; **Output**: equi-spatially interpolated axis points $\tilde{\mathbf{P}}[\mathbf{k}] = (\tilde{x}[k], \tilde{y}[k], \tilde{z}[k])^t$, $k = 1, \cdots, N$. The algorithm is as follows:

1. Using the *spline* with $f'[1]$ and $f'[n]$ approximated by differences, obtain three cubic spline interpolation formulas $f_x(p)$, $f_y(p)$ and $f_z(p)$, corresponding to $x[i]$, $y[i]$ and $z[i]$, $i = 1, \cdots, n$;
2. Set the starting point of the interpolated central axis:

$$\begin{cases} \tilde{x}[1] = x[1], \\ \tilde{y}[1] = y[1], \\ \tilde{z}[1] = z[1]; \end{cases}$$

3. Set k, p, $\tilde{p}[1]$ and δp

$$\begin{aligned} k &= 1, \\ p &= 1, \\ \tilde{p}[1] &= 1, \\ \delta p &= \text{a sufficiently small constant (for example, 0.01);} \end{aligned}$$

4. For $p \leq n$,

$$\begin{cases} \Delta x = f_x(p + \delta p) - f_x(p), \\ \Delta y = f_y(p + \delta p) - f_y(p), \\ \Delta z = f_z(p + \delta p) - f_z(p), \end{cases}$$

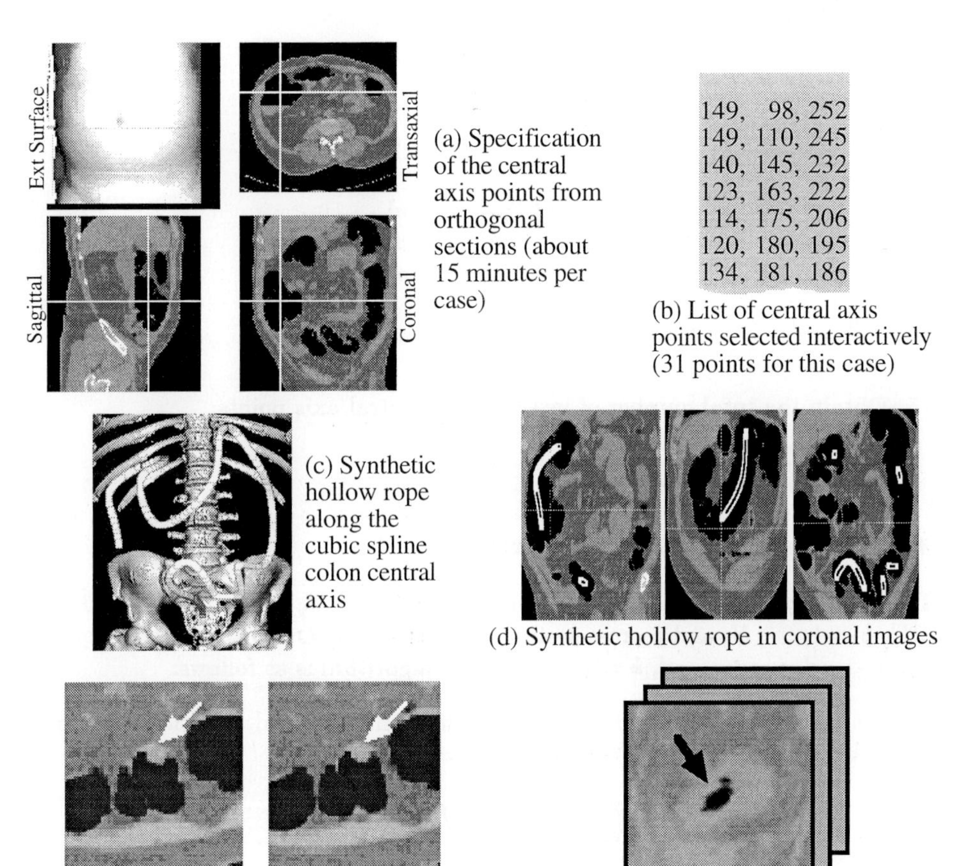

(a) Specification of the central axis points from orthogonal sections (about 15 minutes per case)

(b) List of central axis points selected interactively (31 points for this case)

(c) Synthetic hollow rope along the cubic spline colon central axis

(d) Synthetic hollow rope in coronal images

(e) Cross-sections perpendicular to the colon central axis (arrows show polypoid sigmoid mass. The synthetic rope was turned off)

(f) "Fly-through" in a cine loop (arrow shows focal rectal wall thickening corresponding to the cancer). Colon sections orthogonal to the central axis are displayed sequentially.

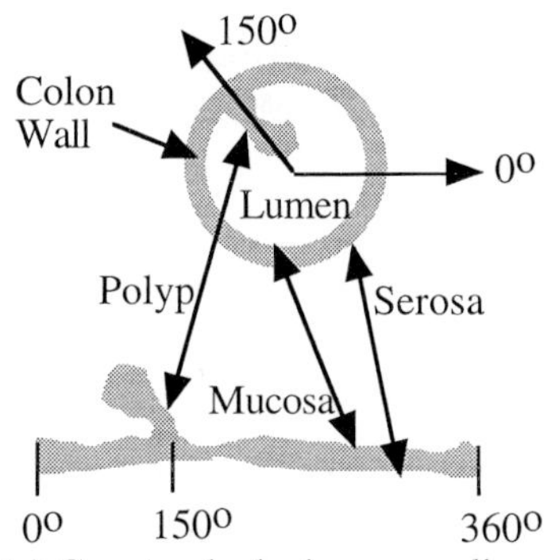

(g) Counterclockwise unraveling of orthogonal colon sections into a flat image volume

(h) Unraveled colon segment slices (arrows denote a polyp)

Fig. 2. Semi-automatic fly-through and unraveling of a GI tract *in vivo*.

$$\begin{aligned} \Delta p &= \frac{\delta p}{\sqrt{\Delta x^2+\Delta y^2+\Delta z^2}}\Delta s, \\ p &= p + \Delta p, \\ k &= k+1, \end{aligned}$$

then

$$\begin{cases} \tilde{x}[k] = f_x(p), \\ \tilde{y}[k] = f_y(p), \\ \tilde{z}[k] = f_z(p), \end{cases}$$

and

$$\tilde{p}[k] = p;$$

5. Obtain the total number of interpolated central axis points:

$$N = k.$$

3.2 Orthogonal Section Formation

Input: $image[i_x, i_y, i_z]$, $i_x = 1, \cdots, M_x$, $i_y = 1, \cdots, M_y$, $i_z = 1, \cdots, M_z$, $\tilde{\mathbf{P}}[k]$, $\tilde{p}[k]$, $k = 1, \cdots, N$, and a cross-sectional dimension d; **Output:** $sections[i, j, k]$, $i = 1, \cdots, d$, $j = 1, \cdots, d$, $k = 1, \cdots, N$. The algorithm is as follows:

1. Set a reference vector $\mathbf{r} = (1, 0, 0)^t$, and compute local directional vectors $\mathbf{L}[k] = (L_x[k], L_y[k], L_z[k])^t$ at $\tilde{\mathbf{P}}[k]$, $k = 1, \cdots, N$:

$$\begin{cases} L_x[k] = f'_x(\tilde{p}[k]), \\ L_y[k] = f'_y(\tilde{p}[k]), \\ L_z[k] = f'_z(\tilde{p}[k]); \end{cases}$$

2. Check the angle between vectors $\mathbf{r}$ and $\mathbf{L}[1]$. If $\alpha <$ a sufficiently small angle (for example, 1°), then set $\mathbf{r} = (0, 1, 0)^t$;
3. Compute horizontal and longitudinal vectors $\mathbf{U}[k] = (U_x[k], U_y[k], U_z[k])^t$ and $\mathbf{V}[k] = (V_x[k], V_y[k], V_z[k])^t$ to define cross-sections orthogonal to the central axis and through $\tilde{\mathbf{P}}[k]$, $k = 1, \cdots, N$:

$$\begin{cases} \mathbf{U}[k] = \mathbf{L}[k] \times \mathbf{r}, \\ \mathbf{V}[k] = \mathbf{L}[k] \times \mathbf{U}[k]; \end{cases}$$

4. Construct $sections[i, j, k]$, $i = 1, \cdots, d$, $j = 1, \cdots, d$, $k = 1, \cdots, N$:

$$\begin{cases} i_x = \lfloor \tilde{x}[k] + (\frac{d}{2} - i) * U_x[k] + (\frac{d}{2} - j) * V_x[k] + 0.5 \rfloor, \\ i_y = \lfloor \tilde{y}[k] + (\frac{d}{2} - i) * U_y[k] + (\frac{d}{2} - j) * V_y[k] + 0.5 \rfloor, \\ i_z = \lfloor \tilde{z}[k] + (\frac{d}{2} - i) * U_z[k] + (\frac{d}{2} - j) * V_z[k] + 0.5 \rfloor, \end{cases}$$

and

$$sections[i, j, k] = image[i_x, i_y, i_z],$$

if $1 \le i_x \le M_x$, $1 \le i_y \le M_y$, and $1 \le i_z \le M_z$.

3.3 GI Tract Unraveling

Input: $image[i_x, i_y, i_z]$, $i_x = 1, \cdots, M_x$, $i_y = 1, \cdots, M_y$, $i_z = 1, \cdots, M_z$, $\tilde{\mathbf{P}}[k]$, $\mathbf{U}[k]$ and $\mathbf{V}[k]$, $k = 1, \cdots, N$, and the cross-sectional dimension d; **Output:** $volume[i, j, k]$, $i = 1, \cdots, c$, where c is a constant related to a reference circle for unraveling (for example, $c = 2d$), $j = 1, \cdots, \frac{d}{2}$, $k = 1, \cdots, N$. The algorithm is as follows.

1. Compute discretize angles $\theta[i]$, $i = 1, \cdots, c$,

$$\theta[i] = \frac{2\pi i}{c};$$

2. Compute radial directional vectors $\mathbf{R}[i,k] = (R_x[i,k], R_y[i,k], R_z[i,k])^t$, $i = 1, \cdots, c$, $k = 1, \cdots, N$:

$$\begin{cases} R_x[i,k] = U_x[k] \cos\theta[i] + V_x[k] \sin\theta[i], \\ R_y[i,k] = U_y[k] \cos\theta[i] + V_y[k] \sin\theta[i], \\ R_z[i,k] = U_z[k] \cos\theta[i] + V_z[k] \sin\theta[i]; \end{cases}$$

3. Determine original voxel coordinates while stepping in j along each radial direction $\mathbf{R}[i,k]$, $i = 1, \cdots, c$, $j = 1, \cdots, \frac{d}{2}$, $k = 1, \cdots, N$:

$$\begin{cases} i_x = \lfloor \tilde{x}[k] + R_x[i,k]j + 0.5 \rfloor, \\ i_y = \lfloor \tilde{y}[k] + R_y[i,k]j + 0.5 \rfloor, \\ i_z = \lfloor \tilde{z}[k] + R_z[i,k]j + 0.5 \rfloor, \end{cases}$$

and construct $volume[i, j, k]$

$$volume[i,j,k] = image[i_x, i_y, i_z].$$

if $1 \le i_x \le M_x$, $1 \le i_y \le M_y$, $1 \le i_z \le M_z$, and $(i_x - \tilde{x}[k])^2 + (i_y - \tilde{y}[k])^2 + (i_z - \tilde{z}[k])^2 \le \frac{d^2}{4}$.

4 Automatic Unraveling In Simulation

Optimally, fly-through and unraveling should be done *fully* automatically. It is evident that this is a quite complicated project. In our automatic tracking of an extracted curvilinear central axis of the GI tract from user specified starting to end points, the next position is estimated at a constant step along the local direction at the starting point, adjusted to the mass center of the section orthogonal to the local direction and through the estimated position, and scaled to have the increment equal to the step length. Then, the local direction is updated. The adjustment process can be iterated for higher accuracy. The iteration process will converge to the ideal position, which is on the central axis and a step away from the previous position. Tracking is continued in this manner until the end point is sufficiently close. The optimal step length corresponds to *the minimum average turning angle* [13].

After the tracking process is finished, a list of ordered central axis points and associated local directions are available. With this information, sections orthogonal to the central axis can be defined as in the study on semi-automatic unraveling. Similarly, each cross-section can be described in a polar coordinate system. Consequently, an unraveled cross-section can be formed in the two polar coordinates. By stacking all these unraveled cross-sections, the GI tract is mapped into a planar image volume, which can be surface/volume-rendered for various elongated displays.

To work in an efficient fashion towards the ideal goal, we designed a mathematical GI tract phantom of 1,000 mm in length, as sketched in Figure 3(a). Elliptic cross-section, linear taper, spherical lesions and symmetric stenosis were incorporated in the design. A software simulator was developed to synthesize the phantom. To mimic a twisted central axis of the GI tract, a mathematical knot was taken from knot theory as the central axis of the GI tract phantom. The straight GI tract phantom was convoluted along the knot as shown in Figure 3(b), and automatically tracked and marked with disks in Figure 3(c). The coordinate system used in unraveling the GI tract phantom is given in Figure 3(d). Figure 3(e) and 3(f) are respectively an isometric three-dimensional view and an internal surface topography of the unraveled GI tract phantom. After linear interpolation for the central axis position and local orientation at the midpoint of each tracking interval, the GI tract cross-sectional features, including the cross-sectional area, average radius, eccentricity and wall thickness, were computed. For example, the average radius feature was plotted in Figure 3(g) with respect to the central axis length for the original straight GI tract phantom and the digitally unraveled counterpart, respectively. Our results show that the lesions and stenosis significantly contributed to the feature values, and the ideal and unraveling results were highly consistent. Actually, the total lengths of the original and the unraveled GI tract phantoms agreed with about 1% error only.

In the automatic unraveling study, we assumed that segmentation of the GI tract and extraction of its central axis can be accurately performed. It is unlikely that the entire GI tract will be filled with contract at the same time. Appropriate bowel preparation would alleviate the problem. Promising segmentation algorithms include adaptive region growing, graph-theory-based and "snake" algorithms. The central axis of a segmented GI tract can be determined using various thinning algorithms. There is still much work to be done in this area.

5 Conclusion

Clearly, unraveling and fly-through visualization of the colonic mucosal surface is feasible. Quantitative analyses of the colonic morphology is readily achievable in this manner. There is significant potential for these techniques to become a screening tool for colon carcinoma.

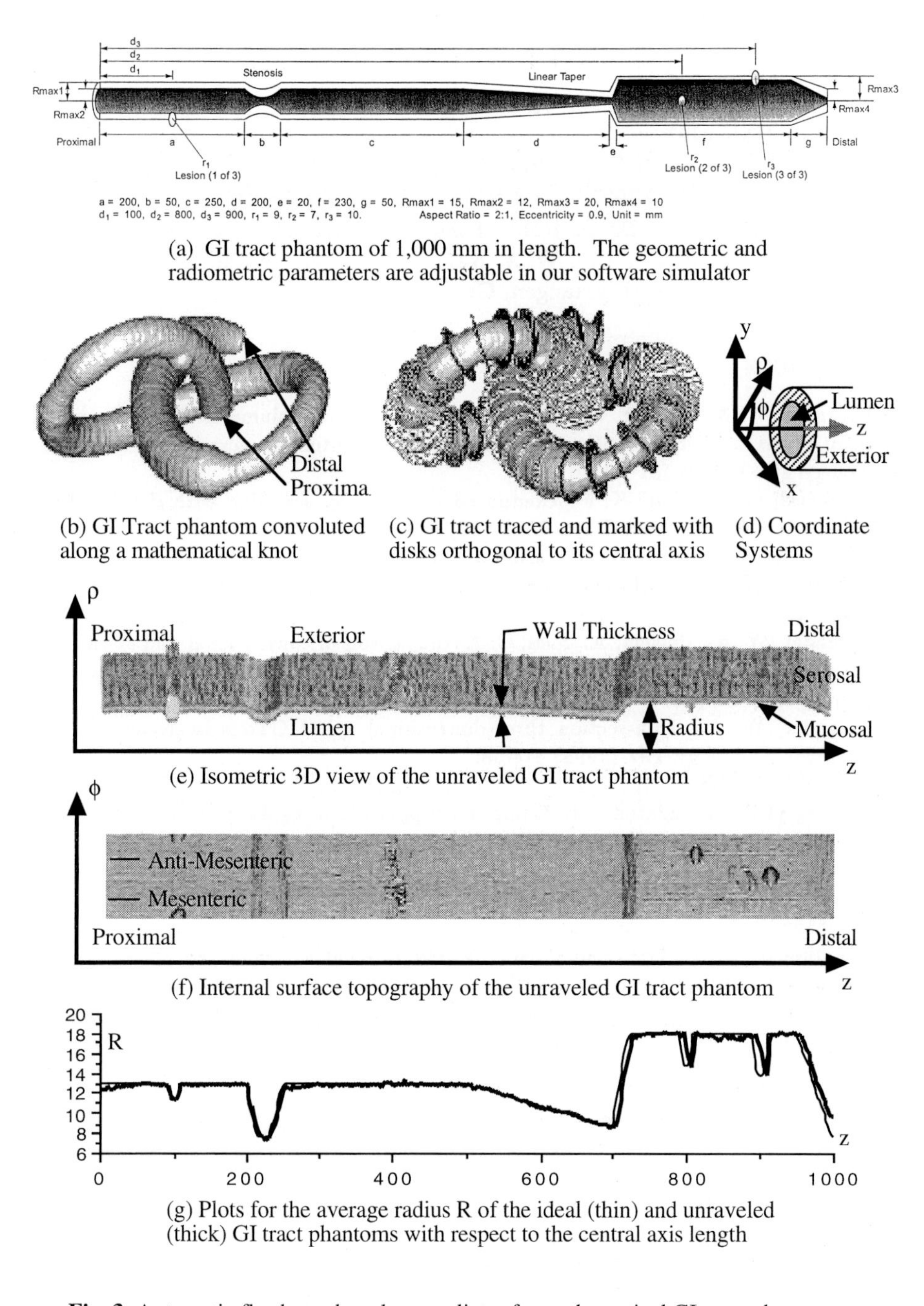

(a) GI tract phantom of 1,000 mm in length. The geometric and radiometric parameters are adjustable in our software simulator

(b) GI Tract phantom convoluted along a mathematical knot

(c) GI tract traced and marked with disks orthogonal to its central axis

(d) Coordinate Systems

(e) Isometric 3D view of the unraveled GI tract phantom

(f) Internal surface topography of the unraveled GI tract phantom

(g) Plots for the average radius R of the ideal (thin) and unraveled (thick) GI tract phantoms with respect to the central axis length

Fig. 3. Automatic fly-through and unraveling of a mathematical GI tract phantom.

Acknowledgments

The authors thank Drs. W. A. Kalender, Z. Li and G. Aliperti for valuable discussions. This work was supported in part by grants from the Whitaker Foundation (The Biomedical Engineering Program) and the National Institutes of Health (NIDCD R03 DC02798 and NIDDKD R29 DK50184). The Analyze^{TM} system was provided by Dr. R. Robb of the Mayo Clinic. The research spiral CT software was provided by Drs. W. A. Kalender, A. Polacin and E. Klotz of the Siemens Medical Systems, Erlangen, Germany.

References

1. Kalender, W. A., Seissler, W., Klotz, E., Vock, P.: Spiral volumetric CT with single-breath-hold technique, continuous transport, and continuous scanner rotation. Radiology **176** (1990) 181–183
2. Crawford, C. R., King, K. F.: Computed tomography scanning with simultaneous patient translation. Med. Phys. **17** (1990) 967–982
3. Wang, G., Vannier, M. W.: Longitudinal resolution in volumetric x-ray CT – Analytical comparison between conventional and helical CT. Med. Phys. **21** (1994) 429–433
4. Kalender, W. A., Polacin, A., Süss, C: A comparison of conventional and spiral CT: An experimental study on detection of spherical lesions. J. of Computer Assisted Tomography **18** (1994) 167–176
5. Kalender, W. A.: Thin-section three-dimensional spiral CT: Is isotropic imaging possible? Radiology **197** (1995) 578-580
6. Vining, D. J., Shifrin, R. Y., Liu, K., Grishaw, E. K., Padhani, A. R.: Virtual reality imaging of human anatomy. In Computer Applications to Assist Radiology (J. M. Boehme, A. H. Rowberg, and N. T. Wolfman, eds., Symposia Foundation) (1994) 703-708
7. Wang, G., Vannier, M. W.: Unraveling the GI tract by spiral CT. SPIE **2434** (1995) 307–315
8. Woodhouse, C. E., Friedman, J. L.: *In vitro* air-contrast-enhanced spiral 3D CT (virtual colonoscopy) appearance of colonic lesions. Radiology **197(P)** (1995) 500
9. Hara, A. K., Johnson, C. D., Reed, J. E., Ahlquist, D. A., Nelson, H., Ehman, R. L., McCollough, C. H., Ilstrup, D. M.: Detection of colorectal polyps by computed tomographic colography: Feasibility of a novel technique. Gastroenterology **110** (1996) 284–290
10. Rubin, G. D., Beaulieu, C. F., Argiro, V., Ringl, H., Norbash, A. M., Feller, J. F., Dake, M. D., Jeffrey, R. B., Napel, S.: Perspective volume rendering of CT and MR images – Applications for endoscopic imaging. Radiology **199** (1996) 321–330
11. Robb, R. A.: Analyze^{TM} Reference Manual. Biomedical Imaging Resource of Mayo Foundation (1995) (version 7.5)
12. Press, W. H., Teukolsky, S. A., Vetterling, W. T., Flannery, B. P.: Numerical recipes in C - The art of scientific computing. Cambridge University Press (1992) (2nd edition) 113-116
13. Wang, G., Vannier, M. W., Skinner, M. W., Kalender, W. A., Polacin, A: Unwrapping cochlear implants by spiral CT. IEEE Trans. on Biomed. Eng. (1996) (in press)

Segmentation of the Visible Human for High Quality Volume Based Visualization

Thomas Schiemann, Jochen Nuthmann, Ulf Tiede, Karl Heinz Höhne

Institute of Mathematics and Computer Science in Medicine (IMDM),
University Hospital Hamburg-Eppendorf, Germany
e-mail: *lastname*@uke.uni-hamburg.de

Abstract. A combination of interactive classification and supersampling visualization algorithms is described, which delivers greatly enhanced realism of 3D reconstructions of the Visible Human data set. Objects are classified on basis of ellipsoidal regions in RGB-space. The ellipsoids are used for supersampling in the visualization process.

1 Introduction

Volume visualization of medical images has become a standard tool, which can be applied for many different purposes. While in some fields, such as planning of craniofacial surgery on basis of CT images, 3D visualization is frequently applied, other fields are still suffering from some difficulties, which prevent a wide use of these methods. For *clinical* applications the main difficulties arise from the segmentation needed for every single case to define the structures to be visualized. Although countless segmentation approaches are existing, none of them solves the problem in general, and the time effort for segmentation usually remains high. Thus in clinical routine many applications, for which 3D visualization might be useful, do not use these methods. The situation changes, if we consider *teaching* applications in radiology or anatomy, for which segmentation has to be done only once for a few typical cases. Nevertheless, these systems are suffering from a lack of realism and detail, because their typical basis are radiological cross sectional images from CT or MRI, which have limited spatial resolution of about 1 mm only, and which do not yield any color values for realistic visualization. Theses difficulties can be overcome, if the high resolution color images of the Visible Human Project [11] are used. These are basically anatomic images, whose important link to radiology is established by corresponding stacks of CT and MRI slices.

Although the availability of the Visible Human images is rather recent, a number of systems for their exploration does already exist. Unfortunately, most of these systems present the images as stacks of orthogonal slices only, which can be browsed through with different techniques. The number of real 3D applications [3, 10] is still limited because of several reasons: On the one hand the vast amount of data imposes problems with computer resources. On the other hand the new type of color images leads to algorithmic problems for high quality volume visualization especially in determination of surface location and orientation.

In this paper we will describe an approach, which overcomes the algorithmical problems of volume based rendering of the Visible Human by object segmentation on basis of ellipsoidal regions in RGB-space, which allows supersampled ray-casting for computation of very realistic images. The proposed visualization method does further provide integrated display of anatomical and radiological images, which is one of the most important properties regarding teaching applications. The usefulness of the method will be demonstrated with images of different parts of the Visible Human, which show a significant improvement of image quality and realism.

2 Method

2.1 Image preprocessing

For our experiments we worked with the Visible Human male. The data are transverse cross-sectional photographic images of a frozen male cadaver with a resolution of 0.33 mm and slice distance of 1 mm. The images come together with two sequences of high-resolution CT images from the fresh and frozen cadaver with resolutions between apx. 0.5 mm and 1.0 mm and slice distances of 1 mm for the frozen and between 1 mm and 5 mm for the fresh cadaver. The fresh cadaver has also been scanned by MRI, resulting in separated stacks of (mainly coronal) images, which have been obtained with three acquisition protocols (T1, T2, PD). The MR images have lower resolution between 1 mm and 2 mm and slice distances up to 5 mm.

Before acquisition of the photographic images, the cadaver was sawed up into different portions. This procedure leads to some shift artefacts in the resulting volume, which are visible as horizontal lines on 3D renderings.

Next to algorithmical problems, which will be addressed later, one of the main difficulties in working with the Visible Human data is their huge size, especially that of the anatomical data. Handling of even small parts of the body in full resolution requires very large disk storage and computer main memory (e.g. 440 Mbyte just for the raw data of the head or 1100 Mbyte for the upper abdomen). We have thus reduced the resolution of the anatomical images to 1 mm^3 by averaging 3×3 pixels on every slice. While this procedure has the advantage, that the resulting data (49 Mbyte for the head or 121 Mbyte for the abdomen) can be processed in high-end workstations, it also introduces new segmentation problems, because partial volume effects are created or much enhanced by the averaging procedure.

A 3D scatterplot of the color distribution of the anatomical images shows, that the components are highly correlated. Thus we also considered the possibility to reduce the RGB-data into 2-channel data by principal components analysis. But finally, we did not choose this procedure, because for later interactive segmentation the original colors of the anatomical images are still needed. Thus they cannot be replaced by the artificial 2-channel data.

The CT and MRI data are coarsely registered with the anatomical data. We improved this registration using an interactive landmark-based tool [7]. The main mapping parameters, which had to be refined, were translation and scaling both within the slices. The other parameters (scaling and translation perpendiclar to the slices and rotation) were already correct. The registration error is reduced to approx. 1 voxel for the frozen CT dataset, but for the images obtained from the fresh cadaver, locally larger registration errors occur due to morphological changes caused by freezing. The registration procedure allows to create combined renderings of different modalities in a single view with object parameters split onto the different modalities.

2.2 Volume based visualization

For the task of 3D visualization of volume data, there are in general three different approaches:

- Rendering of triangular surfaces, which have been obtained from the volume data in preprocessing steps [5].
- Volume rendering by projection of the whole volume under different constraints [4].
- Volume based rendering of surfaces, which have been obtained from previous segmentation [12].

While triangle surface rendering has the advantage, that it is fast, if the number of triangles is kept low, a lot of detail is usually lost during the preprocessing steps for generation of the surfaces. Accurate mapping of surface texture and realistic cutting are difficult to implement. On the other hand pure volume rendering is usually very slow on standard hardware and can in general not create realistic images. Thus we have concentrated on volume based surface rendering, which has the advantage to compute images with a high grade of detail and realism. Although computing times for these volume based methods are still longer than for triangular rendering, this disadvantage will be overcome by future hardware.

In our previous work we have developed a framework for volume based anatomy atlases [2, 9]. One of the key procedures of this system is a high quality volume rendering module, which uses volume based ray casting methods. There are two conditions, which have to be fulfilled in general for the computation of high quality images:

1. The surface shading appears most realistic, when the surface normals are determined by voxel intensity gradients [12]. Thus the segmentation procedure, which precedes visualization, has to define the object borders "reasonably well" at the location of the intensity gradients.
2. The clusters in RGB space defining an object must be expressed in a simple analytical form in order to allow a fast refined classification of supersampled data at ray casting time.

For scalar images (CT, MRI) these prerequisites are fulfilled by determination of an intensity range by a pair of lower und upper threshold values: First the threshold range yields object boundaries at the location of high intensity gradients and second the threshold can be interpolated in sub-voxel-resolution. The strong influence of the threshold specification on the rendering results is shown in figure 1.

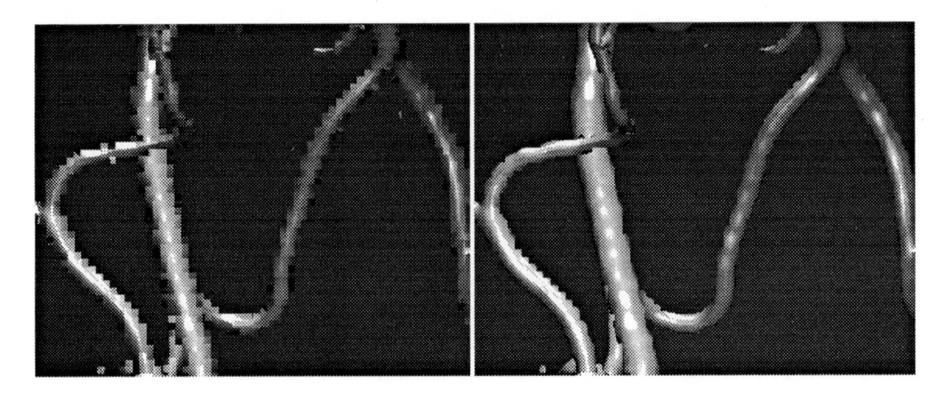

Fig. 1. Influence of the threshold specification on the rendering result. Both images show a volume based rendering of the same voxel set. Left: Disregarding of the threshold leads to blocky surfaces, because even supersampling hits the voxel boundaries only. Right: Regarding the threshold during supersampling produces a smooth and realistic surface appearance.

For volume based rendering of the Visible Human color images, the key problem for computation of high quality images is a generalization of scalar thresholds to a description in RGB-space. Pure enumeration of the RGB-tuples is not practical. The cluster should have a simple formal description for efficient computation during ray-casting. Attempts with a box shaped classification in RGB space gave unsatisfactory results. Since objects in RGB space look rather ellipsoidal (see fig. 3), we have decided to describe the classified regions by ellipsoids in RGB-space. In order to find the sub-voxel surface location it is sufficient to intersect a straight line between two RGB-tuples with the ellipsoid (fig. 2).

Next to the described threshold generalization the visualization system has also been adapted in the following way to handle and take advantage from the Visible Human images:

- Instead of assigning an artificial color, texture mapping is performed by using the RGB-tuple at the surface location for the Phong ilumination model.
- The surface normal is calculated as the normalized sum of the intensity gradients of each RGB-component.
- Rendering parameters can be taken from different data sets (anatomical, CT, or MRI) depending on the object (e.g. bone surface location and normal can be obtained from CT but the color from RGB data).

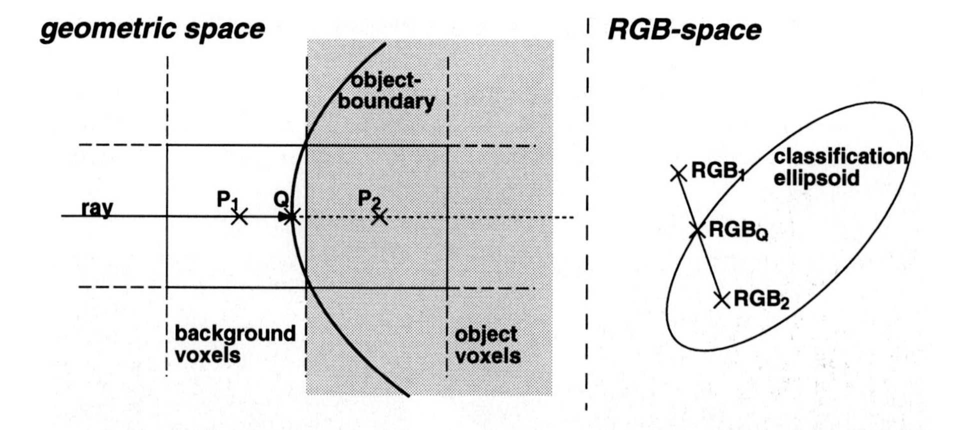

Fig. 2. Using the ellipsoidal description of an object for sub-voxel computation of the surface location: Voxel P_2 with color RGB_2 carries the object label, voxel P_1 with color RGB_1 is outside the object to be visualized. The object surface is assumed to be at that point Q, which devides the line between P_1 and P_2 in the same ratio as RGB_Q devides RGB_1 and RGB_2.

2.3 Segmentation on basis of RGB-ellipsoids

Since the segmentation has to be done in great detail, an automatic segmentation will not be possible and we rather aim at a semi-automatic procedure. The segmentation is adapted from an interactive method, which has proven to be successful for many purposes with scalar images like CT or MRI [1, 6]. The procedure is based on thresholding followed by binary mathematical morphology and connected component labeling. While the latter two steps remain unchanged for color images, the intensity range describing an object is replaced by an ellipsoid in RGB space. Determination of the ellipsoid is done by the following steps (fig. 3):

1. The users outlines a (usually small) typical region of the object, which is to be segmented. This results in a set of color-triples $(RGB)_i$.
2. In most cases the sample set $(RGB)_i$ will contain "wrong" triples, which belong to other structures, but have also been hit during outlining. In order to get rid of these triples, the median rgb-triple μ_{RGB} and the median μ_d of all distances $d(\mu_{RGB} - (RGB)_i)$ are computed. For determination of the ellipsoid only those triples are further regarded, which are closer to μ_{RGB} than an interactively specified multiple f of the median μ_d.
3. The ellipsoid's center is taken from the median of the restricted sample set, the axes directions are taken as the principal axes computed from covariance analysis, and the extent of the ellipsoid is determined such, that all restricted triples are inside the ellipsoid.

During a practical session the first goal is to define an ellipsoid, which covers the object to be segmented. Due to the huge color feature space several outlining

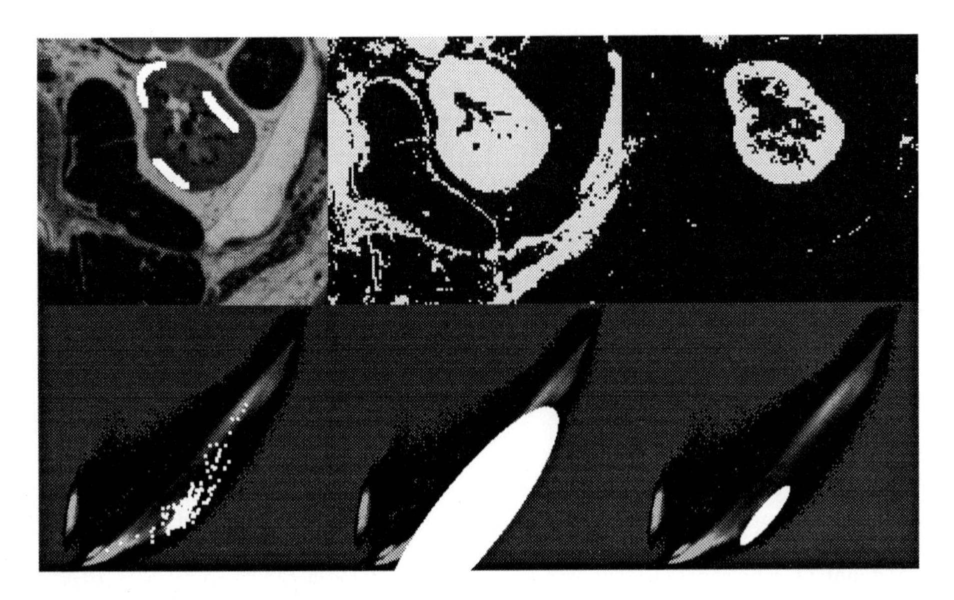

Fig. 3. Interactive classification in RGB-space for the left kidney: The intermediate results are shown in the image space (upper row) and on 2D projections along the blue axis of the 3D scatterplot (lower row). The user first outlines some typical areas of the kidney (left column). A preliminary results is obtained, when all marked triples are used (middle). By restriction to the "substantial" triples the final sharp classification is computed (right).

steps will be neccessary and "wrong" triples will be included in the ellipsoid (fig. 3 left). In this state of the procedure the multiple f is set to an upper limit, so that it has no influence (fig. 3 middle). After that the second goal is to refine the ellipsoid by reducing f (fig. 3 right). In our experiments we found out, that values between 2 and 4 are a good choice for f.

The problem of wrongly marked pixels in the first step is very much reduced, if the full resolution (0.33 mm) images are used. This is due to the fact, that the full resolution images show much sharper object boundaries, which can thus be identified in the marking step much easier. In addition partial volume effects are not very strong on the full resolution images. Thus the cluster of marked RGB-tuples is much sharper compared to procesing on the images with reduced resolution.

Computing times for the ellipsoid itself are negligible, but for fast binary decision, if a triple is inside the ellipsoid or not, we use a binary volume of size 256^3, in which all triples inside the ellipsoid are set. Updating of this volume can take several seconds depending on the size of the ellipsoid.

The ellipsoidal classification in RGB-feature space without further processing steps in image space is generally not sufficient for segmentation of an object, because several structures are classified together. Connected component labeling and binary mathematical morphology are used for a further spatial decomposition of the voxels in geometric space covered by the initial ellipsoid.

Instead of the proposed basic ellipsoidal specification, the segmentation step could also be performed with more sophisticated classification algorithms. However, these do usually not result in easy to formalize descriptions, which we need for fast computations during ray-casting.

The proposed segmentation procedure is successful for all major constituents of the body like brain or abdominal organs. For a detailed subdivision of these structures, e.g. into gyri of the brain, manual segmentation is still needed due to a lack of image properties for this purpose.

3 Results

We have applied our method to several regions of the Visible Human male. The major constituents of the head, the upper abdomen, and the right shoulder have been segmented with the described interactive segmentation procedure. Most structures (e.g. brain, kidneys, muscles, intestine) can be segmented in a few minutes on a workstation DEC 3000 with 256 Mbyte of main memory. Extented organs like the liver with many links to other structures need additional interaction on single slice sequences.

Figure 5 shows a view of the upper abdomen with the kidneys and abdominal vasculature. The dissection planes show the different image modalities obtained in the Visible Human project. The sagittal and upper transverse plane show the anatomical colors, the coronal plane in the back shows the intensities of the fresh CT data set, and the lower transverse plane shows the T1-weighted MRI data. Skin and kidneys are rendered with texture mapping of the anatomical colors, while for the blood vessels artificial colors are chosen in order to enhance their appearance.

Figure 6 shows a view of the muscles of the right shoulder, which have been segmented from the anatomical data. The bone structures have been segmented from the frozen CT data. Also the rendering is performed on the CT data only. Texture mapping is not performed for the bone, but again an artifical color is chosen for enhancement. The shoulder region is hit by several shift artefacts, which are visible as horizontal lines.

Further 3D renderings of the Visible Human's head and abdomen have been published elsewhere [8, 13].

4 Conclusion

We have proposed a new combined approach of interactive classification and volume based visualization of RGB volumes such as those of the Visible Human. The pictures show, that the rendering quality obtained with the described method is superior to that of other approaches published so far.

Our first steps in segmentation for high resolution volume rendering show, that the Visible Human data set represents a new quality of anatomical imaging, when used in a state-of-the-art visualization system. Nevertheless, a huge amount of work still has to be done for detailed segmentation towards a complete 3D anatomical atlas. But unlike clinical imaging, which requires new segmentation for every case, segmentation of the Visible Human has to be done only once. If this work is completed, the results will surely have great impact on many applications of visualization such as education and simulation.

5 Acknowledgements

We are grateful to all members of our department, who have supported this work. Special thanks to Rainer Schubert for anatomical and Uwe Pichlmeier for statistical support. Our thanks go also to Victor Spitzer, University of Colorado School of Medicine, and Michael Ackerman, National Library of Medicine, for providing the Visible Human data set.

References

1. Höhne, K. H., Hanson, W. A.: Interactive 3D-segmentation of MRI and CT volumes using morphological operations. *J. Comput. Assist. Tomogr. 16*, 2 (1992), 285–294.
2. Höhne, K. H., Pflesser, B., Pommert, A., Riemer, M., Schiemann, T., Schubert, R., Tiede, U.: A new representation of knowledge concerning human anatomy and function. *Nature Med. 1*, 6 (1995), 506–511.
3. Hong, L., Kaufman, A., Wei, Y.-C., Viswambharan, A., Wax, M., Liang, Z.: 3D Virtual Colonoscopy. In *IEEE Biomedical Visualization Symposium.* 1996, 26–32.
4. Levoy, M.: Display of surfaces from volume data. *IEEE Comput. Graphics Appl. 8*, 3 (1988), 29–37.
5. Lorensen, W. E., Cline, H. E.: Marching cubes: A high resolution 3D surface construction algorithm. *Comput. Graphics 21*, 4 (1987), 163–169.
6. Schiemann, T., Bomans, M., Tiede, U., Höhne, K. H.: Interactive 3D-segmentation. In Robb, R. A. (Ed.): *Visualization in Biomedical Computing II, Proc. SPIE 1808.* Chapel Hill, NC, 1992, 376–383.
7. Schiemann, T., Höhne, K. H., Koch, C., Pommert, A., Riemer, M., Schubert, R., Tiede, U.: Interpretation of tomographic images using automatic atlas lookup. In Robb, R. A. (Ed.): *Visualization in Biomedical Computing 1994, Proc. SPIE 2359.* Rochester, MN, 1994, 457–465.
8. Schiemann, T., Nuthmann, J., Tiede, U., Höhne, K. H.: Generation of 3D anatomical atlases using the Visible Human. In Kilcoyne, R. F. et al. (Eds.): *Computer Applications to Assist Radiology, Proc. SCAR '96*, Symposia Foundation, Carlsbad, CA, 1996, 62–67.
9. Schubert, R., Höhne, K. H., Pommert, A., Riemer, M., Schiemann, T., Tiede, U.: Spatial knowledge representation for visualization of human anatomy and function. In Barrett, H. H., Gmitro, A. F. (Eds.): *Information Processing in Medical Imaging, Proc. IPMI '93*, Lecture Notes in Computer Science 687, Springer-Verlag, Berlin, 1993, 168–181.
10. Seymour, J.: A New 3D Database and Atlas Based on the Visible Human. In *Second Southeastern Medical Informatics Conference.* Gainsville, FL, 1996.

11. Spitzer, V., Ackerman, M. J., Scherzinger, A. L., Whitlock, D.: The Visible Human Male: A Technical Report. *J. Am. Med. Inf. Ass. 3*, 2 (1996), 118–130.
12. Tiede, U., Höhne, K. H., Bomans, M., Pommert, A., Riemer, M., Wiebecke, G.: Investigation of medical 3D-rendering algorithms. *IEEE Comput. Graphics Appl. 10*, 2 (1990), 41–53.
13. Tiede, U., Schiemann, T., Höhne, K. H.: Visualizing the Visible Human. *IEEE Comput. Graphics Appl. 16*, 1 (1996), 7–9.

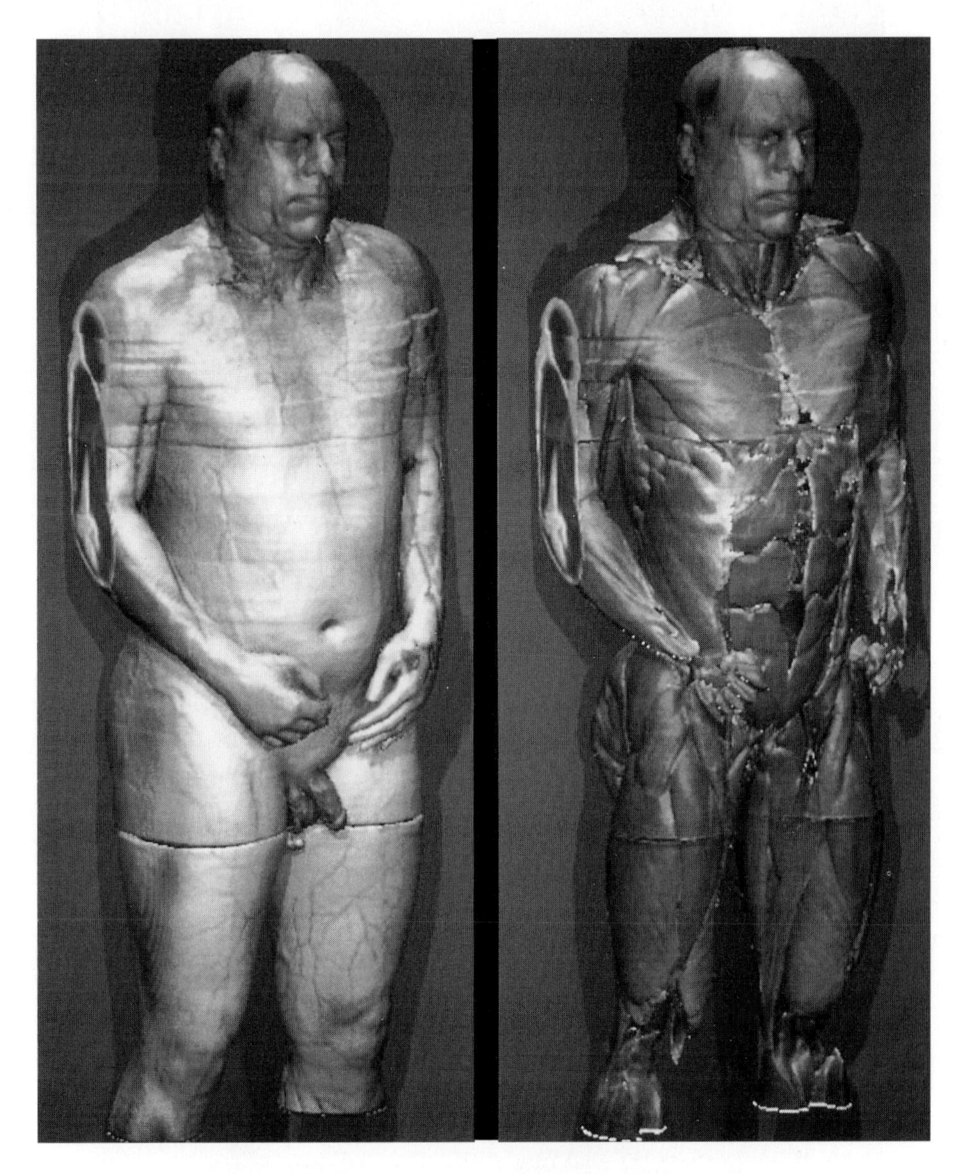

Fig. 4. Views of the torso: Skin surface and muscles.

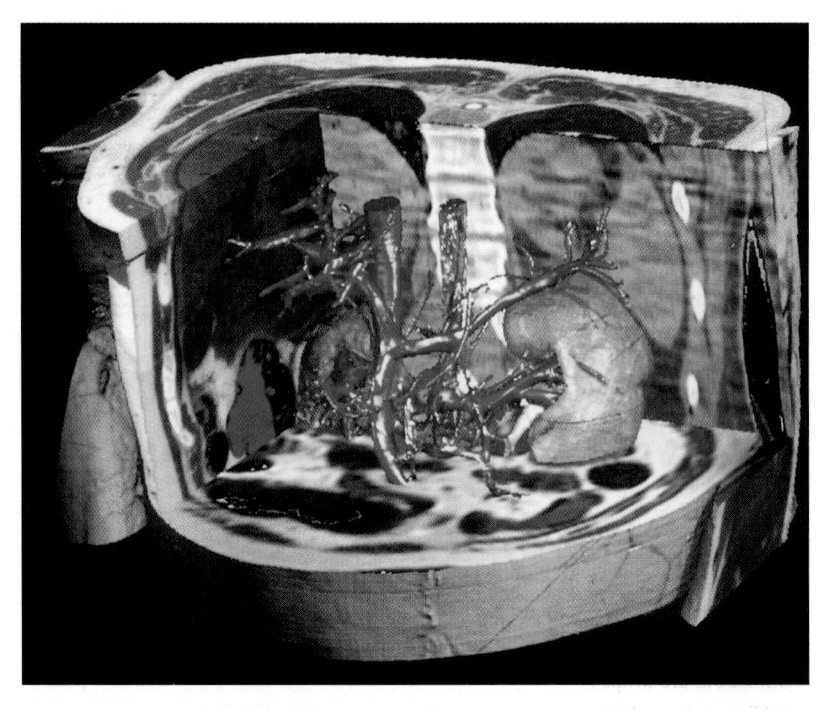

Fig. 5. View of kidneys and vascular structures in their radiological environment.

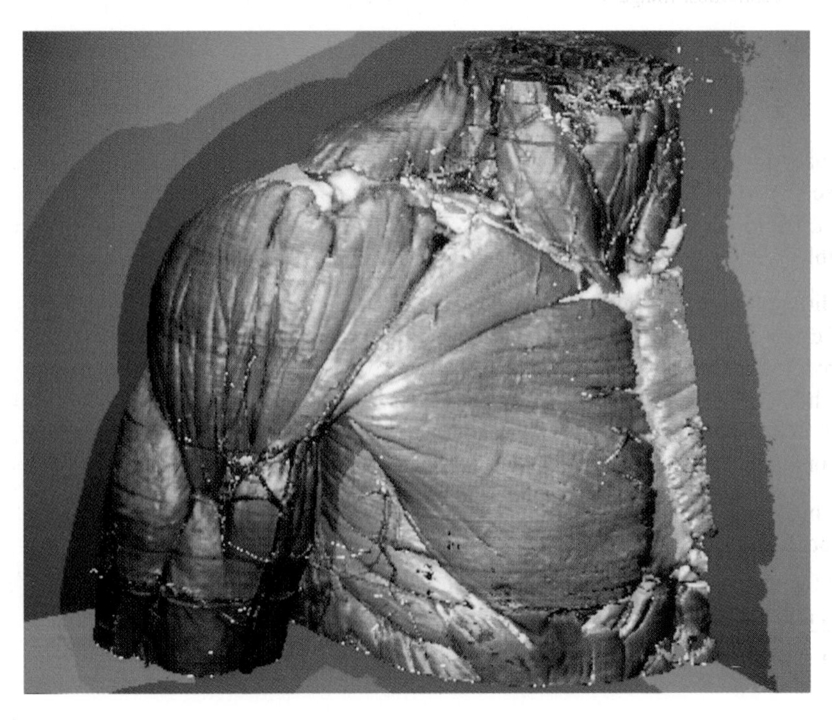

Fig. 6. View of the right shoulder with bones, musculature and some major veins.

Illustrating Anatomic Models – A Semi-Interactive Approach

Bernhard Preim, Alf Ritter, Thomas Strothotte

Institut für Simulation und Graphik, Otto-von-Guericke-Universität Magdeburg
Universitätsplatz 2, D-39106 Magdeburg
e-mail: {bernhard, alf, tstr}@isg.cs.uni-magdeburg.de

Abstract: We present the ZOOM ILLUSTRATOR which illustrates complex 3D-models for teaching anatomy. Our system is a semi-interactive tool which combines an animation mode and an interactive mode. While the animation mode is suitable for beginners, the interactive mode is dedicated to experienced users. The animation is controlled by scripts specifying *what* should be explained in *which level of detail*.

The design of the animation and the interactive components is directed to generate illustrations according to the user's interest. This includes the presentation of text and the parameterization of the rendered image.

Changes on the textual part, interactively requested or generated in an animation, are propagated to the graphics part and vice versa. Thus the display of an explanation results in an adaptation of the corresponding graphical part.

Keywords: Script-Based Animation, Interactive Illustrations, Emphasis Techniques, Image-Text-Relation, Fisheye Zoom Techniques

1 Introduction

In many areas, learning involves the study of complex 3D-phenomena. This is especially true for anatomy, where highly complex structures are studied under different aspects and viewing directions. To get an insight into the spatial structure and to be able to name parts of objects, are important issues in the learning process.

Traditional teaching materials do not support all aspects of this learning process sufficiently. Textbooks (see e.g. Sobotta (1988)) contain valuable drawings but require a lot of flipping through the pages. Numerous illustrations and the text referring to it have to be integrated mentally. Videos do a better job in explaining spatial relations because continuous transitions of different camera positions are possible. The person who explains uses pointers to direct the student to certain parts of the image.

The main problem with these media is the separation of initiative. When studying a textbook the student decides himself what to read and to look at. On the other hand, when watching a video the student can only influence the speed of the presentation.

Based on these observations we developed the ZOOM ILLUSTRATOR to combine techniques from traditional media with the flexibility of a computer system. Our system generates illustrations which are adapted to the parts currently explained. Thus a detailed study of interesting parts is possible while maintaining the context.

This paper describes the scripting language which controls the animation, the interactive mode and mechanisms to switch between them. The interpretation of a script results in commands to present text and to accentuate parts of the underlying 3D-model. To enhance comprehension, smooth transitions for both the display of textual information and changes of the graphics are realized.

The interactive mode allows to ask for explanations which are displayed using Fish-eye Zoom Techniques, and further result in an accentuation of the parts referred to in the image. Emphasis is put on the combination of the interactive mode and the animation mode. While the beginner tends to work with prepared film-sequences the advanced student will shift to the interactive mode. Interactive exploration within an animation as well as access to film-sequences during interactive working are provided.

2 Related Work

Previous work on interactive anatomy teaching is based on scanned images or on 3D-models. The latter are more advanced because they allow flexible transformations.

The leading example in this field is the VOXELMAN™ (see e.g. Pommert *et al.* (1994)). This system exploits volume-models and offers free positioning and the facility to interactively cut off, i.e. literally to operate virtually. Although the images constructed this way are labeled and optionally explained, the coherence of pictorial and textual information is not treated in its own right, meaning that no automatic adaptation takes place in the image due to interaction on the textual part.

While anatomy teaching systems are related in terms of the application area, systems generating illustrated animations are important in terms of the underlying principles.

The generation of animations based on high-level scripts has been described in Zeltzer (1990) as well as in Karp and Feiner (1993). Zeltzer (1990) refers to the level on which an animation is specified as the *task level*, indicating that tasks (communicative intents in the terminology of Zeltzer) are supported. A simple example for a task is "*show object*". To bridge the gap between task level specifications and the low-level commands for an animator (camera movements and geometric transformations) *decomposition rules* are used. These rules are oriented at *filmmaking heuristics* which are generally accepted practices which have evolved during the 90-year-history of cinema. Besides traditional methods, specific techniques from the medium *computer animation* are applied. This includes exploded views and cutaways.

The ESPLANADE-System (Expert System for PLANning Animation, Design and Editing) of Karp and Feiner (1993) exploits a sophisticated planning scheme to fulfill communicative intents. The design of ESPLANADE is guided by the hierarchical structure of traditional films with sequences, scenes and individual shots.

These systems are based on an object-structure within the scene-description. Furthermore, information is available, as to where an object is positioned and what it occludes. This information is used to calculate the transformations necessary to study certain aspects in detail.

Previous work concentrated either exclusively on the graphical part or on the textual part in combination with relatively simple images. The ZOOM ILLUSTRATOR is dedicated to illustrate complex models with a large space of related textual information. This requires flexible mechanisms to emphasize/deaccentuate parts of the image to adapt to a more or less detailed textual description.

3 Design of the Animation Component

This chapter describes the animation techniques developed and the combination of animations with interactive exploration. Our design is guided by educational videos for anatomy. We observed that parts of the model are colored to ensure easy recognizability. It is striking with how much effort the extent of complicated objects is explained. The explanation involves the removal of occluding parts. Pointing devices are used to make clear where an object starts and ends.

3.1 Animation Techniques

From these observations we concluded that to explain an object, it should be clearly recognizable and an appropriate camera position has to be chosen. Although the object of interest must not be occluded, the context should be maintained. Because of the importance of proportions we avoid distortions and selective scaling as a means to accentuate certain parts. From these principles the following techniques are derived:

- *Automatic Transformation*: Positioning the camera so that the object to be explained is visible and appears large enough.
- *Direct Emphasis*: Modifications of an object to direct the viewer towards this object. Colors and textures are important parameters with respect to emphasis. Textures, however, cannot be arbitrarily modified because their use may lead to false impressions of the surface structure.
- *Indirect Emphasis*: Changes of objects to deaccentuate them and thereby to pronounce other objects. This includes the modification of occluding objects, which are removed, clipped or made (semi-)transparent. Furthermore, the colour, especially saturation and brightness, is modified to deaccentuate objects.

Besides these general, dedicated methods for complicated objects are provided:

- *Clarify Shape*: The shape is shown with pointing devices (short: pointers) which are moved from the starting point to endpoint(s) and follow the contour of the object to be explained. This technique is especially helpful for objects with branching structures, like muscles.
- *Separate*: An object is separated, presented in a second window and slowly rotated. The rotation is stopped in some standard viewing directions and in an optimal viewing direction (optimal in terms of visibility and projected size).

3.2 Design of the Data Structures

This section describes the structure of the information necessary to realize the animation techniques listed above. To clarify this description, two terms are explained. An *object* is a part of the geometric model. Each object corresponds to a part of the textual description (with labels and explanations), which is called *node*.

3.2.1 Additional Information about the Graphics

To apply the techniques explained in Section 3.1, more information about the 3D-model than the mere geometry is required. For automatic transformations and indirect emphasis, the following information about the 3D-model is necessary:

- Standardized viewing directions from textbooks, optimal viewing direction
- For a number of precalculated viewing directions: size of the projected object, visibility, position (image coordinates), occluding objects

This information is summarized in the *visibility information*. For each object it contains 64 viewing directions (azimut- and declination-angle are varied in steps of 45 degrees). This data structure is gained by an offline calculation in which the rendered image is analyzed once for each direction. The analysis is performed by casting rays into the scene and evaluating which objects are hit in which order.

The visibility of an object o is a relative value, which is the quotient of the number of rays which hit o first and the overall number of rays which hit o. The system registers not only which objects occlude o in a specific viewing direction, but also how often o_i was hit before object o and thereby to which extent o_i occludes o. This information is evaluated to decide how occluding objects should be handled if o is explained. If o_i occludes o only to a small amount, o_i can be clipped against the bounding box of o. If on the other hand o_i almost totally occludes o it is made semitransparent or removed.

3.2.2 Structural Information on the Textual Descriptions

While the visibility information contains additional information on the geometric model, structural information about the related textual information is needed as well.

This *structure information* assigns a category, for our domain an organ-system, to each node. Furthermore, subcategories can be assigned e.g. to summarize nodes of a region. Although appropriate explanation techniques depend on the specific object, an assignment of animation techniques to categories is useful as a default value. The shape of muscles is often complicated, whereas most bones have a simpler structure. This information is collected in the *structure information* and described below:

1. Membership of nodes to categories/subcategories
2. *Aspects*, under which nodes from a category can be textually explained
3. Hypertextlinks between nodes under a certain *aspect*
4. Default animation techniques for a certain category

The information on memberships of nodes (1) is used to change the properties of all objects of a category. Thus the saturation of all muscles can be reduced in order to recognize one of them better. The *aspects* (2) of a node depend on its organ-system. At least a label and a short explanation are assigned to each object. Additional aspects are possible, e.g. muscles have an aspect "Function". Each aspect is characterized by its name and a value to indicate how detailed this information is. The registration of links (3) is used to evaluate which nodes are conceptually near to the one currently explained. The default techniques (4) specify which animation techniques should be employed. Furthermore it is stated which aspect of textual information will be presented, concerning the explained object or linked objects.

To apply the techniques presented in Section 3.1 at all, the *visibility information* is necessary. *Structural information* gives hints about when to use which technique. One problem is still left out: how to demonstrate the path an object takes. The application of this technique assumes information on starting-points and branches. To define these points from an analysis of the geometry, is complicated and not generally applicable. Therefore an interactive approach was chosen which will be described later on. So far we only assume that a path exists.

3.3 Design of the Scripting Language

Important questions when designing a formal language are *what* should be expressed and *who* should use it. We call the person who writes a script the *author*. The author is assumed to know the content to be illustrated well. The scripting language allows him to specify easily *what* should be explained in *how much detail*.

For this purpose the author does not need to know which data structures are used and relies on the information organized by a *Content Provider*. The role of the author is similar to that of a lecturer who uses a textbook someone else has written.

Our experience with interactive illustrations is that a the connection between rendered images and related text is crucial for comprehension (Preim *et al.* (1995)). From this experience we suppose that a coupling of these media is also essential for the efficiency of educational animations. Therefore we provide constructs to state that changes on the graphical part are accompanied by simultaneous changes of text presentations.

Smooth transitions are required for both, changes on the textual part and on the graphical part. For smooth changes of text presentations Fisheye Zoom Techniques (see Dill *et al.* (1994)) are exploited. They allow continuous scaling of rectangular areas for the presentation of text with one node growing at the expense of others which are automatically scaled down. The text presented is adapted to the space available. Rüger *et al.* (1996) describes the application of these techniques for illustrations.

Scripts are basically sequences of *explain*-statements specifying what should be explained. An optional parameter list specifies the timing and the techniques to apply. An *explain*-statement can be added by an *emphasize*-statement to specify which techniques should be used. A *while*-statement specifies what should happen on the textual part in the meantime. To stress the role of the language to generate commands for the ZOOM ILLUSTRATOR, we call it *Illustrator Control Language (IC-Language)*.

Concerning the timing, we decided that all movements which are connected with a distance (linear or angular in the case of a rotation) are specified in terms of speed so that the timing is relative to this distance. The display of textual information can only be specified as a duration. Relative specifications are superior to absolute ones.

To explain a group of objects, summarizing text is required. For this purpose we designed a *table*-statement and a *title*-statement. The latter serves to list all objects of a (sub-)category under the name of this (sub-)category. Similar, but more flexible is the *table*-statement. It generates a table with descriptions of all objects involved, where aspects can be specified. To define the details of the layout, a *format*-specification is available. The *IC Language* is described in more detail in Preim *et al.* (1996).

3.4 Combination of the Animation Mode with the Interactive Mode

Our previous work concentrated on interactive learning. As already noted in the introduction of this paper, interaction alone is not sufficient, hence we added animation techniques. However, we regard the coupling of interactive and animation techniques as the key for a good learning help. This leads to several interesting questions:

- Which interaction – over and above the video-recorder functionality of spooling – is useful for the viewer of an animation? (interactive interrupts of an animation)
- Which kind of film-sequences can be included to enhance interactive working? (animation during interactive exploring)

The answers to these question will be given later in connection with sample outputs.

4 Architecture

The ZOOM ILLUSTRATOR reads a *Scene Description*, containing a polygonal model, and related *Textual Information* derived from Sobotta (1988). The structure of the *Scene Description* corresponds to that of the *Textual Information*. Furthermore, the *Visibility Information* and the *Structure Information* (recall Section 3.2) are read in. From these sources an *Internal representation* is constructed which guides the execution of the *Animation script*. Figure 1 shows the relation between these components.

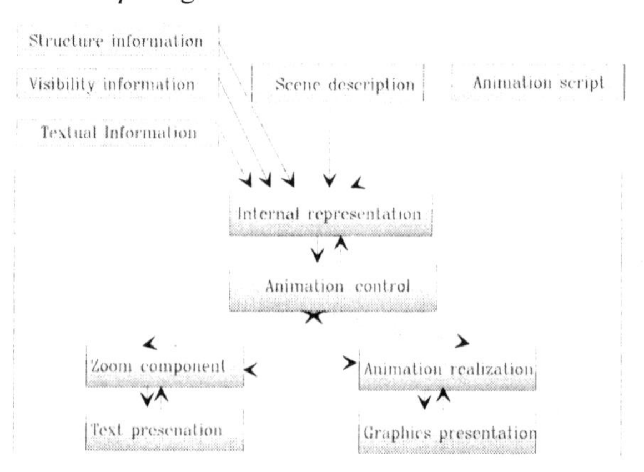

Figure 1: Architecture of the Zoom Illustrator

The ZOOM ILLUSTRATOR can be operated in interactive mode, in animation mode, or in a mixture of both. An illustration (see Figure 2, next page) with an image and textual labels is generated from the specification of the 3D-model and the categories of interest. The details of the layout-generation are described in Preim *et al.* (1995).

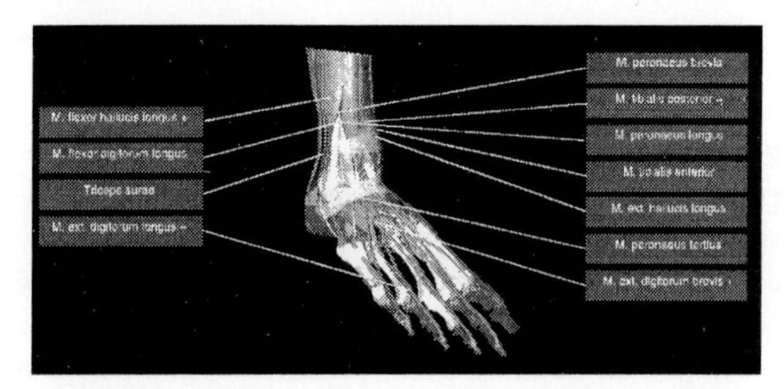

Figure 2: Illustration of a foot with focus on the muscles

The model can be transformed freely while the labels and the reference lines which connect them to the graphics are updated. On the textual part, explanations can be requested which result in zoom steps to accommodate the explanation.

The way the system works does not differ that much between animation mode and interactive working. The main difference is that in the animation mode the *Zoom component* and the *Animation realization* are triggered by the *Animation control* instead of getting "interactive requests" from the user.

5 Implementation

The ZOOM ILLUSTRATOR was developed on Silicon Graphics Indigo² Workstations using the graphics library Open Inventor™. The interaction mechanisms offered and the support for animation predestine Open Inventor for the purpose at hand.

We experimented with a commercially available model with 6000 polygons. It offers enough detail for stepwise exploration but is not too complex to be displayed at interactive rates. The assignment of textual information is based on Sobotta (1988).

For the realization of smooth transitions on the graphical part, the concept of the Inventor Engines is helpful. Engines support animation and simulating object behaviour. Special engines allow the interpolation between initial and final states. Several changes can be specified to happen at the same time. This mechanism is employed for geometric transformations as well as for changes of material properties.

In Section 3.2 we left out how the path to show an object's shape is defined. We developed an interactive tool in which the object to be explained is handled separately. The author marks the starting point, branching points and endpoints resulting in a tree structure of the path to show (see Figure 3). The path contains 3D-points and is therefore independent of the viewing direction.

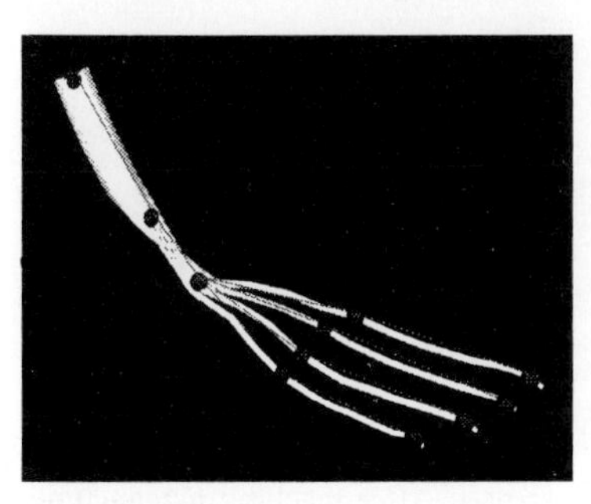

Figure 3: Interactive path definition of a muscle in a separate window

Additional points are derived from the coordinates of that object. These points are extracted near the linear connection of two control points. If the object is strongly curved no coordinates can be found near this connection. In this case the points derived strongly influence the path. With this procedure the surface structure can be well approximated. The author can test the path and modify control points.

Path-specification is not the only interesting detail in the demonstration of an object's shape.

Other important issues are: Which pointers are appropriate? How to move the pointer along the path?

According to our observations, we modelled objects used in videos for anatomy teaching to point at something. These include tweezers, scissors and ballpoints. In addition, we modelled some 3D-arrows. These pointers can be selected via a symbolic name. The movement of a pointer is guided by the current surface normal, that is, the pointer is always perpendicular to the path. The peak of the pointer touches the surface directly. This is in contrast to human pointing gestures, where a gap between the pointer and the object to explain remains. We believe that this gap is due to the fear not to damage the object and is therefore not necessary for our virtual objects.

6 Examples

The left image of Figure 4 shows the start of an animation, while the right image shows modifications to explain a muscle.

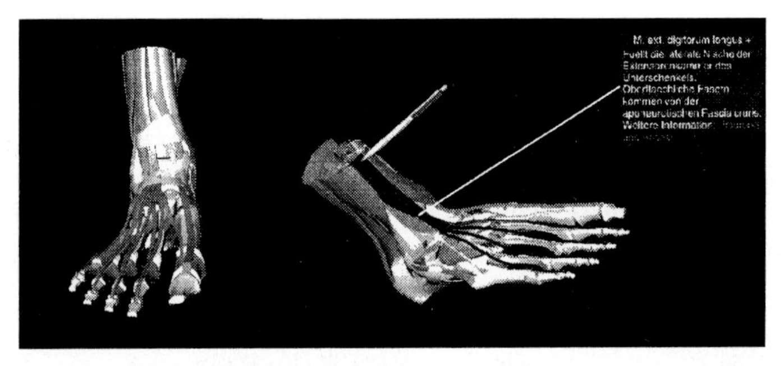

Figure 4: Left image: Original position; Right Image: Modifications to prepare the explanation of the course. The camera position has been adjusted and a ballpoint is directed towards the origin of the emphasized muscle.

Figure 5 presents how the animation continued. The course is explained with arrows, while the ballpoint remains unchanged. An explanation is displayed, which describes the course. Structure information is used to display the labels of related nodes.

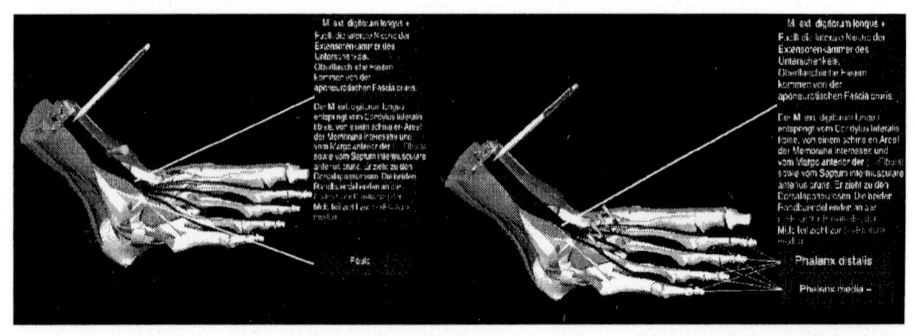

Figure 5: Left image a large arrow demonstrates the upper part. After reaching a branching point, the arrow splits (right image) into several small arrows. As the motion continues, the label of the bone where the muscle starts disappears while others become displayed.

Now let us come back to the questions from Section 3.4: How can an animation – as illustrated by the images of this chapter – be interrupted for interactive exploration? How can interactive exploration be enhanced by film-sequences?

The hypertext links displayed in an explanation can be activated. Such an activation leads to the display of an explanation to this object. Moreover, the viewer is asked whether this object should be explained by an animation, which may result in a sequence as demonstrated in Figure 4 and Figure 5. The rotation can be stopped and the viewer can rotate the explained model freely.

Figure 6 illustrates the answer to the second question. The left image of Figure 6 contains an example from the interactive mode. All muscles are labeled and the user requested an explanation (right part of left image). This causes the other rectangles on this part of the image to be scaled down; their labels are automatically hidden.

The user may rotate the model so that it is clear what is explained. This, however, is a time-consuming trial and error-process. To work more efficiently, the user can start the *explain*-dialog and specify the parameters for a short film (or use the defaults). From this specification an explain-statement of the *IC Language* is interpreted.

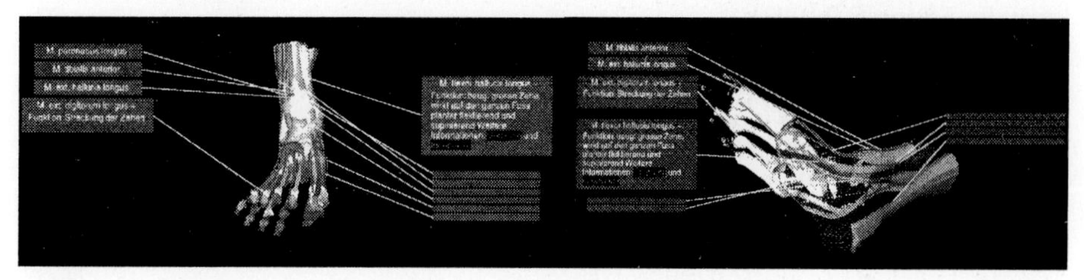

Figure 6: Combination of interactive exploration and animation: Left Image: The user requested an explanation on a muscle (upper right of the left image). After the display of an explanation the user „asks" to transform the model to the optimal viewing direction, rendering a sinew transparent which occludes the muscle.

7 Conclusions and Future Work

This paper describes strategies for illustrating complex 3D-models using a combination of interactive exploration and animation. For the techniques developed geometry data, related visibility information, textual descriptions and information concerning structure are exploited. We developed the IC-Language for specifying an animation. To explain the shape of objects, we developed special pointing techniques.

The ZOOM ILLUSTRATOR focuses on a consistent presentation of rendered images and explaining text. In this paper we extend this approach by introducing an automated adaptation of text attributes to changing parameters of the model display. This allows to transform the model and display text information at the same time. Future work should include "before" - "after" specifications of changes on the graphical part and text presentations.

Our system is based on a huge amount of information concerning the subject to be studied, in this case anatomy. So far this information is collected and updated manually including complex consistency checks. A data base to collect this complex information is desirable as well as facilities to automate data acquisition.

The animation techniques incorporated into the ZOOM ILLUSTRATOR are designed especially for beginners seeking for an overview. Once being familiar with the system, the student composes his or her own combination of interactive and animation features to accomplish the learning task efficiently.

The usability has to be proofed in a field test which has to be conducted in the time to come. Questions to be answered are whether our system covers the main features a student expects and whether our scripting language provides enough flexibility.

References

Dill, J., Bartram, L., Ho, A., Henigman, F. (1994)
"A Continuously Variable Zoom for Navigating Large Hierarchical Networks", *IEEE Conference on Systems, Man and Cybernetics*, November, pp. 386-390

Karp, P., Feiner, S. (1993)
"Automated Presentation Planning of Animation using Task Decomposition with Heuristic Reasoning", *Proc. of Graphics Interface*, Toronto, May, pp. 118-127

Pommert, A., Pflesser, B., Riemer, M., Schiemann, T., Schubert, R., Tiede, U., Höhne, K. H. (1994)
"Advances in Medical Volume Visualization", *Eurographics '94, State of the Art Report*, Eurographics Technical Report, ISSN 1017-4656, pp. 111-139

Preim, B., Ritter, A, Steinicke, G. (1996)
"Gestaltung von Animationen zur Erklärung komplexer Modelle", *Proc. of Fachtagung für Simulation und Animation für Präsentation, Planung und Bildung*, Magdeburg, February, pp. 255-266

Preim, B., Ritter, A., Strothotte, T., Pohle, T., Bartram, L., Forsey, D. R. (1995)
"Consistency of Rendered Images and Their Textual Labels", *Proc. of CompuGraphics,* Alvor, Portugal, December, pp. 201-210

Rüger, M., Preim, B., Ritter, A. (1996)
"Zoom Navigation – Exploring Large Information and Application Spaces", *Proc. of Advanced Visual Interfaces*, Gubbio, Italy, May (to appear)

Sobotta, J. (1988)
Atlas der Anatomie des Menschen, 19. Edition, J. Staubesand (Ed.), Urban & Schwarzenberg, Munich-Vienna-Baltimore

Zeltzer, D. (1990)
"Task-Level Graphical Simulation: Abstraction, Representation and Control", in: Badler, N., Barsky, B., Zeltzer, D. *Making them Move: Mechanics, Control and Animation in Articulated Figures*, pp. 3-33

A Fast Rendering Method Using the Tree Structure of Objects in Virtualized Bronchus Endoscope System

Kensaku MORI[1], Jun-ichi HASEGAWA[2], Jun-ichiro TORIWAKI[1], Hirofumi ANNO[3] and Kazuhiro KATADA[3]

[1] Department of Information Engineering, Graduate School of Engineering, Nagoya University, Furo-cho, Chikusa-ku, Nagoya-shi, 464-01 Japan
E-mail: mori@toriwaki.nuie.nagoya-u.ac.jp
[2] School of Computer and Cognitive Sciences, Chukyo University, 101 Tokodate, Kaizu-cho, Toyotya-shi, 470-03 Japan
[3] School of Health Sciences, Fujita Health University, Dengakugakubo, Kutsukake, Toyoake-shi 470-11 Japan

Abstract. In this report we present a fast rendering method using tree structure of the object and its application to virtualized endoscope system (VBE). VBE enables users to navigate and observe inside the bronchus extracted from 3-D X-ray CT images in real time. Faster rendering is necessary to improve operationability of VBE. The proposed method achieves faster response by rendering only visible branches using the tree structure of the target object (bronchus). This method consists of two steps : (a) extraction of the tree structure of the bronchus by 3-D image processing techniques and (b) dynamic selection of visible branches at the current viewpoint and the view direction. We implemented this method into VBE and evaluated the rendering speed by using real chest X-ray CT images. In the result, improvement of rendering speed was confirmed comparing to our previous system.

1. Introduction

In recent years, the progress of three dimensional imaging equipment enables us to take very precise volumetric images of human body. In the real clinical stage treating large volumes of 3-D images are becoming indispensable. Virtualized endoscope system (VES), which is based on computer graphics and virtual reality techniques, is focused as a new diagnosis method of the target organs [1-7]. This system enables doctors to observe inside the human body constructed from images of each patient.

Our group also has been developing the virtualized bronchus endoscope system (VBE) as one of VES [1,3,4,6]. The most significant feature of our VBE is the interactive operation with real time response. VBE has many functions such as a multiple view window which includes both the outside and the inside view of the bronchus, navigation by real time animation, measurement of feature values of the bronchus, and interactive operations. These functions are all driven by users interactions in interactive mode with real time response. It has also many advantages over the real endoscope as discussed in [6].

In such interactive systems, the response time, especially the rendering speed, is one of the most important elements which determine the system operationability. For example the rendering speed is of critical importance when the user flies through inside the bronchus with observing the bronchus wall. The previous VBE presented in [3] and [4] can render the endoscopic images with the ration of four frames per second by using the triangle surface model and graphics hardware (Triangle surfaces : 66305, Computer : Silicon Graphics Inc. Indigo2 HighIMPACT.). However medical doctors are requesting higher speed for the VBE. A few studies about virtual bronchoscopy have been reported by other research groups [2,5,6,7]. In these researches, however, a lot of endoscopic images were first generated in preprocessing stage and then a movie was made by suitably editing these images. This is because "Volume Rendering" algorithm [8,9], which takes a lot of time to compute one frame, was used for rendering and fast graphics hardware was not used there. There is no implementation of the system which can generate moving sequences of the virtualized endoscopic images in real time with interactive operations except for VBE by authors' group.

Our VBE generates virtualized endoscopic images by rendering triangle surfaces which are constructed from the bronchus area extracted from 3-D X-ray CT images automatically [10,11]. The VBE presents dynamic virtualized bronchus endoscope images by continuously changing the viewpoint and the view direction corresponding to user's operation with fast graphics hardware. Currently the VBE renders all triangle surfaces of the bronchus at each frame. However if it is possible to reduce by an appropriate method the number of triangle surfaces which should be processed in each frame, the system is possible to render endoscopic images faster. In the case of the observation from the viewpoint inside the bronchus by the VBE, the visible area is very limited since the bronchus has tree structure. If we can extract the tree structure of the bronchus, this structure information is applicable to determine parts of the bronchus.

In this paper we propose a method for fast rendering of the virtualized bronchus endoscopic images. This method classifies the surface data of the bronchus into the groups corresponding to each branch. In the stage of interactive rendering, the system presents only the surface data belonging to the branches which are visible from the current viewpoint.

In Section 2 we briefly summarize the basic idea of the proposed method. In Section 3, we describe a processing procedure of the proposed method in detail. Next we implement this method in the virtualized endoscope system and show results of improvement in rendering speed in Section 4. Finally we add brief discussion in Section 5.

2. Overview of fast rendering algorithm

The VBE presented here generates virtualized endoscopic images as follows [1,3,4,6]. First the bronchus area is extracted from 3-D chest X-ray CT images. Second the surface data of the bronchus is generated by Marching Cubes algorithm [12]. Finally this surface data is rendered from an arbitrary viewpoint and a view direction corresponding to user's operation in real time. The triangle surface model is used for rendering. The rendering time of one frame is in proportion to the number of triangle surfaces rendered in a frame. To save rendering time, the number of surfaces included in a target object should be reduced at each frame.

Two major strategies are possible in reducing the number of triangle surfaces.

One is that a group of small triangle surfaces are integrated into one larger triangle surface. This method is called "mesh optimization" [13-17]. Another is that only the visible surfaces are selected in rendering of each frame. The former method may cause deterioration of the image quality and is not suitable for the presentation of medical images, because a medical doctor needs to observe small protrusions and depressions on the wall of the organs. The latter method does not cause deterioration like the former one, but its effect depends on the form of the target organ. To employ this we must develop an efficient procedure which rapidly detects surfaces visible from the current viewpoint and view direction.

When a target object has the tree structure and the inside area of it should be rendered such as in the VBE, the visible branch from one viewpoint is very limited. As an extreme case, for example, when the current viewpoint is located in the right side bronchus and the view direction is towards the bronchus side, the left side bronchus will never be seen. Thus the system does not need to render the left side branches in this situation. The proposed method uses these features for fast rendering. First the tree structure is extracted from the object and all triangle surfaces of the object are classified branch by branch. Second the visible branch corresponding to the current viewpoint and the view direction is automatically determined.

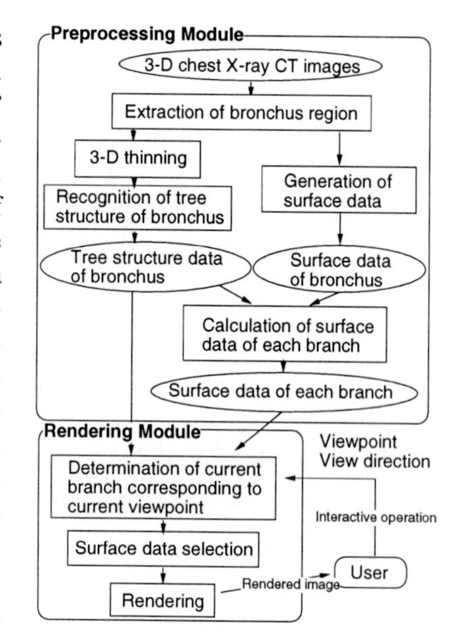

Fig.1 Processing procedure

3. Processing Procedure

3.1 Outline of the procedure

Fig.1 shows the flow of the proposed processing procedure. It consists of two steps; (a) preprocessing and (b) rendering. In the preprocessing step, first we apply the 3-D thinning [18,19] to the 3-D bronchus region in order to extract the medial axis of the bronchus. Second the structure of the bronchus is recognized from the thinned result by computer. This recognized result has the information of branching points and branches. Finally the surface data (= a set of triangular patches) which belongs to each branch is calculated by using the structure recognized above. In the rendering step, the following procedures are performed at each frame. First the current branch which the current viewpoint belongs to is determined automatically. Next the branch which is regarded as visible is selected and the surface of the selected branch is rendered.

3.2 Preprocessing module

(1) Extraction of the bronchus region

The bronchus region is extracted from 3-D chest X-ray CT images by employing the method described in Refs [1,3,4,10,11].

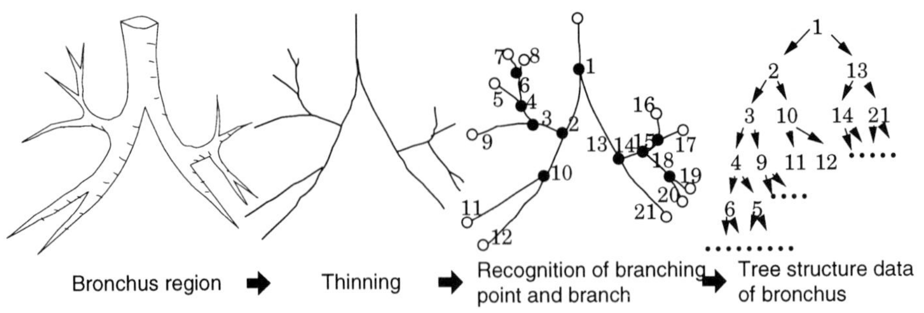

Fig.2 Overview of the recognition of the tree structure of the bronchus

(2) Generation of the surface data of the bronchus

The surface data which represents the bronchus wall is generated from the bronchus region data by the Marching Cubes method [12].

(3) Extraction of the medial axis of the bronchus

The medial axis of the bronchus is extracted from the bronchus region obtained in (1) by employing the 3-D thinning procedure which preserves topology (Fig.2). We employ the thinning algorithm augmented with Euclidean distance transformation [18,19].

(4) Extraction of the tree structure of the bronchus

We extract the tree structure of the bronchus from the thinned result. First the graph representation is derived from the thinned result (Fig.2,3). This graph represents the connective relation between voxels. Each voxel of the thinned line figure is used as the node. The arc of the graph represents the connection between voxels. Second each node is classified into three types, the connecting point (exactly two arcs connected), branching point (more than two arcs meet), and the end point (only one arc is connected) (Fig.3). Each node has the information of its position, nodes connected to it, and the path to the connected nodes on the thinned result. Next all the connecting points are deleted from the graph. Small loops at branching points and small length of branches (which are caused by noise in the original image) are also deleted. Only branching points and the end points are left. The tree structure of the node is obtained from this result by considering the node located at trachea as the root node. By this process, each node is given its parent node number and child node number. Finally the connection structure of each branch is determined from this node structure. Each branch has the information of the branch number of itself, its path (represented by the voxels on the thinned line figure), the parent branch number, and the child branch number. We can easily know the branch number located at one level ahead or behind of the current branch from this tree structure (Fig.3).

(5) Calculation of the surface data which belongs to each branch

The surface data which belongs to each branch is calculated as follows (Fig.4). The minimum distance is calculated between the center of the each triangle patch and the branch path. If the distance is lower than a given threshold value, this patch is assigned

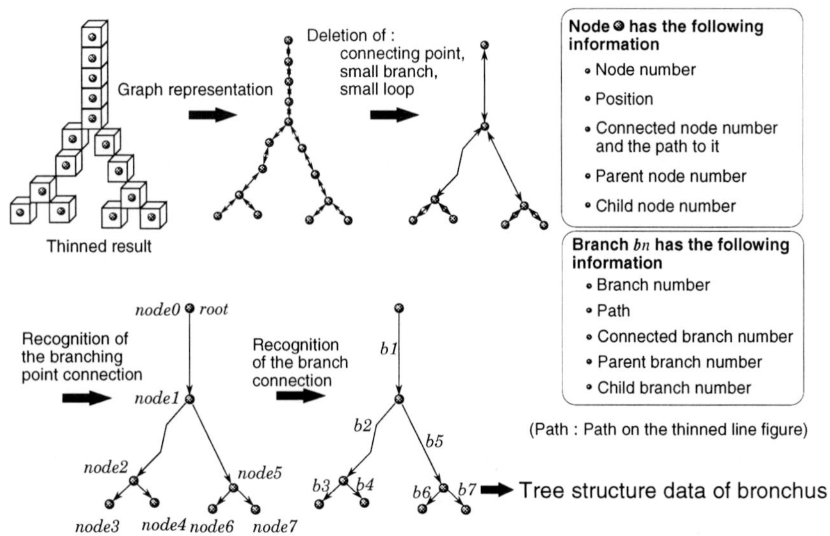

Fig.3 Recognition method of the tree structure of the bronchus region from 3-D thinned result

to the branch. A set of triangle patches assigned to the same branch is regarded as the surface data corresponding to the branch. Duplicate classification (a patch is assigned to more than one branch) is permitted. If there exist some surfaces which do not belong to any branch, those are treated as the surfaces which should be always rendered.

3.3 Rendering Module

(1) Determination of the current branch corresponding to the current viewpoint

The branch corresponding to the current viewpoint is determined by using the branch structure data recognized in the preprocessing module (Fig.5). The distance between the viewpoint and the each branch is calculated. The branch which is closest to the viewpoint is derived as the branch of the current viewpoint. Furthermore the current observation direction, either of <from central to peripheral> or <from peripheral to center>, is decided by the angle between the direction of the current branch and the current view direction. The branch direction is defined as the direction to the end point from the start point of the branch.

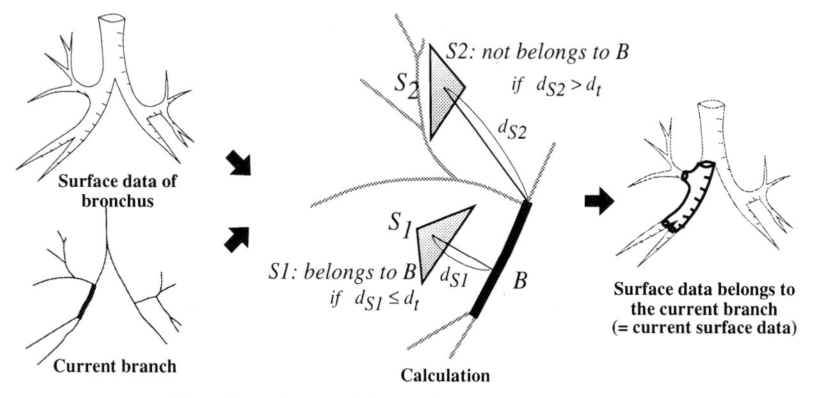

Fig.4 Calculation of the surface data of each branch

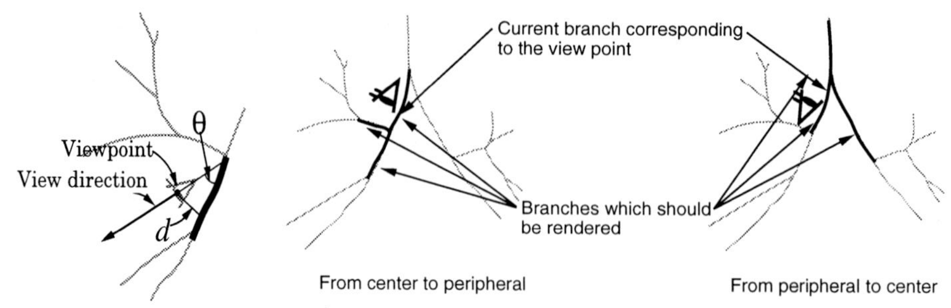

Fig.5 Decision of the current branch corresponding to the current viewpoint and the current direction in the tree structure

Fig.6 Selection of the branch which should be rendered. The selection pattern is changed corresponding to the current view direction.

(2) Selection of the current surface data

The surface data corresponding to the current branch (the current surface data) is selected as shown in Fig.6. This selection is performed for each branch. If the view direction is <from center to peripheral>, two branches, the current branch and the one with the level next to it are adopted. If the view direction is <from peripheral to center>, three branches, the current branch, the branch higher by one level and the one lower by one level are extracted. Only the surface data which belongs to the selected branches are rendered.

(3) Rendering

The system renders only the surface data selected by the above-stated procedure and the surfaces which are not classified into any branches. Thus the fast rendering is realized, since the above process means that only the visible surface from the current viewpoint and the current view direction is selected.

4. Experiments

The proposed rendering method was incorporated into the VBE [4,6]. Seven cases of real 3-D chest X-ray CT images were processed by this system. Original images were taken by helical (spiral) CT scanner. The image size is 512x512, the number of slices is from 50 to 118, the width of X-ray beam is 2mm or 5mm, and the reconstruction pitch is 1mm or 2mm. Parts of original slices are shown in Fig.7. Fig.8 presents the result of the extraction of the bronchus region, its thinned result, and the result of the branch structure recognition. Fig.9 demonstrates the result of the surface data selection when the viewpoint and the view direction were changed inside the bronchus by user's operation. It is known from this result that the surface data was appropriately selected corresponding to the current viewpoint and the view direction, and that rendered endoscopic images are of satisfying quality.

Furthermore we evaluated the rendering performance of the proposed method and the previous method. Fig.10(a) is a plot of the frame rate in rendering an image sequence along an observation path (track of the viewpoint in the inside space of the bronchus). This path starts from the trachea, passes through the appropriate peripherals of both the right and the left bronchus, and returns to the trachea. Fig.10(b) gives the change in the number of the selected triangle surfaces. These results confirm that the rendering performance is significantly improved by the proposed method. The maximum improvement in performance is fifteen times, the minimum is two times and the average is four times that of the previous method. The computer used here is Silicon Graphics Inc. Indigo2 HighIMPACT (R4400 250MHz, MainMemory 256MB).

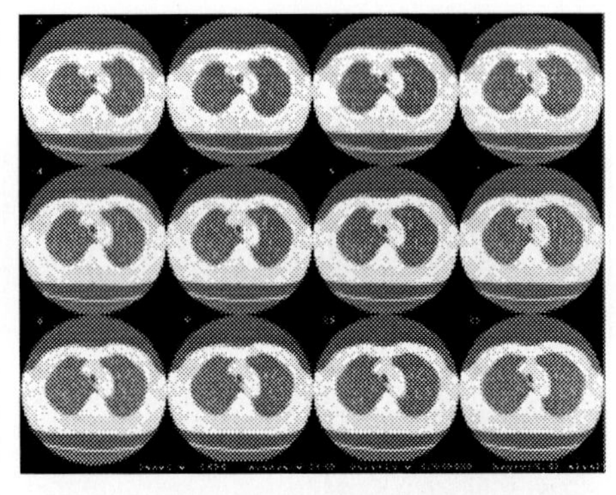

Fig.7 An example of an original image

5. Discussion

(1) Effectiveness of the proposed method : From the results shown in Section 4, we can see that the proposed method improves the rendering speed without deterioration of the image quality. This method is characterized by the point that the tree structure of the bronchus is automatically extracted and is effectively used for fast rendering. Generally the number of polygons contained in the surface data is extremely large in visualization of medical 3-D images. The deterioration of the quality in rendered images is not desired because presented images are used for the diagnosis of human body. Therefore the proposed method of fast rendering is also very effective in these points.
(2) Advantages of the fast rendering : The response time of the interactive system is a critical factor in general. In particular the response time determines operationabilty and practical usefulness of the system in such a case of the virtualized bronchus endoscope system which performs all of the rendering, observation, navigation, operation, and measurement in real time. The proposed method is regarded as one of the basic techniques for improvement of the system operationability. The rendering speed becomes much more important in the virtual reality environment where the rendering is performed according to the user's posture with the head mounted display and the three dimensional position sensor. The proposed method is also effective in that situation.

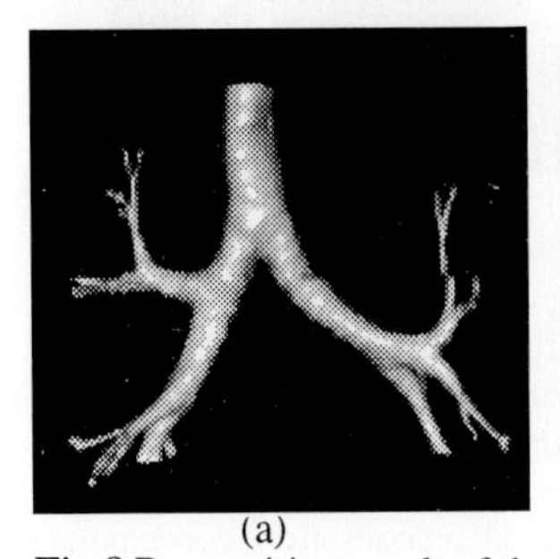

(a)

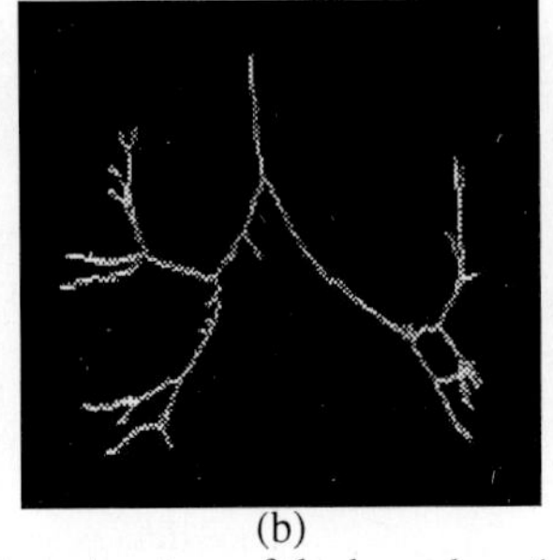

(b)

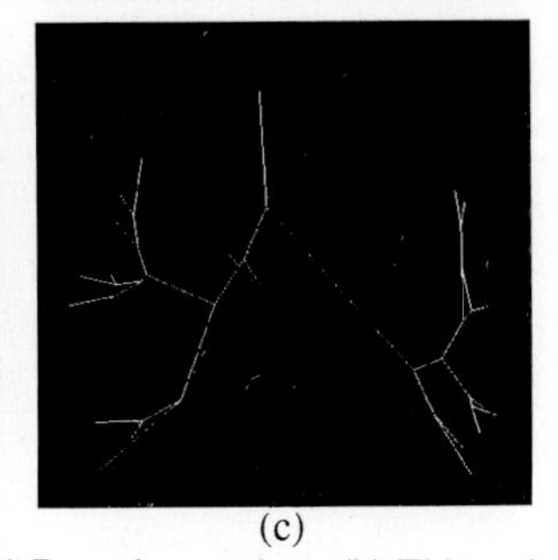

(c)

Fig.8 Recognition result of the tree structure of the bronchus (a) Bronchus region (b) Thinned result, and (c) Structured result

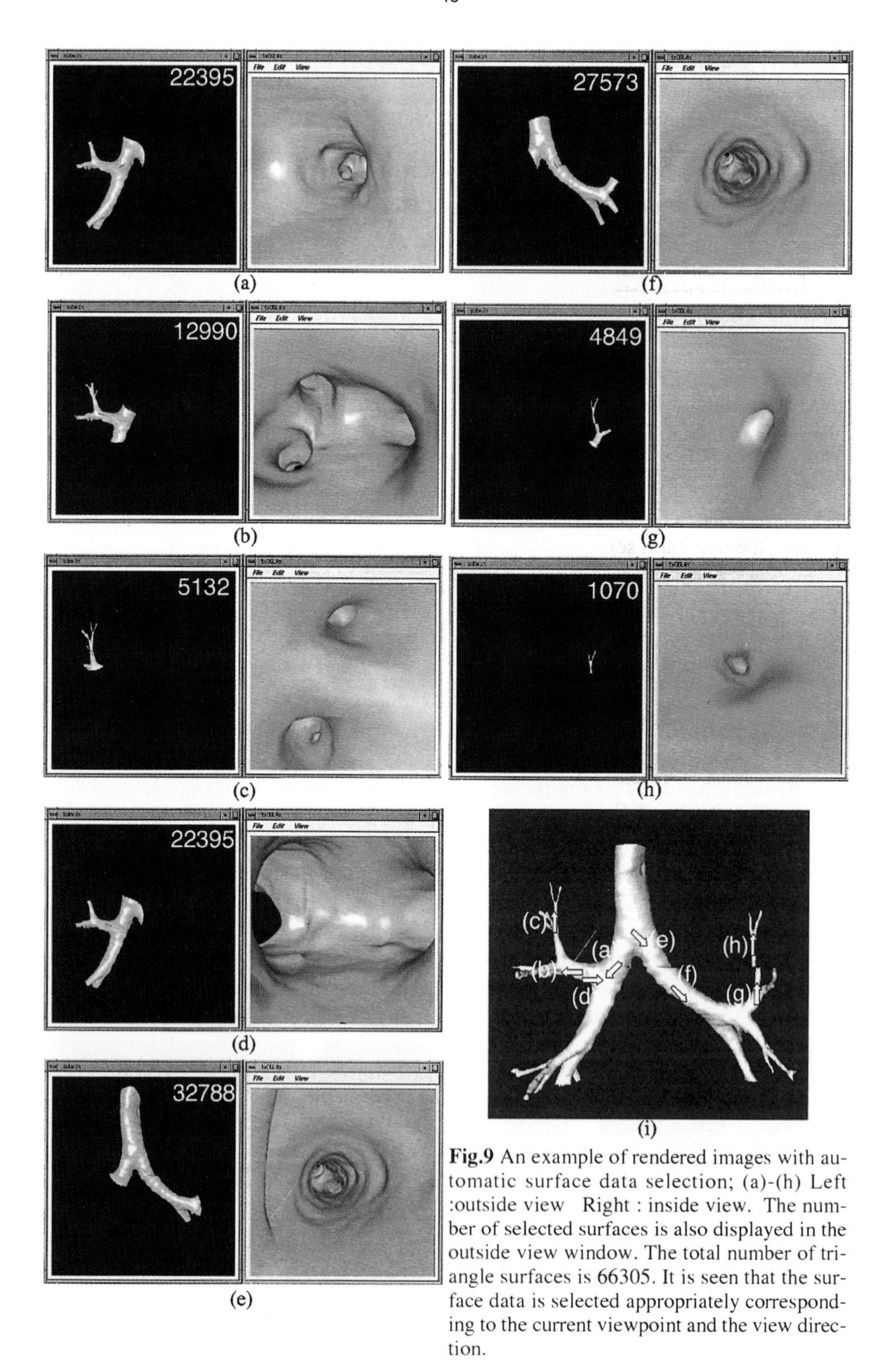

Fig.9 An example of rendered images with automatic surface data selection; (a)-(h) Left :outside view Right : inside view. The number of selected surfaces is also displayed in the outside view window. The total number of triangle surfaces is 66305. It is seen that the surface data is selected appropriately corresponding to the current viewpoint and the view direction.

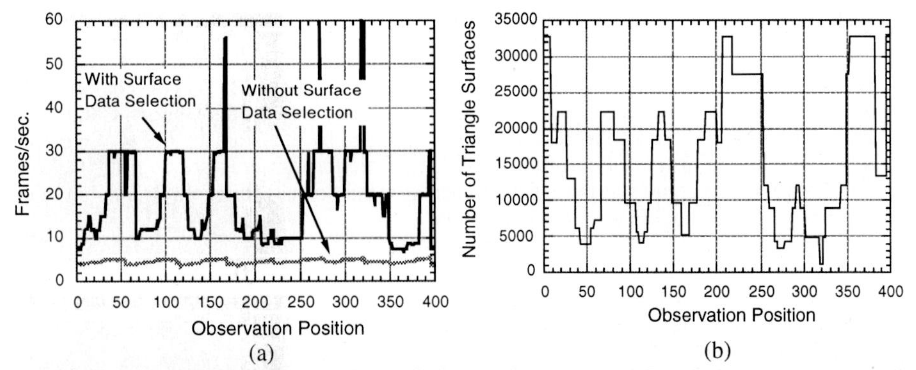

Fig.10 The result of improvement in rendering performance. (a) Rendering rate per second. Horizontal axis means the observation point number along the observation path. Vertical axis means the number of frames which are rendered in a second. (b) The number of selected surface data at each observation point. The total number of triangle surfaces is 66305.

(3) Improvement in the strategy of surface data selection : This method selects the surface to be rendered by each branch. When the view point passes through the branching point, the unnatural selection of the surface data may occur. Further improvement is desired in the way of selecting the surface data in the neighborhood of the branching point.

(4) Combination of another method : Because only the needed branch is rendered for saving computation time, there is unnatural phenomenon that the branch which was rendered in the previous frame suddenly disappears with the movement of the viewpoint. To eliminate this type of unnaturalness, we are now developing the method which renders the branch with the two level ahead by using rough surface data. This rough data consists of relatively small number of the triangle surfaces. This is acceptable because the precise data is not needed for rendering the branch far from the current viewpoint. We expect more natural rendering by this method.

(5) Use of the result of classification of the bronchus branch : The classified data of the bronchus branch is applicable to other purposes, since the classified branch corresponds to the anatomical bronchus branch. For example, each branch name used in anatomy can be also determined by using these classification results and the tree structure data recognized in preprocessing stage. In fact, the branch is named by the explicit rule in anatomy, and such names are found easily by using these data. Implementing this method in VBE, it is possible to display the anatomical name of the currently observing branch in the VBE image.

6. Conclusion

This paper proposed a method for fast rendering of the object which has tree structure and its application to VBE. We confirmed the significant improvement in the rendering speed experimentally. Future problems are the evaluation of the rendering speed with larger number of clinical cases and the improvement of the surface data selection algorithm in the neighborhood of the branching points.

Acknowledgments

Authors thank to Dr. H.Natori (Sapporo Medical College, Japan), Dr.H.Takabatake (Hokkaido Keiaikai Minami-ichijo Hospital, Japan) for their useful suggestions, Dr.M.Ikeda (Nagoya Univ., Japan) for providing one of the original CT images, and colleagues of author's laboratory for useful discussion. Parts of this research were supported by the Grant-In-Aid for Scientific Research from the Ministry of Education, the Grant-In-Aid for Cancer Research from the Ministry of Health and Welfare, Japanese Government and the Science Research Promotion Fund from Japan Private School Promotion Foundation.

References

1. K.Mori, J.Hasegawa, J.Toriwaki et al. : A method to extract pipe structured components in three dimensional medical images and simulation of bronchus endoscope image, Proc. of 3D Image Conf. '94 pp.269-274 (1994-07)
2. D.J.Vinning, R.Y.Shitrin, E.F.Haponik et al. : Virtual Bronchoscopy, Radiology, **193(P)**, p.261, Supplement to Radiology (RSNA Scientific Program) (1994-11)
3. K.Mori, J.Hasegawa, J.Toriwaki et al. : Automated Extraction and Visualization of Bronchus from 3D CT Images of Lung, in N.Ayache ed., Proc.of CVRMed'95, Lecture Notes in Computer Science, **905**, pp.542-548, Springer-Verlag, Heiderberg (1995-04)
4. K.Mori, J.Hasegawa, J.Toriwaki et al. : Bronchus Endoscope Simulation System Based on Three Dimensional X-ray CT Images (Virtualized Bronchus Endoscope System), Japanese Journal of Medical Electronics and Biological Engineering, **33**, 4, pp.343-351 (1995-12)
5. B.Geiger and R.Kikinis : Simulation of Endoscopy, in N.Ayache ed., Proc.of CVRMed'95, Lecture Notes in Computer Science, **905**, pp.542-548, Springer-Verlag, Heiderberg (1995-04)
6. K.Mori, A.Urano, J.Hasegawa et al. : Virtualized Endoscope System - an Application of Virtual Reality Technology to Diagnostic Aid, IEICE* Trans. of Information and System, **E79-D** (in printing)
7. G.Rubin, C.Beaulieu, V.Argiro et al. : Perspective Volume rendering of CT and MR Images : Applications for Endoscopic Imaging, Radiology, **199**, 2, pp.321-330 (1996-5)
8. M.Levoy : Volume Rendering -Display of Surfaces from Volume Data, IEEE Computer Graphics & Applications, **8**, 3, pp.29-37 (1988-05)
9. R.A.Drebin. : Volume Rendering, Computer Graphics, **22**, 4, pp.65-74 (1988-08)
10. K.Mori, J.Hasegawa, J.Toriwaki et al. : Automated Extraction of Bronchus Area from Three Dimensional X-ray CT Images, IEICE* Technical Report, **PRU93-149** (1994-03)
11. K.Mori, J.Hasegawa, J.Toriwaki et al. : Recognition of Bronchus in Three Dimensional X-ray CT Imgaes with Applications to Virtualized Bronchoscopy System, Proc. of 13th ICPR (1996-08, in printing)
12. W.E.Lorensen and H.E.Cline : Marching Cubes - A High Resolution 3D Surface Construction Algorithm, Computer Graphics, **21**, 4, pp.163-169 (1987-07)
13. G.Turk : Re-Tiling Polygonal Surfaces, Computer Graphics, **26**, 2, pp.55-64 (1992-07)
14. W.J.Schroeder, J.Zarge, and W.E.Lorensen : Decimation of Triangle Meshes, Computer Graphics, **26**, 2, pp.65-70 (1992-07)
15. H.Hoppe, T.DeRose, T.Duchamp et al.: Mesh Optimization, Computer Graphics, **27**, 2, pp.19-26 (1993-07)
16. A.Gueziec : Surface Simplification with Variable Tolerance, Proc. of MRCAS'95, pp.132-139 (1995-11)
17. A.Kalvin, R.Taylor : "Superfaces : Polygonal Mesh Simplification with Bounded Error", IEEE Computer Graphics and Applications, **16**, 3, pp.64-77 (1996-05)
18. T.Saito and J.Toriwaki : A Sequential Thinning Algorithm for Three Dimensional Digital Pictures Using the Euclidean Distance Transformation, Proc. of the 8th Scandinavian Conf. on Image Analysis, pp.507-516 (1995-07)
19. T.Saito, K.Mori, and J.Toriwaki : A Sequential Thinning Algorithm for Three Dimensional Digital Pictures Using the Euclidean Distance Transformation, Trans. of IEICE*, **J79D-II** (1996, in printing)

(*IEICE : Japanese Institute of Electronics, Information and Communication Engineering)

An Optimal Ray Traversal Scheme for Visualizing Colossal Medical Volumes

Asish Law and Roni Yagel

Department of Computer and Information Science
The Ohio State University
Columbus, Ohio

Abstract. Modern computers are unable to store in main memory the complete data of high resolution medical images. Even on secondary memory (disk), such large datasets are sometimes stored in a compressed form. At rendering time, parts of the volume are requested by the rendering algorithm and are loaded from disk. If one is not careful, the same regions may be (decompressed and) loaded to memory several times. Instead, a coherent algorithm should be designed that minimizes this thrashing and optimizes the time and effort spent to (uncompress and) load the volume. We present an algorithm that divides the volume into cubic cells, each (compressed and) stored on disk, in contrast to the more common slice-based storage. At rendering time, each cell is allocated a queue of rays. For a sequence of images, all rays are spawned and queued at the cells they intersect first. Cells are loaded, one at a time, in front-to-back (FTB) order. A loaded cell is rendered by all rays found in its queue. We analyze the algorithm in detail and demonstrate its advantages over existing ray casting volume rendering methods.

1 Introduction

There has been a growing interest in visualizing extremely large medical databases, one classic example being the Visible Human, comprising of more than 30 gigabytes. This database was created by the National Library of Medicine's Visible Human Project, with the intention of creating anatomical atlases. This voxelized human provides a new level of educational value to anatomical visualization. Real time or interactive software rendering of such large databases still looks like an unattainable goal. A more immediate requirement is to reduce the rendering time as far as possible, so that the frame generation time decreases. Several attempts have been made to "conquer" the visible human, some notable ones been described in [5][6][10][12][14]. In [6], an attempt has been made to develop a comprehensive virtual environment, including efficient segmentation and realistic ray tracing of the volume. In [14] the Visual Human is used as a basis for a comprehensive medical atlas. Palmer et al., [12] have gained speed by implementing a parallel volume renderer on clusters of shared-memory multiprocessors. Apart from using parallel computers, some researchers have taken a different route to improve the performance of the rendering algorithm by suggesting coherent algorithms, as in [1][4].

Similarly, direct rendering of compressed volumes have also gained attention [3][7][11][15].

We present an algorithm that divides the volume into cubic cells, each (compressed and) stored on disk, in contrast to the more common slice-based storage. At rendering time, each cell is allocated a queue of rays. For a sequence of images, the front-to-back (FTB) order of cells is determined. For all frames that share the same FTB order, all rays are spawned and queued at the cells they intersect first. Cells are loaded, one at a time, in FTB order. A loaded cell is rendered by all rays found in its queue. The end result is that the costly operation of decompressing and loading a cell into memory is done once for all the frames that share the same FTB order.

The coherent ray casting method proposed here can be used in three ways. First, it can be used to render compressed volumes by explicitly decompressing small blocks of the volume (cells in our case). The images produced in this manner are exactly the same as would be produced by a direct volume renderer. The algorithm is independent of the compression technique used, thus allowing higher compression ratios. The only assumption is that the compression is lossless [2] and that the corresponding decompression routine is available at render-time. Second, it can be used in conjunction with the methods for direct rendering of compressed volumes (e.g., [11]) to exploit coherency to a far higher degree. The image quality will be the same as guaranteed by the respective algorithms. Finally, the algorithm can simply be used advantageously to improve cache efficiency, for volumes that do fit in main memory.

In the next section, we describe our method in more detail and some optimizations we implemented (Section 2.2). We also propose an extension (Section 2.3) that will enable us to render a cell exactly once even when the images do not share the same FTB order. In Section 3, we thoroughly analyze our current implementation and suggest our conclusions in Section 4.

2 Method

Volumes are compressed because they do not fit in secondary or primary memory. Efficient compression techniques [15] and direct volume visualization of compressed volumes [3][7][11] have been suggested. These algorithms generally compromise with accuracy, but with an *a priori* knowledge of the behavior of the data, very efficient schemes can be developed. In this paper, we take a different approach to visualize compressed volumes. We believe that the pain and effort taken in acquiring high resolution and high quality data, such in the Visible Human Project, cannot be compromised at rendering time by employing lossy compression techniques. Our algorithm preserves the accuracy of the direct volume renderer by employing lossless compression [2] (or no compression) while trying to optimize rendering time.

The idea of thrashless volume rendering was first proposed for multicomputers by Westermann and Augustin [16], and later extended to multi-frame thrashless ray-casting by Law and Yagel in [8]. In this paper, we extend the scope of this novel approach to the visualization of compressed or colossal volumes on uniprocessor machines. The algorithm described in [8] preserves the thrashless property across all the frames that share the same FTB order. In the next section, we briefly explain the idea of multi-frame thrashless volume rendering, followed by a proposal (in Section 2.3) to extend the method to work for arbitrary sets of frames, i.e., frames that do not share the same FTB order of cells. The net effect of this proposed feature will make the algorithm limited only by the memory required to store the frames since every cell will be decompressed and fetched exactly once for *all* images.

The only requirement of our algorithm is that volume is divided into cells that can fit in the machine's main memory. Each cell is optionally compressed and stored onto a disk. We assume that the corresponding decompression routine is available at rendering time. Compressing cells instead of the whole volume may lead to lower compression ratios [2]. Alternatively, very large volumes can also be divided and stored, as uncompressed cells, on distributed remote disks. The algorithm basically renders the original data, but uses an efficient ray traversal scheme to ensure that each cell is decompressed exactly once for a number of frames. It takes advantage of coherency between slowly changing frames to eliminate thrashing. In Section 2.3, we propose an extension to this method, so that the thrashless property is maintained even for arbitrary jumps of the screen during the animation.

2.1 Multi-Frame Thrashless Volume Rendering Revisited

In this section, we briefly describe the idea of multi-frame thrashless ray-casting. Refer to [8] for a detailed discussion. During preprocessing, the volume is divided into equal-sized cells and stored on a remote disk, possibly in a compressed form. The size of a cell is fixed at the size of the main memory. Before rendering begins, the list of cells is ordered in a front-to-back (FTB) manner depending on the position and orientation of the screen. This list is referred to as FTBL (*Front-To-Back-List*). For example, Figure 1 shows a 24^2 2D raster divided into 36 2D cells of size 4^2 voxels each. The number in each cell indicates its position in the FTBL for any screen position in region I, when viewing towards the center of the volume. We also determine the first cell entered by each ray and push the ray into the queue associated with that cell.

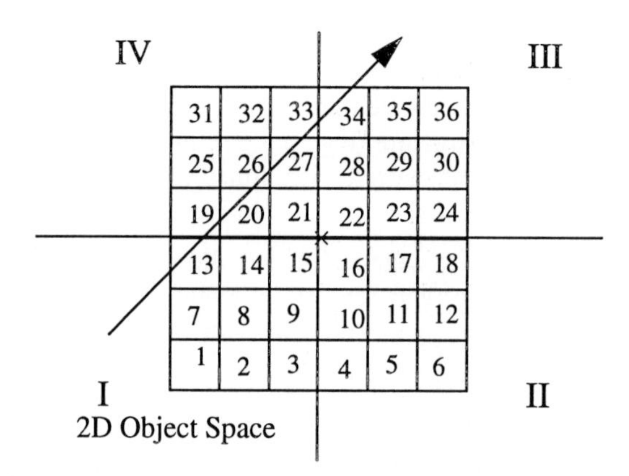

FIGURE 1. For all viewing positions in region I, and when viewed towards the center of the volume, the FTB order of the cells are as shown. ╳ denotes the center of the volume. A ray is also shown which traverses cells 13, 19, 20, 26, 27, 33, 34, in order.

When rendering begins, the first cell in FTB order is (decompressed and) fetched from disk. This cell is referred to as the *active cell*. Since the cell size is the same as the size of main memory, only one cell can be active at a time. Also, as the cells are processed in FTB order, a cell becomes active at most once during the generation of a frame. All the rays associated with the queue of this active cell are advanced until they exit the cell. These rays are then dequeued from the currently active cell and, in turn, are queued into the list of the respective cells they enter. Figure 1 shows an example of a ray which is initially queued into the list of cell 13, then queued and dequeued from cells 19, 20, 26, 27, 33, 34 (in order) as these cells become active and inactive. Rays which either exit the volume or accumulate enough opacity during their traversal, are considered done, and are not queued into the list of any cell. Once all the rays in a cell's queue are traced, the active cell is removed from main memory, and the next cell in FTB order is brought in. In this manner, the rest of the cells in the volume are (decompressed and) fetched from disk one at a time and all the rays associated with the fetched cell (now the active cell) are advanced accordingly.

The algorithm as described so far is thrashless within a single frame. We can extend the method so that such coherency can be exploited across a number of *similar* frames also. By similar frames, we mean those screen positions for which the FTB order remains the same. The algorithm can now be followed for updating all such frames in the same pass. A frame number is attached to each ray to be processed. All the rays for all frames which are associated with the currently active cell, are advanced through the cell before the active cell is given up. This provides an algorithm which is thrashless across several frames. The set of all the frames with the same FTBL is referred to as a *phase*.

2.2 Enhancements

Before proposing a couple of major extensions to the original algorithm, we describe a few minor enhancements we have incorporated for tuning it to perform well with compressed volumes. As a preprocessing step, we break the volume into cells, and each cell is marked empty if none of the voxels in it are occupied. This allows us to totally skip all the empty cells (not even decompress or read) during rendering. For even better performance, the empty cells can be combined during pre-processing and the whole volume may be put into a hierarchical structure (e.g., octree). In addition, early ray-termination for opaque volumes saves the decompression times of cells lying totally behind an opaque object.

In order to save ray and queue storage space, we allocate queues only for non-empty cells. Therefore, when a ray intersects (enters) an empty cell, we look for the next cell rather then queue it there. This saves some time in pushing and popping rays from queues but, more importantly, rays that eventually hit the background go through a series a ray-cell intersection calculations without ever being allocated memory and without ever being queued.

2.3 Extension to Arbitrary Frame Animation

The method of multi-frame thrashless ray-casting described above can be extended to work for any arbitrary frame sequence with a minor extension. Firstly, we realize that the FTB order of cells, which remains unaltered for a set of frames, is also the BTF (back-to-front) order for a different set of frames. As a 2D example, the order of cells shown in

Figure 1 conforms with the FTB order for all screen positions in region I, while looking towards the center of the volume. Similarly, the same order is maintained as a BTF order for all screen positions in region III, while looking towards the center of the volume. The drawback of the latter is that early-ray termination can no longer be applied, and so all the cells in the volume have to traversed. Working with colossal or compressed volumes, it may be worthwhile to stick to this drawback than to decompress a cell more than once.

For parallel viewing of a 3D volume, there are eight FTB orders of cells, one for each octant. With the above extension, we can divide the stream of frames into four sets, frames with same FTB or BTF falling in the same set. This reduces the method to just four phases, implying that a cell in the volume would be decompressed at most four times for generating all the frames in any animation sequence.

Now, we propose another extension which reduces the number of times a cell is decompressed to exactly once, irrespective of the order of frames. It is evident from the description of the original rayfront method that, at any time, each ray maintains exactly one segment of traversed volume, starting from the entering point in the volume to the current sampling point along the ray. This claim holds as long as cells are processed in an FTB or a BTF order.

We claim that if two segments of each ray are maintained, then there is a thrashless order for any arbitrary set of frames. This implies that for a single order of cells (which is not necessarily FTB), all the frames in the animation can be generated by decompressing a cell exactly once in the whole process.

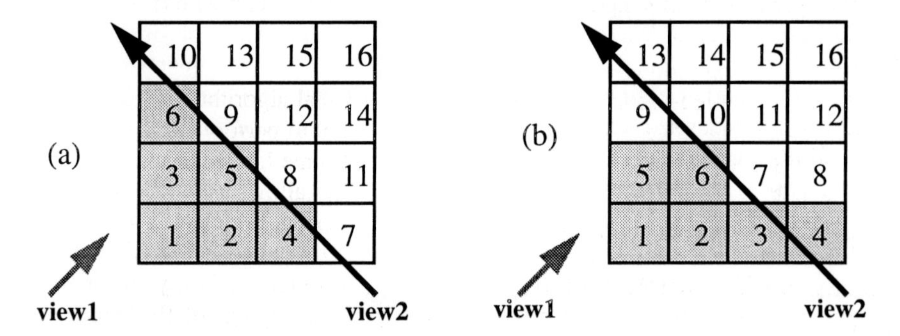

FIGURE 2. (a) An FTB order for the case where the eye in the left bottom side, at 'view1' (gray arrow). That order requires the maintenance of arbitrary number of ray segments for some other eye positions such as 'view2' (black arrow). (b) An FTB order that requires at most two ray segments for any ray orientation.

Let us consider the FTB order as shown in Figure 2a. When some cells (i.e. 1-6 in Figure 2a) have been rendered for 'view1', the image from some 'view2' will have to save the information of several rays segments (e.g., in Figure 2a, for cell 4, 5, and 6) so that when other cells (8 and 9) are rendered, their results could be properly composited in FTB order. However, there is one order, the one that scans row by row and plane by plane, as shown in Figure 2b, that will require maintaining at most two ray segments for each ray (e.g., in Figure 2a, one segment for cells 3 and 4, and one segment for cell 6). We call this method *ZZ-buffer* to portray the correspondence between Z-buffer and ray-casting.

In normal ray casting, a single segment of the ray is maintained, and so is in Z-buffer. In our method, two segments are maintained corresponding to two Z values in a ZZ-buffer.

3 Results

The algorithm was implemented on a single 90 MHz R8000 processor of Silicon Graphics Power Challenge. Our machine, located in The Ohio Supercomputer Center, has 16 processors, 2 Gbytes main memory (8-way interleaved), and 4 Mbytes secondary cache. In this section, we thoroughly investigate all aspects that influence the performance and behavior of our algorithm.

3.1 Timings for Non-Compressed Volumes

We experimented with various types of objects and gathered timings by varying a number of parameters: volume size, cell size, opacity, illumination vs. pure compositing, and most importantly, the number of frames across which the thrashless property was maintained. The specifications for six different volumes are given in Table 1. As means of comparison, we have also shown the raw rendering times (i.e., times for each of these datasets. "Simple-128" is a 128^3 volume consisting of two almost transparent spheres and an almost transparent cube. "SOD-128" is another 128^3 volume consisting of electron densities, and was obtained from UNC Chapel Hill volume repository. The volumes "Capsid-256" and "Capsid-512" each consist of 252 (almost transparent) spheres. The spheres are organized in the shape of the molecular structure of a capsid, taking the shape of a dodecahedron. "Head-256" is a 256^3 volume obtained from CT-scan also taken from the UNC Chapel Hill volume repository. "VH" is the cryosection of the head, taken from the Visible Human dataset. We took only 250 slices of the head which originally took a total of 1.3 Gbytes. The images were cropped to size 600x700, and quantized to one color band (grayscale), yielding a 113Mb dataset. Since we save our data in a compressed form large data sets such as Capsid-512 and VH occupy fraction of disk space – 1MB and 16Mb, respectively.

Table 1 shows the raw times for rendering non-compressed volumes, as signified by the almost insignificant time taken for reading in the cells.

TABLE 1. The volumes, along with reading, rendering, and total times (in seconds).

	Volume Size (in Mbytes)	Cell Size	Opacity	Screen Size	Render Time (secs.)	Read Time (secs.)	Total Time (secs.)
Simple-128	2	8^3	0.01	181^2	3.57	0.11	3.68
SOD-128	2	8^3	1.00	181^2	3.08	0.19	3.27
Capsid-256	16	16^3	0.01	362^2	17.84	0.16	18.00
Head-256	16	16^3	1.00	362^2	12.01	0.10	12.11
Capsid-512	128	32^3	0.01	724^2	49.74	0.07	49.81
VH	113	64^3	1.00	700^2	78.00	1.12	79.12

3.2 Effect of Cell Size

As a preprocessing step, the volume is divided into cells. Each cell is marked empty if none of the voxels in it is occupied. Needless to say, a smaller cell size will result in more empty cells in the volume and vice versa. The advantage of having smaller cell size is that a better portion of the empty space can be skipped. On the other hand, extraneous computation is involved with smaller cells, offsetting some of its advantages. Entry and exit points have to be calculated for each cell along with the extra time needed for reading a larger number of cells one at a time. Most of the volumes show optimal timings in the vicinity of cell size 16^3 or 32^3.

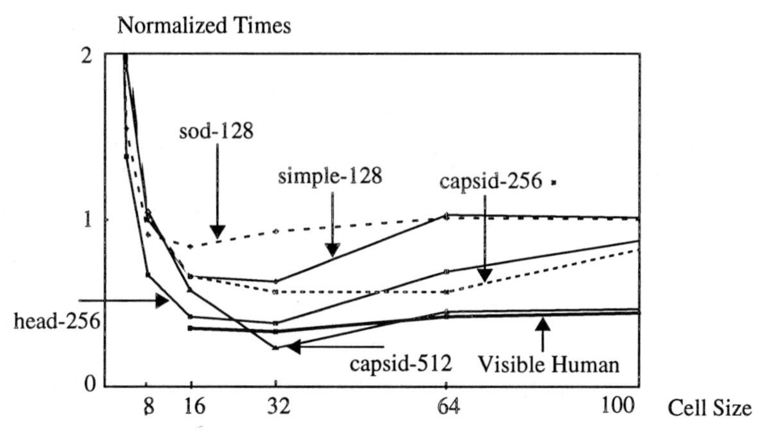

FIGURE 3. Effect of cell size for different datasets.

Table 2 shows the timings obtained for various cell sizes. In Figure 3, the times are normalized with respect to the time taken to render the complete volume as a single cell, i.e., a cell with the size of the volume itself. For large datasets we did not measure for small cell sizes due the immense number of files required.

TABLE 2. Total times (reading + rendering) in seconds, as a function of Cell Size. The minimum times are shown in bold.

	2^3	4^3	8^3	16^3	32^3	64^3	128^3	256^3	512^3
Simple-128	20.3	7.1	3.7	2.4	**2.3**	3.8	3.7	--	--
SOD-128	14.5	5.5	3.3	**3.0**	3.3	3.6	3.6	--	--
Capsid-256	164.5	55.4	27.6	18.0	**15.4**	15.4	27.6	27.1	--
Head-256	112.5	39.7	19.3	12.1	**10.8**	19.9	29.5	28.8	--
Capsid-512	--	--	230.5	127.6	**49.8**	97.6	106	208	218.7
VH	--	--	--	65.9	**62.9**	79.1	86.1	85.2	163.5

3.3 Cost of Overheads

If one has access to memory that can store the whole dataset, then there is no need for the maintenance and management of cells. To verify the additional cost of this overhead, we compared the performance of our algorithm with a ray tracer that stores all the data in the memory. Table 3 shows the cost of our algorithm compared with a normal ray caster. The increase in times will be compensated by the amount of time saved when a number of frames are generated in each phase (see Section 3.4). A phase includes all the frames for which a cell is decompressed only once. The tides change when decompression times are also included in the total animation time. With decreasing size of memory, or equivalently, with increasing volume size, our method will show much better speedup than normal ray casting of compressed volumes. The average frame generation times are also comparable. The number of cells used for timings in the second column of Table 3 are the same as used in Table 1. From Table 3, we see that our algorithm closes on the performance of a normal ray-caster for larger volumes or for larger cells.

TABLE 3. Degradation factor of our algorithm as compared to a normal ray-caster

	Time with volume as a single cell (normal ray-casting)	Time with several cells (our algorithm)	Degradation Factor
Simple-128	3.69	6.32	1.71
SOD-128	3.58	5.43	1.52
Capsid-256	27.10	36.00	1.33
Head-256	28.78	37.66	1.31
Capsid-512	218.70	242.56	1.11
VH	85.18	87.36	1.03

3.4 Effect of Simultaneous Multi-Frame Rendering of Compressed Volumes

The term *Phase* is used to denote a collection of frames for which the thrashless property is preserved. Table 4 shows the variation in total rendering time for generating 20 frames. All the frames had the same FTBL of cells. Most of the time was spent in decompressing the cells to be read in. This time is amortized over all the frames as the number of frames in a phase (FPP) increases. On the extremes, each cell is decompressed 20 times when FPP is 1, while a cell is decompressed only once when FPP is 20. The graph in Figure 4 shows the speedup gained for different volumes for various FPPs between 1 and 20. The occupancy of the SOD volume is higher than the other volumes, signifying a higher number of non-empty cells, thus leading to more time spent for decompressing. This is the reason for the higher gain for this densely occupied volume. For higher density volumes, even higher speedup can be expected. For sparser volumes, the speedups is lower.

The speedups shown in Figure 4 are relative to the case where no thrashing happens whatsoever, i.e., for the case when FPP equals 20. For example, the rendering for sod-128 is about 9 times slower with an FPP of 1 compared to an FPP of 20. It must be stressed here that even with FPP of 1, the thrashless property is preserved across individual frames. Normal ray casting algorithms, when applied to such large datasets, are unable to maintain the thrashless property even across rays. It will be highly impractical to use normal ray casting with these datasets. Our algorithm will show speedups of enormous magnitude when compared with normal ray casting, especially in cases when main memory (cache) is extremely limited compared with the size of the complete volume..

TABLE 4. Average rendering times (decompressing + reading + rendering) in seconds, as a function of number of frames in a phase. Total number of frames = 20

	1	2	4	5	10	20
Simple	33.31	19.02	11.85	10.41	7.55	6.13
SOD	89.67	48.19	27.24	23.03	14.48	10.25
Capsid-256	69.63	46.89	35.34	33.02	28.40	25.90
Head-256	43.74	29.60	22.53	21.09	18.23	16.72
Capsid-512	92.60	76.10	67.38	65.47	61.33	58.87
VH	262.40	175.36	131.50	122.65	104.68	95.52

4 Discussion

In this paper we have presented a method for efficiently handling the rendering of colossal datasets. Our method is based on subdivision into cells. At rendering time, cells are brought, one at a time, to main memory and rendered from several viewpoints, so that a sequence of images can be generated without loading the same cell more than once. The only limit is the memory required to save the multiple frames. This is rather demanding since each frame is a 2D array of rays (each ray requesting ~40 bytes) rather the an RGBα tuples (4 bytes).

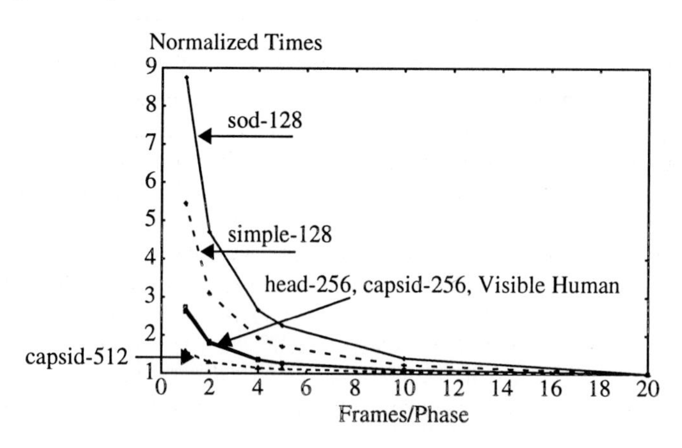

FIGURE 4. Speedups achieved for different volumes as FPP is varied between 1 and 20. Total number of frames remains the same at 20.

The main advantages of our approach is that it exploits the benefits of both object-order methods (for the FTB order) and image-order methods (for early ray termination and other optimizations). It provides, for the first time, a completely thrashless method for rendering of an arbitrary set of views.

Although we didn't measure cache utilization, we believe that the emphasized localization of memory accesses typical to our method also enhances cache hit ratio. The timings compare well with those reported for direct rendering of compressed volumes; in addition our algorithm does not compromise with accuracy.

Acknowledgments

We thank the Ohio Supercomputer Center for allowing us to use their facilities. This research was partially supported by DARPA BAA 92-36, the National Science Foundation CDA 9413962, US Department of Energy DE-PS07-95ID113339, and General Electrics.

5 References

1. J. Arvo, "Space-Filling Curves and a Measure of Coherence". *Graphics Gems II*, Chapter 1.8, pp. 26-30.
2. J.E. Fowler, and R. Yagel, "Lossless Compression of Volume Data", *Proceedings of The 1994 Symposium on Volume Visualization*, Washington, DC, October 1994, pp. 43-50.
3. M.H. Ghavamnia and X.D. Yang. "Direct Rendering of Laplacian Pyramid Compressed Volume Data". Proceedings of *Visualization 1995*, Atlanta, Georgia, pp. 192-199.
4. S. Green, D. Paddon. "Exploiting Coherence for Multiprocessor Ray Tracing", *IEEE Computer Graphics and Applications* 9, (6), pp. 12-26, November 1989.
5. L. Hong, A. Kaufman, Y.C. Wei, A. Viswambharan, M. Wax, Z. Liang. "3D Virtual Colonoscopy". Proceedings of *Biomedical Visualization '95*, Atlanta, October 30, pp. 26-32.
6. J. Kerr, P. Ratiu, M. Sellberg. "Volume Rendering of Visible Human Data for an Anatomical Virtual Environment". Chapter 44 in *Health Care in the Information Age*, H. Sieburg, S. Weghorst, K. Morgan (eds.), IOS Press and Ohmsha, 1996.
7. R. Knittel. "High-Speed Volume Rendering Using Redundant Block Compression". Proceedings of *Visualization 1995*, Atlanta, Georgia, pp. 176-183.
8. A. Law and R. Yagel. "Multi-Frame Thrashless Ray Casting with Advancing Ray-Front", to be published in Proceedings of *Graphics Interface '96*, Toronto, Canada.
9. M. Levoy. "Efficient Ray Tracing of Volume Data". *ACM Transactions on Graphics 9*, July 1990, pp. 245-261.
10. W.E. Lorensen. "Marching Through the Visible Human". Proceedings of *Visualization 1995*, Atlanta, Georgia, pp. 368-373.
11. P. Ning and L. Hesselink. "Fast Volume Rendering of Compressed Data". Proceedings of *Visualization 1993*, pp. 11-18.
12. M.E. Palmer, S. Taylor, B. Totty. "Interactive Volume Rendering on Clusters of Shared-Memory Multiprocessors". Proceedings of *Parallel Computational Fluid Dynamics '95*, Pasadena, CA.
13. I.E. Sutherland, R.F. Sproull, R.A. Schumacker. "A Characterization of Ten Hidden-Surface Algorithms", *Computing Surveys*, Vol. 6, No. 1, March 1974, pp. 1-55.
14. U. Tiede, T. Schiemann, K.H. Hohne. "Visualizing the Visible Human". *IEEE Computer Graphics and Applications*, Vol. 16, No. 1, January 1996, pp. 7-9.
15. R. Westermann. "Compression Domain Rendering of Time-Resolved Volume Data". Proceedings of *Visualization 1995*, Atlanta, Georgia, pp. 168-175.
16. R. Westermann, S. Augustin. "Parallel Volume Rendering", Proceedings of *International Parallel Processing Symposium*, 1995, pp. 693-699.
17. L. Westover. "Footprint Evaluation for Volume Rendering". Proceedings of *Siggraph '90*, Vol. 24, 1990, pp. 367-376.

A Direct Multi-Volume Rendering Method. Application to Visual Assessment of 3-D Image Registration Algorithms

Jean-José Jacq and Christian Roux

Département Image et Traitement de l'Information
Telecom Bretagne, Technopôle de Brest-Iroise,
Laboratoire de Traitement de l'Information Médicale
B.P 832 - 29285 Brest Cédex - France
E-mail: JJ.Jacq@enst-bretagne.fr, Christian.Roux@enst-bretagne.fr

Abstract. We present a new method for direct volume rendering of multiple 3-D functions using a density emitter model. This mainly consists of the definition of merging rules that are applied on emitter clouds. This work aims at obtaining visual assessment of the results of a 3-D image registration algorithm.

1. Introduction

A new approach to the 3-D image global registration problem - volume-to-volume or surface-to-volume - which operates directly at the voxel level, has been recently proposed [1, 2]. This algorithm does not make use of an explicit surface model and works on a non-segmented data basis. In order to obtain visual assessment of the results, we have developed a specific visualization tool. Among the two 3-D rendering approaches for scientific visualization - Surface Rendering (SR) and Direct Volume Rendering (DVR) - the latter naturally fits in with our underlying purpose for rendering: like the registration algorithm, this rendering approach works at the voxel level. The rendering algorithm needed consists of an extension of standard DVR and projects multiple 3-D functions modeled as sets of fuzzy surfaces that overlap partially.

A new trend in the graphics research community is Voxel Graphics [3]: the key idea is to use the voxel as the basic building block; complex operations can then be reduced to operations on groups of voxels [4]. To achieve SR realism, Voxel Graphics attemptes to transpose global illumination models - ray tracing and radiosity - in this voxel context. In the medical imaging area, a multi-modal volume visualization tool has been described in [5]; it offers the possibility of choosing and tuning visualization strategies (SR or DVR) on an object-by-object basis. In this paper, despite of a similar multi-volume objective, we are only concerned with direct volume rendering; we can therefore make use of a simpler and more unified model. It should be noted that the two papers quoted above [3, 5] do not explicitly address the case of simultaneous direct rendering of fuzzy surfaces over two (or more) overlapping volumes.

Beyond the immediate purpose of our rendering application, this paper aims at assessing the completeness of the Blinn physical model concerning scientific visualization of volume data (especially biomedical data). Provided that all parameters are expressed at physical model level (including end user interaction), extension of standard DVR to the multi-volume case becomes quite simple. This is why section 2 describes the well-known background theory of standard DVR, along with an unified physical formulation: this model simply merges two major aspects of medical volume rendering: fuzzy classification [6] and fuzzy surface rendering [7] within an explicit formulation of the common underlying density emitter model as expressed in [8]. Section 3 shows that this explicit physical modeling permits us to deal easily with the multi-volume rendering case by means of simple and general *ad hoc* rules. Section 4 demonstrates application of the rendering algorithm to visual assessment of medical dataset matching, in particular bone structure registration (volume-to-volume) and matching of a surface to a fuzzy bone sub-structure (surface-to-volume).

2. Background

2.1 Density Emitter Radiation Transport Model

Since we are not looking for some realistic rendering effect, we directly express the problem within a simple and efficient physical model - instead of a realistic one, whose applicability would, in turn, necessitate strong simplifications. Although using various types of formulation, all DVR algorithms implicitly rely on some adaptation of the simple density emitter model introduced by Blinn [9]. In such a model, the volume dataset is described as an inhomogeneous cloud formed by virtual spherical particles (density emitters) and acts as a glowing gel. Let us consider the transmission of a monochromatic ray of light through a slab, centered at ray abcissa t, whose thickness dt is of the same order of magnitude as the diameter of a particle. The model considers only two possibilities: either the ray (R) hits a particle and is completely absorbed (the ray does not experience any diffusion process) or the ray (R') is transmitted without any attenuation. In this model, σ will denote the equivalent elementary surface area that the particle offers to a monochromatic ray modelled as a geometric line.

An elementary reflecting surface is attached to each particle; the normal direction N_t is computed from the local gradient of the 3-D physical density function. Only primary rays coming from the light source can interact with the particle micro-reflector (the particles are said to have a low albedo). Moreover the source of light is directly visible at *each* particle location; intensity of the light source is transmitted following $\boldsymbol{L}_t$ without occlusion nor dispersion untill the current section t. Its only interaction occurs by reflection and diffusion when colliding with a micro-reflector associated with a particle located in the section t. This unrealistic behavior is indicated by the term 'virtual' particle.

We have first to express the amount of light emited by the particle located at t along the view ray $\boldsymbol{R}$ toward the observer. Its analytical expression refers to the choice of a local illumination model. Like in SR algorithms, this amount is defined by some reflection function $Ph(t,\boldsymbol{S}(t),\boldsymbol{R},\boldsymbol{N}_t,\boldsymbol{L}_t)$ usually obtained from the Phong model; in this model, the components of the parameter vector $\boldsymbol{S}(t)$ are respectively the ambient coefficient, the diffuse coefficient, the specular coefficient, and the specular exponent. In this local model the amount of ambient light is context-independent and can also be seen as a particle luminescent emission component. In order to differentiate particle material membership, each material is assigned its own illumination parameter $\boldsymbol{S}$. Geometric parameters associated with the surface are only of interest at the evaluation step of the local illumination; therefore, the amount of light emitted by a particle along the view ray $\boldsymbol{R}$ may be written as $\phi(\boldsymbol{S}(t)) = Ph(t, \boldsymbol{S}(t), \boldsymbol{R}, \boldsymbol{N}_t, \boldsymbol{L}_t)$.

Ray propagation is linked to the particle *population density* $\rho(x,y,z)$. Considering the particle elementary model described above, it is obvious that only particles belonging to the view ray contribute to the final amount of light arriving at the observer location. Therefore, the radiative transport problem can be solved by a one-dimensional integration process and will involve one major parameters: the path *transmittance* θ denoting the fraction of the incident light transmitted - its dual parameter is the *opacity* $\alpha = 1 - \theta$. The transmittance is expressed as a Poisson distribution $\theta = \exp(-\tau)$ of the path *optical density* τ - the optical density is defined as the number of particles included in the cylinder of section area σ described by the ray segment.

Numerical solving of the transport equation along ray $\boldsymbol{R}$ is done through uniform sampling of the ray path. With a simplified notation (the index i stands for a sampling interval δt_i centered on t_i), the resulting global contribution of n samples of ray $\boldsymbol{R}$ becomes:

$$B = \sum_{i=1}^{n} I_0(i) \left(\prod_{j=1}^{i-1} 1 - \alpha(j) \right), \tag{1}$$

with $\alpha(i) = 1 - \exp\left(-\sigma\rho(t_i)\,\delta t_i\right)$ and $I_0(i) = \alpha(i)\,\phi\left(S(t_i)\right)$. The former relation can be reformulated in a recursive way, following a Front-To-Back (FTB) or Back-To-Front (BTF) order - the 'Front' position being at the observer. The FTB formulation needs the introduction of a second relation concerning the *cumulative opacity* associated with the first i samples already accounted for; it represents the progressively accumulated contribution of each voxel. The DVR algorithm relies on this recursive relation. Considering computation, the FTB order is more efficient since the calculations can be stopped when the cumulative opacity is close to one; thus, we do not consider the BTF order.

Since the section σ is a constant parameter of the model - particles only differ in their micro-reflector surface properties - it will no longer appear in the sequel; to avoid the need of a new variable, the product $\sigma\rho$ will be directly denoted as ρ and will be considered as the *optical density per unit length*; in [10] such a parameter is termed as the *differential opacity*.

2.2 Mapping of Physical Properties

Applying the discrete radiative transport equation defined by relation (1) first necessitates mapping of local physical properties within each sample i encountered along the ray at location t_i; these properties are the optical density per unit length $\rho(t_i)$ and the intensity of the emitter toward the observer $\phi(\boldsymbol{S}(t_i))$. As the mapping problem is not concerned with the relative ordering of the samples, these two local parameters will be simply expressed as ρ and $\phi(\boldsymbol{S})$ in the remaining part of this subsection.

Going back to the framework of the physical model, it can be noted that the fundamental quantities involved (ρ and ϕ) correspond basically to mean values, and can be seen as resulting from a linear combination of the properties of m different substances potentially involved in the volume element surrounding a sample point. This is why it is relevant to perform a fuzzy classification of the volume element [6] by assigning to each volume element a *membership value* μ_i associated with each of the m materials in the assumed mixture. In the case of medical imaging, fuzzy classification can be performed assuming that each voxel can be a mixture of a restricted number of different materials. Provided the acquisition system operates with a sufficient sampling grid resolution, most biomedical dataset voxels will contain at most two types of material [6]. Each of the m virtual materials defined *a priori* is assigned a series of four attributes discussed below.

Model of the Distribution Function of Material. An accurate fuzzy classification can be made available at the cost of an additional 3-D function issued from a fuzzy classifier applied to the coding range of the samples. If some *a priori* knowledge of the material distribution over the dynamic range of the acquisition system is available and if $f(x,y,z)$ is a scalar function, one can easily obtain a rough fuzzy classification through a lookup table describing the *kernel* and *support intervals* of each material. However, a complex dataset still needs a computed fuzzy classification.

Physical Density. The *physical density* $\rho_{0,i}$, leading to a scalar 3-D function $\rho_0(x,y,z)$ is required to apply the gradient operator. A sample physical density is expressed through its fuzzy membership as:

$$\rho_0 = \sum_{i=1}^{m} \mu_i \rho_{0,i} \quad . \tag{2}$$

However, if $f(x,y,z)$ is a scalar function, the requirement of knowledge of this parameter can be avoided and $\rho_0(x,y,z)$ could then be equated to $f(x,y,z)$. The normal N and the relative magnitude G are expressed as:

$$N = \vec{\nabla}\rho_0 \Big/ \left|\vec{\nabla}\rho_0\right| \quad , \quad G = \left|\vec{\nabla}\rho_0\right| \Big/ \left|\vec{\nabla}\rho_0\right|_{\max} \quad , \tag{3}$$

where $\vec{\nabla}$ is the gradient operator and $\left|\vec{\nabla}\rho_0\right|$ is the contrast of the physical density. In the generic case, numerical implementation of the ∇ operator is based on finite differences and accounts for anisotropies. In case of noisy dataset, in order to make up for lack of accuracy, the gradient can be obtained from three additional channels storing an off-line computation of the vector components.

Optical Depth per unit Length ρ_i. Since we aim at selecting arbitrary materials that will appear in the rendering process, this virtual parameter is no longer dependent on the related physical density $\rho_{0,i}$; the optical depth per unit length of the material mixture is then expressed as:

$$\rho = \sum_{i=1}^{m} \mu_i \rho_i \quad . \tag{4}$$

To avoid the requirement of having the end users directly specifying ρ_i (which takes values in an unbounded interval) , one could, as proposed in [10], let them express the more intuitive *opacity per unit length* α_i of material i; this bounded parameter would then - following inversion of the opacity expression - be internally decoded into an optical depth per unit length through a relation of the form:

$$\rho_i = \log\left(1/1-\alpha_i\right) \quad . \tag{5}$$

Reflection Model Parameter Sets. The *reflection model* parameter sets S_i associated with each material - these may consist of the coefficient n-tuple <ambient, diffuse, specular, specular exponent> usually associated with the Phong model. Given the viewing parameters and the surface normal N, one could compute the emitter intensity $\phi(S_i)$. As we have made the assumption that all of the particles contained in the volume sample exhibit the same properties, we have to define a common parameter set $\boldsymbol{S}$ in order to compute an equivalent emitter intensity $\phi(\boldsymbol{S})$. Instead of defining some mixing rules, it is more suitable to choose one of the parameter sets associated with the two substances belonging to a voxel under study: all the particles contained in the voxel receive the reflection parameters of the material which represents the major part of the optical depth:

$$\boldsymbol{S} = \boldsymbol{S}_j \text{ with } j \text{ such that } \mu_i \rho_i \leq \mu_j \rho_j \;\forall\; i \in [1, m] \;. \tag{6}$$

In order to obtain a projection that enhances structural information, the homogeneous matter light contribution is overlooked and the emitters located in the neighborhood of those voxels showing noticeable gradient magnitude are focused on; this is the so called *fuzzy surface rendering* technique [7]. It is simply based on multiplying optical density per unit length by the relative gradient magnitude G (see relation (3)). Taking into account this *effective optical density per unit length*, the opacity related to the sample interval is $\alpha = 1 - \exp(-G\rho\,\delta t)$. It can be noticed that the former relation does not require that the relative gradient magnitude G be normalized. Therefore, the presence of the structural information can be enhanced by setting $\left|\vec{\nabla}\rho_0\right|_{\max}$ to an intermediate value of contrast.

Ray acceleration requires some *a priori* knowledge of the properties of the samples that will be encoutered by the ray path. Voxel sites which exhibit an effective opacity lower than a threshold $\alpha_{\min}$ are removed from consideration; the resulting view independant binary map of active voxels is obtained by testing the inequality:

$$\left|\vec{\nabla}\rho_0\right| \geq -\frac{\left|\vec{\nabla}\rho_0\right|_{\max}}{\rho\, e_{\max}} \log\left(1 - \alpha_{\min}\right), \tag{7}$$

with $e_{\max} = \max(e_x, e_y, e_z)$ being the worst resolution component of the dataset. The acceleration approach we retain is based on Ray Acceleration by Distance Coding (RADC) [11]. This consists simply of transforming the binary map into a distance map - this is obtained through an anisotropic version of a fast algorithm [12] computing an exact Euclidian distance. Finally, any inactive voxel will contain an encoded distance to its nearest active voxel and will enable space leaping.

3. Extension to the Multi-Volume Rendering Case

From the geometric point of view, the use of an image-order projection such as raycasting, leads by itself to the multi-volume extension of the rendering algorithm presented in the former section. In order to obtain full graphics capabilities, raycasting must be done through a standard rendering pipeline. A local coordinate system is centered and aligned with the bounding box of each 3-D function; a 3-D image coordinate system, centered on the sample grid origin, is attached to the previous one and permits direct indexing of the sample on the regular grid.

Since, in the general case each volume considered may have its own specific modality, we will consider as many modalities as volumes - however, in a given pratical case a modality may often be redirected through pointer duplication to a more restricted common set of modalities predefined by the user. The classification and modeling knowledge associated with an M-volume rendering will be termed as:

$$\left\{ m(j), \left\{ \rho_{0,i}^j,\ \rho_i^j, \mu_i^j, \boldsymbol{S}_i^j,\ i = 1, \cdots, m(j) \right\},\ j = 1, \cdots, M \right\}, \tag{8}$$

with:

M, # of modalities (= # of volumes);
$m(j)$, # of materials in modality j;
$\rho_{0,i}^{j}$, physical density of material i in modality j;
ρ_{i}^{j}, optical density per unit length of material i in modality j;
μ_{i}^{j}, fuzzy membership function associated with material i in modality j;
$\mathbf{S}_{i}^{j}$, parameter vector describing the surface properties of the reflectors associated with material i in modality j.

Unlike the mono-volume case, the direct multi-volume rendering process does not perform an on-line compositing (through FTB recursive form of relation (1)); the process is segmented into three major steps that are applied successively: 1) collecting traces of the individual intersection signals; 2) resampling and merging of the signals collected; and 3) integration (compositing) of the resultant signal.

The first step consists of a passive ray traversal of the world space - i.e. separate resampling along intersection of the view ray with each volume enclosure and computation of the sample properties - and results in an unordered set of P ($\leq M$) aligned segments that potentially overlap. As in the mono-volume generic case, each function is resampled according to an interval equal to half the extension of a function voxel along the view ray. Knowledge of the vectorial sample $[\rho, \phi]$ set distributed among the P segments results from the application of relations (2), (4), and (6) (their extensions to the context defined by relation (8) are straightforward) and can be termed as:

$$\left\{l(p),\, \delta t^{p},\, t_0^{p},\, t_1^{p},\, \left\{\phi^{p}(\mathbf{S}(k)),\, \rho^{p}(k),\ k = 1,\cdots,l(p)\right\},\, p = 1,\cdots,P\right\}, \qquad (9)$$

with:

P, # of segments ($\leq$ # of volume);
$l(p)$, # of samples along segment p;
δt^{p}, sample thickness along segment p;
t_0^{p}, t_1^{p}, abscissa of both extremities of segment p along the view ray;
$\phi^{p}(\mathbf{S}(k))$, intensity of an emitter belonging to sample k of segment p;
$\rho^{p}(k)$, effective optical depth per unit length in sample k of segment p.

It should be noted that relation (9) requires knowledge of G and $\mathbf{N}$; this implies computation of gradient vector within each sample of each segment.

The next step aims at producing an equivalent and single series of contiguous slabs of density emitter gel; this will come down to the single-volume case and will enable one to compute their overall contribution toward the observer. The abcissae of the extremities and the slab thickness of the resulting segment are defined by:

$$t_0 = \min_{p=1,\cdots,P} t_0^{p};\ t_1 = \max_{p=1,\cdots,P} t_1^{p};\ \delta t = \min_{p=1,\cdots,P} \delta t^{p}\ . \qquad (10)$$

In order to fill in their homologous cells into the final segment, each of the P segments must be resampled following the uniform grid defined by relation (10). To avoid rendering artifacts, one must take care of this resampling step: the explicit physical model we adopt leads to this operation being performed in an intuitive and simple way; the resampling process must reconstruct new voxels along the initial segment so that its overall optical depth remains unchanged:

$$\sum_{k} \rho^{p}(k)\, \delta t^{p} = \sum_{k'} \rho^{p}(k')\, \delta t\ . \qquad (11)$$

This can be obtained through linear gathering of the two portions of contiguous samples that project in the same new sample. Equivalent emitter intensity of the resulting mixture is obtained through weighting of the intensities of the two emitter types by their respective number of emitters.

Given all the knowledge on the P segments within the same sampling grid (as defined by relation (10)), one has to propose some merging rule. In fact, this merging step is an application-dependent part of the rendering process. We introduce here three *ad hoc* rules - like minimization, maximization, and averaging - which are of interest in relation to our specific goal, but are endowed with more general properties. In order to fill in undefined portions of the ray trace on $[t_0, t_1]$, sample contents of the output merging segment are initialized by vacuum properties. The three basic merging rules are:

$$\begin{cases} \rho(k) = \max\limits_{p=1,\cdots,P} \rho^p(k) \text{ and } \phi(S(k)) = \phi^{p_1}(S(k)) \text{ with } p_1 = \arg\left(\max\limits_{p=1,\cdots,P} \rho^p(k)\right); \\ \rho(k) = \min\limits_{p=1,\cdots,P} \rho^p(k) \text{ and } \phi(S(k)) = \phi^{p_2}(S(k)) \text{ with } p_2 = \arg\left(\min\limits_{p=1,\cdots,P} \rho^p(k)\right); \\ \rho(k) = \frac{1}{P}\sum\limits_{p=1}^{P} \rho^p(k) \text{ and } \phi(S(k)) = \sum\limits_{p=1}^{P} \rho^p(k)\ \phi^p(S(k)) \Big/ \sum\limits_{p=1}^{P} \rho^p(k)\ . \end{cases} \quad (12)$$

They stand respectively for maximum, minimum, and average of the contributions.

As a special case, the two first non-linear rules exhibit effects similar to those of spatial fuzzy set operations (union and intersection). More generally, applying rules issued from fuzzy set theory on density emitter clouds would require to express differential opacity ρ as opacity per unit length α; as this parameter is bounded over [0, 1.], it could act as a fuzzy set membership. Following relation (5), fuzzy set operators could be applied through the general scheme:

$$\rho \longrightarrow \left.\begin{matrix} \cdots \\ \alpha = 1 - \exp(-\rho) \\ \cdots \end{matrix}\right\} \xrightarrow{\text{Fuzzy logic operators}} \alpha' \longrightarrow \rho' = \log\left(\frac{1}{1-\alpha'}\right).$$

In such a way, other fuzzy set operators - difference, complement, algebraic sum and algebraic product - could be implemented. As an example, in such a scheme, complement ρ^c of the density ρ vould be $-\log(1-\exp(-\rho))$.

Following the merging rule, one must take care of the optimization rules that are applied while collecting the segments. For example, in order to prevent an early termination of the ray traversal within a specific volume, accumulated opacity should be processed only for the maximization merging rule.

The last step deals with ray integration along the final segment obtained from the merging step. This is performed in the same way as in mono-volume rendering through the FTB recursive form of relation (1) - in this final step, early termination of ray integration process is alway enabled.

4. Application to Visual Assessment of a 3-D Registration Algorithm

Two kinds of elastic registration tasks are illustrated in the application described: volume-to-volume and surface-to-volume. All of them involve one or two distinct CT volume datasets with voxel resolution ranging from 0.66 to 0.75 mm. Although our implementation allows for perspective effects, projections presented here are nearly parallel.

Fig. 1 shows applications of our rendering scheme to the visual assessment of volume-to-volume global elastic registration. It illustrates the recovery of a deformation between two states of a non-segmented volume, which is the distal part of a humerus bone. The simulated transform corresponds to a helicoidal deformation with compression along the axis of the helix, with major distortions on the side of the femural head. Here, two basic merging rules, maximizing and minimizing, are of interest; the first one permits union of fuzzy surfaces whereas the second rule permits intersection of fuzzy surfaces. Union is of major interest when merging unregistered or uncorrelated datasets. Conversely, as the dataset morphologies are correlated, intersection of the fuzzy surfaces is an accurate way to estimate the elastic registration: holes should appear whenever registration fails. Moreover, color labelling of the datasets gives a differential view of the relative positions of the fuzzy surfaces.

The surface-to-volume applications make use of the same optimization procedure in order to fit a geometric model over a raw dataset. Visual assessment of the fitted surface requires one to be able to insert geometric object into the world space. Instead of doing hybrid rendering, we want to insert the geometric object in the space so that the same underlying physical model may apply. Since our specific geometric model enables analytic computation of the Euclidian distance $d(x,y,z)$ in each locus of the space, the fitted model is inserted into the world space through a distance-based fuzzy voxelization (the exterior of the object is assumed to have a negative distance). The blending function is based on a hyperbolic tangent given by:

$$f(x,y,z) = 255\,\frac{1+\tanh(\beta\, d(x,y,z))}{2},$$

with β being a slope parameter controlling the sharpness of the voxelized object (in the results presented, β is set to 1).

Fig. 2 depicts the modeling of the diaphysal part of the radius medullary cavity - this part corresponds to approximately one third of the long bone structure in its central part. Medullary cavity enhancement is obtained through pre-processing of the CT dataset. Such a structure does not exhibit well-defined boundaries and can be accurately visualized as a fuzzy surface. The voxelization of the fitted geometric model is performed at the resolution of the raw dataset. Fig. 3 shows that simple Voxel Graphics capabilities can also be achieved by our rendering scheme; it gives an *in situ* visualization of the computed model in the openned (two halves) cortical structures of the original bone. In Fig. 2 and 3 input data only consist of two datasets that are internally cloned; each of the resulting four objects are cropped through validation of specific Volumes Of Interest, and moved according to symmetrical affine transformations.

Conclusion

We have presented a direct multi-volume rendering framework based on fuzzy surface modeling. We have shown that this framework is very useful in rendering and comparing structural information in different 3-D functions that partially overlap one another. From a more general point of view, this unified rendering scheme encompasses various issues of medical imaging and can be used as a general imaging framework.

Our experience is that DVR of fuzzy surfaces involves a non-critical parameterization step; with regard to end-user interaction, this should be another major advantage of this rendering approach. The amount of relevant information included in the projection depends on the fuzzy classification step; this critical step is upstream and still necessitates theoretical and experimental studies. As fuzzy surface rendering can directly express results coming from the fuzzy classification theory, it should give strong basis to the visualization of multi-modal datasets. Giving end-user ability to show through intrinsic uncertainty of medical datasets is also a major topic of this rendering scheme. Furthermore, the rendering scheme can include geometric objects (or surface models obtained from usual segmentation of the data acquisition) without any modification, by means of fuzzy voxelization (see Fig. 2 and 3), giving intrinsic possibilities of surface rendering without the artifacts usually encountered in the case of surface rendering. Compared to an iso-surface projection, fuzzy surface rendering does not significantly increase rendering time, provided that the data exhibit a high level of coherence.

In our application, the local merging rules applied on the emitters are scalar; however, they can be extended to the vectorial case and it may highlight more complex local interactions between structures belonging to different functions.

References

1. Jacq, J.J., Roux, C.: Registration of non-Segmented 3-D Images using a Genetic Algorithm. *Computer Vision, Virtual Reality and Robotics in Medecine. Lecture Notes in Computer Sciences **905**,* Springer (Apr. 1995) 205-211
2. Jacq, J.J., Roux, C.: Registration of 3-D Images by Genetic Optimization. *Pattern Recognition Letters*, Vol **16**, No **8** (Aug. 1995) 823-841
3. Sobierajski, L., Kaufman, A.E.: Volumetric Ray Tracing. In *Proc. of the IEEE/ACM Volume Visualization'94 Symposium*, Washington, D.C. (Oct. 1994) 11-18
4. Gallagher, R.S.: Visualization: the look of reality. *IEEE Spectrum*, Vol. **31**, no **11** (Nov. 1994) 48-55
5. Zuiderveld, K.J., Viergever, M.A.: Multi-modal Volume Vizualization using Object-Oriented Methods. In *Proc. of the IEEE/ACM Volume Visualization'94 Symposium*, Washington, D.C. (Oct. 1994) 59-66
6. Drebin, R.A., Carpenter, L., Hanrahan, P.: Volume Rendering. *Computer Graphics* (proc. SIGGRAPH) , Volume **22**, No. **4**, Atlanta (Aug. 1988) 65-74
7. Levoy, M.: Display of Surfaces from Volume Data. *IEEE Computer Graphics & Applications* (May 1988) 29-37
8. Sabella, P.: A Rendering Algorithm for Visualizing 3D Scalar Data Fields. *Computer Graphics*, Vol. **22**, No. **4** (Aug. 1988) 51-58
9. Blinn, J.: Light Reflection Functions for Simulation of Clouds and Dusty Surfaces. *Computer Graphics*, Vol. **16**, No. **3** (Jul. 1982) 21-29
10. Wilhelms, J., Van Gelder, A.: A Coherent Projection Approach for Direct Volume Rendering. *Computer Graphics*, Vol. **25**, No. **4** (Jul. 1991) 275-284
11. Zuiderveld, K.J., Koning, A.H.J., Viergever, M.A.: Acceleration of ray-casting using 3-D distance transforms. In *Proc. Visualization in Biomedical Computing*, Vol. **1808** (Oct. 1992) 324-325
12. T. Saito, T., Toriwaki, J.I.: New algorithms for Euclidian distance transformation of an n-dimensional digitized picture with applications. *Pattern Recognition*, Vol. **27**, no. **11** (May 1994) 1551-1565

Acknowledgement. CT scan data were provided by le Centre Hospitalier Universitaire de Brest, France, and we thank Drs. Ch. Lefèvre, D. Colin, and E. Stindel.

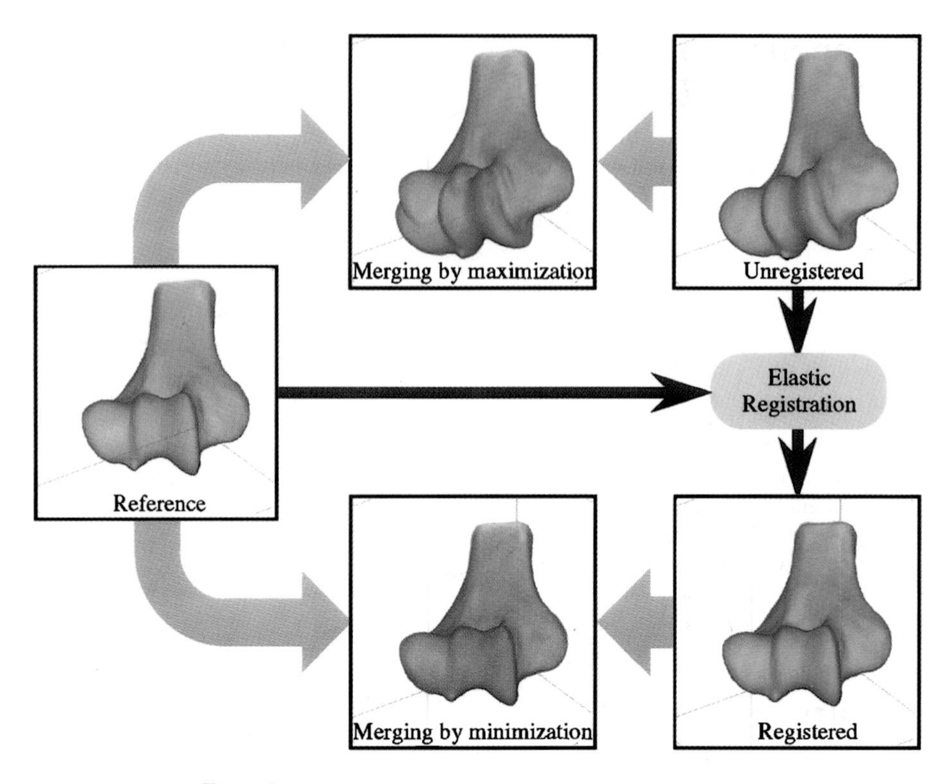

Fig. 1. Recovering an elastic deformation (humerus bone)

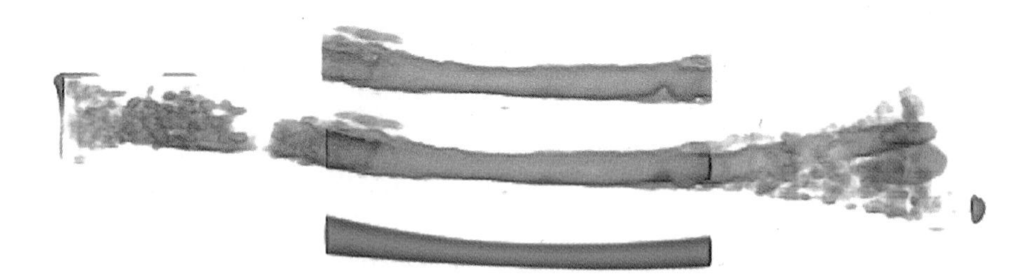

Fig. 2. Merging of the medullary cavity and its fitted model (radius bone)

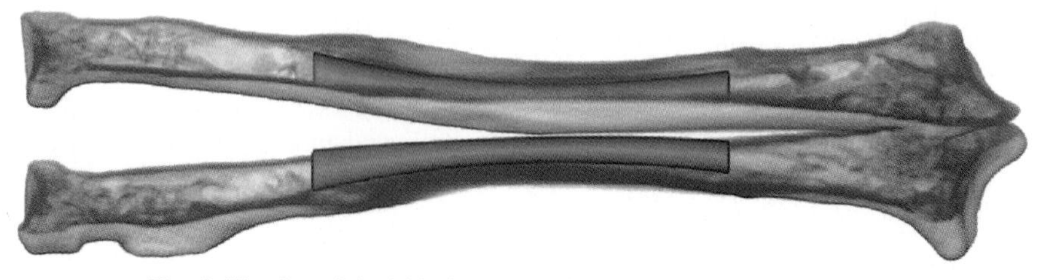

Fig. 3. Fitted model within its openned cortical bone (radius bone)

Visualization of Tissue Elasticity by Magnetic Resonance Elastography

A. Manduca, R. Muthupillai, P. J. Rossman, J. F. Greenleaf, R. L. Ehman

Mayo Clinic and Foundation
Rochester, MN 55905 USA

Abstract. A newly developed magnetic resonance imaging technique can directly visualize propagating acoustic strain waves in tissue-like materials [1, 2]. By estimating the local wavelength of the acoustic wave pattern, quantitative values of shear modulus can be calculated and images generated that depict tissue elasticity or stiffness. Since tumors are significantly stiffer than normal tissue (the basis of their detection by palpation), this technique may have potential for "palpation by imaging," with possible application to the detection of tumors in breast, liver, kidney, and prostate. We describe the local wavelength estimation algorithm, study its properties, and show a variety of sample results.

1 Introduction

A new technique which can directly visualize propagating acoustic strain waves in materials subjected to harmonic mechanical excitation has recently been developed [1, 2]. This technique, termed magnetic resonance elastography (MRE), presents the unique opportunity of generating medical images that depict tissue elasticity or stiffness. This is significant because palpation, a physical examination that assesses the stiffness of tissue, is far more effective at detecting tumors than any current imaging modality, but is restricted to parts of the body that are accessible to the physician's hand. MRE shows promise as a potential technique for "palpation by imaging," with potential applications in the detection of tumors in breast, liver, kidney, and prostate.

The MRE technique incorporates: (i) conventional phase and frequency encoding gradients, (ii) harmonic motion-sensitizing magnetic gradient waveforms, and (iii) synchronous trigger pulses which are fed to a waveform generator driving an electromechanical actuator coupled to the surface of the object being imaged, inducing phase-locked shear waves. MRE images have been acquired which provide snapshots of propagating shear waves (200 - 1100 Hz) in gel phantoms and organ specimens. Wave patterns with displacement amplitudes of less than 200 nanometers are clearly observable. Additionally, these images allow the calculation of regional mechanical properties. In particular, the shear modulus μ of a material is given by $\nu^2\lambda^2\rho$, where ν is the frequency of excitation, ρ the density of the medium, and λ the wavelength of the shear wave in the material. Thus, estimation of the local wavelength (or frequency) of the shear wave propagation pattern at each point in the image allows one to quantitatively calculate local values of shear modulus across the image and generate an image mapping tissue elasticity. These "elastograms" or "stiffness images" clearly depict areas of different elastic moduli in test phantoms, and the calculated values correlate well with moduli calculated independently by mechanical means [1].

2 Local Wavelength Estimation

The classical concept of frequency is not well defined for non-stationary signals, but is so useful that many attempts have been made to define some analogous quantity. The instantaneous frequency, for instance, is commonly defined as the rate of change in phase of the analytic extension of a real signal, though this definition leads to some difficulties and alternatives have been proposed [3]. Many ways to derive local frequency estimates have been proposed, with windowed Fourier transforms and Gabor transforms being the most well-known. Recently, Knutsson et al. [4] described an algorithm which estimates local signal or image frequencies by combining local estimates of instantaneous frequency over a large number of scales, derived from filters which can be considered to be lognormal quadrature wavelets and are a product of radial and directional components.

The filters defined in [4] have a radial component of the form $R_i(\rho) = e^{-C_B \ln^2(\rho/\rho_i)}$ (a Gaussian on a logarithmic scale), where C_B expresses the relative bandwidth and ρ_i is the central frequency. This function is illustrated for six different central frequencies ρ_i in Fig. 1. The directional component has the form $D_k(\hat{u}) = (\hat{u} \cdot \hat{n}_k)^2$ if $(\hat{u} \cdot \hat{n}_k) > 0$ and $D_k(\hat{u}) = 0$ otherwise (where $\hat{n}_k$ is the filter directing vector). This component thus varies as $\cos^2(\phi)$, where ϕ is the angle difference between $\hat{u}$ and the filter direction $\hat{n}_k$, and the filter is non-zero in the positive $\hat{n}_k$ direction only. Along the orientation direction, the filter profiles correspond to the radial component alone, as in Fig. 1.

It is shown in [4] that an isotropic estimate of signal strength, that is local both spatially and in the frequency domain, is obtained by summing the magnitudes of the outputs of orthogonally oriented filters of this kind at one scale, and that the magnitude of the ratio between two such filters at different scales can be interpreted as a narrow band estimate of instantaneous frequency. By combining the outputs from two or more sets of filters which differ only in center frequency ρ_i, it is possible to produce a local frequency estimate. If the filters have the appropriate bandwidth relative to the ratio of the two central frequencies [4], a particularly simple case results: the ratio of the two responses times the geometric mean of the two central frequencies is the local frequency estimate. This estimate will work well only if the signal spectrum falls within the range of the filters, but a wide range estimate can be obtained by using a bank of filters and performing a weighted summation over the different filter pairs, with the weighting factor correponding to the amount of energy encountered by each filter pair.

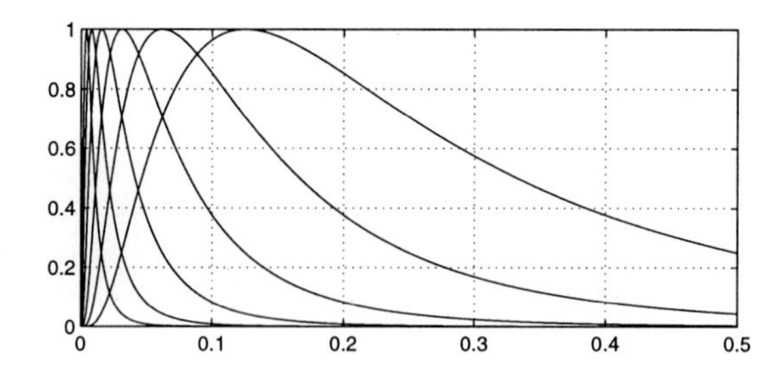

Fig. 1. Filter profiles along the orientation direction for six filters with central frequencies spaced an octave apart from 1/256 to 1/8. The bandwidth parameter is $C_B = 1/(2 \ln 2)$.

3 Results on Test Data

We have applied this algorithm to test images and to MRE data using (as default) a bank of 6 filter sets, with each set containing both x and y axis oriented filters. Knutsson et al. [4] recommend spacing the central frequencies of such a filter bank an octave apart, which is both a natural choice and one which works well in practice. All our images were 256 x 256 (though some of the figures below are truncated to save space), and we spaced the central filter frequencies an octave apart from 1/256 to 1/8, as shown in Fig. 1. Calculations were also carried out with a bank of 11 filter sets, spaced from 1/256 to 1/8 half an octave apart, as an experiment in noise suppression (see below). Fig. 2 shows the results of the algorithm on two test images. The results are displayed using a full gray scale, but the quantitative values in each region correctly match the frequencies used to generate the data set. Vertical profiles across the local frequency estimate images are shown in Fig. 3. The transition region is fairly sharp, and the transition between the two regions is essentially complete one half wavelength into each region. This results in a sharper transition in the image with shorter wavelengths, as is evident in Figs. 2d and 3; the algorithm is able to utilize the sets of filters at widely different scales to effectively select the appropriate scale for analysis.

To test the effects of noise on the local frequency estimates, we added simple zero-mean Gaussian noise with a standard devation of 10 to the test data in Fig. 2a (whose original values ranged from 0 to 255). The noisy image and the results of the algorithm are shown in Fig. 4, and vertical profiles are shown in Fig. 5. The original algorithm now yields estimates that are slightly higher than expected, due to the high frequencies in the noise influencing the averages. Furthermore, these estimates themselves are fairly noisy (see solid line in Fig. 5). There is no "correct" answer in this situation, since the underlying signal is not purely a single frequency. However, in practice we would consider it desirable if the algorithm was minimally influenced by such noise and continued to give smooth estimates close to the noise-free case. This can be accomplished by modifying the algorithm slightly, to use 11 filter sets spaced half an octave apart (and with a reduced bandwith of $C_B = 1/\ln 2$) rather than the default values. This modified

Fig. 2. (a) A test image with two regions of different frequencies (4 and 8 cycles/image respectively), (b) the local frequency estimate, (c) a second test image with two regions of higher frequency (8 and 16 cycles/image), (d) its local frequency estimate, and (e) the first image with Gaussian noise added (analyzed later). Note how sharply the frequency estimate changes in the transition region; the transition between the two estimates is essentially complete one half wavelength into either region in both cases.

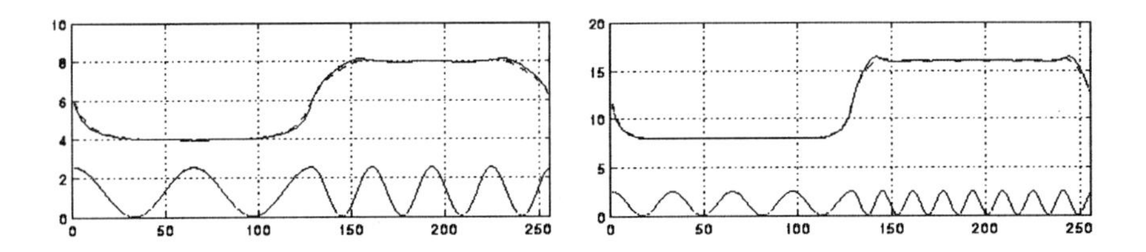

Fig. 3. (a) Left: (Top, solid line) A vertical profile along fig. 2b, calculated with the default 6 filter sets. (Top, dashed line) The same profile calculated with 11 filter sets spaced half an octave apart. (Bottom) A vertical profile along fig. 2a, the original data. The left half of the data set (lower half in fig. 2a) has a frequency of 1/64 or 4 cycles across the image, while the right half has a frequency of 1/32 or 8 cycles across the image. (b) Right: A similar graph for vertical profiles along figs. 2c and 2d. The frequencies are 8 and 16 cycles/image in the left and right halves respectively.

filter bank clearly gives more accurate, smoother estimates in Fig. 4.

The modified filter bank covers essentially the same frequency range, with the central frequencies ranging from 1/256 to 1/8, but the larger number of more precise estimators allow it to estimate a better average frequency at any point. Conversely, viewed in the spatial domain, each of the new fiters has a larger spatial extent (since it has a smaller frequency bandwith), so the new filters are sampling more of space and therefore obtaining better averages over the noise. This implies, however, that the resolution of the modified filter bank should not be as sharp, since the filters are sampling larger spatial regions. This is indeed the case in Fig. 3; the modified filters do have a slightly larger transition region. However, the difference is surprisingly small, and one can argue that even in the noise-free case the modified filters might be better since they avoid the slight ringing evident for the standard filter bank in Fig. 3b.

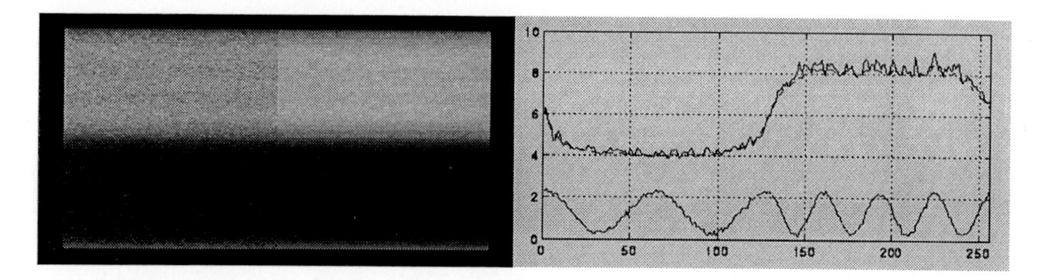

Fig. 4. Left: (a) The local frequency estimate of the noise-added image in fig. 2e with the default 6-band filter bank. (b) The local frequency estimate with the 11-band filter bank. Right: (Top, solid line) A vertical profile along fig. 4a, calculated with the 6-band filter bank. (Top, dashed line) The same profile along fig. 4b, calculated with the 11-band filter bank. This estimate is much smoother and closer to the desired values of 4.0 and 8.0. (Bottom) The vertical profile along fig. 2e, the original data.

4 Results on MRE Data

Fig. 5 shows the results of the MRE technique on a gel phantom which contains two cylindrical plugs of different stiffness than the background material. The standard MRI magnitude image (not shown) of this object is simply a uniform field. The mechanical motion is perpendicular to the image plane, as is the motion-sensitizing magnetic gradient oscillation. Planar shear waves are generated at the top and propagate down the image. The bright areas in the acoustic strain wave image indicate motion in phase with the magnetic gradient, while the dark areas indicate motion out of phase. The actual displacement values of the motion at any point can be calculated, and correlate well with optical deflection measurements [1, 2]. The wavelength of the shear waves changes as it encounters the plugs of differing stiffness, with the wavelength getting longer in the stiffer plug and shorter in the less stiff plug. Applying the algorithm described above, and converting from frequency to wavelength, a local wavelength estimate (LWE) image is obtained, which clearly delineates the two plugs. Fig. 6 shows the results of imaging an excised porcine kidney. The cortex of the kidney can be seen to be stiffer than the medulla.

5 Conclusion

The Knutsson et al. [4] algorithm appears to be very suitable for local wavelength estimation on MRE data, and for the generation of elasticity images. Since the shear modulus μ of a material is given by $\nu^2\lambda^2\rho$, where ν is the frequency of excitation, ρ the density of the medium, and λ the wavelength of the shear wave in the material, the local wavelength measures at each point can be converted to quantitative measures of shear modulus (a mean density of $1.0g/cm^3$ is an appropriate assumption for tissue-equivalent gels and soft tissues). Such values calculated in test phantoms correlate well with shear moduli calculated independently by other means [1]. Palpation, a physical examination that assesses the

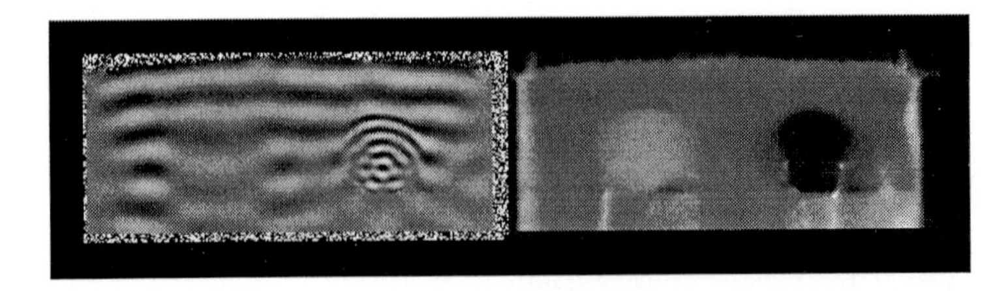

Fig. 5. (Left) MRE image of a tissue-simulating gel object containing two embedded gel cylinders with differing stiffness. Shear wave excitation at 250 Hz (perpendicular to the image plane) with an amplitude of 5.0 mm was applied to a wide contact plate on the surface. Planar shear waves are propagating down the image. The cylinder on the left is stiffer than the surrounding gel, resulting in an increased wavelength; conversely, the cylinder on the right is less stiff and shows a shorter wavelength and refraction effects. (Right) The local wavelength estimate of the image. The plugs stand out clearly from the surrounding gel.

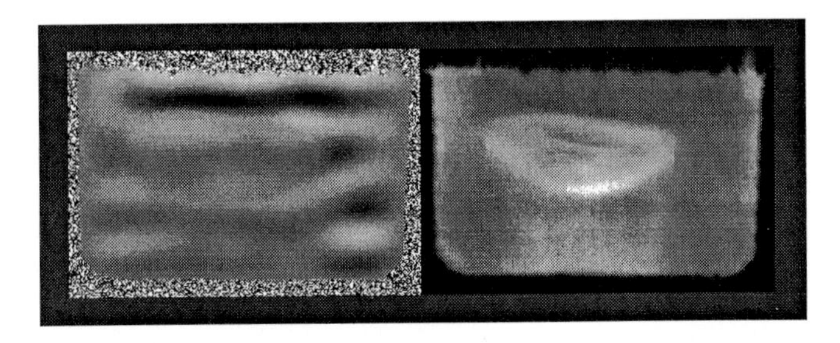

Fig. 6. An MRE image of an excised porcine kidney and the local wavelength estimate. The cortex (outer region) of the kidney is stiffer than the medulla (inner region).

stiffness of tissue, has historically been an effective method of detecting tumors, but is restricted to parts of the body that are accessible to the physician's hand. MRE shows promise as a possible technique for "palpation by imaging," with potential applications in the detection of tumors in breast, liver, kidney, and prostate.

References

1. Muthupillai R., Lomas D., Rossman P., Greenleaf J., Manduca A., Ehman R.: Magnetic Resonance Elastography by Direct Visualization of Propagating Acoustic Strain Waves. Science **269** (1995) 1854–1857
2. Muthupillai R., Rossman P., Lomas D., Greenleaf J., Riederer S., Ehman R.: Magnetic Resonance Imaging of Transverse Acoustic Strain Waves. Submitted to Journal of Magnetic Resonance Imaging (1996)
3. Boashash B.: Estimating and Interpreting the Instantaneous Frequency of a Signal. Proc IEEE **80** (1992) 520–568
4. Knutsson H., Westin C., Granlund G.: Local Multiscale Frequency and Bandwidth Estimation. Proceedings of the IEEE Intl Conf on Image Processing **1** (1994) 36–40

Accurate Vessel Depiction with Phase Gradient Algorithm in MR Angiography

Romhild Hoogeveen, Chris Bakker, and Max Viergever

Imaging Center, University Hospital Utrecht, Heidelberglaan 100,
3584 CX Utrecht, The Netherlands

Abstract. Current MRA techniques do not always accurately depict vascular anatomy, particularly in areas of disturbed flow. Various reasons cause a wrong delineation of vessel boundaries. A phase contrast (PC) based post-processing operation, the phase gradient (PG), is introduced to detect phase fluctuations indicating flow. By means of numerical, phantom and in vivo experiments, it is shown that PG angiograms give better impressions of (stenotic) vessels and of their diameters for both laminar and disturbed flow.

1 Introduction

A variety of magnetic resonance angiography (MRA) techniques have been introduced in clinical practice for the depiction of vascular structures. Still, established techniques like time of flight (TOF) or phase contrast (PC) are unable to produce a geometrically correct visualization of the vascular anatomy in the presence of complex flow [1, 2, 3], and when precise vessel wall location is concerned [4], which makes the detection of stenoses[5, 6] a challenging topic of research. Problems with PC MRA comprise: 1) signal loss caused by intravoxel phase dispersion due to high order motion or disturbed flow, 2) sensitivity to the velocity sensitivity V_{enc} which must be set before acquisition higher than the maximum velocity present (if not, phase aliasing will occur), 3) partial volume effects resulting in a lower signal intensity at the vessel wall which may lead to an underestimation of the vessel diameter and 4) non-linearity of the MIP projection.

We introduce a new phase-sensitive, PC-based, post-processing technique, the *phase gradient* (PG), so as to to overcome the problems of (PC) angiography concerning a good visualization of the vessel anatomy. In this study, the potentials of PG images towards MR angiography are demonstrated. We will show that PG images are less V_{enc}-dependent and that a better delineation of vessels can be achieved than with standard PC images. This will be investigated numerically, in vitro and in vivo. Furthermore, a comparison of signal losses in laminar and disturbed flow in PG and conventional PC angiograms will be evaluated.

2 Theory

2.1 Phase Gradient

PC methods rely on a change of spin phase induced by motion: $\phi = \gamma \int \overline{x}(t) \cdot \overline{G}(t)dt$ where γ is the proton gyro-magnetic ratio, $\overline{x}(t)$ the position of the spins and $\overline{G}(t)$ the magnetic gradient applied. Typical PC acquisitions assume first or second order motion to be present, but we allow *any* motion to occur, which should be detectable by a change of phase. By applying a gradient operation $d\phi/ds$ to the phase of the magnetic signal, we detect transitions of static spins to moving spins, so that vessel walls rather than the highest flow yield a high signal. The gradient operator has proven to detect vessel walls [7]. The PG operation produces the highest output when performed perpendicular to the direction of the flow. The performance of the PG operator is illustrated in figure 1. If V_{enc} is set high, no phase wraps will occur (a), which results in a CD profile as in in (b) and a PG profile as in (c). Note that for continuous signals, the PG profile has two extreme values exactly positioned at the vessel walls. If V_{enc} is lowered, phase wraps will occur, as seen in (d). The CD profile is no longer 'valid' and produces stripe-like artifacts. However, the PG profile (f) is not affected by these phase wraps since it is a differential operation. In the case of disturbed flow ((g)-(i)), the phase is unstable (g) because the signal vanished due to intravoxel phase dispersion. Complex subtraction (h) yields a low signal due to the small magnitude signal of the flow sensitive and flow compensated datasets. However, the PG detects any change of phase and produces a signal within the constricted lumen (i). The one-dimensional PG $\nabla\phi_s[\tilde{f}(\overline{x})]$ in direction s at point $\overline{x}$ can be calculated from the real and imaginary parts of the reconstructed MR-image $\tilde{f}(\overline{x})$:

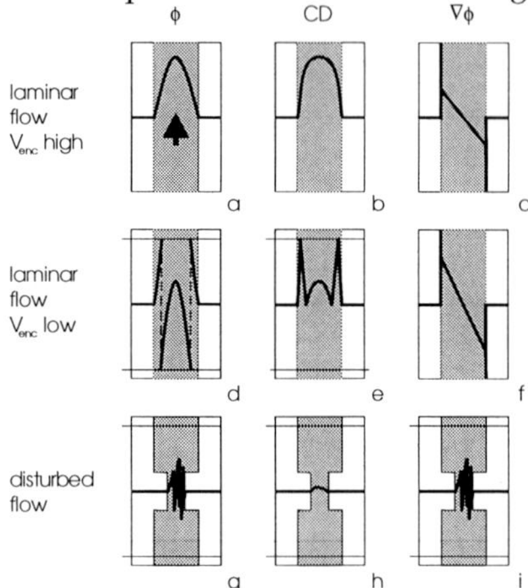

Fig. 1.: *PG performance compared to complex-difference. (a)(d)(g) Phase of profile within lumen, (b)(e)(h) complex difference signal within lumen, (c)(f)(i) PG signal within lumen. (a)(b)(c) Laminar flow V_{enc} high, (d)(e)(f) laminar flow V_{enc} low, (g)(h)(i) disturbed (stenotic) flow. PG is not affected by phase wraps and still produces signal in areas of disturbed flow.*

$$\nabla\phi_s[\tilde{f}(\overline{x})] = \frac{d\phi[\tilde{f}(\overline{x})]}{ds} = \frac{Re[\tilde{f}(\overline{x})]\cdot\delta Im[\tilde{f}(\overline{x})]/\delta s - Im[\tilde{f}(\overline{x})]\cdot\delta Re[\tilde{f}(\overline{x})]/\delta s}{Re^2[\tilde{f}(\overline{x})] + Im^2[\tilde{f}(\overline{x})]} \tag{1}$$

The real and imaginary derivatives are calculated in k-space since the signal $\tilde{f}(\overline{x})$ is completely determined by its Fourier coefficients and since differentiation in the spatial domain is equivalent to multiplying with a linear imaginary vector in the frequency domain. Phase wraps can occur in phase images when V_{enc} is not properly dimensioned [8]. Since the PG is a differential operator it will be clear that phase wraps do not influence PG images. Consequently, a *dynamic range extension* can be achieved by lowering V_{enc}. Susceptibility and field inhomogeneity artifacts can be reduced by subtracting the reconstructed PG images of two differently flow encoded datasets:

$$\nabla\phi'_s[\tilde{f}(\overline{x})] = \nabla\phi_s[\tilde{f}_{dataset1}(\overline{x})] - \nabla\phi_s[\tilde{f}_{dataset2}(\overline{x})] \tag{2}$$

For vessel wall location determination, formula (2) is applied. This one-dimensional PG can be expanded to a two-dimensional and three-dimensional PG when directionally independent visualization of vessels is needed:

$$\nabla\phi'_{s,t}[\tilde{f}(\overline{x})] = \sqrt{\nabla^2\phi'_s[\tilde{f}(\overline{x})] + \nabla^2\phi'_t[\tilde{f}(\overline{x})]} \tag{3}$$

$$\nabla\phi'_{s,t,u}[\tilde{f}(\overline{x})] = \sqrt{\sum\nolimits_{v=s,t,u}\sum\nolimits_{w=s,t,u}\nabla^2\phi'_w[\tilde{f}_v(\overline{x})]} \tag{4}$$

with s, t and u an orthogonal frame and $\tilde{f}_s$, $\tilde{f}_t$, and $\tilde{f}_u$ three orthogonally flow encoded datasets. A MIP of the PG images produced by formula (3) or (4) contains high signal intensities not only at the vessel walls but also in the middle of the lumen since the projection rays have penetrated the vessel wall at least once.

2.2 Sampling

In this section formulas are presented as one-dimensional. They can easily be extended to 2D and 3D. It can be shown that a discrete MR image can be obtained by sampling a continuous image $\tilde{f}(x)$, obtained by convolving a point-spread function $q(x)$ with the original image $f(x)$:

$$\tilde{f}(x) = (f * q)(x) \qquad with \qquad q(x) = \frac{2\pi}{N} e^{jxN} \frac{1 - e^{-jx2\pi}}{1 - e^{-jx2\pi/N}} \tag{5}$$

N is the number of samples. Since, during convolution with $q(x)$, no phase-aliasing is desired, we can deduce that the phase change per pixel (=1/N) should obey $\frac{\Delta\phi}{\Delta pixel} \leq \pi$. In practice, (1) amplifies noise, so we need to pre-smooth the real and imaginary components of the MR-image a little. We found that prefiltering by means of convolution with a 3D Gaussian with σ=0.5 pixel produced a good trade-off between smoothing and aliasing:

$$\tilde{f}'(x) = (f * q * r)(x) \qquad with \qquad r(x) = \frac{1}{\sqrt{2\pi\sigma^2}} e^{-\frac{x^2}{2\sigma^2}} \tag{6}$$

From formula (6) the phase $\phi[\tilde{f}'] = Arg[\tilde{f}']$ and the PG $\nabla\phi_s[\tilde{f}']$ of an MR-image can be calculated, given a continuous image f, σ, and the number of samples.

3 Materials and Methods

3.1 Computer Simulations

Computer simulations were performed to investigate the influence of V_{enc} on the PG profile across a vessel with laminar flow. The PG profiles were calculated by applying formula (6) to a mathematical laminar flow model for different diameters and different values of V_{enc}. The conventional complex-difference (CD) profile was also calculated[9]. To assess the accuracy of diameter estimation by means of the PG and CD profiles, we investigated three measures: 1) PG Max: distance between the two maxima of the absolute value of the PG profile, 2) PG FWHM: distance between the two outer points at half the maximum (Full Width at Half Maximum) of the PG profile, 3) CD FWHM: the FWHM of the CD profile.

3.2 Phantom Studies

In our phantom studies, we investigated the influence of V_{enc} on the PG and the CD images. A 5.5 mm diameter tube with a 2 mm diameter constriction was put in a water filled box. A blood mimicking fluid was guided through the tube with a constant flow rate of 10.0 ml/s (42 cm/s and 318 cm/s in the wide and narrow parts of the tube) by a computer-controlled pump. Standard non-triggered 3D PC acquisitions were performed on a 0.5T whole-body MR system with a spoiled gradient-echo sequence, TR/TE/α=47/28/15°, bandwidth 9250 Hz, matrix size 256x256, 16 slices, 1mm^3 voxels. We studied in-plane flow with the foldover-direction perpendicular to the flow. Only a flow-sensitivity in the direction of the main flow was measured, with varying V_{enc}: 100, 50, 20, 10, 5 and 2.5 cm/s. MIPs of the CD images were made and compared with the MIPs of the reconstructed PG images. From the MIPs, the signal-to-noise ratio (SNR) of the wide part of the phantom was calculated by dividing the mean signal of the MIPs in the unconstricted part of the tube by the standard deviation of the background signal. Similarly, the SNR was calculated for the constriction.

3.3 In Vivo Studies

Preliminary in vivo studies were done to show the potentials of PG images in MRA. The carotid arteries of healthy volunteers were scanned using a standard 3D PCA acquisition on a 1.5T imager: 256x256 matrix, FOV 256x256 mm, 32x1mm coronal slices, TR/TE/α=19/11/15°, receiver bandwidth 50kHz, flow sensitivity in feet-head direction only, V_{enc}=20 cm/s, smaller than the average velocity of 35 cm/s. The 2D PG reconstruction (formula (3)) was used. Bone, air and the jugular vein were segmented manually and removed to be able to compare PG MIPs with CD MIPs. A similar acquisition was performed on the Circle of Willis with V_{enc}=20cm/s and flow sensitivity in all directions. PG images were calculated using the 3D PG reconstruction (formula (4)).

4 Results

4.1 Computer Simulations

Because of space limitations we refer to our poster for more detailed results. The computer simulations show that the conventional diameter measure (CD FWHM) underestimates the real vessel width, especially for large diameters. For $\phi_{max} < \pi$, the maximum error is about 1.5 pixel at a diameter of 8 pixels. The PG Max measure has a comparable performance. If we compare the PG FWHM measure with the ideal diameter, we find a very good correlation. The maximum error is less than 0.5 pixel. For $\phi_{max} > \pi$ (realized by lowering V_{enc}) the CD reconstruction shows phase aliasing, resulting in smaller diameter estimations, whereas the PG can be calculated until the phase across one pixel exceeds π. In all cases studied, the PG FWHM measure gives a better estimate of the vessel diameter than the PG Max and the CD FWHM measure. Increasing ϕ_{max} results in a better PG Max measure since the extreme values of the PG are more closely situated to the vessel wall. The error in the PG FWHM measure, however, increases slightly when ϕ_{max} is increased, caused by the more spread-out profile of the PG for small vessels.

4.2 Phantom Studies

In the phantom studies, ϕ_{max} was controlled by a selection of V_{enc} with respect to the maximum velocity in the phantom tube. MIPs of the CD and PG images with different values of V_{enc} are shown in figure 2 (a) and (b). The flow is upwards. The flow proximal to the constriction is close to laminar. Unlike the PG MIPs, the CD MIPs show a complete loss of signal in and distal to the constriction, where the flow is disturbed. The diameter of the wide part of the tube is strongly V_{enc}-dependent and varies significantly with intensity scaling. The PG MIPs show an improved lumen delineation for all values of V_{enc}. The flow proximal to the constriction shows a constant diameter for V_{enc}=100, 50 and 20 cm/s. For lower values of V_{enc} the PG MIPs still give a good impression of the anatomy of the vessel both in laminar and disturbed flow. Figure 2(c) shows the signal-to-noise ratios of the PG and CD images in the wide and constricted part of the phantom. The PG plots show an important increase of the SNR in both parts of the phantom, especially for low values of V_{enc}. The range for which the PG angiograms produce high SNRs is much broader than for conventional angiograms.

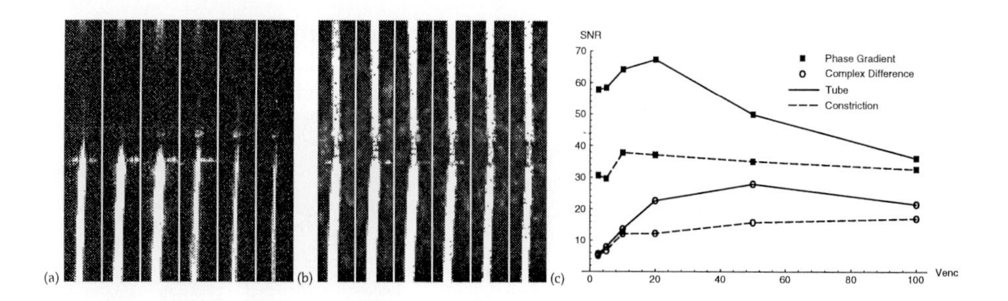

Fig. 2. *MIPs of CD(a) and PG (b) images of the flow phantom with steady flow (42cm/s) in the wide part for V_{enc}=100, 50, 20, 10, 5 and 2.5 cm/s (left to right). SNR in the wide part of the tube and in the constriction as a function of V_{enc} (c)*

4.3 In Vivo Studies

Figure 3 shows MIPs of the CD and PG images of the carotid bifurcation of another healthy volunteer. As indicated by the arrows, the PG MIP suggests larger diameters than the CD MIP, in accordance with the phantom results. Since the average velocity in the common and internal carotid arteries is higher than V_{enc}, the highest intensities are seen in those vessels in the PG MIP. Areas of laminar flow, e.g. proximal to the bifurcation in the common carotid artery and distal to the bifurcation in the internal carotid artery, will benefit from the phase dynamic range extension and will therefore yield a higher PG signal than around the bifurcation.

5 Discussion and Conclusions

Laminar and disturbed flow results in a predictable or unpredictable change of MR phase signal. In both cases, such this change of phase can be detected by means of the PG operator as a post-processing technique on a conventional PC dataset. In case of normal (laminar) flow, a velocity dynamic range extension is achieved as long as $\Delta\phi/\Delta pixel$ does not exceed π. Because of the dynamic range extension we are able to depict a large number of vessels with only one setting of V_{enc} without the need of multiple acquisitions and without the penalty of longer scan time! For disturbed flow profiles, the phase distribution within a pixel exceeds π. The phase and PG are not well defined and diameter measures cannot be given. Nevertheless, since the PG detects any change of phase, disturbed flow is also visualized. The PG MIPs of the flow phantom studies show signal present in the tube constriction, nearly independent of V_{enc}. When producing PG angiographic images, the major difference with conventional phase contrast angiography is the setting of V_{enc}, which is less critical and should be set equal to the *lowest* velocity to be depicted rather than the *maximum* velocity present in the imaging volume. However, we found that V_{enc} is practically bounded to a lower limit of about 5 cm/s, which means that very slow flow cannot easily be visualized by means of PG images. It is possible to accurately detect vessel walls from PG images confirmed by the numerical simulations. It was shown that the PG FWHM measure offers the best diameter estimates. PG images show less variation in diameter when changing the window and level settings compared to CD images. This was seen in both the in vitro and the in vivo studies. PG MIPs show an important increase in SNR compared to CD MIPs, because 1) the PG operation enhances signal at vessel walls and 2) PG images give signal when flow is disturbed and 3) PG images benefit from the dynamic range extension when V_{enc} is set low with respect to the maximum velocity present. PG images cannot easily be used for vessels in or near regions of air and bone since the PG gives high output in these areas. Suppression of venous flow is difficult and image segmentation is commonly needed before visualization.

To conclude, PG images can be calculated from real and imaginary images, without phase-wrap complications. PG images show enhanced intensities at vessel walls and in regions of phase dispersion, but also in areas of low signal magnitude, like air and bone. Compared to conventional PC images, a broader range of velocities can be depicted with one setting of V_{enc}, owing to the differential nature of the PG, which only puts a limit on the maximum change of phase per pixel. It is also shown that PG MIPs produce higher signal-to-noise ratios than conventional PC angiograms and give a better impression of the vessel wall location and vessel diameters for both laminar and disturbed flow.

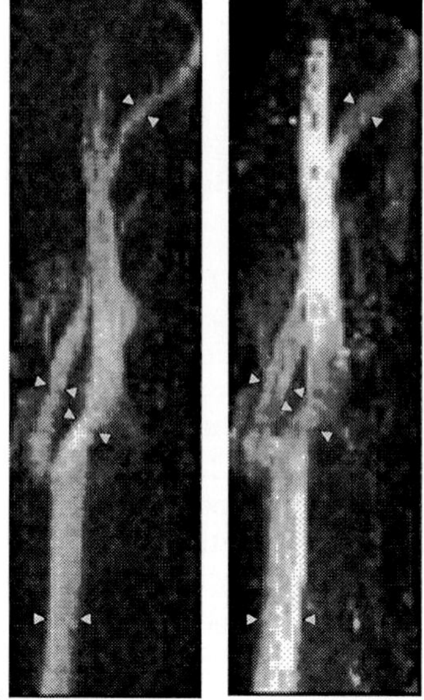

Fig. 3.: *MIPs of (a) CD images and (b) segmented PG images of the carotid bifurcation for V_{enc}=20 cm/s.*

References

1. Evans AJ, Richardson DB, Tien R, MacFall JR, Hedlund LW, Heinz ER, Boyko O, Sostman HD. Poststenotic signal loss in MR angiography: effects of echo time, flow compensation and fractional echo. American Journal of Neuroradiology 1993; 14:721–729.
2. Oshinski JN, Du DN, Pettigrew RI. Turbulent fluctuation velocity: The most significant determinant of signal loss in stenotic vessels. Magnetic Resonance in Medicine 1995; 33:193–199.
3. Urchuk SN, Plewes DB. Mechanisms of flow-induced signal loss in MR angiography. Journal of Magnetic Resonance Imaging 1992; 2:453–462.
4. Anderson CM, Saloner D, Tsuruda JS, Shapeero LG, Lee RE. Artifacts in maximum-intensity-projection display of MR angiograms. American Journal of Radiology 1990; 154:623–629.
5. Bowen BC, Quencer RM, Margosian P, Pattany PM. MR angiography of occlusive disease of the arteries in the head and neck: current concepts. American Journal of Röntgen Ray Society 1994; 162:9–18.
6. Patel MR, Klufas RA, Kim D, Edelman RR, Kent KC. MR Angiography of the carotid bifurcation: artifacts and limitations. American Journal of Röntgen Ray Society 1994; 162:1431–1437.
7. Hoogeveen RM, Bakker CJG, Viergever MA. Improved visualisation of vessel lumina using phase gradient images. volume 1, page 578. Proc. SMR/ESMRMB, 3rd and 12th annual meeting, 1995.
8. Lee AT, Pike GB, Pelc NJ. Three-point phase-contrast velocity measurements with increased velocity-to-noise ratio. Magnetic Resonance in Medicine 1995; 33:122–126.
9. Polzin JA, Alley MT, Korosec FR, Grist TM, Wang Y, Mistretta CA. A complex-difference phase-contrast technique for measurements of volume flow rates. Journal of Magnetic Resonance Imaging 1995; 5:129–137.

Visualization and Labeling of the Visible Human™ Dataset: Challenges & Resolves

R. Mullick and H. T. Nguyen

Center for Information Enhanced Medicine (CIeMed)
Institute of Systems Science, Nat. Univ. of Singapore, SINGAPORE 119597.
EMail: {rakesh,htn}@iss.nus.sg

Abstract. The Visible Human™ is a vast resource, which few methods can fully explore and analyze. We aim to completely label and volume display it; here we discuss the main obstacles, and our means for overcoming them. Our segmentation into over 350 anatomical regions spans the various body systems. Volume rendering by a memory-independent, partitioned, color and translucency scheme gives realistic human images, with many possibilities for medical education and "edutainment".

Keywords Human Anatomy, Volume Rendering, Visualization, Visible Human, Segmentation

1 Introduction

The Visible Human™ dataset (VHD) from the National Library of Medicine [1, 2] is the much-awaited standard for medical image visualization, analysis, and registration. The dataset consists of mutually registered MRI, CT, and high resolution cryogenic macrotome sections of a male ($0.33mm \times 0.33mm \times 1.0mm$) and female ($0.33mm \times 0.33mm \times 0.33mm$) cadaver. Numerous research and commercial sites are developing tools to browse, visualize, and label this gigantic 3D dataset. Although details of segmentation methods used at various centers are not available in public domain, it appears that most groups have based their labeling of the VHD on one of the RGB channels of the macrotome sections. A majority of the groups report usage of manual methods to achieve this monumental task. The sheer magnitude of the data offers numerous challenges in management, visualization and analysis. We have embarked on the task of completely labeling this data into the various anatomical regions, and have developed visualization techniques to volume render it in its entirety. Our approach allows the user to independently specify the various rendering parameters, and relative translucency, of each segmented anatomical structure (Figure 3). We present an

* The authors thank Tim Poston and Raghu Raghavan for their valuable technical and editorial comments. We are grateful to Ms. Jin Xiaoyang and all the NUS students who enthusiastically segmented the data with great patience and precision. We also express our gratitude to Chun Pong Yu and Pingli Pang who developed the keyframe animation, and built and integrated the GUI for the rendering system. Thanks to Seema Mullick for her support in segmenting data and in matters related to design aesthetics. This work would not have been possible without the support of the National Institutes of Health for the Visible Human Project.

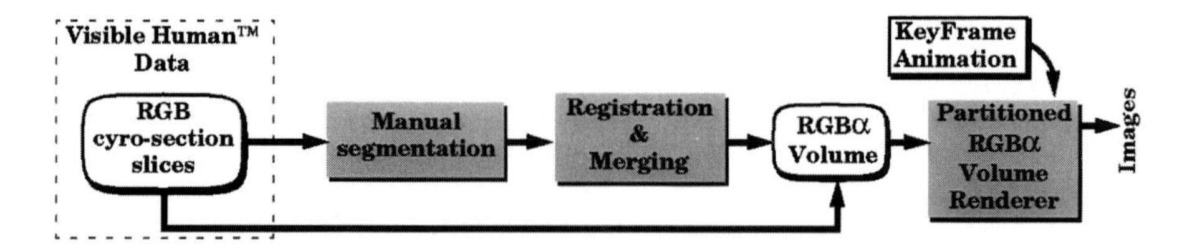

Fig. 1: Information Flow

overview of our methods for segmentation and visualization of this data. The challenges faced at each step of this process are discussed in the relevant sections, followed by proposed ways to resolve them.

2 Information Flow

We have used the full resolution RGB for segmentation of the VHD section data. The labeling was done manually by drawing Bézier curves for each anatomical region in the image. Corresponding slices of the segmented(α)-volume and photo RGB sections were then registered to generate the RGBα volume, which was used for all volume rendering. An overview of the information flow is presented in Figure 1.

3 Segmentation & Labeling

Anatomical regions in each RGB section of the data were outlined manually by students from the Anatomy, Medicine, and other science departments at the National University of Singapore, under the supervision of CIeMed staff. For each structure, the student sketched out one or more Bézier curves, such that the interior of the region corresponded exactly to that of the anatomical feature. This representation of the labeling has infinite resolution and offers easy scalability of any structure, with an option to generate its geometric model. Figure 2(a) shows the liver region outlined for slice 1500 of the dataset. Thereafter, a labeled image is generated for each slice using the control points and label of each contour. This is done in two steps: (i) A connected raster version of each curve is drawn into a blank (or partially labeled) image; (ii) then, an efficient region filling algorithm is used to fill/erase the interiors of each curve with a region-specific label represented by an intensity value. The region filling algorithm robustly handles special cases such as regions with single or multiple hollow interiors.

The stability of the region filling algorithm comes from a two-mode filling technique. Once the closed contours corresponding to each label are drawn into an image, the region filling algorithm begins by employing the traditional parity check method[3] in a left-to-right raster scan. Then, at points of possible ambiguity, defined as boundary locations and pixels adjacent to the boundary, the algorithm employs a "point in/outside polygon" test in order to correctly label the underlying region. The result of labeling the slice in Figure 2(a) is presented in Figure 2(b).

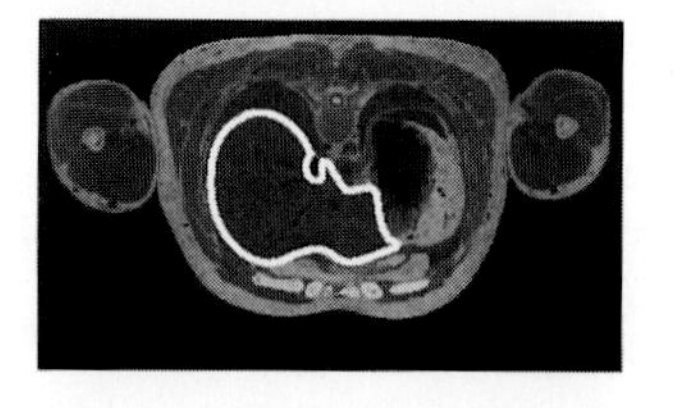

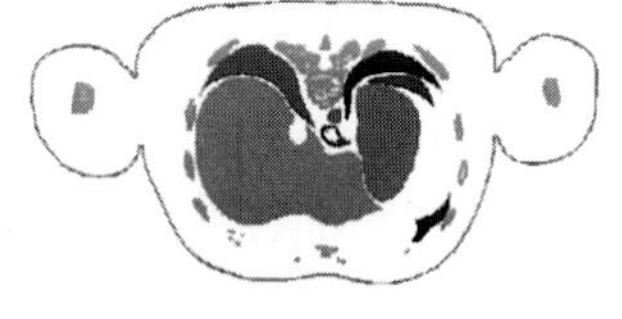

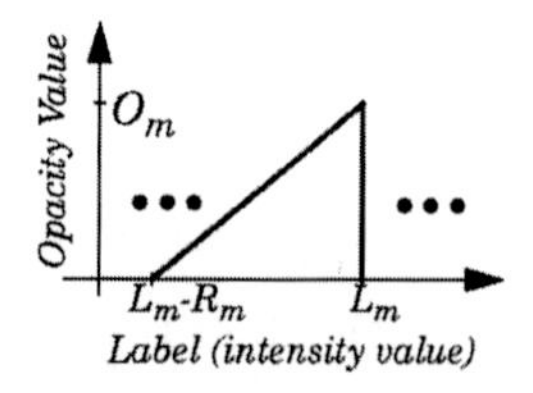

Fig. 2: (a) Manual feature extraction from RGB section data (liver outlined); (b) Labeled image; (c) Opacity value determination.

Although our segmentation of this dataset is an ongoing process, we have labeled all the visible anatomical structures comprising the primary human systems. They are: (i) Central Nervous; (ii) Respiratory; (iii) Digestive; (iv) Excretory; (vi) Skeletal; (vii) Circulatory; (viii) Endocrine; and (ix) Sensory organs. A Java Applet to view the labeled VH data is available on WWW at `http://ciemed.iss.nus.sg/research/brain/IIBA/JavaAtlas/VHD.html`.

A majority of the difficulties of this phase were linked to one main issue, the sheer size of this dataset. To flawlessly store, modify, and manage this data requires significant amounts of time, human resources, and hardware. In order to minimize some of this burden, an in-house interactive volume browser was developed. The browser facilitates efficient resource management, with easy search options for volume data and associated anatomical structures. The magnitude of this dataset also pushes the hardware memory requirements far beyond their limits, allowing few ways to simultaneously segment or visualize a particular chunk of data.

4 Volume Rendering

We have developed a new volume rendering technique called Partitioned RGBα Volume Rendering of Segmented Data (ParVo) to visualize the segmented photo images of the VHD. The unique features of this approach are:

- size independent translucent volume rendering, with an option to render a specified block of slices one at a time,
- unified rendering of RGB color channels and the α-channel (labeled volume),
- removal of possible aliasing effects during interpolation of α-channel using a constrained 3D blur operation.

The volume rendering method is based on object space projection rendering. This basic approach has been modified to resample all voxels to planes transverse to the viewing direction, then project them onto the viewing plane in back-to-front or front-to-back order using a color blending equation [4]. A detailed description of this method is given in [5, 8]. The following paragraphs highlight the qualitative enhancements to this method.

1. A special 3D blurring operation is applied to the α-channel to remove aliasing effects due to a round-off methodology used in interpolation during resampling. Each anatomical structure represented by a specific intensity value

($L_m, m \in$ [1, Number of regions]), is allowed a unique range (R_m) for the blur operation. The blur operation at a voxel v_{xyz} is carried out only if the $n \times n \times n$ neighborhood surrounding this voxel has only one label (excluding the background). This operation averages the transformed values (a_{xyz}) of the labels (v_{xyz}) in a $n \times n \times n$ neighborhood: $\bar{v}_{xyz} = \frac{\sum_i \sum_j \sum_k a_{ijk}}{n^3}$, where $a_{ijk} = \begin{cases} L_m - R_m, \text{if } v_{ijk} = 0 \\ L_m, \text{otherwise} \end{cases}$. The transformation limits the output of the blur operation to fall within the specified range R_m.This step is applied to the α-volume only once prior to rendering.

2. An opacity of range between 0 and 1 is assigned to each label. To smooth out the aliasing effect at the surface, opacity within the blurring range (R_m) of the specified label is linearly ramped from 0 to the object's opacity value (O_m). (See Figure 2(c))
3. Up to four light sources can be specified, with independent control of their location, intensity, color, attenuation, and specular parameters based on the Phong lighting model.
4. The surface normal for shading and lighting is approximated based on the gray-scale value computed from the R, G, and B values using the equation $g = 0.299R + 0.587G + 0.114B$. Four methods of gradient calculation are provided: central difference, adaptive gray-level, Zucker-Hummel[7], and 3D Sobel edge detector which is a 3D extension of the 2D Sobel operator [6].
5. Each channel (R, G, and B) is blended, shaded, and lighted individually.

A voxel based volume of 1-*mm* resolution of VHD was created from photo images (as rendered in Figure 3) containing a total of ($586 \times 340 \times 1878 \times [3+2+1]$) 2.24 Gbytes where each voxel consists of three bytes for RGB, a two byte label, and one byte for the grey channel. Hard disk requirements ($\approx$17 Gbytes) limited us in creating and thus rendering the entire dataset in full resolution. With a volume dataset of this magnitude, it is not practical to require that the entire volume be resident in memory. Hence a divide-and-conquer approach called Partitioned Volume Rendering [5] was adopted. PaVe4-16 was originally designed for multi-processor systems, but has been applied quite naturally and efficiently to a single processor architecture (ParVo [8]) to handle large datasets. ParVo renders n slices of data (n depends on the available memory and memory required to contain each slice) at a time, and merges (blends) each of them in order to form a final image. The slice, and hence partition, ordering depends on the viewing direction. ParVo has allowed us to render the 1-*mm* resolution multi-Gbyte VHD volume on an SGI Indigo-2 with 32 MB of main memory in $<$ 3 hours (*time estimate for unoptimized code*).

5 Results & Future Directions

A rendition highlighing many of the over 350 segmented anatomical structures is presented in Figure 3. This translucent view (15° roll) of the human form clearly depicts the relative and absolute position of bodily organs with respect to the skin surface and the skeletal system. Most of the images presented here have been rendered with a single light source, Zucker-Hummel gradient estimate, and back-to-front compositing.

Figure 5 is a detailed visualization of skin of the whole body rendered from a 586x340x1878 sized RGB volume. An isolated illustration of the pelvis region is depicted in Figure 4. This was rendered from a $1mm$-cubic voxel RGB volume of size 320x192x300. An unshaded unlighted visualization of the neuroanatomy is illustrated in Figure 7. Pseudo-color volume rendering of the blurred α-volume for the digestive and cardio-pulmonary systems is shown in Figure 6 and 8 respectively. (Additional images and animations can be obtained via the WWW site at **http://ciemed.iss.nus.sg/research/human_av/human_av.html**)

Naming of each skeletal tissue and labeling of the muscular system is also being completed. Other more efficient methods and fidelity enhancement techniques for rendering this type of data are under investigation.

Transformation of the raw VHD into structured geometric objects is vital for all uses beyond passive display. For a simulated catheter to explore its arteries, the arteries' geometry must be known, before it can be translated into force walls that limit the catheter's motion. For the VHD head to smile realistically, the muscles must be mapped and transformed into biomechanical models that contract and bunch up according to suitable virtual force laws and nerve impulses. Transforming the VH into an electronic 'crash dummy' depends on passive biomechanics that follow the segmentation into bone, sinus, soft tissue and so on. All such things require, before any progress can be made, an effective segmentation that responds to discontinuity in any data channel (red, green, blue or scan density), and efficient manipulation of the mass of data involved. (Recall also that succeeding data sets will be far larger.) The toolkit described here is a key element in CIeMed's movement beyond visualization, to actively modeled 3D medicine.

References

1. Spitzer, V. *et al.*: The Visible Human Male: A Technical Report. Journal of American Medical Informatics Association (1995)
2. http://www.nlm.nih.gov/extramural_research.dir/visible_human.html
3. Pavlidis, T.: Algorithms for Graphics and Image Processing. Computer Science Press, Rockville, MD, (1982).
4. Levoy, M.: Display of surfaces from Volume Data. IEEE Computer Graphics and Applications. **8**.5 (1988) 29–37
5. Nguyen, H. T., Srinivasan R.: PaVe4-16: A Distributed Volume Visualization Technique. TR94-135-0 Technical Report, Inst. of Sys. Science, National Univ. of Singapore, Singapore (1993)
6. Jain, A. K.: Fundamentals of Digital Image Processing. Prentice Hall, Englewood Cliffs, NJ (1989)
7. Monga, O., Deriche, R., Malandain, G., Cocquerez: Recursive filtering and edge closing: 2 primary tools for 3D edge detection. Image and Vision Comp., **9**.4 (1991)
8. Nguyen, H. T., Mullick R., Srinivasan, R.: Partitioned Volume Rendering (ParVo): An efficient approach to visualizing large datasets. Manuscript in preparation.

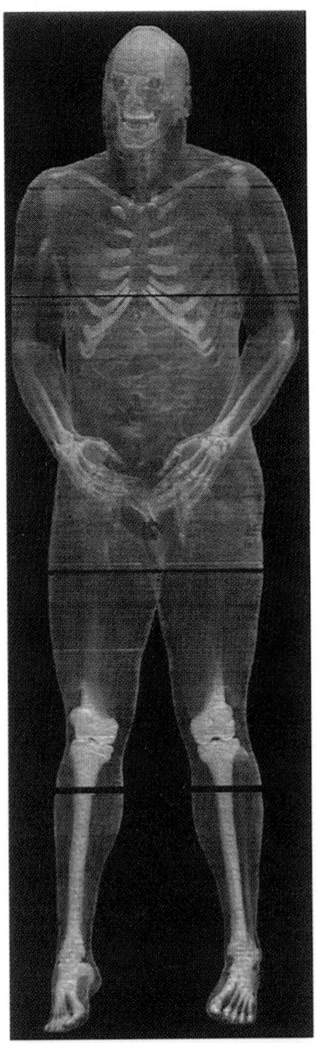

Fig. 3: Transluscent viz. of labeled VH Dataset.

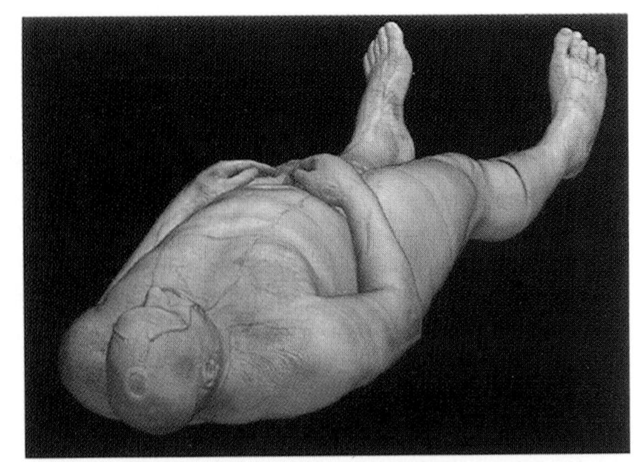

Fig. 5: Whole-body/Skin visualization.

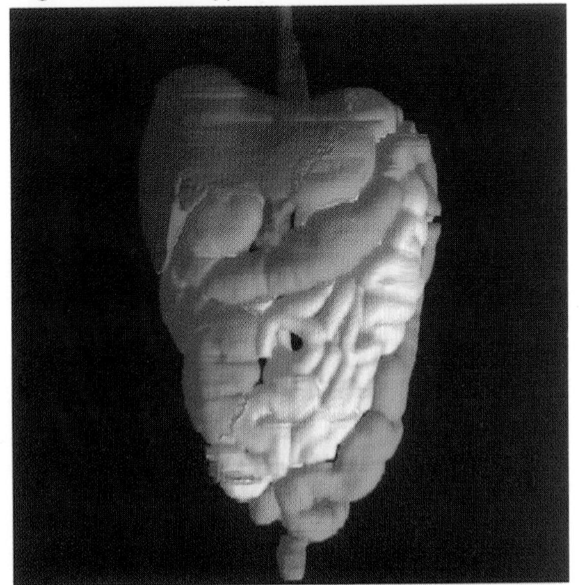

Fig. 6: Pseudo-color labeled visualization of blurred α-volume of digestive system.

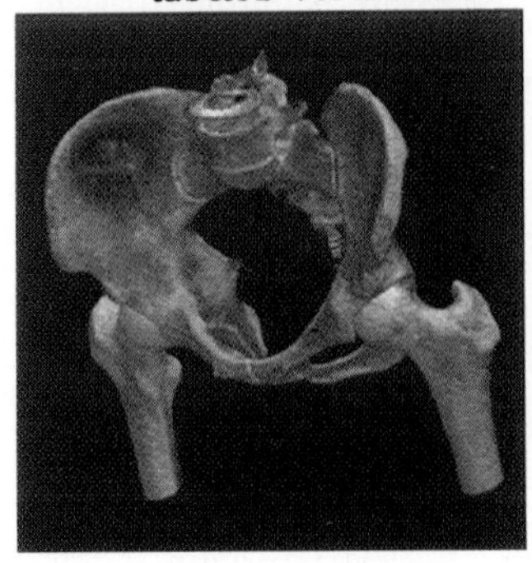

Fig. 4: Pelvis Structure

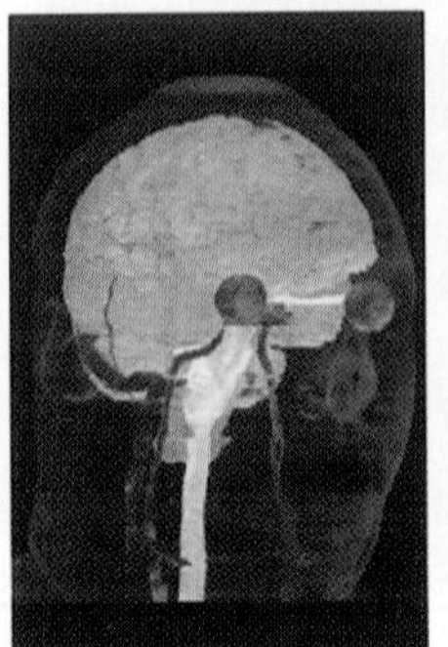

Fig. 7: CNS System

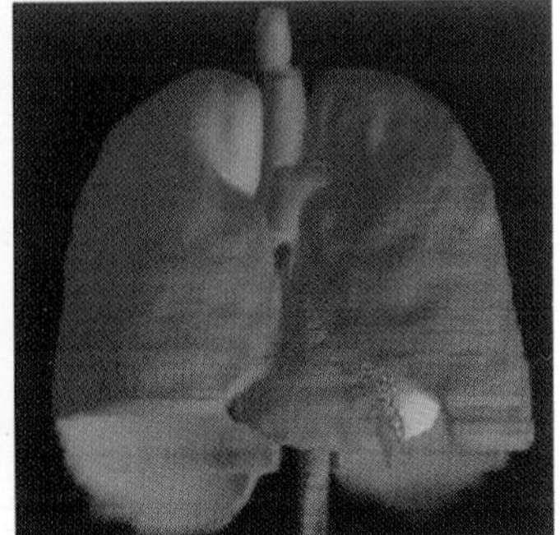

Fig. 8: Thoracic Region

Rebuilding the Visible Man

John E. Stewart[1], William C. Broaddus[1], and James H. Johnson[2]

[1]Division of Neurosurgery and Department of Biomedical Engineering
[2]Department of Anatomy
Virginia Commonwealth University / Medical College of Virginia, Richmond, Virginia 23298

Abstract. This paper describes a collection of techniques designed to create photo-realistic computer models of the National Library of Medicine's Visible Man. An image segmentation algorithm is described which segments anatomical structures independent of the variation in color contained in the anatomical structure. The generation of pseudo-radiographic images from the 24-bit digital color images is also described. Three-dimensional manifold surfaces are generated from these images with the appropriate anatomical colors assigned to each surface vertex. Finally, three separate smoothing algorithms -- surface, normal, and color, are applied to the surface to create surprisingly realistic surfaces. A number of examples are presented which include solid surfaces, surface cutaways, and mixed opaque and translucent models.

1 Introduction

"The Visible Man" is the name given to the male cadaver of the National Library of Medicine's Visible Human Project [1]. This project involved the medical imaging and physical sectioning of a male and female cadaver. We have focused our attention on the Visible Man data primarily because these were the data made first available to the public. The Visible Man was imaged using both Magnetic Resonance (MR) and Computed Tomography (CT) imaging techniques. MR images were spaced at 5mm intervals while CT images were spaced at only 1mm intervals. Pixel size of these image sets varied depending on the field of view of the scanner. CT images were obtained before and after the cadaver was frozen. Color digital (24-bit) images were taken after each 1 mm section of the cadaver was removed. These images are 2048 x 1216 pixels with a pixel size of 1/3 mm. The memory required to store all image data for the visible man is approximately 15 gigabytes.

There are currently a large number of institutions, both public and private, which are looking at possible uses for these data. Most of these are focused around educational uses such as interactive anatomy atlases or three-dimensional (3D) surfaces for virtual surgery. A basic requirement of these applications is that each image set be segmented to identify the anatomical structures contained within the image. 3D surfaces can then be constructed using these segmentations. In order to achieve the maximal visual effect, each surface should be colored using the anatomical colors contained in the 24-bit image data and shaded as realistically as possible. Techniques which can be used to achieve these goals are the topic of this paper.

2 Image Segmentation

The first step in creating 3D surfaces is segmentation. Segmentation of color digital images offers a unique challenge to those who are used to dealing with CT and MR grayscale images. Twenty-four-bit color images contain three 8-bit variables at each pixel -- red, green, and blue. The approach to segmenting these images is to create a new image which combines these three variables into a single variable at each pixel. The resulting image would then be analogous to a CT or MR image which could be segmented via thresholding. The challenge is to create a grayscale image from the digital color images which has high-intensity pixel values for the anatomical object of interest and lower-intensity pixel values elsewhere. This is particularly difficult when one realizes that most anatomical structures are made up of multiple colors that may be quite different from one another. Techniques which perform segmentation based on a range of red, green, or blue values will often produce segmentations which include much more than the object of interest. Other techniques based on color gradients may divide an anatomical structure into two parts if two colors are present in the structure.

The choice was made to look at the colors in the image as existing in a 3D Cartesian coordinate system, with red increasing along the x-axis, green increasing along the y-axis, and blue increasing along the z-axis (Fig. 1). Each axis has a range of 0 to 255 creating a *3D color volume* capable of representing every color in the 24-bit digital color images. A 1-byte variable is allocated for each point in this volume resulting in a total storage requirement of 16.7 megabytes. The values contained in these 1-byte variables will be used to transform the 24-bit color images into 8-bit grayscale images.

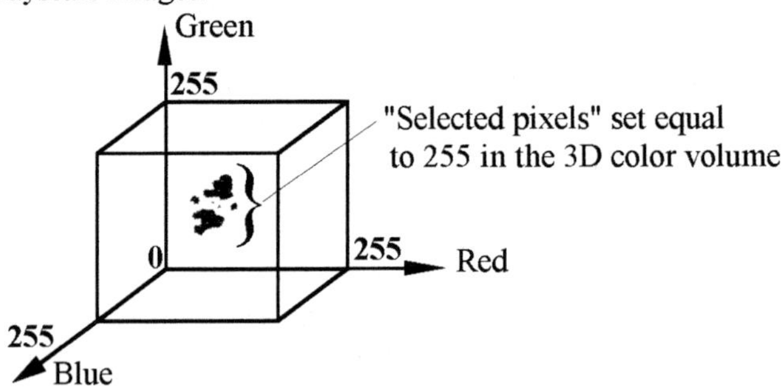

Fig. 1. 3D color volume.

All 1-byte variables in the color volume are initialized to a value of 0. A collection of colors is then interactively selected from the Visible Human Project 24-bit color images. These "selected pixels" belong to the anatomical structure to be segmented. A software system developed at Virginia Commonwealth University entitled *IsoView* [2] allows a user to perform this task. It is not necessary to meticulously identify a large number of pixels at this stage. A small group of pixels is often sufficient to select a structure. A value of 255 is placed in the 1-byte variables of the 3D color volume for all colors represented by the selected pixels. All other values in the volume remain equal to 0. If the selected pixels were drawn opaque, a set of small surfaces would appear in the 3D color volume (Fig. 1). Note that typically there are multiple, disconnected surfaces.

A dilation algorithm is then applied to the 3D color volume in order to fill in all 1-byte variables with a non-zero number. This algorithm begins with the 1-byte variables in the color volume which equal 255 but border a 1-byte variable equal to 0. These bordering variables are set equal to 254. This process is repeated for those

1-byte variables which equal 254 such that the bordering 1-byte variables which equal 0 are set equal to 253. Each iteration of this dilation algorithm adds a layer on top of the original selected pixels. The result is that every color in the 3D color volume has a value indicating its distance from the colors of the selected pixels.

A 1-byte pseudo-radiographic grayscale image is generated for each 24-bit color image by using the 3D color volume as a lookup table. The colors in close proximity to the selected colors are of high intensity while those distant from the selected color are of low intensity. This allows for image segmentation of the anatomical object of interest by thresholding. The threshold is interactively adjusted until contours wrap around the anatomical structure of interest in the image. If the segmentation does not capture the entire structure, the currently segmented pixels plus additional pixels manually selected are used as the selected pixels as described above. The grayscale images are then regenerated using the adjusted 3D color volume and a new threshold. The segmentation of the images generated in this step can then be saved to a file. This file contains an index to each image segmented and a 1-bit variable set to 1 for pixels above the threshold and 0 for pixels below the threshold.

3 Segmentation of Bone

Segmentation of bone is performed in a different manner than that of other anatomical structures. Because bone is well delineated on CT images, the CT images of the Visible Man replace the 1-byte grayscale images described above. This requires that the CT images be registered with the 24-bit digital color images such that there is a one-to-one correspondence between pixels in both sets of images. Fortunately, there is a CT image of the Visible Man taken in the same plane as almost every color image. Unfortunately, these images are not registered with the color images and do not have the same pixel size as the color images. To remedy this problem, a separate program was written which reads in the CT images, finds the pixel size, resamples the CT images using bilinear interpolation, and outputs a new image with the same pixel size as the 24-bit color images. The x and y offset of the CT images are read from the command line allowing the new CT images to be shifted to register with the 24-bit color images. This program allows a user to manually register the CT images with the color images. The registered CT images are then stored for future use.

4 Surface Creation

Surfaces are created using the software system IsoView [2]. This system uses the Marching Cubes [3] and Border Case Comparison [4] algorithms to create 3D manifold surfaces made up of triangles. IsoView runs on a Silicon Graphics Indigo2 with an Extreme graphics engine. Because the Extreme graphics engine does not support texture mapping, a color is stored at each vertex to be used later in rendering. This color comes from the 24-bit digital color images and is determined at the time the vertex is created. Figure 2 illustrates a typical voxel triangulation as defined by Marching Cubes. The color assigned to this vertex is the color of the "on" voxel vertex. This is demonstrated by the upper and lower case letters in Fig. 2. Note that the edge vertex is the same letter as the on voxel vertex of that edge. Linear interpolation of the red, green, and blue colors of the image is not appropriate since this would result in a mixing of the colors outside the surface with those inside the surface. A single 4-byte variable stores the color and alpha value. The alpha value will be used later to permit surface transparency.

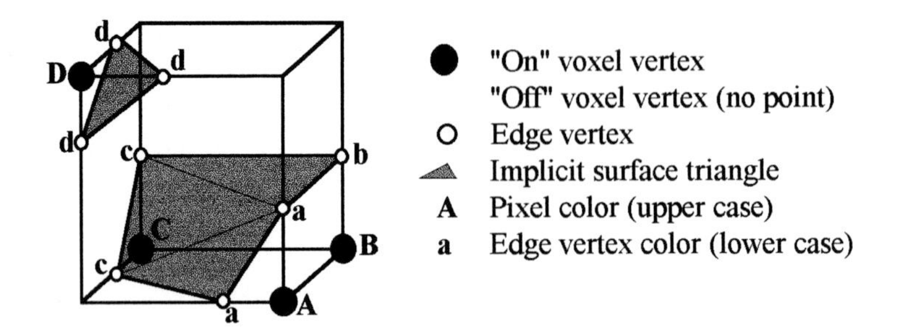

Fig. 2. Marching Cubes triangulation showing edge vertex color assignment.

Surface normals are computed by taking an area-weighted average of the triangle normals surrounding a vertex. These normals are then normalized to unit length and stored at the vertex to permit surfaces to be Gouraud shaded.

5 Smoothing

It became apparent soon after creating the first few 3D surfaces from the Visible Man that some degree of smoothing is necessary if these surfaces were to appear lifelike. After experimenting with a number of different algorithms, three separate types of smoothing were incorporated into IsoView -- surface smoothing, normal smoothing, and color smoothing. All three types are essentially Laplacian smoothers with constraints applied to them. All smoothing is vertex centered rather than triangle centered. The surface and normal smoothers are constrained by the angle formed between the central vertex normal and the surrounding normals. If this angle exceeds 45 degrees, smoothing is not performed for this vertex. This constraint preserves the appearance of surface edges. The color smoother is constrained by the difference between the red, green, or blue index of the central vertex and the surrounding vertices. If either of these indices differs by more than 25 from the central vertex indices, the vertex is not smoothed. This constraint permits large changes in surface color to be preserved.

The number of iterations for each type of smoothing is determined by the user. The results of the smoothing are displayed immediately to permit the interactive selection of the most effective smoothing criteria.

6 Results

Figures 3 and 4 are representative 3D surfaces of the Visible Man created using the techniques described above. Each figure is a composite of a number of surfaces created from the Visible Man data. The skull seen in these figures is created from both the registered CT images and the 24-bit color images. The entire skull is made up of 881,000 triangles and requires 30 seconds to generate on a Silicon Graphics Indigo2 Extreme with a 150 MHz processor and 192 Mbytes of RAM. The muscle and brain surfaces are generated from the 24-bit color images using the pseudo-radiographic images created from these 24-bit color images.

Figure 3 demonstrates the use of cut-away surfaces to display the internal anatomy of the Visible Man. These cutaways are generated by simply limiting the range of pixels which can be used to create the 3D manifold surfaces. Transparency can also be used to display the internal anatomy through solid surfaces. Figure 4 demonstrates the internal anatomy of the head with the skin made 50% transparent. This is accomplished by setting the alpha value of the 32-bit triangle vertex color to 128.

7 Conclusion

The Visible Human Project has provided an amazing (and somewhat overwhelming) amount of anatomical data which can be used for a multitude of applications. In this report, we have focused on the generation of photo-realistic 3D anatomical surfaces for the purposes of education. Although these surfaces clearly illustrate the human anatomy, they are too complex to be rendered at interactive speeds with typically available hardware. Recent efforts have been aimed at surface simplification (decimation) which should provide a means of interactively manipulating these surfaces in 3D. The ultimate goal of this project is to create a database of 3D surfaces which can be used for medical education, surgical simulation, and finite-element modeling.

8 Acknowledgments

This work was supported in part by the Jeffress Memorial Trust and the Virginia Commonwealth University / Medical College of Virginia M.D./Ph.D. Fund.

References

1. Ackerman, M.J.: The Visible Human Project. J. Biocomm. vol. 18 no. 2 (1991) pp. 14

2. Stewart, J.E.: IsoView: An Interactive Software System for the Construction and Visualization of Three-Dimensional Anatomical Surfaces. Va. Med. Q. vol. 121 no. 4 (1994) pp. 256

3. Lorensen, W.E. and Cline, H.E.: Marching Cubes: A High Resolution 3D Surface Construction Algorithm. Comput. Graphics. vol. 21 no. 4 (1987) 163-169

4. Stewart, J.E., Samareh, J.A., and Broaddus, W.C.: Border Case Comparison: A Topological Solution to the Ambiguity of Marching Cubes. IEEE Comp. Graphics, Submitted.

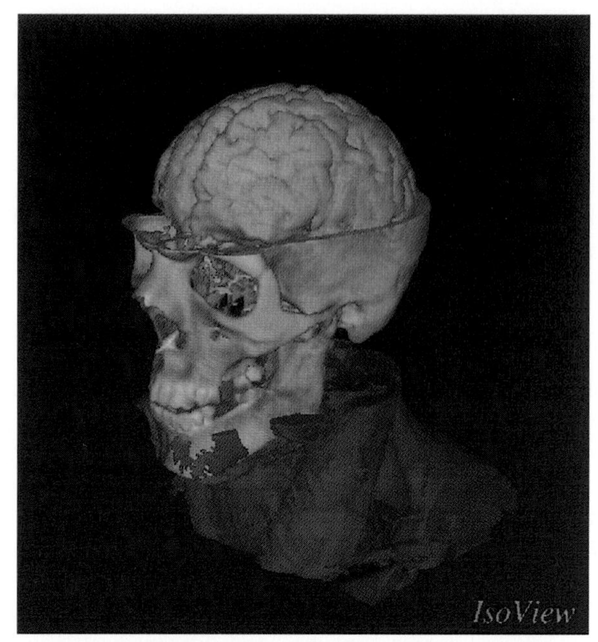

Fig. 3. Brain, skull and muscle composite picture.

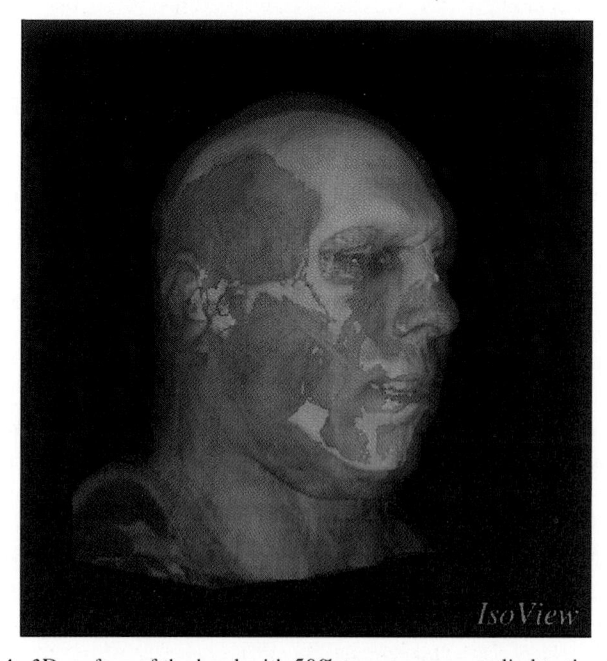

Fig. 4. 3D surface of the head with 50% transparency applied to the skin.

Image Quality Improvements in Volume Rendering

Jeroen Terwisscha van Scheltinga[1], Marco Bosma[2],
Jaap Smit[2] and Steven Lobregt[1]

[1] Philips Medical Systems Nederland B.V., Best, The Netherlands
[2] University of Twente, Enschede, The Netherlands

Abstract. This paper presents some methods to improve the image quality obtained with volume rendering. By computing the opacity, color and gradient of each sample point directly at the sample position, the image quality has improved over methods which compute these values at the voxel positions. A new method for calculating the gradient is presented. These improvements result in small details becoming clearly visible. It also allows high zoom rates without generating blurry images. The opacity is corrected for the sample rate, allowing a consistent translucency setting.

1 Introduction

Volume rendering generates images directly from three-dimensional range data, for example medical datasets from CT or MRI scanners. The ray casting method of Levoy [1] is well known. This method (pre-)calculates opacity, color and gradients at voxel locations, and uses interpolation to obtain the values at sample locations.

Our method, which we called super resolution [2], [3], is based on Levoy's, but produces higher quality images, because of different ways to calculate the sample opacity, color and gradient directly at sample locations.

The sample rate along rays is often constant. But when an image is magnified, the sample rate should increase to avoid artefacts. Then the opacity should be adjusted because more samples are taken in the same region.

2 Opacity and Color Calculation

Levoy first performs classification to find the opacity at voxel locations, after which interpolation gives the opacity at the sample location. The color is also first computed at the voxel locations, and then interpolated to the sample location.

Another possibility is to first interpolate the grey-values to find the grey-value at the sample location, after which classification will provide the opacity. This results in a better image quality, because now the classification is performed at the correct position.

These processes are illustrated in Fig. 1. An original object (a) is sampled on a 16×16 grid (b). Levoy then applies classification (c), after which interpolation gives the final opacity (d). Our method first applies interpolation (e), and then classification (f). The Levoy method results in a blurred image with blocking artefacts, while super resolution results in (almost) the original image.

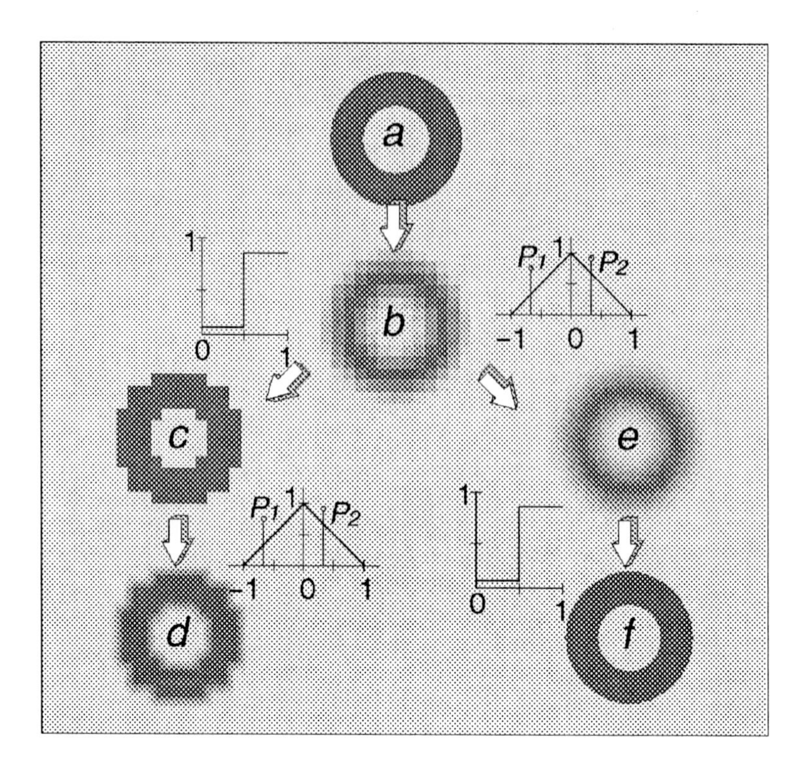

Fig. 1. Opacity calculation with the Levoy method (*abcd*), and super resolution (*abef*)

Likewise, the color can be computed directly at the sample locations. Therefore it is essential to have the grey-value gradient available at arbitrary positions. This will be discussed in the next section.

3 Gradient Calculation

The surface normal used to perform shading is approximated by the grey-value gradient. In existing methods by Levoy [1] and Höhne [4], the gradient is often computed at a voxel location using the central difference method. Interpolation of these gradients at the voxel grid is used by Zuiderveld [5] to provide the gradient at the sample location.

A better method is to use the intermediate difference method. Here two neighboring voxels are subtracted to give a gradient component halfway these voxels (Fig. 2). This results in three new grids (one for each component) which are shifted half a voxel in each corresponding component direction. An interpolation for each component is used to calculate the gradient at the sample location.

Note that on exact voxel locations the two methods give the same results. The advantage of the intermediate difference method is that the gradient is calculated using a one voxel distance instead of two for the central difference method. This results in a better image quality, especially in regions with higher frequencies.

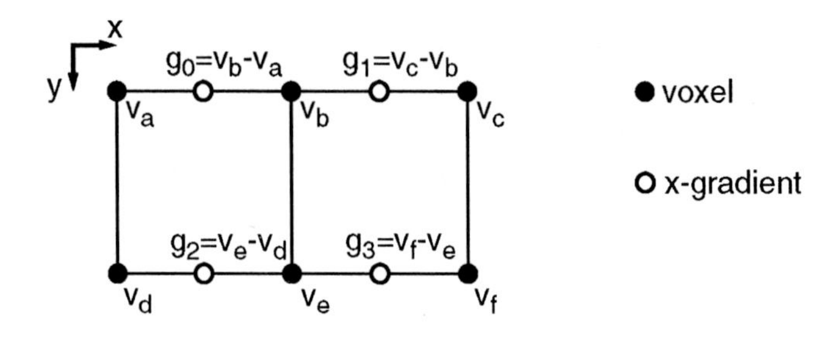

Fig. 2. Intermediate difference gradient grid for the x-component

4 Image Quality Comparison

To demonstrate the advantage of our super resolution method, the image quality obtained with both methods are compared. Therefore a cone with 3 slits of 2, 1 and 0.5 voxels wide is used. The image is rendered with a high zoom factor, to clearly show the differences.

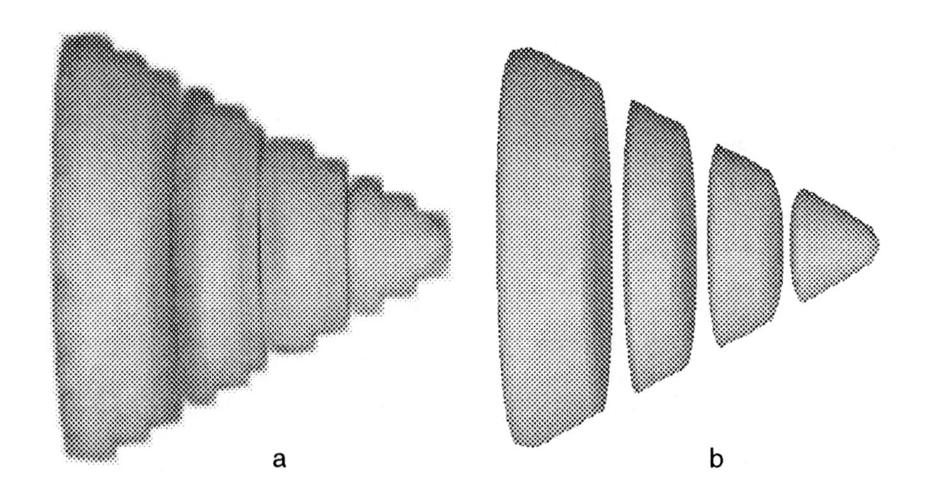

Fig. 3. Cone rendered with the Levoy method (a), and super resolution (b)

Fig. 3 shows the result of the Levoy method (a), and of super resolution (b). In (a) the image is blurry and individual voxels can clearly be recognized, furthermore the slits are hardly visible. In (b) the image is much sharper and the individual voxels have disappeared, while the slits are clearly visible. So the image quality has improved over the existing methods, because small details become sharp and clearly visible.

5 Opacity Correction for Sample Rate

It is desirable to be able to change the sample rate along a ray, because a high sample rate is computationally expensive, while a low sample rate gives lower quality images. When zooming in, a higher sample rate is normally needed to prevent the occurrence of unnecessary artefacts.

Simply changing the sample rate will result in an altered image. This can be understood by looking at the formula for the color of an image pixel. The color $C(i,j)$ of each image pixel for N samples along a ray is given by (see [1]):

$$C(i,j) = \sum_{k=1}^{N} \left\{ \alpha(i,j,k) \cdot C(i,j,k) \cdot \prod_{l=1}^{k-1} (1-\alpha(i,j,l)) \right\} \tag{1}$$

With $C(i,j,k)$ the color and $\alpha(i,j,k)$ the opacity of a sample point.

If the sample rate is increased, the total translucency of a region is reduced, because more samples are taken in the same region. So it is necessary to adjust the sample opacity values to the used sample rate.

If the sample rate is increased by a factor n with regard to a reference sample rate, then for each position on a ray the total translucency of the previous part of the ray must be the same, so:

$$\prod_{l=1}^{k} (1-\alpha_r(l)) = \prod_{l=1}^{n \cdot k} (1-\alpha_n(l)) \tag{2}$$

If we assume the opacity of a region is constant (α_r for the reference sample rate), this results in the corrected opacity α_n:

$$\begin{aligned} 1-\alpha_r &= (1-\alpha_n)^n \\ \alpha_n &= 1-\sqrt[n]{1-\alpha_r} \end{aligned} \tag{3}$$

This opacity correction, introduced in [2], allows the translucency to be defined in translucency per unit distance, so it is independent of the voxel size and thus always consistent, even between datasets. This makes it possible to do a low quality but fast rendering at a low sample rate, and a high quality but slower rendering at a high sample rate, without changing the translucency of a region.

6 Results

Fig. 4 shows an image of a MRA dataset with an AVM in the head, generated using the super resolution method. The image is sharp and even clearly shows small vessels without blocking artefacts. The shading is correct, even for small vessels with a size approaching the voxel distance.

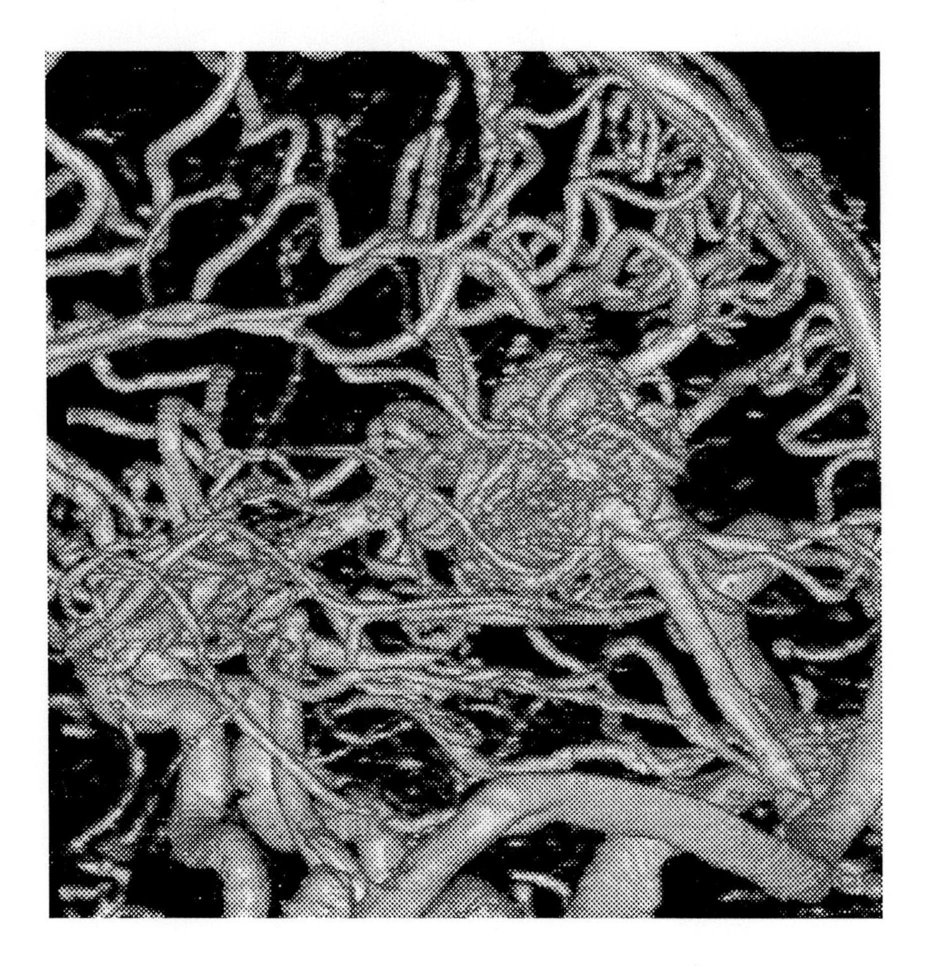

Fig. 4. MRA dataset showing an AVM in the head

7 Conclusions

The image quality obtained with volume rendering has improved over existing methods. The super resolution method, which calculates the opacity and color directly at sample positions instead of at voxel positions, greatly improves the image quality, because images become sharp and small details clearly visible. The calculation of the grey-value gradient is improved by using the intermediate difference method, which results in a better shading of the images. The opacity correction for the sample rate makes it possible to adjust the sample rate without changing the translucency of a region.

References

1. Levoy, M.: Display of Surfaces from Volume Data. IEEE Computer Graphics and Applications **8**(3) (1988) 29-37
2. Bosma, M., Terwisscha van Scheltinga, J.: Efficient Super Resolution Volume Rendering. Master's Thesis University of Twente (1995)
3. Bosma, M., Smit, J., and Terwisscha van Scheltinga, J.: Super Resolution Volume Rendering Hardware. Proceedings of the 10th Eurographics Workshop on Graphics Hardware. (1995) 117-122
4. Höhne, K.H., Bomans, M., Pommert, A., Riemer, M., Schiers, C., Tiede, U. and Wiebecke, G.: 3D Visualization of tomographic volume data using the generalized voxel model. The Visual Computer. **6** (1990) 28-36
5. Zuiderveld, K.J.: Visualization of Multimodality Medical Volume Data using Object-Oriented Methods. Thesis Universiteit Utrecht (1995)

Real-Time Merging of Visible Surfaces for Display and Segmentation

Robert P. Grzeszczuk[1] and Charles A. Pelizzari[2]

[1] Departments of Radiology and
[2] Radiation Oncology, the University of Chicago

Abstract. A technique has been developed which combines the high image quality of volumetric rendering with the interactivity associated with surface rendering by texture mapping pre-rendered images onto depth map-based partial models of the surface. An original real-time algorithm is presented which allows us to combine multiple overlapping visible surfaces free of occlusion artifacts associated with poorly estimated fragments. In virtual exploration, decoupling the rendering process from display can result in improved interactivity and navigability without a corresponding loss of image quality. Visible surfaces can also be used as an aid in segmentation for applications requiring object models and to easily combine volumetrically defined structures with purely geometric information (e.g., computed isodose surfaces).

1 Introduction

In many visualization tasks a scene is repeatedly rendered from a number of similar viewpoints. In the conventional approach each new view is rendered from a full geometric model (i.e., a mathematical description such as a set of polygons or voxels). However, such methods place a limit on the scene complexity and/or rendering quality which is particularly acute in real time applications like architectural walk-throughs and virtual endoscopic procedures. A number of solutions have been proposed that avoid the problems associated with exploring large environments by abandoning the concept of the scene model altogether. Instead, each new view of the environment is obtained from a set of sparsely acquired pre-rendered images via interpolation [2] [1] [3]. The technique we propose here is representative of this class of solutions in that it uses image-based description of the scene for the sake of rendering. However, it is not a true view interpolation technique as 3D information from partial surfaces is used explicitly to resolve occlusion ambiguities via re-projection.

The aim of the present work is to create a visualization tool that provides high quality renditions of volumetric data sets at interactive speeds and at the same time aids in the task of segmenting complex 3D structures. Several views of an object are rendered volumetrically, producing high quality images in a relatively long time. The range images (or Z buffers) produced as a side effect of the process are retained in order to reconstruct the visible surface. These two images (i.e., the rendered view and the range image) are combined via texture mapping to arrive at a partial 3D surface model. By decoupling the rendering and display processes we arrive at an object that retains the high visual quality characteristic of a volumetrically-rendered image, at the same time offering interactivity typical of a surface-based rendering resulting in increased navigability in large environments. In addition, the partial surface provides a geometric model description that is missing from volumetrically rendered views. By combining several such partial models we can build complete surfaces which fully enclose structures of interest. These in turn can be used for data set segmentation. Finally, partial surface models are suitable for combining volumetrically rendered structures with purely geometric ones (e.g., dose distributions, models of segmented structures, models of radiation beams).

During the process of volumetrically rendering a tomographic data set, rays are traced from the eye point through the volume, and visual attributes such as color and opacity are accumulated along each ray. This process terminates when enough opacity along the ray path is accumulated. It is fairly straightforward to record the termination point for each ray. Such a depth map image is commonly referred to as a Z-buffer. Figure 1 shows an example of a rendered image and the corresponding depth map. Because the ray termination points together with the model/view transformation matrix define coordinates on the portions of surfaces in the rendered image that are visible from the eye point, we can use the depth map to re-create a partial model of an object (called the visible surface) as shown in Figure 2a. Clearly, the partial surface itself is a fairly poor visualization of the model and by no means comparable to traditional surface-based methods; this is is the result of quantization and uncertainty as to the actual ray termination. In order to enhance the rendering quality we may texture map the rendered image onto the surface. The resulting hybrid rendering (Figure 2b) can be manipulated rapidly, as is typical of purely geometric rendering schemes, while retaining the high quality of the volumetrically rendered image.

Naturally, a visible surface model derived from a depth map as described above is only suitable for viewing the object from viewpoints that are sufficiently close to that of the original rendering (which we call the reference view), otherwise visibility artifacts occur (Figure 2c). However, partial surfaces from other views (Figure 2d) can be combined in registration in order to produce a unified view of the scene. One possibility here is to build a full model of the object, which calls for resolving occlusion issues and "zippering" adjacent views together [4]. However, such an approach, being time consuming, has to be performed off-line in a pre-processing step. Furthermore, in some situations (e.g., virtual endoscopic procedures) the resulting full models can potentially grow quite large. The rendering labor in such methods scales with the size of the entire model, rather than with the number of images near the current viewpoint. This can unnecessarily strain the rendering process, possibly requiring data base culling with associated performance or display quality penalties. The alternative presented here is to combine only the partial models that sufficiently contribute to the current view. This, however requires that mutual occlusion issues be resolved in real time.

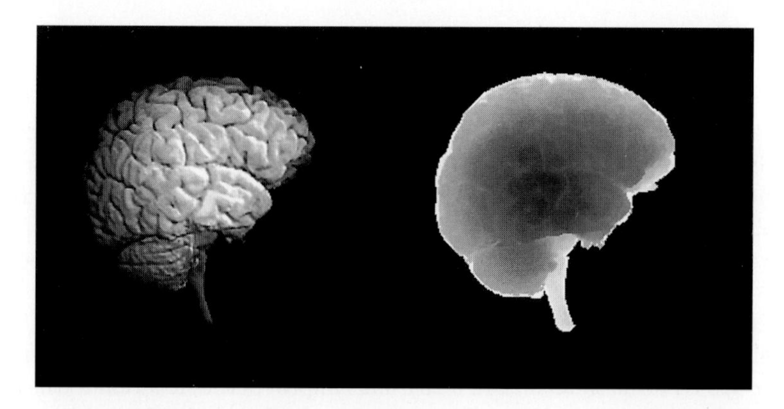

Fig. 1. (a) A volumetric rendition of an MRI of a human brain (b) the corresponding range map (bright pixels are far away).

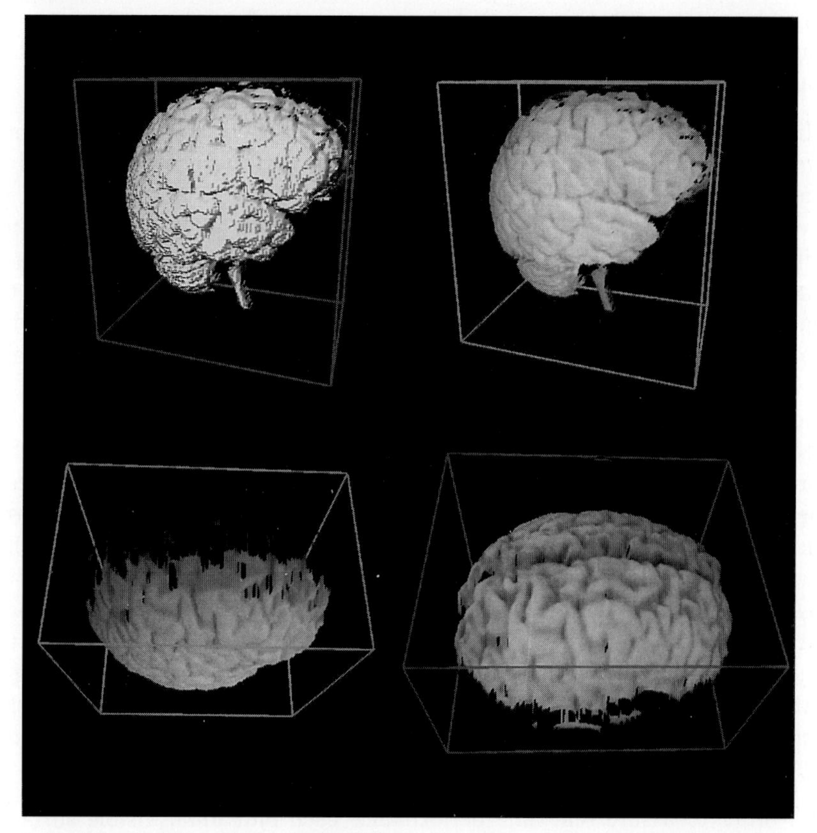

Fig. 2. (a) The visible surface obtained from the range map in Figure 1b (b) the visible surface texture mapped with the image from Figure 1a (c) a single visible surface is an accurate description of the model only within a limited scope (d) another visible surface can be used for near-vertex views.

2 Real-time merging of visible surfaces

The visible surface approximates the original model with varying degree of accuracy. Small polygons that have normals nearly parallel to the view vector of the reference view are likely to represent the model well while large polygons with normals that point away are poor estimates. This should be taken into account when several partial models are merged together because poorly estimated polygons of one surface may occlude parts of the objects that were clearly visible in another view. One easy solution is to simply discard all grazing polygons (in our implementation a polygon is considered grazing if the dot product of its normal and the eye vector is smaller than 0.20). This improves the situation to the extent that no gross errors are present. Figure 3c depicts the result of jointly displaying the two visible surfaces shown in Figures 3a,b after all grazing polygons were discarded. Note, that pieces of the lateral surface (yellow) can be seen on the forehead, even though this area is better seen in the frontal view (green). This straightforward form of joint display can result in loss of small surface detail and misalignment of textures, affecting the quality of display.

In order to address this problem we resolve conflicts between well and poorly estimated polygons in favor of the former. For that purpose we sort all the triangles by their quality as measured by the dot product of their normal and the eye vector of the reference view. Then the triangles are binned into a small number (2-16) of classes based on their ranking. During the display phase the triangles are drawn starting with the best class using stencil planes to assure that members of an inferior class do not overwrite the previously drawn triangles (i.e., a fragment passes a stencil test only if its label is less than or equal to the value already stored in the stencil planes). This procedure, however, has to be applied selectively: it is possible for a poor triangle to properly occlude a good triangle as long as they do not represent the same piece of the surface. Therefore, at the initial stage of the display process all the triangles (regardless of their quality) are rendered into the Z buffer with the near clipping plane moved back a little (znear=0.05) and color planes disabled. Effectively, this fills the Z buffer with values that are pushed back a little. Subsequently, the near plane is reset (znear=0.0) and drawing is re-enabled. With a Z buffer thus prepared re-drawing the scene into the frame buffer will result in only the polygons that were visible or "nearly" visible (i.e., occluded by a nearby fragment) passing the depth test. This allows us to display the quality-sorted list using stencil masking without running a risk of occluded high quality triangles shining through lower quality triangles that are indeed in front of them.

In order to assure real-time performance each individual visible surface is quality-sorted in the pre-processing stage. The lists are merge-sorted during the display phase with little overhead. This allows us to add (delete) partial surfaces as they come in (disappear from) the view.

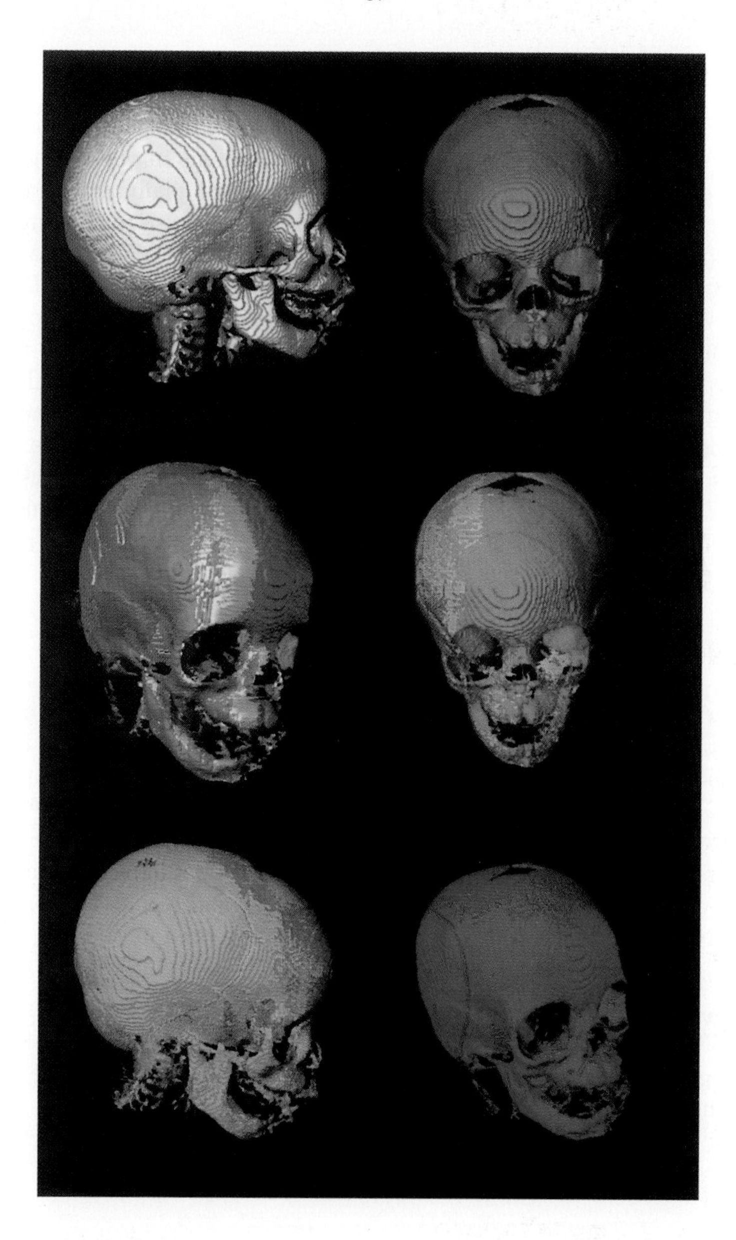

Fig. 3. (a) A lateral visible surface (b) a frontal visible surface (c) both surfaces displayed together without regard for occlusion artifacts–poorly estimated fragments from a lateral view (yellow) show up in areas where frontal surface (green) is a better choice and vice versa (d,e) surfaces merged with the new algorithm; note that the lateral (frontal) parts of the model are made up mostly of the lateral (frontal) visible surface (f) merged surfaces colored by the quality of the surface estimate–well estimated polygons are drawn in blue, poorly estimated in red.

3 Visualization-aided segmentation

Although the superior visualization available using volume rendering appears attractive in several contexts, there are many situations in which a geometric model of an object is required. In radiation treatment planning, for example, once an acceptable beam configuration is found, we may wish to compute organ-specific dose statistics such as dose volume histograms (DVH), tumor control probability (TCP) or normal tissue complication probability (NCTP). Definition of which voxels lie within the object of interest is essential in order to perform such calculations, thus a segmented model of the organ of interest is required. Using the present technique, segmentation of an object requires identification of a region of interest containing it on a modest number of rendered views, which may be done manually or automatically. There are no requirements on view directions, and for any given object the set of rendered views which most clearly visualize it may be chosen. The user is free to choose the viewpoints which result in best definition of the object. Depending on the complexity of the object, from three to ten views may be required to give adequate definition.

4 Conclusion

The superior image quality available from volume rendering may be combined with interactivity normally associated with surface rendered displays by use of texture mapping onto depth maps generated during the rendering process. A number of such partial models may be merged in real time for view interpolation increasing navigability during virtual environment exploration without sacrificing the quality of the display. A segmentation method based on identification of visible objects in volume rendered scenes, and reconstruction of their surfaces from the corresponding depth maps has also been developed.

Acknowledgements

Jason Leigh and Matthew Szymanski are gratefully acknowledged for their insight in the uncharted territories of SGI Inventor class extensions.

References

1. S. Chen and L. Williams. View interpolation for image synthesis. In *Proceedings of ACM SIGGRAPH '93*, pages 279–289, 1993.
2. A. Lippman. Movie maps: an application of the optical videodisk to computer graphics. In *Proceedings of ACM SIGGRAPH '80*, pages 32–43, 1980.
3. L. McMillan and G. Bishop. Plenptic modeling: an image-based rendering system. In *Proceedings of ACM SIGGRAPH '95*, pages 39–46, 1995.
4. G. Turk and M. Levoy. Zippered polygon meshes from range images. In *Proceedings of ACM SIGGRAPH '94*, pages 311–318, 1994.

Cortical Peeling: CSF/Grey/White Matter Boundaries Visualized by Nesting Isosurfaces

Colin J. Holmes[1], David MacDonald[2], John G. Sled[2] Arthur W. Toga[1] and Alan C. Evans[2]

[1] Laboratory of NeuroImaging, Dept. of Neurology, UCLA, Los Angeles, CA, USA,
[2] Montreal Neurological Institute, McGill University, Montreal, Quebec, Canada

Abstract. One computational tool of particular interest in the analysis of MR images of the cerebral cortex has been the automatic extraction of surfaces. Recently, enhanced MR images have been produced whose quality presented the opportunity to extract several of these surfaces from the same volume. We present preliminary data using this approach to visualize the boundaries of the cortex from the dura to the white matter. A series of isosurfaces were extracted from a signal–enhanced T1 MRI. The lower MR-intensity isosurfaces featured cortical vessels and dura, in contrast to higher intensity isosurfaces, in which the deeper structure of the superficial gyri became apparent. Two surfaces were chosen, one representative of the outer cortical surface and the other of the inner grey/white boundary, and their separation was computed to construct a "cortical thickness map." The use of nested isosurfaces with high quality MR images can differentially segment surface features and reveal the underlying structure of gyri and sulci, aiding anatomical delineation, and may ultimately result in the automatic extraction of cortical thickness maps.

1 Introduction

The vast majority of recent anatomical and functional MRI investigations have been focused on the human cerebral cortex. One method of analysis of volumetric MR data applied to the cortex has been the automatic extraction of surfaces and surface features [6, 2]. These techniques often rely on a cardinal feature of the dataset such as an intensity isovalue. For cortical surface extraction, this isovalue is generally the subjectively–determined best intermediate value between the cerebrospinal fluid (CSF) and the grey matter. Theoretically, by choosing values that are incrementally higher or lower than this optimal value, isosurfaces could be extracted that lie just superficial or just deep to the true cortical surface. Similarly, by choosing values that approach that of white matter, surfaces could be extracted that approximate the grey-white interface. Indeed, by visualizing a series of such surfaces in sequence, the thickness of the cortex could be "peeled" off, revealing the underlying white matter. More significantly, given two surfaces approximating the grey/CSF and grey/white boundaries, the cortical thickness could be inferred by determining their separation.

Of course, the validity of this proposed procedure would rely heavily on the use of extremely high quality data. Typically, isosurface extraction is complicated by noise within the data, and the partial volume effect inherent to the MRI process. Furthermore, in order for a single isovalue to suffice across an entire volume, the tissue intensity must be homogenous across the same space. MR images, however, typically suffer from signal intensity non-uniformity (RFI), which must be corrected.

The purpose of this study was to implement the nested isosurface strategy on the best available data. We used a posthoc registration technique [4] to produce an MRI volume of high signal, low noise, and enhanced image quality, which was then automatically corrected for RFI, and extracted nested isosurfaces using a previously described surface matching algorithm [6]. The sequence of extracted surfaces was inspected for its anatomical usefulness and the perpendicular distance between a superficial (grey/CSF) and a deep (grey/white) surface was computed to generate a cortical thickness map.

2 Methods

*High Resolution Image.*To produce an MRI volume with an extremely high signal to noise ratio, 27 individual high contrast [4] T1 volumes were acquired from a single subject. These volumes were automatically registered to a standard, Talairach-based space [3] where they were intensity–averaged, (for an expected signal to noise ratio increase of 5.2x).

RFI Correction. Although a number of post-processing methods for the compensation of field non-uniformity exist [1, 5, 7], the method we used is an extension to that of Wells *et al.* [9]. This method iterates between segmentation and estimation of the corrupting field, with intensity non-uniformity modelled as a smoothly varying multiplicative field. We have enhanced this model by incorporating spatial priors (in the form of tissue probability maps: TPM) in both the training of the classifier and segmentation of the volume. Initially, the skull is stripped (using a stereotaxic mask) and the remaining voxels are segmented into CSF, grey matter, and white matter using their intensity and the TPM for each class. The corrected intensity of each voxel is estimated from a weighted average of the mean intensities for the respective classes. The ratio of this estimate to the actual intensity is used to estimate the non-uniformity field (NUF) at that location. The NUF is then smoothed by a filter whose region of support is restricted to the region of the brain mask. Iteration proceeds by correcting the original volume using the NUF estimate and repeating the process of classifying and filtering. After convergence, thin plate splines are fit to the NUF to extrapolate it to regions outside the brain mask.

Nested Isosurface Extraction. Surfaces were extracted using a previously described algorithm [6] that continuously deforms a spherical mesh surface (i.e. 'shrink-wrapping'), to match a target boundary defined by a threshold value. The algorithm operates in a multi-scale fashion, so that progressively finer surface detail is extracted at finer scale representations of the data. Initially, both surface mesh and 3-D intensity field are coarsely sampled and the latter heavily blurred. The initial surface, composed of as few as 4096 polygons, is extracted rapidly, but expresses only the gross shape of the cortex. After several finer scale steps, the data are sampled at the original 1.0 mm intervals, resulting in a surface consisting of 100-150k polygons. To produce nested isosurfaces, we applied this surface extraction algorithm at incrementally increasing threshold values (5000 unit steps ranging from 85,000 to 300,000 units).

3 Results

Initial Image Quality. Averaging 27 individual scans produced a single MRI volume whose enhanced signal made it possible to observe fine structure in the thalamus and brainstem. The homogeneity of the cortical grey matter was greatly enhanced, as was that of the white matter. Surface vessels and connective tissue sheets (eg. the dura) became quite apparent and continuous from slice to slice. Example slices from untreated and enhanced volumes are presented in Figure 1. In addition to the increased anatomical detail, however, the averaging process revealed intensity inhomogeneity normally concealed by noise. The RFI correction allowed this artefact to be removed through the production of a smoothly varying field that mapped the visible inhomogeneity. The effect of the correction is demonstrated in Figure 2.

Nested Isosurfaces. We produced a series of surfaces that, when viewed in sequence, give the impression that the layers of the meninges, CSF and cortex are "peeled" away from the underlying white matter. The lower intensity surfaces contained features consistant with the dura, dural sinuses, and dural (as well as superficial cerebral) vessels. Higher intensity isosurfaces lay deeper in the gray substance, with no surface vessels visible. As the isosurface intensity approached that of the white matter, the extracted surface lay closer to the grey/white boundary, and the deeper structure of the superficial gyri became apparent. The extraction did not, however, appear as uniform as expected after the RFI correction. The parietal and especially temporal white matter became apparent earlier than that of the frontal lobes. While it may be exciting to speculate that these differences may be due to the myeleoarchitectonic properties of the various lobes, at this early stage it is more likely that the RFI correction was inadequate. Examples of individual views of these surfaces are presented in colour plate Figure 3 (a,c,e: view from the left; b,d,f: view from above; a,b: 140k; c,d: 185k; e,f: 275k units) and mpeg movies of all the intermediate surfaces are available via the World Wide Web (http://www.mni.mcgill.ca/).

Cortical Thickness Map. By selecting the surfaces empirically "closest" to the true CSF-grey and grey-white boundaries, we hoped to be able to extract a cortical thickness map by finding the perpendicular distance between the two surfaces. Figure 4a (on the colour plate) presents a preliminary example of this process, with the cortical thickness map of Von Economo [8] for comparison (fig. 4b). The cortical thickness is displayed as a colourscale superimposed on the outer (CSF-grey) surface, with darker colours (blue) implying thinner cor-

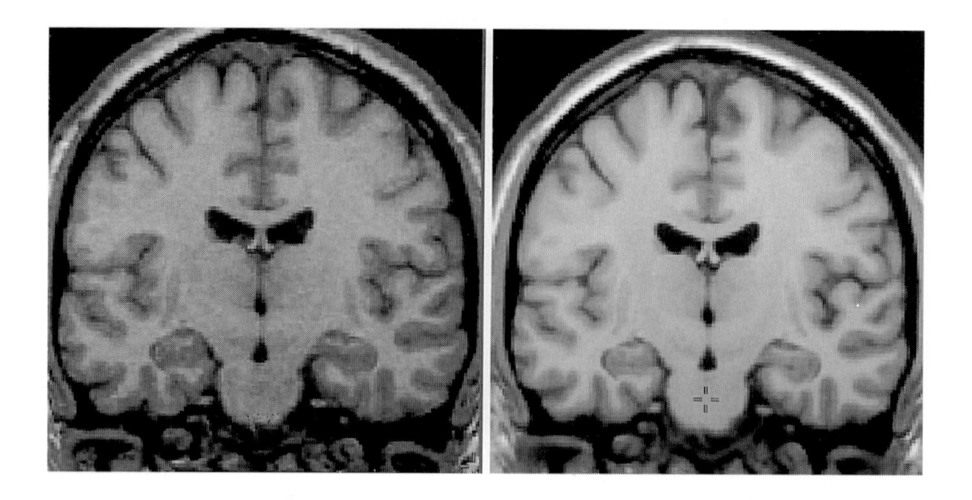

Fig. 1. Representative slices from a single 1.0mm^3 T1 MRI (left) and a composite of 27 individual scans from the same subject after registration and averaging (right). Note the great reduction in noise apparent in the averaged volume, and the homogeneity of the MR intensity in the grey and white matter relative to the noisy tissue in the single scan.

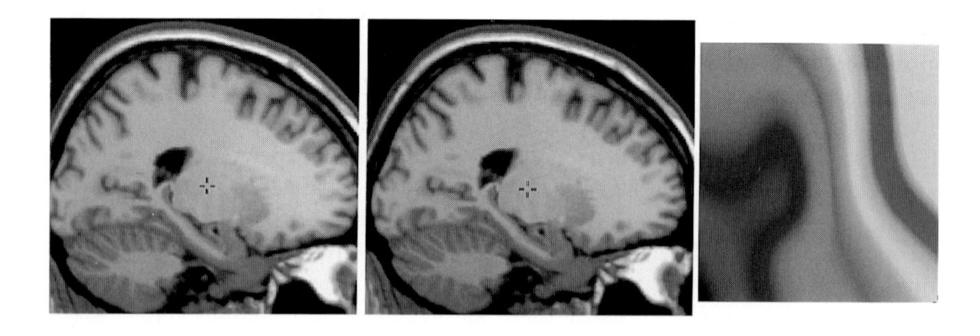

Fig. 2. Sagittal slice from the high resolution T1 before (left) and after (center) RFI correction, with the extracted field for the same slice (right). Note that the decrease in the intensity of the white matter from rostral (right of each image) to caudal (left of each image) within the initial slice is absent in the corrected slice, and obvious in the corrective field.

tex, and brighter colours (green–white) meaning thicker cortex. The colour scale ranges from 0 to 6 mm in thickness. While superficially similar to the thickness map of vonEconomo (fig. 4b, where heavier stipple reflects thicker cortex) , the map generated here suffers from the gradient limitations of the surface extraction. It can only be valid where the two surfaces run parallel, and as the outer surface does not descend into the depths of the sulci, the thickness map must be constrained to the crests of each gyrus. Nevertheless, this preliminary computation gives great hope that with more elaborate automatic surface extraction schemes, the automatic generation of thickness maps may be possible.

4 Summary and Conclusion

Through posthoc averaging we produced a MRI volume whose characteristics permitted the extraction of nested isosurfaces. The extracted surfaces tracked the changing features of the grey/CSF and grey/white interfaces as the isovalue target was incrementally increased, and the generated thickness map was superficially similar to that of von Economo. With improved RFI correction and surface extraction techniques, it ought to be possible to produce reliable cortical thickness maps using nested isosurfaces. *This work was supported by the Canadian MRC (*SP-30*), the McDonnell-Pew Cognitive Neuroscience Program, the NSF (*BIR 93-22434*), the NLM (*LM/MH05639*), the NCRR (*RR05956*) and the Human Brain Project, jointly funded by the NIMH and NIDA (*P20 MH/DA52176*).*

References

1. Axel, L., Costantini, J., Listerud, J.; Intensity correction in surface-coil MR imaging America Journal of Roentgenology **148** (1987) 418–420
2. Carman, G.J., Drury, H.A., Van Essen, D.C.; Computational methods for reconstructing and unfolding the cerebral cortex. Cerebral Cortex **5** (1995) 506–517
3. Collins, D.L., Neelin, P., Peters, T.M., Evans, A.C.; Automatic 3D inter-subject registration of MR volumetric data in standardized Talairach space. Journal of Computer Assisted Tomography **18(2)** (1994) 192–205
4. Holmes C.J., Hoge R., Collins L., Evans A.C.; Enhancement of T1 MR images using registration for signal averaging. 2nd International Conference on Functional Mapping of the Human Brain (1996)
5. Lim, K.O., Pfefferbaum, A.; Segmentation of MR Brain Images into Cerebrospinal Fluid Spaces, White and Gray Matter Journal of Computer Assisted Tomography **13(4)** (1989) 588–593
6. MacDonald D., Avis D., Evans A.C.; Multiple surface identification and matching in magnetic resonance imaging. Proc SPIE **2359** (1994) 160–169
7. Meyer, C., Bland, P.H., Pipe, J.; Retrospective Correction of Intensity Inhomogeneities in MRI IEEE Transactions on Medical Imaging **14** (1995) 36–41
8. von Economo C.F.; The Cytoarchitectonics of the Human Cerebral Cortex. Milford Press, London, 1929
9. Wells, W.M., grimson, W.E.L., Kikinis, R.; Statistical Intensity Correction and Segmentation of MRI Data Proceedings of the SPIE. Visualization in Biomedical Computing **2359** (1994) 13–24

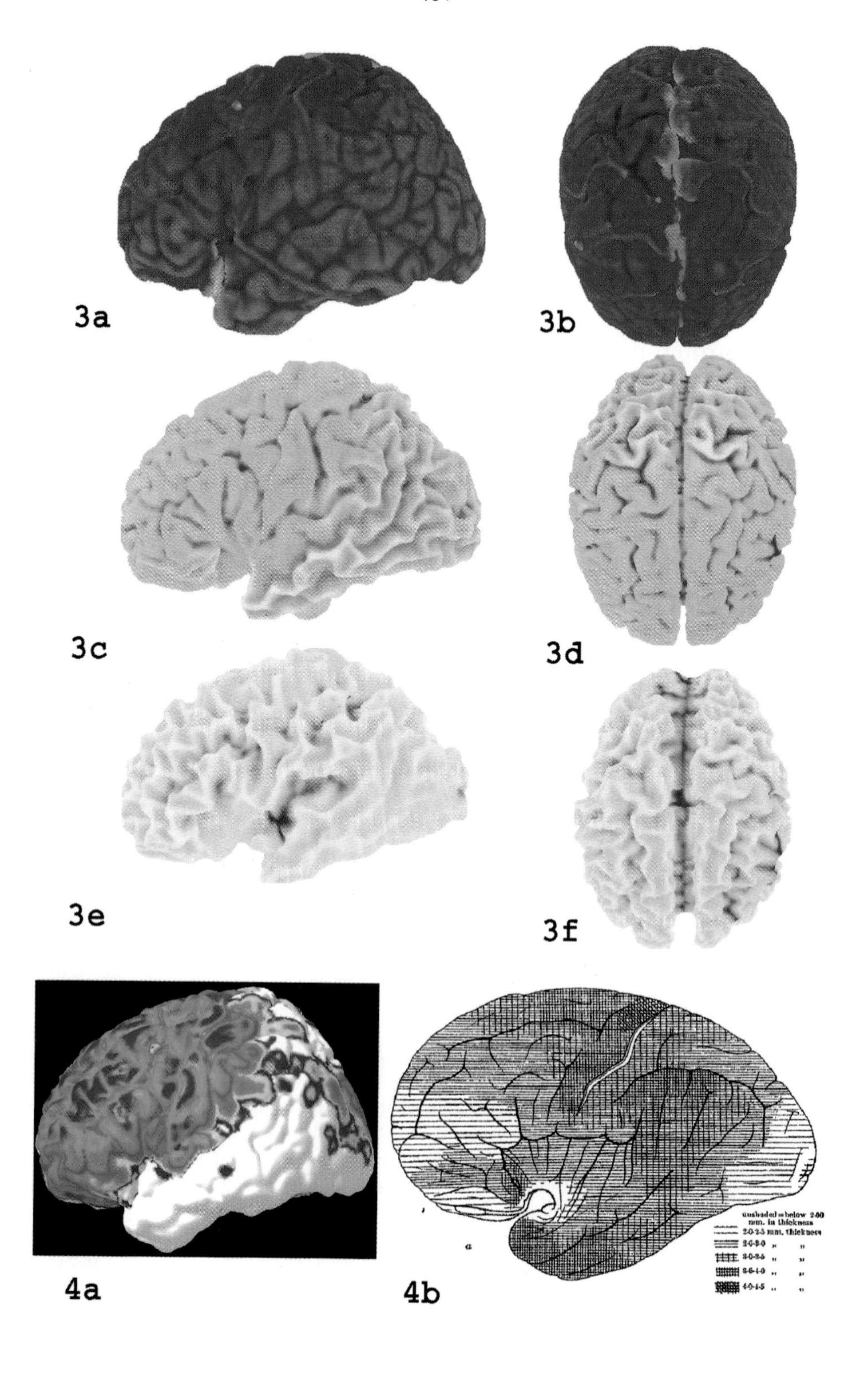
3a
3b
3c
3d
3e
3f
4a
4b
unshaded = below 2·00 mm. in thickness
2·0-2·5 mm. thickness
2·6-3·0 „ „
3·0-3·5 „ „
3·6-4·0 „ „
4·0-4·5 „ „

Image Processing

Enhancement of MR Images

Régis Guillemaud[1,2*] and Michael Brady[2]

[1] Clinical Neurology, Radcliffe Infirmary,
[2] Robotics Research Group, Dept of Engineering Science
University of Oxford, Oxford OX1 3PJ, UK.

Abstract. We propose a modification of Wells' et. al. technique for bias field estimation and segmentation of MR images. Replacement of the class *other* that includes all tissue not modeled explicitly by Gaussians with small variance by a uniform probability density, and amending the EM algorithm appropriately, gives significantly better results. The performance of any segmentation algorithm is affected substantially by the number and selection of the tissue classes that are modeled explicitly, the corresponding defining parameters, and, critically, the spatial distribution of tissues in the image. We present an initial exploration of the application of minimum entropy to choose automatically the number of classes and the associated parameters that give the best output.

1 Introduction

Many applications of magnetic resonance (MR) imaging require that the data be segmented into subsets corresponding to different tissue classes. Ideally, for any given set of MR imaging parameters, the intensity values of voxels of a tissue class should be constant, or, in view of the partial volume effect, correspond to a Gaussian distribution with small standard deviation. For this reason, the goal of restoration algorithms is to yield an output image that is piecewise constant, and which can then easily be segmented.

Following [1], $I_{measured}(Y) = I_{restored}(Y) \times B(Y) + N(Y)$, where $I_{measured}(Y)$, is the given image intensity at voxel Y; $I_{restored}(Y)$ is the image intensity to be "restored" by the segmentation algorithm; $B(Y)$ is the slowly varying bias field; and $N(Y)$ is the image noise. There have been numerous approaches to estimating the bias field in order to remove it to restore the image. Space precludes a comprehensive survey, for which see [2].

This paper may be viewed as a contribution toward refining [1] , in order to overcome a number of the problems we have encountered using it. In the next section, we recall the features of a conventional statistical classifier in order to set up the notation for a brief review of Wells' et. al. segmenter/classifier in Section 3. The first of our contributions is in Section 4, in which we combine Wells' et. al. technique for tissue classes whose intensity values are well approximated by a small variance Gaussian distribution with a Bayesian decision to model the tissue

* permanent address from June 1996 : LETI(CEA-Technologies Avancées)
17, avenue des Martyrs, F 38054 Grenoble, FRANCE

class "other". Results on simulated data (originally inspired by conversations with G. Gerig), and on MR brain images are presented. We observe that the performance of any segmentation algorithm, in particular that of Wells et. al. (and our refinements of it) is affected substantially by the *selection* of classes that are modeled explicitly (for example, as Gaussians), the corresponding defining parameters, and, critically, the spatial distribution of tissues in the image. It is reasonable to suppose that the first two of these could be decided automatically, for example by knowing the MR imaging parameters; but the latter is often more problematical : in Section 6 we suggest choosing the parameters that give minimum entropy.

2 Conventional Statistical Classifier

The goal is to estimate the multiplicative bias field and so restore the image to be piecewise constant. We begin with the conventional pattern recognition approach, in which a set of classes are chosen and fixed in advance, and each voxel is assigned to a class depending only on its intensity level. Associate the jth tissue class with the probability distribution on intensities Γ_j. Typically, Γ_j is modeled by a Gaussian distribution with mean μ_j and variance σ_j^2: $p(Y_i|\Gamma_j) = G_{\sigma_j}(Y_i - \mu_j)$, where G_{σ_j} is the scalar Gaussian distribution with variance σ_j^2, and Y_i is the ith voxel of the image volume. Next, a stationary probability $p(\Gamma_j)$ is defined for each class Γ_j so that the sum of all the probabilities over the classes equals one. This leads to a class-independent intensity distribution on the image that is a mixture of Gaussian populations (one population for each tissue class):

$$p(Y_i) = \sum_{\Gamma_j} p(Y_i|\Gamma_j)p(\Gamma_j) \tag{1}$$

Classification consists of estimating the likehood $p(Y_i|\Gamma_j)p(\Gamma_j)$ of each class for each voxel i, finally accepting, for example, the class whose likehood is maximum. The segmentation is the mapping of this classification over the image. If (i) the assumption of piecewise constant intensity value holds; (ii) each class has a known mean and variance; and (iii) the bias field is very small, then this maximum likelihood estimator can give very good results. For real MRI, however, (ii) does not hold, since there is generally at least one class which has a large variance, typically to represent those voxels that are either not of interest or not expected. Similarly, (iii) often does not hold (and is the main motivation behind [1]). Note that in the above approach, all the voxels are considered as independent samples drawn from a population; no spatial constraint is used in the segmentation.

3 Bias Field Correction

Wells et. al. [1] suggested modifying the maximum likelihood approach developed in the previous section in order to apply the EM (*expectation-maximisation*) algorithm. If we ignore $N(Y)$ for the moment (or pre-filter to remove it), we can convert the equation $I_{measured}$ to a summation by taking logarithms. The logarithm of the bias field is then denoted $\beta(Y_i)$, more succinctly β_i. A conventional Gaussian classifier (as described above) is used to classify the data Y_i after correction for the (currently) estimated bias field $\beta_i : p(Y_i|\Gamma_j, \beta_i) = G_{\sigma_j}(Y_i - \mu_j - \beta_i)$. Plugging this into Equation 1, and pretending that the voxels in the image are independent, we have:

$$p(Y|\beta) = \prod_i [\sum_{\Gamma_j} p(Y_i|\Gamma_j, \beta_i) p(\Gamma_j)] \tag{2}$$

Wells et. al. suggest modeling the slowly-varying bias field by a n-dimensional zero-mean Gaussian prior probability density $p(\beta) = G_{\Gamma_{\psi_\beta}}(\beta)$. Next, Bayes' rule is used to obtain the posterior probability of the bias field, given the observed intensity data.

Substituting Equations 2 in Bayes' Equation yields a new form for the posterior probability of the bias field leading to the following pair of coupled equations:

$$\mathbf{W}_{ij} = \frac{p(Y_i|\Gamma_j, \beta_i) p(\Gamma_j)}{\sum_{\Gamma_j} p(Y_i|\Gamma_j, \beta_i) p(\Gamma_j)} \tag{3}$$

$$\beta = \mathbf{H}(Y - \mathbf{W}U) \tag{4}$$

The first of these is the probability that a bias corrected voxel i belongs to a particular tissue class j. In the second, Y is a stacked vector of all the voxels, $U = (\mu_1, \mu_2, ...)^T$ is the vector comprising the mean tissue class intensities, and $\mathbf{H}$ is usually implemented as a low-pass filter, and is here represented as a matrix. The EM algorithm is applied to Equations 3 and 4.

The outputs from the algorithm are: a bias field estimate $\beta(Y)$, and a MRI image segmented by maximum likelihood tissue class probabilities. Given the bias field, it is easy to compute the restored MR image $I_{restored}(Y)$, simply by dividing $I_{measured}(Y)$ by the inverse log of $\beta(Y)$. Our experiments confirmed good performance in many cases, but revealed a number of problems, however, initially on breast MR images, then on simulated images, finally on brain images. Our main conclusions from these experiments are: (i) the method requires that the MR intensity distribution be modeled as a Gaussian mixture; (ii) it is straightforward to compute a sample mean and (small) sample variance for large, well-defined regions, typified by grey matter and white matter; however (iii) other small "regions" consist of several different types of tissue, and when they are modeled as a Gaussian, the variance is large and the mean may be unrepresentative of any constituent class (the multi-population problem). This is typified in brain images by "CSF", which may in fact comprise CSF, pathologies, and "other" small regions of tissue. Such small regions can cause a disproportionately large distortion to the computation of the bias field. This is because the estimation of the bias field for a voxel is computed from the difference between the intensity at the voxel and the assumed mean value. If the mean value is unrepresentative of the class, this may (and does) introduce errors into the *global* estimation of low-pass filtered bias field.

We illustrate these points in a number of examples computed using an implementation of Wells' et. al. algorithm. First, we present a simulated image consisting of vertical strips, Fig.1(a). The first two strips were chosen to simulate white and grey matter structures (the intensity values were 150 and 120 respectively). The third strip is chosen to be typical of what would be assigned to CSF, and has intensity 10. In accordance with the model presented above, the image was corrupted with a multiplicative bias field that is sinusoidal ($\beta(Y) = (6 + \sin(2\pi \times 3.0(y/ymax)))/7$) and the resulting measured image is shown in Fig.1(b).

The Wells et. al. algorithm was applied to the image with different sets of parameters. The number of tissue classes (3) and the number of iterations (5) remained constant throughout. First, the true mean intensities were used for each tissue class. The first and the last segmentation, the bias field and the corrected image are shown in Fig.1 (c), (d), (e) and (f). The bias field was removed and the segmentation is nearly perfect. Then an incorrect mean intensity value (40) is assigned to the CSF class. The result after the bias correction is similarly shown in Fig. 1 (g), (h), (i) and (j). The estimated bias field is clearly erroneous: one strip of "grey matter" has been completely lost. As well, there is a clear artifact at the boundary between "grey matter" and "CSF" in the restored image.

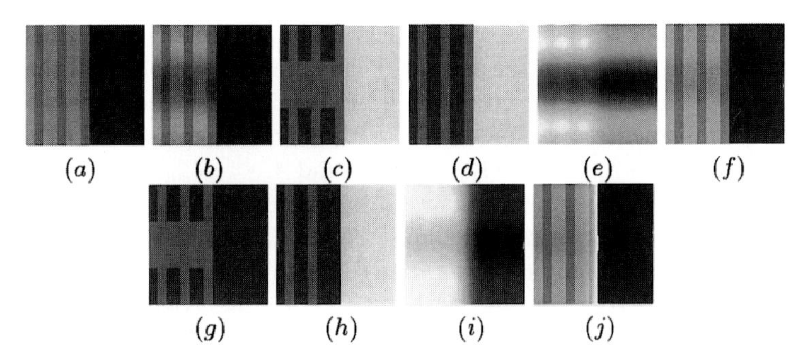

Fig. 1. Simulation results. See text for details.

Similarly, axial proton density density slices were obtained using a General Electric Signa 1.5 Tesla clinical MR imager. Fig.2 (a) and (e) show two typical slices for the same patient. In slice (a), we estimated the intensity parameters for four classes: air, white matter (plus CSF), grey matter, and the tissue surrounding the cortex. We then applied the EM algorithm to correct for the bias and to segment *both* slices (a) and (e). Fig.2(c),(d),(g) and (h) shows the segmentations after the first iteration (no bias correction), and after 5 iterations. In addition, Fig. 2 (b) and (f) show the slices corrected for bias at the end of the process. The segmentation is much poorer for slice (e) than for slice (a). For instance, the grey matter is missing from the bottom left side of the brain. It seems that the tissue surrounding the brain in slice (a) is slightly different from that in slice (e). Therefore the class "other tissue", defined in slice (a), has a wrong value for this region. This affects the bias correction and the segmentation.

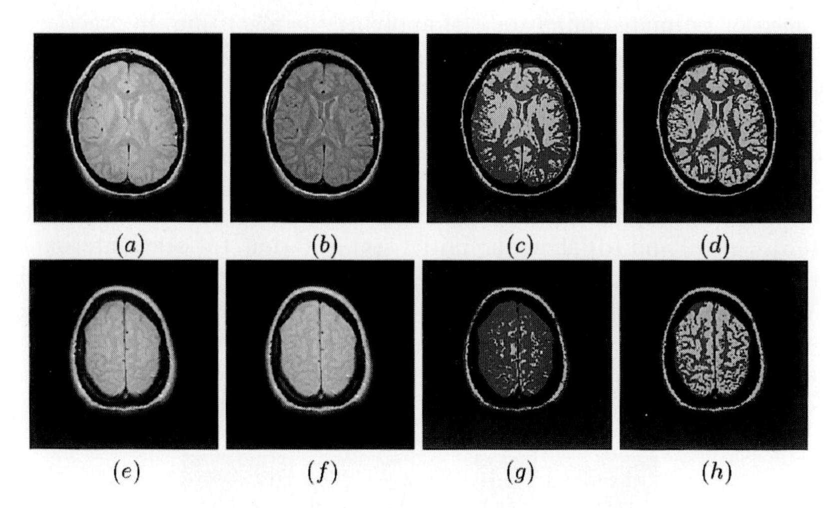

Fig. 2. Results of applying Wells' et. al. algorithm to brain MR. See text for details.

4 Modified EM Algorithm

The modification we propose to the method of Wells et. al. is to introduce a tissue class Γ_{other} with a *non-Gaussian* probability distribution. The new extra class is intended to gather all the pixels which are far from the Gaussian distributions of the identified tissue classes. We conservatively assume that the probability density is uniform over the set of intensities of voxels inside the region, and is zero outside. In practice, the a priori probability of a voxel intensity being assigned to the class Γ_{other} is small. Otherwise, tissue classes are modeled by Gaussian distributions with small variance, as in the two previous sections. The resulting combination of tissue classes is a mixture of Gaussian densities with small variance and a uniform density with typically small probabilities for each intensity.

We now revisit the EM algorithm proposed by Wells et. al. For the Gaussian classes, we retain the intensity conditional probability density given the bias field. For the class Γ_{other}, the conditional probability density is a constant, and so the derivative of the distribution with respect to the bias is zero. The new grey level probability density on the image becomes:

$$p(Y_i|\beta_i) = \sum_{\Gamma_j \text{ is Gaussian}} p(Y_i|\Gamma_j)p(\Gamma_j) + \lambda . p(\Gamma_{other}). \quad (5)$$

where λ is a constant. This new density is now used in Equation 3, but where $p(Y_i|\Gamma_j, \beta_i)$ can be a Gaussian distribution or a constant. Equation 4 has the same form as before but the component of U corresponding to the class Γ_{other} is zero. That is, *the bias is only estimated with respect to the Gaussian classes*. We note that the mean and variance of each Gaussian tissue class must be defined or estimated in advance of applying the algorithm. In practice, this does not present a problem, since so long as the acquisition protocol remains fixed over a series of scans, there is no reason to change the parameter values.

We present results first for the simulated data, where we are interested primarily in finding the regions of "white" and "grey" matter. As before, we associate Gaussian densities with these two tissue classes, but use a uniform density for the rest. Fig.1 (a) and (b) show the original measured data (as before), while Fig.3 (a), (b) (c) and (d) show the initial segmentation, the segmentation after 5 iterations, and the corresponding estimated bias field and the corrected data. The result is extremely good. Note that we have not defined any parameters for the "CSF" region, so it is impossible to make any errors. Consequently the approach with the new statistical segmenter should be more robust.

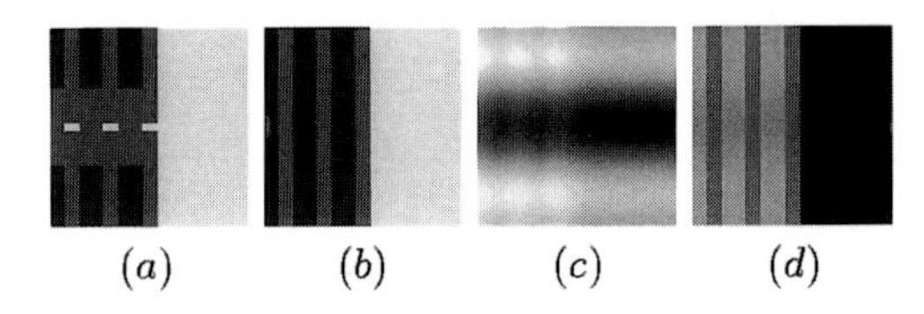

Fig. 3. The modified algorithm applied to simulated data. Refer to the text for details.

The second experiment repeats that in Fig.2. Again, we defined two tissues classes (white and grey matter) supplemented by a uniform density class. We used the same parameters for the white and grey matter classes as previously. Fig.4 shows the results obtained with the modified algorithm. The initial segmentations are similar to the Wells et. al. The final segmentation of slice (a) (Fig.4 (d)) is very close to the segmentation with the Wells et. al. method, but the final segmentation of slice (e) (Fig.4 (h)) is very different at the bottom of the image, and is a more accurate repesentation of the brain structure. The change in tissue type around the cortex, from slice (a) to (e), does not affect the segmentation as it did previously, reaffirming the robustness of the modified algorithm.

5 Limitation

The EM algorithm sometimes converges to a local minimum that is far from the desired result. We illustrate this in Fig.5(a), which shows a variant to the simulated image shown earlier, but this time with just a single "grey matter" strip and a large "white matter" region. As before, the image was multiplied

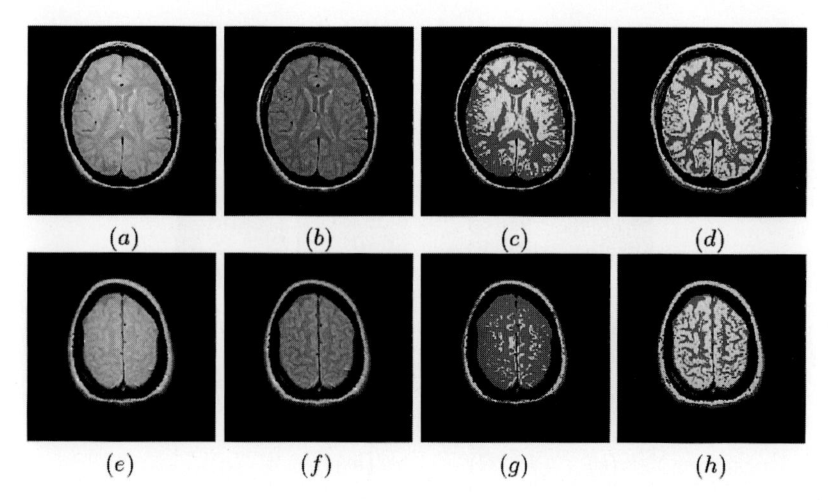

Fig. 4. The modified EM algorithm applied to proton density brain data. Refer to the text for details.

by a sinusoidal bias field to give the measured image Fig.5(b). We applied the Wells' et. al. algorithm and our modified version, in both cases using three tissue classes ("grey matter", "white matter", and *other*). The results of applying the Wells et. al. algorithm are shown in Fig. 5 (c), (d), (e) and (f), and those for the modified algorithm in Fig.5 (g), (h), (i) and (j)). In both cases, the algorithm converges to a solution that is close to the initial segmentation. However, (and we return to this point in the next section), when we used two tissue classes ("white matter" and the uniform density class *other*), the modified segmenter gave good results (see Fig.5 (k), (l), (m) and (n)). We also get good results for the modified algorithm in the case that there are many "grey matter" strips and three tissue classes are assumed ("grey matter", "white matter", and *other*).

In general, the two class bias correction problem is quite straightforward and the EM algorithm nearly always gives good results, so long as the class models are correct with respect to the data. However, the complexity of the problem increases sharply with three tissue classes. In particular, in that case the EM algorithm often has problems with large uniform regions for which it cannot reliably estimate the bias field, and ends up splitting the large region into many subregions of different tissue classes. It is the relative lack of signal *changes* that make it difficult to estimate the bias field reliably. On the other hand, if the image is a complex pattern of adjacent regions (as is often the case in the brain) then there is more information to anchor the solution to the bias estimation. It is at the boundaries between regions that the the optimal solution of the bias is achieved by smoothing, and that the maximum spreading of the consequent constraint occurs. This is why the results presented by Wells et. al. are good for brain data but far less good on the latter simulations. The modified algorithm makes it possible to use a smaller number of classes, which simplifies the estimation problem, and in turn increases the likelihood of good

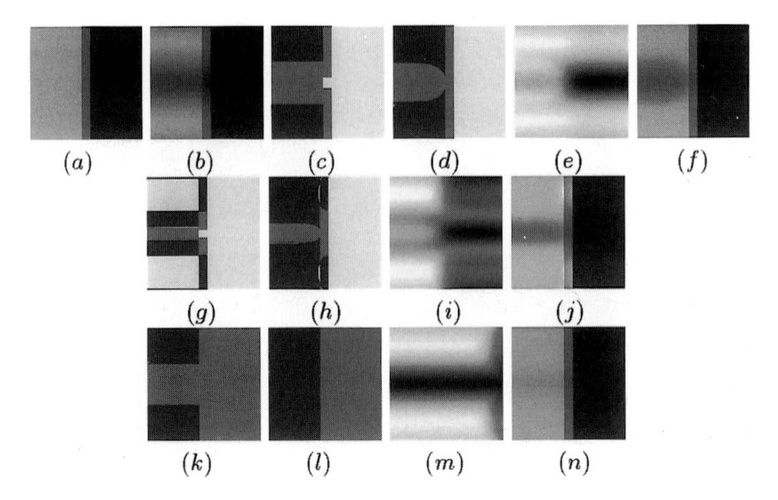

Fig. 5. Simulated data to show the effect of varying the number of modeled classes and the parameters of each modeled class. Refer to the text for details.

results. In cases where it gives poor results, for example the three tissue model discussed in this section, a different number of models, or the same number of models but with different parameters, can give excellent results. The next section presents an initial foray into automatically choosing the number of classes and their parameterisation.

6 Choosing Classes and their Parameters

According to the MR model underlying our approach and that of Wells et. al. a "perfect" restored image would have one grey value per tissue class, resulting in an intensity histogram consisting of a small number of single intensity spikes. The effect of the bias field is to spread such a spike over a possibly large number of intensity values, and to overlap those corresponding to different tissue types. According to this view, bias correction corresponds to a transformation of histograms to one that most closely approximates the "ideal" few spikes. Conventionally, entropy is used to assess the degree of departure from such an ideal. If $hist(g; I)$ is the intensity histogram of an image I, its entropy is defined to be:

$$E(I) = - \sum_{\text{intensity} g} hist(g; I).log[hist(g; I)].$$

We propose to use entropy as a measure of the quality of a restored image for a given choice of classes and parameters, then identify the best selection as that giving the minimum entropy. This choice is equivalent to choosing the restored image that minimises the number of bits needed to code it [3, 4]. It is necessary to constrain the set of output images for which we compute the entropy, otherwise the minimum would be the image that is everywhere zero! We note that the bias field has the effect of limiting image contrast by decreasing intensity values. Therefore, we insist that the bias correction *amplifies* intensities, hence amplifies image contrast.

In principle, one expects to compute the entropy on the histogram taken over the entire image, since this should be the best approximation to the probability density. However, it is important to avoid voxels outside the region of interest that could act as outliers and skew the histogram density function, hence the entropy. For example, coronal MR images often include a part of the neck of the patient. For this reason, we use a region of interest that is close to the region we want to segment. The region of interest could be defined interactively by the user, or it could be computed from the segmentation given by the bias field correction algorithm. In the latter case, given the restored MR image and the associated segmentation, we extract the region of interest (eg the largest connected component gives the white and grey matter for brain MR), then measure the entropy for that region. The region extracted depends on the body part being imaged and can involve anatomical knowledge. Our current method is a search for that set of classes and,for each, the intensity value $\mu^j \in [\mu^j_{min}, \mu^j_{max}]$ of the mean intensity, that overall gives the lowest entropy.We plan next to introduce the region of interest computation into the estimation scheme and to investigate the application of minimisation techniques for non-linear functions.

We begin with a simulated image, typical of the sort that causes problems for Wells' et. al. algorithm, and which we have already used in Figure 5. The multiplicative bias field was sinusoidal. In this case, the region of interest was the whole image. Fig.6(a) shows the restored image that results from two classes ("white matter" and *other*). The minimum entropy (plotted in Fig.6(b)) found by the program was 3.15, the tissue parameter for "white matter" was calculated to be 148.41, which is close to the actual value of 150. The restored image is close to the original. Figure 6(c) shows the result found by the program for three classes ("grey matter", "white matter", and *other*). In this case, the minimum entropy is 3.47, and while the estimated parameters are quite good (147 for "white matter" instead of 150, and 77 for "grey matter" instead of 80), the restored image is relatively poor. The entropy is plotted in Fig.6(d), in which the plateau corresponds to parameters for which the final bias amplification was not greater than 0.99. Fig.(e) is an enlargement of the Fig.(d) in the region of the mimimum of entropy.

In the case of proton density brain images, we have estimated the minimum entropy solution for two tissues classes (grey matter, white matter) plus *other.* The region of interest in Fig. 6 (f) is a presegmented region corresponding mainly to white matter and grey matter (plus CSF in this case). The entropy is plotted in Fig. 6 (h) and (i), in which the plateau corresponds to parameters for which the final bias amplification was not greater than 0.99. Fig.(f) and (g) show the corrected brain image and the segmentation corresponding to the minimum entropy solution.

In practice, we find that the partial derivative of entropy with respect to class parameters is quite small. For a given MR scanner, and a given acquisition

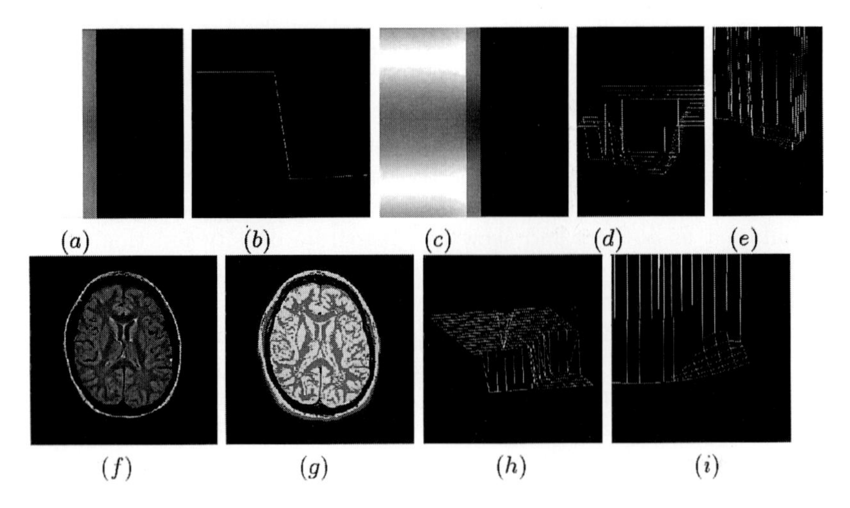

Fig. 6. Application of entropy mininisation. See text for details.

protocol one can use approximately the same parameter sets for restoration of all acquisitions. Indeed, one may regard parameter estimation as a calibration step for MRI. This is a line of investigation that we propose to pursue further.

Acknowledgments

The work was carried out on an MRC grant G9220033. We thank Tim Crow, Sebastine Gilles, Ralph Highnam, Patrick Marais, Alison Noble, and Andrew Zisserman for comments. We have benefited enormously from conversations on this topic with Guidi Gerig, Sandy Wells, and Gabor Szekely. The brain images shown in the paper were supplied by Dr. Lynn de Lisi of SUNY Stony Brook.

References

1. W. M. Wells III, W.E.L Grimson, R. Kikinis, and F. A. Jolesz, "Adaptive Segmentation of MRI Data", in *Proceedings of CVRMed'95*, 1995, vol. 905 of *Lectures Notes in Computer Science*, pp. 59–69, Springer Verlag.
2. William A. Wells III, *Statistical Object Recognition*, PhD thesis, 1992.
3. Y. G. Leclerc, "Constructing simple stable descriptions form image partitioning", *International Journal of Computer Vision*, vol. 3, 1989.
4. S. C. Zhu, T. S. Lee, and A. L. Yuille, "Region competition: unifying snakes, region growing, energy/bayes/mdl for multi-band image segmentation", in *Int. Conf. Computer Vision*, 1995, pp. 416–423.

Radiometric Correction of Color Cryosection Images for Three-Dimensional Segmentation of Fine Structures

Jorge Márquez and Francis Schmitt

Dept. Images, Ecole Nationale Supérieure de Télécommunications,
46, rue Barrault, 75634, Paris Cedex 13, France

Abstract. We address the problem of correction of radiometric inhomogeneity present in the color images of the anatomical cryosections from the Visible Human Project (VHP) data base. We devised an adaptive correction that is propagated along a series of parallel slices, taking advantage of spatial coherence between consecutive slices. No blurring is introduced, and fine detail and texture are respected. Results of 3D segmentation of fine blood vessels on the corrected volume are presented.

1 Introduction

Many laboratories have long experience with three-dimensional reconstruction of structures from data collected from transmitted, emitted or reflected radiation (x-ray computer and emission tomography, nuclear magnetic resonance imaging, ultrasound, confocal and electron microscopic imaging, etc.). This is not always the case with traditional, physical, cross-section reconstruction, which is lengthy and requires bloc fixation and destruction of the specimen. Even if physicians are familiar with anatomic slices of the human body, until very recently there where no large color data sets to reconstruct whole organs for anatomy atlases.

Cryosection techniques have been used in the Visible Human Project (VHP) [1] of the National Library of Medicine to make a high resolution volume of anatomical information available to scientists. 1871 axial-plane slices spaced at 1mm have been sampled at a resolution of 0.33 mm, with 2048x1216 pixels of 24 bits. Three sets of images corresponding to the three channels RGB have been recorded. The three channels are spatially well registered and a signal of one byte per pixel and per channel is available. The complete data file corresponds to 14 Gigabytes. Computer tomography and magnetic resonance imaging sets are also available.

This paper deals with the problem of radiometric inhomogeneities that we have found on the physical color images from the VHP male body. We consider in the following our data subset or Region Of Interest (ROI) which is the cropped sub-volume that encloses both lungs. This ROI consists of 180 cross sections of 832x640 pixels of 24 bits. Figures in this paper illustrate only a subset of 40 images, at half reduced resolution, to show better the inhomogeneity problem.

2 Problem Description

In order to perform three-dimensional reconstructions of the bronchial blood vessels or any other structures in volume data from cross-sections, these images must be spatially registered, but they must also be *radiometrically* registered. The VHP color images are correctly registered in rotation and translation, except for some images in which translation offsets of a few pixels had to be corrected. To ensure radiometric registration, each color image was digitized by the VHP team including a gray scale test card, to allow for *gamma homogenization* [2] of a stack of color images.

We have observed that even after a separate gamma correction of each image of the data subset by using the accompanying gray scale, there remain some radiometric discontinuities between slices which affect the three-dimensional segmentation of small blood vessels and bronchi. The origin of these inhomogeneities is unknown, but they seem to be unrelated to acquisition artifacts. They could be evidence of an irregular oxidation of the sliced surface of the frozen body between two image acquisitions. These acquisitions could have been unevenly spaced in time, or taken under different temperature conditions; but these are just hypothesis.

Full radiometric registration and homogeneity implies some kind of continuity in localization of histogram main features for small, medium and large regions in each image. Radiometric inhomogeneities can then be easily visualized by coronal or sagittal re-sampling of the data subset.

Let I_{ORG} be the original color data subset and I_γ the subset obtained by gamma correction, with linearization, using the gray scale provided for each image. Fig. 1-a shows a coronal slice sub-image from I_{ORG}; each of its lines corresponds to one of the successive axial-plane images. Fig. 1-b shows the corresponding re-sampled slice from I_γ.

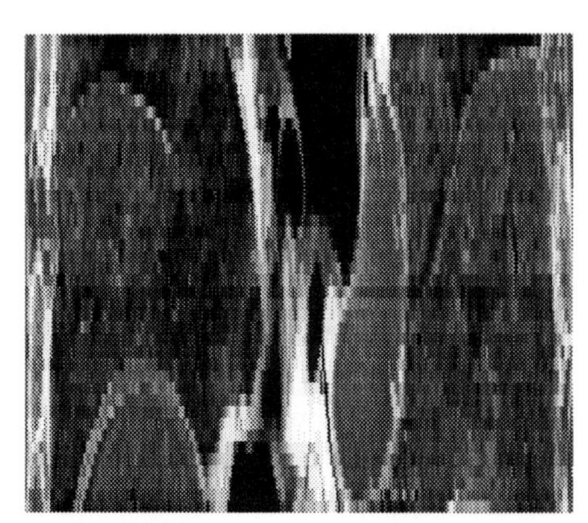
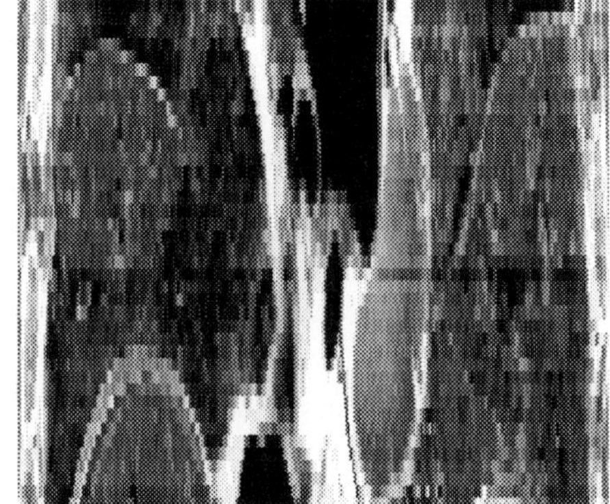

Fig. 1. A coronal re-sampled sub-image of a), the original volume set, and b), after gamma correction

Even if some homogenization has taken place, we may observe that the most important remaining horizontal discontinuities can not be interpreted as just a color offset between two successive acquisitions.

3 Adaptive Correction Methods and Results

To correct for these discontinuities, we have adopted the following first order autoregressive model:

$$I_i(x,y) = I_{i-1}(x,y) + \Delta_i(x,y) \tag{1}$$

where $I_i(x,y)$ is the RGB vector of pixel (x,y) in slice i, and $\Delta_i(x,y)$ is an offset vector that comprises at least two kinds of radiometric discontinuities that may arise between two axial-plane slices: one is the texture and structure features (*innovations*) which must be preserved, and the second kind consists of some regional variations caused by an unknown physical process which has changed the color of the tissue before digitization. In some imaging modes as MRI [4], this component of $\Delta_i(x,y)$ can be further separated in inter- and intra-slice intensity variations, but some a priori knowledge is required for an explicit formulation.

In order to correct the second kind of inhomogeneities, we have taken advantage of spatial coherence. First, a difference color image, $\delta I_i(x,y)$ is obtained by subtraction between two consecutive cross-sections. The coherence hypothesis consists in having local histograms of $\delta I_i(x,y)$ centered at RGB=(0,0,0) for all (x,y). Even if fine structure is very complex, two consecutive thin cross-sections are similar enough to expect that, *locally*, $\delta I_i(x,y)$ will have an histogram centered at (0,0,0). "Locally" means that histograms are measured over small sized regions, the minimum size being that required for a meaningful statistics of coherent pixels. Furthermore, secondary peaks in histogram correspond to innovations, which represent the end or beginning of a different tissue region between two slices. Any shift of the central mode from (0,0,0) implies a radiometric discontinuity of the second kind in that pixel.

Thus, we have performed a local adaptive correction, for each color channel, which is propagated along the images $I_i(x,y)$. Figure 2-a shows a difference color image in luminance, plus a global offset. The dominant gray value corresponds to "ground" (0,0,0), and darker or lighter pixels represent inhomogeneities. A local histogram is obtained for each pixel of the difference color image $\delta I_i(x,y)$; the pixel neighborhood consists of several hundred pixels. The main peak position is measured, and its shift from (0,0,0) is calculated. This shift constitutes the local coherence deviation between the two images. A partial correction is made on the second image, a new image is subtracted from the corrected one, and the process is repeated, until all images of the ROI have been processed. A second pass is performed backwards, to ensure a minimal deviation from original undistorted values, while rejecting inhomogeneities which are not innovations. We found that a partial correction of 60% in one direction and another of 60% of the residual difference in the opposite one (making a total of 90% of the original difference), perform better than a full 100% correction in one single pass, which produces some color artifacts. These artifacts depend of the selection of the starting slice for propagation, but a proper choice can be devised, using homogeneity testing techniques [3].

Figure 2-b shows the same re-sampled slice as in Fig. 1-a, after homogenization; where horizontal artifacts have almost disappeared. We call the new color data subset I_{HOMOG}. Other coronal and sagittal image re-samplings of I_{HOMOG} show a similar performance of the correction.

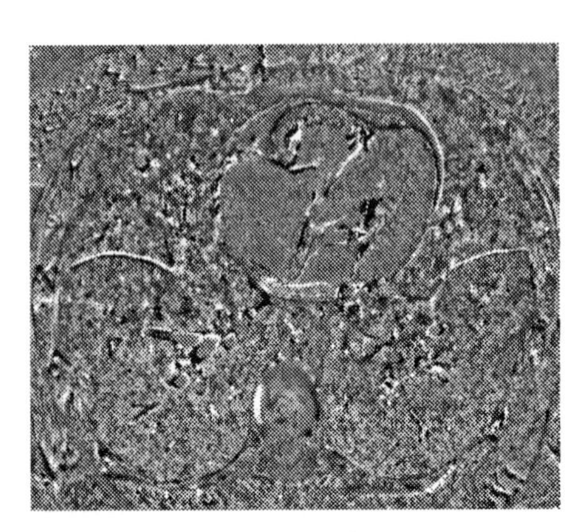

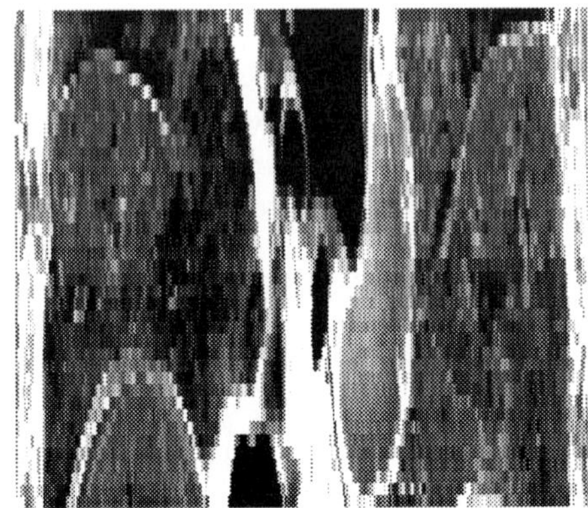

Fig. 2. a) Difference image between two cross sections; b) coronal re-sampled image of the volume set after adaptive homogenization

Let I_{GRAY} be the set of gray level images, taking the luminance from I_{HOMOG}. Using local thresholding on I_{GRAY}, we have obtained a binary volume I_{BIN}, where fine blood vessels may be isolated for three-dimensional reconstruction. A 3D back-to-front rendering [5] view of I_{BIN} is shown in Fig. 3-a. An improper neighborhood of analysis and other parameters affect homogenization results. In order to evaluate the performance of 3D segmentation of I_{HOMOG} with different homogenization parameters, we use an axial *minimum intensity projection* (MIP) image (see [6] and [7]) of I_{ORG}, which allows to visualize the projection of most blood vessels present in the sub-volume. Such a MIP image is shown in Fig. 3-b.

4 Discussion and Conclusion

An important advantage of the adaptive correction method is that no blurring is introduced in I_{HOMOG}, and fine detail and texture are respected, because no low-pass filtering is done, and pixel modification consists of local offsets, which depend only on the coherence mismatch between regions of two adjacent images. For the same reasons, our algorithm does not filter out smooth variations along the treated volume, and local thresholding or region growing techniques are still

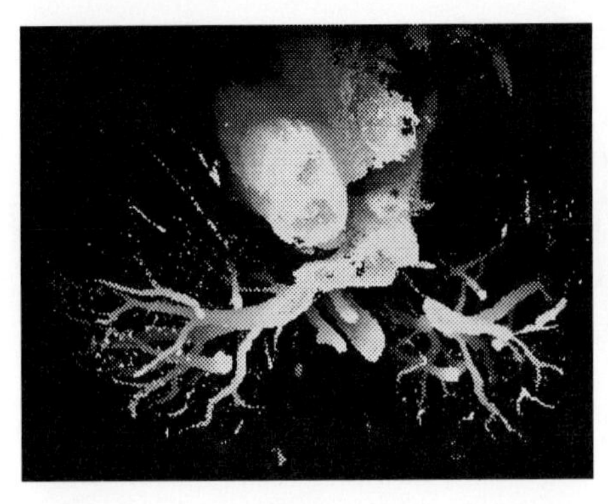

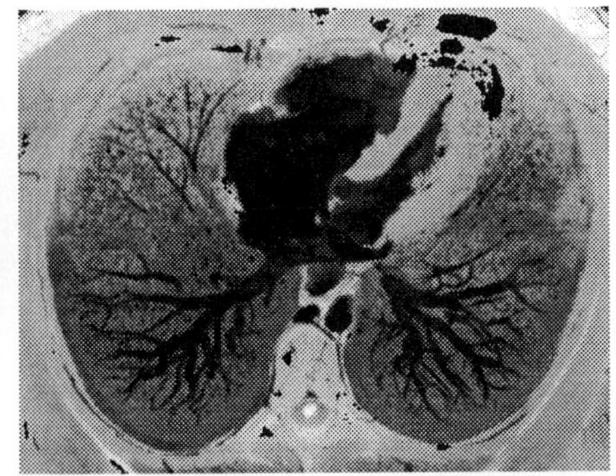

Fig. 3. a) Three-dimensional rendering of blood vessels detected in the radiometric corrected sub-volume; b) axial minimum intensity projection image of the original ROI

needed in order to segment fine structures such as the blood vessels. Classical segmentation problems such as noise, sub-sampling, partial volume, etc, remain to be solved.

Remaining inhomogeneities still visible in Fig. 2-b could be smoothed out, if larger neighborhoods of lung tissue were analyzed. This is not possible in the presence of important innovations and parenchyma other than the lung tissue. In these regions it appears that radiometric distortions are different from those in the lungs.

The use of local region information and coherence between two images suggests performing a preliminary segmentation of innovations during image correction. Furthermore, large blood vessels and heart, stomach and other structures should be masked out beforehand, by contour extraction, to improve homogenization of lung tissue and three-dimensional segmentation of small details. Our present work deals with these issues.

Acknowledgments. We thanks Mr. Michel Ackerman and the National Library of Medicine for providing the VHP data set.

References

1. Ackerman M.J.: *The Visible Human Project*, National Library of Medicine; World Wide Web site **http://www.nlm.nih.gov** (1995).
2. Wyszecki, G., Styles, W.S.: *Color Science: Concepts and Methods, Quantitative Data and Formulae,* Second Edition, John Wiley & Sons, New York (1982).
3. Wu. Z.: Homogeneity Testing for Unlabeled Data: A Performance Evaluation. CVGIP: Graphical Models and Image Processing Vol. 55: **5** (1993) 370–380
4. Zijdenbos, A., Dawant, B., Margolin, R.: Intensity Correction and its Effect on Measurement Variability in MRI. Computer Assisted Radiology, edited by Lemke H. et al. Proceedings of the Int. Symp. on Comp. and Comm. Sys. for Im. Guided Diag. and Therapy (1995) 216–221

5. Frieder, G., Gordon, D., Reynolds, R.: Back-to-Front Display of Voxel-Based Objects. IEEE Computer Graphics and Applications **5** (1985) 52–60
6. Udupa, J.K.: 3D Visualization of Images. Technical Report No. MIPG196. Medical Imaging Processing Group. University of Pennsylvania. (1993) 52–62
7. Napel, S., Marks, M.P., Rubin, G.D., Dake, M.D., McDonell, C.H., Song, S.M., Enzmann, D.R., Brooke, R.J.: CT Angiography with Spiral CT and Maximum Intensity Projection. Radiology **185** (1992) 607–610

3D Reconstruction of Cerebral Vessels and Pathologies from a Few Biplane Digital Angiographies

Laurent Launay[1,2], Eric Maurincomme[2], Pierre Bouchet[1], Jean-Laurent Mallet[1] and Luc Picard[3]

[1] CRIN - CNRS, B.P. 239, F-54506 Vandœuvre-lès-Nancy CEDEX, France
[2] General Electric Medical Systems, 283 rue de la Minière, B.P. 34, F-78533, Buc CEDEX, France
[3] Service de Neuroradiologie Diagnostique et Thérapeutique, CHU, Hôpital Saint-Julien, F-54037 Nancy CEDEX, France

Abstract. 3D reconstructions of cerebral vessels and Arteriovenous Malformations from six digital angiography sequences are compared for three iterative methods, one of them being new. The interest of Multiplicative ART is underlined, and the effects of a criterion minimizing the total density are discussed.

1 Introduction

Three-dimensional images of cerebral vessels are known to be very useful to the diagnosis of pathologies such as aneurysms or arteriovenous malformations (AVMs). 3D X-ray angiography systems have been built where a volumic image of the vessels is reconstructed from a sequence of projections acquired while the acquisition system is rotating around the patient [1]. Unfortunately, such reconstructions cannot be made on standard acquisition systems, because of the high speed of rotation which is needed. Nevertheless, the traditional biplane systems usually allow the positioning of the imaging chain in almost any incidence and the simultaneous acquisition of two orthogonal sequences. Moreover, in the case of a complex pathology, radiologists often acquire two or three of such injected sequences with different angulations in order to understand its three-dimensional structure.

These remarks motivate investigations on the ability of performing 3D reconstructions from a few number of angiographic biplane static sequences. Due to the low number of projections available (from 4 to 8), one cannot expect an image quality as good as with rotational systems. On the other hand, many projection images are available in every incidence. This allows to select the same particular phase of the contrast agent propagation in all acquired series. Thus many reconstructed volumes can be computed to give a sort of 3D sequence of the injection (sometimes called 4D image, the fourth dimension being time). Such reconstructions are particularly interesting for interventional neuroradiology procedures and for stereotactic angiography examinations performed before radiotherapy irradiation of arteriovenous malformations.

Some authors have shown that three-dimensional reconstructions can be obtained from only two or three projections, by use of a prior model of the vessels [2, 3]. Due to the very diffuse aspect of certain AVMs (see Fig. 1), we only investigate fully tomographic algorithms, and we restrict ourselves to discrete iterative methods. After a brief description, ART, MART and a new algorithm called DSI are compared on different reconstructions of two patient vasculatures.

2 Iterative Algorithms

In the algebraic formulation of the reconstruction problem, a linear system $\mathbf{Y} = \mathbf{Hf}$ is to be resolved where the unknown is $\mathbf{f}$, N-dimensional vector representing the 3D image. We denote P the number of X-ray projection images, and vector $\mathbf{Y}$ represents all the M pixels of these projections. Matrix $\mathbf{H}$ is the projection operator, whose j^{th} line is denoted $\mathbf{h}_j$.

2.1 Algebraic Reconstruction Technique (ART)

In ART [4], one iteration corresponds to the action of one row of matrix $\mathbf{H}$ (i.e. one X-ray equation) on volume $\mathbf{f}$, and one cycle is completed when the M rows have been processed. The iterative formula for a ray j and a voxel i is:

$$f_i^{(n+1)} = f_i^{(n)} + \lambda^{(n)} \frac{y_j - \mathbf{h}_j \cdot \mathbf{f}^{(n)}}{||\mathbf{h}_j||^2} h_{ji} \tag{1}$$

In the multiplicative version (MART), the update formula is:

$$f_i^{(n+1)} = f_i^{(n)} \cdot \left(\frac{y_j}{\mathbf{h}_j \cdot \mathbf{f}^{(n)}} \right)^{\lambda^{(n)} h_{ji}} \tag{2}$$

When the system is consistent (i.e. a solution exists), ART converges to the solution having minimum variance, which is also the one having minimum quadratic norm [5], and MART converges to the maximum entropy solution [6]. If data are noisy, good reconstructions can be obtained by stopping the iterations after a few cycles, the optimal number being deduced experimentally.

When treating a large amount of projection data, MART tends to give almost the same solution as ART. Nevertheless, the two algorithms can give very different results for few-view reconstructions, as detailed in the next section.

2.2 Regularization Methods

The method we present here is based on the regularization of the inverse problem with quadratic terms, one of them being a smoothness criterion. It was called DSI because the formulation is inspired from a generic Discrete Smooth Interpolation algorithm proposed by Mallet [7] and mainly used in geometric modelling. We state the reconstruction problem as finding $\mathbf{f}^*$:

$$\mathbf{f}^* = \text{Argmin}\left\{ R(\mathbf{f}) + \varpi^2 \left\|\mathbf{Y} - \mathbf{H}\mathbf{f}\right\|^2 + \varpi_d^2 \left(\sum_i f_i\right)^2 \right\} \tag{3}$$

$R(\mathbf{f})$ quantifies the roughness of the volume and is defined by:

$$R(\mathbf{f}) = \sum_{k=1}^{N} \left| \sum_{\alpha \in \Lambda(k)} f_\alpha - p f_k \right|^2 \tag{4}$$

where p is the number of neighbours for a given connectivity (usually $p = 6$) and $\Lambda(k)$ is the set of the neighbouring voxels of voxel k. If $\mathbf{f}$ is positive, the last term in (3) is proportional to $||\mathbf{f}||_{L_1}^2$; it models a prior information specific for vessels reconstruction: the solution is supposed to be close to a null volume, in the L_1-norm sense.

When setting to zero the derivative of the criterion (3) with respect to $\mathbf{f}$, a linear system is obtained. Its resolution by a Gauss-Seidel method leads to a voxel by voxel algorithm [8], whose basic iterative formula for a voxel α is:

$$f_\alpha^{(n+1)} = \frac{\Gamma_\alpha^{(n)} + \varpi^2 \left(P f_\alpha^{(n)} + \sum_i h_{i\alpha} \left(y_i - \mathbf{h}_i \mathbf{f}^{(n)}\right)\right) - \varpi_d^2 \sum_{j \neq \alpha} f_j}{(p^2 + p) + \varpi^2 P + \varpi_d^2} \tag{5}$$

$$f_\beta^{(n+1)} = f_\beta^{(n)} \quad \text{for } \beta \neq \alpha \tag{6}$$

where $\Gamma_\alpha^{(n)}$ stands for

$$\sum_{k \in \Lambda(\alpha)} \sum_{\substack{\beta \in \Lambda(k) \\ \beta \neq \alpha}} f_\beta^{(n)} - 2p \sum_{k \in \Lambda(\alpha)} f_k^{(n)} \tag{7}$$

By adding a relaxation coefficient λ and a positivity constraint:

$$f_\alpha^{(n+1)} = \max\left\{ \lambda \hat{f}_\alpha^{(n+1)} + (1 - \lambda) f_\alpha^{(n)}, 0 \right\} \tag{8}$$

where $\hat{f}_\alpha^{(n+1)}$ is given by (5).

The special form of $R(\mathbf{f})$ ensures uniqueness of the solution to (3) and the convergence of the algorithm to this solution (a general proof can be found in [7]). Moreover, the coefficients ϖ and ϖ_d can be normalized so that each constraint has a comparable weight in the iterative formula. Thus, we define ω and ω_d by:

$$\omega = \frac{\varpi^2 P}{p^2 + p} \quad \text{and} \quad \omega_d = \frac{N \varpi_d^2}{N_r (p^2 + p)} \,. \tag{9}$$

Where N_r is the approximate number of voxels intersected by each ray.

3 Comparison on Vessels and Pathologies

This section presents the comparison of the previously described algorithms on cerebral vascular image data of two different patients. Three positions of a bi-plane acquisition system furnished six Digital Subtraction Angiography (DSA) sequences corresponding to regularly disposed angulations, for a rotation around the patient axis. While dataset A is normal, a very large AVM is present in dataset B, which disturbs very much the blood flow and makes surrounding vessels difficult to see (Fig. 1).

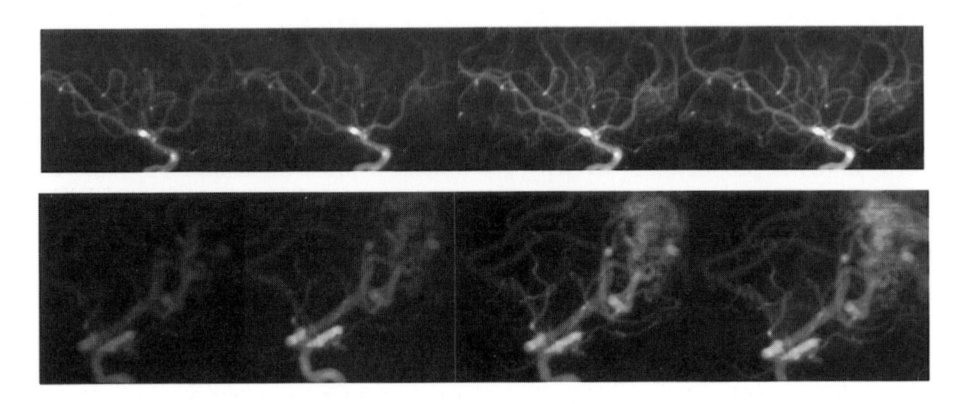

Fig. 1. Part of a DSA sequence of patient A (up) and patient B (down).

We compared ART, MART, and DSI for a particular injection time of dataset A (Fig. 2). We define DSI1 as DSI with ($\omega = 1.5, \omega_d = 0$) and DSI2 as DSI with ($\omega = 1.5, \omega_d = 1.5$). Images are displayed in Maximum Intensity Projection (MIP), which consists in visualizing on the screen among all the voxels composing the ray, the one with maximum density. This display mode is known to reinforce the contrast for small vessels. The first interesting result is that with six projections, ART and MART give very different reconstructions. Due to the highly underdetermined nature of the system, the entropic behaviour of MART gives a better result than the small variance solution of ART, which produces streaking artifacts. DSI1 gives a locally smooth reconstruction, but even on MIP views, it is sometimes difficult to distinguish between vessels and the background fluctuations. Furthermore, less vessels can be seen than with MART. With DSI2, the contrast on small vessels is increased and more vessels can be recognized. It shows that the regularizing effect of the L_1-norm criterion is favourable to few-view vessels reconstruction.

The other way to visualize a 3D vascular reconstruction is to compute an X-ray reprojection of the volume. When comparing MART reconstruction and DSI2 reconstruction on MIP views, it seems that MART is better because the background is more uniform. On the other hand, as shown in Fig. 3, DSI2 appears to be superior in X-ray-like reprojection. Furthermore, while MIP images are

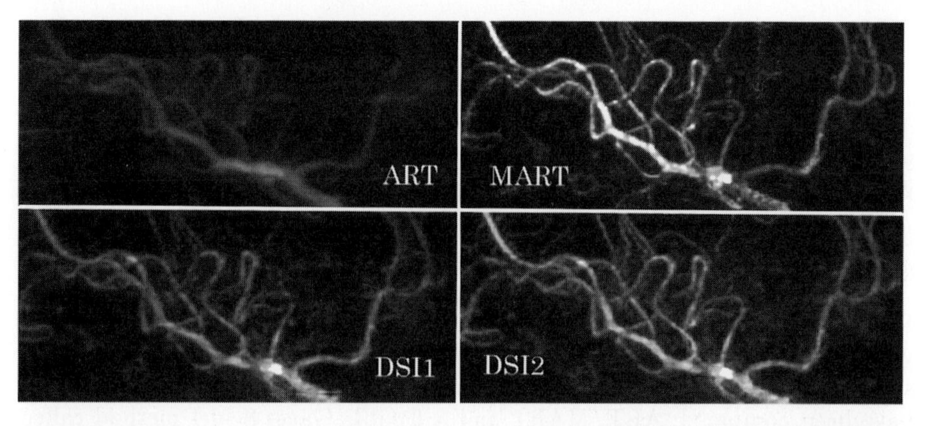

Fig. 2. Comparison of ART, MART, DSI1 and DSI2 reconstructions for dataset A, MIP rendering.

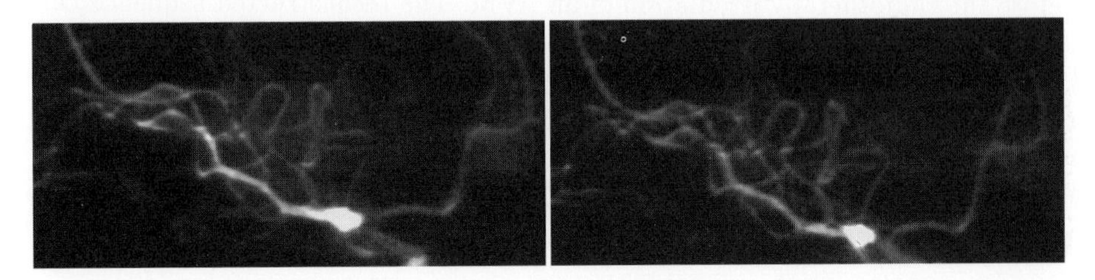

Fig. 3. Comparison of MART (left) and DSI2 (right) reconstructions for dataset A, X-ray like rendering.

usually preferred, in cases such as dataset B they lead to a very noisy aspect of the AVM nidus (see Fig. 4), which is better understood in X-ray reprojection.

Another difference between DSI and MART is the sensitivity to measurement errors. If a part of a projection image has very low density (because of a collimator or an occluding object), MART will give more artefacts than DSI, because of its multiplicative nature. From the computational point of view, one

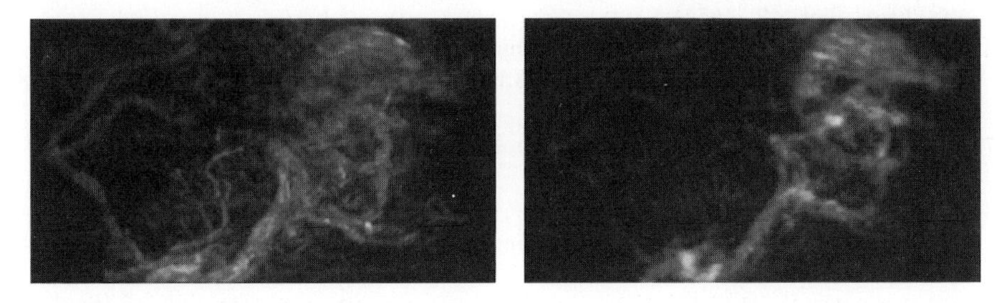

Fig. 4. Comparison of MIP (left) and X-ray like (right) rendering for a 3D reconstruction of dataset B.

DSI iteration is almost as time consuming as one cycle of ART or MART. More memory is required for DSI, since original and current projections need to be stored in addition to the current volume. Nevertheless, the difference is not very important for few-view reconstructions. The number of iterations we used is 6, in order to be sure that convergence is reached. This could possibly be improved by optimizing the relaxation coefficient and choosing a better initial volume.

4 Conclusion

We have experimented that a three-dimensional injected sequence of cerebral vessels and Arteriovenous Malformations can be reconstructed from six digital subtraction angiography sequences. Three iterative reconstruction algorithms have been compared: ART, MART and a flexible Gauss-Seidel method called DSI.

Such reconstructions are interesting in the case of a stereotactic examination before the radiotherapy irradiation of an AVM. The reconstructed sequence can then be merged to a MR or a CT stereotactic volume to help the radiotherapy planning. Interventional neuroradiology procedures could also benefit from such reconstructions; matching with another modality being this time achieved by use of registration methods [9].

References

1. Trousset, Y., Picard, C., Ponchut, C., Romeas, R., Campagnolo, R., Croci, S., Scarabin, J., Amiel, M.: 3D X-ray angiography: From numerical simulations to clinical routine. In Int. Meeting on Fully 3D Image Reconstruction in Radiology and Nuclear Medicine, Aix les Bains, France (1995) 3–11
2. Coatrieux, J.-L., Garreau, M., Collorec, R., Roux C.: Computer vision approaches for the three-dimensional reconstruction of coronary arteries: Review and prospects. Crit. Rev. in Biomed. Eng., **22(1)** (1994) 01–38
3. Pellot, C., Herment, A., Sigelle, M., Horain, P., Maitre, H., Peronneau, P.: 3D reconstruction of vascular structures from two X-ray angiograms using an adapted simulated annealing algorithm. IEEE Trans. on Med. Im. **13(1)** (1994) 48–60
4. Herman, G. T.: Image Reconstruction from Projections: the Fundamentals of Computerized Tomography. Academic Press, New York (1980)
5. Herman, G. T., Lent, A., Rowland, S. W.: ART: Mathematics and Applications. J. of Theor. Biol. **42** (1973) 1–32
6. Lent, A.: A convergent algorithm for maximum entropy image restoration, with a medical x-ray application. In Image Analysis and Evaluation, Soc. of Photo. Sci. and Eng. (1977) 249–257
7. Mallet, J.-L.: Discrete smooth interpolation in geometric modelling. Computer Aided Design **24(4)** (1992) 177–191
8. Launay, L., Bouchet, P., Maurincomme, E., Berger, M.-O., Mallet, J.-L.: A flexible iterative method for 3D reconstruction from X-ray projections. In Int. Conf. on Patt. Recog. (1996) (to appear)
9. Feldmar, J., Ayache, N., Betting, F.: 3D-2D projective registration of free-form curves and surfaces. INRIA Research report **2434** (1994)

Robust 3-D Reconstruction and Analysis of Microstructures from Serial Histologic Sections, with Emphasis on Microvessels in Prostate Cancer

Paul A. Kay, Richard A. Robb Ph.D., David G. Bostwick MD, Jon J. Camp

Mayo Clinic and Foundation, Rochester MN, 55905, USA

Abstract. We have developed a technique to reconstruct 3-D microstructures from serial histologic section, and demonstrated the robustness of the method on microvessels associated with prostate cancer and bile ducts in liver specimens. The method is critically based on image coregistration techniques, particularly two dimensional surface matching between extracted image features. The reconstruction paradigm allows 3-D regions of information in several locations throughout a block of tissue to be obtained at different magnifications while maintaining proper spatial relationship to the original data. The reconstructed objects have a generally unmapped and unknown topography. Visualizing anatomic microstructures in 3-D and analyzing their topography will help physicians to better understand and correctly interpret changes in disease.

1. Introduction and Background

New diagnostic tests, such as serum testing for PSA, have increased the percentage of prostate cancers detected[1]. Because of the varying malignant potential of prostate cancers, this increase in diagnostic incidence highlights the need for accurate stratification of prostate cancer patients into appropriate therapies. This stratification is difficult because the metastatic potential of prostate cancer cannot be assessed prior to histologic confirmation [1]. Several measures of the pathologic state of prostate cancer have been proposed [2,3] to facilitate stratification of patients into either a specific therapy or into a period of "watchful waiting". These measures include prostate tumor size and microvessel density, both of which have been shown to be indicators of metastatic potential in prostate cancer [2-6]. However, *there is no currently reliable technique* to measure these potential markers *in vivo*. If such measures were available, patients who have cancer with low probability of metastasis may forgo aggressive treatment and its associated risks. The role of microvessel density as a marker for the metastatic potential of a prostate tumor could, therefore, significantly contribute to improved outcomes at reduced cost. However, fundamental knowledge about the spatial and functional relationships between microvessels and the prostate tumor need to be obtained before such benefits can be realized.

The process by which a tumor induces the growth of nutrient microvessels is called angiogenesis [5]. Since unlimited tissue growth is one characteristic of cancer, an increase in the number of microvessels occurs in many types of cancerous tissue [5]. Pathologists can measure the microvessel density which is proportional to the number of small nutrient vessels. Increases in microvessel density have been shown to correlate with the metastatic potential of prostate cancer [2, 4-6]. Some researchers have proposed that there is a differential distribution of microvessels in prostate tumors, with an area of highest density (called the "hot spot") which correlates with the metastatic potential of the tumor [4,6]. However, the volumetric distribution of microvessels and their 3-D patterns of growth have not been studied. If the pattern of density can be shown to be consistent over the entire volume, a random core sample of prostate tissue could determine the metastatic potential of the tumor. But a "gold standard" 3-d study is required to confirm the efficacy of such a useful procedure.

To be able to rigorously study microvessels associated with prostate cancer, the microvessels must be reconstructed from 2-D serial sections and analyzed in consistent spatial relationship to the cancer and normal tissue surrounding the tumor. Several methods [7,8] of aligning serial 2-D images have been used previously including user-induced fiducial markers (for example laser holes), interactive operator alignment and computer automated methods of alignment. Fiducial marking is a frequent method used in alignment but requires locating the area to be reconstructed before slicing a tissue block with the negative consequence that the markers must be in the field of view. And in many cases the area of reconstruction can not be defined before processing. The most common method for serial section realignment is interactive operator alignment [7-9]. In this method one section is digitized and the next section is overlayed on a video screen. The user adjusts the section until they are aligned. There are inherent and significant operator errors and fatigue factors involved with this method [8,9]. Some work has been done on automated realignment methods [8] and these techniques have proven useful, but they are generally slow (~10 matches per hour). We have developed an automated method to accurately and rapidly co-register sequential 2-D slices using a surface matching algorithm [10] applied to features extracted from the images.

2. Method

Others investigators [2] have used 2-D image analysis methods to count the number of microvessels in specific microscope fields. We analyzed the 3-D density and growth patterns of the microvessels by using color stains of serial sections (4μ thick) of excised tissue following retropubic prostatectomy. The sequential sections are differentially stained with antibodies to factor VIII - related antigen isolated on the endothelial cells of the vessels. The chromogen used in the factor VIII stain was amino ethyl carbazole (AEC) which produces a red reaction. The counterstain used was a light hematoxylin which provides good color contrast to the vessels. The sections are acquired with a slow-scan CCD camera attached to a microscope. Color separation of the microvessels is accomplished using multispectral classification [11]. 3-D alignment of the sequential digitized slices is accomplished using a novel technique developed in our laboratory [12]. Registration and reconstruction is a non-trivial task due to the absence of internal landmarks that can serve as natural fiducial points. Registration is accomplished using a special combination of image processing filters and the surface matching algorithm [10] in ANALYZE™.

Our method of registering sequential 2-D images follows these steps: 1) Relevant feature are extracted from the images. Simple thresholding produces prominent objects and Sobel filtering followed by thresholding produces an enhanced edge image. 2) The feature images are processed using math morphology and component analysis to eliminate noise and small structures and retain only the large, prominent features that can be located consistently from slice to slice. Edge images from consecutive sections are shown in Figure 1. 3) The extracted features are then serially matched using a chamfer-distance surface matching algorithm (ANALYZE™). The quality of the surface match is demonstrated in Figure 2. This technique is relatively fast producing approximately 100-150 matches per hour on a modern workstation.

3. Results

A volume rendering of microvessels associated with prostate cancer (Fig. 3a) and normal tissue (Fig. 3b) produced by this method are shown. Analysis of a small focus of prostate cancer (> 1 mm^3) using this technique [12] is summarized in Table 1. The vessel density in the region of the tumor was approximated by counting the pixels within the different classified objects and subregions of the tissue. The number of pixels representing vessels within the tumor was computed to be twice as large as the number outside of the tumor, suggesting a significantly higher vascularity in tumor compared to normal tissue. This finding supports similar observations by others studying 2D microvessel density patterns [2,4-6].

Table 1

V1 = Total pixels / region	= 19,966,500
V2 = Tumor pixels / region	= 2,884,716
V3 = Vessel pixel / region	= 195,836
V4 = Vessel pixels / tumor	= 60,027
V4/V2 = ratio of vessels in tumor	= .021
V3/V1 = ratio of vessels in sample	= .0098

This method has also been used to align serial section of liver tissue to reconstruct bile ducts. A volume rendering of bile ducts is shown in Figure 4. For microscopic bile ducts, a branching pattern that had only been implied from 2-D sections can now being visualized and analyzed rigorously for the first time.

4. Discussion

Significant problems associated with 3-D analysis of serial histologic sections include uneven staining from section to section and missing or damaged sections. We have overcome these problems during the tissue preparation phase by carefully recording the location of missing or damaged sections. The sections are stained using a VENTANA (Tucson AZ) automated instrument to reduce inconsistent staining through a tissue block. The co-registration technique was faithful in matching non-consecutive sections when a damaged or missing section was found. This is due in part to the thin spacing of the sections. The structures tend to vary slowly over slice thickness. When missing sections in the tissue where encountered, they where placed in the reconstruction as blank slice so to preserve the 3-D relationships of the vessels.

We will explore methods to quantify the microvessel patterns in the prostate tumors. We will measure vessel density, vessel spacing, branching patterns, tortuosity and surface area of the vessels. Tree analysis methods within ANALYZE™ will be utilized to extensively study the microvessels branches. We will study variations within the tumors of all of these measures to quantify the 3-D microvessel patterns.

Rigorous three-dimensional visualization and analysis of histologic structures will be a powerful method for investigating, monitoring and quantifying disease patterns. The methods developed in our research constitutes an "enabling" technology which will allow pathologists and histologists to directly visualize and analyze features of microstructures in 3-D, which currently can only be deduced from 2-D samples of these features. The method requires a minimum of specialized equipment and can be readily reproduced by others.

5. Acknowledgments

We thank Dr. Jurgen Ludwig and Dr. Nick LaRusso for allowing us to use liver images from there biliary anatomy study. We thank the Mayo histology sectioning and immunostaining labs for the many hours of work on this project. We also thank the staff of the Biomedical Imaging Resource for their multi-faceted contributions to this work.

References

1. Garnick MB, "The dilemmas of prostate cancer", *Scientific American*, pp 72-81, April 1994.
2. Brawer MK, Deering RE, Brown M, Preston SD, Biglor SA, "Predictors of pathologic stage in prostatic carcinoma: The role of neovascularity", *Cancer*, vol 73(3), pp 678-687, 1994.
3. McNeal JE, Bostwick DG, Kindrachuk RA, Redwine EA, Freiha FS, Starney RA, "Patterns of progression in prostate cancer", *Lancet*, vol I, pp 60-63, 1986.
4. Weidner N, Carroll PR, Flax J, Blumenfeld W, Folkman J, "Tumor angiogenesis correlates with invasive prostate cancer", *Am J Path*, vol 143(2), pp 401-409, 1993.
5. Fidler IJ, Ellis LM, "The implications of angiogenesis for the biology and therapy of cancer metastasis", *Cell*, vol 79, pp 185-188, 1994.
6. Siegal JA, Yu E, Brawer MK, "Topography of Neovascularity in Human prostate carcinoma" *Cancer*, vol 75, pp 2545- 2551, 1994.
7. Kreite A., ed., *Visualization in Biomedical Microscopies: 3-D Imaging and Computer Applications*, VCH, 1992.
8. Salisbury JR, Whimster WF, "Progress in computer-generated three-dimensional reconstruction" *Journal of Pathology*, vol. 170, pp 223-227, 1993
9. Rydmark M, Jansson T, Berthold C.-H., Gustavsson T, "Computer Assisted realignment of light micrograph images from consecutive sections series of cat cerebral cortex", *Journal of Microscopy*, vol. 165, pp 29-47, 1992.
10. Jiang HJ, Robb RA, Holton KS, "A new approach to 3-D registration of multi-modality medical images by surface matching", *Proceeding of Visualization in Biomedical Computing*, vol 2, 1992.
11. Robb RA, Hanson DP, Karwoski RA, Larson AG, Workman EL, Stacy MC, ANALYZE™: A comprehensive, operator-interactive software package for

multidimensional medical image display and analysis, Computerized Medical Imaging and Graphics, vol 13(6), pp 433-454, 1989.

12. Kay PA, Robb RA, Bostwick DG, Leske DA, Camp JJ, Three-dimensional visualization of microvessels in prostate cancer, SPIE Proceeding of Medical Imaging 1995, San Diego, CA, Feb. 1995.

Figures

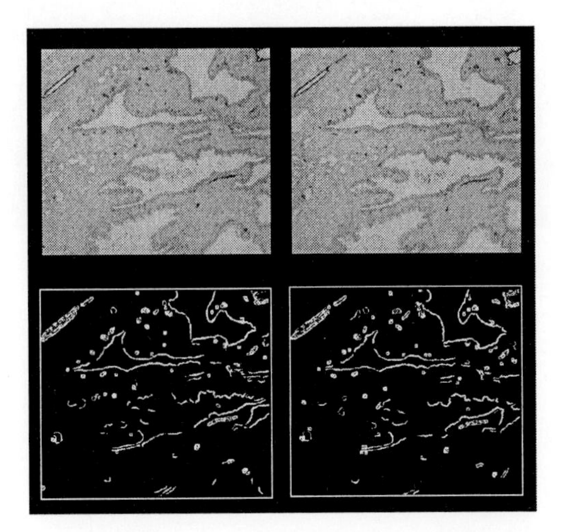

Fig. 1. Two consecutive sections digitized sections (top) and their respective edge image (bottom).

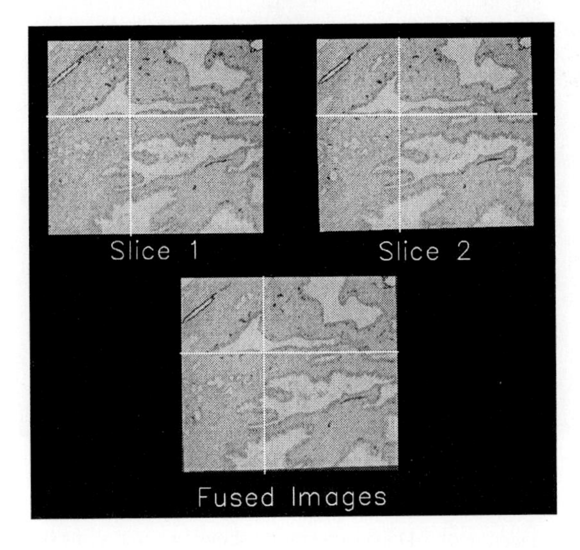

Fig. 2. Matched consecutive pair of sections. The lower image is a fused image of both sections. The registration marker lies on the same spot in both images.

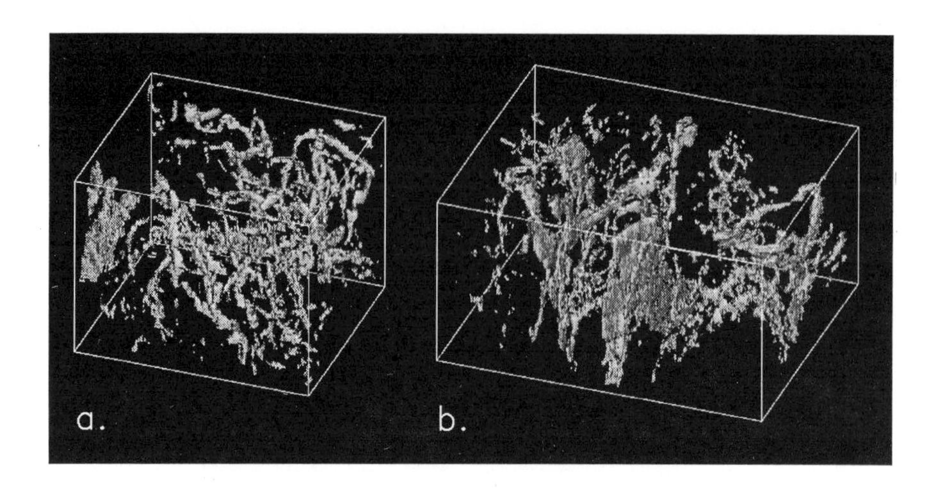

Fig. 3. Volume rendering of 3-D microvessels: a) Associated with Prostate Cancer. b) Associated with normal prostate tissue.

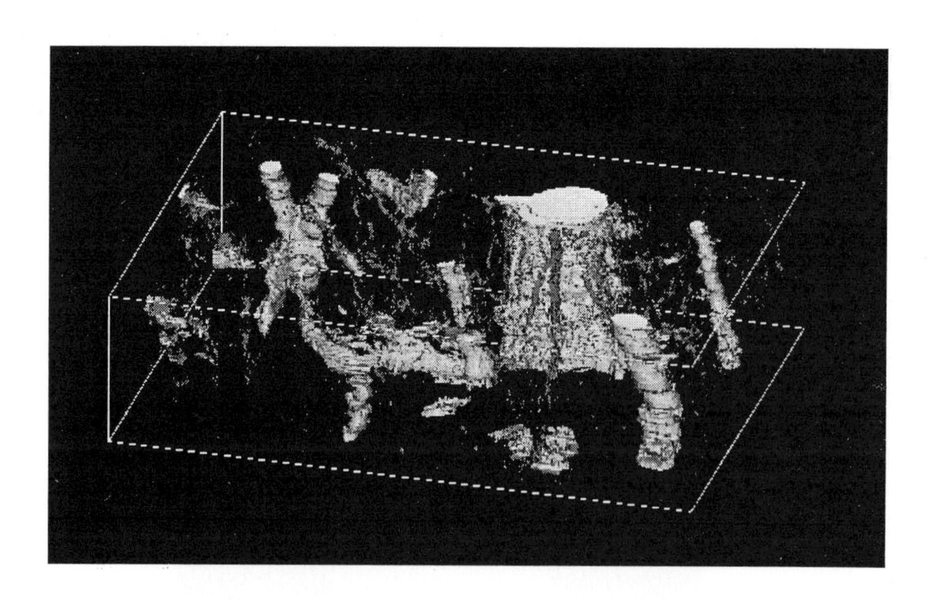

Fig. 4. Volume rendering of 3-D liver specimens showing portal veins (lighter) and bile ducts.

An Extensible MRI Simulator for Post-Processing Evaluation

Remi K.-S. Kwan*, Alan C. Evans, and G. Bruce Pike

McConnell Brain Imaging Centre, Montréal Neurological Institute, McGill University, Montréal, Québec, Canada, H3A 2B4.

Abstract. An extensible object-oriented MRI simulation system is presented. This simulator uses first-principle modelling based on the Bloch equations to implement a discrete-event simulation of NMR signal production. A model of the image production process, incorporating noise and partial volume effects, has also been developed based on first-principles. This is used to generate realistic simulated MRI volumes, based on a labelled data set, for the evaluation of classification algorithms and other post-processing routines.

1 Introduction

One of the strengths of MRI is the versatility with which contrast can be obtained for images of soft tissue in the brain. However, while this ability makes MRI a flexible tool for clinical diagnosis, the multi-dimensional nature of the image parameter space often confounds the optimal selection of acquisition parameters. With the increasing need for post-processing and quantitative analysis of medical images, it has become useful to evaluate the effects of imaging parameters on image quality with respect to the performance of post-processing algorithms. With this application in mind, an extensible MRI simulator has been developed.

By necessity, simulations are an approximation to reality. The benefit of using simplified models is their tractability and the ease with which they may be controlled. Of course, models should not be oversimplified at the expense of predicting key behaviour, but instead they should be tailored to the application at hand. The approach taken has thus been to develop an extensible system that allows for the incorporation of increasingly complex models depending on the particular simulation study.

2 Simulation Design

In order to flexibly support future applications, the simulator makes use of several models organized according to principles of object-oriented design [2, 7]. The base models in Fig. 1 represent objects often used to describe MRI systems

* This work has been supported by a fellowship from the Natural Sciences and Engineering Research Council of Canada (NSERC).

and define the basic interface for interaction between models required to produce simulated images. As more sophisticated or specialized simulations are needed, new models may be introduced by inheritance from these base models without affecting the functionality of the rest of the simulator model.

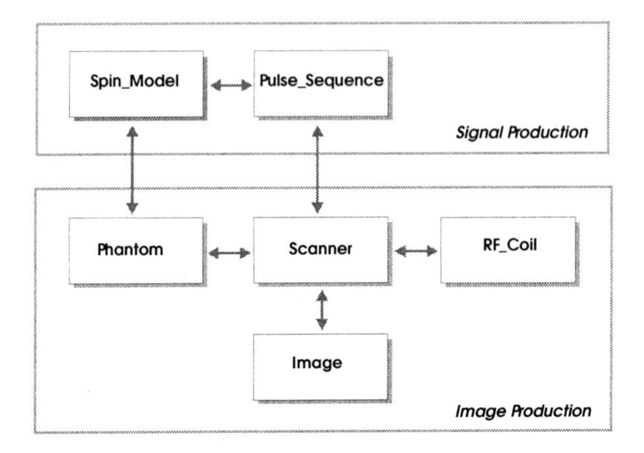

Fig. 1. An overview of the principal models in the MRI simulator and their interactions.

The simulator is organized around two major tasks. The first task, signal production, is concerned with the computation of tissue intensities from a Pulse Sequence model and a model of tissue magnetization properties encapsulated in a Spin Model. The second task, image production, is described by the Scanner, Phantom, and RF Coil models which incorporate the calculation of noise and partial volume effects with the results of the signal production simulation to generate realistic MRI volumes.

2.1 Signal Production

The simulation of NMR signal production proceeds as a discrete-event simulation of a pulse sequence's effects on magnetization state. The timing of events, such as RF pulses, are described by the Pulse Sequence model. When a pulse sequence is applied each event triggers a message to be sent to a Spin Model.

The Spin Model encapsulates the current state of tissue magnetization dynamics and its behaviour under the influence of events such as RF pulses, gradient fields, or relaxation. The Spin Model interface defines services that must be implemented in all inherited models, yet implementation details are hidden so that the behaviour of the system is dependent on the particular model used in the simulation.

A simple model of magnetization dynamics is described by the Bloch equations [1],

$$\frac{d\mathbf{M}}{dt} = \gamma\left(\mathbf{M} \times \mathbf{B}\right) - \frac{\left(M_x\hat{\mathbf{x}} + M_y\hat{\mathbf{y}}\right)}{T_2} - \frac{\left(M_z - M_0\right)\hat{\mathbf{z}}}{T_1} \tag{1}$$

Using the operator formalism introduced by Ernst and Anderson [4], in its simplest form, the effect of an RF pulse will then be a rotation of the magnetization vector, $\mathbf{M}$, with the transitions between events modelled by relaxation given by the Bloch equations. Another useful model is the isochromat spin distribution model introduced by Hahn [5], which can be used to model more complex phenomena such as the emergence of stimulated echoes.

2.2 Image Production

The results of the signal production simulation are used by the image production models to generate MRI volumes. The Scanner model is responsible for coordinating the various components in the simulation and interfacing to the Pulse Sequence model of the NMR signal production simulation.

The spatial distribution of tissues in the object to be imaged, and their intrinsic parameters, are described by the Phantom model. In order to generate realistic MRI brain phantoms for post-processing evaluation, the input to the simulator is a 3D brain model with labelled grey matter, white matter, and cerebrospinal fluid, which is generated from a volumetric MRI data set. Given this labelled volume of tissue type and voxel position, signal intensities computed using the signal production simulation are then mapped to produce a pseudo-MRI volume. Partial volume effects can then be evaluated depending on the sampling resolution specified by the pulse sequence.

Signal reception is modelled by the RF Coil model. This includes effects such as noise and spatial RF inhomogeneities. Different coil models allow images to be simulated under ideal noiseless conditions or with a noise level dependent on imaging parameters such as slice thickness and receiver bandwidth. A noise model that takes into account image parameters has been derived based on [6], with white Gaussian noise added to the in-phase and quadrature components of the received signal. This results in images with Rician or Rayleigh distributed statistics.

3 Results

The MRI simulator has been implemented in C++ and versions are currently running on a Silicon Graphics workstation and a Linux based PC. In Fig. 2, the effects of a dual spin-echo pulse sequence on a population of isochromat spin elements, with a Lorentzian off-resonance distribution, has been simulated. The NMR signal production simulation thus allows pulse sequences to be studied, or animations of magnetization evolution to be visualized.

Figure 3 shows a simulated transverse slice through the lateral ventricles. The simulation was produced by acquiring data on a Philips Gyroscan 1.5 T scanner at the Montréal Neurological Institute, estimating tissue parameters, and generating a labelled brain volume for input. The right image in Fig. 3 shows a simulated T2-weighted image generated from the late echo of a dual

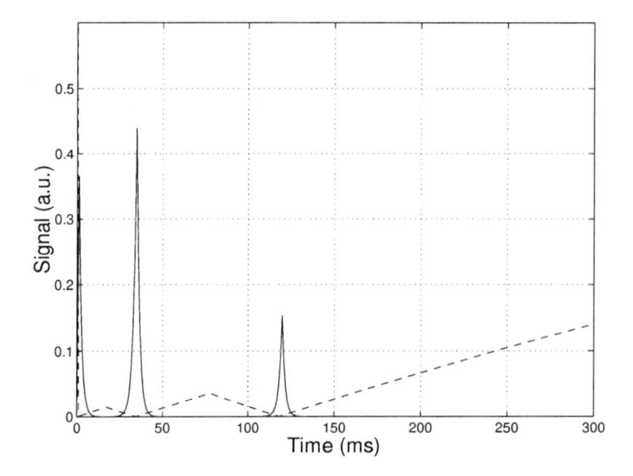

Fig. 2. A trace of a simulated dual spin-echo sequence. Solid line is transverse magnetization. Dashed line is longitudinal magnetization. Note how the isochromat distribution model predicts the exponential lineshape of the echoes, as well as the echo intensities.

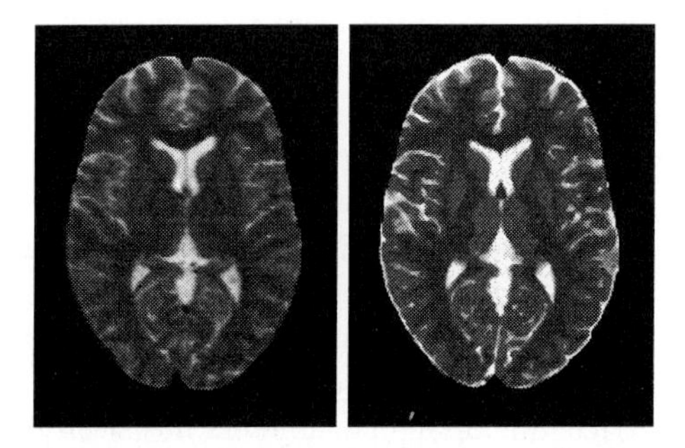

Fig. 3. An MRI image acquired from the late echo of a dual spin-echo pulse sequence (left) and a simulated image generated with the same imaging parameters (right).

echo spin-echo sequence (TR=3300 ms, TE=35 ms, 120 ms). For comparison, a real image acquired using the same acquisition parameters is shown on the left.

One shortcoming of this method, is the presence of sharp tissue boundaries in simulated images, due to the discrete nature of the labelled data set. In principle, this is remedied by specifying labelled volumes at a higher resolution than that required of simulations; however this quickly becomes computationally intensive. Alternative approaches which are being investigated include the use of fuzzy labelled data sets and interpolation schemes to improve the realism of the partial volume computation.

One application of the simulator is the evaluation of varying pulse sequence parameters on image contrast. In Fig. 4 a range of contrast is obtained, from a T1-weighted image on the left to a T2-weighted image on the right, by varying the repetition time in a spin-echo sequence. This ability to predict image contrast can be used to evaluate intensity based post-processing routines, and should be particularly useful given the increased use of automatic multispectral segmentation techniques in MRI.

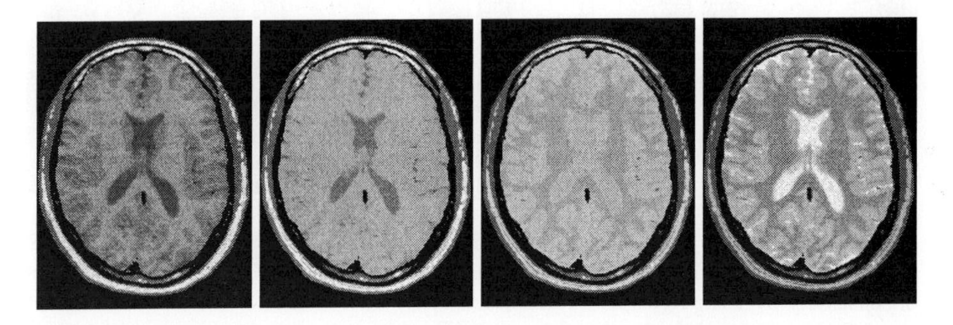

Fig. 4. Example of varying TR in a simulated spin-echo sequence. From left to right: TR = 500 ms, 1000 ms, 2000 ms, and 4000 ms, all with TE = 50 ms.

The simulator also allows other image parameters to be varied. The effect of varying slice thickness is shown in Fig. 5. Note the improvement in the signal-to-noise ratio of the images as thicker slices are acquired, at the expense of partial volume effects. One of the benefits of the simulation over real acquisitions is that it is also possible to independently control these effects. Partial volume could therefore be studied independently of any signal-to-noise variations.

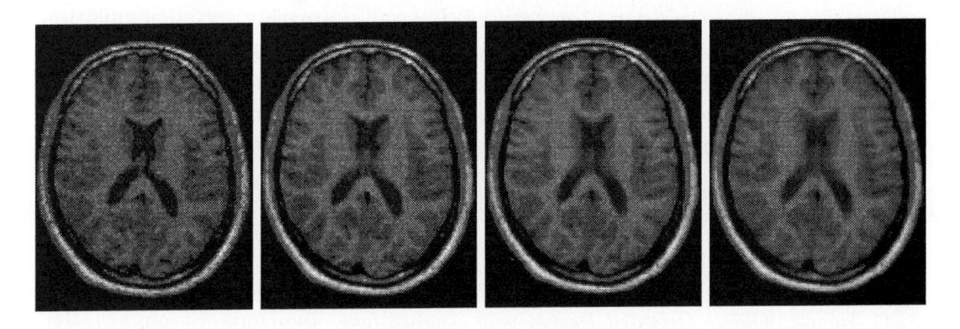

Fig. 5. Example of varying slice thickness in a simulated spoiled FLASH sequence (TR = 18 ms, TE = 10 ms, Flip angle = 30°). From left to right: 1 mm, 3 mm, 5 mm, and 9 mm slice thicknesses.

As a final example, Fig. 6 shows the effect of RF inhomogeneity on a minimum distance classifier [3]. These simulations can be used to evaluate a classifier's susceptibility to RF inhomogeneity, as demonstrated by the white-matter classification in Fig. 6. The ability to generate ideal uncorrupted data also provides an objective method of evaluating the performance of RF correction algorithms, designed to overcome the effect of RF inhomogeneity on classification.

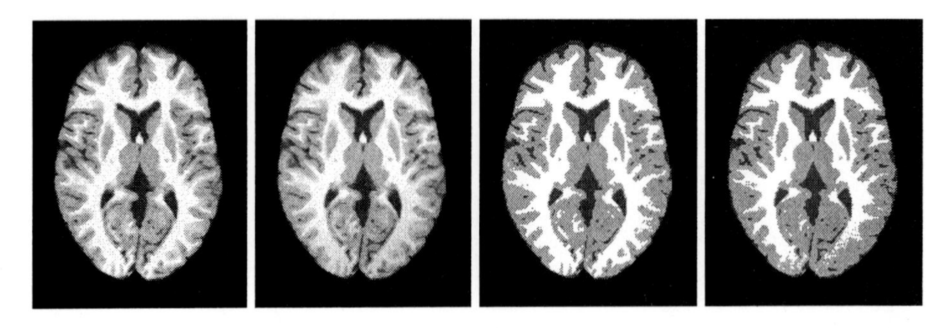

Fig. 6. Simulation of RF inhomogeneity and classification results. A realistic RF bias field has been used with an approximate sensitivity range of 20% and a maximum loss of sensitivity over the right posterior region. From left to right: Simulated ideal image (no RF inhomogeneity), simulated image (RF inhomogeneity), classified ideal image, classified RF corrupted image.

4 Conclusion

In order to support the evaluation of post-processing algorithms, an extensible MRI simulator has been developed. The design of the simulator uses object-oriented methods to ensure that models of NMR signal production and the MRI image production process can be extended depending on the requirements of specific simulation studies.

Using classified labelled volumes, obtained from high-resolution MRI data, realistic brain phantoms can be generated. Results comparing actual acquisitions to simulated images, along with examples demonstrating potential applications, have been shown. Similar images can serve as the basis for the generation of a database of images for evaluating classification algorithms or other image-processing procedures, as well as the validation of quantitative analysis studies.

References

1. F. Bloch. Nuclear Induction. *Physical Review*, 70:460–474, Oct. 1946.
2. G. Booch. *Object Oriented Design with Applications.* Benjamin/Cummings, Redwood City, CA, 1991.
3. R. O. Duda and P. E. Hart. *Pattern Classification and Scene Analysis.* Wiley, New York, 1973.
4. R. Ernst and W. Anderson. Application of Fourier Transform Spectroscopy to Magnetic Resonance. *Review of Scientific Instruments*, 37(1):93–102, Jan. 1966.
5. E. Hahn. Spin Echoes. *Physical Review*, 80(4):580–594, Nov. 1950.
6. D. L. Parker and G. T. Gullberg. Signal-to-noise efficiency in magnetic resonance imaging. *Medical Physics*, 17(2):250–257, Mar/Apr 1990.
7. J. Rumbaugh, M. Blaha, W. Premerlani, F. Eddy, and W. Lorenson. *Object-Oriented Modeling and Design.* Prentice-Hall, Englewood Cliffs, NJ, 1991.

Compensation of Spatial Inhomogeneity in MRI Based on a Parametric Bias Estimate

Christian Brechbühler, Guido Gerig, and Gábor Székely

Communication Technology Laboratory, ETH-Zentrum, CH-8092 Zurich, Switzerland
email: brech(gerig,szekely)@vision.ee.ethz.ch

Abstract. A novel bias correction technique is proposed based on the estimation of the parameters of a polynomial bias field directly from image data. The procedure overcomes difficulties known from homomorphic filtering or from techniques assuming an initial presegmented image. The only parameters are a set of expected class means and the standard deviation. Applications to various MR images illustrate the performance.

1 Introduction

A major obstacle for segmentation by multiple thresholding or by multivariate statistical classification [1, 2, 3] is insufficient data quality. Beside corruption by noise, MR image data are often radiometrically inhomogeneous due to RF field deficiencies. This variability of tissue intensity values with respect to image location severely affects visual evaluation and also segmentation based on absolute pixel intensities. The effects of radiometrical inhomogeneities to the subsequent segmentation have been carefully discussed by Kohn et al. [4].

Beside various methods [1, 2], homomorphic filtering is often used to separate the low-frequency bias field from the higher frequencies of the image structures [5, 6]. Wells [7] proposed an expectation-maximization (EM) algorithm to achieve an interleaved bias correction/statistical segmentation. The algorithm iterates between two steps, the E-step for calculating the posterior tissue probabilities and the M-step for estimating the bias field. Wells presented excellent results on double-echo spin-echo data and on surface-coil MR images. Own experiences showed that the initialization is a critical step. The procedure either has to start with a good estimate of the bias field or with a relatively good initial segmentation. In the case of a step edge image function, for example, a large part of the signal can be incorrectly classified due to a bias distortion. With no prior assumptions about the bias field the EM cannot recover the correct classification.

Based on the experience with several inhomogeneity correction schemes we developed a new approach. It was our intention to separate the estimation of the parameters of the bias field from the statistics of the homogeneous image regions. Our model is based on the observation that the pixel values of a corrupted image still belong to one single class, but are moved away from their original values which were clustered around a common mean. Therefore, we try to correct for the inhomogeneity by "pushing" each pixel to a value which is near to one of the

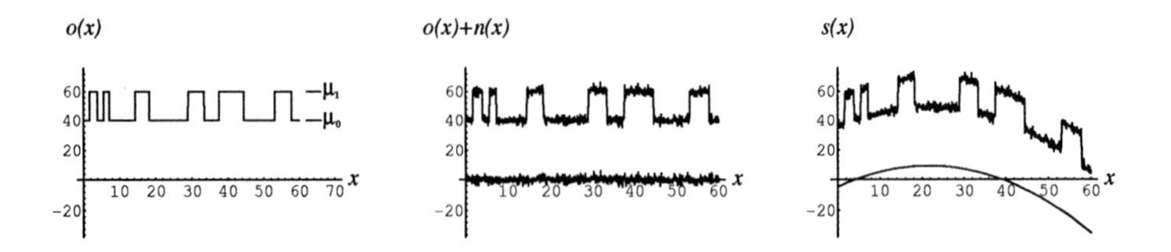

Fig. 1. Signal formation in 1D. The original signal $o(x)$ (*left*) is superimposed by noise $n(x)$ (*middle*) and a smooth bias field $b(x)$(*right*, lower curve), yielding the signal $s(x)$ (*right*, upper curve).

predefined class means. We estimate the parameters of a parametric bias field model without requiring a preliminary segmentation. This seems suitable if we consider an intensity-based segmentation of the image data into a small set of classes.

2 Formation of the Signal

The measurement volume contains several kinds of tissues or other substances (like cerebral gray and white matter, cerebro-spinal fluid, bone, muscle, fat, air), which are labelled $k = 0, 1, \ldots$. The ultimate goal of our efforts is a determination of the correct class k for each pixel of the data set, i.e. the *segmentation* of the image data.

The mathematical model of the signal formation assumes that each tissue type k has a specific value μ_k of the property being measured (e.g., proton density). The idealized original signal $o(\mathbf{x})$ consists of piecewise constant regions, each having one of the values μ_k. Biological tissues have interior structures, which we model by an additive noise term n_{bio}. Characteristics of the measuring device lead to intermediate values in the border region between tissue classes. This effect can be modeled as a convolution with a small kernel $h(\mathbf{x})$. The measuring device adds statistical noise n_{MR} and and a systematic bias $b(\mathbf{x})$. The bias field $b(\mathbf{x})$ is partly caused by imperfections of the RF field and partly induced by the patient. The bias field is usually very smooth across the whole data set, i.e. it varies slowly and has only very low frequency components. To summarize, the measured signal $s(\mathbf{x})$ is modeled to be formed as follows.

$$s(\mathbf{x}) = (o(\mathbf{x}) + n_{bio}(\mathbf{x})) \star h(\mathbf{x}) + b(\mathbf{x}) + n_{MR}(\mathbf{x}) \ , \tag{1}$$

where $o(\mathbf{x}) = \mu_{k(\mathbf{x})}$ and $k(\mathbf{x})$ is the tissue class found at location $\mathbf{x}$. An intensity-based segmentation of the corrupted signal will be severely hampered.

When $h(\mathbf{x})$ is neglected, and n_{bio} and n_{MR} joined into one term, which leads to

$$\tilde{s}(\mathbf{x}) = o(\mathbf{x}) + b(\mathbf{x}) + n(\mathbf{x}) \ . \tag{2}$$

To introduce smoothness, we take $b(\mathbf{x})$ from a family of very smooth functions. We choose them to be linear combinations of smooth basis functions. The m parameters $\mathbf{p}$ specify one particular function from the family. The estimated bias field $\hat{b}$ takes $\mathbf{p}$ as a further argument. Lacking information about the specific MR equipment, we select products of Legendre polynomials in x and y as the basis functions f_{ij} for the estimated bias field. The image coordinates are scaled such that each x_j varies from -1 to 1. With $m = (m' + 1)(m' + 2)/2$, we have

$$\hat{b}(\mathbf{x}, \mathbf{p}) = \sum_{i=0}^{m'} \sum_{j=0}^{m'-i} p_{ij} P_i(x) P_j(y) \,,$$

3 Bias Estimation Method

If $\hat{b}(\mathbf{x}, \mathbf{p})$ is a good estimate for the real bias field $b(\mathbf{x})$, we can subtract it from the measured signal and, using (2), get

$$\tilde{s}(\mathbf{x}) - \hat{b}(\mathbf{x}, \mathbf{p}) = o(\mathbf{x}) + b(\mathbf{x}) - \hat{b}(\mathbf{x}, \mathbf{p}) + n(\mathbf{x}) \approx o(\mathbf{x}) + n(\mathbf{x}) \,. \tag{3}$$

In pixels not disturbed by nearby boundaries, we can expect the *difference to be close to one of the* μ_k, only disturbed by $n(\mathbf{x})$. An evaluation or energy function must honor this desirable situation.

The dissimilarity between two general sequences, say u_i and v_i, is often assessed by summing the squares of their differences.

$$e = \sum_i (u_i - v_i)^2 \; = \; \sum_i \text{square}(u_i - v_i) \tag{4}$$

The function "square", which has a minimum at 0, assesses the individual differences. *Least squares* techniques vary some parameters $\mathbf{p}$ that influence u or v to minimize e. various kinds of valley-shaped functions besides "square" can be considered, e.g., a negated Gaussian or

$$valley(d) = -\frac{\sigma^2}{d^2 + \sigma^2} \; = \; -\frac{1}{1 + (\frac{d}{\sigma})^2} \,. \tag{5}$$

Supplying $d - \mu$ as an argument will shift the position of the minimum from 0 to $d = \mu$, and σ controls the width of the valley. An energy function of the form $e(\mathbf{p}) = \sum_i valley\,(u_i - v_i\,(\mathbf{p}))$ will yield a small value when the "model" v is well fitted to the "samples" u. Outliers have little influence on the position of the minimum if *valley* is used instead of "square".

The noisy signal $o(\mathbf{x}) + n(\mathbf{x})$ likely takes values close to one of the μ_k. An overall energy function $e_0(d)$ must yield a small value when d is close to μ_0 *or* to μ_1 *or* to μ_2 etc. It will have local minima at the values μ_k.

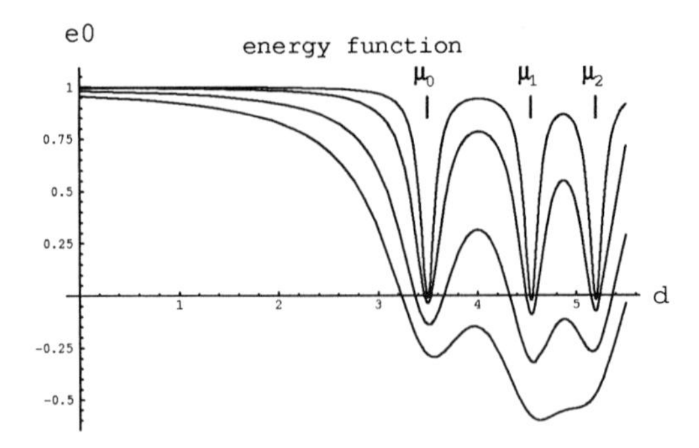

Fig. 2. Plot of energy functions representing the class mean values $\mu_k (\mu_0 = 3.49, \mu_1 = 4.54, \mu_2 = 5.20)$ for $\sigma = 0.05, 0.1, 0.2$ and 0.3, from top to bottom.

We combine several *valleys* into a function e_0 which has a small value when d is close to any of the μ_k. Summing the individual *valley* functions is just one possibility[1].

$$e_0(d) = \sum_k valley(d - \mu_k) \tag{6}$$

The function $e_0(d)$ allows the bias-corrected signal $o(\mathbf{x}) + n(\mathbf{x})$ to take one out of several values with equal goodness of the fit. $e_0(d)$ takes the role of an *OR* function for the set of possible μ_k.

Summing the local energy term e_0 over the whole data set yields the global goal function e.

$$e(\mathbf{p}) = \sum_{x_0} \cdots \sum_{x_{n-1}} e_0(\tilde{s}(\mathbf{x}) - \hat{b}(\mathbf{x}, \mathbf{p})) = \sum_{\mathbf{X}} \sum_k valley(\tilde{s}(\mathbf{x}) - \hat{b}(\mathbf{x}, \mathbf{p}) - \mu_k)$$

The parametes $\mathbf{p}^*$ that minimize $e(\mathbf{p})$ yield the optimal bias field estimate, $\hat{b}(\mathbf{x}, \mathbf{p}^*)$. The estimation is thus driven by a force that pushes each pixel value close to some class mean μ_k.

The optimization is carried out using the discrete taboo search (TS) technique [8]. This is an effective method for finding the optimum in a high-dimensional search space without getting trapped in local minima. It takes a few minutes to find the optimal bias field estimate on a SPARCstation 4. There are alternatives to taboo search, like Genetic Algorithms, Simulated Annealing or Newton and Quasi-Newton schemes. We did not yet explore other optimization techniques for the bias estimation method.

[1] Other possibilities include $\min_k valley(d - \mu_k)$ and $\prod_k valley(d - \mu_k)$. In the case of the product, the minimal values $valley(0)$ should be 0.

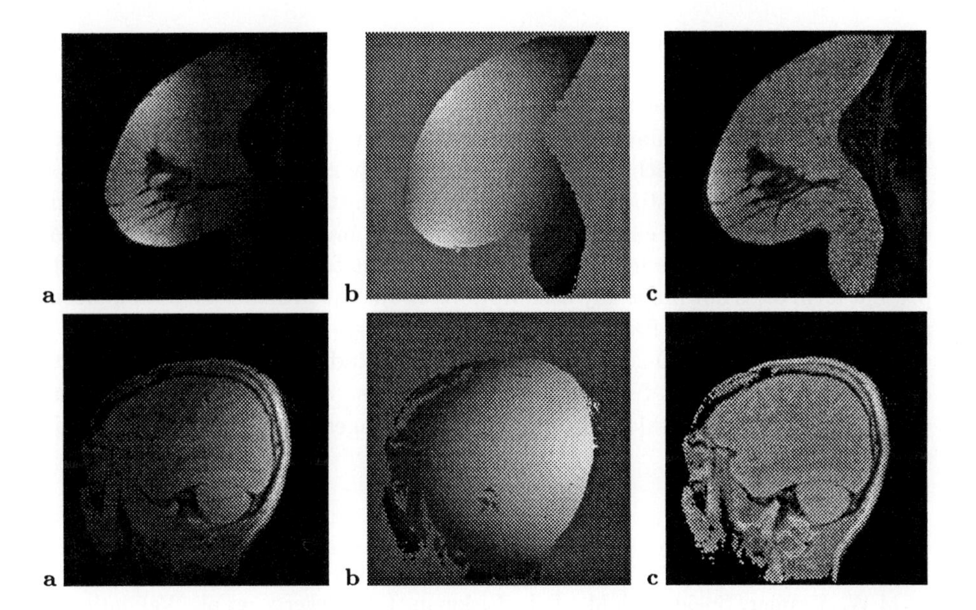

Fig. 3. Inhomogeneity correction of MR surface coil images. *Top row:* breast image(courtesy of Mike Brady, Oxford); *bottom row:* head (courtesy of Ron Kikinis and Sandy Wells, Brigham and Women's Hospital). Figure **a** shows the original image, **b** the polynomial fit of the bias field within the ROI, and **c** the bias-corrected image.

4 Results

Bias correction is highly needed for surface coil MR imaging because the large drift does not allow a sufficient visual inspection or a segmentation by windowing. Figure 3 (top) illustrates one slice of an MR mammography volume data set. A region of interest was marked as a mask for the bias correction. We chose a polynomial degree of five and two categories to represent the dark linear structures and the surrounding large bright region. The bias field estimation was based on a the logarithmic values of the original data.

The original surface coil MR image (Fig. 3, bottom) shows a large drift from left to right, impeding visual inspection and segmentation. Bias field estimation was calculated on a logarithmic version of the image data. A polynomial degree of two turned out to be sufficient to correct for the bias corruption.

5 Conclusions

The bias correction as presented in this paper estimates the parameters of an inhomogeneity field model by nonlinear optimization. The parametric description of the bias field also allows the inclusion of prior knowledge about the type of inhomogeneity by choosing an appropriate model. The method is based on clear assumptions about the imaging process and about the observed scene. Good local energy contributions of each pixel are obtained by bias-corrected pixel values which are close to one of the predefined mean values. The energy formulation as proposed herein simulates a logical OR function, as we do not know in advance the category of each pixel. The global energy minimum is achieved for a correction with a bias estimate which completely flattens the image and assigns each pixel to one of the categories without relying on a presegmented image or on the existence of large homogeneous regions. The fact that the method in its present form does not make any assumptions about the presence of homogeneous patches leaves room for further extensions.

The extension to 3-D is straightforward by just adjusting the dimensionality of the polynomial basis functions. Further, we also see possibilities to extend the method to treat double-echo spin-echo MR images or spectral image data in general.

References

1. M.W. Vannier, Speidel Ch.M., D.L. Rickman, L.D. Schertz, et al. Validation of Magnetic Resonance Imaging (MRI) Multispectral Tissue Classification. In *Proc. of 9th Int. Conf. on Pattern Recognition, ICPR'88*, pages 1182–1186, November 1988.
2. K.O. Lim and A.J. Pfefferbaum. Segmentation of MR brain images into cerebrospinal fluid spaces, white and gray matter. *Journal of Computer Assisted Tomography*, 13:588–593, 1989.
3. H.E. Cline, W.E. Lorensen, St.P. Souza, F.A. Jolesz, R. Kikinis, G. Gerig, and Th.E. Kennedy. 3D surface rendered MR images of the brain and its vasculature. *Journal of Computer Assisted Tomography*, 15(2):344–351, March 1991.
4. M.I. Kohn, N.K. Tanna, G.T. Herman, S.M. Resnick, P.D. Mozley, R.E. Gur, A. Alavi, R.A. Zimmerman, and R.C. Gur. Analysis of brain and cerebrospinal fluid volumes with MR imaging. Part I. Methods, reliability, and validation. *Radiology*, 178:115–122, January 1991.
5. B.M. Dawant, A.P. Zjidenbos, and R.A. Margolin. Correction of intensity variations in MR images for compuer-aided tissue classification. *IEEE Transactions on Medical Imaging*, 12(4):770–781, 1993.
6. Charles R. Meyer, Peyton H. B land, and James Pipe. Retrospective correction of Intensity Inhomogeneities in MRI. *IEEE Transactions on Medical Imaging*, 14(1):36–41, March 1995.
7. William Wells, Ron Kikinis, and Ferenc A. Jolesz. Statistical intensity correction and segmentation of magnetic resonance image data. In Richard A. Robb, editor, *Proceedings of the Third Conference on Visualization in Biomedical Computing VBC'94*, volume 2359, pages 13–24. SPIE, October 1994.
8. Colin R. Reeves. *Modern Heuristic Techniques for Combinatorial Problems*. Blackwell Scientific Publications, 1993.

Analysis of Coupled Multi-Image Information in Microscopy

St. Schünemann[1], Ch. Rethfeldt[2], F. Müller[2], K. Agha-Amiri[2], B. Michaelis[1], W. Schubert[2]

Otto-von-Guericke University Magdeburg
[1] Institute for Measurement and Electronics
[2] Institute for Medical Neurobiology
P.O. Box 4120, D-39016 Magdeburg, Germany

Abstract. Series of microscope recordings of cells, labeled for many different molecules, contain important biological information. The correct interpretation of those coupled multi-images, however, is not directly possible. This paper introduces a procedure for the analysis of higher-level combinatorical receptor patterns in the cellular immune system, which were obtained using the fluorescence multi-epitope-imaging microscopy. The cell recognition and the classification algorithm with an artificial neural network is described.

1 Introduction

Series of microscope recordings, i.e. N images of the same region of interest with different object markings, are of special interest in biomedical research. Because those coupled multi-microscope images may contain not directly visible combinatorical information, the correct biological interpretation is difficult.

The immuno-fluorescence microscopy method is based upon antibody-antigen reactions to make cellular locations of molecules visible. For such purposes antigens are labeled with special fluorochromes [1]. Under exposition of ultraviolet light such fluorochromes allow the topographical localization of receptor epitopes at cell surfaces and the estimation of epitope concentrations. But generally the method is limited by the number of simultaneous incubations of antigens labeled by different fluorochromes and the problems of spectral optical separation.
The advanced fluorescence multi-epitope-imaging microscopy [2, 3] is a repetitive method which provides a tool for the localization of random numbers of different receptor epitopes at single cell surfaces. It leads to special preparation techniques, the automation of immuno-chemical routines in strong time regimes.

The goal of the fluorescence multi-epitope-imaging microscopy method is the quantitative analysis of higher-level combinatorical receptor patterns present at the cell surface of immune cells, i.e. the detection of the existing combination patterns under normal and pathological conditions. Assuming a binary molecular cell-pattern, 2^N combinations are theoretically possible. Truly distinct concentrations of antigens in the cells produce different gray levels during the recording of images. Therefore the analysis algorithm has to find clusters of combined concentrations and make a considerable biological interpretation possible.

In the following the cell recognition and classification algorithm is described.

2 Image Preprocessing and Cell Recognition

The biological data samples are characterized by a phase contrast image and N fluorescence image sequences, depending on the number of considered receptor epitopes. Fig. 1-2 (Appendix 1) show selected short image sequences of cellular distributions of human surface antigens, such as CD7, CD38, CD45 and CD71, for normal mononuclear blood leukocytes in comparison to an experimentee with sporadic Amyotrophic Lateral Sclerosis (ALS). The main task is the detection of multiple marked cells, which supports the exploration of higher-level differentiation in the immune system.
Fig. 3 shows the strategy for the image preprocessing and cell recognition.

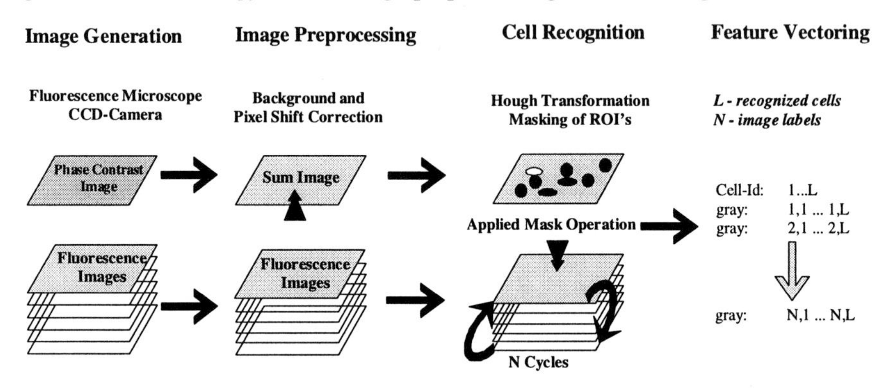

Fig. 3: Image preprocessing and cell recognition of N sequential fluorescence images

The first task for the *Image Preprocessing* is a correction of the N image layers to achieve the best possible geometrical overlay, in the case of the diffuse shining fluorescent cell pattern. The occurred pixel shifts result from systematic errors, caused by changes in the optical refraction for different spectral channels, and from small mechanical and thermal movements during long time operations of the microscope. All images are background subtracted and edge enhanced. Furthermore, a contrast stretching unifies the gray level dynamics of the images.

The *Cell Recognition* step uses a modified Hough-Transformation-Technique [4] for recognition of circular objects applied to the sum image of N fluorescence channels. Because of the good circle approximation of the fluorescent blood leukocytes, the recognition rate is in the order of 90 percent. The output of the Hough Transformation leads to arrays of coordinates and radii for L recognized cells, which define the regions of interests (ROIs). The final shape tuning of each ROI takes into account the deviation from the ideal circle shape. The individual cell geometries are estimated by applying simple growing algorithms. This mask matrix contains the geometrical information of L labeled ROIs.

The mask matrix allows a fast access to the ROIs of N fluorescence image sequences and the calculation of gray level statistics for the *Feature Vectoring*. Using this approach an individual cell is characterized by an array of N determined gray values. The analysis is confined to simple averaged gray level statistics, which leads to a (L, N)-dimensional data set for the neural network processing.

3 Classification of the Feature Vectors

The determined feature vectors $\boldsymbol{v}$ represent a N-dimensional feature space $\boldsymbol{v} \in \mathbf{R}^N$, which contains the desired molecular combination pattern. The theory of Self-Organizing Feature Maps (SOFM) provides an efficient method for reduction and cluster analysis of the high-dimensional feature spaces onto two-dimensional arrays of representative weight vectors [5].

In the *training phase* the weight vectors $\boldsymbol{w}(x,y)$ of the neurons are trained for a suitable representation of the feature space. In each training step i, a feature vector $\boldsymbol{v} \in \mathbf{R}^N$ is chosen from the learning set with a certain probability. Each neuron calculates its Euclidean distance to the presented vector:

$$\boldsymbol{d}_v^w(x,y) = \sqrt{\sum_{k=1}^{N}\left[v_k - w_k(x,y)\right]^2} \tag{1}$$

The weight vectors of the winner neuron (the neuron featuring the minimal distance), and its coupled neighbors are updated according to the following rule:

$$\boldsymbol{w}_{i+1}(x,y) = \boldsymbol{w}_i(x,y) + \alpha_i\ \varphi_i(x,y)\ [\boldsymbol{v} - \boldsymbol{w}_i(x,y)] \tag{2}$$

in the direction of the feature vector $\boldsymbol{v}$. The learning rate α determines the magnitude of the update and the neighborhood function φ describes the effective training region on the SOFM. The left side of Fig. 4 demonstrates the training phase.

The problem is that the smoothing property of the SOFM leads to a relative insensitivity to clusters of low feature density [6]. But, in the case of the combination patterns of extracted human antigens the clusters of low feature density represent significant qualities of the distribution [7]. For this reason a modification of the neighborhood function were developed, which more strongly differentiates the degree of coupling of the neighboring neurons with the winner neuron [8, 9].

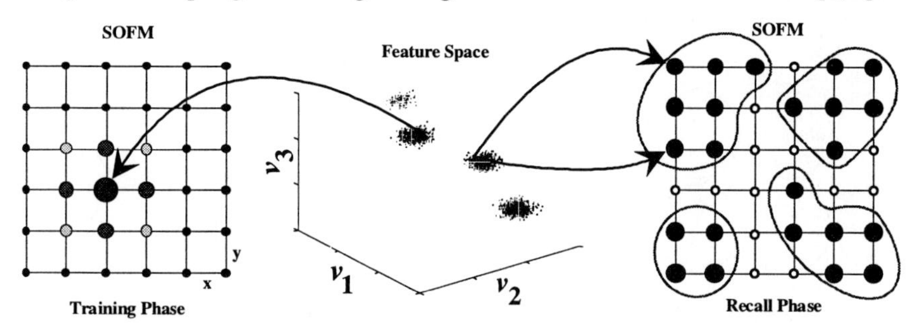

Fig. 4: Reduction and cluster analysis with the SOFM

In the *recall phase* only the neuron of the trained SOFM with the shortest Euclidean distance between the feature vector and the weight vector turns into the state of excitation. In this phase no manipulation of the neuron weights occurs. The activated neuron represents the membership in a cluster. Feature clusters can be represented by several neurons (right side of Fig. 4). A few of the neurons are never or rarely activated during the recall phase. Such neurons separate the clusters on the SOFM.

4 Application of the Algorithm

The performance of the described algorithm using the image correction, cell recognition and the cluster analysis with the modified algorithm of the SOFM for a similarly reduced four-dimensional feature space for several combination patterns of the human antigens: CD7, CD38, CD45 and CD71 in the blood (see Fig. 1-2) are presented in Fig. 5 [9]. Normally in this application the feature space is 15 or higher dimensional.

The different gray level distributions of the extracted cells in the N fluorescence images are used for the classification. The determined feature vectors for a healthy and an experimentee suffering from sporadic Amyotrophic Lateral Sclerosis (ALS) are mixed to train a SOFM with 5*5 neurons. The recall phases of the normal and pathological conditions are shown separately.

The distinct frequencies shown in percentages and the different existing combination patterns for healthy and experimentees suffering from ALS are recognizable from the Fig 5. Each existing combination of the fluorochrome labeled human antigens, i.e. each cluster of the used distribution, is stored in a separate neuron of the SOFM while maintaining the topographics in $\mathbf{R}^N$. The neuron weights of the SOFM are reference vectors of the desired clusters and make a considerable biological interpretation possible.

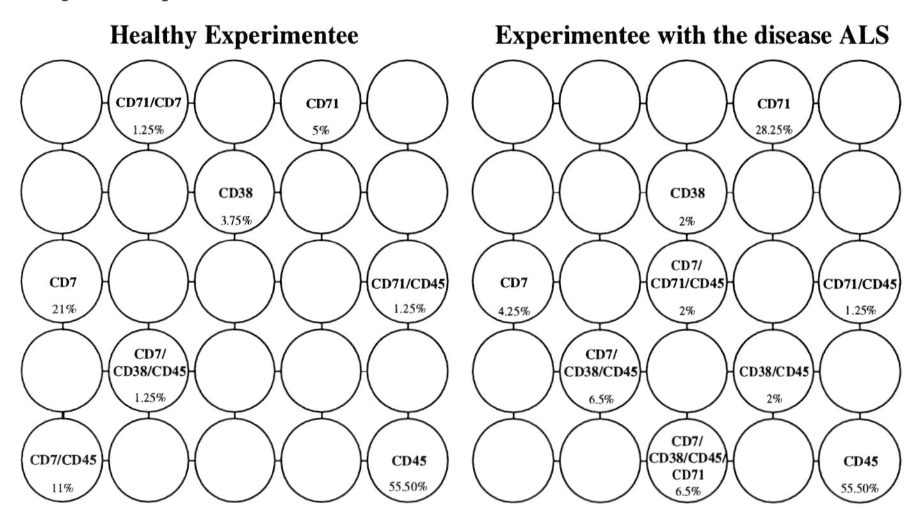

Fig. 5: SOFM for several combination patterns of human antigens

In case of an adequate number of training sets with representive feature vectors the cluster analysis, i.e. the determination of the number and position of the clusters, need not be repeated. Provided that the measurement conditions for the fluorescence microscopy remain unchanged, the training phase is omitted. Small deviations could be corrected by means of system calibration. The SOFM with the trained neuron weights can be used for the classification and detection of the desired combination patterns.

5 Conclusion

The advantage of the introduced algorithm is that the detection and classification of the molecular combination patterns of human antigens from multiple marked images is possible without a-priori knowledge, i.e. without previous definition of classes and borderlines. The algorithm is applied to the detection of human antigens in the blood, muscular tissue and organs with up to twenty different cell markings and several features, such as gray levels and contours of the cells. In further works different cell regions have to be included into the analysis.

The described algorithm is also suitable for other high-dimensional classification problems in biology and medicine with multiple markings of objects that are approximate circles. The used classification algorithm is universally applicable.

In the foreseeable future the algorithm of cell extraction will become independent from cell contours. The expansion of the classification algorithm to cluster-dependent multi-layer structures of SOFM in connection with the definition of several hyper-ellipsoids seems also reasonable for a further improvement.

This work was supported by the DFG/BMBF grant (Innovationskolleg 15/A1).

References

1. Methods of Immunological Analysis, Cells and Tissues, Volume 3, Ed. by R.F. Masseyff; W.H. Albert; N.A. Staines, VCH-NewYork, 1993
2. W. Schubert: Antigenetic determinants of T lymphocyte αβ receptors and other leucocyte surface proteins as differential markers of skeletal muscle regeneration: detection of spatially and timely restricted patterns by MAM microscopy, Eur. J. Cell Biol. 62, pp. 395-410, 1993
3. W. Schubert, H. Schwan: Detection by 4-parameter microscopic imaging and increase of rare mononuclear blood leucocyte types expressing the γ receptor (CD16) for immunoglobulin G in human sporadic amyotrophic lateral sclerosis (ALS), Neurosci. Lett. 198, pp. 29-32, 1995
4. J. Sklansky: On the Hough Technique for Curve Detection, IEEE Trans. on Comp. 27 (10), pp. 923-926, 1978
5. T. Kohonen: Self-Organizing Maps, Springer Series in Information Sciences, Springer, New York 1995
6. K. Obermayer, et.al.: Statistical-Mechanical Analysis of Self-Organization and Pattern Formation During the Development of Visual Maps, Physical Review A, Vol. 45(10), pp. 7568-7589, 1992
7. Ch. Rethfeldt, et.al.: Multi-Dimensional Cluster Analysis of Higher-Level Differentiation at Cell-Surfaces, Proc. 1. Kongress der Neurowissenschaftlichen Gesellschaft, p.170, Spectrum Akademischer Verlag, Berlin 1996
8. St. Schünemann, B. Michaelis: A Self-Organizing Map for Analysis of High-Dimensional Feature Spaces with Clusters of Highly Differing Feature Density, Proc. 4th European Symposium on Artificial Neural Networks, pp. 79-84, Bruges 1996
9. St. Schünemann, et.al.: Analysis of Multi-Fluorescence Signals Using a Modified Self-Organizing Feature Map, accepted for: The Int. Conf. on Artificial Neural Networks, July 16 - 19, 1996, Bochum, Germany

Appendix 1

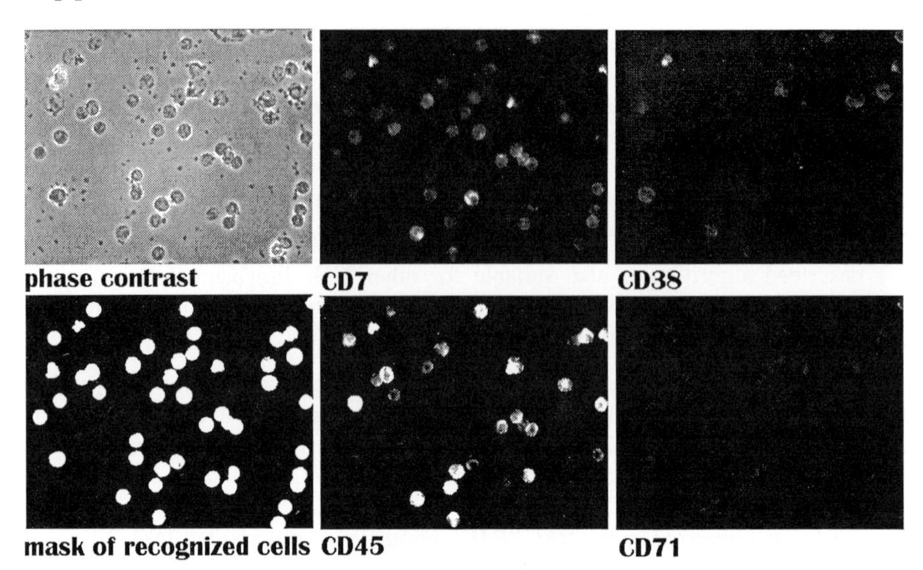

Figure 1: Phase contrast image, the corresponding recognition mask and fluorescence images of fluorochrome labeled CD-antibody-cell reactions for normal human blood leukocytes

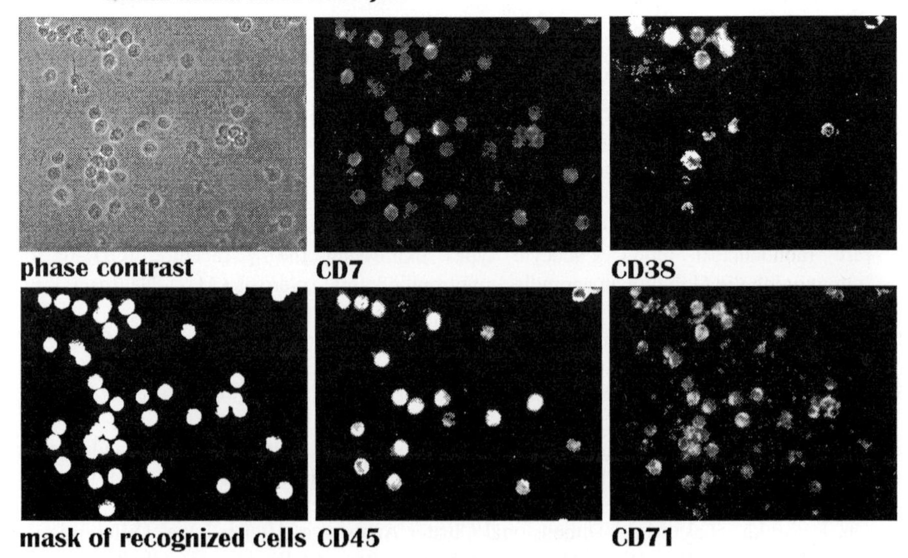

Figure 2: Phase contrast image, the corresponding recognition mask and fluorescence images of fluorochrome labeled CD-antibody-cell reactions for human sporadic amyotrophic lateral sclerosis (ALS)

Bias Field Correction of Breast MR Images

Sébastien Gilles[1], Michael Brady[1], Jérôme Declerck[2], Jean-Philippe Thirion[2] and Nicholas Ayache[2]

[1] Robotics Research Group, Dept of Engineering Science, Oxford University
[2] Epidaure group, INRIA, Sophia-Antipolis, France

Abstract. We present a method to automatically estimate and remove the bias field of MR images where there is a single dominant tissue class. Assuming that a multi-class image is corrupted by a multiplicative, low-frequency bias field, the method evaluates the bias field on a single tissue class, and extends it to the whole image. The algorithm works iteratively, interleaving tissue class domain and bias field estimation using B-spline.

1 Introduction

Many applications of magnetic resonance (MR) imaging require that the data is segmented into subsets corresponding to different tissue classes. Ideally, images are piecewise constant, each corresponding to a tissue class. As Figure 1 shows in practice, the ideal image is corrupted, here by a large smoothly varying bias field. Following [4], $\hat{I}(\mathbf{X}) = I(\mathbf{X})B(\mathbf{X}) + N(\mathbf{X})$, where $\hat{I}(\mathbf{X})$ is the observed image intensity at voxel $\mathbf{X}$, $I(\mathbf{X})$ is the restored image intensity, assumed to be piecewise (roughly) constant, $B(\mathbf{X})$ is the smoothly varying bias field and $N(\mathbf{X})$, is the voxel noise, supposed to be zero-mean Gaussian. The goal is to estimate I given $\hat{I}$.

One way to do this is to precede each MR image acquisition by acquiring an image of a phantom. Unfortunately, underlying to this idea is the assumption that the bias field is independent of the placement of the coil, which it is not. An alternative approach might be to develop a physics-based model of the coil, comprising, say, its geometry, material composition, shape, and the voltage field, in order to estimate $B(\mathbf{X})$ "from first principles". This appears to be difficult if not analytically intractable. For these reasons, most authors have tackled the restoration of $I(\mathbf{X})$ from the standpoint of image processing, directly from the given image $\hat{I}(\mathbf{X})$. For example, [1] propose correcting for intra-slice intensity variations after fitting a 2D surface to reference points obtained by a neural network classifier. [3] selects reference points automatically and deals with 3D images. An alternative approach has been introduced by [4], based on the EM algorithm, that iteratively interleaves estimation of the bias field and probabilities of voxels to belong to tissue classes, until convergence. The key-point of the method is the observation that *bias field estimation and segmentation are coupled*, as is the technique that we develop in this paper. See [2] for improvements and analysis of the method.

However, Wells' technique often gives poor results when applied to images where there are only two classes, one of which is relatively small and/or is subject to partial volume effects. A typical example of this sort is breast MR. Figure 1 shows a slice of a typical original breast MR image supplied to us by the *Institut Gustave Roussy*, Kremlin-Bicêtre, France. Observe the large bias field caused by C-type receiver coil that surrounds the breast.

Temporarily ignoring $N(\mathbf{X})$, the Equation for $\hat{I}(\mathbf{X})$ can be converted to a summation by noticing that $I(\mathbf{X}) = const$ in the dominant class D. Choosing $const = 1$ [1], we get $\hat{I}(\mathbf{X}) = B(\mathbf{X}) \quad$ for $\mathbf{X}$ in D. The idea is to estimate $B(\mathbf{X})$ as the B-spline minimizing

$$\sum_{i \in D} (\hat{I}(\mathbf{X}) - B(\mathbf{X}))^2 + \text{smoothness constraints}$$

To do this, we interleave estimate of B given D, and D given B.

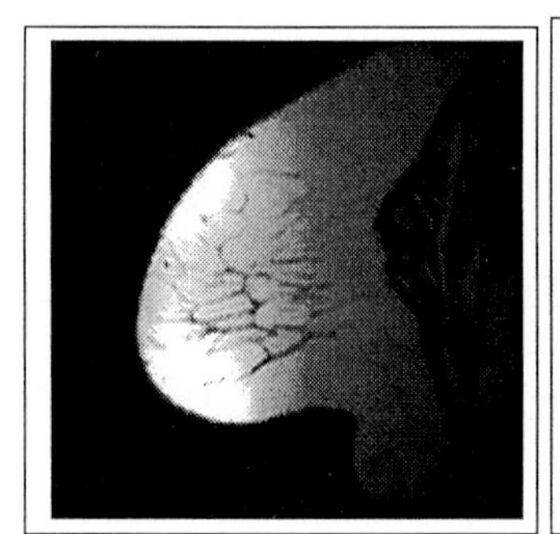

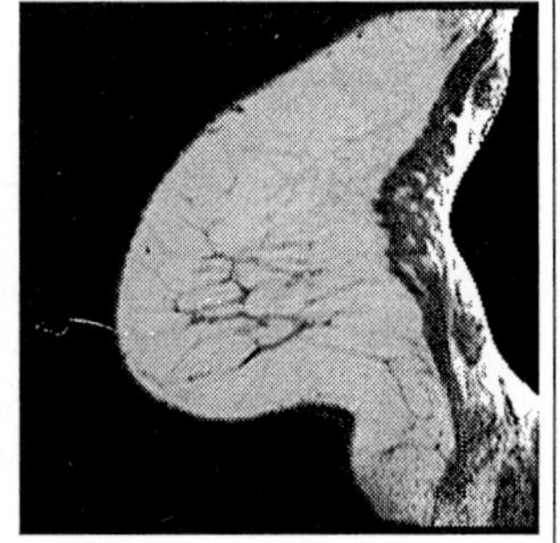

Fig. 1. *Left:* (a) Original breast MRI (slice 1) with coil effect. *Right:* (b) Same image after correction. The right part of the image (chest bones) has been badly restored because it is not part of the breast, and has therefore not been taken in account by the method.

2 The new method

The images to which our method applies have a single dominant tissue class D, with another, much smaller, set of typically "thin" structures. Following [3, 1], we develop an *explicit* model of the bias field. We follow [4] in iteratively updating both the bias field estimate and the segmentation that results. We choose to model the bias field by a B-spline estimated from an iteratively updated subset D_0 of the dominant class.

The algorithm is as follows (k indexes the number of loops):

[1] Actually, we may choose any value for *const* , because a scale-factor will be included in the spline.

1. Initialise the estimate of the dominant class $\hat{D}_0$. For breast MR images, this is easily and reliably done by thresholding the voxel data. Alternatively, it can be computed using a robust technique. Set $k = 0$.
2. Randomly select a set of points in $\hat{D}_k$, and compute a B-spline $B_k(\mathbf{X})$ approximating them.[2]
3. Divide the observed image by $B_k(\mathbf{X})$, so that the output image intensity equals 1.0 on $\hat{D}_k$.
4. Scan the neighbourhood of $\hat{D}_k$ for points that can be added to it, using the method described below. Merge the additional points with $\hat{D}_k$ to form $\hat{D}_{k+1}$. If there are no further points that can be added to $\hat{D}_k$, stop.
5. Set $k \leftarrow k + 1$, and return to step 2.

Figure 2 shows various stages of the algorithm, using an image cross section to make things clearer.

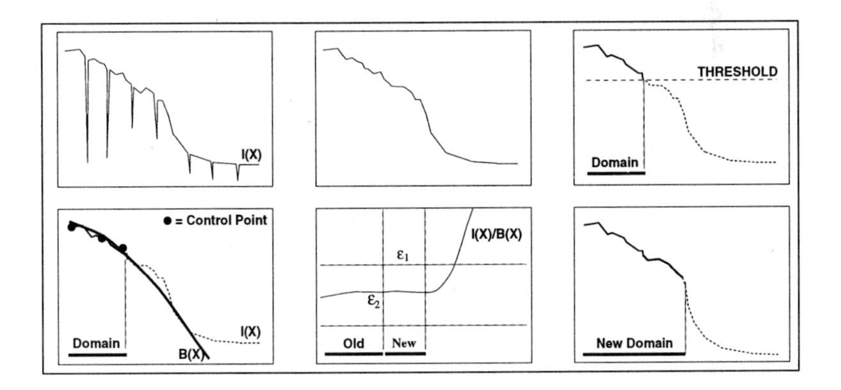

Fig. 2. 1) The initial cross section (the sharp dips correspond to glands and milk ducts). 2) The same cross section after a morphological closing. 3) Thresholding. 4) Reference points have been selected automatically, and used to approximate the function with a B-spline. 5) Thresholding on $I(\mathbf{X})/B(\mathbf{X})$ gives a new set of points that can be used to update the B-spline. 6) The new domain.

Updating $\hat{D}_k$: Recall that at the kth iteration of the algorithm we have a B-spline $B_k(\mathbf{X})$ approximating $\hat{I}(\mathbf{X})$ on $\hat{D}_k$. For efficiency, but with no sacrifice in reliability, we use the following criterion for accepting points into $\hat{D}_{k+1}$:

$$\text{if } \mathbf{X}_0 \in \hat{D}_k \text{ AND } ||\mathbf{X}_0 - \mathbf{X}|| \leq \epsilon_1 \text{ AND } |\frac{\hat{I}(\mathbf{X})}{B_k(\mathbf{X})} - 1| \leq \epsilon_2 \Longrightarrow \mathbf{X} \in \hat{D}_{k+1} \quad (1)$$

We chose ϵ_1 and ϵ_2 experimentally, but the results indicated very little dependance on these parameters. This criterion is useful because testing the mem-

[2] Typically, the B-spline $B_k(\mathbf{X})$ is a good approximation of $\hat{I}(\mathbf{X})$ only on $\hat{D}_k$

bership of points to class $\hat{D}_k$ after having divided the image by the bias field estimation $B_k(\mathbf{X})$ amounts to selecting an interval along the intensity-axis.

B-spline approximation: Since the bias field is assumed $\mathcal{C}^2$-differentiable, we use cubic splines. Since the spline is 2- or 3-dimensional, we use tensor-product B-splines. We make the control points the nodes of a regular mesh. A feature of the B-spline implementation that we have used is that it does not require that at least one data point lies between each pair of nodes. The bias field, modeled as a bicubic B-spline can be written (here in two dimensions for simplicity):

$$B(x,y) = \sum_{i=0}^{n_x-1} \sum_{j=0}^{n_y-1} \alpha_{ij} B_i^x(x) B_j^y(y)$$

where n_x (n_y) is the number of control points in the x (y) direction and B_i^x is the ith B-spline basis function.

The best $B_k(\mathbf{X})$ is computed using regularised least-squares. We balance two constraints: conformance to the data and smoothness, where *Conformance to the data* is:

$$J_{position}(B) = \sum_{k=1}^{N} \left(\sum_{i=0}^{n_x-1} \sum_{j=0}^{n_y-1} \alpha_{ij} B_i^x(x) B_j^y(y) - I(\mathbf{X}_k) \right)^2 \qquad (2)$$

and *smoothness* is:

$$J_{smooth}(B) = \rho_{smooth} \int\int_{\mathbb{R}^2} \left(\frac{\partial^2 B}{\partial x^2}^2 + \frac{\partial^2 B}{\partial y^2}^2 + 2 \times \frac{\partial^2 B}{\partial x \partial y}^2 \right) \qquad (3)$$

where ρ_{smooth} is a weighting coefficient. We minimise the sum of the two terms. The extension to 3D is immediate. The only issue that needs to be addressed concerns the anisotropy of breast MRI slices. As the spline is slowly-varying, a scale factor should be applied to the z-coordinate.

Prefiltering: To further increase the likelihood that the initial set of points selected all belong to the dominant class D, we prefilter the data, by morphologically closing the image using a structuring element that is a circle in two dimensions, or a sphere in three dimensions. We use morphological prefiltering because it preserves discontinuities. More precisely, linear filtering inevitably forms linear combinations of voxels belonging to different classes, worsening the partial volume effect and acting contrary to the anatomical segmentation sought. The diameter of the structuring element is chosen to be larger than the diameter of blood vessels. The pre-filtering acts like a low-pass filtering, but, because morphological closing is non-linear, (i) dark fine structures (like blood vessels) disappear, because they are composed of smaller zones than the structuring element, and darker than the dominant class, and (ii) the dominant class topology is preserved if the circle's diameter is small compared to the characteristic variations in the length of the bias field.

3 Results

We show results first for a synthetic image. We simulated a 2-class image corrupted by a bias field. The first class is dominant and covers the whole image, with thin dark structures. The bias field was modelled as a sum of 2D Gaussians: the standard deviations σ_1 and σ_2 are typically set to about the half of the image length (or width). In all experiments, the spline was modelled using a grid of 8×8 control points, and about 10 iterations ($\sim$ 2 minutes) sufficed to get good results (Mean error less than 0.3 between the B-spline and the bias field)

Figure 3a shows 100 random lines with varying intensities superimposed. Figure 3b shows the resulting corrected image.

The method was then applied to a 3D breast volume made of 10 adjacent slices, leading to both the estimated bias field and a breast tissue extraction. Figure 1a shows a T1-weighted, sagittal image (TR=300ms, TI=11ms). The coil effect is clear, causing over a 50 % magnitude variation within the breast tissue class. Figure 1b shows the corrected image.

Note that the restoration quality reduces when far from the coil poles: as the coil locally increases the SNR, its effect will be more important on pole regions. Given the slow-varying assumption, the spline is locally planar. That is, dividing the original image by it will not modify the SNR, causing higher standard deviation zones to be observed on the amplification zone's outskirts. Elevation maps from the original image and its associated bias field are shown in figure 4.

4 Discussion

The assumptions required for the method described in this paper are (i) the ideal intensity is, ignoring random noise, piecewise constant; (ii) there is a tissue class that covers a wide area in the image and that we call the *dominant class*;(iii) the real intensity is corrupted by a multiplicative slow-varying field; (iv) the bias field is $\mathcal{C}^2$-differentiable; and (v) some pixels of the dominant class are known (a heuristic generally allows automatic retrieval). No assumptions are made about (i) the ideal intensity distribution; (ii) the imaging process; or (iii) intra/Inter-patient invariance. We emphasize as well that our algorithm requires no pre-segmentation.

Acknowledgements

We thank Daniel Vanel and Anne Tardivon of the the *Institut Gustave Roussy*, Kremlin-Bicêtre, France for supplying us with the images. The work reported here was carried out while JMB was a Professeur Invité at INRIA Sophia Antipolis.

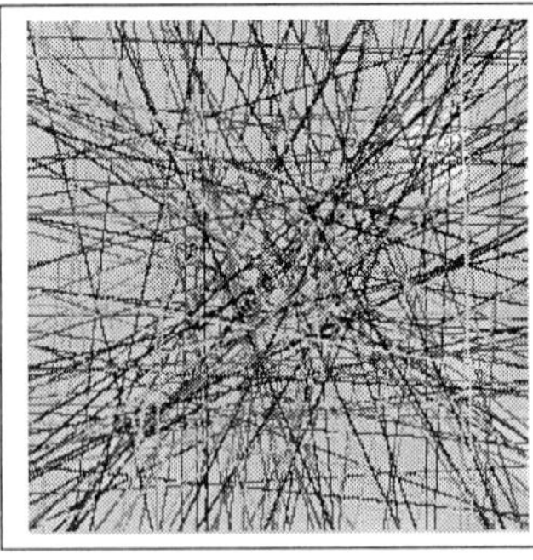

Fig. 3. *Left image*: (a) Synthetic image with 100 random lines. *Right image*: (b) Same image after correction

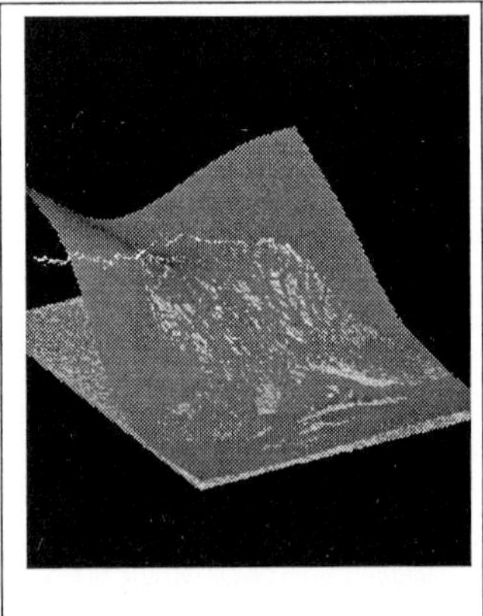

Fig. 4. *Left image*: Surface rendering of slice 1. *Right image*: Processed bias field superimposed

References

1. B. M. Dawant, P. Zijdenbos, and R. A. Margolin. Correction of Intensity Variations in MR Images for Computer-Aided Tissue Classification. *IEEE Transactions on Medical Imaging*, 12(4):770–781, Dec. 1993.
2. R. Guillemaud and M. Brady. Enhancement of MR Images. Sept. 1996.
3. C. R. Meyer, P. H. Bland, and J. Pipe. Retrospective Correction of MRI Amplitude Inhomogeneities. In N. Ayache, editor, *Computer Vision, Virtual Reality and Robotics in Medicine*, pages 513–522, Apr. 1995.
4. W. M. Wells III, W. Grimson, R. Kikinis, and F. A. Jolesz. Adaptive Segmentation of MRI Data. In *Proceedings of CVRMed'95*, volume 905 of *Lectures Notes in Computer Science*, pages 59–69. Springer Verlag, 1995.

Segmentation

Region-Growing Based Feature Extraction Algorithm for Tree-Like Objects

Yoshitaka Masutani, Ken Masamune and Takeyoshi Dohi

The University of Tokyo Graduate School of Engineering,
Division of Precision Machinery, 7-3-1 Hongo Bunkyo-ku Tokyo 113, JAPAN

Abstract. To overcome limitations of the conventional 'toward-axis' voxel-removal way of thinning operations, a new 'along-axis' style of algorithm was developed for topological information acquisition of tree-like objects like vascular shapes based on region-growing technique. The theory of mathematical morphology is extended for closed space inside binary shapes, and the 'closed space dilation' operation is introduced as generalized form of region growing. Using synthetic and clinical 3D images, its superior features, such as parametric controllability were shown.

1 Introduction

Recently biomedical image processing is moving toward description and modeling for superior analysis and simulation beyond just visualization.[1,2] One of the final goals of the description of internal organs is to express inter-organ spatial linkage and its functional interaction. Vasculature is a good example and an important organ for such spatial and functional linkage. The main feature of tree-like objects such as vessels are their bifurcation structure and their radii.[3,4] In intravascular surgery, for instance, such structural description of vascular shape can give far more useful information like a path length, bifurcation angle and etc., instead of vascular shape information as a set of voxels.

Such topological information acquisition of vessels is carried out conventionally by thinning[5] operations both in 2D and 3D, which remove surface voxels (or pixels) toward the medial axes, preserving its topology[6]. However, they seem not to be optimal because of including some of following limitations:

Low robustness: Thinning operation often yields undesired branches caused by noise shape along the border. Conventionally, such branches are

removed according to their length after structure acquisition.
Size uncontrollability: This attribute is more important than robustness as insensitivity for surface noise. As shown in Fig.1, bifurcation structure extraction is essentially subjective, because of human judgment. Therefore, we need to define the minimum size of branches numerically. The parameter gives a meaning of threshold between branch shape and noise.

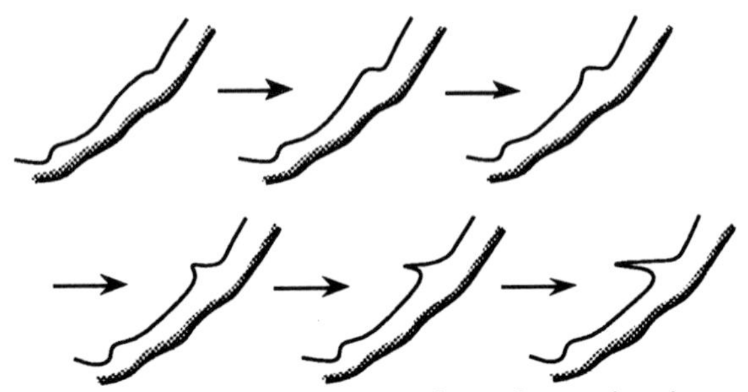

Fig.1. Subjectivity in topology determination
-How do you define the existence of a 'branch'?-

Lack of irregular shape manageability: If vasculature contains irregular shapes like aneurysms which are unlike tubes, thinning is not proper method to acquire its feature as tree-like object. We have shown the possibility and limitation of separation of such shape using a method based on mathematical morphology.[7] Though recent efforts especially based upon multiscale expression[8] of skeleton, or medial axis show possibility to overcome the former two limitations, irregularity management seems to be difficult to achieve.

Above limitations of thinning operations are supposed to appear from its toward-axis and voxel-removal way of medial axis acquisition of the structure. In this paper, we propose a new, along-axis style of algorithm as one of the solutions based on the development of mathematical morphology and region growing, and show some fundamental experiments using synthetic and clinical images.

2 Theory and Method

In this chapter, we describe an algorithm to extract structural information of tree-like binary shape and its theory based upon mathematical morphology.[9] Therefore, a binary image is required after certain enhancement[10] and segmentation of original pictures like.

Region-growing[11] is usually used for segmentation of spatially connected and homogenous regions according to grayscale pixel values. An important characteristic of region-growingis flow simulation starting from a seed point. The extension of region growing in our method can be categorized as a morphological operation called 'closed-space dilation' inside a binary object.

Shape Decomposition By Closed Space Dilation: Closed space dilation is performed to decompose a binary shape S into pieces called indexed components X_1, X_2, ..., X_n. The starting point w which is the seed of region-growing, can be considered as X_0.
The first computed component X_1 is defined as

$$X_1 = f_w(S \cap (w \bullet B)), \tag{1}$$

where f_w is *connection filter* of a point(or a region)w, B is structuringelement (growing kernel) and $\bullet$ is conventional dilation operation. The X_0 can be found by raster scanning of the volume. Generally, a sphere with proper radius as described later is used for structuring element.
he n th component(n>1) can be expressed as

$$X_n = f_{X_{n-1}}\left(\left(S \cap \left(\left(\bigcup_{i=0}^{n-1} X_i\right) \bullet B\right)\right) - \bigcup_{i=0}^{n-1} X_i\right). \tag{2}$$

The *connection filter* f characterizes this closed space dilation and is defined

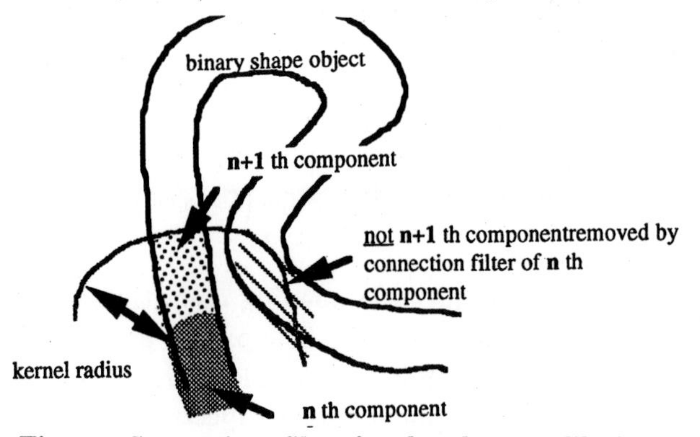

Fig.2. *Connection filter* for closed space dilation

to remove regions which have no contact with a region. As shown in Fig.2, only such regions as spatially connected to the n-1 th component can be part of the n th component. With this theory, the shape dilation of usual region-growing algorithm can be regarded as a special case of closed space dilation where kernel radius is 1.

After this decomposition of the binary shape into several components, the clusters are defined. If n th component exists spatially apart, they are regarded as other cluster (Fig.3). Simultaneously, inter-cluster linkage information is encoded. For the final step, the linkage information is used as topological information of the shape, with clusters as its nodes.

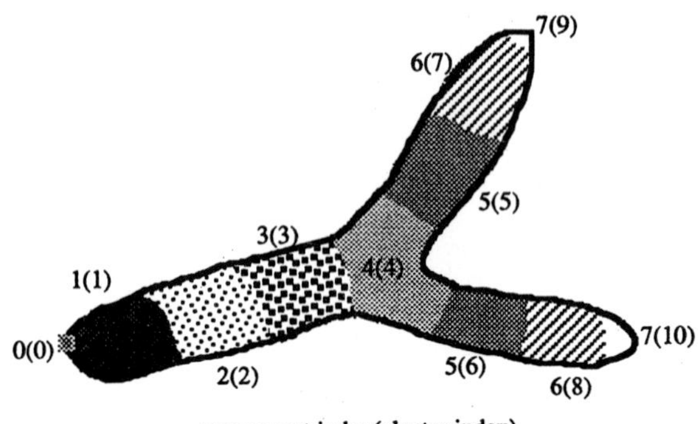

Fig.3. Conversion of components to clusters

Taking into consideration that vessels are almost round pipes, spherical kernel is adequate. For selection of proper kernel size, global pattern spectrum[4] derived from the original development of Maragos[12] is helpful to estimate the thickness of object. It can be regarded as the radius distribution of branches using morphological opening operation. In pattern spectrum inspection for this purpose, a spherical kernel of which radius equals one is adequate.
The n th component of global pattern spectrum for object S with structuring element B is expressed as

$$GPS_S(n,B) = V\left[(S \circ nB) - (S \circ (n+1)B)\right] \quad (n \geq 0), \tag{3}$$

where *V[ ]* is volume of shape, o is opening operation and nB is n times dilation of structural element by itself. The proper growing kernel radius is the average diameter of branches not to yield the artifact described in the discussion part. Examples of global pattern spectrum are shown in the next chapter.

Webbed signal analysis and shape segmentation: After decomposition of a binary shape into connected clusters, analysis for the parameters of this network must be carried out. The parameters are position, volume and number of connected clusters. The cluster's center of gravity is a good approximation for node position. Volume is helpful to estimate its thickness and is important to join similar clusters with segmenting the whole object into groups. In the next chapter, examples are shown.

3 Results

Some simple binary 3D images were synthesized for fundamental analysis. A clinical study was also performed with an image data set of 3D X-ray CT.

Synthetic Image #1(a) and #1(b) <Fig.4>
A vessel #1(a) and relatively complicated vasculature #1(b) were segmented to clusters with different gray values. Global pattern spectrum of kernel radius 1 is also shown for each shape.
Synthetic Image #2 (#1(a) with 2% surface noise) <Fig.5>
2% of spatial noise voxels were added on the surface of a vessel #1(a). Less branches were obtained by larger kernel size. It shows the size controllability of this method.
Synthetic Image #3 (#1(a) with an aneurysm)<Fig.6>
An aneurysm on a branch was simulated. As shown in the webbed signal, volume of clusters which contain aneurysm is high. According to thresholding of these volume values, the aneurysm shape can be separated.

Clinical Image #1 (Cerebral arteries) <Fig.7-10>
Cerebral arteries and gigantic AVM(Arterio-VenousMalformations) were imaged using contrast medium by 3D X-ray CT. Those regions were extracted as a binary shape. The pattern spectrum indicates the average radius of vascular tubes is about 2.0 or 3.0. Therefore kernel size was set to 6. Using cluster volume thresholding, the AVM was separated and left regions as thin vascular objects are reconstructed using pipes. The result seems to express completely the whole structure of vasculature without no manual interactive compilation such as noisy branch removal was performed.

4 Discussion

In comparison with other methods, major features for topology acquisition of this method are simplicity, and parametric controllability of

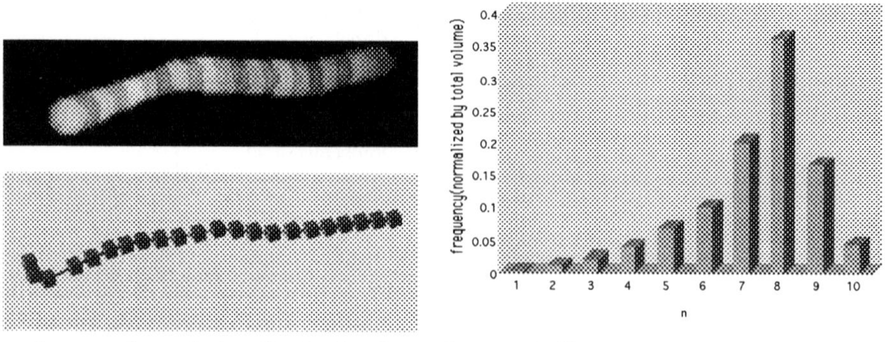

clusters (kernel size 7), obtained topology and Global Pattern Spectrum

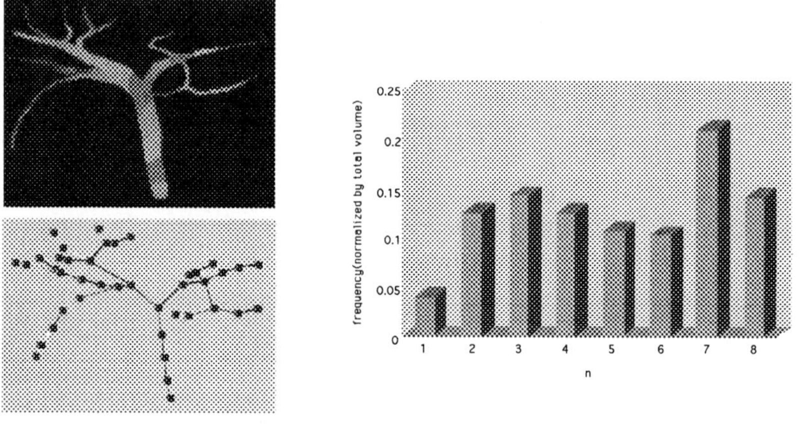

clusters (kernel size 14), obtained topology and Global Pattern Spectrum

Fig. 4. Synthetic Image #1(a) and #1(b)

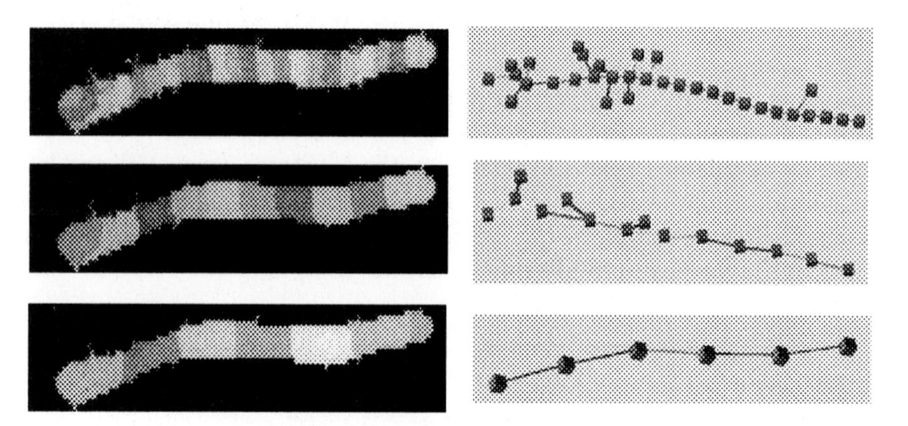

Fig. 5. Synthetic Image #2
kernel sizes are upper: 7, middle: 15, lower 24

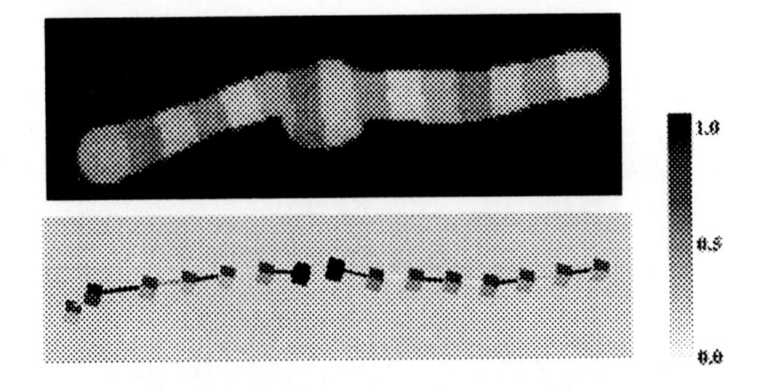

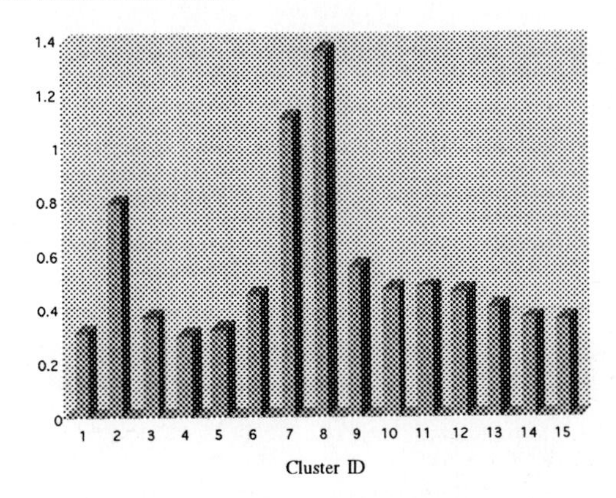

clusters (kernel size 10) and cluster volumes

Fig. 6. Synthetic image #3

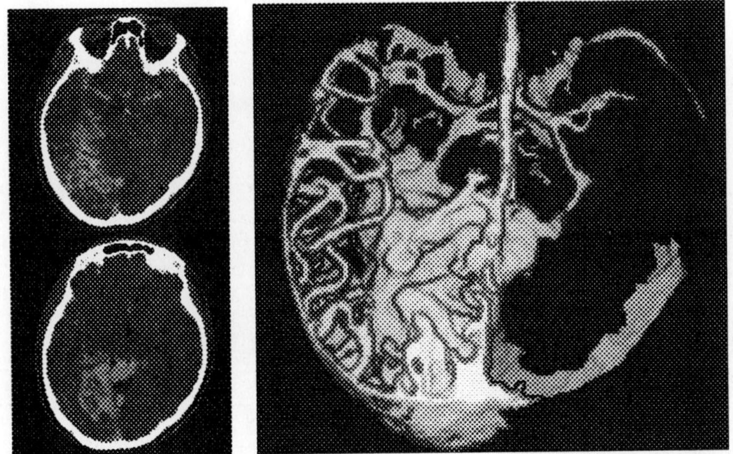

Fig. 7. Part of the clinical image data set (X-ray CT) and segmented vascular tubes connected to AVM region

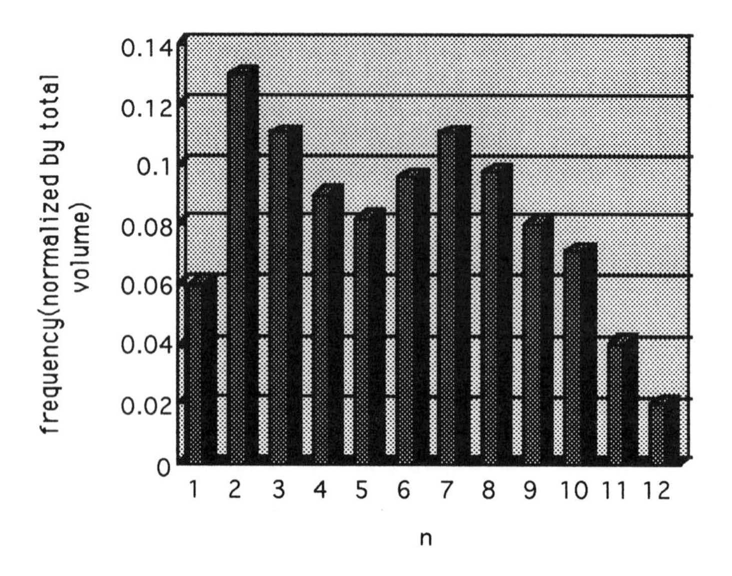

Fig. 7 (contd.). Pattern spectrum of the whole object

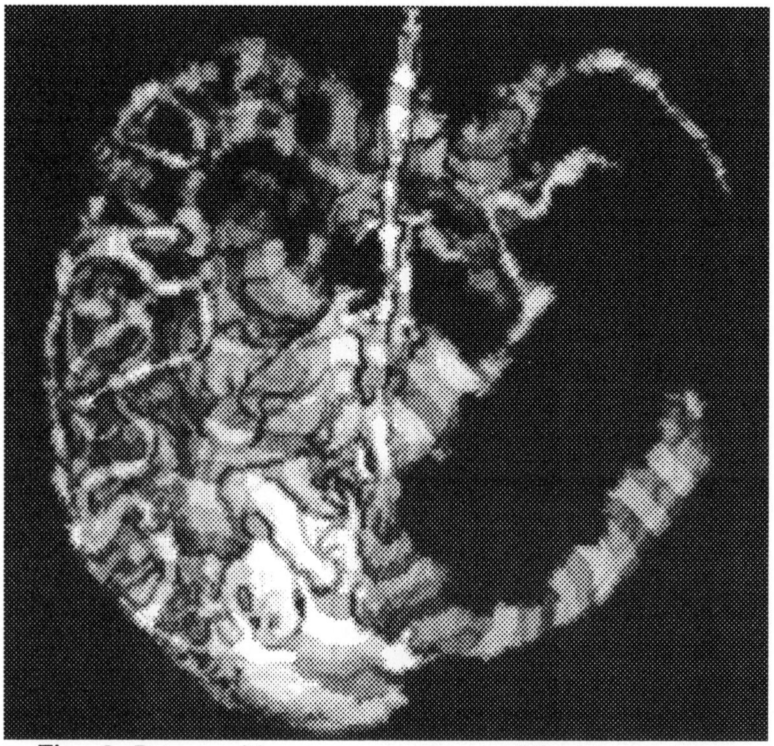

Fig. 8. Decomposition to components in different gray tones

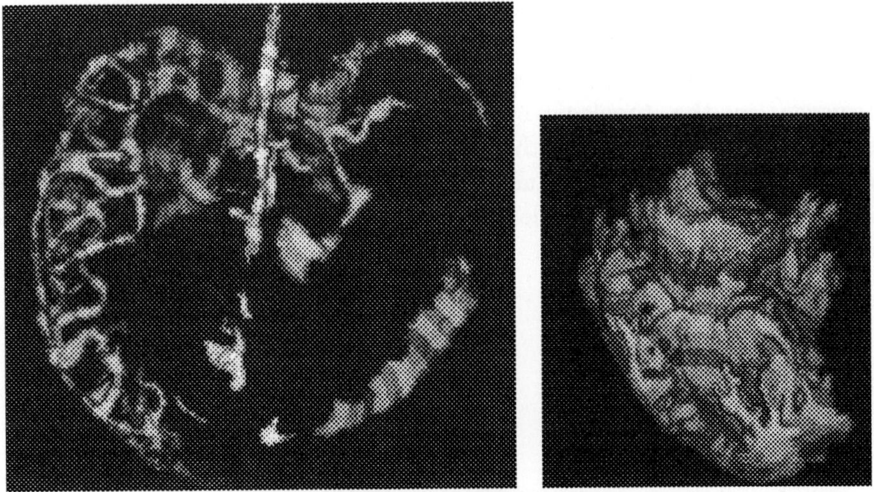

Fig. 9. Separated vascular tubes(left) and AVM(right) according to cluster volumes

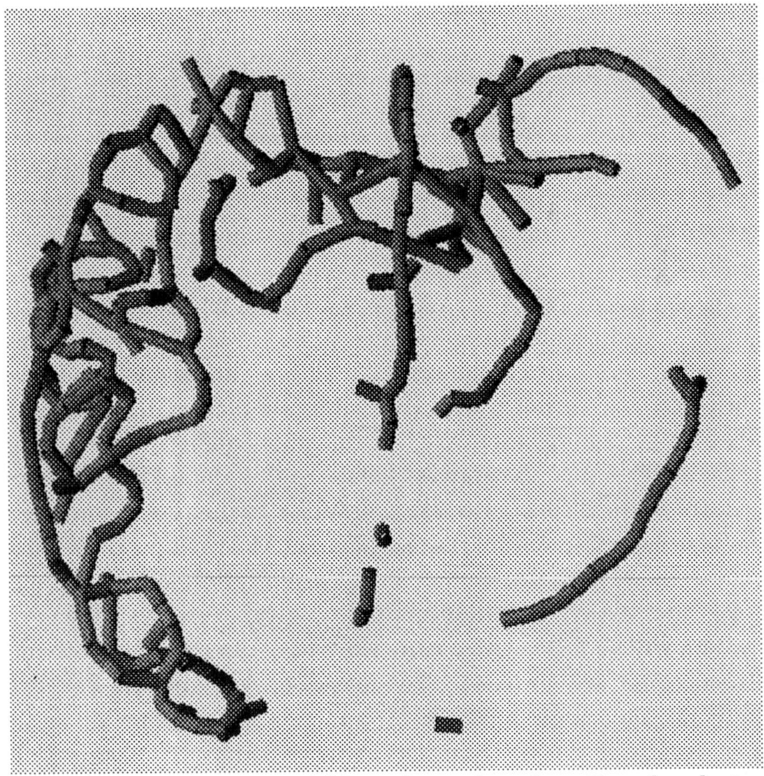

Fig. 10. Selective topology acquisition of vascular tubes. Vascular clusters were replaced with round pipe.

results by selection of kernel shape and size. However, there exist potential complexity in selection of seed. Node position depends on the selection of the seed position. It is affected also by shape irregularity and bifurcation. However, in the topological sense, the extraction never fails. Like conventional thinning algorithms, the location of each node should be regarded just as a rough estimation. Regarding fine center estimation and radii estimation, deformable models like SNAKES[13] using grayscale information are superior.

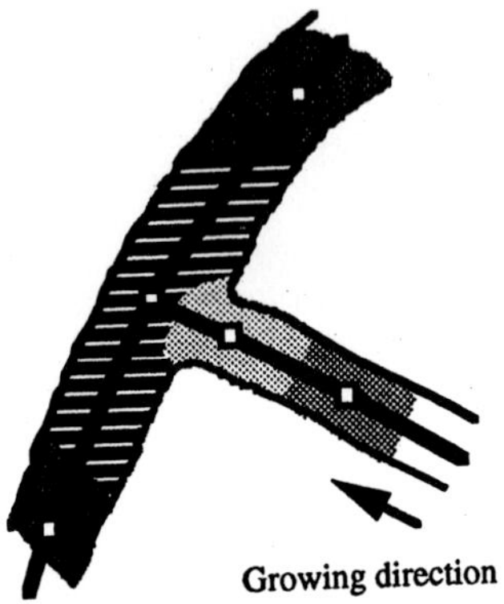

Fig.11. The "Hammer-head effect" -Volume irregularity at bifurcation point-

Webbed signal analysis makes it possible to detect and separate shape irregularity. Except for the clusters in bifurcation point or at a steep corner, a cluster shape is a cylinder, inter-node distance is constant and its volume is uniform if the radii are similar. If kernel size is not adequate, some artifact are observed. Therefore, cluster volume is not enough for estimation of irregularity. Combined methods with the number of connected branches and their angle value are necessary. As shown in Fig. 11, the bifurcation point with a too big kernel, the cluster volume is maximally increased twice of an adjacent straight branch. We call this phenomenon as "Hammer-head effect". It is obvious that the growing kernel radius should be the diameter of pipes. Global pattern spectrum was helpful for estimation of average radius of tubes. However, if irregularshapes like the AVMin our clinical study are contained in the object, the size must be carefully decided.

As we just started fundamental analysis and studies, further investigation will be followed for more theoretical analysis and other clinical applications. And we propose other possible application. For example, asymmetric kernels might be useful for analysis of directional distribution of branches. Furthermore this technique can also be used for 2D instead of 3D.

5 Summary

As an extension of mathematical morphology, a new notion of closed space dilation was introduced and a new algorithm was developed for feature extraction of tree-like binary shapes like vasculature. Using synthetic and clinical images, its fundamental characteristics were investigated.

References

1. Robb R. A., Three-dimensional biomedical imaging -principles and practice-, VCH Publishers, Inc. (1995)
2. Höhne K. H., et al. A new representation of knowledge concerning human anatomy and function, Nature medicine, vol. 1, **6**, (1995) 506-511
3. Székely G., et al. Structural description and combined 3-D display for superior analysis of cerebral vascularity from MRA. Proc. of visualization in biomedical computing, (1994) 272-281
4. Masutani Y., et al., Development of interactive vessel modelling system for hepatic vasculature from MR images, Medical & biological engineering and computing **33** (1995) 97-101
5. Tsao Y. F., et al., A parallel thinning algorithm for 3-D pictures, Computer graphics and image processing **17**, (1981) 315-331
6. Ma C. M. On topology Preservation in 3D Thinning, CVGIP: Image understanding, vol. 59, **3**, (1994) 328-339
7. Masutani Y., et al., Quantitative vascular shape analysis for 3D MR-Angiography using mathematical morphology, Proc. of CVRMed'95, (1995) 449-454
8. Morse B. S., et al. Multiscale medial analysis of medical images, proc. of IPMI '93, (1993) 112-131
9. Serra J., Introduction to mathematical morphology, Computer vision, graphics and image processing **35**, (1986) 283-305
10. Vandermeulen D., et al. Local filtering and global optimisation methods for 3D Magnetic Resonance Angiography (MRA) image enhancement. Proc. of visualization in biomedical computing, (1992) 274-288
11. Foley J. D., et al. J. Computer Graphics: Principles and practice. 2nd Ed. Addison_wesley Publishers CP., Reading MA, (1990)
12. Maragos P., Pattern spectrum and multiscale shape representation, IEEE Trans. on pattern analysis and machine intelligence vol. 11, **7** (1989) 701-716
13. Kass M., et al., SNAKES: Active contour models, Int. J. comput. vision 1, (1987) 321-331

Scale-Space Boundary Evolution Initialized by Cores

Matthew J. McAuliffe[1], David Eberly[2], Daniel S. Fritsch[1],
Edward L. Chaney[1], Stephen M. Pizer[1]

[1]University of North Carolina, Chapel Hill, NC 27599, USA
[2]SAS Institute Inc. Cary, NC 27513, USA

Abstract. A novel interactive segmentation method has been developed which uses estimated boundaries, generated from *cores*, to initialize a scale-space boundary evolution process in greyscale medical images. Presented is an important addition to core extraction methodology that improves core generation for objects that are in the presence of interfering objects. The boundary at the scale of the core (BASOC) and its associated width information, both derived from the core, are used to initialize the second stage of the segmentation process. In this automatic refinement stage, the BASOC is allowed to evolve in a *spline-snake*-like manner that makes use of object-relevant width information to make robust measurements of local edge positions.

1 Introduction

1.1 Motivation

Radiotherapy is the process of applying high energy radiation to specific locations in the body to destroy diseased tissue. To ensure that a lethal dose of radiation is applied to the tumor volume and that minimal dose is applied to healthy tissue, three-dimensional (3D) radiotherapy treatment planning (RTP) is performed. In one RTP approach, *virtual simulation* [1], a 3D graphical model of the patient is created from computed tomography (CT) or magnetic resonance (MR) images and is used to design a treatment protocol that includes information necessary to position the patient and to aim the shaped radiation beams. The first step in the virtual simulation process involves the segmentation of critical structures, including the tumor volume and sensitive surrounding tissues, in the 3D dataset. In typical clinical practice, a radiation oncologist employs a computer application [2] to manually or semi-automatically outline the critical structures of the patient's anatomy in each slice of the dataset. The contours are then used to create 3D models of the anatomy that are input to the virtual simulation program. The 3D models are used both as visual aids to target the radiation beams and for computing volume weighted metrics, such as dose-volume histograms.

1.2 Overview

Clinically available segmentation methods are labor intensive and subject to intra- and inter-user variability. This paper describes the research of a two stage process to segment, on a slice-by-slice basis, objects in greyscale medical images to facilitate more reproducible segmentation for eventual use in a virtual simulation process. The first stage computes medial structures of objects, called cores, directly from image intensities with little human interaction. A core defines the middle and approximate width of the object and is extracted via image measurements in a manner that provides invariance to rotation, translation and zoom [3]. Discussed in this paper is an important addition to core extraction methodology that improves core generation for objects that are in the presence of interfering objects. Derived from and isomorphic to the core, the boundary at the scale of the core (BASOC) is an approximation to the actual boundary of the object. In addition, the BASOC includes a width parameter that varies according to the local width of the object. The BASOC and its associated width information are used to initialize the second stage of the segmentation process. In this automatic refinement stage, the BASOC is allowed to evolve, in a process similar to *spline-snakes* [4,5], where the scale[1] and shape of the aperture used to measure the local geometric properties along the boundary change according to the local object width.

2 Cores and Boundary Evolution

2.1 Overview of the segmentation process

An overview of the two-stage segmentation process diagrammed in Figure 2.1. In the first stage, the user places a circular cursor near the center of the object and adjusts the diameter of the cursor such that it approximately engages opposing object boundaries. A single click of the mouse button is used to stimulate core generation. The BASOC for each core is then calculated. The BASOC (x, y, r), calculated directly from the core, is an excellent initial ordered set of points for the boundary evolution process. All image measurements that combine to form the energy function of the boundary evolution process are made at relevant scales derived from the width parameter of the BASOC. Using a relevant scale to make these measurements greatly increases the likelihood that the BASOC will converge to the desired boundary. Each BASOC, for an object, is allowed to evolve and a union of the BASOCs is then found. The boundary of the union is the localized boundary that defines the object.

---

[1] For the purposes of this paper, scale refers to the standard deviation of a Gaussian kernel.

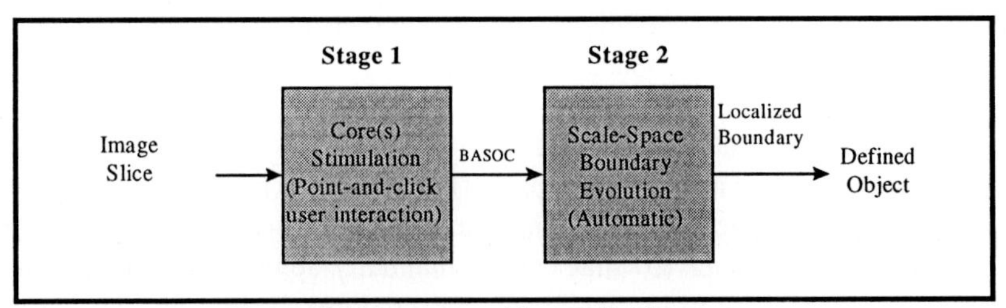

Figure 2.1 Two-stage object definition process.

2.2 Medialness and Cores

In a previous paper, [3], we have characterized the fundamental mathematics of cores and have described how they are extracted from a scale-space of measurements called *medialness*. To summarize, medialness is a measure that describes the degree that an image location appears to be the middle of a figure of specified radius. Medialness is calculated by linking boundaries measured at a scale proportional to that radius.[3]. More specifically, medialness is defined as $\boldsymbol{M(x,y,r)} = \boldsymbol{I(x,y)} \otimes \boldsymbol{K(x,y,\sigma)}$ where $\boldsymbol{K}$ is a scale normalized medialness kernel, $\boldsymbol{I}$ is the actual image intensities and $\otimes$ indicates convolution.

A core is a compact representation of multiscale geometric structure that indicates an object's middle and approximate width (see Figure 2.2a). For a 2D image, a core is defined as a 1D ridge in medialness, defined on $R^2 \times R^+$ where R^2 is the set of all positions (x,y) and R^+ is the set of all widths r.

2.2.1 Medialness Kernels

It is fundamental that medialness kernels yield a maximum response when a level curve of the kernel is tangent to the boundary of the object at multiple loci (see Figure 2.2b).

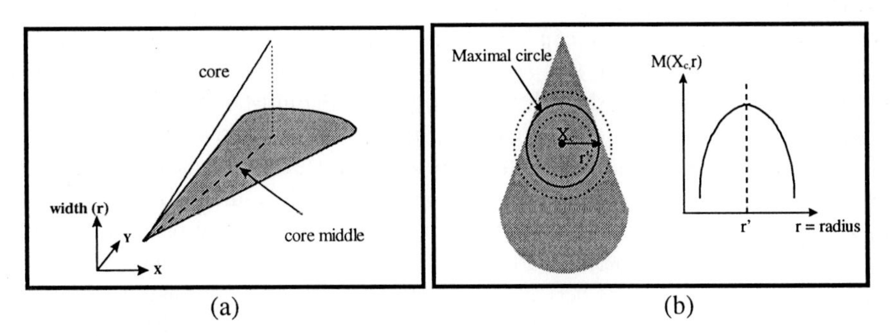

Figure 2.2 (a) Example core of a drop-shaped object. The core middle is the projection of the core into the image plane. (b) A properly designed medialness kernel must yield a maximum response when the kernel tangentially osculates the boundary of the object.

Medialness kernels can be generally classified as either *local* or *multilocal.* A scale normalized Laplacian of the Gaussian, $\mathbf{K(x, y, \sigma) = -\sigma^2 \nabla^2 G(x, y, \sigma)}$ [6], is an example of a local medialness kernel. Advantages include its relative insensitivity to noise and efficient computational properties. Integration of derivative measurements at a radial distance r and at a scale (σ) proportional to r ($\sigma = \rho r$) [7] is an example of a multilocal medialness kernel. This medialness kernel has the advantage of having a tunable parameter (ρ) that can be manually adjusted to compensate for differing amounts of noise in images. Reducing ρ reduces the interference caused by neighboring objects but increases sensitivity to noise as seen in Figure 2.3.

Medialness kernels currently being investigated have proven to be inadequate to the task of object definition in CT image slices. These medialness kernels are subject to interference effects when the "stimulated" object has close neighboring objects (see Figure 2.3). Moreover, neighborhood interference can be so great that it is impossible to stimulate the desired core. Therefore, improvements to the medialness generation process is required to be able to robustly calculate cores in CT images. We define the term *adaptive medialness*, a multilocal medialness, to represent the kernel that addresses the insufficiencies of the other medialness kernels.

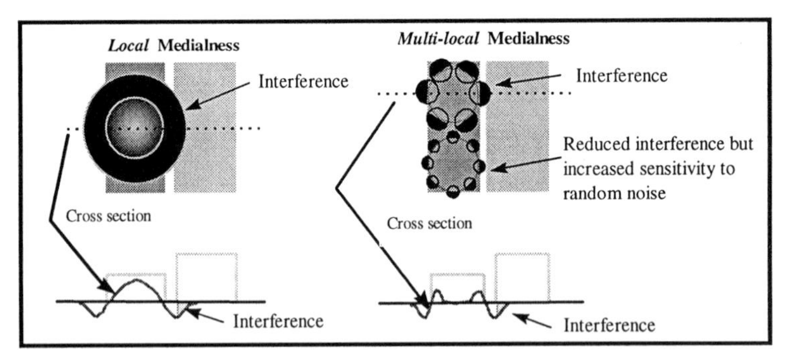

Figure 2.3 Examples showing how local and multi-local medial interference can affect medialness calculations.

Before describing adaptive medialness in detail, a short discussion of boundariness operators, fundamental to the accurate calculation of adaptive medialness, is required. The Gaussian and its derivatives possess the attributes needed to build a medialness scale-space for core generation. In addition, because the Gaussian family of operators inherently have smoothing properties, measurements using them are typically insensitive to noise that is proportionally less then the scale of the measurement operator. These properties make the first derivative of the Gaussian a logical boundariness operator for general image processing and medialness generation. We use the first derivative of a bivariate Gaussian to make boundariness measurements. The use of a bivariate Gaussian as opposed to a radially symmetric Gaussian is key to the process of adaptive medialness and is discussed in more detail in the next section. Equation 1 is a 2D bivariate Gaussian.

$$G(\vec{x},\sigma)=\frac{1}{2\pi(\det\Sigma)^{1/2}}e^{-1/2\vec{x}^T\Sigma^{-1}\vec{x}} \tag{1}$$

where the covariance matrix $\Sigma = R^T DR$ and

$$R=\begin{bmatrix}\cos\theta & -\sin\theta\\ \sin\theta & \cos\theta\end{bmatrix} \tag{2}$$

$$D=\begin{bmatrix}\sigma_x{}^2 & 0\\ 0 & \sigma_y{}^2\end{bmatrix} \tag{3}$$

and σ_x, σ_y are free variables that describe the amount of blurring in the x and y direction, respectively. The angle, θ, describes the rotation.

2.2.2 Adaptive Medialness

Adaptive medialness, at a point, is the sum of boundariness measurements around a circle of radius R (see Figure 2.4a) At every sampled position on the circle of radius r, the shape of the boundariness operator is optimized to reduced interference and maintain relative insensitivity to noise. The net effect is a more accurate medialness and thus a core that better reflects the shape of the object. Moreover, the BASOC, calculated from the core, is also more accurate and therefore a better initialization to the boundary evolution stage.

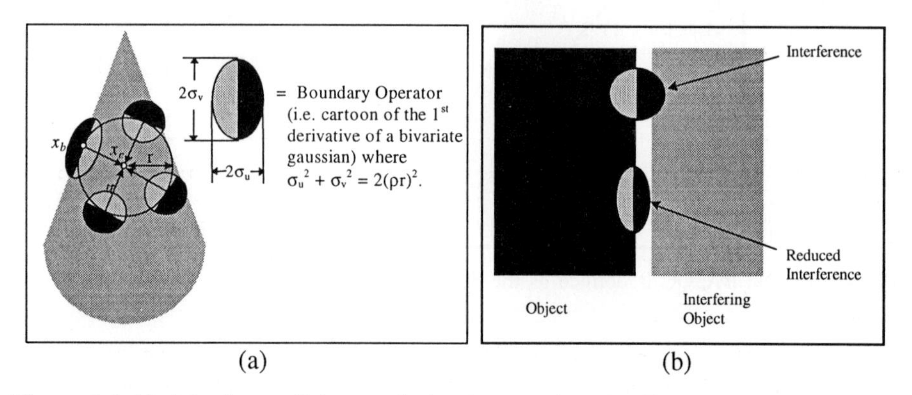

Figure 2.4 (a) Adaptive medialness calculated at a point. (b) Illustrative example of the advantage of using bivariate Gaussian derivative. Reduced interference while maintaining relative insensitivity to random noise.

Mathematically, medialness at a point is described by

$$M(\vec{x}_c,r)=\sum_{u\in S^1}B(\vec{x}_c-r\vec{u},\vec{u}) \tag{4}$$

Boundariness is described by

$$B(\vec{x}_b,\vec{u}) = \underset{(\sigma_u,\sigma_v)}{MAX}[F(\vec{x}_b,\sigma_u,\sigma_v,\vec{u})] \tag{5}$$

where

$$F(\vec{x}_b,\sigma_u,\sigma_v,\vec{u}) = D_u G(\vec{x}_b,\sigma_u,\sigma_v,\vec{u}) \otimes I(\vec{x}_b) \tag{6}$$

and $D_u G(\vec{x}_b,\sigma_u,\sigma_v,\vec{u})$ is the derivative of a bivariate Gaussian oriented in the $\vec{u}$ direction and convolved with the image $I(\vec{x}_b)$. The search for the maximum response over σ_u/σ_v of $F(\vec{x}_b,\sigma_u,\sigma_v,\vec{u})$, where $\sigma_u^2 + \sigma_v^2 = 2(\rho r)^2$, is key in minimizing the effects of neighborhood interference while maintaining relative insensitivity to random noise (see Figure 2.4b).

2.2.3 Boundary at the scale of the core (BASOC)

The boundary at the scale of the core is defined as the boundary of an object whose medial axis [8] is the core middle (see Figure 2.5). The BASOC is directly calculated from the core and represents an approximation to the actual boundary of an object.

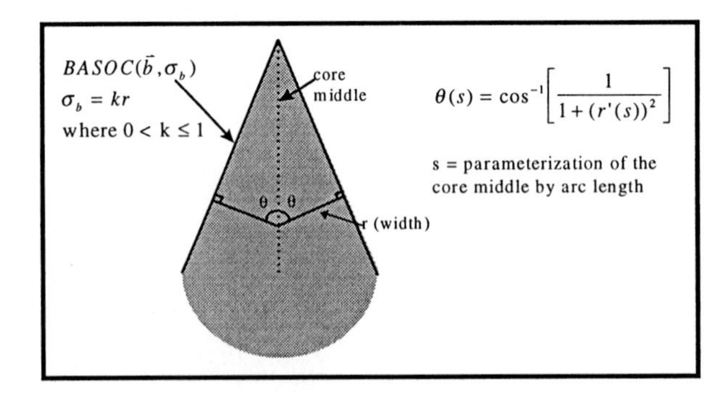

Figure 2.5 The BASOC is defined as the boundary of an object whose medial axis is the core middle.

More specifically, consider the parameterization of the core by arclength (s) which corresponds to a set of points $(\vec{c}(s), r(s))$ where $\vec{c}(s)$ is the core in image space and $r(s)$ is the corresponding radius associated with the core. The BASOC points are calculated by moving from the core middle $\vec{c}(s)$ at an angle $\theta(s) = \cos^{-1}[1 / (1 + (r'(s))^2)]$ with the core, by a distance r (by definition these rays are perpendicular to the BASOC). With every core point there exists two BASOC points that are assigned a scale value σ_b ($\sigma_b = kr$). Image measurements used to control the evolution of the boundary, discussed in the next section, are made at scale σ_b where $0 < k \leq 1$.

2.3 Boundary Evolution

Boundary evolution is a broad area of research that has been adapted in various ways to perform object definition in greyscale medical images. These methods normally start from an initial approximation of an object's boundary and evolve to a more accurate boundary for the purposes of object definition. Kass et al [9] presented the active contours "*snake*" paradigm that minimizes an energy function using a parameterized model that balances the internal energy of a spline (bending energy) and external energy stored in the image. This model and similar models' requirement of a good starting contour is a drawback in clinical practice if significant human labor is required The next section discusses how we address this problem.

2.3.1 Boundary Localization

Approximate object boundaries provided by the BASOC are not suitable for object representation for RTP. However, the BASOC can be generated with minimal user interaction and is an excellent initialization to the boundary evolution process. Moreover, knowledge of the width of the object at every point along the BASOC provides, at a minimum, an upper bound on the scale of the boundariness aperture σ_b with which to make geometric measurements of the image. The width information is critical to the evolution from the BASOC to the final localized boundary. Figure 2.6 illustrates the importance of making measurements at a relevant scale.

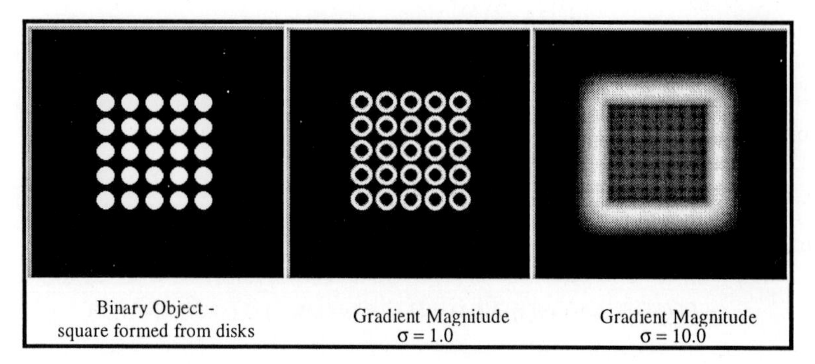

Figure 2.6 Simple example of the importance of scale when making measurements of the image. Measurements made at small scale are sensitive to noise (i.e., the disks) while measurements using large scale operators minimizes the effects of noise. Alternatively, if the circles were the objects of interest, making large scale measurements uninteresting.

While boundariness measurements using large scale operators are less sensitive to small scale noise they can be sensitive to neighboring objects or the adjacent edges of the same object. This interference reduces the accuracy of the measurement and thus the accuracy of the boundary. Conversely, small scale measurements are less sensitive to objects but are more sensitive to noise.

2.3.1.1 Scale-space Boundary Localization

Let $C(s){:}[0,1] \rightarrow R^2$ be a parameterized curve where $C_0(s) = \text{BASOC}$ (see Figure 2.7). The boundary, $C(s)$, is fit to a B-spline and points along the curve are allowed to move in a normal direction in an effort to maximize boundariness of each point. Similar to the medialness generation, boundariness operators are formed from the first derivative of a bivariate Gaussian oriented normal to the curve. At every sampled position on the curve, the shape of the boundariness operator is optimized to reduced interference and maintain relative insensitivity to noise. The net effect is an accurate boundary localization. The curve stops evolving when the change in total energy (boundariness) of the curve goes to zero.

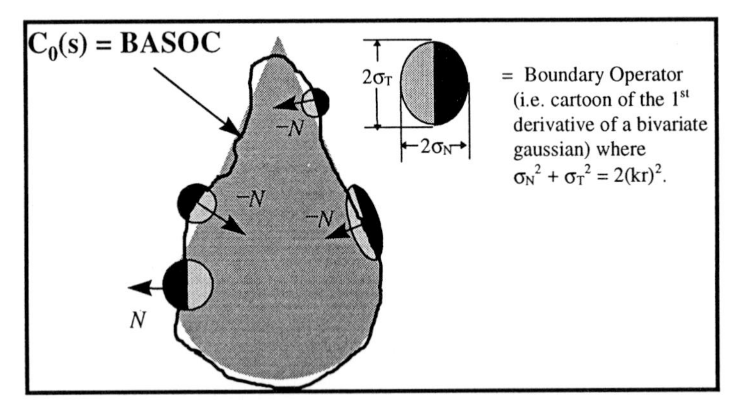

Figure 2.7 The BASOC and boundariness operators (with varying scales, again, derived from the core) used to control the scale-space evolution of the boundary. The boundary is allowed to move normal to the curve to maximize boundariness.

Mathematically, boundaryiness at a point is described by

$$B(C(s,t), \vec{N}(s)) = \underset{(\sigma_N, \sigma_T)}{MAX} [F(C(s,t), \sigma_N(s), \sigma_T(s), \vec{N}(s))] \tag{7}$$

where

$$F(C(s,t), \sigma_N(s), \sigma_T(s), \vec{N}(s)) = D_N G(C(s,t), \sigma_N(s), \sigma_T(s), \vec{N}(s)) \otimes I(\vec{x}) \tag{8}$$

The search for the maximum response of $F(C(s,t), \sigma_N(s), \sigma_T(s), \vec{N}(s))$, where ${\sigma_N}^2(s) + {\sigma_T}^2(s) = 2(kr(s))^2$, is, again, key in reducing the effects of neighborhood interference while maintaining insensitivity to noise, and thus, improving localization of the boundary.

3 Results

We have applied the object definition process to a number of medical images, two of which are shown in Figure 3.1 and Figure 3.2. The first figure is composed of two images, both of the same CT image. On the left image are the cores for the liver, and kidneys, calculated from adaptive medialness, and the associated BASOCs. On the right are the localized boundaries. The second figure is composed of two images, both of the same MRI image of the head. On the left image are the cores for the cerebellum, brain stem and ventricle, all calculated from Laplacian medialness, and the associated BASOCs. Again, on the right are the localized boundaries.

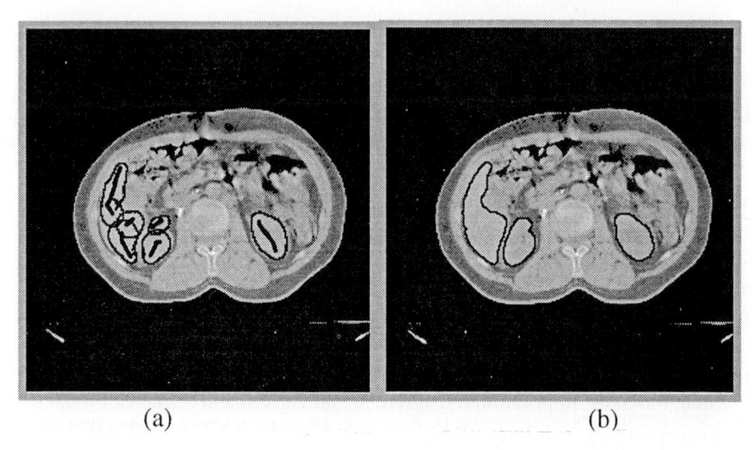

(a) (b)

Figure 3.1 Enhanced (for display purposes, cores where calculated from the original intensities) CT image (a) showing both the core middles, generated using adaptive medialness, and the BASOCs for the liver and both kidneys. (b) the localized boundary after boundary evolution from the respective BASOCs where k =1/4.

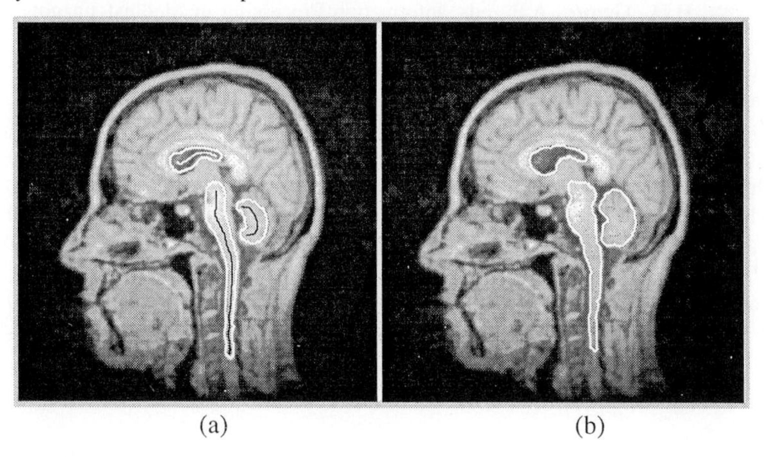

(a) (b)

Figure 3.2 MRI image (a) core middles and BASOCs generated from Laplacian of Gaussian medialness for the cerebellum, brain stem and ventricle, (b) localized boundary evolved from the respective BASOCs where k=1/4.

4 Conclusions

Adaptive medialness is an addition to core extraction methodology that improves core generation for objects that are in the presence of interfering objects. It is planned to use this method of core extraction to build shape-based probabilistic models of normal human anatomy. In addition, we believe that the BASOC provides an excellent initialization for a *spline-snake* boundary evolution process. Future work includes using the BASOC in an implicit parameterized boundary evolution scheme where the BASOC is a contour embedded as a level set of some function [10-13].

References

1. Sailer S.L., Chaney E.L., Rosenman J.G., Sherouse G.W., and Tepper J.E.: Three Dimensional Treatment Planning at the University of North Carolina, Seminars in Radiation Oncology **2(4):1(4)** (1992) 267-273
2. Tracton G, Chaney EL, Rosenman JG, Pizer SM,: MASK: Combining 2D and 3D Segmentation Methods to Enhance Functionality, Math. Methods Med. Imaging III, F.L. Bookstein et al eds, SPIE Proc. Volume **2299** (1994) 98-109
3. Pizer S.M., Eberly D, Morse B.S., Fritsch D.S.: Zoom-Invarient Vision of Figural Shape: The Mathematics of Cores. University of North Computer Science Technical Report, TRN96-004 (1996). Submitted for publication.
4. Leitner F., Margue I., Lavallee S., Cinquin P.: Dynamic Segmentation: Finding the Edge With Snake-splines. International Conference on Curves and Surfaces Proc, (1990) 1-4
5. Cohen L.: Auxiliary Variables for Deformable Models. ICCV **5** (1995) 975-980
6. Fritsch D.S.: Registration of Radiotheraphy Images using Multiscale Medial Descriptions of Images Structure. Ph.D. Dissertation, Department of Biomedical Engineering, University of North Carolina, 1993
7. Morse B.S., Pizer S.M., Liu A.: Multiscale Medial Analysis of Medical Images. In: Barrett, H.H., Gmitro, A.F., eds. Information Processing in Medical Imaging, Lecture Note in Computer Science 687. Berlin: Springer-Verlag; (1993) 112-131
8. Blum H., Nagel R.N.: Shape Description Using Weighted Symmetric Axis Features. Pattern Recognition **10** (1978) 167-180
9. Kass M., Witkin A., Terzopoulos D.: Snakes: Active Contour Models. Int J Comp Vision **1** (1987) 321-331
10. Osher S.J., Sethian J.A.: Fronts Propagation With Curvature Dependent Speed: Algorithms Based on Hamilton-Jacobi Formulation. J. of Comp. Physics **79** (1988) 12-49
11. Caselles V., Kimmel R., Sapiro G.: Geodesic Active Contours. ICCV **5** (1995) 694-699
12. Malladi R., Sethian J., Vemuri B.C.: Shape Modeling with Front Propagation: A Level Set Approach, IEEE-PAMI (1995) **17** 158-175
13. Whitaker R.T.: Algorithms for Implicit Deformable Models. ICCV **5** (1995) 822-827

Fast, Accurate, and Reproducible Live-Wire Boundary Extraction

William A. Barrett and Eric N. Mortensen
Brigham Young University

Abstract. We present an interactive tool for efficient, accurate, and reproducible boundary extraction which requires minimal user input with a mouse. Optimal boundaries are computed and selected at interactive rates as the user moves the mouse starting from a user-selected seed point. When the mouse position comes in proximity to an object edge, a "live-wire" boundary snaps to, and wraps around the object of interest. Input of a new seed point "freezes" the selected boundary segment, and the process is repeated until the boundary is complete. Data-driven boundary cooling generates seed points automatically and further reduces user input. On-the-fly training adapts the dynamic boundary to edges of current interest.

Using the "boundary snapping" technique, boundaries are extracted in one-fifth of the time required for manual tracing, but with 4.4 times greater accuracy and 4.8 times greater reproducibility. In particular, *inter*observer reproducibility using boundary snapping is 3.8 times greater than *intra*observer reproducibility using manual tracing.

1 Introduction

Due to the wide variety of image types and content, fully automated segmentation is still an unsolved problem, while manual segmentation is tedious and time consuming, lacking in precision, and impractical when applied to extensive temporal or spatial sequences of images. However, most current computer based techniques require significant user input to specify a region of interest, initialize or control the segmentation process, or perform subsequent correction to, or adjustment of, boundary points. Thus, to perform image segmentation in a general and practical way, intelligent, interactive tools must be provided to minimize user input and increase the efficiency and robustness with which accurate, mathematically optimal contours can be extracted.

Previous segmentation algorithms have incorporated higher level constraints, but still use local boundary defining criteria which increases susceptibility to noise [1,2]. Others have incorporated global properties for robustness and to produce mathematically optimal boundaries, but most of these methods use one-dimensional (1-D) implementations which impose directional sampling and searching constraints to extract two-dimensional (2-D) boundaries, thus requiring 2-D boundary templates, as with snakes [3-7]. In addition, many of the former techniques exhibit a high degree of domain dependence, and usually still require extensive user interaction to define an initial set of boundary points or region of interest, or to modify results and thereby produce accurate boundaries. Finally, these methods are often very computational or make use of iterative algorithms which limits high level human interactivity as well as the speed with which boundaries can be extracted.

More recent boundary definition methods make use of active contours or snakes [8-11] to "improve" a manually entered rough approximation. After being initialized with a rough boundary approximation, snakes iteratively adjust boundary points in parallel in an attempt to minimize an energy functional and achieve an optimal boundary. The energy functional is a combination of internal forces such as boundary curvature and distance between points and external forces like image gradient magnitude and direction. Snakes can track frame-to-frame boundary motion provided the boundary hasn't moved drastically. However, active contour models follow a pattern of initialization followed by energy minimization; as a result, the user does not know what the final boundary will look like when the rough approximation is input. If the resulting boundary is not satisfactory, the process must be repeated or the boundary must be manually edited.

While snakes and the boundary snapping technique presented here both require user interaction and make use of similar boundary features and cost functions to find optimal boundaries, the two methodologies differ in several significant ways. First, snakes iteratively compute a final optimal boundary by refining a single initial boundary approximation, whereas the live-wire tool interactively *selects* an optimal boundary segment from potentially *all* possible minimum cost paths. Secondly, live-wire boundaries are piece-wise optimal (i.e. between seed points), providing a balance between global optimality and local control, whereas snakes are globally optimal over the entire contour. This piece-wise optimality also allows for path cooling and inter-object on-the-fly training. Third, snakes are typically attracted to edge features only within the gravity of an edge's gradient energy valley whereas the live-wire boundary can snap to strong edge features from arbitrary distances since the live-wire search window is the entire image. Fourth, the smoothing term for live-wire boundaries is data-driven (i.e.computed from external image gradient directions), whereas the active contour smoothing term is internal, based on the contour's geometry and point positions. Finally, the laplacian zero-crossing binary cost feature seems to have not been used previously in active contour modes.

Interactive optimal 2-D path selection is what makes boundary snapping work and what distinguishes it from all previous techniques. As a practical matter, boundary snapping typically requires less time and effort to accurately segment an object than it takes to manually input an initial approximation to the object boundary. Live-wire boundary snapping for image segmentation was introduced initially in [12-13]. This paper reports on several significant enhancements to the original technology, including:

1. A new laplacian cost function and cursor snap for boundary localization.
2. Interleaved boundary selection and wavefront expansion.
3. On-the-fly training for dynamic adaptation to the current edge.
4. Data-driven boundary cooling for automated generation of seed points
5. Application to full color images.

In particular, this paper reports on the application of the boundary snapping technology to medical images and its performance in terms of the *speed*, *accuracy*, and *reproducibility* with which boundaries can be extracted.

2 Boundary Snapping

With boundary snapping, boundary detection is formulated as a graph searching problem [4] where the goal is to find the optimal path between a start node and a set of goal nodes, where pixels represent nodes and edges are created between each pixel and its 8 neighbors. Thus, in the context of graph searching, boundary finding consists of finding the globally optimal path from the start node to a goal node. The optimal path or boundary is defined as the minimum cumulative cost path from a start node (pixel) to a goal node (pixel) where the cumulative cost of a path is the sum of the local costs (or edge link) on the path.

2.1 Local Costs

Since an optimal path corresponds to a segment of an object boundary, pixels (or links between neighboring pixels) that exhibit strong edge features should have low local costs and vice-versa. The local cost function is a weighted sum of the component cost functionals for each of the following features. Features include the gradient magnitude, f_G, and the laplacian zero-crossing, f_Z.

Letting $l(\boldsymbol{p},\boldsymbol{q})$ represent the local cost for the directed link (or edge) from pixel $\boldsymbol{p}$ to a neighboring pixel $\boldsymbol{q}$, the local cost function is

$$l(\boldsymbol{p},\boldsymbol{q}) = \omega_G \cdot f_G(\boldsymbol{q}) + \omega_Z \cdot f_Z(\boldsymbol{q}) \quad (1)$$

where each ω is the weight of the corresponding feature function. Empirical default values for these weights are $\omega_G = 0.5$ and $\omega_Z = 0.5$.

The gradient magnitude feature, $f_G(\boldsymbol{q})$, provides a "first-order" positioning of the live-wire boundary. However, so that high image gradients will correspond to low costs, the gradient magnitude, G, is scaled and inverted using an inverse linear ramp function.

The laplacian zero-crossing feature, $f_Z(\boldsymbol{q})$, is a binary edge feature used for boundary localization. That is, it provides a "second order" fine-tuning or refinement of the final boundary position. The output, I_L, of the convolution of the image with a laplacian edge operator is made binary by letting $f_Z(\boldsymbol{q}) = 0$ where $I_L(\boldsymbol{q}) = 0$ or has a neighbor with a different sign, and $f_Z(\boldsymbol{q}) = 1$ otherwise. If $I_L(\boldsymbol{q})$ has a neighbor with a different sign, of the two pixels, the one closest to zero represents the zero-crossing or the position to localize the boundary. Thus, $f_Z(\boldsymbol{q})$ has a low local cost (0) corresponding to good edges or zero-crossings and a (relatively speaking) high cost (1) otherwise. This results in single-pixel wide cost "canyons" which effectively localize the live-wire boundary at point $\boldsymbol{q}$.

2.2 Boundary Detection as Graph Searching

Boundary finding can be formulated as a directed graph search for an optimal (minimum cost) path. Nodes in the graph are initialized with the local costs described above. Then, a user-selected seed point is "expanded," meaning that its local cost is summed into its neighboring nodes. The neighboring node with the minimum cumulative cost is then expanded and the process continues producing a "wavefront" which expands in order of minimum cumulative cost such that for any node within the wavefront, the optimal path back to the seed point is known.

Unlike some previous approaches to optimal boundary detection [3-7], this approach allows two degrees of freedom in the search, producing boundaries of arbitrary curvature and complexity. Previous approaches also require a fixed goal node/pixel which must be specified before the search begins, whereas selection of a goal node or "free point" within the wavefront expansion allows any goal node to be specified interactively with the resulting boundary immediately available by following pointers (i.e.boundary points) back to the seed point. In addition, we expand nodes in order of minimum cumulative cost, creating a dynamic wavefront which expands preferentially in directions of highest interest (i.e. along edges). This has several advantages: (1) it almost always allows path expansion to keep pace with path selection so that (2) interactively selected paths *are* available and (3) it overcomes the limitations of previous approaches which use dynamic programming where nodes must be expanded in fixed stages and therefore require up to n iterations through the cost matrix for paths of length n. Details of our algorithm are reported in [14].

2.3 Interactive "Live-Wire" Segmentation Tool

Once the optimal path pointers are generated, a desired boundary segment can be chosen dynamically via the free point specified by the current cursor position. Interactive movement of the free point by the mouse cursor causes the boundary to behave like a live wire as it erases the previous boundary points and displays the new minimum cost path defined by following path pointers from the free point back to the seed point. By constraining the seed point and free points to lie near a given edge, the user is able to interactively "snap" and "wrap" the live-wire boundary around the object of interest. Figure 1 shows how a live-wire boundary segment adapts to changes in the free point (cursor position) by latching onto more and more of the coronary artery. Specifically, note the live-wire segments corresponding to user-specified free point positions at times t_0, t_1, and t_2. Although Figure 1 only shows live-wire segments for three discrete time instances, live-wire segments are actually updated dynamically and interactively (on the fly) with each movement of the free point.

When movement of the free point causes the boundary to digress from the desired object edge, input of a new seed point prior to the point of departure effectively "ties off" or freezes the boundary computed up to the new seed point and reinitiates the boundary detection starting from the new seed point. This produces a wavefront expansion emanating from the new seed point with a new set of optimal paths to *every* point within the wavefront. Thus, by selecting another free point with the mouse cursor, the interactive live-wire tool is simply *selecting* an optimal boundary segment from a large collection of optimal paths.

Since each pixel (or free point) defines only one optimal path to a seed point, a minimum of two seed points must be deposited to ensure a closed object boundary.

Two seed points are sufficient to provide a closing boundary path from the free point if the path (pointer) map from the first seed point of every object is maintained during the course of an object's boundary definition. The closing boundary segment from the free point to the first seed point eliminates the need for the user to manually close off the boundary.

Placing seed points directly on an object's edge is often difficult and tedious. If a seed point is not localized to an object edge then spikes results on the segmented boundary at those seed points. To facilitate seed point placement, a cursor snap is available which forces the mouse pointer to the maximum gradient magnitude pixel within a user specified neighborhood. The neighborhood can be anywhere from 1x1 (resulting in no cursor snap) to 15x15 (where the cursor can snap as much as 7 pixels in both x and y). So that the responsiveness of the live wire and the associated interactivity is not encumbered, a cursor snap map is *precomputed* by encoding the (x,y) offset from every pixel to the maximum gradient magnitude pixel in the neighborhood. Thus, as the mouse cursor is moved by the user, it snaps or jumps to a neighborhood pixel representing a "good" static edge point.

3 On-The-Fly Training

On occasion, a section of the desired object boundary may have a weak gradient magnitude relative to a nearby strong gradient edge. Since the nearby strong edge has a relatively low cost, the live-wire segment snaps to it rather than the desired weaker edge. This is demonstrated in the Cine' CT scan of the heart in Figure 2(a) where the live-wire segment cuts across the corner of the ventricle and in Figure 2(b) where the live-wire segment snaps to the epicardial-lung boundary surface, the edge of greater strength, rather than following the endocardial edge.

Training allows dynamic adaptation of the cost function based on a previous sample boundary segment that is already considered to be "good" (i.e. between the two red seed points in Figure 2(c)). Since training is performed "on-the-fly" as part of the boundary segmentation process trained features are updated interactively as an object boundary is being defined. This eliminates the need for a separate training phase and allows the trained feature cost functions to adapt *within* the object being segmented (Figure 2(a)) as well as between objects in the image (Figure 2(b)). Thus, training overcomes the problems in Figures 2(a) and 2(b) with *no additional seed points.*

Training is facilitated by building a distribution of features, notably image gradients, from the training segment. As is done with the gradient feature itself, the distribution is inverted and scaled so that relatively higher gradients, in particular those associated with the "good" training segment, have lower costs. To allow training to adapt to slow (or smooth) changes in edge characteristics, the trained gradient magnitude cost function is based only on the most recent or closest portion of the currently defined object boundary. A monotonically decreasing weight function (either linearly or gaussian based) determines the contribution from each of the closest t pixels. The training algorithm samples the precomputed feature maps along the closest t pixels of the edge segment and increments the feature distribution element by the corresponding pixel weight to generate a distribution for each feature involved in training. Since training is based on learned edge characteristics from the most recent portion of an object's boundary, training is most effective for those objects with edge properties that do not change radically as the boundary is traversed (or at least change smoothly enough for the training algorithm to adapt).

The training length is typically short (in the range of 32 to 64 pixels) to permit local dependence (prevent trained features from being too subject to old edge characteristics) and thereby allow it to adapt to slow changes.

4 Data-Driven Path Cooling

As described in Section 2.3, closed object boundaries can be extracted with as few as two seed points. However, more than two seed points are often needed to generate accurate boundaries. Typically, two to five seed points are required for boundary definition but complex objects may require many more. Even with cursor snap, manual placement of seed points can be tedious and often requires a large portion of the overall boundary definition time.

The idea of path cooling was introduced in Section 2.3. Namely, that depositing a new seed point by fixing the free point causes the previous boundary segment to cool or freeze, meaning that the previous segment is now fixed and is no longer part of the live-wire boundary - the live-wire boundary is now rooted at the new seed point. Since this kind of path cooling requires the user to manually input new seed points with a click of the mouse, it would be preferable to have seed points deposited, and associated boundary segments freeze automatically, as a function of image data and path stationarity. Thus, automatic seed point generation is the motivation behind data-driven path cooling.

Automatic seed point generation relieves the user from placing most seed points by automatically selecting a pixel on the current active boundary segment to be a new seed point. Selection is based on "path cooling" which in turn relies on path coalescence. Though a single minimum cost path exists from each pixel to a given seed point, many paths "coalesce" and share portions of their optimal path with paths from other pixels. Due to Bellman's Principle of Optimality, if any two optimal paths from two distinct pixels share a common point or pixel, then the two paths are identical from that pixel back to the seed point. This is particularly noticeable if the seed point is placed near an object edge and the free point is moved away from the seed point but remains in the vicinity of the object edge.

Though a new path is selected and displayed every time the mouse cursor moves, the paths are typically all identical near the seed point and only change local to the free point. As the free point moves farther away from the seed point, the portion of the active "live-wire" boundary segment that does not change becomes longer. New seed points are generated automatically at the end of a stable segment (i.e. that has not changed recently). Stability is a function of (1) time on the active boundary and (2) path coalescence (number of times the path has been drawn from distinct seed points). This measure of stability provides the live-wire segment with a sense of “cooling.” Since the degree to which paths will coalesce over a given interval is a function of the underlying data (noise, gradient strength, variability in geometry and brightness of object boundary), we say that path cooling is data-driven. Pixels that are on a stable segment sufficiently long will freeze, automatically producing a new seed point.

Path cooling was used to extract the boundaries of the coronary vessel in Figures 1 and 3, automatically generating additional seed points in both cases. Both cooling and training were used in Figure 3.

5. Results

Figures 1 through 4 illustrate the application of live-wire boundary extraction for a variety of medical image types: Figure 1 (X-ray projection angiography), Figure 2 (CT), Figure 3 (MRI) and Figure 4 (a color photograph (mid-thigh) from the Visible Human Project). Table 1 shows the times and the number of seed points required to extract boundaries in these images.

Figure	Anatomy	Time (seconds)	Seed Points	Training Used	Cooling Used
1	coronary (right side)	2.02	2	No	No
	coronary (left side)	3.50	3	No	No
2	left ventricle	3.71	2	Yes	No
3	brain	2.30	3	Yes	Yes
4	thigh muscle A	6.40	5	No	No
	thigh muscle B	1.33	2	No	No
	thigh muscle C	1.31	2	No	No
	thigh muscle D	3.24	3	No	No
	Average:	2.98	2.75		

Table 1: Times to extract boundaries for anatomy contained in Figures 1-4 with number of seed points used or automatically generated.

As shown in Table 1, the average time required to extract the boundaries shown in Figures 1-4 was just under three seconds with an average of close to three seed points needed. (Recall that 2 seed points are required.) A more detailed study shows that, for an experienced user, the time required to extract boundaries with the live-wire tool is roughly 4.6 times less than for manual tracing. In other words, the average time required to manually trace each boundary in Figures 1-4 would be about 13.7 seconds. However, although live-wire boundary extraction is only four to fives times faster, it is worth pointing out that live-wire boundaries are also much more accurate and reproducible, as shown below. In other words, to get the same kind of accuracy and reproducibility with manual tracing, many times more effort would need to be expended.

Figure 4 demonstrates application of the live-wire tool to a full-color image - one in which the separations between muscle groups are subtle, but visually noticeable. This is a classic example of where traditional edge-following or region-growing schemes would have difficulty due to weak gradients, touching objects, and similar color or contrast. However, the live-wire tool snaps to these boundaries of separation quite easily.

Figure 5 graphically compares the live-wire boundary accuracy with manual tracing. The graph shows the average time and accuracy from a study where 8 untrained users were asked to define the boundaries of five objects. Each user spent a few minutes becoming familiar with the live-wire tool as well as a manual tracing tool before defining the boundaries of the 5 objects. Each boundary was defined multiple times by each user with both the live-wire and the manual tracing so that inter- and intra- observer reproducibility could also be measured (Figure 6).

The graph in Figure 5 shows that only 20% of the manually defined boundary points corresponded to the "true" boundary positions as determined by an expert, whereas 87% of the boundary points determined with the live-wire tool matched those chosen by the expert. In other words, the accuracy obtained with the live-wire tool was 4.4 times (87/20) greater. On the other hand, 52% of the manually specified boundary points were within 1 pixel of the position chosen by the expert while 93% of the live-wire boundary fell within that range. A range of up to 4 pixels was chosen since the curves asymptote and connect at that point.

A similar study was performed to compare reproducibility for both the live-wire and the manual trace tools. Eight users extracted 5 object boundaries 5 times with the live-wire tool and the same object boundaries 3 times with the manually tracing tool (we couldn't pay them enough to do it more). The results are reported in Figure 6. As the graph shows, for a given user (intra-observer), slightly over 95% of the same points were consistently chosen over the 5 repeated trials using the live-wire tool. What is even more significant is that virtually the same percentage was obtained when all live-wire boundaries were pooled (inter-observer) demonstrating *dramatically* that the boundaries are virtually identical regardless of which user is performing the task. This is in striking contrast to the inter- and intra-observer reproducibility shown for the manual tracing tool. (High variability for manual tracing is a well known problem). In particular, the inter-observer reproducibility is roughly 4.8 times (95/20) better for live-wire boundaries at the zero-pixel tolerance, with the curve asymptoting quickly to 100% for higher error tolerances. Perhaps the most striking result is that the *inter-observer* reproducibility using the live-wire tool is 3.8 times (95/25) greater than the *intra-observer* reproducibility at the zero error tolerance level. This, of course, shows that we will have much better consistency with the live wire tool regardless of who is performing the boundary extraction than we would just having a single person doing the boundary extraction manually. When taken in conjunction with the accuracy and timing results, this is significant indeed.

6. Conclusions and Future Work

Live-wire boundary extraction provides an accurate, reproducible and efficient interactive tool for image segmentation. In fact, and in sharp contrast to tedious manual boundary definition, object extraction using the live wire is almost fun. The live-wire tool is intuitive to use and can be applied to black and white *or* color images of arbitrary complexity. There are many rich extensions of this work, including (1) making use of the weighted zero-crossings in the Laplacian to determine boundary position at the sub-pixel position and (2) extension of the graph search and application of the live-wire snap and training tools to spatial or temporal image sequences.

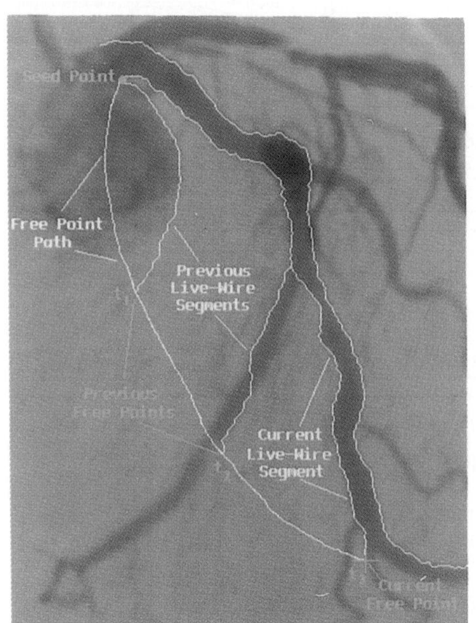

Figure 1. Continuous snap-drag of live wire to coronary edge (shown at times t_1, t_2, and t_3). A boundary is completed in about 2 seconds.

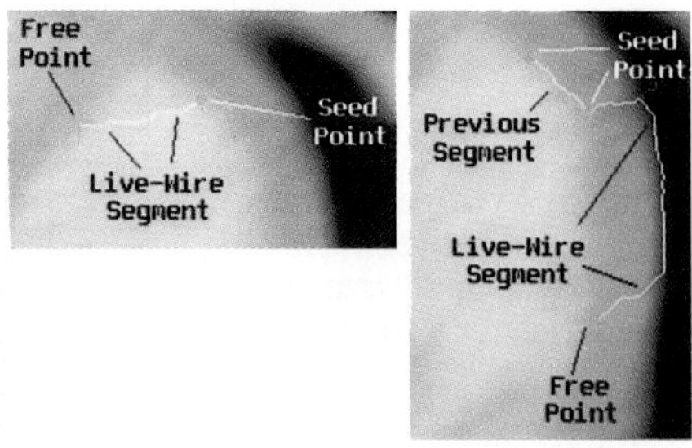

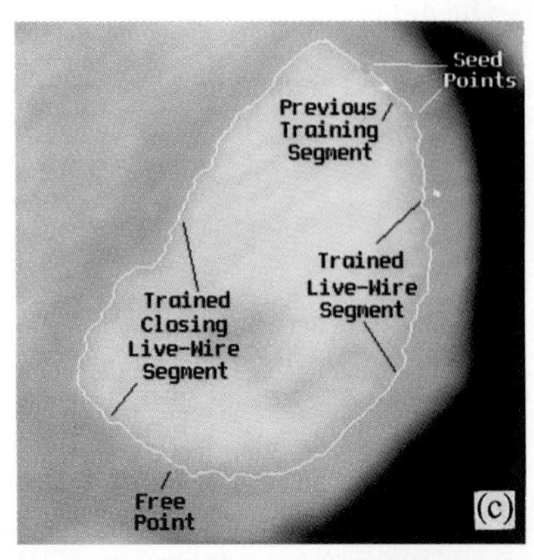

Figure 2. CT slice - mid heart. *Without* training live wire "cuts" corner, (a), or snaps to lung boundary, (b). *With* training, (c), correct boundary is followed.

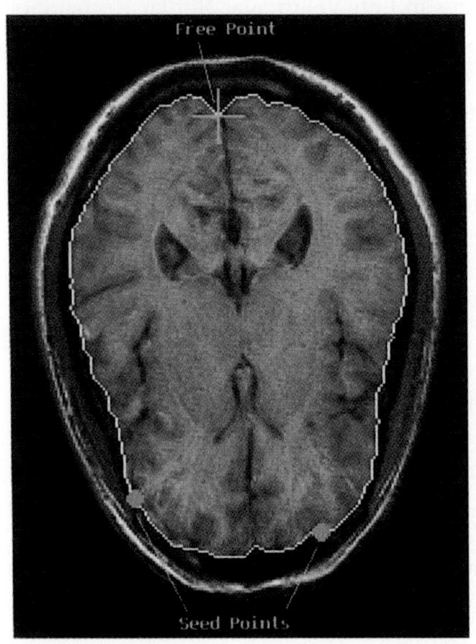

Figure 3. MRI brain boundary extracted in 2.3 seconds using cooling and on-the-fly training.

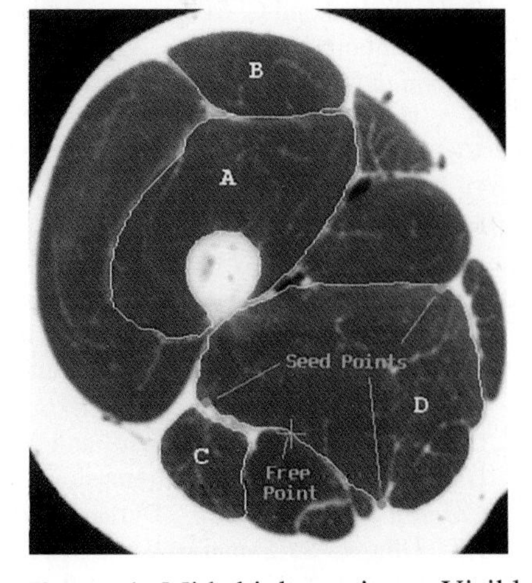

Figure 4. Mid-thigh section - Visible Human project. Live wire separates muscle groups in seconds, although they touch, are of similar color and contrast, and have weak gradients at junctures.

Figure 5. Average accuracy comparison between live-wire and manually traced boundaries for 8 users.

Figure 6. Average reproducibility comparison between live-wire and manually traced boundaries for 8 users.

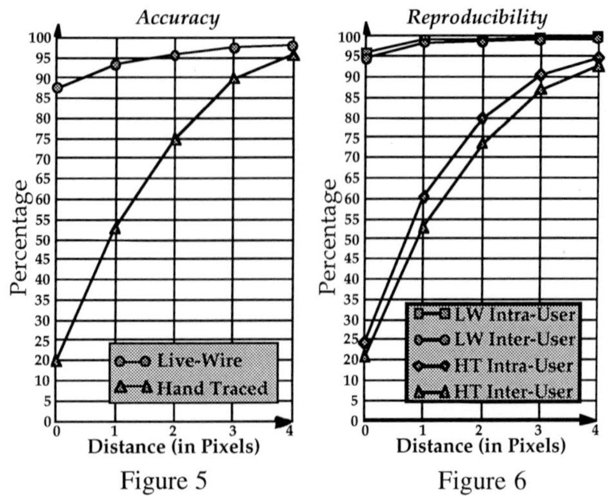

References

1. J.F. O'Brien and N.F. Ezquerra, "Automated Segmentation of Coronary Vessels in Angiographic Image Sequences Utilizing Temporal, Spatial and Structural Constraints," *VBC'94*, SPIE Vol. 2359, pp. 25-37, October, 1994.
2. M. Gleicher, "Image Snapping," *Comp. Graphics SIGGRAPH '95 Proc.* pp. 191-198.
3. Y. P. Chien and K. S. Fu, "A Decision Function Method for Boundary Detection," *Computer Graphics and Image Processing*, Vol. 3, No. 2, pp. 125-140, June 1974.
4. A. Martelli, "An Application of Heuristic Search Methods to Edge and Contour Detection," *Communications of the ACM* Vol.19, No. 2, pp. 73-83, February 1976.
5. D.L. Pope et. al., "Dynamic Search Algorithms in Left Ventricular Border Recognition and Analysis of Coronary Arteries," *IEEE Computers in Cardiology*, pp. 71-75, 1984.
6. D. H. Ballard and C. M. Brown, *Computer Vision*, Prentice Hall, 1982.
7. U. Montanari, "On the Optimal Detection of Curves in Noisy Pictures," *Communications of the ACM*, Vol. 14, No. 5, pp. 335-345, May 1971.
8. A.A. Amini et. al., "Using Dynamic Programming for Solving Variational Problems in Vision." *PAMI*, Vol. 12, no. 2, pp. 855-866, Sept., 1990.
9. D. Daneels, et al., "Interactive Outlining: An Improved Approach Using Active Contours," *SPIE Proceedings of Storage and Retrieval for Image and Video Databases*, vol. 1908, pp. 226-233, Feb. 1993.
10. M. Kass et. al., "Snakes: Active Contour Models," *Proceedings of the First International Conference on Computer Vision*, London, England, pp. 259-68, June 1987.
11. D.J. Williams and M. Shah, "A Fast Algorithm for Active Contours and Curvature Estimation," *CVGIP: Image Understanding*, vol. 55, no. 1, pp. 14-26, Jan. 1992.
12. Eric Mortensen, Bryan Morse, and William Barrett: "Adaptive Boundary Detection Using 'Live-Wire' Two-Dimensional Dynamic Programming," *IEEE Computers in Cardiology, pp. 635-638,* Durham, North Carolina, October, 1992.
13. J.K. Udupa, S. Samarasekera, W.A. Barrett: "Boundary Detection Via Dynamic Programming," *Visualization in Biomedical Computing '92*, pp. 33-39, Chapel Hill, North Carolina, October, 1992.
14. Eric Mortensen and William Barrett: "Intelligent Scissor for Image Composition," *Computer Graphics (SIGGRAPH '95 Proceedings),* pp. 191-198.

Automatic Segmentation of Cell Nuclei from Confocal Laser Scanning Microscopy Images

A. Kelemen[1,2], G. Székely[1], H-W. Reist[2] and G. Gerig[1]

[1]Federal Institute of Technology, CH-8000 Zürich, Switzerland
[2]Paul Scherrer Institute, F2, CH-5232 Villigen PSI, Switzerland

Abstract. In this paper we present a method for the fully automatic segmentation of cell nuclei from 3D confocal laser microscopy images. The method is based on the combination of previously proposed techniques which have been refined for the requirements of this task. A 3D extension of a wave propagation technique applied to gradient magnitude images allows us a precise initialization of elastically deformable Fourier models and therefore a fully automatic image analysis. The shape parameters are transformed into invariant descriptors and provide the basis of a statistical analysis of cell nucleus shapes. This analysis will be carried out in order to determine average intersection lengths between cell nuclei and single particle tracks of ionizing radiation. This allows a quantification of absorbed energy on living cells leading to a better understanding of the biological significance of exposure to radiation in low doses.

1 Introduction

Single traversals of protons or α particles through the nucleus of an individual cell are particularly relevant for understanding the biological significance of exposure to ionizing radiation in low doses, both with respect to risk assessment and to mechanistic studies. The ultimate goal of our study is to develop and evaluate a new method to relate selected biological endpoints to the physical parameters of low dose rate irradiation of cells and cultured mouse pre-implantation embryos by reconstructing single tracks through individual cells. By comparing the experimental results with microdosimetric models in terms of track structure, our understanding of the biological consequences of low doses can be improved.

Newly developed experimental methods [1] can combine the possibility of irradiating more than a thousand cells simultaneously with an efficient colony-forming ability and the capacity of localization of a particle track through a cell nucleus together with the assessment of the energy transfer. The cell chamber combined with the attached detector permits to localize a track through a cell nucleus by digital superposition of the image containing the tracks with that of the cells.

To assess the amount of energy deposition by particles traversing the cell nucleus the intersection lengths of the particle tracks have to be known. Intersection lengths can be obtained by determining the 3D surface contours of the irradiated cell nuclei. Confocal laser scanning microscopy using specific DNA fluorescent dye (fourochrome) offers a possible way for the determination of the 3D shape of individual cell nuclei. Unfortunately, such experiments cannot be performed on living cells due to the toxicity of the dye under irradiation.

One solution to this problem can be provided by building a statistical model of cell nucleus shapes, allowing the determination of the average intersection lengths as function of the distances of the tracks from the centre or from the contour of the cell nucleus visible on 2D microscopic images.

In order to build such a statistical model, a large number of cell nuclei have to be identified and segmented from confocal laser scanning microscopy images. The present paper describes a method, to perform this 3D segmentation task in an automatic manner in order to create a solid basis for the statistical model building. In a first step the raw CLSM images are restored using the measured point spread function of the optical system as described in Section 2. Section 3 describes an efficient method for the localization of the cell nuclei in the CLSM images using combined anisotropic wave propagation and diffusion. As the localization method also provides size information about the detected elliptic models, the resulting ellipsoid can be used as an initialization for an elastically deformable parametric surface model, leading to a segmentation of the cell nuclei and a parametric description of their shapes. This process is described in Section 4. The proposed procedure is demonstrated with the automated segmentation of the nucleus of type *v79* cells.

2 Reconstruction of confocal laser scanning microscope (CLSM) images

Detection of the cell nucleus requires a reliable assessment of the surface contour of that nucleus. Laser scanning microscopy provides a set of confocal sections through a living cell in real time. However, each section has many details obscured by blurred light from parts of the cell that are out of focus. Image acquisition in confocal laser scanning microscopy is being investigated in order to establish restoration of the obtained images. Three-dimensional optical blurring of the image acquisition system is characterized by the continuous point spread function (PSF), the image of a point like source. Considering the image acquisition as a linear and shift-invariant system, the observed images can be expressed mathematically as a 3D-convolution of the original fluorescent image and the PSF. The PSF can be used to improve the quality of the image data.

Point spread functions of different objectives were determined using subresolution fluorescent beads [2]. Four different restoration algorithms were examined. It was shown that linear restoration techniques, such as Wiener-filter, perform well at good signal to noise ratios. At high noise levels, however, it became clear that non-linear restoration techniques give superior results to those obtainable by linear filters. Investigating non-linear methods, reconstructions by constrained iterative deconvolution were tested. We implemented a modification of the Jansson-van Cittert [3, 4, 6] method of successive deconvolution.

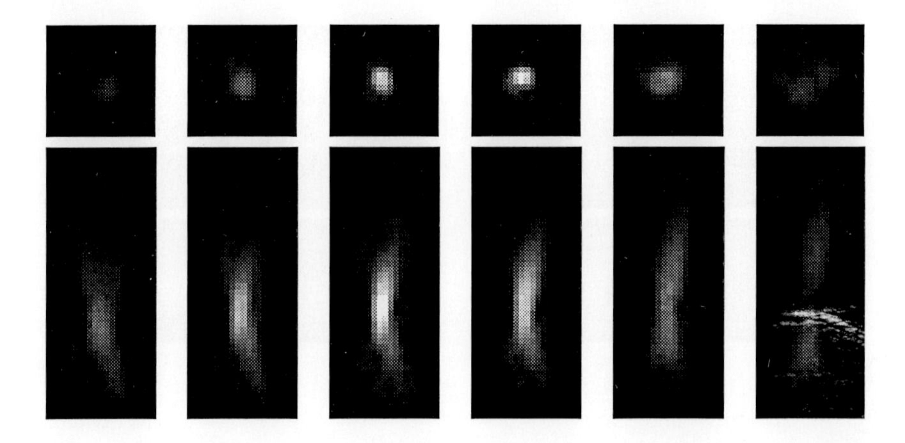

Figure 1: Measured point spread function for the Zeiss oil-immersion 63x objective with *numerical aperture*=1.3. For the measurement we used subresolution fluorescent beads of 0.175 μm diameter. **Top row**: x-y sections with section separation = 0.78 μm, **bottom row**: x-z sections with section separation = 0.1 μm.

This approach provided good performance even at a high noise level. A second non-linear algorithm which was originally developed by Gold converges about five times faster than the Jansson-van Cittert method, but it shows poor performance on noisy images. There is a trade-off between computation time and accepted noise. A further approach is being investigated which is based on a conjugate gradient algorithm proposed by Carrington [5]. This algorithm is expected to give restoration with high resolution.

Additional compensation is performed for other aberrations caused by mismatched indices of refraction of lenses and objects studied. This results in an image that is scaled along the axial direction. In case of small numerical aperture there is also an intensity drop of up to 40% every 10 μm.

3 Finding centres of ellipsoids using the wave-equation

Using the wave equation on gradient magnitude images has been proposed to detect axial [16] or rotational [15] symmetries in unsegmented gray-scale images. It has been demonstrated, that the combined use of the wave and the diffusion equation offers an efficient alternative to Hough-transformation to detect circles and to compute the 'symmetric axis transform' (SAT) of different objects. Figure 3 illustrates this process. At first, a gradient image is computed to find object boundaries in the image, and used as initial value for the wave propagation process. In the case of a circle two wavefronts will be created propagating into opposite directions. After a certain number of iterations the inner wavefront reaches the centre of the circle. This can easily be detected, since the

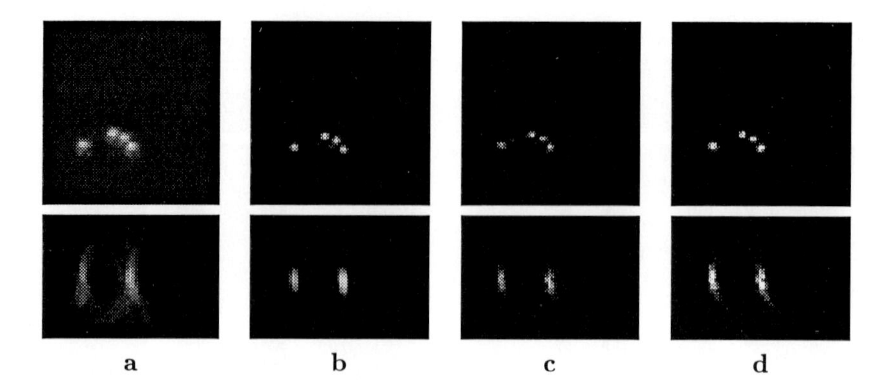

Figure 2: Restoration examples of different algorithms. The 3-D data of fluorescent beads shown in **a** was restored with the Wiener-filter (**b**), Jansson-van Cittert (**c**) and Gold (**d**) methods. **Top row**: x-y slices, **bottom row**: x-z slices.

wavefronts superpose with maximal intensity there. The constant propagation speed and the number of iterations needed to reach maximum intensity permits the computation of the radius of the circle.

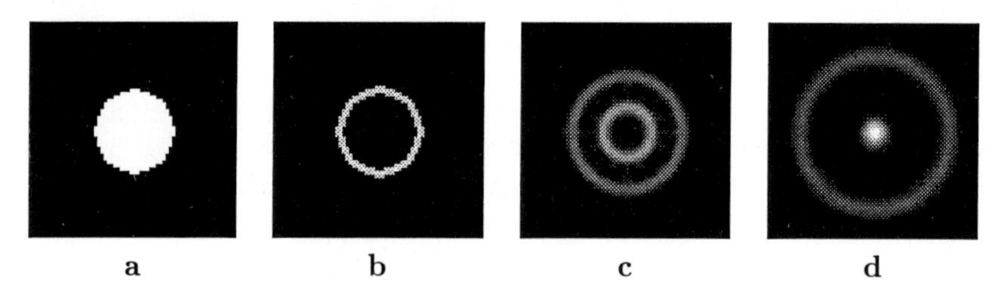

Figure 3: Detection example using the wave-equation. Image **a** shows the original image, **b** the gradient magnitude, **c** and **d** the wave propagation after 7 and 17 iterations, respectively.

We extended this procedure for 3D processing in order to recognize spheres or ellipsoids in volume data. The wave-equation is given by

$$\frac{\partial^2 u(\mathbf{x},t)}{\partial t^2} = c_w^2 \cdot \Delta u(\mathbf{x},t)$$

where Δ denotes the Laplacian-operator and c_w the propagation speed of the wave. The wave propagation has been combined with the diffusion-equation

$$\frac{\partial u(\mathbf{x},t)}{\partial t} = c_d \cdot \Delta u(\mathbf{x},t)$$

in order to make the process more stable and less sensitive to noise. The diffusion equation acts like the gravity on a surface wave in fluids, it makes the wavefront smoother and wider after each iteration step. If we set $\Delta x = \Delta y = \Delta z = 1$, the approximate solution of the mixed differential equation can be computed by the following algorithm:

```
repeat
    for all pixels
        laplace = ℒ ⋆ u
        u_t += Δt · c_w^2 · laplace
    for all pixels
        u += Δt · (u_t + c_d · laplace)
```

where $\mathcal{L}$ represents a small $3 \times 3 \times 3$ filter mask approximating the Laplacian operator.

In the case of confocal laser scanning microscopy images the sampling distance along the z-direction is different from those along the x- and y-directions resulting in strongly anisotropic data sets. In such cases the images either have to be axially rescaled, or an anisotropic Laplacian operator has to be calculated which allows for a different propagation speed along the axial direction. Considering that a rescaling of 3D images is a memory- and time-consuming process, we have chosen the more elegant second way. An anisotropic Laplacian operator has been developed which is able to overcome the problem caused by non-isotropic sampling. At the same time it allows to detect flat (elliptical) cell nuclei as shown in Figure 4.

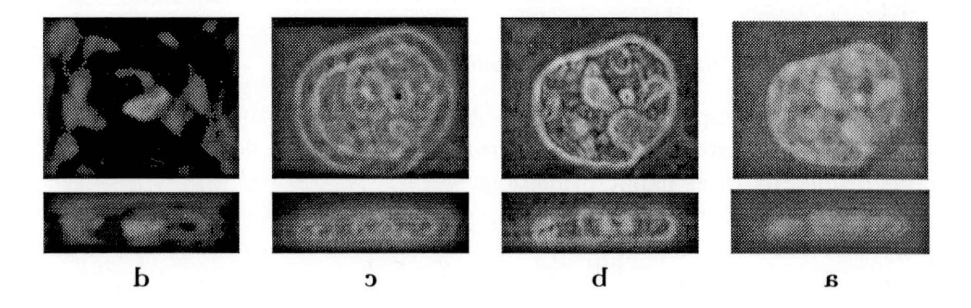

Figure 4: Detection of the centre of a cell nucleus (type v79). **a:** original gray-value images, **b:** edge detection, **c:** wave propagation after 10 and **d:** after 60 iteration steps. **Top row:** x-y sections, **bottom row:** x-z sections through volume data.

4 Segmentation of the cell nucleus

Elastically deformable surface models [7, 8] have been proven to be efficient tools for the segmentation of 3D images if a reasonable initialization can be provided. Parametric techniques as flexible Fourier surface models proposed by Staib and Duncan [9, 10] offer a convenient way for the elastic deformation of the ellipsoid

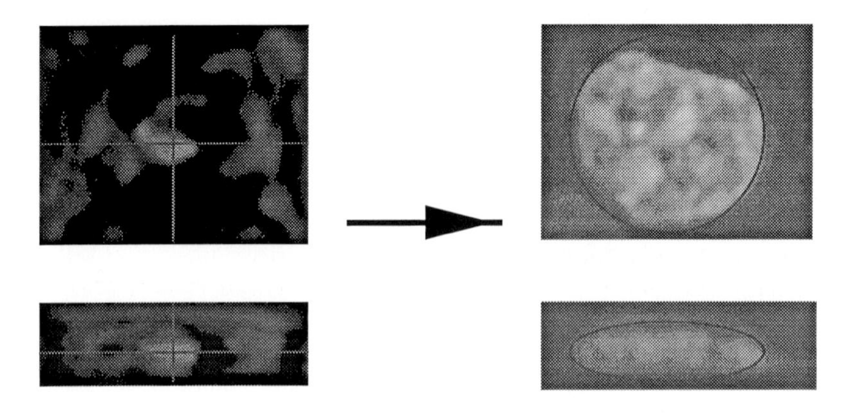

Figure 5: The detected center, the known propagation speed and number of iterations permit a determination of the surface of the cell nucleus by an ellipsoidal approximation. This ellipsoid is then used for the initialization of the elastic matching process.

resulting from the initialization step. The deformation is achieved by refining the parameters in a parameter space defined by the spherical harmonics as basis functions. In the following parts we first summarize the mathematics of 3D Fourier descriptors followed by a short description of the parametric deformable model applied for the image segmentation.

4.1 Parametric description of closed 3D surfaces

The surface of a simply connected 3D-object is parametrized by two variables, the θ and ϕ polar parameters, ranging from 0 to π and from 0 to 2π, respectively, and is defined by three explicit functions

$$\mathbf{r}(\theta,\phi) = \begin{pmatrix} x(\theta,\phi) \\ y(\theta,\phi) \\ z(\theta,\phi) \end{pmatrix}.$$

If we select the spherical harmonic functions (Y_l^m denotes the function of degree l and of order m, see [14]) as a basis, the coordinate functions can be written as

$$\mathbf{r}(\theta,\phi,\mathbf{p}) = \sum_{k=0}^{K} \sum_{m=-k}^{k} \mathbf{c}_k^m Y_k^m(\theta,\phi) \quad \text{, where} \quad \mathbf{c}_k^m = \begin{pmatrix} c_{x}{}_k^m \\ c_{y}{}_k^m \\ c_{z}{}_k^m \end{pmatrix}$$

We restricted the expansion to the first $K+1$ terms.

When the free variables θ and ϕ run over the whole sphere (e.g. $\theta = 0\ldots\pi, \pi = 0\ldots 2\pi$, the point $\mathbf{r}(\theta,\phi)$ moves over the whole surface of our object. The surface is described by the parameters

$$\mathbf{p}(c_{x}{}_0^0, c_{y}{}_0^0, c_{z}{}_0^0, c_{x}{}_1^{-1}, c_{x}{}_1^0, c_{x}{}_1^1, c_{y}{}_1^{-1}, c_{y}{}_1^0, c_{y}{}_1^1, c_{z}{}_1^{-1}, c_{z}{}_1^0, c_{z}{}_1^1, \ldots c_{x}{}_K^{-K} \ldots c_{z}{}_K^K)^T$$

The initializing ellipsoid surface can be described by the first four terms of the coordinate functions and can be easily calculated based on information from the initialization step.

4.2 3D Fourier Snakes

Elastically deformable contour and surface models as proposed by Witkin et.al. [13] locate the position of a curve or surface by minimization of an energy function compromising between the attraction to image features and internal deformation energy. The internal energy is responsible for the regularization of the resulting boundary. In a parametric notation, we search for optimal parameters of the boundary function $\mathbf{r}(\theta, \phi)$ which minimize the total energy

$$E(\mathbf{p}) = E(\mathbf{r}(\theta, \phi, \mathbf{p})) = E_I(\mathbf{r}(\theta, \phi, \mathbf{p})) + E_D(\mathbf{r}(\theta, \phi, \mathbf{p})).$$

By varying $\mathbf{p}$, the surface deforms itself to minimize the image energy

$$E_I(\mathbf{r}) = \int\int_A P(\mathbf{r}(\theta, \phi, \mathbf{p}))dA,$$

searching for an optimal position in the image described by the image potential $P(\mathbf{r}(\theta, \phi, \mathbf{p}))$. A typical choice takes $P(\mathbf{r}(\theta, \phi, \mathbf{p}))$ equal to the negative magnitude of the image gradient:

$$P(\mathbf{r}(\theta, \phi, \mathbf{p})) = -|\nabla I(\theta, \phi, \mathbf{p}))|,$$

where I is the smoothed image. The deformation term $E_D(\mathbf{r})$ serves as a regularization force. It restricts the expansion and bending of the snakes. In the case of Fourier surface models Staib and Duncan [9, 10] proposed an energy model which makes use of the normal direction $\mathbf{n}(\theta, \phi, \mathbf{p})$ of the parametrized surface and of the image gradient:

$$E_I(\mathbf{r}) = \pm \int\int_A \nabla I(\theta, \phi, \mathbf{p})) \cdot \mathbf{n}(\theta, \phi, \mathbf{p})dA.$$

The sign of the energy $E_I(\mathbf{r})$ decides whether bright objects are segmented on dark background or vice versa. The polarity of the boundary can be neglected by using the absolute value of the dot product in the integration term.

We applied the Fourier snake algorithm for the segmentation of the CLSM images. To segment the cell nuclei we used spherical harmonics up to degree 4 (75 descriptors), initialized by the 12 descriptors (degree 1) of the ellipsoid obtained from the wave equation process. Figure 6 shows the result of the segmentation on an example of the nucleus of a type v17 cell. A conjugate gradient technique has been applied for the optimization, which took 10 minutes on a SUN Sparc10 workstation. The resulting parametric models can be normalized to be invariant to translation, rotation, scaling and the starting point of the parametrization, resulting in an invariant descriptor of the shape of cell nuclei [11, 12].

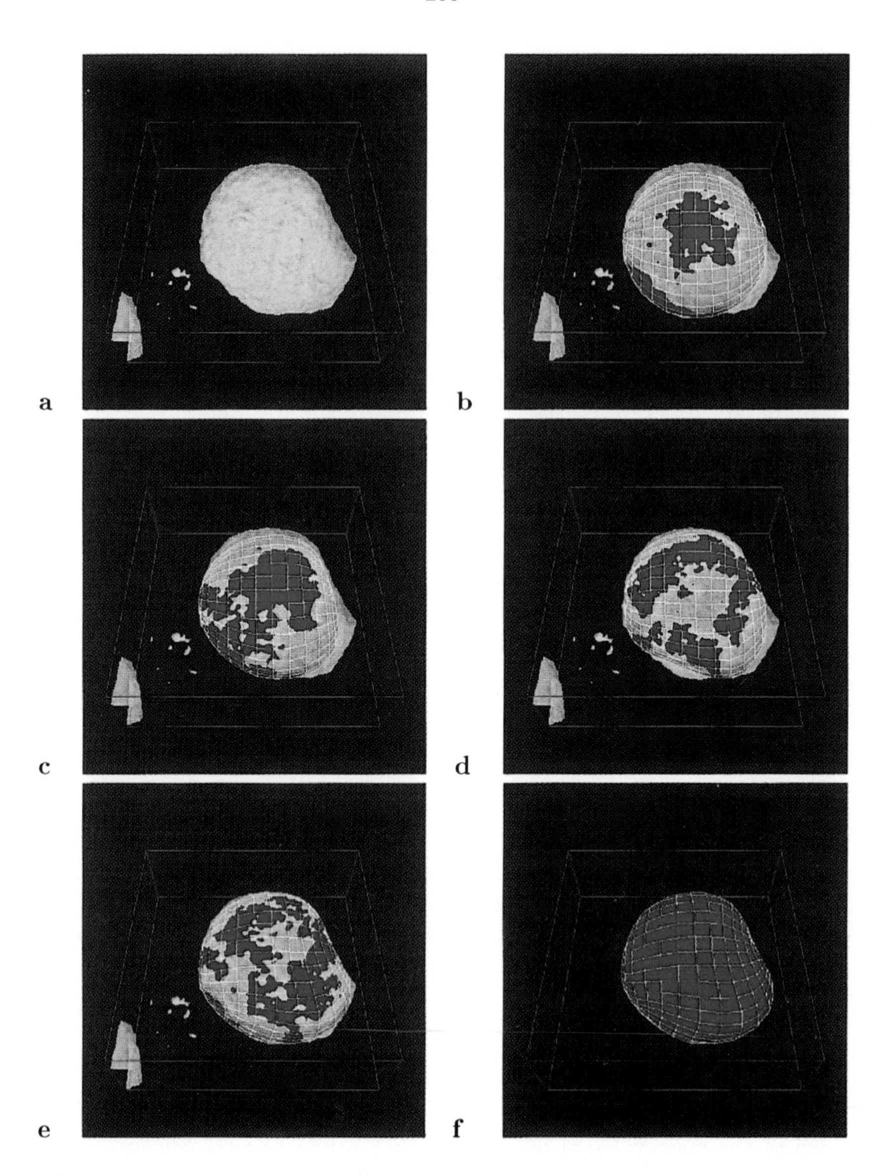

Figure 6: Segmentation example of a v79 cell nucleus; **a:** the manually segmented surface of the cell nucleus; **b:** initial ellipsoid determined by the wave-equation; **c** to **e:** iteration steps of the parameter optimization; **f:** the final result of the segmentation. The half transparent yellow color denotes the sought surface of the cell nucleus and dark blue represents the deformable 3D model.

5 Validation

Two different methods are planned to validate the accuracy of the segmentation results. In the first one we are embedding fluorescent micro-spheres of $10 \mu m$ diameter into a medium of similar refractive index to those of the spheres preventing lens effects caused by mismatched refractive indices. After taking confocal images of the spheres, the segmentation procedure described above will be carried out. The aberration of the result from the known spherical shape will provide us with an accuracy measure.

While the above method gives us a first approximation about the accuracy of the complete process the real experimental conditions are very different from that of the validation experiment. We plan to perform measurements under more realistic circumstances by comparing results from the evaluation of confocal images with independent light microscopy images. Since distortions caused by confocal imaging mainly occur along the z-axis, we intend to produce thin layers by axial slicing of frozen cell cultures and compare light microscopy images of them with the axial sections of 3D confocal data sets.

6 Conclusions

In this paper we presented a method for the fully automatic segmentation of cell nuclei from 3D confocal laser microscopy images of synchronous cells in the G_0/G_1 cell cycle stage. Previously proposed techniques have been reimplemented and refined for the requirements of this task. Different modules have been integrated into a fully automatic segmentation system. A new concept based on wave/diffusion equation has been developed for initializing the elastically deformable Fourier models. The resulting system allows us the simultaneous detection of many cell nuclei in the same 3D image without any manual intervention as required by the experimental conditions.

We presently apply the implemented procedures for the segmentation on a large number of CLSM images together with 2D microscopy data of the same cells. The normalized 2D contours and 3D cell nucleus boundaries will be used for the building and analysis of the statistical shape model of the cell nucleus which should enable us to estimate their thickness from the 2D contour data. This study will also provide information about the statistical variation of the model limiting the accuracy of the approximated intersection length.

References

[1] H. W. Reist et. al.: A new method to study low dose radiation damage, *Radiation Protection Dosimetry, Vol. 61*, No. 1-3, pp. 221-224, (1195)

[2] Peter J. Shaw and David J. Rawlins: The point-spread function of a confocal microscope: its measurement and use in deconvolution of 3-D data, *Journal of Microscopy, Vol 163*, pp. 151-165, (1991)

[3] Jose-Angel Conchello and Eric W. Hansen: Three-dimensional reconstrucion of noisy confocal scanning microscope images, *SPIE Vol 1161*, pp. 279-285, (1989)

[4] Jose-Angel Conchello and Eric W. Hansen: Enhanced 3-D reconstruction from confocal scanning microscope images 1: Deterministic and maximum likelihood reconstructions (1990)

[5] W. A. Carrington, R. M. Lynch, E. D. W. Moore, G. Isenberg, K. E. Fogarty, F. S. Fay: Superresolution Three-Dimensional Images of Fluorescence in Cells with Minimal Light Exposure, *SCIENCE, Vol. 268*, pp. 1483-1487, (1995)

[6] D. A. Agard, Y. Hiraoka, P. Shaw, J. W. Sedat: Fluorescence Microscopy in Three Dimensions, *Methods in Cell Biology, Vol. 30*, pp. 353-377. Academic Press, San Diego, (1989)

[7] Terzopoulos, D., Witkin, A. and Kass, M.: Symmetry-Seeking Models and 3D Object Reconstruction. *Int. J. Comp. Vision* 1, 3, 211–221 (1988)

[8] Cohen, I., Cohen, L.D. and Ayache, N.: Using Deformable Surfaces to Segment 3D Images and Infer Differential Structures. *CVGIP: Image Understanding* 56, 2, 242–263 (1992)

[9] L. H. Staib and J. S. Duncan: Boundary finding with parametrically deformable models. *IEEE PAMI 14*, 11, pp. 1061-1075, (1992)

[10] L. H. Staib and J. S. Duncan: Deformable Fourier models for surface finding in 3D images. *in Proc. VBC'92 Conf.*, pp. 190-194, (1992)

[11] C. Brechbühler, G. Gerig and O. Kübler: Surface parametrization and shape description., *Proc VBC'92 Conf.*, pp. 80-89, (1992)

[12] C. Brechbühler, G. Gerig and O. Kübler: Parametrization of closed surfaces for 3D shape description. *CVGIP: Image Understanding*, 61, 2, pp. 154-170, (1995)

[13] M. Kass, A. Witkin and D. Terzopoulos: Snakes: Active contour models, *Int. J. Comp. Vision*, 1, 4, pp. 321-331, (1988)

[14] W. Greiner and H. Diehl: Theoretische Physik – Ein Lehr- und Übungsbuch für Anfangssemester, Band 3: Elektrodynamik, 61-65, Verlag Harri Deutsch, Zürich und Frankfurt am Main, (1964)

[15] K. Hanahara and M. Hiyane: A Circle-Detection Algorithm Simulating Wave Propagation, *Machine Vision and Applications*, 3, pp. 97-111, (1990)

[16] G. L. Scott, S. C. Turner and A. Zisserman: Using a mixed wave/diffusion process to elicit the symmetry set, *Image and Vision Computing*, vol 7, pp. 63-70, (1989)

Evaluation of Segmentation Algorithms for Intravascular Ultrasound Images

Carolien J. Bouma[1], Wiro J. Niessen[1], Karel J. Zuiderveld[1], Elma J. Gussenhoven[2], Max A. Viergever[1]

[1] Imaging Center, University Hospital Utrecht, The Netherlands
[2] Erasmus University Rotterdam, The Netherlands

Abstract

A large number of paradigms to segment 30 MHz intravascular ultrasound (IVUS) subtraction images is evaluated. A fully automated active contour model, initialized by thresholding and preceded by filtering with a large scale median filter is a good alternative for manual delineation of both pre-dilation and post-dilation images. Results obtained with this automated method are even better than manual segmentation, which is a major step forwards in IVUS image analysis.

1 Introduction

One prerequisite for routine clinical use of intravascular ultrasound (IVUS) is the rapid availability of diagnostically relevant data such as lumen size and stenotic index. To extract these data from the IVUS images, (automated) lumen boundary detection is necessary, and this detection has to be both quick and reliable. Only few studies to obtain the lumen boundaries have been reported.

In first studies lumen boundary detection of IVUS images was performed manually [1, 2], by thresholding [3, 4] or by region growing methods [5]. These studies did not pay attention to the quality of the segmentations. In more recent studies contour models were developed to define the lumen boundary [8, 6, 7]. None of these methods is fully automated. Sonka *et al.* [8] obtained in vitro IVUS images which could be properly segmented in 105 out of 127 images. In some of these images minor editing was required. The quality of the segmentations was assessed by visual inspection. In the contour model developed by Li *et al.* [6, 7] a good agreement between segmented images and true dimensions was found in a circular phantom. Until now, neither of these methods has been validated in vivo.

The purpose of our study is to investigate whether it is possible to automatically determine the lumen boundary from IVUS images, with a quality similar to manual segmentation. Conventional IVUS images obtained by 30 MHz ultrasound probes contain a high contribution of backscatter of blood in the lumen, which precludes automatic segmentation of the vessel lumen. While the backscatter of blood can be reduced by flushing the lumen with saline, this procedure presents a potential risk for the patient. We therefore use subtraction of a series of IVUS images obtained at one location in the vessel, a method originally proposed by Pasterkamp *et al.* [9]. In the resulting subtraction images, the vessel wall is suppressed while the lumen highlights.

We have applied 80 combinations of smoothing techniques and (semi-)automatic object definition methods to the IVUS subtraction images, and compared the results with manual tracings of the lumen boundary produced by four experienced observers. The results have been evaluated in 15 femoral artery images.

2 Data acquisition

We use a 30 MHz 4.2 F flexible IVUS catheter (Du-MED, Rotterdam, The Netherlands). The axial resolution of the system is 0.1 mm and the lateral resolution is better than 225 μm at a depth of 1 mm. To obtain accurate relative distance over the trajectory of interest, the catheter is led through a small displacement sensing device [10] which can obtain an accuracy of 0.1 mm.

Images stored on sVHS are digitized using an IBM PC-based framegrabber (512 $\times$ 512 pixels with 256 grey-levels). Data acquisition of subtraction images is performed off-line with TCL-image software (Multihouse, The Netherlands). Subsequent analysis is performed on a Hewlett Packard 9000/720 UNIX workstation.

Subtraction images are obtained by subtracting two consecutive images in time at one position (determined by the displacement sensing device). A total of 20 subtraction images, obtained at the same position in the blood vessel, is averaged and the resulting image is used for further processing.

For the present study 15 IVUS recordings obtained from the femoral artery of 12 patients are used. Of these, 7 images are obtained before balloon dilation (pre-dilation) and 8 images are obtained after balloon dilation was performed (post-dilation).

3 Segmentation

The segmentation paradigm consists of three steps. First, the catheter is replaced by lumen-equivalent texture (Fig. 1b). This step is necessary since in the subtraction images the area of the catheter appears as a dark spot while it does belong to the lumen (Fig. 1a). Second, the image is smoothed. This step is necessary because there is a large amount of noise in the images which - if not filtered out - hinders proper object definition (Fig. 1c). And third, the lumen is delineated by an object definition method (Fig. 1d). Combinations of different filters and different object definition methods result in a total of 80 different segmentation methods. The abovementioned steps are explained in more detail below.

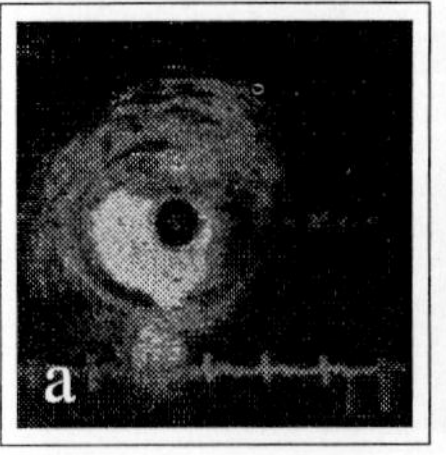

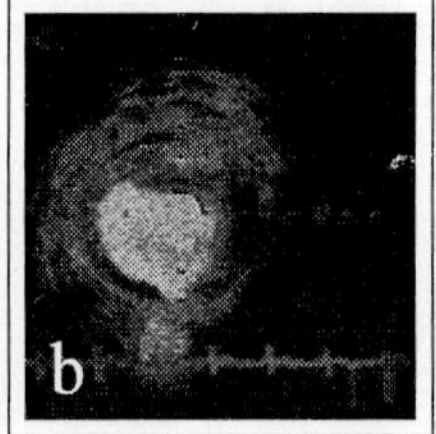

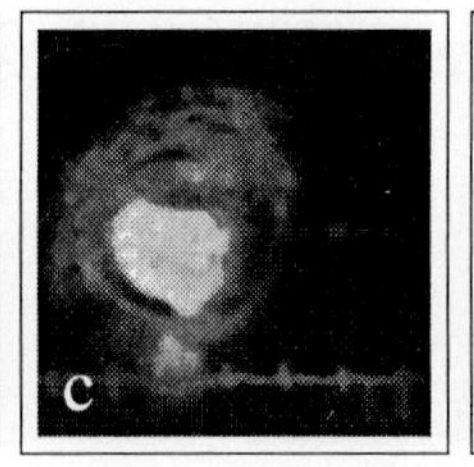

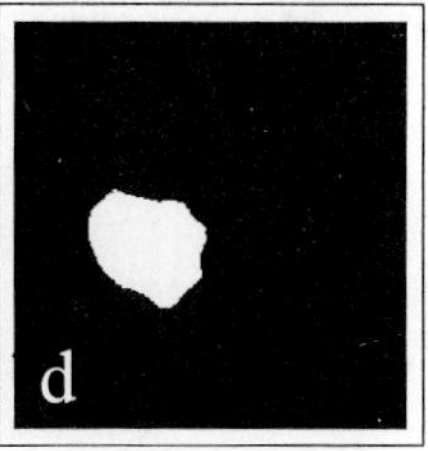

Figure 1: An example of the subsequent steps of a segmentation paradigm. a. An original IVUS subtraction image. The lumen highlights while the vesselwall is suppressed. A dark area in the lumen represents the catheter. b. The catheter is automatically filled with lumen texture. c. Smoothing of the image with a median filter with an edge length of 7 pixels. d. Object definition by means of a dynamic contour model. The initial contour is obtained by thresholding.

3.1 Filling of the catheter

Not only the catheter, but also an area between the catheter and the lumen does not contain lumen-equivalent texture. We therefore designed an automatic algorithm to fill these areas with lumen texture. First, we automatically determine the mean grey-value of lumen pixels. Out of 16 radial regions surrounding the catheter, which are large enough to be able to perform statistical analysis, and small enough that the region with the highest mean value contains only lumen pixels, we select the region with the highest mean grey-value. Secondly, a threshold, which defines whether a pixel is classified as a lumen pixel, is set to the mean grey-value minus two times the standard deviation in that region. Empirical evidence shows that this criterion performs well. Thirdly, we use this threshold to determine which pixels, besides the catheter pixels, should be filled with lumen texture. We therefore search radially in a small area around the catheter. When we meet lumen in a direction, the pixels between the edge of the catheter and the pixel with lumen are labeled. After having smoothed the resulting object (catheter plus labeled pixels) by means of morphological operations, and upon removing discontinuities between the lumen and the catheter by use of tangent information, we have obtained all pixels to be filled with lumen equivalent texture (see Fig. 2). As a final step these pixels are randomly filled out of an array containing all pixels in the dataset which are classified as lumen texture pixel. This method of catheter filling results in a texture somewhat different from lumen texture. For our purpose, segmentation of the lumen, this is not important because the images are smoothed, which strongly reduces the difference between lumen texture and texture in the catheter area (see Fig. 3). This heuristic procedure is fully automatic and yields satisfactory results in all of our tested cases; the segmentation errors obtained in the neighborhood of the catheters are within the variations of the manual segmentations.

3.2 Smoothing

To obtain images suitable for (semi-)automated segmentation, noise reduction is performed by filtering the input images. Three different filters, a linear Gaussian filter, a median filter, and a non-linear diffusion operator (Euclidean Shortening Flow, ESF) [11, 12, 13, 14] are used (Fig. 3), each at three or four different scales (*i.e.* width of the filter window around a pixel). Convolution with the Gaussian filter models an isotropic

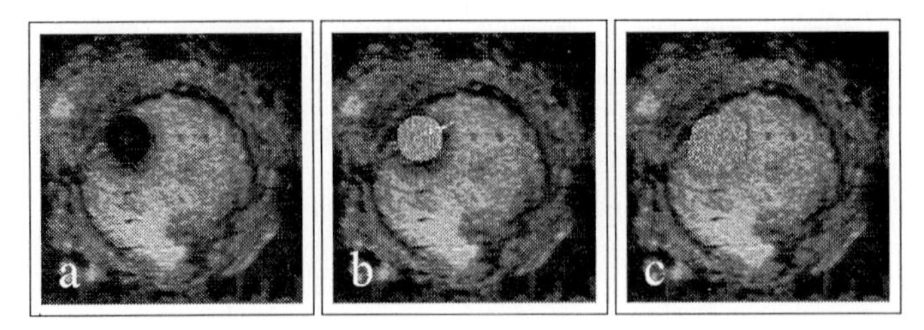

Figure 2: Typical result of the procedure to fill the catheter with lumen texture. a. Original subtraction image. b. Filling of only the catheter. A small dark area surrounding the catheter remains. c. Filling of the small area surrounding the catheter. Discontinuities at the transitions between lumen and vessel wall are removed by use of tangent information. A sufficiently smooth transition between the lumen and the catheter is finally achieved at the lumen borders.

linear smoothing process, *i.e.* blurring occurs equally in all directions, independent of the image structure present. In contrast, non-linear diffusion as proposed by Catté et al. [11] and Alvarez et al. [12] only smoothes in the local tangent direction; no blurring in the gradient direction occurs. In this way a high level of smoothing can be obtained in the lumen, with a minimal amount of smoothing between lumen and vessel wall. The median filters also smoothes while preserving edges, by replacing a pixel with the median of pixels in the filter window.

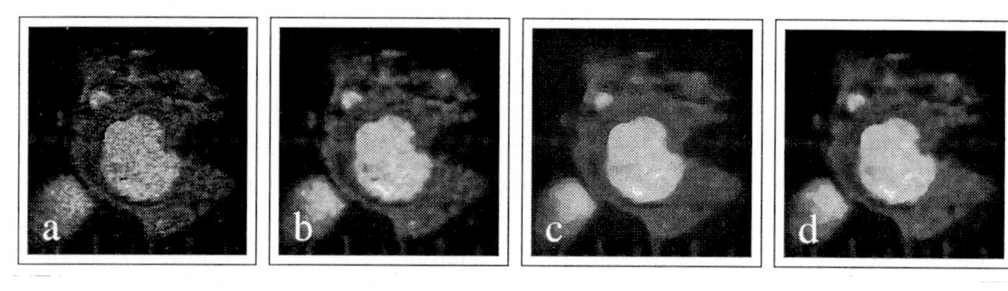

Figure 3: Examples of different preprocessing filters. a. original image. b. linear Gaussian filter with a spatial scale of 1 pixel, c. Euclidean Shortening Flow with 4 iterations, d. Median filter with window length of 5 pixels.

3.3 Object definition methods

The first four object definition methods are combinations of grey-level thresholding and a morphological operation. In order to obtain an independent threshold range, the threshold value chosen is situated in the middle of the second and the third peak of the grey value histogram. The first peak in the histogram represents the background of the image, the second peak represents the vessel wall and the third peak represents the moving particles of the image, *i.e.* the lumen.

Segmentation by thresholding gives some artifacts, mostly containing of small clusters of pixels. These small artifacts disappear with morphological filtering [15, 16]. Four different morphological operators are used: 1. Closing (dilation followed by erosion); 2. Closing followed by opening (opening is erosion followed by dilation); 3. Opening; and 4. Opening followed by closing. For all morphological operations, a square structuring element of 3 × 3 pixels was used (3 pixels is equal to 0.2 mm, the resolution of the imaging device). Morphological filtering preserves size and shape of the luminal regions, while small artifacts (smaller than the resolution of the imaging system) disappear. Remaining regions that do not belong to the lumen, can be simply removed by a connecting technique starting from the catheter.

Region growing, the fifth (semi-automated) method studied, requires the initialization of a seedpoint at the place of the catheter. This method performs threshold based segmentation by connecting all pixels within a specified threshold range of the seed pixel, and then automatically draws a trace around the connected region. The threshold range can be adapted to arrive at a best - by visual inspection - lumen definition.

In the sixth object definition method, the discrete dynamic contour model [17], a human operator outlines a rough initial shape of the lumen contour. From the initial shape, the dynamic contour model actively modifies the shape, until it approximates the desired contour. The contour model requires an edge-image as input which we calculate by means of the squared gradient magnitude for the smoothed images at a number of scales.

The dynamic contour model is a semi-automatic object definition method. It can be made fully automatic by taking the output of a threshold segmentation as input of the active contour model. The resulting automated method is the seventh object definition method that is evaluated in the course of this study.

4 Validation

To validate the (semi)-automated segmentation methods, the results should be compared with the real lumen shape. Because the actual shape is not known, a 'gold standard' is obtained from manual segmentations.

Manual segmentation of the lumen is performed on original subtraction images, *i.e.* without filtering operations, by four observers. Each observer traced the contour of the lumen twice with a period of one week between the tracings. The observers are one physician, two medical biologists and one clinical physicist, all having a long experience with IVUS images.

The manual segmentation results show a large variability between different observers and sometimes also between the two traces of one observer, so a single manual segmentation can not act as a standard. To obtain a 'gold standard' the average shape of the manual segmentations is calculated using shape-based interpolation [18, 19]. This technique uses the shapes of the binary object in each manual segmentation to calculate so-called distance scenes; interpolation of the distance scenes from the other manual segmentations is then used to calculate the average object shape.

The results of all segmentation methods, including the manual segmentations, are subsequently compared to the reference images. To this end the squared distance images of all segmentation methods are calculated as described by Zuiderveld [20]. The contour of the segmented lumen is used as the object, and the squared distance of each pixel in the image to the contour is calculated. The squared distance is used as a measure for the quality of the segmented image, because small deviations in segmentation over a large part of the lumen will not have a large influence in clinical decision making, but large deviations at a particular place, for example missing a dissection, can significantly affect the decision. Subsequently, the squared distance of each contour pixel in a segmented image to the nearest contour pixel in the reference image is obtained using the distance image. Finally, the mean squared distance (*msd*) of the segmentation result is calculated from these values.

For quantitative assessment of the results pre-dilation and post-dilation images are treated separately, because variations found in the manual segmentations are much larger for post-dilation images than for pre-dilation images. These larger variations for post-dilation images may be ascribed to the more irregular lumen boundaries in most of these images as a result of balloon dilation. The validation method assigns as a quality measure to a segmentation method the average *msd* of that method over all images. This average *msd* is compared to the average *msd* of all manual segmentations over all images.

5 Results

The manual segmentations show a large variability between different observers and occasionally also between the two traces of one observer (Table 1). The mean of the *msd* of all observers is 1.17 (range 0.54 - 1.63) for the images of vessel segments which are not balloon dilated (pre-dilation) and 10.86 (range 1.53 - 36.3) for the images of balloon dilated vessel segments (post-dilation).

	pre-dilation		*post-dilation*	
observer	msd *(per observer)*	msd *(all observations)*	msd *(per observer)*	msd *(all observations)*
1a	*1.30 (0.78–2.36)*		*36.3 (1.18–127)*	
1b	*1.23 (0.53–2.99)*		*25.5 (0.63–152)*	
2a	*0.54 (0.25–0.86)*		*1.64 (0.50–6.59)*	
2b	*0.80 (0.37–1.20)*	*1.17*	*1.53 (0.46–3.19)*	*10.86*
3a	*1.48 (0.60–2.22)*		*10.1 (1.50–52.2)*	
3b	*1.27 (0.46–3.62)*		*6.31 (0.78–28.4)*	
4a	*1.07 (0.59–2.03)*		*3.01 (0.75–13.0)*	
4b	*1.63 (0.82–3.95)*		*2.55 (0.63–11.4)*	

Table 1: Msd's of manual segmentations for different observers for the 7 pre-dilation images and the 8 post-dilation images.

The results for all (semi-) automated segmentation methods are shown in tables 2 and 3. For both pre-dilation and post-dilation images nearly all methods using thresholding or region growing for object definition give worse results than segmentation methods using the contour model. Segmentation methods using region growing give slightly better results for pre-dilation images than methods using thresholding. For post-dilation images, methods using thresholding give slightly better results than methods using region growing for object definition. For pre-dilation images only two segmentation methods using region growing for object definition are better than manual segmentation. None of the methods using thresholding is better than manual segmentation. The methods using a contour model for object definition give better results than manual segmentation in 9 out of 15 methods when the initial contour is obtained manually and in 10 out of 15 when the initial contour is obtained by thresholding. Smoothing with a median filter results in the best segmentations. For post-dilation images nearly all methods using thresholding or region growing and all methods using the contour model for object definition are better than manual segmentation.

filters	*pre-dilation mean square distance (range)*				
	Close	*CloseOpen*	*Open*	*OpenClose*	*region grow*
L0.8	*1.75 (0.50–4.67)*	*1.70 (0.47–4.83)*	*2.07 (0.53–6.36)*	*1.88 (0.49–5.82)*	*1.46 (0.53–2.69)*
L1.0	*1.52 (0.47–3.73)*	*1.47 (0.44–3.84)*	*1.55 (0.46–4.24)*	*1.51 (0.43–4.12)*	*1.46 (0.49–2.90)*
L1.5	*1.50 (0.43–3.71*	*1.47 (0.43–3.52)*	*1.49 (0.43–3.55)*	*1.47 (0.43–3.55)*	*1.31 (0.49–3.21)*
ESF4	*1.44 (0.47–3.57)*	*1.41 (0.47–3.61)*	*1.68 (0.49–4.84)*	*1.64 (0.49–4.76)*	*1.31 (0.53–3.73)*
ESF8	*1.58 (0.43–4.30)*	*1.56 (0.43–4.34)*	*1.61 (0.54–4.75)*	*1.61 (0.54–4.75)*	*1.14 (0.44–2.77)*
ESF16	*1.83 (0.67–5.11)*	*1.83 (0.67–5.11)*	*1.83 (0.69–5.30)*	*1.82 (0.69–5.26)*	*1.28 (0.40–3.30)*
Med3	*1.66 (0.65–3.81)*	*1.63 (0.62–4.02)*	*1.93 (0.70–5.28)*	*1.81 (0.70–4.73)*	*1.75 (0.80–4.76)*
Med5	*1.49 (0.59–3.54)*	*1.48 (0.57–3.58)*	*1.53 (0.57–3.64)*	*1.49 (0.57–3.58)*	*1.41 (0.56–4.30)*
Med7	*1.50 (0.54–3.51)*	*1.50 (0.54–3.57)*	*1.52 (0.54–3.65)*	*1.51 (0.54–3.60)*	*1.19 (0.45–3.19)*
Med9	*1.50 (0.52–3.48)*	*1.50 (0.52–3.48)*	*1.53 (0.52–3.51)*	*1.52 (0.52–3.48)*	*1.07 (0.51–2.54)*

filters	*post-dilation mean square distance (range)*				
	Close	*CloseOpen*	*Open*	*OpenClose*	*region grow*
L0.8	*7.69 (1.69–19.4)*	*6.65 (1.56–13.1)*	*9.55 (1.59–17.5)*	*7.71 (1.52–14.5)*	*10.5 (1.68–30.1)*
L1.0	*6.42 (1.74–11.2)*	*6.69 (1.72–12.9)*	*8.13 (1.80–15.1)*	*7.08 (1.71–13.5)*	*10.3 (1.60–30.9)*
L1.5	*9.47 (1.59–32.5)*	*9.82 (1.57–34.2)*	*6.53 (1.60–14.1)*	*9.54 (1.57–32.4)*	*9.34 (1.37–32.0)*
ESF4	*11.1 (1.46–40.5)*	*5.97 (1.41–15.6)*	*7.99 (1.99–18.5)*	*7.37 (1.94–16.4)*	*9.36 (1.05–29.5)*
ESF8	*5.66 (1.16–14.9)*	*5.50 (1.18–13.4)*	*6.38 (1.43–11.7)*	*6.08 (1.38–12.0)*	*8.68 (0.78–31.9)*
ESF16	*7.65 (1.22–24.5)*	*7.66 (1.22–24.5)*	*5.44 (1.30–11.6)*	*5.41 (1.30–11.3)*	*7.34 (0.92–20.5)*
Med3	*11.2 (1.89–42.0)*	*10.8 (1.64–36.6)*	*9.21 (1.79–18.5)*	*11.5 (1.72–35.9)*	*11.7 (2.39–30.5)*
Med5	*10.7 (1.40–36.7)*	*10.0 (1.18–32.1)*	*7.28 (1.20–14.4)*	*6.92 (1.20–12.4)*	*6.30 (1.04–17.4)*
Med7	*9.19 (1.13–31.7)*	*9.32 (1.16–32.7)*	*9.65 (1.17–32.8)*	*9.19 (1.16–31.6)*	*9.58 (0.96–32.4)*
Med9	*8.61 (0.91–30.7)*	*8.33 (0.95–28.3)*	*8.40 (0.91–28.3)*	*8.37 (0.95–28.4)*	*9.27 (0.84–35.2)*

Table 2: Mean square distances to reference images of the thresholding and region growing methods. L0.8, L1.0, L1.5 = linear Gaussian filter with a spatial scale equal to 0.8, 1.0 and 1.5 pixels respectively, ESF4, ESF8, ESF16 = Euclidean Shortening Flow with 4, 8 and 16 iterations respectively. Med3, Med5, Med7, Med9 = Median filter with a window length of 3, 5, 7. and 9 pixels respectively.

For pre-dilation images the best segmentation result is obtained by filtering the image with a median filter with a window length of 9 pixels, followed by measurement of the image gradient at scale 1.5, and subsequent object definition by a contour model with a manual initial contour. When the initial contour is obtained from thresholding the results are slightly worse, but this method is nonetheless preferred, because it is completely automated.

For post-dilation images the smallest *msd* is obtained by filtering the image with a median filter with a window length of 9 pixels, followed by measurement of the image gradient at scale 1.0, and subsequent object definition by a contour model with an initial contour obtained from thresholding.

filters	*pre-dilation mean square distance (range)*		*post-dilation mean square distance (range)*	
	contour	*thr + contour*	*contour*	*thr + contour*
G2.5	*1.09 (0.43–2.12)*	*1.14 (0.44–2.14)*	*4.79 (0.87–11.1)*	*4.80 (1.01–11.4)*
G3.0	*1.12 (0.39–1.97)*	*1.11 (0.47–1.94)*	*5.66 (1.10–19.7)*	*4.69 (1.14–12.0)*
G3.5	*1.21 (0.42–2.35)*	*1.17 (0.46–2.31)*	*5.49 (1.02–18.1)*	*4.93 (1.24–13.1)*
ESF4 G1.0	*1.63 (0.45–5.06)*	*1.25 (0.54–2.51)*	*8.78 (1.23–44.1)*	*4.33 (0.83–9.75)*
ESF4 G1.5	*1.13 (0.39–2.24)*	*1.21 (0.59–2.27)*	*9.23 (0.97–47.1)*	*4.36 (1.03–10.7)*
ESF8 G1.0	*1.21 (0.49–2.76)*	*1.26 (0.56–2.53)*	*9.03 (0.83–46.5)*	*4.66 (0.92–11.2)*
ESF8 G1.5	*1.07 (0.34–2.11)*	*1.16 (0.43–2.47)*	*8.92 (0.99–45.8)*	*4.17 (1.02–9.70)*
ESF16 G1.0	*1.61 (0.51–4.94)*	*1.28 (0.48–2.90)*	*7.99 (0.95–38.4)*	*4.63 (1.11–9.81)*
ESF16 G1.5	*1.32 (0.39–3.01)*	*1.27 (0.45–2.79)*	*7.54 (1.22–36.1)*	*4.43 (1.24–10.3)*
Med5 G1.0	*1.03 (0.49–1.93)*	*1.17 (0.44–2.47)*	*8.29 (0.81–40.1)*	*4.85 (0.85–11.6)*
Med5 G1.5	*1.12 (0.48–2.09)*	*1.11 (0.57–2.28)*	*8.30 (0.95–41.3)*	*4.60 (0.89–11.3)*
Med7 G1.0	*1.12 (0.35–2.14)*	*1.13 (0.48–2.19)*	*7.26 (0.70–32.8)*	*4.65 (0.76–11.5)*
Med7 G1.5	*1.10 (0.45–1.92)*	*1.03 (0.48–2.10)*	*7.20 (0.84–31.7)*	*4.43 (0.99–11.4)*
Med9 G1.0	*1.03 (0.45–2.01)*	*1.08 (0.46–2.10)*	*7.63 (0.83–34.2)*	*3.58 (0.97–11.3)*
Med9 G1.5	*0.98 (0.38–2.04)*	*1.02 (0.44–2.02)*	*7.34 (0.82–33.7)*	*3.99 (0.91–9.47)*

Table 3: Mean square distances to reference images of the contour methods. The initial contour is obtained from manual editing in the first and third column, and from thresholding in the second and fourth column. G1.0, G1.5 G2.5, G3.0 G3.5 = linear Gaussian squared gradient with a spatial scale equal to 1.0, 1.5, 2.5, 3.0 and 3.5 pixels respectively, ESF4, ESF8, ESF16 = Euclidean Shortening Flow with 4, 8 and 16 iterations respectively. Med5, Med7, Med9 = Median filter with window length of 5, 7, 9 pixels respectively. An initial blurring with a Euclidean Shortening Flow followed by measurement of the image intensity gradient was indicated for example as ESF4 G1.0.

6 Discussion

Almost all IVUS studies detecting lumen boundaries were performed on original IVUS images. In IVUS images obtained from 20 MHz systems there is little backscatter of blood in the lumen, and lumen segmentation can be performed on original IVUS images by edge detection methods or thresholding if the entire vessel wall is visible. In the present study a 30 MHz ultrasound catheter is used. The high frequency causes a blood-scatter effect, which sometimes precludes discrimination between lumen and vessel wall in these IVUS images. To obtain better discrimination between lumen and vessel wall Li et al. [21] averaged several subsequent images. Using this method they found an intra-observer variability of 8.8% in lumen boundary detection in single-frame images and 3.6% when images were averaged. Thus averaging increases discrimination between lumen and vessel wall. In other studies, flushing with saline was performed [22, 2] or the vessel was injected with contrast media [4, 23] for better discrimination between lumen and vessel wall.

In the present study, subtraction images of IVUS are used for segmentation of the vessel lumen. The most important advantage of subtraction images is that dissections of the vessel wall can easily be detected; this is even true for dissections which can not be detected in flushed images or averaged images [24]. On the other hand, the subtraction method has several limitations. First, by use of subtraction images information about the vessel wall will be lost. This problem can be solved by overlaying the lumen contour as obtained from a subtraction image onto the original IVUS image. Secondly, the subtraction technique will not work when the vessel pulsates, unless the images used for subtraction are triggered. Fortunately, severely atherosclerotic vessels, which are the majority of the vessels studied by IVUS, pulsate only to a limited extent. Finally, the interventional procedure will take more time, because several ultrasound images should be obtained at one position for measurement of one subtraction image. This disadvantage cannot be remedied and, consequently, is the most significant of all.

7 Conclusion

The large variations in manual segmentation show that manual segmentation of ultrasound images is not straightforward. In this study it is shown that out of a large number of smoothing techniques and object definition methods a median filter with a rather large edge length (7 or 9 pixels) combined with a contour model with an initial contour obtained from thresholding gives the best results for both pre-dilation and post-dilation images. This fully automated method is at least a good alternative for manual segmentation and is possibly even preferable to manual segmentation. Support of the latter conjecture requires an evaluation with a substantially larger number of observers and a larger number of images. The purpose of the present study was to show that automated lumen representation of IVUS images was possible, so as to pave the way for clinical use of IVUS. This objective has been accomplished.

Acknowledgments
This research is supported by the Technology Foundation of the Netherlands (STW), the Netherlands Heart Foundation and the Interuniversity Cardiology Institute, the Netherlands. The authors would like to thank dr. Koen Vincken for his part in the software development.

References

[1] S. E. Nissen, J. C. Gurley, C. L. Grines, D. C. Booth, R. McClure, M. Berk, C. Fischer, and A. N. DeMaria, "Intravascular ultrasound assessment of lumen size and wall morphology in normal subjects and patients with coronary artery disease," *Circulation*, vol. 84, pp. 1087–1099, 1991.

[2] D. W. Losordo, K. Rosenfield, A. Pieczek, K. Baker, M. Harding, and J. M. Isner, "How does angioplasty work? serial analysis of human iliac arteries using intravascular ultrasound," *Circulation*, vol. 86, pp. 1845–1858, 1992.

[3] K. M. Coy, J. C. Park, M. C. Fishbein, T. Laas, G. A. Diamond, L. Adler, G. Maurer, and R. J. Siegel, "In vitro validation of three dimensional intravascular ultrasound for the evaluation of arterial injury after balloon angioplasty," *Journal of the American College of Cardiology*, vol. 20, pp. 692–700, 1992.

[4] K. Rosenfield, D. W. Losordo, K. Ramaswamy, J. O. Pastore, R. E. Langevin Jr, S. Razvi, and J. M. Isner, "Three-dimensional reconstruction of human coronary and peripheral arteries from images recorded during two-dimensional intravascular ultrasound examination," *Circulation*, vol. 84, pp. 1938–1956, 1991.

[5] K. Rosenfield, J. Kaufman, A. M. Pieczek, R. E. Langevin, P. E. Palefski, S. A. Razvi, and J. M. Isner, "Human coronary and peripheral arteries: On-line three-dimensional reconstruction from two-dimensional intravascular US scans," *Radiology*, vol. 184, pp. 823–832, 1992.

[6] W. Li, J. G. Bosch, Y. Zhong, H. van Urk, E. J. Gussenhoven, F. Mastik, F. van Egmond, H. Rijsterborgh, J. H. C. Reiber, and N. Bom, "Image segmentation and 3d reconstruction of intravascular ultrasound images," *Acoustical Imaging*, vol. 20, pp. 489–496, 1993.

[7] W. Li, C. von Birgelen, C. Di Mario, E. Boersma, E. J. Gussenhoven, N. van der Putten, and N. Bom, "Semi-automatic contour detection for volumetric quantification of intracoronary ultrasound," IEEE Computers in Cardiology, (Los Alamitos), pp. 277–280, IEEE Computer Society Press, 1994.

[8] M. Sonka, X. Zhang, M. Siebes, R. R. Chada, R. McKay, and S. M. Collins, "Automated detection of wall plaque borders in intravascular ultrasound images," in *proceedings of SPIE*, vol. 2168, (Bellingham), SPIE Press, 1994.

[9] G. Pasterkamp, M. S. van der Heiden, M. J. Post, B. M. ter Haar Romeny, W. P. T. M. Mali, and C. Borst, "Discrimination of the intravascular lumen and dissections in a single 30 MHz US image: Use of "confounding" blood backscatter to advantage," *Radiology*, vol. 187, pp. 871–872, 1993.

[10] E. J. Gussenhoven, A. van der Lugt, M. van Strijen, L. Wenguang, H. Kroeze, S. H. K. The, F. C. van Egmond, J. Honkoop, R. J. G. Peters, P. de Feyter, H. van Urk, and H. Pieterman, "Displacement sensing device enabling accurate documentation of catheter tip position," in *Intravascular Ultrasound* (J. Roelandt, E. Gussenhoven, and N. Bom, eds.), vol. 143, pp. 157–163, Dordrecht: Kluwer Academic Publishers, 1993.

[11] F. Catté, P.-L. Lions, J.-M. Morel, and T. Coll, "Image selective smoothing and edge detection by nonlinear diffusion," *SIAM Journal on Numerical Analysis*, vol. 29, no. 1, pp. 182–193, 1992.

[12] L. Alvarez, P.-L. Lions, and J.-M. Morel, "Image selective smoothing and edge detection by nonlinear diffusion. II," *SIAM Journal on Numerical Analysis*, vol. 29, no. 3, pp. 845–866, 1992.

[13] W. J. Niessen, B. M. ter Haar Romeny, and M. A. Viergever, "Numerical analysis of geometry-driven diffusion equations," in *Geometry-Driven Diffusion in Computer Vision* (B. M. ter Haar Romeny, ed.), vol. 1 of *Computational Imaging and Vision*, pp. 393–410, Dordrecht: Kluwer Academic Publishers, 1994.

[14] W. J. Niessen, B. M. ter Haar Romeny, L. M. J. Florack, and M. A. Viergever, "A general framework for geometry-driven evolution equations," *International Journal of Computer Vision*, 1996. In print.

[15] J. Serra, *Image analysis and mathematical morphology*. San Diego: Academic Press, 1988.

[16] R. M. Haralick, S. R. Sternberg, and X. Zhuang, "Image analysis using mathematical morphology," *IEEE Transactions on Pattern Analysis and Machine Intelligence*, vol. 9, pp. 532–550, 1987.

[17] S. Lobregt and M. A. Viergever, "A discrete dynamic contour model," *IEEE Transactions on Medical Imaging*, vol. 14, pp. 12–24, 1995.

[18] S. P. Raya and J. K. Udupa, "Shape-based interpolation of multidimensional objects," *IEEE Transactions on Medical Imaging*, vol. 9, pp. 32–42, 1990.

[19] G. T. Herman, J. Zheng, and C. A. Bucholtz, "Shape-based interpolation," *IEEE Computer Graphics and Applications*, vol. 12, pp. 69–79, 1992.

[20] K. J. Zuiderveld, *Visualization of Multimodality Medical Volume Data using Object-Oriented Methods*. PhD thesis, Utrecht University, March 1995.

[21] W. Li, E. J. Gussenhoven, Y. Zhong, S. H. K. The, H. Pieterman, H. van Urk, and K. Bom, "Temporal averaging for quantification of lumen dimensions in intravascular ultrasound images," *Ultrasound in Medicine and Biology*, vol. 20, pp. 117–122, 1994.

[22] S. H. K. The, E. J. Gussenhoven, Y. Zhong, F. van Egmond, H. Pieterman, H. van Urk, G. P. Gerritsen, C. Borst, R. A. Wilson, and N. Bom, "Effect of balloon angioplasty on femoral artery evaluated with intravascular ultrasound imaging," *Circulation*, vol. 86, pp. 483–493, 1992.

[23] B. M. Ennis, D. M. Zientek, N. T. Ruggie, R. A. Billhardt, and L. W. Klein, "Characterization of a saphenous vein graft aneurysm by intravascular ultrasound and computerized three-dimensional reconstruction," *Catheterization and Cardiovascular Diagnosis*, vol. 28, pp. 328–331, 1993.

[24] G. Pasterkamp, M. S. van der Heiden, M. Post, C. Borst, E. J. Gussenhoven, H. Pieterman, H. van Urk, and N. Bom, "Discrimination of the intravascular lumen and dissections in single intravascular ultrasound images using subtraction, averaging and saline flush," *Ultrasound in Medicine and Biology*, vol. 21, pp. 149–156, 1995.

Fast Algorithms for Fitting Multiresolution Hybrid Shape Models to Brain MRI *

B. C. Vemuri[1], Y. Guo[2], S. H. Lai[4] and C. M. Leonard[3]

[1] Department of Computer & Information Science & Engg.
[2] Department of Electrical & Computer Engg.
[3] Department of Neuroscience
University of Florida, Gainesville, Fl. 32611
[4] Siemens Corporate Research, Princeton, NJ

Abstract. In this paper, we present new and fast algorithms for shape recovery from brain MRI using the multiresolution hybrid shape models introduced in Vemuri and Radisavljevic. We present three new computational methods for model fitting to data and demonstrate the efficiency of our methods via experiments on brain MRI data.

1 Introduction

Modeling shapes is an important and integral part of computer graphics as well as computer vision. In computer graphics, modeling shapes is the key ingredient of shape synthesis while in computer vision, it is needed for shape reconstruction and shape recognition from sensed data. Many shape modeling schemes have been proposed in the graphics as well as vision literature [1, 10, 2]. In this paper, we will be concerned with shape modeling with a view toward facilitation of shape recovery from 3D data for computer vision/medical imaging applications. The main thrust of the paper is in developing fast numerical algorithms for use in model fitting to brain MRI.

A hybrid modeling scheme dubbed, "deformable superquadrics" was introduced by Terzopoulos et al., [10]. These models have the advantage of combining the descriptive power of lumped and distributed parameter models and thus, simultaneously satisfying the requirements of shape reconstruction and shape recognition. However, these hybrid models do not exhibit a smooth transition in the number of parameters required for their description. For example, the two extremes of the spectrum are occupied by lumped models at one end and distributed models at the other. The former requiring, few parameters and the later requiring a large number of parameters for their description. Also, the lumped parameter models allow only for global deformations while the distributed parameter models allow for local deformations. The modeling scheme described in [10] involves a superposition of the lumped and distributed parameters to completely characterize the plethora of shapes occupying the spectrum. Although,

* This work was supported in part by the Whitaker Foundation and the NSF grant ECS-9210648.

these hybrid models can be used to generate both local and global shape deformations, they *do not depict a smooth transition between these types of deformations.* In addition, there is *no smooth transition in the number of parameters* that can capture descriptions of the range of generated shapes.

Recently, Vemuri & Radisavljevic [11] introduced multiresolution hybrid models that uses an orthonormal wavelet basis [7]. These hybrid models have the property of being able to smoothly scale up or down the range (from local to global) of possible deformations and the number of parameters required to characterize them. These properties are inherited by virtue of the multiresolution representation. Depending on the application, one may choose a set of wavelet coefficients at a particular coarse-level from the multi-resolution representation of the hybrid model and augment the global (lumped) parameters of the model. This augmented set of "global" (lumped) parameters define a larger class of shapes than those characterized by purely lumped parameters. This enhanced class of shapes characterized by a few global parameters were used as prior information in a probabilistic shape estimation framework.

Like any deformable surfaces, these multiresolution hybrid models can also be interactively or non interactively manipulated to mold into desired shapes via the use of externally applied forces. *In this paper, external forces will be derived from volumetric medical image data (MRI brain scans) and the model will be forced into desired anatomical shapes in the MRI data of a human brain.* This process of molding the model via externally applied forces is also known as model fitting. *In this paper, we will focus on developing very fast numerical algorithms for model fitting.* The process of fitting the model to the data is achieved numerically. In [11], a gradient descent scheme was used where as, a first order Euler scheme was employed in fitting the dynamically deformable superquadric model in [10]. Both these methods are slow and can prove to be unstable for large step size in each iteration. To overcome this problem, *we develop a new preconditioned nonlinear conjugate gradient (PNCG) algorithm that is fast, stable and yields a locally optimal solution in very few iterations.* We compare the performance of the model fitting with PNCG against that with gradient descent (GD) and the nonlinear conjugate gradient (NCG) algorithms. The experimental results show that our algorithm outperforms both GD and NCG.

The hybrid shape model comprises of global and local shape parameters. The global parameters consist of : nine parameters to describe the superquadric (5 for the shape, 2 for bending and 2 for tapering), three for translation and four for rotation (quaternion). The local part is represented by a triangular finite element discretization. In the PNCG, we use the Hessian of the energy function as a preconditioner. However, we do not use the exact Hessian but only an approximation to it namely, the diagonal elements of the Hessian matrix. Since the analytical calculations for the Hessian of the global parameters of the model are very complex, we have developed a semi-numerical approximation of the Hessian for the part of the preconditioner corresponding to the global parameters of the model. For the local part, we have developed three solution techniques. Two of them use preconditioning while the third does not and instead involves the use of the alternating direction implicit (ADI) method.

The rest of the paper is organized as follows: In the next section, we briefly describe the multiresolution hybrid shape model and introduce the model fitting problem. In section 4, we describe the nonlinear conjugate gradient technique to solve the model fitting problem numerically. It also contains a description of the Hessian based preconditioning technique while section 4.2 describes a wavelet-based preconditioning method. In section 4.3 we describe the derivation leading up to the use of the ADI method. In section 5, we present some preliminary experimental results and conclude in section 6.

2 Multi-resolution Geometry of the Modeling Scheme

In this section, we will briefly describe the geometry and the construction of a multi-resolution wavelet basis for the modeling scheme. For a detailed description of the embedding of the deformable superquadric geometry in a wavelet basis, we refer the reader to Vemuri et al.,[11].

2.1 Geometry of the Deformable Superquadrics

The deformable superquadrics are closed surfaces in space whose intrinsic coordinates $\mathbf{u} = (u, v)$ are defined on a domain Ω. The positions of points on the model in an inertial frame of reference are given by a vector-valued function $\mathbf{x}(\mathbf{u}) = (x_1(\mathbf{u}), x_2(\mathbf{u}), x_3(\mathbf{u}))^T$. In a model-centered coordinate frame, the position vector $\mathbf{x}$ becomes $\mathbf{x} = \mathbf{c} + \mathbf{R}\mathbf{p}$, where $\mathbf{c}$ is the location of the center of the model, and the rotation matrix $\mathbf{R}$ specifies the orientation of the model-centered coordinate frame. Hence, $\mathbf{p}(\mathbf{u})$ denotes the canonical positions of points on the model relative to the model frame. We further express $\mathbf{p}$ as the sum of a reference shape $\mathbf{s}(\mathbf{u})$ and a displacement $\mathbf{d}(\mathbf{u})$ namely, $\mathbf{p} = \mathbf{s} + \mathbf{d}$.

For the parameterized reference shape $\mathbf{s}(\mathbf{u})$ we use the superquadric with a bending and tapering deformation (as in Bajscy et al., [1]), due to it's attractive global shape characterization. The reference shape $\mathbf{s}$ is given by, $\mathbf{s} = \mathcal{B}(\mathcal{T}((\tilde{\mathbf{s}}))$. Where, $\mathcal{B}$ is a bending deformation and $\mathcal{T}$ is a tapering deformation (see [1] for details). The parametric equation of a superquadric $\tilde{\mathbf{s}}$ is given by

$$\tilde{\mathbf{s}} = a \begin{pmatrix} a_1 C_u^{\epsilon_1} C_v^{\epsilon_2} \\ a_2 C_u^{\epsilon_1} S_v^{\epsilon_2} \\ a_3 S_u^{\epsilon_1} \end{pmatrix} \tag{1}$$

Where, $-\pi/2 \leq u \leq \pi/2 \ -\pi \leq v \leq \pi$, and $S_w^{\epsilon} = sgn(sin\, w)|sin\, w|^{\epsilon}$ and $C_w^{\epsilon} = sgn(cos\, w)|cos\, w|^{\epsilon}$. Here, $0 \leq a_1, a_2, a_3 \leq 1$ are aspect ratio parameters, $\epsilon_1, \epsilon_2 \geq 0$ are "squareness" parameters and $a \geq 0$ is a scale parameter. This defines the geometry of our underlying reference shape namely, the superquadric. We collect all these shape parameters along with the two bending and two tapering parameters into a vector $\mathbf{q}_s$. We will denote $\mathbf{c}$ by the vector $\mathbf{q}_c$ and use $\mathbf{q}_\theta$ to denote the vector of rotational coordinates of the model in a quaternion representation (see [10]).

We represent the displacement vector function $\mathbf{d}$ in a wavelet basis. Thus, $\mathbf{d} = \Phi S \mathbf{q}_d$ is completely determined by a set of wavelet coefficients collected into the vector $\mathbf{q}_d$, with S corresponding to the discrete wavelet transform and Φ containing nodal interpolation functions. Thus, the state of the model is fully expressed by the vector $\mathbf{q} = \left[\mathbf{q}_c^T, \mathbf{q}_\theta^T, \mathbf{q}_s^T, \mathbf{q}_d^T\right]^T$.

The above described multiresolution model is embedded in a Bayesian framework to facilitate incorporation of specific prior information about anatomical shapes of interest. The prior model in this case consists of statistics (mean and variance) on the global parameters of the model namely, $\left[\mathbf{q}_c^T, \mathbf{q}_\theta^T, \mathbf{q}_s^T, \mathbf{q}_{dg}^T\right]^T$. The local parameters are free to deform away from the mean due to the influence of data forces and are converted into a probability distribution over expected shapes via statistical mechanics techniques. We combine the prior model with a data acquisition model and determine the maximum a posteriori (MAP) estimate using the Bayes' rule. The MAP estimate is the same as the deterministic solution if the covariance matrix is chosen to be the stiffness matrix $\mathbf{K}$ described earlier. For more details on the probabilistic formulation, we refer the reader to [11].

3 Numerical Solution

The numerical problem involves minimizing minimizing the total energy $E = E_p + E_D$ for which we have developed several numerical schemes that form the focus of the rest of the paper. Where, $E_p = 1/2[(\mathbf{q} - \bar{\mathbf{q}})^T K_p (\mathbf{q} - \bar{\mathbf{q}})]$ and

$$E_D(\mathbf{D}, \mathbf{q}) = \begin{cases} \frac{1}{2}\sum_i \beta(\mathbf{D}_i - \mathbf{x}(\mathbf{q}, \mathbf{u}_i))^2 & \text{spring energy} \\ \int_\Omega \beta\, |P(\mathbf{x}(\mathbf{q}, \mathbf{u}))| & \text{image potential} \end{cases} \tag{2}$$

A gradient descent method was used in [11] which amounts to iteratively solving,

$$\mathbf{q}^{k+1} = \mathbf{q}^k - \Lambda \frac{\partial E(\mathbf{q})}{\partial \mathbf{q}} \tag{3}$$

where Λ is a diagonal matrix containing step sizes and consequently effecting the speed of convergence. Partial derivatives of the prior model and data energies defined above can be now written as $\frac{\partial E_p}{\partial \mathbf{q}} = K_p(\mathbf{q} - \bar{\mathbf{q}})$, $\frac{\partial E_D}{\partial \mathbf{q}} = \mathbf{f}_\mathbf{q}$. By substituting into the equation for gradient descent 3, we obtain an expression for iterative updating of the state vector $\mathbf{q}^{(k+1)} = \mathbf{q}^{(k)} + \Lambda\left(\mathbf{f}_q^{(k)} - K(\mathbf{q}^{(k)} - \bar{\mathbf{q}})\right)$

While the above iterative equation will eventually converge to the correct estimate, in practice it may be unacceptably slow. In the following sections, we present algorithms for solving this problem very efficiently.

4 Preconditioned Nonlinear Conjugate Gradient Algorithm

The conjugate gradient (CG) algorithm is well suited for solving quadratic optimization problems. The nonlinear version can be used for finding the locally optimal solution of a nonlinear nonconvex function. In this paper, we develop a preconditioned nonlinear conjugate gradient algorithm and apply it to the energy minimization discussed in the previous section.

Given any continuous function $f(\mathbf{x})$ with gradient $f'(\mathbf{x}) = [\frac{\partial}{\partial x_1} f(\mathbf{x}) \ \frac{\partial}{\partial x_2} f(\mathbf{x})$ ...... $\frac{\partial}{\partial x_n} f(\mathbf{x})]^T$, we can use Nonlinear Conjugate Gradient Algorithm to minimize it. The *residual* $\mathbf{r_i} = -f'(\mathbf{x_i})$ indicates how far we are from the correct value that minimizes $f(\mathbf{x})$. If $\mathbf{d_0}, \mathbf{d_1}, \ldots\ldots, \mathbf{d_{n-1}}$ are a set of orthogonal *search directions*, α_i and βi are the step sizes, and we choose a preconditioner M that approximates f'' and has the property that $M^{-1}r$ is easy to compute [6]. The outline of the Preconditioned nonlinear conjugate gradient method is as follows:

1. Let $\mathbf{r_0} = -f'(\mathbf{x_0})$.
2. Calculate a preconditioner $M_0 \approx f''$.
3. $\mathbf{d}_0 = M_0^{-1}\mathbf{r}_0$.
4. Find α_i that minimize $f(\mathbf{x_i} + \alpha_i \mathbf{d_i})$.
5. $\mathbf{x_{i+1}} = \mathbf{x_i} + \alpha_i \mathbf{d_i}$.
6. Calculate the preconditioner $M_{i+1} \approx f''$.
7. $\mathbf{r_{i+1}} = -f'(\mathbf{x_{i+1}})$.
8. $\beta_{i+1} = \dfrac{\mathbf{r_{i+1}^T M_{i+1}^{-1} r_{i+1}}}{\mathbf{r_i^T M_i^{-1} r_i}}$ *or*

$$\beta_{i+1} = \max\left(\frac{\mathbf{r_{i+1}^T}(\mathbf{M_{i+1}^{-1} r_{i+1}} - \mathbf{M_i^{-1} r_i})}{\mathbf{r_i^T M_i^{-1} r_i}}, 0\right).$$

9. $\mathbf{d_{i+1}} = \mathbf{r_{i+1}} + \beta_{i+1}\mathbf{d_i}$.

In step 8, using the first expression for β_{i+1} leads to the *Fletcher-Reeves* method, and using the second formula leads to the *Polak-Ribiére* method. In step 4, the function $f(\mathbf{x} + \alpha\mathbf{d})$ is approximately minimized by setting $\alpha = -\frac{f'^T \mathbf{d}}{\mathbf{d}^T f'' \mathbf{d}}$. Where, $f''(\mathbf{x})$ is the *Hessian matrix*, in our case, it is:

$$H = \begin{bmatrix} \frac{\partial^2 E}{\partial q_1 \partial q_1} & \frac{\partial^2 E}{\partial q_1 \partial q_2} & \cdots & \frac{\partial^2 E}{\partial q_1 \partial q_n} \\ \frac{\partial^2 E}{\partial q_2 \partial q_1} & \frac{\partial^2 E}{\partial q_2 \partial q_2} & \cdots & \frac{\partial^2 E}{\partial q_2 \partial q_n} \\ \cdots\cdots & \cdots\cdots & \cdots\cdots & \cdots\cdots \\ \frac{\partial^2 E}{\partial q_n \partial q_1} & \frac{\partial^2 E}{\partial q_n \partial q_2} & \cdots & \frac{\partial^2 E}{\partial q_n \partial q_n} \end{bmatrix}$$

When it's too expensive to compute the full Hessian matrix, it is reasonable to use the diagonal of the Hessian as a preconditioner. However, if the initial guess is too far from a local optimum, the diagonal elements may not all be positive. Since a preconditioner should be positive definite, non-positive diagonal entries will violate this property and hence can not be allowed. A conservative approach in this case is to use the identity matrix as a preconditioner ($\mathbf{M} = \mathbf{I}$) i.e., not to precondition.

4.1 Computation of the Approximate Hessian

From the expression of Hessian matrix, it can be seen that the elements in one row of Hessian matrix, say row i can be obtained by taking the partial derivative with respect to q_i of the vector $\frac{\partial E}{\partial \mathbf{q}}^t$, i.e, the ith row is in the form:

$$\frac{\partial}{\partial q_i}[\frac{\partial E}{\partial q_1} \frac{\partial E}{\partial q_2} \cdots \frac{\partial E}{\partial q_n}] = \frac{\partial}{\partial q_i}[\frac{\partial E}{\partial \mathbf{q}}]$$

For example, let's consider $\mathbf{q}_\theta = [v1\ v2\ v3\ s]^T$ then, (Note that q_i refers to an element of $\mathbf{q_c}, \mathbf{q}_\theta, \mathbf{q_s},\ or\ \mathbf{q_d}$).

$$\frac{\partial}{\partial v_1}[\frac{\partial E}{\partial v_1}\ \frac{\partial E}{\partial v_2}\ \frac{\partial E}{\partial v_3}\ \frac{\partial E}{\partial s}]$$

$$= (\frac{\partial E}{\partial x_1}\ \frac{\partial E}{\partial x2}\ \frac{\partial E}{\partial x_3}) \begin{bmatrix} \frac{\partial}{\partial v_1}(\frac{\partial x_1}{\partial v_1}) & \frac{\partial}{\partial v_1}(\frac{\partial x_1}{\partial v_2}) & \frac{\partial}{\partial v_1}(\frac{\partial x_1}{\partial v_3}) & \frac{\partial}{\partial v_1}(\frac{\partial x_1}{\partial s}) \\ \frac{\partial}{\partial v_1}(\frac{\partial x_2}{\partial v_1}) & \frac{\partial}{\partial v_1}(\frac{\partial x_2}{\partial v_2}) & \frac{\partial}{\partial v_1}(\frac{\partial x_2}{\partial v_3}) & \frac{\partial}{\partial v_1}(\frac{\partial x_2}{\partial s}) \\ \frac{\partial}{\partial v_1}(\frac{\partial x_3}{\partial v_1}) & \frac{\partial}{\partial v_1}(\frac{\partial x_3}{\partial v_2}) & \frac{\partial}{\partial v_1}(\frac{\partial x_3}{\partial v_3}) & \frac{\partial}{\partial v_1}(\frac{\partial x_3}{\partial s}) \end{bmatrix}$$

$$+ (\frac{\partial^2 E}{\partial x_1^2}\ \frac{\partial^2 E}{\partial x_2^2}\ \frac{\partial^2 E}{\partial x_3^2}) \begin{bmatrix} \frac{\partial x_1}{\partial v_1} & & \\ & \frac{\partial x_2}{\partial v_1} & \\ & & \frac{\partial x_2}{\partial v_1} \end{bmatrix} \begin{bmatrix} \frac{\partial x_1}{\partial v_1} & \frac{\partial x_1}{\partial v_2} & \frac{\partial x_1}{\partial v_3} & \frac{\partial x_1}{\partial s} \\ \frac{\partial x_2}{\partial v_1} & \frac{\partial x_2}{\partial v_2} & \frac{\partial x_2}{\partial v_3} & \frac{\partial x_2}{\partial s} \\ \frac{\partial x_3}{\partial v_1} & \frac{\partial x_3}{\partial v_2} & \frac{\partial x_3}{\partial v_3} & \frac{\partial x_3}{\partial s} \end{bmatrix}$$

Where x_1, x_2, x_3 are the world coordinates in which the model is described.

Similarly, we can get $\frac{\partial}{\partial v_2}[\frac{\partial E}{\partial v_1}\ \frac{\partial E}{\partial v_2}\ \frac{\partial E}{\partial v_3}\ \frac{\partial E}{\partial s}], \ldots, \frac{\partial}{\partial s}[\frac{\partial E}{\partial v_1}\ \frac{\partial E}{\partial v_2}\ \frac{\partial E}{\partial v_3}\ \frac{\partial E}{\partial s}]$. Since we require only the diagonal entries of the Hessian, only the following elements need to be computed, $\frac{\partial}{\partial v_1}(\frac{\partial E}{\partial v_1}), \frac{\partial}{\partial v_2}(\frac{\partial E}{\partial v_2})$, $\frac{\partial}{\partial v_3}(\frac{\partial E}{\partial v_3})$ and $\frac{\partial}{\partial s}(\frac{\partial E}{\partial s})$.

4.2 Preconditioning in Wavelet Basis

Alternatively, one may precondition the estimation of local and global parameters separately. For the global part, we can use the same procedure as above however, for the local part which involves the vector $\mathbf{q_d}$, we can construct the stiffness matrix of this local part and find its eigen values in closed form. This stiffness matrix happens to be a Toeplitz matrix and we can find its spectrum by finding its eigen values and eigen vectors. We can use the spectral characteristics of this matrix to modulate a chosen wavelet basis to construct the preconditioner. If M represents the modulation matrix, a spectral approximation of the $\mathbf{K}$ matrix in a wavelet basis (see [6] for details), then the preconditioner is defined as $\mathbf{P} = \mathbf{WMW}^t$ where $\mathbf{W}$ is the wavelet transform and W^t is its inverse. In our implementation, the diagonal entries of the $\mathbf{M}$ matrix at a particular resolution j are of the same value and are determined by taking the average of the spectral values of $\mathbf{K}$ at the resolution j.

4.3 ADI Method for the Local Parameter Estimation

The alternating direction implicit (ADI) method [5] can be used for updating the local displacement instead of using a preconditioning scheme. We have proved that this algorithm takes $O(N)$ time to converge for the local part. Here, N is the number nodes in the models triangular finite element discretization. Due to the presence of the north and south poles, we need special matrix machinery to solve the linear system that results from the discretization. To this end, we use the Schur complement formula [9] to the linear system with the system matrix containing the contributions from the north and south poles. The stiffness matrix $\hat{\mathbf{K}} \in \Re^{(N+2)\times(N+2)}$ for our problem can be written as a (2×2) block matrix with $\hat{\mathbf{K}}_{11} = \mathbf{K}$ described below, $\hat{\mathbf{K}}_{22}$ a (2×2) diagonal block containing the coefficients corresponding to the north and south poles. $\hat{\mathbf{K}}_{12} = \hat{\mathbf{K}}_{21}$ consists of the interaction between the poles and other nodes.

The component $\hat{\mathbf{K}}_{11}$ of the stiffness matrix can be expressed as the sum of three tensor products namely

$$\mathbf{K} = w_1(\mathbf{A} \otimes \mathbf{I} + \mathbf{I} \otimes \mathbf{A}) + w_0 \mathbf{I} \otimes \mathbf{I} \tag{4}$$

In the above equation, $\mathbf{A}$ is an $(n \times n)$tridiagonal and Toeplitz matrix. A has $2s$ on the diagonal and $-1s$ on the two sub-diagonals. Matrix $\mathbf{I}$ is an $(n \times n)$ identity matrix. $\otimes$ represents the kronecker product. $\mathbf{K}$ is an $(N \times N)$ matrix, where $N = n^2$. We know that the matrix $\mathbf{A}$ is an SPD matrix. Equation 4 can be further simplified to

$$\mathbf{K} = \mathbf{C} \otimes \mathbf{I} + \mathbf{I} \otimes \mathbf{C} \tag{5}$$

where $\mathbf{C} = w_1\mathbf{A}+(w_0/2)\mathbf{I}$. It is obvious that the matrix $\mathbf{C}$ is also SPD for positive w_0 and w_1. In addition, the condition number for the matrix C is bounded by a constant.

Thus, we can write the linear system $\mathbf{Kx} = \mathbf{b}$ (that results from treating the local part independent of the global part and taking the gradient and setting to zero) as a matrix equation by using equation 5 as follows:

$$\mathbf{CX} + \mathbf{XC} = \mathbf{D} \tag{6}$$

Where, $\mathbf{D}$ is an $(n \times n)$ representation of the n^2 vector $\mathbf{b}$. This equation is a Lyapunov matrix equation. The ADI method [5] can be applied to solve this matrix eqn. Since the matrix A' has a bounded condition number (can be proved from that fact that the condition numbers for $\mathbf{B}$ and $\mathbf{C}$ are bounded), the number of iterations required in ADI for a given error tolerance is also bounded by a constant. Therefore, the solution to the linear system $\mathbf{Kx} = \mathbf{b}$ within a given error tolerance can be obtain in O(N) operations.

5 Model Fitting Results

From an application point of view, our goal is to apply the multi-resolution shape modeling scheme described in previous sections for recovering the 3D shape of cortical and subcortical structures such as the Cerebellum, gyrus and hippocampus, from brain MRI data. The shapes will be accurately quantified by a parameter vector that can be used in computing shape measures such as volume. Volumetric analysis of structures contributes to understanding of various neurological disorders such as dyslexia, specific language impairment etc. (Leonard, et al., [4]). Nontrivial shapes, low contrast images and high noise levels in this application domain prompted us to take advantage of our statistical modeling scheme by incorporating object-specific information into our prior model. To facilitate construction of such specialized prior models we divide shape recovery process into a training phase and an operational phase. In the training phase human supervision via interaction and manipulation is needed to provide good training samples, while in the operational phase the model-fitting process becomes fully autonomous and relies heavily on statistics accumulated in prior model.

We now present two examples depicting the experiments with various numerical algorithms for model fitting process to a cerebellum and a skull from an MR brain scan. The data set consisted of 30 and 45 slices respectively, with each slice of 1.25mm thickness. Figure 1, illustrates our results for a **cerebellum**. From left to right and top to bottom, the displayed images are, (a) a sagittal section of the brain showing a region of interest (containing a cerebellum) indicated in a box, (b) a slice of the initialized 3D shape model superimposed on the corresponding slice from the original data; super-imposition of a slice from the fitted model on the corresponding slice from the original data,(c) using 22 iterations of a gradient descent scheme (d) using 22 iterations of a nonlinear conjugate gradient algorithm (e) using 18 iterations of the diagonal Hessian preconditioned conjugate gradient algorithm (PNCG) and (f) is the 3D fitted model using PNCG. The iteration time for each method was fixed to the convergence time for the PNCG scheme. As is evident, the convergence is far superior in the PNCG than the other schemes for a fixed time. The initialized model was subject to externally applied forces synthesized from a 3D edge based potential obtained by computing edges in 3D using the recursive filtering discussed in [8].

A similar arrangement of results is shown in figure 2. The 3D shape in this example was a human skull. The CPU time was fixed to the convergence time for the PNCG (= 160 seconds on a sparc-10). In this fixed time, gradient descent executed 11 iterations, NCG executed 6 and PNCG executed 16 iterations. Once again the convergence results obtained for the PNCG are far superior to the other methods.

6 Conclusion

In this paper, we introduced several numerical schemes to solve the nonlinear optimization involved in model fitting to data. The models were multiresolution deformable superquadrics introduced in [11]. We developed a fast preconditioned nonlinear conjugate gradient (PNCG) algorithm and compared its performance on two test cases involving a cerebellum and a skull from an MR brain scan. In both the cases the PNCG outperformed GD and NCG algorithms. We also

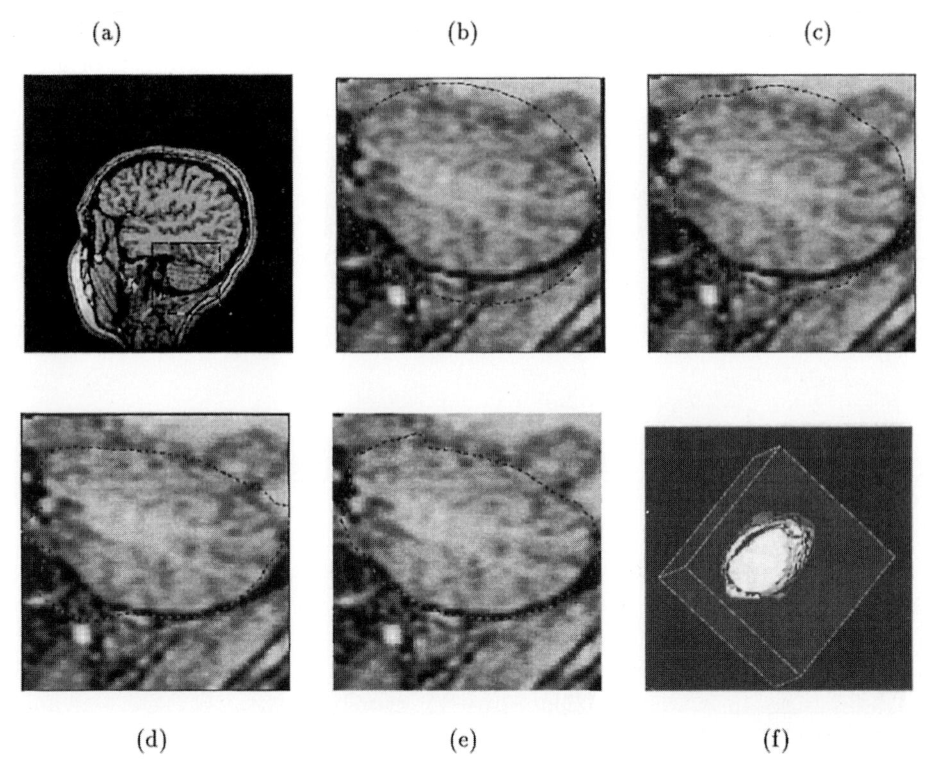

Fig. 1. Convergence results for (c) GD, (d) NCG, (e) PNCG on a Cerebellum, a slice of which is shown in (a) with corresponding model initialization in (b). (f) Depicts the 3D fitted model using PNCG.

proposed three new methods for estimating the local model parameters and our future efforts will be focused on more experimentation with these methods on MRI data.

References

1. R. Bajscy and R. Solina : Three-dimensional object representation revisited. IEEE First Conference on Computer Vision. (1987) 231–240
2. L. D. Cohen and I. Cohen : Deformable models for 3D medical images using finite elements and balloons. IEEE Conference on Computer Vision and Pattern Recognition. **Ill** (1992)
3. S. Geman, and D. Geman : Stochastic relaxation, Gibbs distribution, and Bayesian restoration of images. IEEE Transactions on Pattern Analysis and Machine Intelligence. **6(6)** (1984) 721–724
4. C. M. Leonard, C. A. Williams, R. D. Nicholls, O. F. Agee : Angelman and Prader-Willi Syndrome: A Magnetic resonance imaging study of differences in cerebral structure. American Journal of Medical Genetics. (1992) (in press)
5. S. H. Lai and B. C. Vemuri : An $O(N)$ iterative solution to the Poisson equation in low-level Vision problems.IEEE Conference on Computer Vision and Pattern Recognition **CVPR'94** (1994) 9–14.

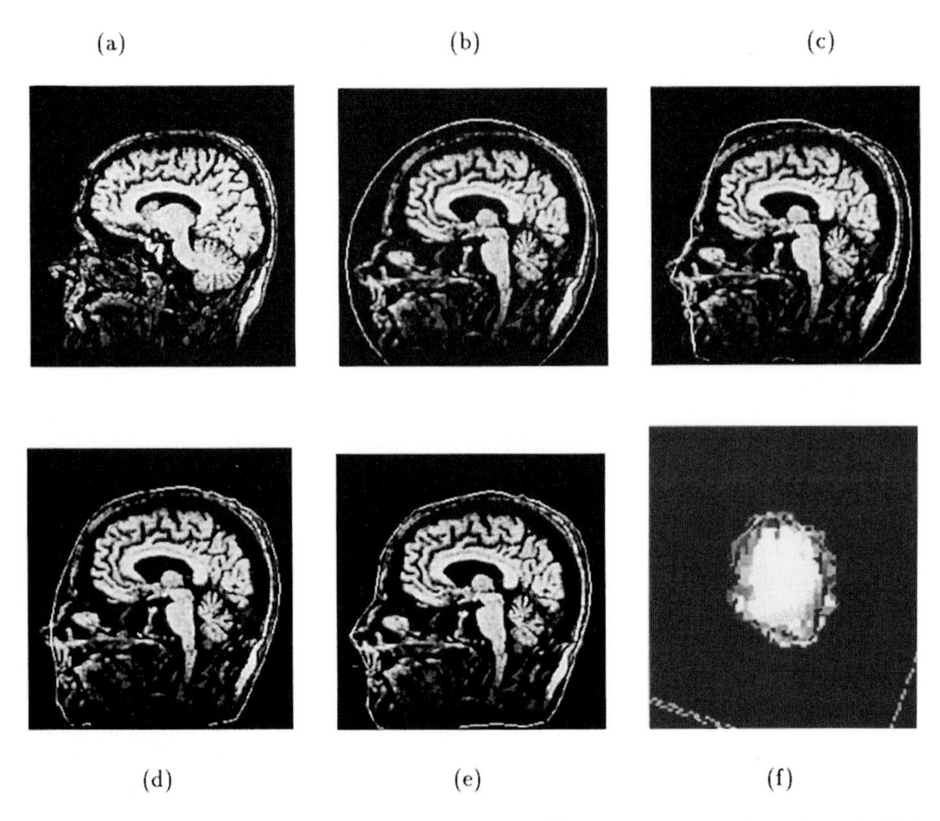

Fig. 2. Convergence results for (c) GD, (d) NCG, (e) PNCG on a skull, a slice of which is shown in (a) with corresponding model initialization in (b). (f) Depicts the 3D fitted model using PNCG.

6. S. H. Lai and B. C. Vemuri : Physically-based adaptive preconditioning for early vision. IEEE Workshop on Physics-based Modeling in Computer Vision. (1995)
7. S. G. Mallat : A theory of multi-resolution signal decomposition: the wavelet representation. IEEE Transactions on Pattern Analysis and Machine Intelligence. **11(7)** (1989) 674–693.
8. A. Monga, R. Deriche : 3D edge detection using recursive filtering: Application to scanner images. Proc. of IEEE Conference on Computer Vision and Pattern Recognition. (1989) 28–35.
9. D. V. Ouellette: Schur complements and statistics. Linear Algebra and its Applications. **36** (1981) 187–295.
10. D. Terzopoulos and D. Metaxas : Dynamic 3D models with local and global deformations: Deformable superquadrics. IEEE Transactions on Pattern Analysis and machine Intelligence. **13(7)** (1991) 703–714.
11. B. C. Vemuri and A. Radisavljevic : Multiresolution stochastic hybrid shape models with fractal priors. special issue of ACM Transactions on Graphics on Interactive Sculpting. **13** (1994) 177–207.

Computer Aided Screening System for Lung Cancer Based on Helical CT Images

Keizo Kanazawa [1], Mitsuru Kubo [1], Noboru Niki [1]
Hitoshi Satoh [2], Hironobu Ohmatsu [3], Kenji Eguchi [3], Noriyuki Moriyama [3]

[1] Dept. of Information, Univ. of Tokushima, JAPAN
[2] Medical Engineering Laboratory, Toshiba Corporation, JAPAN
[3] National Cancer Center, JAPAN

Abstract : In this paper, we describe a computer assisted automatic diagnosis system for lung cancer that detects tumor candidates at an early stage from helical CT images. This automation of the process reduces the time complexity and increases the diagnosis confidence. Our algorithm consists of analysis and diagnosis sections, and detects regions of lung tumor based on image processing techniques and medical knowledge. We have applied our algorithm to 450 patient's data for mass screening. The results show that our algorithm detects lung cancer candidates successfully.

1 Introduction

Lung cancer is known as one of the most difficult cancers to cure, and the number of deaths that it causes is generally increasing. Detection of lung cancer in its early stage can be helpful for medical treatment to limit the risk. One of the measures is a mass screening process for lung cancer. As a conventional method for the mass screening process, chest X-ray films have been used for lung cancer diagnosis. However, small lung cancers at early stages are difficult to detect because of overlapping of bone and organ shadows. One of the proposed techniques that assist detection uses helical CT which makes a wide range measurement of the lung in a short time. We expect that the proposed technique will increase diagnostic confidence because the helical CT information has 3D cross section images of the lung. However, mass screening based on helical CT images leads to a considerable number of images for diagnosis. This time-consuming fact makes it difficult for utilization in the clinic. The automation of this process reduces the time complexity and increases diagnostic confidence. To increase the efficiency of the mass screening process, we are developing an algorithm for the automatic detection of lung cancer candidates based on helical CT images.

As input data for the automation process, we use helical CT images with specified measurement conditions (beam width 10mm, table speed 20mm/sec, tube voltage 120kV, tube current 50mA and scan duration 15seconds). Images are reconstructed at 10mm increments by using an 180° linear interpolation algorithm and a matrix size of 512×512 pixels. These conditions are used to measure all the lung area at one breath stop. As these measurement conditions are rougher than general clinic measurement conditions and their 3D information is of low confidence, we apply our diagnostic algorithm to each image.

2 Diagnostic Algorithm

In the lung area, the CT values of lung cancer are similar to those of pulmonary blood vessels, so it is difficult to separate them using only their CT values. We thus extract the regions which have high CT values in the lung area such as the pulmonary blood vessels and tumors, and we detect the tumor candidates from the extracted regions based on defined diagnostic rules. Our diagnostic algorithm consists of the two following sections :

Analysis Section : We extract the necessary regions for diagnosis and analyze these regions based on image processing techniques.

Diagnosis Section : We define the diagnosis rules to detect suspicious shadows based on those features, and we detect only lung cancer shadows using these rules.

2.1 Analysis Section

Extraction and Analysis of the Lung Area

The original image is transformed to a binary image using the thresholding algorithm. However, this thresholding technique excludes lung boundary with high CT values from the real lung area, so we follow a correction process for making up such lost parts based on the curvature of the lung boundary. We classify each pixel on the lung boundary into three types, either a concave, convex or smooth point based on its curvature. And we perform following two steps.

(1) We connect two points $\mathbf{P_1}$, $\mathbf{Q_1}$ as shown in **Fig.1**(a), where $\mathbf{P_1}$ and $\mathbf{Q_1}$ are $\mathbf{n_1}$ pixels away from the concave point $\mathbf{A_1}$ and $\mathbf{A_n}$.

(2) We connect **A** and **B** as shown in **Fig.1**(b) if the ray $\mathbf{P_2A}$ or $\mathbf{Q_2A}$ crosses the lung boundary within the distance d, where $\mathbf{P_2}$ and $\mathbf{Q_2}$ are $\mathbf{n_2}$ pixels away from the convex point **A**.

This method replaces the boundary line with rapidly changing curvature with a straight line.

The organs in the lung are seen differently according to the position of the lung in each slice image. So, we classify the lung into four sections based on the area and shape of the extracted lung regions as shown in **Fig.2**. In the diagnosis section, the diagnositic rules are defined using these classification results.

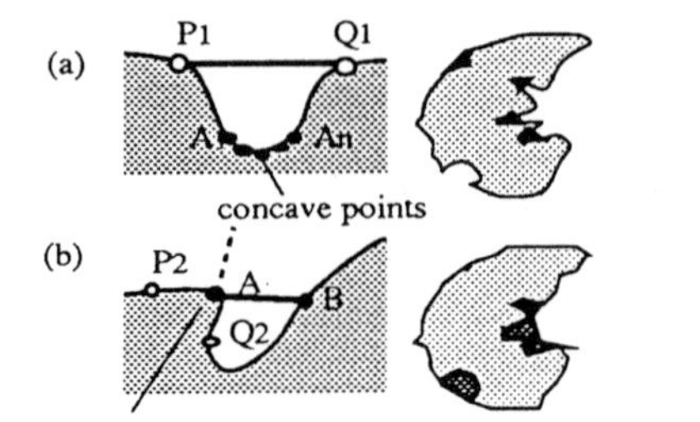

Fig.1. Correction of the lung boundary.

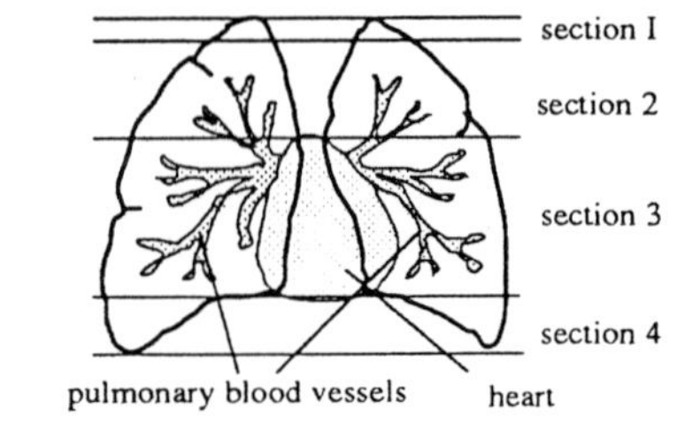

Fig.2. Section classification of the lung.

Extraction and Analysis of Pulmonary Blood Vessels

Structures in the lung, such as blood vessels and tumors, are separately given by a segmentation process using the fuzzy clustering method [4]. We apply the fuzzy clustering method to the histogram of the pixel values within the lung area, and we separate the lung area into two classes. we then apply the gray-weighted distance transformation [5] to the segmented image. After which we apply the threshold algorithm to the distance transformed image to eliminate the pixels lower than a threshold distance value. Finally, we apply an inverse distance transformation. Results of this extraction process are shown in **Fig.3**. In **Fig.3**(c), the yellow regions are the extracted blood vessels and tumor regions as the candidate regions for diagnosis. From here, we call these regions CR (Candidate Region).

Most of the extracted CR are blood vessel regions, we distinguish between tumors and blood vessels according to features of the CR. Here, we give attention to shape, gray value and the position of each CR. We give serial numbers to all CRs, and calculate the six features, Area **F**a(**n**), Thickness **F**t(**n**), Circularity **F**c(**n**), Glay level **F**g(**n**), Variance of values **F**v(**n**) and Position **F**p(**n**) for each CR. Where, **n** is the serial number of each CR, **F**a(**n**) is the number of the pixels, **F**t(**n**) is the maximum gray-weighted distance, **F**c(**n**) is the rate of occupation inside the circumscribed circle of each CR, **F**g(**n**) is the average CT numbers of pixels, **F**v(**n**) is the variance of the CT numbers, and **F**p(**n**) is the minimum distance between the circumscribed circle's center of each CR and the lung wall.

2.2 Diagnosis Section

Here we describe the diagnosis rules which detect suspicious shadows as candidate tumors utilizing six features. We refer to the following diagnostic medical knowledge;

Knowledge 1: The shape of lung cancer is generally spherical, and it is seen as a circle on the cross section. However the shape of blood vessels running horizontally is oblong.

Knowledge 2: The thickness of the blood vessel becomes smaller as its position is nearer the lung wall, however the thickness of the lung cancer is generally larger than the ordinary thickness of the blood vessels at each position.

Knowledge 3: As the periphery blood vessels are too small to be seen in the helical CT image and are difficult to be recognized, shadows contacting the lung wall are generally tumors,

Knowledge 4: The values of blood vessels are generally higher than cancer when running vertically in helical CT images using these measurement conditions.

Knowledge 5: The gray values of pixels of the cancer region are comparatively uniform.

We classify the diagnosis rules into three types. 1) we define RULE 1 to eliminate the CR which is not obviously suspicious. 2) we define RULE 2 to detect the tumors which do not have contact to the lung wall. 3) we define RULE 3 to detect the tumors which contact the lung wall. We define the diagnostic rules as follows:

[RULE 1] : Object for elimination.

We eliminate the current CR if the following conditions are satisfied.

(1) If **Fa(n)** is very small, we eliminate this CR.
(2) If **F**c(**n**) is very small, we eliminate this CR as a blood vessel region based on knowledge 1.
(3) If **F** g(**n**) or **F** v(**n**) are very high, we eliminate this CR as blood vessel or calcification based on knowledge 4, 5.
(4) If bone exists at the location of the current CR in the previous or next slice image, we eliminate this CR as a shadow of partial volume effect.

[**RULE 2**] : Detection of tumor candidates in the case of noncontact with the lung wall.

In this rule, we quantify the likelihood of the lung cancer based on the extracted features for each CR. Here we determine the three quantities of likelihood as follows;

Roundness : We think that the likelihood of lung cancer for each CR depends on the circularity **F** c(**n**). Roundness is in proportion to the circularity **F** c(**n**).

Size : The area generally becomes larger with distance from the lung wall. We assume the basic area which is varied according to the position **Fp(n)**, and we define the size as the ratio of **Fa(n)** to the assumed area.

Brightness : Brightness becomes smaller as the CT value **F** g(**n**) and the variance **F** v(**n**) become large compared with the average value of lung cancer.

Furthermore, we define the synthetic diagnosis index as follows :

$$\text{Diagnosis index} = \text{Roundness} \times \text{Size} \times \text{Brightness}$$

We class the current CR as a tumor candidate, if this index is over a threshold value.

[**RULE 3**] : Detection of tumor candidates in the case of contact with the lung wall.

The shadows contacting the lung wall are ordinarily suspicious as described in knowledge 3, however these shadows involve the artifact values which are caused by the partial volume effect. The tumors have a convex shape inside the lung, the artifacts are oblong along the lung wall, and the gradient around the tumors is higher than the artifacts. Therefore to distinguish between tumors and artifacts we define two indexes of convexness and contrast as follows :

Convexness : For the three pixel dilation area outside the current CR, we count the number of pixels which are higher than the average value **F** g(**n**), and similarly count the number of pixels which are lower than the **F** g(**n**). We define the convexness as ratio of these numbers.

Contrast : We define the contrast as the difference between the average values of the current CR and the inner dilatation area outside this CR.

Fig.4 shows descriptions of RULE3. If both indexes are higher than threshold values, we class this CR as a tumor candidate.

When utilizing this algorithm, at first RULE 1 is applied to all CRs, then, either RULE 2 or RULE 3 is applied to the remainder CRs. Selection of the rules depends on whether a CR contacts the lung wall or not. In **Fig.3**(f), our diagnosis results show tumor candidates as red regions.

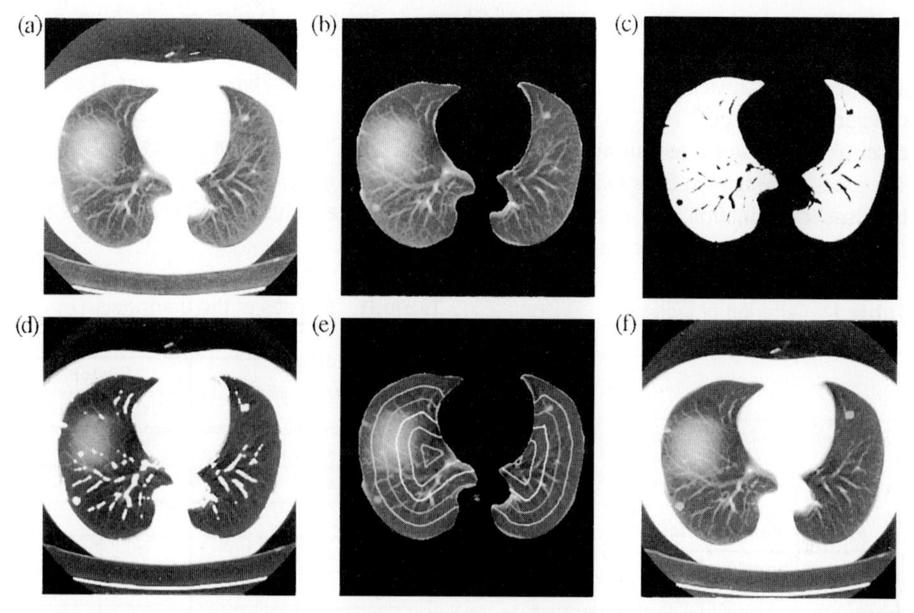

Fig.3. Results of diagnostic process: (a) Original image; (b) Extraction of the lung area; (c) Segmentation result using the fuzzy clustering; (d) Extraction of the pulmonary blood vessels; (e) Contour map of the distance from the lung wal; (f) Detection result.

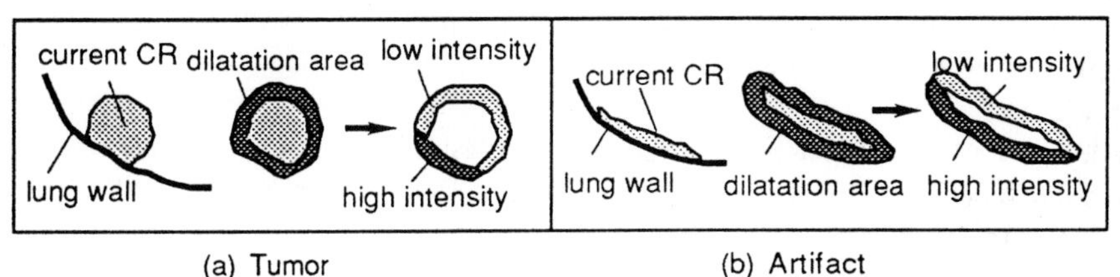

(a) Tumor (b) Artifact

Fig.4. Descriptions of RULE 3

3 Experimental Results

The diagnostic algorithm is applied to the helical CT images of 450 cases (total: 15,750 images). All data are diagnosed by three expert physicians with the criterion as shown in **Table.I.** Our diagnostic algorithm aims to detect not only the lung cancer but other suspicious regions as tumor candidates, and our system separates all detected tumor candidates into two clusters according to their likelihood of cancer as follows :

Cluster 1 : The suspicious shadows of lung cancer

Cluster 2 : The shadows of other tumors.

In **Table.II**, we show the comparative results between our algorithm and three physicians. Here we separate each tumor candidate into six groups according to the judgment and the number of the physician who detected it. As a result, all the judgment E tumors (group

I, II, III) are detected perfectly (100%) as Cluster 1. Judgment D tumors which are judged by three physicians (group IV) are detected at 90% as Cluster 1, and all remainder cases are detected as Cluster 2. The tumors of group V are detected at 76% as Cluster 1, and also considering Cluster 2 at 87%. Regarding false positive cases, there is an average of 2.4 cases about Cluster 1 and an average of 7.2 cases about Cluster 2 for one case (35 slices). The cases of judgment E include three tumors which were diagnosed as definite lung cancers by close examination, our system detected such tumors as Cluster 1. These results show that our algorithm is helpful in lung cancer diagnosis.

Table.I Criterion of judgment

Judgment E	sure malignant lesion
Judgment D	probably malignant lesion
Judgment C	non malignant lesion
Judgment B	normal case

Table.II Results of the diagnosis

	Judgment	Number of physicians	Detections		
			physician	System	
				Cluster1	Cluster2
I	E	3 physicians	1	1	0
II		2 physicians	2	2	0
III		1 physician	7	7	0
IV	D	3 physicians	10	9	1
V		2 physicians	38	29	4
VI		1 physician	168	98	36
Total			226	146	41

References

[1]T.Iinuma,Y.Tateno,T.Matsumoto,S.Yamamoto, M.Matsumoto, "Preliminary Specification of X-ray CT for Lung Cancer Screening (LSCT) and its Evaluation on Risk-Cost-Effectiveness", NIPPON ACTA RADIOLOGICA, Japan, vol.52, no.2, pp.182-190, 1992

[2]J.Hasegawa,K.Mori,J.Toriwaki,H.Anno,K.Katada, "Automated Extraction of Lung Cancer Lesions from Multi-Slice Chest CT Images by Using Tree-Dimensional Image Processing", Trans.IEICE, Japan, vol.J76-D-II, no.8, pp.1587-1594, 1993

[3]M.L.Giger, K.T.Bae, H and MacMahon, "Computerized Detection of Pulmonary Nodules in computed Tomography Images",Invest Radiol,vol.29,no.4,pp.459-465,1994

[4]M.M.Trivede, J.C.Bezdek, "Low-Level segmentation of aerial images with Fuzzy clustering", IEEE Trans.Syst., Man. & Cybern., SMC-16, 4, pp.589-59, 1986

[5]J.Toriwaki, A.Fukumura, T.Maruse, "Fundamental Properties of the Gray Weighted Distance Transformation", Trans.IEICE,Japan,vol.J60-D,no.12,pp.1101-1108, 1977

[6]K.Kanazawa, N.Niki, H.Nishitani, H.Satoh, H.Omatsu, N.Moriyama, "Computer Assisted Diagnosis of Lung Cancer Using Helical X-ray CT", IEEE Computer Society Press, pp.261-267,1994.

[7]K.Kanazawa, K.Kubo, N.Niki, H.Satoh, H.Ohmatsu, K.Eguchi, N.Moriyama: "Computer Assisted Lung Cancer Diagnosis Based on Helical CT Images", Image Analysis Applications and Computer Craphics, Lecture Notes in Computer Science, Springer-Verlag, pp.323-330,1995.

Unsupervised Regularized Classification of Multi-Spectral MRI

D. Vandermeulen, X. Descombes, P. Suetens, G. Marchal

Laboratory for Medical Imaging Research, ESAT-Radiology
U. Z. Gasthuisberg, Radiology, Herestraat 49, B-3000 Leuven, Belgium

Abstract. In this paper[1], we present a framework for automatic classification of multi-spectral MRI data. We propose a novel clustering method using a contextual hypothesis which succeeds in discriminating largely overlapping component distributions. The initial classifications obtained for each channel in a multi-spectral study independently are subsequently merged by minimizing a Minimum Description Length (MDL) criterion trading the data-fit accuracy for simplicity of the model (number of classes). In a final stage, we use a refined Markovian prior to regularize the final segmentation. This prior preserves fine structures and linear shapes as opposed to the typically used Ising or Potts MRF priors. This work represents work-in-progress. Results on a limited number of data are presented.

1 Introduction

In this paper, we propose a framework for unsupervised segmentation of MRI images. A completely unsupervised procedure is yet only feasible in restricted situations where the imaging protocol is well defined and does not change over time. Alternatively, we define a procedure that contains a minimal number of input parameters. Furthermore, small changes in the value of these input parameters should only result in small and predictable changes in the final segmentation. In the proposed framework, we define segmentation as pixel or voxel classification whereby each pixel/voxel is assigned a label with no other semantic meaning (such as grey matter, bone, white matter...) but indicating similarity in grey value. As such, it amounts to unsupervised classification or clustering. Although promising in some applications, existing unsupervised clustering procedures such as K-means and Fuzzy C-Means (FCM) require a good automatic initialisation of the position of the cluster centers. Moreover, most of the clustering approaches applied to image classification take only the first order statistics into account. Often these procedures do not succeed in discriminating overlapping component distributions. Therefore, we propose to take contextual information into account, based on the hypothesis that neighbors of a pixel are expected to belong to the same class as the central pixel. It can be demonstrated that modes of the conditional mean variance (variance of the average of the neighbours of a

[1] This paper is a short version of a technical report KUL/ESAT/MI2/9608 which can be obtained from the first author (e-mail: dirk.vandermeulen@uz.kuleuven.ac.be).

pixel s, excluding s, as a function of the grey value x_s of pixel s) separate the individual component distributions. These modes are extracted using a a Scale Space Analysis (SSA) procedure [4, 1]. Although this approach can be extended to n-D histograms, $n \geq 1$, it soon becomes computationally intractable. Instead, we perform the contextual clustering on each channel independently and define a combined segmentation as the intersection of the individual channel segmentation maps. This will often result in an over-segmentation of the image. We subsequently merge the regions using a Minimum Description Length (MDL) criterion which trades data fitting for complexity (in casu the number of classes or components) of the model. Finally, we regularize the solution using Markovian priors in a Maximum A Posteriori Bayesian classification. Instead of using the classical binary Ising model or its m-ary extension, the Potts model, as priors that impose smoothness of the segmentation map, we apply a more refined prior, called the DMPS-model, first presented in [5] which better preserves fine structures and detail.

2 Methods

2.1 Automatic clustering

Consider a set of objects $X = \{x_1, ..., x_n\}, x_i \in \mathbb{R}^n$. A clustering is a partitioning of X into subsets $C_j, j = 1, k$ such that $C_r \cap C_s = \emptyset$ for $r \neq s$, $C_r \neq \emptyset, \forall r$ and $\cup_{j=1,k} C_j = X$. In this context the elements x_i are vectors whose components are the grey levels of the n channels (or spectra) corresponding to pixel i. The objective of the unsupervised clustering procedure is to group the x_i into clusters based on a similarity criterion such that elements belonging to the same cluster are more similar than elements belonging to different clusters. The clustering procedure outputs the cluster centers, an estimation of scatter parameters (class conditional distribution parameters) and possibly also an a priori probability for each cluster.

We use the Fuzzy C-Means (FCM) clustering algorithm as a reference unsupervised method. Our experiments (see [6] and comments by Clarke et al.[2, p.349]) indicate that the FCM results are, in general, badly conditioned on the position of the initial cluster centers. An alternative approach [4, 1] models the histogram as a sum of Gaussians (normal mixture distribution). These methods are based on a scale space analysis of the 1-D histograms. The histogram is convolved with a series of Gaussian kernels of gradually increasing width. Each convolution refers to a resolution at a coarser scale. At each resolution zero-crossings of the second derivative are computed. These points correspond to inflection points that split the grey level set into several intervals defining classes and transitions between classes. This first order statistics histogram based method fails when the conditional class distributions overlap to such an extent that the modes as well as the zero-crossings disappear. Using a **contextual hypothesis** (*the neighbors of a pixel belong to the same class ("with high probability") as the considered pixel*), we can recover both distributions as is demonstrated in [6]. We first compute the *conditional mean variance* $\Sigma^2(x_s)$ of the average μ_s of the neighbors of s, excluding s, as a function of x_s, the grey level of s. The typical shape of this curve has more or less constant steps, corresponding to non-overlapping classes, separated by maxima corresponding to mixture intervals. We use the SSA principle to detect modes of the conditional mean variance. Pairs of zero-crossings of the second derivative define an interval corresponding to mixtures between two different classes.

2.2 Merging multi-channel classifications

Although the approach presented above can be extended to n-D histograms, $n \geq 1$, it soon becomes computationally intractable. Instead, we perform the 1-D clustering on each channel independently and define a combined segmentation as the intersection of the individual channel segmentation maps. This will often result in an over-segmentation of the image. We subsequently merge the regions using a Minimum Description Length (MDL) criterion which trades data fitting for complexity of the model. The data-fitting model is defined by the log-likelihood function whereas the complexity of the model is represented by the number of its parameters (see [6] for details).

2.3 MAP Classification using MRF regularizing priors

The classification resulting from section 2.2 amounts to a Maximum Likelihood classification[2]. A pixel s is classified based on its grey value $x_s^k, k = 1, K$ (K = number of input channels) only. A typical problem with these local procedures is the appearance of isolated or wrongly classified pixels. Spatial homogeneity of neighbouring pixel labels is not taken into account. However, by incorporating this additional assumption we can arrive at a regularized solution in a Maximum A Posteriori (MAP) Bayesian framework.

Let $X^1, ..., X^n$ be the data, representing the different input channel images, and Y the segmented image. The MAP segmentation maximizes the a posteriori probability $P(Y|X^1, ..., X^n)$. The known quantities are the a priori model $P(Y)$ and the data term $P(X^1, ..., X^n|Y)$. Using the Bayes rule we get $P(Y|X^1, ..., X^n) = \frac{P(X^1,...,X^n|Y)P(Y)}{P(X^1,...,X^n)} \propto P(X^1, ..., X^n|Y)P(Y)$

Local homogeneity constraints (neighbouring labels are expected to be similar) are easily coded using the framework of Markov Random Fields (MRF): $P(Y) \propto \exp - \sum_{c \in C} V_c(y_s, s \in c)$, with Y the segmented image and C the set of cliques induced by the MRF neighbourhood system. Homogeneous label regions are often modeled using the Potts model (the m-ary extension of the binary Ising model): $V_c(y_s, s \in c) = \sum_{s \in c, s' \in c} \delta(y_s, y_{s'})$ It can be shown that this model favours objects with a low edgelength-to-surface ratio. An improved model, the DMPS-model first proposed in [5], better preserves fine structures and linear shapes in images. This model seems very well adapted to medical images. Indeed, fine structures such as the interface between white and grey matter or CSF

[2] The only prior used in the MDL procedure is to minimize the number of clusters

can be erased or at least spoiled using a Potts model. The DMPS-model is defined by explicitly classifying clique configurations according to noise, lines and objects, associated to 0, 1 and 2-dimensional patterns, respectively. Reference[6] explains the definition of this extended model for the 2-D binary case and its generalization to a 3D m-ary model.

3 Results

Because of page limitations we just report on a single exemplary case of real MRI data (a proton-density/T2-weighted MR image pair). Figure 1 shows the results of the contextual clustering procedure, the MDL-based merging and the MRF-based regularisation using a conventional Potts model and a refined DMPS prior model. For a more extended discussion see [6].

4 Discussion

The Contextual Clustering method can be considered as fully non-supervised. There is one parameter though, the variance of the Gaussian convolution kernel, which depends on the desired precision of the clustering. Moreover, this parameter is also related to the SNR of the input data. For lower SNR, we need a broader kernel and vice versa. A quantitative relationship has not been established but is probably feasible.
The drawback of this method is the slight lack of precision in the parameter estimation due to the contextual hypothesis. Nevertheless, this fact influences the result only when the images contain a lot of edges. Moreover, this can be corrected during the regularization procedure using a MRF-prior.

The results after MDL-based merging are better than those obtained with a multi-channel/spectral fuzzy C-means algorithm. Furthermore, the method does not need an initialization of the cluster centers. There is one parameter (λ), weighting the likelihood function and the term penalizing a large number of classes, in the MDL-criterion. A theoretical value can be obtained based on information theory [3]. In our experiments, we set $\lambda = 400$.

The DMPS-model seems to be an improvement over the Potts model in case of fine structures, especially for low SNR. The DMPS-model is more complicated and more time-consuming, however.

5 Conclusion

In this paper, we have proposed a framework for un-supervised classification or clustering of multi-channel/spectral data. This framework has been applied to the classification of brain structures on multi-spectral MRI. The complete procedure is basically un-supervised apart from the parameters discussed in section 4.

A requirement for this framework to succeed is the assumption of spatial stationarity or translation invariance of the cluster parameters (class conditional

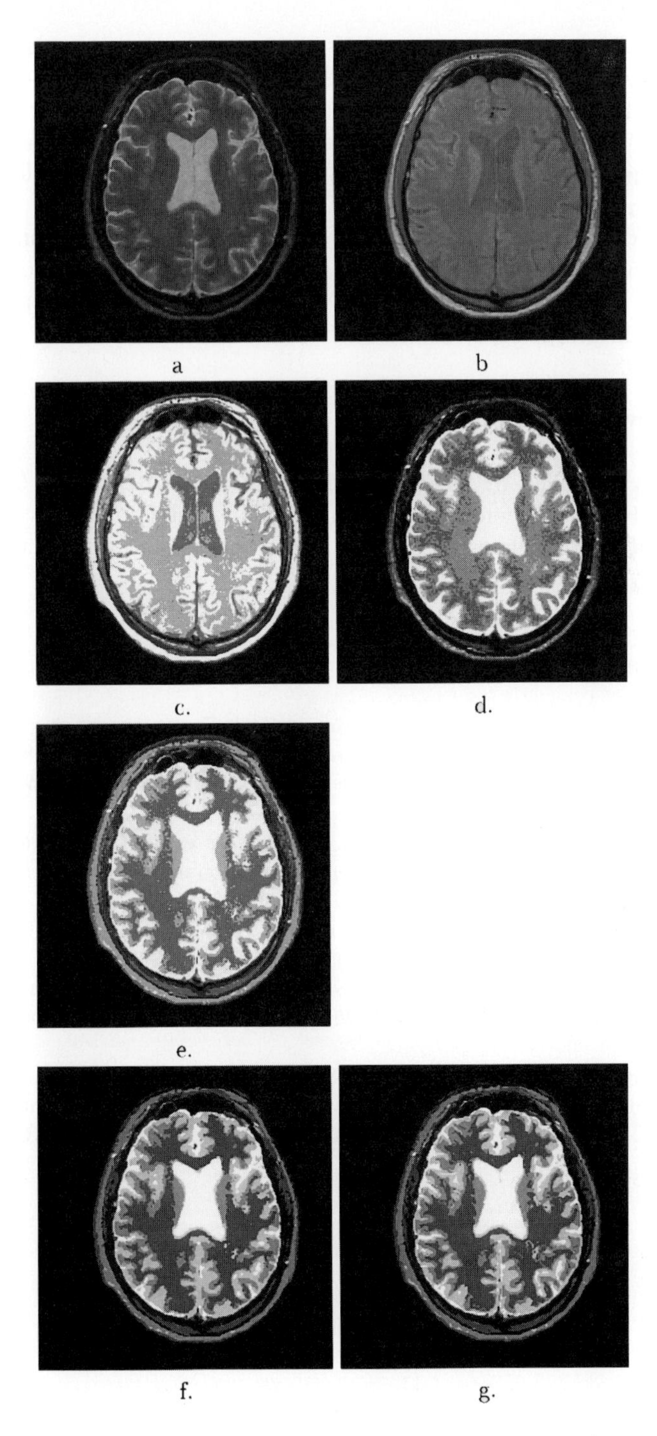

Fig. 1. **a**: T2-weighted MR; **b**: proton density MR; **c**: Contextual clustering for (**a**); **d**: Contextual clustering for (**b**); **e**: MDL-based merging of (**c**) and (**d**); **f**: regularisation of (**e**) using a MRF-Potts model; **g** idem using the refined MRF-DMPS model

distributions). It is well known that MRI data in particular are subject to spatial inhomogeneity distortions due to, a.o., antenna imperfections. Therefore, as a first step of the framework we need to include a method for correcting these stationarities. Supervised as well as non-supervised methods have been reported in the literature (see a.o. the papers by Guillemaud et al., Brechbuhler et al. and Gilles et al. in these proceedings).

We acknowledge that our results are only indicative and far from complete. More quantitative results need to be obtained using software phantom data and a database of expert labelings of "real" MRI-data. As such, this paper reported only on work in progress.

6 Acknowledgements

This research is funded by research grants number 3.0115.92 and 9.0033.93 of the National Fund for Scientific Research (NFWO), Belgium. This work was also financed by a junior fellowship (Xavier Descombes, F/94/153) of the Reseach Council, K.U.Leuven, Belgium.

References

1. M.J. Carlotto. Histogram analysis using a Scale-Space approach. *IEEE trans. Pattern Analysis and Machine Intelligence*, 9(1):121–129, January 1987.
2. L.P. Clarke and all. MRI segmentation: methods and applications. *Magnetic Resonance Imaging*, 13(3):343–368, 1995.
3. Z. Liang, R.J. Jaszczak, R.E. Coleman. Parameter estimation of finite mixtures using the EM algorithm nad information criteria with application to medical image processing. *IEEE Trans. on Nuclear Science*, 39:1126–1133, 1992.
4. A. Goshtasby, W.D. O'Neill. Curve fitting by a sum of Gaussians. *CVGIP: Graphical Models and Image Processing*, 56(4):281–288, 1994.
5. X. Descombes, J.F. Mangin, E. Pechersky, M. Sigelle. Fine structures preserving model for image processing. In *Proc. 9th SCIA 95 Uppsala, Sweden*, pages 349–356, 1995.
6. D. Vandermeulen, X. Descombes, P. Suetens, and G. Marchal. Unsupervised Regularized Classification of Multi-Spectral MRI. Technical Report KUL/ESAT/MI2/9608, ESAT/MI2, Katholieke Universiteit Leuven, February 1996.

Using an Entropy Similarity Measure to Enhance the Quality of DSA Images with an Algorithm Based on Template Matching

Thorsten M. Buzug, Jürgen Weese, Carola Fassnacht and Cristian Lorenz

Philips Research, Technical Systems Hamburg,
Röntgenstraße 24-26, 22335 Hamburg, Germany
buzug@pfh.research.philips.com

Abstract. The reduction of motion artifacts arising in DSA requires registration of mask- and contrast image prior to subtraction. An algorithm has been developed consisting of a) partitioning an interactively chosen region-of-interest (ROI) and exclusion of low-contrast partitions, b) assignment of homologous landmarks or control points, c) estimation of parameters of an affine transformation and application of the transformation on the mask image (inside the ROI), and d) subtraction of contrast- and corrected mask image. For assigning homologous landmarks, we use the entropy as similarity measure and compare the results to other frequently used measures.

1 Introduction

To enhance the quality of digital subtraction images (DSA) patient motion during image acquisition must be corrected. This is usually done by a registration procedure of mask and corresponding contrast image prior to subtraction. Often only a part of the image is of clinical interest and, therefore, a region-of-interest (ROI) is interactively selected. After this selection the registration procedure should be fully automated. An important constraint is the computational efficiency of the algorithm to be implemented on digital angiography systems for clinical routine. The main problem of the automatic registration procedure is that the ROI always contains the vascular structures of interest contrasted with opaque dye. That makes both images to be compared dissimilar and the used similarity measure must cope with these grey-value distortions. We present a 4-step algorithm which meets the above requirements.

2 DSA Algorithm Using Entropy Similarity Measure

Subdivision of region-of-interest. The interactively defined region-of-interest in the contrast image is divided into quadratic, non-overlapping templates of a given size. Due to the well known fact from signal-processing theory that for matched filters the so-called *correlation gain* depends on the product of template size and signal bandwidth [1], only those templates should be incorporated in our algorithm which have sufficient contrast variation. We propose an exclusion technique using an estimate of the mask-image contrast variation calculated via the entropy $h = -\Sigma\, p_g \log p_g$, where p_g is the fraction of pixels with grey-value g and the sum runs from 0 to g_{max}, the maximum grey-value of the mask image. Templates with a high contrast show a broad histogram whereas templates with a low contrast show a narrow distribution of grey-values. Therefore, only those templates having an entropy that exceeds a certain threshold are taken into account.

Assigning homologous landmarks by template matching. The shift of the templates in the mask image with respect to the contrast image is determined by minimizing the entropy of the template after subtraction of the corresponding region in the contrast image. Let g_c, g_m be the grey-values of the contrast and the mask image, x,y the coordinates of the images, and r,s the (horizontal and vertical) shift of the template, respectively. Using these quantities the entropy measure is defined in a 3-step procedure [2]. Firstly, subtract images (only inside the template): $d_e(x,y,r,s) = g_c(x,y) - g_m(x+r,y+s)$. Secondly, calculate the grey-value histogram of the difference and normalize it according to $\Sigma p_g = 1$, where p_g is the fraction of pixels in image $d_e(x,y,r,s)$ with grey-value g and the sum runs from g_{min} to g_{max}, the minimum and maximum grey-value of the difference image, respectively. And, thirdly, calculate the entropy:

$$H(r,s) = -\sum_{g=g_{\min}}^{g_{\max}} p_g \log p_g \tag{1}$$

For the optimal shift, the subtraction image shows low contrast variation whereas in the case of misregistration the contrast variation is high. Therefore, in the case of misregistration, a broad histogram is expected. A perfect registration leads to a peaky histogram, containing peaks corresponding to the homogeneous background, and further peaks caused by the contrasted vessels that cannot (and should not) vanish after subtraction. Hence, the entropy function is minimized yielding shift values of the motion between corresponding templates of mask and contrast image. The template centers together with the resulting shifts define a set of homologous landmarks (or control points, not meant as anatomical landmarks) which is used to estimate the parameters of an affine transformation.
In order to illustrate the result of template matching, we applied the algorithm to an image pair showing the vessel tree of a transplanted kidney. The images were acquired using a Philips Integris V3000 system. Fig. 1a/b shows the contrast and mask image with the set of homologous landmarks (dark crosses in the contrast image, bright crosses in the mask image). The local shift of each landmark is indicated in the contrast image by lines attached to the crosses. It can be recognized that only those landmarks have been selected which show sufficiently high contrast variation in the corresponding template.

Estimation of affine transformation. The set of homologous landmarks resulting from the template matching is used to estimate the parameters of an affine transformation relating the points of the mask image to the corresponding points in the contrast image inside the region-of-interest. As there are much more than three homologous landmarks, this is an overdetermined problem and a singular value decomposition is used to produce a solution that is the best result in the least-squares sense [3]. The affine transformation can cope with distortions like translation, rotation, scaling and skewing. Higher-order polynomes or elastic approaches may be used [4] to further improve the quality of DSA, especially for more complex distortions.

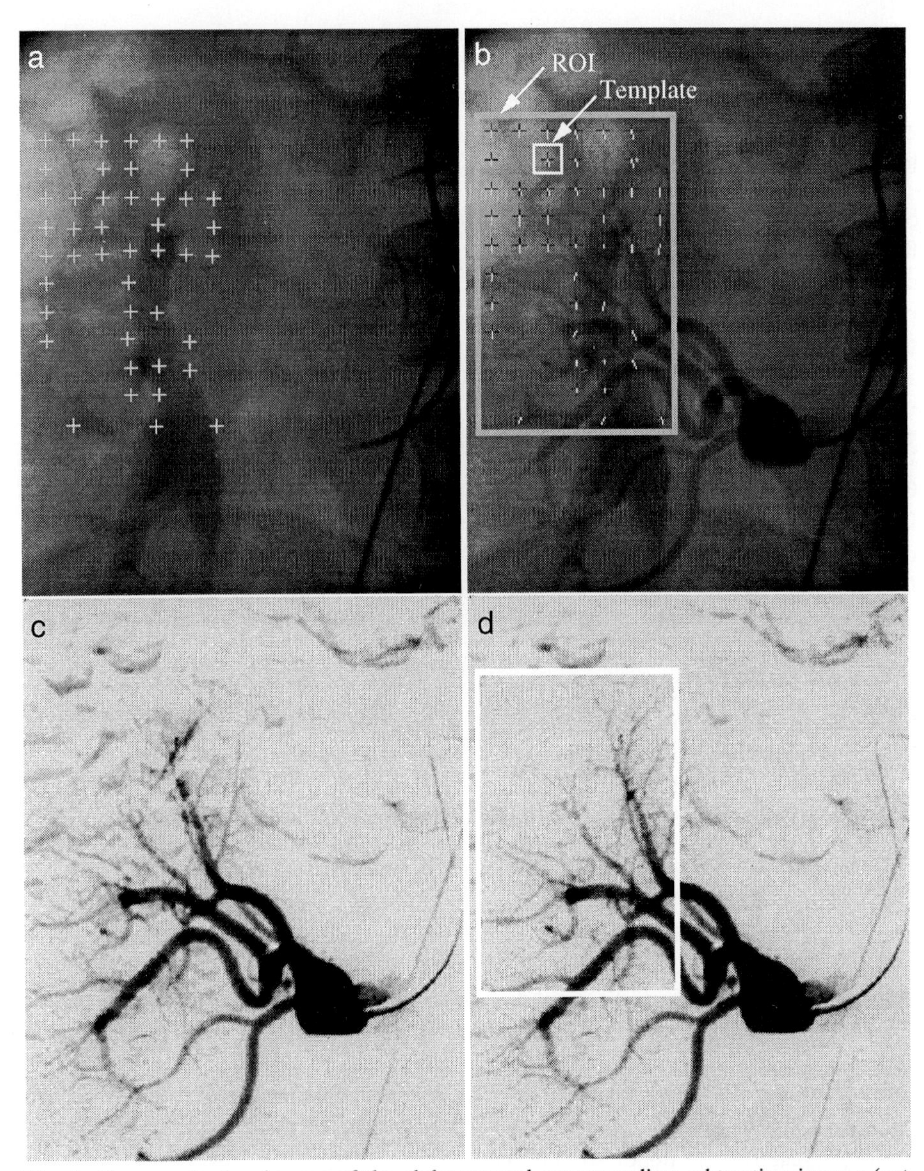

Fig. 1. X-ray projection images of the abdomen and corresponding subtraction images (cutouts of 1024x1024-pixel images can be seen). a) Mask image: The catheter is visible on the right-hand side. b) Contrast image: Opaque dye is injected into the vessel tree of a transplanted kidney. Mask and contrast image are overlaid with the calculated set of homologous landmarks (or control points), indicated as white and dark crosses, respectively. Additionally, lines are attached to each landmark in b) to visualize the motion-vector field in the contrast image. As an example one of the (40x40-pixel) templates is drawn as a white square inside the indicated region-of-interest. c) Uncorrected DSA-image: Smaller vessels are covered by cloud-like artifacts that arise from moving air bubbles of the intestine (motion is caused by breathing and probably peristaltic motion). d) Corrected DSA-image: Affine correction of patient motion inside the indicated region-of-interest.

A general problem is, however, that distortions are inherently 3-dimensional. There are 3-dimensional patient motions leading to distortions which cannot be described by a position dependent shift in the 2-dimensional projection images. Therefore, it is not expected that DSA registration algorithms can ever be developed to the extent that they remove all motion artifacts [5].

Subtraction of images. In a last step, contrast and mask image are subtracted. Within the subtraction the affine transformation is used to compensate the geometric distortions. As an example the algorithm presented has been applied to the image pair of fig. 1a/b. In fig. 1c/d the uncorrected DSA-image and the image corrected inside the region-of-interest are compared. The example clearly demonstrates that the artifacts produced by the patient motion can be reduced significantly.

3 Comparison of Similarity Measures

Of course the question arises whether the entropy measure is better adapted to the DSA problem than other well-known similarity measures such as e.g. cross-correlation function [6], cross-structure function [7], and deterministic sign change (DSC) [8,9]. For that reason these similarity measures are briefly described and compared for another example in fig. 2a/b. The *cross-correlation function* is defined as:

$$G(r,s) = \frac{1}{NM} \sum_{x=1}^{N} \sum_{y=1}^{M} (g_c(x,y) - m_c)(g_m(x+r,y+s) - m_m) \tag{2}$$

where m_c and m_m are the mean grey-values of contrast and mask image, respectively. It can be calculated very fast via the Wiener-Khintchine theorem, but has the disadvantage that it exists only, if the images are spatially stationary. To overcome this problem the *cross-structure function* can be applied. It is defined as

$$S(r,s) = \frac{1}{NM} \sum_{x=1}^{N} \sum_{y=1}^{M} (g_c(x,y) - g_m(x+r,y+s))^2 \tag{3}$$

and exists also for data with drifts. For spatially stationary images the structure function and correlation function are equivalent. Both measures give excellent results if the images to be compared are not distorted. This is demonstrated in fig.2c/d which shows the normalized similarity measures for pixel shifts $r,s \in [-15,15]$. An optimum is expected at $(r,s)=(0,0)$. Unfortunately, both measures will fail, if a structure is overlaid to one image as for DSA (fig.2g/h). In those situations the *deterministic sign change (DSC)* yields much better results. It is defined in three steps: Firstly, add a periodic pattern of depth δ to one of the images to be compared, i.e. $g'_c(x,y)=g_c(x,y)+\delta$ if $x+y$ is even and, $g'_c(x,y)=g_c(x,y)-\delta$ if $x+y$ is odd. Secondly, subtract images $d_d(x,y,r,s)=g'_c(x,y)-g_m(x+r,y+s)$. And, thirdly, evaluate the criterion $D(r,s)$: *Number of sign changes in* $d_d(x,y,r,s)$ *scanned line-by-line*. Fig. 2e/i shows the success of this measure. The disadvantages in DSC are that the image depth δ is not known a priori and the mean value of the images to be compared must be adapted. However, even if these additional parameters are estimated, the objective function may not be smooth (see fig. 2i), and, therefore, be hard to optimize. In contrast, the entropy measure yields reliable results with a smooth basin also for very dissimilar templates as can be seen in fig. 2j.

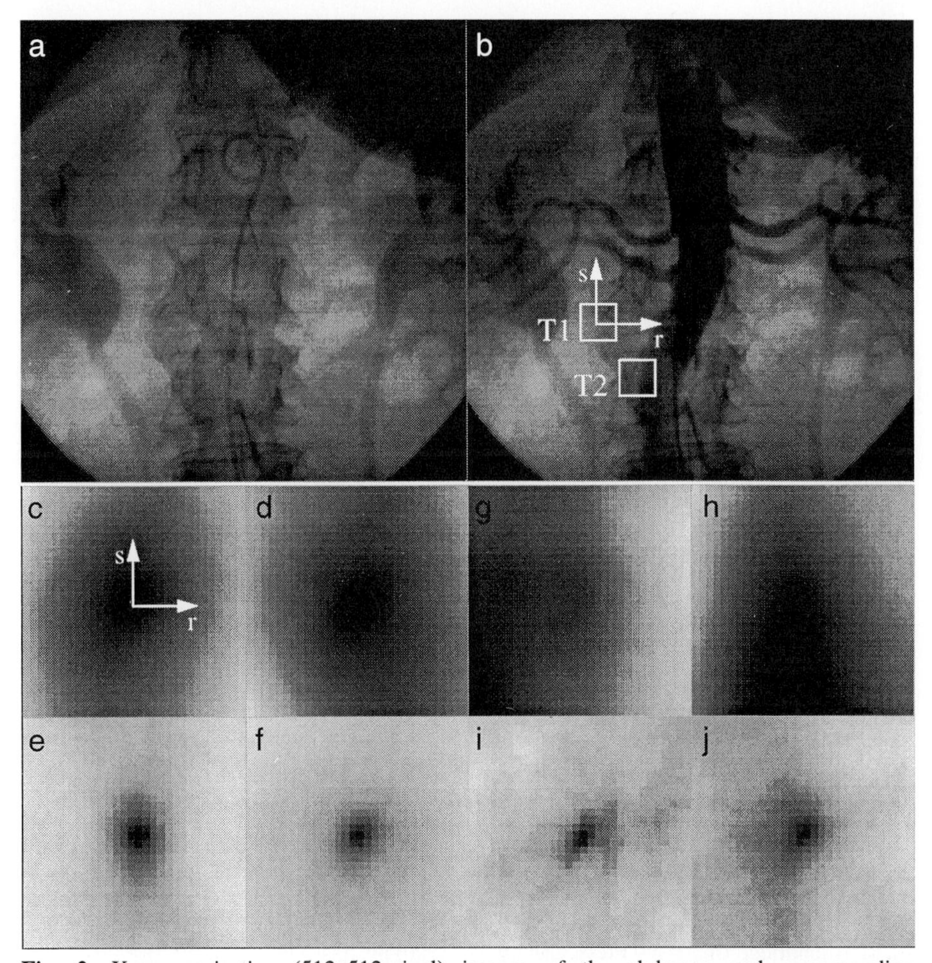

Fig. 2. X-ray projection (512x512-pixel) images of the abdomen and corresponding evaluations of different objective functions for selected templates. a) Mask image: The aorta catheter is visible over the vertebra. b) Contrast image: Opaque dye is injected into the aorta. c) Cross-correlation function G(r,s). d) (Inverted) cross-structure function -S(r,s). e) Deterministic sign change D(r,s). f) (Inverted) entropy -H(r,s). Fig. c) to f) refer to the (40x40-pixel) template T1 indicated in the contrast image. g) to j) Respective similarity measures for the (40x40-pixel) template T2 indicated in the contrast image. S(r,s) and H(r,s) are inverted to allow for a visual comparison of all similarity measures. All functions are normalized to grey-values in the range [0,255]. The origins of the similarity plots are placed in the center of each graphic, as indicated in fig. c). For template T1, where we have no grey-value distortion, all 4 similarity measures yield a conform result for the optimal shift value: (r,s)=(0,0). The entropy measure shows the sharpest optimum with a large, smooth attractive basin. Template T2 is chosen such that approximately one half of it's area includes grey-values from the contrasted aorta. In this situation cross-correlation and cross-structure function, fig. g) and h), respectively, yield false results. We obtain the correct result from the DSC- and entropy measure, that succeed for such dissimilarities. While the DSC shows relatively rough surface, the entropy is again smooth which simplifies an optimization.

4 Conclusion

We presented a 4-step algorithm to enhance the quality of DSA images that are degraded by patient motion. To compensate this motion a point-based registration is employed using homologous landmarks extracted by template matching from the mask and corresponding contrast image. The entropy has been proposed as exclusion criteria in the selection process of landmarks and also as similarity measure in the template-matching procedure. The examples of this paper demonstrate that this measure is especially suitable for image pairs which are dissimilar due to structures overlaid to one image. In this sense the entropy measure can be considered a data-driven approach adapted to DSA.
So far an affine transformation has been used to correct for the motion artifacts. We think, however, that the result can be further improved using elastic transformations. Furthermore, the algorithm may be applied to other images as e.g. to CT/CTA which is the 3D analogous to the DSA problem. For this purpose the algorithm is straightforwardly extended to 3D.

Acknowledgments

The authors would like to thank Dr. L. J. Schultze Kool, University Hospital Leiden, for providing us with the data sets. The algorithm was implemented on an experimental version of the EasyVision workstation from Philips Medical Systems and we would like to thank ICS (EasyVision/EasyGuide) Advanced Development, Philips Medical Systems, Best, for helpful discussions.

References

1. R. Gagliardi: Introduction in Communications Engineering. (John Wiley & Sons, New York, 1978).
2. T. M. Buzug and J. Weese: Improving DSA images with an automatic algorithm based on template matching and an entropy measure. In: Proc. of CAR'96 (Elsevier, Paris, 1996).
3. W. H. Press, B. P. Flannery, S. A. Teukolsky and W. T. Vetterling: Numerical Recipes in C. (Cambridge University Press, Cambridge, 1990) pp. 534-539.
4. K. Rohr, H. S. Stiehl, R. Sprengel, W. Beil, T. M. Buzug, J. Weese and M. H. Kuhn: Point-Based Elastic Registration of Medical Image Data Using Approximating Thin-Plate Splines. In: Proc. of the 4th Int. Conf. on VBC'96, Lecture Notes in Computer Science (Springer, Berlin, 1996).
5. J. M. Fitzpatrick, J. J. Grefenstette, D. R. Pickens, M. Mazer and J. M. Perry: A system for image registration in digital subtraction angiography, in: Information processing in medical imaging. C. N. de Graaf and M. A. Viergever (eds.), (Plenum Press, New York, 1988) p. 415.
6. M. Yanagisawa, S. Shigemitsu and T. Akatsuka: Registration of locally distorted images by multiwindow pattern matching and displacement interpolation: The proposal of an algorithm and its application to digital subtraction angiography, in: 7th Int. Conf. of Pattern Recognition (IEEE Computer Society Press, Silver Spring Md., 1984) p. 1288.
7. E. O. Schulz-DuBois and I. Rehberg: Structure function in lieu of correlation function. Appl. Phys. **24** (1981) 323.
8. A. Venot and V. Leclerc: Automated correction of patient motion and gray values prior to subtraction in digitized angiography. IEEE Trans. on Med. Im. **4** (1984) 179.
9. K. J. Zuiderveld, B. M. ter Haar Romeny and M. A. Viergever: Fast rubber sheet masking for digital subtraction angiography. SPIE **1137**, Science and Engineering of Medical Imaging (1989) 22.

Automatic Segmentation of the Brain in MRI

M. Stella Atkins and Blair T. Mackiewich

School of Computing Science, Simon Fraser University Burnaby, British Columbia, Canada V5A 1S6

Abstract. This paper describes a robust fully automatic method for segmenting the brain from head MR images, which works even in the presence of RF inhomogeneities. It has been successful in segmenting the brain in every slice from head images acquired from three different MRI scanners, using different resolution images and different echo sequences. The three-stage integrated method employs image processing techniques based on anisotropic filters, "snakes" contouring techniques, and a-priori knowledge. First the background noise is removed leaving a head mask, then a rough outline of the brain is found, and finally the rough brain outline is refined to a final mask.

1 Introduction

Magnetic Resonance Imaging (MRI) provides detailed 3D images of living tissues, and is used for both brain and body human studies. Data obtained from MR images is used for detecting tissue deformities such as cancers and injuries. Many uses of head MR images require that the brain tissue is isolated from the head and scalp, for example, in computer-assisted segmentations of brain tissues [5], and in registration of MR images with emission images [9].

This paper describes a new fully automatic method for segmenting the brain from the head in MR images. The key to any automatic method is that it must be robust, so that it produces reliable results on every image acquired from any particular MR scanner using any echo sequence. This method is so robust, that it successfully was able to segment the brain in every slice of head images from 3 different MRI scanners, using different resolution images and different echo sequences. The method works in the presence of radio frequency (RF) inhomogeneity and it addresses the partial volume effect in a consistent reasonable manner.

2 Method

2.1 Overview

We use a 3-stage method to segment the brain. First we remove the background noise, leaving a head mask, then we generate an initial brain mask, and finally we refine the mask for the final segmentation. Each stage has been implemented using the WiT visual programming environment [1], which aids prototype development and enables experimentation [2]. The first stage, *Segment Head*, uses intensity histogram analysis to remove background noise and provide a head mask defining the head. The second stage, *Generate Initial Brain Mask*, produces a mask that approximately identifies the intracranial boundary. It uses nonlinear anisotropic diffusion of the T2-weighted data set and spatial information provided by the head mask to identify regions corresponding to the brain. The nonlinear anisotropic diffusion effectively counters RF inhomogeneity by attenuating non-brain regions. With the initial brain mask as a seed, the third step, *Generate Final Brain Mask*, locates the intracranial boundary using an active contour model algorithm. The active contour model algorithm consistently tracks the edge of the brain, even in the presence of partial volume effects.

The methodology is described in more detail below, and in full detail in [7].

2.2 Segment Head

The head mask is generated using the method suggested by Brummer at al. to determine the "best" threshold level for removing background noise in PD-weighted MR images [3]. By assuming the MR scanner produces normally distributed white noise, they show that background noise in the reconstructed PD-weighted MR volumes has a Rayleigh distribution. The subtraction of the best fit Rayleigh curve from the volume histogram produces a bimodal distribution from which a minimum error threshold can be determined.

Fig. 1 shows the results of automatically thresholding an MR volume using this method. The binary image in Fig. 1(b) produced by thresholding the volume contains speckle outside the head region and there are misclassified regions within the head. This "noise" is easily removed using standard morphological operations.

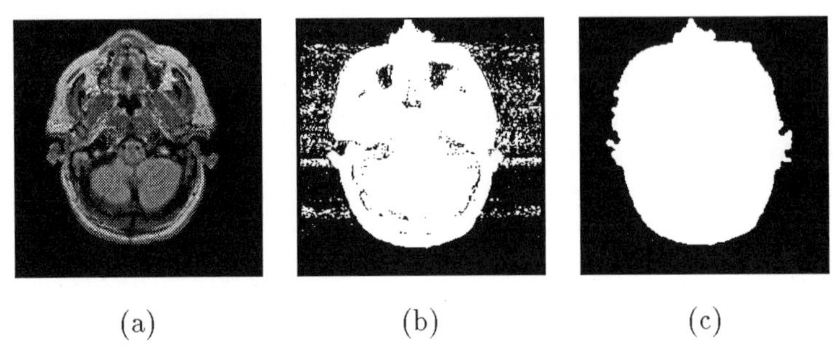

(a) (b) (c)

Fig. 1. A head mask produced using automatic thresholding and morphology. (a) The original PD image slice. (b) Initial head mask. (c) Final head mask after morphology.

2.3 Generate Initial Brain Mask

Generate Initial Brain Mask is a three step process. It first uses 2D nonlinear anisotropic diffusion to attenuate non-brain regions in each MRI slice. Next, it automatically thresholds the diffused MR volume to produce a binary mask.

Finally, misclassified, non-brain regions are removed from the binary mask using morphology and spatial information provided by the head mask.

Nonlinear Anisotropic Diffusion: Nonlinear anisotropic diffusion filters are iterative, "tunable" filters introduced by Perona and Malik [8]. Others have used such filters to enhance or smooth MR images; we use them to sharpen the brain edges so we can attenuate the brain using one filter for the whole volume.

The discrete diffusion process consists of updating each pixel in the image by an amount equal to the flow contributed by its four nearest neighbors. [1] We use 2D nonlinear anisotropic diffusion to attenuate non-brain regions in each MRI slice[2].

Automatic Threshold: Once each T2-weighted MRI slice has been diffused to attenuate non-brain voxels, we segment the brain using a single automatic threshold. The threshold level is determined by fitting a gaussian curve to the histogram of the diffused volume, then choosing a level at two standard deviations below the mean [3]. Fig. 2 shows a slice of the binary mask produced by the threshold.

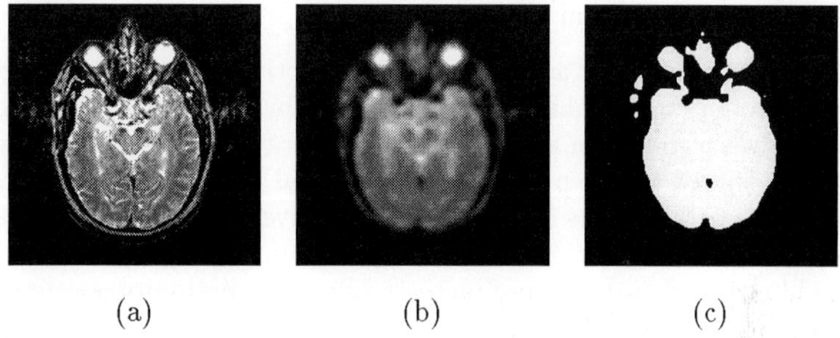

(a) (b) (c)

Fig. 2. A binary mask produced by automatic thresholding. (a) The original T2 image slice. (b) The diffused image slice. (c) The corresponding binary mask.

Mask Refinement: The binary mask produced by automatic thresholding contains misclassified regions, such as the eyes (see Fig. 2(c)). These regions are removed from each slice using morphology and spatial information provided by the head mask. First, holes are filled within each region of the mask. Next, binary erosion is performed to separate weakly connected regions. After erosion, regions whose centroids fall outside a bounding box defined by the head mask

[1] Eight nearest neighbors can be used if the flow contribution of the diagonal neighbors is scaled according to their relative distance from the pixel of interest [4]. Anisotropic data can be handled similarly.

[2] We also attempted to attenuate non-brain tissues using 3D nonlinear anisotropic diffusion over the entire MRI volume. 3D diffusion, however, increased the partial volume effect, blurring the edges of the brain [7].

are eliminated. Finally, binary dilation is performed on the remaining regions to return them to their original size. The results of these steps are shown in Fig. 3.

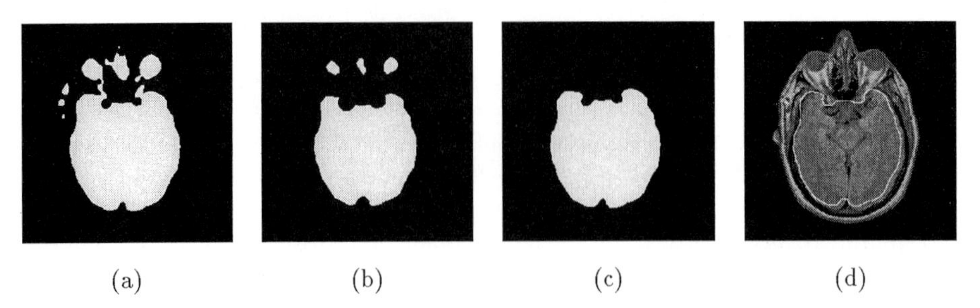

(a) (b) (c) (d)

Fig. 3. Elimination of misclassified regions from the initial binary mask produced by thresholding in Fig. 2(c), (a) after regions have been filled, (b) after binary erosion, (c) after elimination of non-brain regions and binary dilation. (d) shows the initial brain mask overlaid on the PD-weighed slice.

2.4 Generate Final Brain Mask

Given the initial brain mask as a seed, we use an active contour model algorithm, extended from the "snakes" algorithm introduced by Kass et al. [6], to locate the boundary between the brain and the intracranial cavity. The algorithm deforms the contours defined by the perimeter of the initial brain mask to lock onto the edge of the brain. The points in the contour iteratively approach the intracranial boundary through the solution of an energy minimization problem.

A combination of four energy functions based on continuity, gradient, intensity and balloon forces enables the active contour model algorithm to detect the intracranial boundary in all image slices using the same relative energy weightings. The diffusion process ensures that the initial brain mask falls completely inside the brain. Therefore, the adaptive balloon force aides in cases where the initial brain mask is poor. The intensity energy term helps the active contour algorithm produce consistent results where partial volume effects are particularly severe. Finally, computing the gradient energy term from the diffused T2-weighted volume greatly stabilizes the active contour algorithm because the gradient derivatives are small in the diffused volume [7].

Fig. 4 illustrates the result of applying our active contour model algorithm to the MR slice shown in Fig. 3. The parameters were experimentally chosen to produce good results, and we found that the same parameters were suitable for many different data sets.

3 Results

Fig. 5 shows the intracranial contour detected automatically by our algorithm overlaid on selected slices of a PD-weighted MRI data set. The PD and T2 data

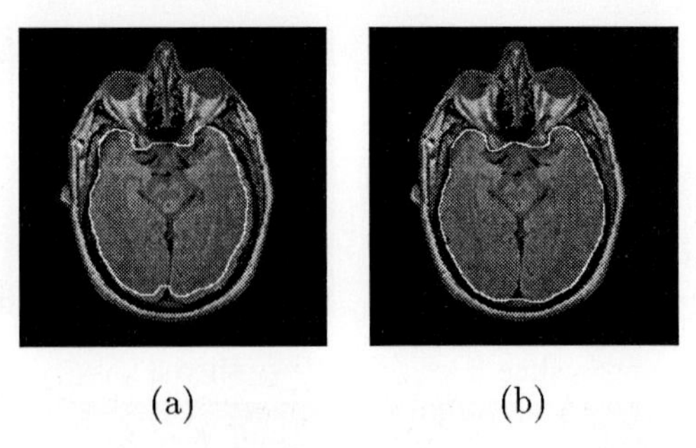

(a) (b)

Fig. 4. Refinement of the intracranial contour. (a), the contour defined by the perimeter of the initial brain mask. (b), the intracranial contour detected using the active contour model algorithm.

sets were acquired axially on a GE 1.5 Tesla MRI scanner, with repetition time TR = 2000 ms, and echo times of 20 ms and 100 ms respectively. The data consists of 28 slices with 256 * 256 8-bit pixels per slice. The pixel size is 0.9375 mm^2, and the slice thickness is 5 mm.

This data was transferred to us "upside-down" with the eyes at the bottom of the slices. The only modification to the program to deal with this, was to change the parameters in the second step which remove blobs whose centroid could not possibly be brain tissue.

The intracranial boundary shown in Fig. 5 is accurate in all slices, except for an apparent slight overshoot at the top in slices 4,5 and 6. Careful examination shows that the partial volume effect in these slices is dealt with in a consistent way, and the apparent overshoot is not a problem for subsequent analysis of the brain data.

Our algorithm produced comparable results for many other data sets from two other scanners [7]. In all cases, our algorithm successfully detected the intracranial boundary without user interaction and without parameter modification.

Acknowledgment This research has been partially funded by the Natural Sciences and Engineering Research Council of Canada. The authors wish to thank Dr. Don Paty, Dr. David Li, Dr. Alex MacKay, Dr. K. Whittall, Andrew Riddehough and Keith Cover of the University of British Columbia MS/MRI Group and to Charlie DeCarli from the NIH, for providing MRI data on which the algorithms were tested.

References

1. T. Arden and J. Poon. *WiT User's Guide*. Logical Vision Ltd., Burnaby, Canada, October 1993. Version 4.1.
2. M.S. Atkins, T. Zuk, B. Johnston, and T. Arden. Role of visual languages in developing image analysis algorithms. In *Proceedings of the IEEE Conference on Visual Languages*, pages 262–269, St. Louis, MO, October 1994. IEEE.
3. Marijn E. Brummer, Russell M. Mersereau, Robert L. Eisner, and Richard R. J.

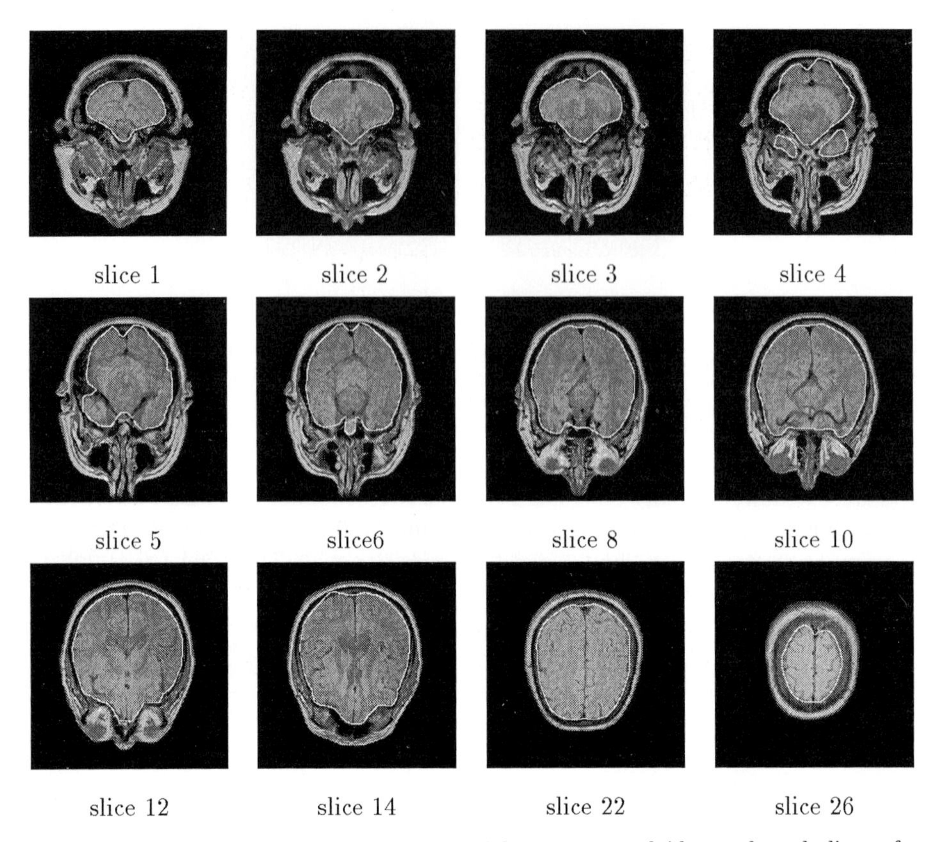

Fig. 5. An automatically detected intracranial contour overlaid on selected slices of a PD-weighed MRI scan.

Lewine. Automatic detection of brain contours in MRI data sets. *IEEE Transactions on Medical Imaging*, 12(2):153–166, June 1993.

4. Guido Gerig, Olaf Kübler, Ron Kikinis, and Ferenc A. Jolesz. Nonlinear anisotropic filtering of MRI data. *IEEE Transactions on Medical Imaging*, 11(2):221–232, June 1992.
5. B. Johnston, M.S. Atkins, B. Mackiewich, and M. Anderson. Segmentation of multiple sclerosis lesions in intensity corrected multispectral MRI. *IEEE Transactions on Medical Imaging*, 15(2):154–169, April 1996.
6. Michael Kass, Andrew Witkin, and Demetri Terzopoulos. Snakes: Active contour models. *International Journal of Computer Vision*, pages 321–331, 1988.
7. Blair Mackiewich. Intracranial boundary detection and radio frequency correction in magnetic resonance images. Master's thesis, Simon Fraser University, Computer Science Department, Burnaby, B.C., August 1995.
8. Pietro Perona and Jitendra Malik. Scale-space and edge detection using anisotropic diffusion. *IEEE Trans. on Pattern Analysis and Machine Intelligence*, 12(7):629–639, July 1990.
9. R.P. Woods, S.R. Cherry, and J.C. Mazziotta. Rapid automated algorithm for aligning and reslicing PET images. *J. Comp. Ass. Tom.*, 16:620–633, 1992.

Image Analysis Using Modified Self-Organizing Maps: Automated Delineation of the Left Ventricular Cavity Boundary in Serial Echocardiograms

Marek Belohlavek[1], Armando Manduca[2],
Thomas Behrenbeck[1], James B. Seward[1], and James F. Greenleaf[2]

[1] Division of Cardiovascular Diseases and Internal Medicine,
[2] Department of Physiology and Biophysics, Mayo Clinic
and Mayo Foundation, Rochester, Minnesota 55905

Abstract. An algorithm is described for delineation and three-dimensional (3D) reconstruction of the left ventricular (LV) cavity boundary from echocardiographic images. The algorithm combines advanced image analysis and neural network techniques with a priori knowledge about LV shapes. Minimal user interaction is required to initiate the process which results in computer-generated outlines of the LV. Laboratory tests with dog hearts compare the algorithm to a conventional method based on endocardial outlines by an expert. LV volume and shape comparisons are assessed.

1. Background

Kohonen introduced self-organizing maps (SOM) in the early 1980s and employed them to various recognition tasks such as texture or speech analysis [1, 2]. Briefly, the SOM is an unsupervised neural network which maps input data, defined by input vectors, from an n-dimensional space onto a lattice of nodes in an m-dimensional space (where $m \leq n$) in an ordered fashion. The geometry of the lattice is fixed (rectangular, hexagonal, etc.), and the nodes have neighborhood relationships defined on the lattice. Each node has an initial associated internal state vector which can be considered to be a position in the n-dimensional input space. During training (self-organization) the internal state vectors automatically fit themselves to the probability distribution of the input vectors while preserving local neighborhood relations, i.e., nearby points in the input space tend to map to nearby nodes on the lattice. The starting value of the internal state vectors may be random or may approximate some (a priori known) target shape, such as LV shape. The self-organizing process can be intentionally influenced by assigning weights to individual input vectors which control the attraction of the SOM nodes.

2. Introduction

Reproducible determination of the endocardial boundary in echocardiographic images is clinically important and has been extensively examined through various filtering and edge detection techniques [3–5], albeit with only moderate success, as ultrasound cardiac images are typically prone to noise, dropout, and other artifacts and thus more difficult to process. Manhaeghe applied the SOM algorithm to estimation of the LV myocardial midwall in emission tomograms in 1994 with good results [6].

The purpose of this study was to explore the capability of a modified SOM algorithm combined with advanced image processing techniques to determine the LV cavity boundary (i.e., endocardium) in serial echocardiographic images.

Custom image processing software was combined with a modified SOM algorithm to automatically delineate the endocardium in serial echocardiographic tomograms. The processing software was designed to reduce image noise and classify the probability of the presence of the endocardium by enhancing cavity-to-wall transition regions. The SOM algorithm was applied to the processed images to recognize the transition regions associated with the LV cavity boundary and label it by distributing individual SOM nodes. The original Kohonen algorithm was modified so that the nodes no longer have the freedom to distribute themselves anywhere in the image during the self-organizing process. Instead, a trainable statistical algorithm has been developed which gains knowledge about LV shapes. This algorithm bounds SOM node distribution to a specific region of interest (ROIs) along the boundary for each node.

3. Methods

3.1. Image Collection

Six dog heart specimens were pressure perfused and fixed with formalin. Their true volumes were measured by introducing a thin latex sheath into the LV and filling it with known amounts of water. The volumes ranged from approximately 30 to 100 ml. Each specimen was placed in a water bath and scanned using a Hewlett Packard 77020A ultrasound system with a 3.5 MHz transducer located adjacent to the apex of the specimen in a rotation apparatus. The transducer was rotated in 22.5° increments about its imaging axis over a 180° span to obtained a series of 8 long-axis tomograms characterizing the LV.

3.2. Manual Delineation of the Endocardium Representing a Conventional Approach

Endocardial contours were outlined manually in each of the 8 tomograms (Fig. 1) by a person trained in echocardiography. Outlines were reconstructed into a 3D solid cast. Volume of the cast was calculated as a sum of cast voxels with known size.

3.3. Image Filtering and Enhancement of a Cavity-to-wall Transition Region

Histogram equalization and stretching to a full 8-bit gray-scale range was performed in each tomogram to facilitate distinction of the blood pool (compare Fig. 2(1) with

Fig. 2(2)), i.e., LV cavity appears anechoic (dark); myocardium appears echogenic (bright). The interface between the blood pool and myocardium was termed cavity-to-wall transition region (CWTR). It encompasses the endocardium and epicardium. We employed a 9x9 median filter (Fig. 2(3)) to remove noise spikes which can be misinterpreted as endocardium. A rank-based operator, modified from [7], which weights the separation of the highest and lowest density values within a sliding 9x9 window, was used to assign higher pixel intensities to locations determined as the CWTR and lower intensities to the cavity and myocardium (Fig. 2(4)).

The next necessary step was to recover LV cavity boundary from the CWTR through a higher order analysis, such as logical distribution of interconnected nodes.

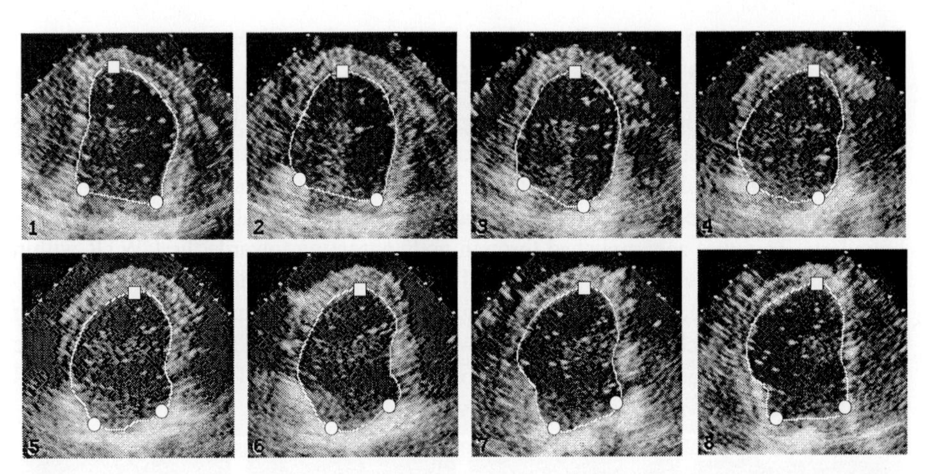

Fig. 1 A set of 8 tomograms representing the heart volume with superimposed manual outlines. Histogram equalization and stretching to a full 8-bit range enhanced definition of edges. Circles approximate mitral valve hinge points (mitral valve leaflets were removed during specimen preparation), squares denote the approximate LV apex in each tomogram. Manual delineations could be compared with those generated by computer (Fig. 3).

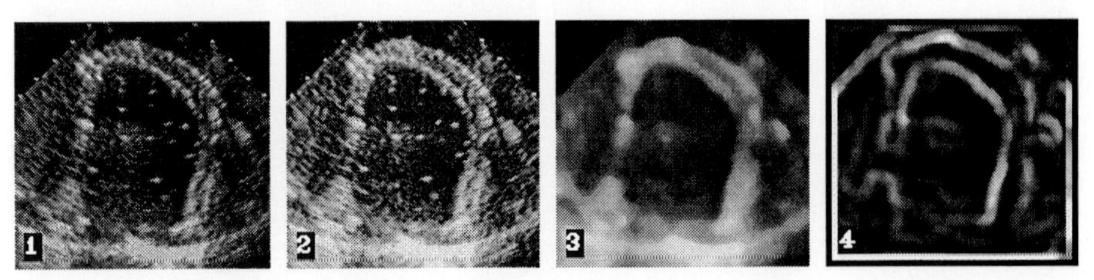

Fig. 2 A cross-sectional tomogram of the dog heart. (1) Original gray scale image. (2) Histogram stretching to a full 8–bit range of gray and equalization enhances distinction between the blood pool (dark) and muscle regions (bright). (3) Filtering with 9x9 median filter removes most of the noise spikes while preserving edges. (4) Processing with a 9x9 rank-based operator (see section 3.3) brightens cavity-to-wall transition regions (CWTR).

3.4. Determination of the LV Cavity Boundary

LV shape and size vary from heart to heart and in the beating heart with the cardiac cycle. To approximate LV boundary within the CWTR, an experimental knowledge database was created using rotational long-axis cardiac scans of 10 sample hearts: Endocardial borders were outlined manually by an expert familiar with echocardiography who identified the apex (AP) and 2 mitral valve (MV) hinge points in each tomogram. The AP and MV points are labelled for illustration in Fig. 1 by squares and circles, respectively. The training images were used to compute deformable "average outline" and its variability for each tomogram.

3.5. Computer-driven LV Boundary Outline using Shape Knowledge and a Modified SOM Algorithm

We used 17 SOM nodes in each tomogram as LV contour markers. The nodes were initially distributed along the "average outline" fitted into AP and MV points in each tomogram, defined interactively. A self-organizing process was employed to redistribute the nodes from their "average" positions to locations mapping the LV boundary. This was accomplished through node attraction towards the CWTR within limits of the known variability of each outline. This prevented distribution of nodes to other than endocardial areas. The self-organization process consisted of 500 training cycles per node which required only a few minutes on a SPARCS20 workstation. The process was designed with a 3D neighborhood relationship of nodes which guided node distribution in regions with a poorly defined boundary.

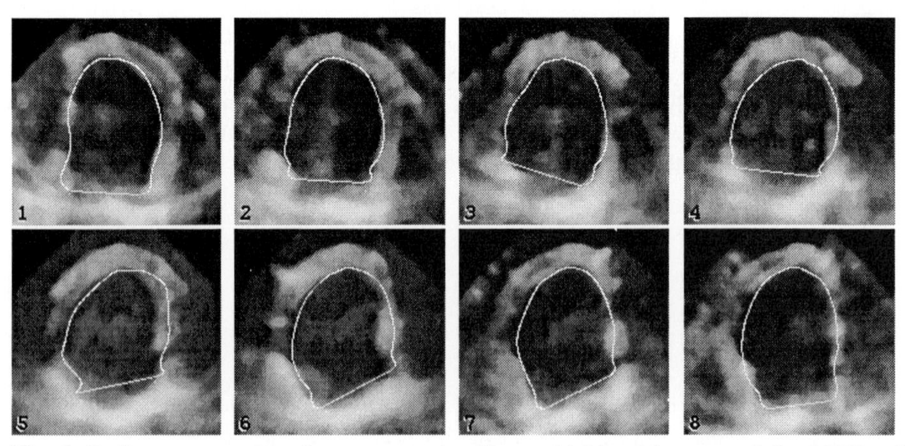

Fig. 3 Outlines generated by the computer algorithm and initiated from user defined MV and AP labels (not visible here, they were removed during algorithm execution).

4. Results

4.1. Manual Versus Computer-generated LV Shapes

Figure 1 shows a series of enhanced tomograms with manual outlines. These outlines were compared to those generated by the computer algorithm (Fig. 3). There are general shape similarities, but also obvious differences, especially in the mid-sections of the outline in Fig. 3. These differences, however, did not influence accuracy of measured volumes.

4.2. Manual Versus Computer-generated LV Volumes (Table 1)

	Volume (ml)		
Heart	Reference	Manual outline	Computer outline
1	103.6	101.2	102.1
2	27.6	28.5	26.4
3	86.3	86.1	83.4
4	90.2	93.5	93.2
5	53.0	55.9	53.2
6	62.8	66.2	63.2

Table 1 LV volumes in 6 dog heart specimens. True (reference) volumes are compared to those estimated from manual and computer-generated outlines.

There was an excellent correlation ($R = 0.99$, $p = 0.001$, SEE = 1.3 – 2.4 ml) and no significant difference by a paired t test at a 95% confidence level in all three paired combinations of measurements. The results reflect the independency of volume from corresponding shape, i.e., high correlation in volumes may not necessarily infer exact shape similarity.

5. Discussion

This is the first known application of a modified SOM algorithm in combination with custom image preprocessing to determine LV cavity surface from ultrasound images in three dimensions and generate data for LV volume measurements.

Trainable a priori knowledge of the likely LV shape represents an important feature of the algorithm. It allows a quick initial estimation of LV configuration based upon elastic fitting of the "average outline" into the AP and MV labels placed by the user. The database for this particular experimental study was built from 10 echo tomographic series. The shape knowledge can be augmented by adding more expert-outlined contours. It can be anticipated, however, that for clinical application, the diastolic and systolic shapes should be separated for their substantially different geometry, particularly in irregularly shaped hearts (myocardial infarction, etc.).

Important positional information during the self-organization process is conveyed through 3D relationships among nodes. This information provides the clues for node placement in image regions with suboptimal definition of the endocardial boundary.

6. Conclusions

We have developed a novel algorithm for LV cavity delineation in echocardiographic images and accurate calculation of its volume.

The limitations include: 1) the requirement for an initial interactive placement of AP and MV points and 2) differences in resulting shapes with respect to manual outlines. However, intra- and inter-individual variability, typical for manual outlines of echocardiographic images, made establishment of a reference standard difficult.

The strengths of the algorithm are in: 1) automated and fast determination of the LV cavity boundary after the interactive placement of the AP and MV points, 2) 3D capability of the algorithm to include a database of LV shapes, and 3) robustness, i.e., the resistance against node escaping through image shadows.

7. Acknowledgments

This work was supported in part by grants HL 41046 and HL 52494 from the National Institutes of Health

8. References

1. Kohonen, T.: Self-organized formation of topologically correct feature maps. Biol. Cybern. **43** (1982) 59–69.
2. Kohonen, T.: *Self-organization and Associative Memory*. Springer-Verlag, New York, NY (1984) 2nd ed. 1988.
3. Chu, C., Delp, E., Buda, A.: Detecting left ventricular endocardial and epicardial boundaries by digital two-dimensional echocardiography. IEEE Trans. Med. Imag. **7** (1988) 81–90.
4. Brotherton, T., Pollard, T., Simpson, P., DeMaria A.: Classifying tissue and structure in echocardiograms. IEEE Eng. Med. Biol. (1994) 754–760.
5. Coppini, G., Poli, R., Valli, G.: Recovery of the 3–D shape of the left ventricle from echocardiographic images. IEEE Trans. Med. Imag. **14** (1995) 301–317.
6. Manhaeghe, C., Lemahieu, I., Vogelaers, D., Colardyn, F.: Automatic initial estimation of the left ventricular myocardial midwall in emission tomograms using Kohonen maps. IEEE Trans. Patt. Anal. Machine Intell. **16** (1994) 259–266.
7. Pitas, I., Venetsanopoulas, A.: Edge detectors based on nonlinear filters. IEEE Trans. Patt. Anal. Machine Intell. **8** (1986) 1893–1921.

Analyzing and Predicting Images Through a Neural Network Approach

L. de Braal[1], N. Ezquerra[1], E. Schwartz[1], C. D. Cooke[2], and E. Garcia[2]
[1] College of Computing, Georgia Institute of Technology, Atlanta, GA 30332
[2] Radiology Department, Emory University, Atlanta, GA 30322

Abstract. A neural network approach has been developed to predict diagnostic image information to assist in the assessment of coronary artery disease. The predicted information represents the redistribution (or reversibility) of perfusion in the myocardium. A multilayer, backpropagation neural network is trained to predict the redistribution information from two other types of images: stress perfusion and myocardial thickening using SPECT imaging. The significance of this approach is two-fold: (i) the predicted reversibility information obviates the additional acquisition of delayed images (with the patient at rest), and (ii) the neural network approach represents a novel way with which to analyze and predict images from other images. This paper presents the methods that underlie the approach, and discusses the most recent experimental results that demonstrate its viability.

1. Introduction and Rationale

The interpretation and visualization of cardiovascular nuclear medicine images remains a challenging, information-intensive process. A recent development in this diagnostic process is the integration of myocardial thickening as a possible indicator of myocardial viability. Myocardial thickening information, used in conjunction with stress perfusion information, can serve as a measure of the redistribution of perfusion in viable (but infarcted) myocardial tissue. This paper discusses how a neural network approach can be used to analyze and process the thickening and stress perfusion information to predict the associated perfusion redistribution information.

The rationale for this approach draws from previous biomedical experiments. Reports have shown that, in dogs, thickening of left ventricular myocardium during systole is reflective of myocardial perfusion [1][2].

A dynamic feature of a normal left ventricular myocardium is its thickening during systole by approximately 50%. With myocardial infarction, wall motion abnormalities are observed and systolic wall thickening is dramatically reduced. Furthermore, clinical researchers have shown [3] that when using Sestamibi Tc-99m, the assessment of systolic wall thickening in humans can provide information about tissue viability and myocardial perfusion. Since the predictive value of (absent) systolic wall thickening has been reported to be 100%, this thickening information can therefore represent an indication of myocardial functioning. If myocardial thickening is measured simultaneous to assessing myocardial perfusion at peak stress, then determination of ischemia, scar and viable myocardium in a single setting would be possible. The clinical importance of this finding would be significant, since aspects such as time, cost and morbidity could be considerably reduced. One difficulty, however, is that the patterns of the rate of thickening are difficult to visualize. Also, the visualization and interpretation of the thickening and stress information together represent a relatively new and complex visualization problem. To address these considerations, machine learning methods appear to offer a viable and computationally efficient approach to assist in the analysis, processing, and interpretation of this complex information, as discussed subsequently.

2. Machine Learning through Artificial Neural Networks

Machine learning methods have evolved as a way with which to create systems that can discern complex image patterns in a manner that is relatively accurate, generalizable, computationally efficient, and robust with respect to noise and data variations. Artificial neural networks ("connectionist systems"), in particular, contain algorithms that can be trained to extract patterns from complex images and subsequently discern similar patterns in newly encountered imagery [4][5][6].

Thus, an artificial neural network (ANN) may enable prediction of redistribution perfusion, based on the images associated with myocardial thickening and stress perfusion. Based on this hypothesis, a connectionist system has been constructed, a schematic overview of which is given in Figure 1. As shown in the figure, the redistribution information predicted by the ANN would serve to create and replace an actual resting perfusion image that presently requires a second clinical test. With this in mind, we have designed, implemented and trained an ANN to accept images associated with stress and thickening as input, and provide redistribution information as output.

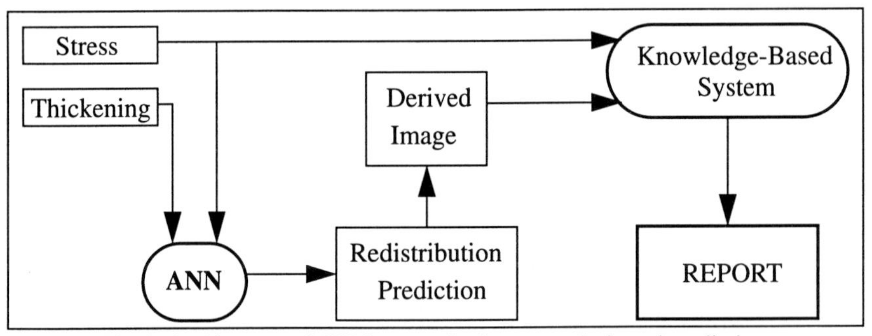

Fig. 1. Architecture of ANN system for image analysis and prediction.

As Figure 1 suggests, the image derived from the ANN-based processing is introduced to a knowledge-based system that further interprets both the stress and predicted image information and provides an overall diagnostic interpretation. This knowledge-based system is discussed in [7][8], and only aspects of this system that are relevant to the present discussions will be included herein. The thrust of the present paper is thus placed on the prediction of imagery through connectionist methods.

In creating the connectionist system, several configurations of input and output data formatting, and ANN topology have been investigated, implemented, and tested. The data consist of thickening and perfusion studies of 109 real patient cases, using Tc-99m.

For the purposes of training and testing the ANN, it was experimentally determined that partitioning the data into regions associated with individual perfusion defects (rather than associating a data set with one complete patient case) yielded optimal results. Thus, the data were represented as perfusion defects, where each defect consists of a closed region with reduced stress perfusion levels equal to or exceeding 2.5 standard deviations from normal levels [7]. This resulted in a data set containing 211 test cases, with one case for each defect found in the original 109 patient cases.

Of the 211 test cases, about 80 percent were randomly selected for training purposes, with the remaining 20 percent left for testing. Consistency and reliability of the study were improved by repeating each ANN training and test cycle 4 additional times, each time using another 20 percent of the complete data set for testing purposes. This approach had the result that, for each network topology, five different networks were trained and tested. The results of testing the five networks were used to calculate the overall performance for all 211 cases, without violating the ANN evaluation requirement that test cases are not allowed to be part of the training data set.

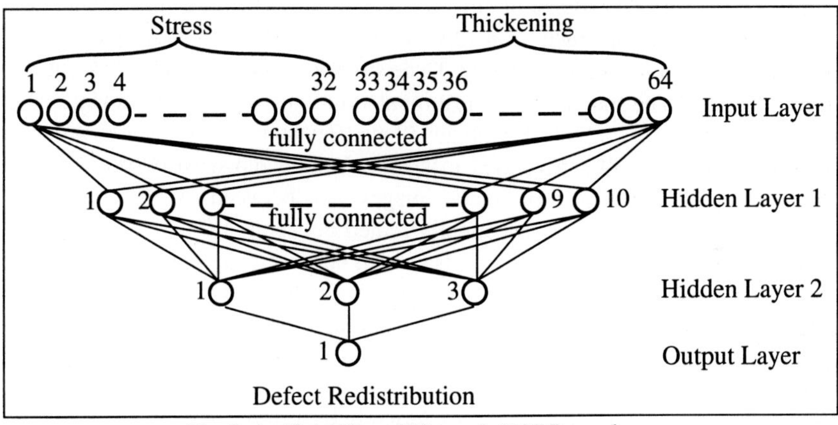

Fig. 2. Artificial Neural Network (ANN) topology.

The ANN topology that has yielded optimal training and testing results thus far has the configuration illustrated in Figure 2. The input layer consists of 64 nodes, made up of 32 values to describe each of the thickening and stress perfusion images. Since the layers are fully connected, no specific ordering of the 32 regions is required other than that the ordering should be the same for all data sets.

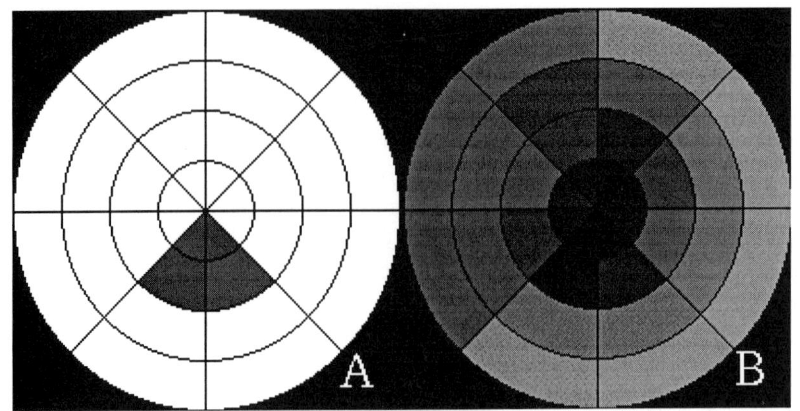

Fig. 3. Examples of input images: (A) stress perfusion and (B) thickening, where each polar map is subdivided into 32 regions.

Each set of 32 values is derived from image information represented in polar format, as illustrated in Figure 3. The polar maps are sampled into 8 angular regions and 4 regions describing the distance from the apical center, giving rise to the 32 regions encompassing the perfusion and thickening information.

Figure 3 shows a polar image of stress perfusion in (3a), and a similar polar image providing thickening information in (3b), where both are subdivided into 8x4 polar representations. These images provide 32 descriptors each, yielding an input with a total of 64 values. The specific numerical input values are 32 standard deviations from normal (for perfusion) and 32 percentage values (for thickening). Prior to processing, these input values are first truncated to span a limited range (between 0.0 and 8.0 standard deviations for stress perfusion images, and between -325% and 600% for thickening percentages), and subsequently scaled between -1.0 and 1.0 to normalize the range of values. This preprocessing is needed because there is a large difference in value ranges between stress perfusion standard deviations and thickening percentages, and the chosen ANN configuration requires similar value ranges as inputs for proper behavior. Outside of a defect area, the stress perfusion input is set to the lowest possible value: -1.0.

The topology is a fully connected, feed-forward, back-propagation network with two hidden layers of 10 and 3 nodes, as shown in Figure 2. A standard sigmoid transfer function and delta learning rule were used. The output consists of one single node, providing a value indicating the average redistribution perfusion associated with the defect area in question. A bias connected to all processing elements of hidden and output layers is used to check if the input and output are scaled sufficiently to let the network focus the calculations on relating the changes in input and output to network weights (instead of improperly allocating resources to scale the values, which could also introduce inaccuracies). The test values for the output have been truncated between 0.0 and 5.0 standard deviations, and are then scaled between the values of 0.20 and 0.80, representing a limitation in useful output values for a backpropagation network in which a sigmoid transfer function is used.

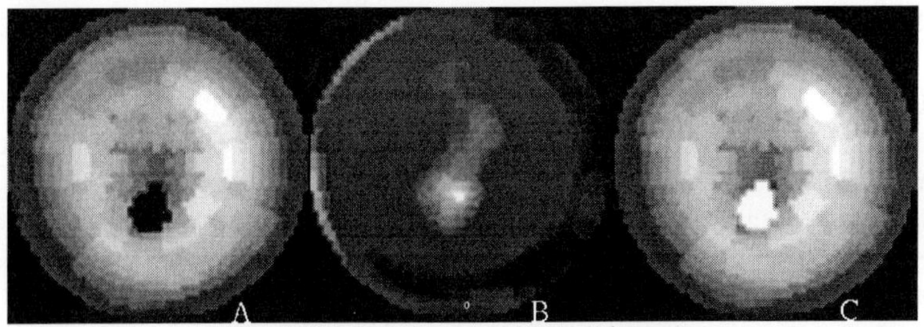

Fig. 4. Example input and output images:
(A) stress perfusion, (B) thickening and (C) predicted redistribution.

This ANN has been successfully trained, using 100,000 cycles. Each case of 80/20 percent training on an SGI indigo (100Mhz, R4000, 80MB internal) workstation requires approximately 25 minutes. An example of how the ANN analyzes the input information that enables the creation of a redistribution image is shown in Figure 4, using data from an actual patient case. The stress perfusion (4a) and thickening (4b) information are segmented into 32 regions each (see 3a and 3b) for each defect area, and subsequently used as input to the trained network. The ANN provides one value as output, indicating the reversibility for that defect area. This single value is then substituted in the defect area of the original stress image as a reversed area (4c) if the value exceeds a threshold of 0.30, a value determined by experimentation.

3. Evaluation

To obtain a measure of the performance of the developed ANN configuration, the 211 defect area test cases were used in a five step 80% training, 20% testing cycle, each time using a different 20% for the testing. This results in 211 tests total. The output of each of these tests was compared to the actual reversibility information, using an experimentally determined threshold (0.30) to create a separation between defects showing reversibility or not.

The results obtained experimentally in predicting the occurrence of reversibility using the ANN-based method described above, based on the tests conducted with the 211 defect area data set, are as follows:

Sensitivity: 74%
Specificity: 55%
Overall Accuracy: 72%

These preliminary results show the viability of this approach. At present, different and more complex data configurations and ANN topologies are being explored. Current research thrusts center on improving the predictive performance in terms of accuracy, reliability, and the ability to make reversibility predictions with greater quantitative granularity.

References

1. Gallagher P, Matsuzaki M, et al. "Effect on Exercise in the Relationship between Myocardial Blood Flow and Systolic Wall Thickening in Dogs with Acute Coronary Stenosis", Circ Res: S2. 1983: 716
2. Gallagher P, Matsuzaki M, et al. "Regional Myocardial Perfusion and Wall Thickening during Ischemia in Conscious Dogs", American Journal of Physiology, 1984: 16:H727
3. Ziffer J, La Pidus A, Alazraki N, Folks R, and Garcia E. "Predictive Value of Systolic Wall Thickening for Myocardial Viability Assessed by Tc-99m Sestamibi Using a Count Based Algorithm", 40th Annual Scientific Session, American College of Cardiology, 1991
4. Hertz J, Krogh A, Palmer R. "Introduction to the Theory of Neural Computing", Addison-Wesley, Redwood City, CA 1991
5. Maren A, Harston C, Pap R. "Handbook of Neural Computing Applications", Academic Press, London, 1990
6. Muller B, Reinhardt J. "Neural Networks: An Introduction", Springer-Verlag, Berlin, Germany, 1990
7. Ezquerra N, Mullick R, et al. "PERFEX: An Expert System for Interpreting 3D Myocardial Perfusion", Expert Systems With Applications, Vol. 6, pp. 459-468, 1993
8. Garcia E, Herbst M, Cooke D, et al. "Knowledge-Based Visualization of Myocardial Perfusion Tomographic Images", In Conference on Visualization in Biomedical Computing, 1990: 157-161

An Artificial Intelligence Approach for Automatic Interpretation of Maxillofacial CT Images

M. Alcañiz[1], V. Grau[1], C. Knoll[2] M. C. Juan[1], C. Monserrat[1] and S. Albalat[3]

[1] Departamento de Ingeniería Gráfica, Universidad Politécnica de Valencia, SPAIN
[2] Technische Universitat Erlangen GERMANY
[3] Faculty of Medicine and Dentistry of Valencia. Valencia. SPAIN

Abstract. In this paper we present an automatic system for segmentation and recognition of relevant tissues in maxillofacial CT images. The system allows to dynamically validate anatomical information of these structures. Our procedure differs from previous attempts in its use of advanced low level segmentation operators and specific knowledge bases that embody knowledge about tissue characteristics, not about specific anatomical structures or organs. System results tested on CT images from five patients running on a PC-based hardware are very promising both in accuracy and processing time. The developed system has applications in dental implantology, allowing the optimization of surgery 3D planning in low cost PC-based workstations.

1 Introduction and Backgrounds.

One of the first medical disciplines which embraced three-dimensional (3D) imaging was craniofacial surgery. 3D imaging is particularly useful in this area due to the complexity of cranial and facial skeletal structures [4]. In recent years, several advances have been made in the field of craniofacial surgery treatment prediction with 3D computer tomographic skull models [2]. In dental implant surgery, planning has almost exclusively been based in information provided by coronal dental CT scans and panoramic images. These techniques give the clinician only a bidimensional vision of the reception zones for implant surgery. The possibility of 3D visualization and navigation through the zone on which the implants will be placed allows to establish the correct implant sites, implant size, and precise drilling and insertion directions, so that the implant can be firmly placed and surrounded by cortical bone on all sides. Three-dimensional display is important because the complete understanding of a treatment plan requires comprehension of the spatial relationships among anatomical structures and prosthetics devices [1]. It is interesting for the clinician that, in these 3D computer models, anatomical structures can be manipulated individually, thus it is necessary to segment the CT images. The nature of this type of images makes them impossible to segment using standard techniques such as thresholding. Manual segmentation, as we have experienced, will never be used routinely by the clinicians, due to the large time cost associated. In this situation, it is absolutely necessary to develop an automatic segmentation system. In spite of the great efforts oriented in this direction, at the present time nobody has yet obtained a totally automatic and reliable segmentation system for CT images. In general, we could think that the general segmentation methods are not enough to obtain a correct segmentation of the images. It is necessary to apply to each case the "a priori" knowledge of the anatomical structures that appear in each image. This way, the segmentation process can be guided, making it specially adapted to each type of image, to obtain much better results.

2 Characteristics of CT Maxillofacial Images.

In our case, the problem is especially difficult, due to the nature of the images (figure 1). The system must distinguish the following anatomical structures and tissues: cortical and espongiosa bone, teeth, maxillary sinus and possible implants. The main problem is that the boundaries separating tissues from their surroundings are not always clear; especially in the case of cortical and espongiosa bone. It is also difficult, in many cases, to separate teeth and implants from cortical bone, as their gray levels are very similar.

3 System Description.

We have implemented a knowledge-based medical image interpretation system for the segmentation of maxillofacial CT images. Our procedure differs from previous attempts in its use of advanced low level segmentation operators and specific knowledge bases that embody knowledge about tissue characteristics, not about specific anatomical structures or organs. The system is designed in the form of a low level segmentation module (LLS) and a brain module (figure 2). The LLS module carries out a low level segmentation, implementing the techniques known as watersheds and region growing. The output of this block is an oversegmented image, formed by very small subregions. The purpose of this block is to convert the initial data volume (a large number of individual pixels) in a smaller number of higher-level structures (regions), which are more easily managed by the brain module. The brain module condenses the knowledge that the clinician possesses about CT maxillofacial images. The basic control structure is that of a production system, in which specific knowledge about independent low level properties of an image is formulated into condition-action rules. The system process matches rules in the brain module against the data originated in the LLS module, so that when a match occurs, the rule fires. This triggers an action that is performed by that process, and usually involves modification of the data. In this way, the segmentation process can be guided, making it specially adapted to each type of image, to obtain much better results. This module acts mainly by joining adjacent regions that correspond to the same tissue, and only in special cases a region can be resegmented through a call to the low level algorithms (figure 2).

4 Low Level Segmentation (LLS) Module.

This block takes as input directly the CT image, and converts it in a predetermined number of small regions. To achieve this segmentation, we use a two-step process. Initially we apply the immersion algorithm to find the watersheds [6] in the image, in order to extract edges in the image, along with a great number of false contours due to noise. This algorithm produces a highly oversegmented image. In second place, a specially designed region growing algorithm is applied that reduces the number of regions thus preserving the important edges. For the region growing, we use a combination of edge detection and clustering approaches [5]. The regions are joined in order, beginning with the ones that give a lower value of a "cost function". Each time a joining is carried out, the costs affected are recalculated. The cost function used in the process has the following form:

$$Cost = C_1 + \alpha C_2$$

where C1 is the difference in gray level of the regions 1 and 2, while C2 represents the "edgeness" of the pixels that form the frontier between the two regions, and is calculated as the sum of the gradient values of those pixels. In this way, we combine the formation of homogeneous regions and the correct detection of the edges. The α coefficient is set empirically. The joining process is stopped only when the number of regions equals a predetermined value. It is important that this value is adequately set, for a high number of regions means much more work for the expert system, and a low one would make some important contours disappear. We try to choose always a conservative value, to make the minimum number of resegmentations.

4.1 Resegmentation.

The low level segmentation algorithms implemented are used always as the first step of the segmentation, but not only in this case. During the high level segmentation, it is possible for the expert system to re-segment a part of the image. This resegmentation would be carried out if it is impossible to label a region with enough confidence, or if a strange structure has been found. Errors of the two kinds are detected by the expert system, which turns the image back to the low level block for resegmentation. Resegmentation can be done of a rectangular area (if some regions have been joined to background) or of a single region (if this region has grown too much). This step is achieved through a call from the expert system to the low level segmentation algorithms.

5 Brain module - high level segmentation (HLS).

5.1 Knowledge Implementation.

In the brain module, we can distinguish three types of knowledge bases (KB): general KBs, which contain constructs to store results provided by low level segmentation modules and other knowledge bases; anatomical KBs, which contain knowledge about the different anatomical regions, and tissue KBs, which embody general knowledge about the tissues.

object.tkb This knowledge base contains significant classes, objects and properties to store resulting data. A class called "region" is characterized mainly by 6 properties: area, length of its boundary, position of the centroid, average gray level, shape and a boolean value that represents if the region is neighbor to background. All those properties are scale-invariant through a normalization process. The idea is to increase the confidence level for labeling a region if the different conditions are satisfied, as in [3].

global.tkb As a result of LLS, a list of all existing global regions in the image is passed to the expert system. A global region (anatomical region) is created by unifying all existing regions determined by LLS, which form connected regions, excluding the background region. In this case, there are three global regions of interest: the maxillary region and the left and right condyle regions (fig. 1).

condyle.tkb This knowledge base is loaded twice in case of upper jaw images, to label tissues of the two condyle areas. These areas are characterized by a closed outer belt of cortical surrounding a darker region of espongiosa.

maxillar.tkb Possible tissues in maxillary zone are cortical, espongiosa, tooth, implant and air (sinus). Criteria based in gray level and shape are used as a first step. Constraints are applied: for example, espongiosa regions do not touch air regions: there should always be a cortical region between them. To distinguish teeth from implants, its inner nerve is detected as a dark central point.

5.2 3D High Level Segmentation.

Now HLS will be continued correcting results of all cuts by extending the segmentation process to 3 dimensions. The comparison of labeled regions with their neighbors in near above and below cuts of the CT-image series leads to a reduction of mislabeled regions and a higher consistency of the complete 3-dimensional segmentation. Using this knowledge base, possible positions of teeth and implants are stored, as well as the numbers of the lowest and highest cuts where each tooth or implant appears. This way, all teeth and implants are correctly detected in all cuts.

6 Results.

We have applied the above system to several complete CT-scan sets, the one presented in fig. 3 consisting in 14 images obtained of a patient's upper jaw. The images were taken by a General Electric Pace Plus CT (General Electric,

Milwaukee, USA), with a collation of 1mm-thickness at slice intervals of 2 mm. We have tested the system in a Pentium CPU, with 32 Mb RAM under Windows NT operating system. Figure 3(a) shows the results for one of the higher cuts of the CT image set. Cortical, espongiosa and air (corresponding to maxillary sinus) have been correctly segmented. In figure 3(b), a lower cut has been segmented. Two new "tissue" types appear: teeth and implants. We appreciate here how the expert system is able to segment correctly quite different images. This is because of the modular structure of the high level segmentation module. We see also that the system correctly differentiates between teeth and implants, although those two elements have very similar gray level and shape. In this image, we also appreciate that, in some implants and teeth, portions of the cortical bone have been identified as if they were part of the teeth or implant. This problem will be corrected with a more precise identification of the shapes.

7 Conclusion.

We have shown that the results of our automatic segmentation system are very accurate, valid for a 3-dimensional reconstruction with only minor manual corrections. For this purpose we have developed a segmentation editor. The modularity of the knowledge bases makes them reusable in other body locations: for example, rules defined for the condyle regions (thin cortical outside espongiosa) could be used almost in any image where we want to differentiate cortical and espongiosa bone, as for example in femur CT, where an accurate reconstruction of the cortical volume would be vital for prosthesis design. As future works we are actually implementing new knowledge bases, that, added to the ones we already have, make our system applicable to other anatomical human locations.

References

1. **Hemmy D. C.** "Future directions in 3D imaging." In: Udupa J.K. and Herman, G.T. (eds). 3D imaging in medicine. Boston: CRC Press, 1991: 331-341.
2. **Lambrecht J.** "3-D modeling technology in oral and maxillofacial surgery." Berlin. Quintessence Books. 1995.
3. **Levine M., Shaheen S.** "A modular computer vision system for picture segmentation and interpretation." IEEE Transactions on Pattern Analysis and Machine Intelligence. 1981.
4. **Luka B, Brechtelsbauer D, Gellrich NC, Konig M.** "2D and 3D CT reconstructions of the facial skeleton: an unnecessary option or a diagnostic pearl?." Int J Oral Maxillofac Surg. 24: 76-83. 1995
5. **Pavlidis, T, Liow, Y.** "Integrating Region Growing and Edge Detection". IEEE Transactions on Pattern Analysis and Machine Intelligence, pp 225-233. March 1990.
6. **Vincent, L, Soille, P.** "Watersheds in Digital Spaces: An Efficient Algorithm Based on Immersion Simulations". IEEE Transactions on Pattern Analysis and Machine Intelligence, pp 583-598. June 1991.

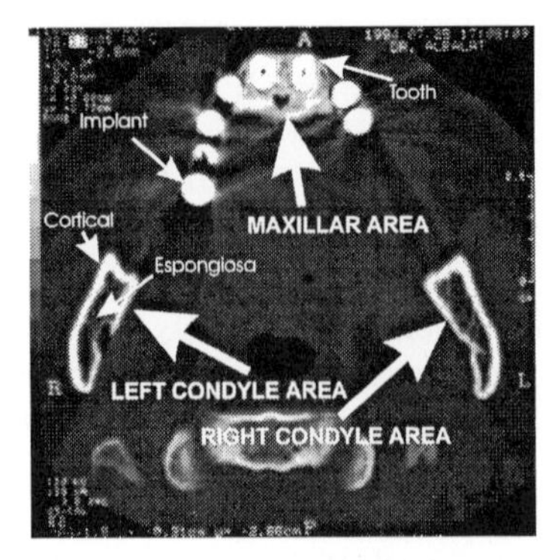

Fig. 1. Upper maxillar CT image

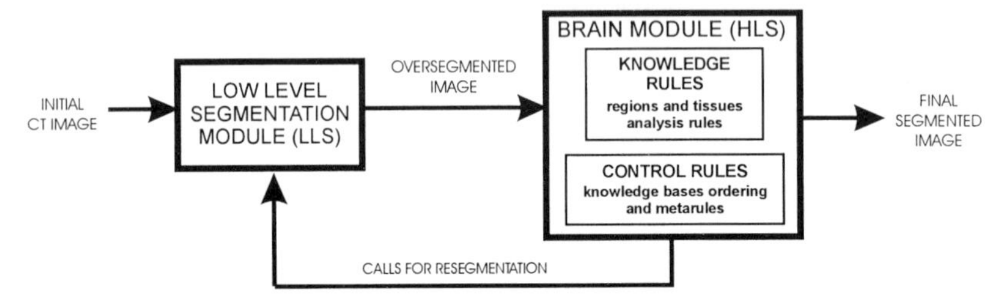

Fig. 2. Block diagram of the whole system. The two modules are executed in order, with the possibility of back calls to the low level segmentation module

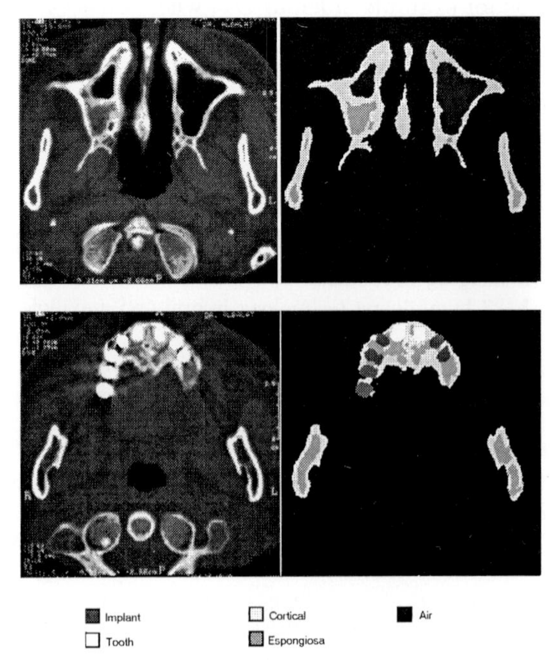

Fig. 3. Results of the whole segmentation process, showing original and segmented images.

Registration

Fast Fluid Registration of Medical Images

Morten Bro-Nielsen[1,2] and Claus Gramkow[1]

[1] Dept. of Mathematical Modelling
Technical University of Denmark, Bldg. 321
DK-2800 Lyngby, Denmark
[2] 3D-Lab, School of Dentistry, Univ. of Copenhagen,
Nørre Alle 20, DK-2200 Copenhagen N, Denmark
e-mail: bro@imm.dtu.dk WWW http://www.imm.dtu.dk/~bro

Abstract. This paper offers a new fast algorithm for non-rigid Viscous Fluid Registration of medical images that is at least an order of magnitude faster than the previous method by Christensen et al. [4]. The core algorithm in the fluid registration method is based on a linear elastic deformation of the velocity field of the fluid. Using the linearity of this deformation we derive a convolution filter which we use in a scale-space framework. We also demonstrate that the 'demon'-based registration method of Thirion [13] can be seen as an approximation to the fluid registration method and point to possible problems.

1 Introduction

Non-rigid registration of two medical images is performed by applying global and/or local transformations to one of the images (which we will call the template T) in such a way that it matches the other image (the study S). It is important to understand that the aim of the transformation is to map the template *completely* onto the study in such a way that information from the template can be applied to the study as well. A very important application of non-rigid registration is using an electronic atlas to segment a study image. Only if the transformation maps the template completely onto the study can the atlas be used to infer conclusions about the contents of the study. In practice this means that the transformation must accommodate both very complex and large deformations.

Bajcsy et al. [1, 10] were the first to demonstrate volumetric non-rigid registration of medical images. Building on initial work by Broit [2], they modelled the template image as a linear elastic solid and deformed it using forces derived from an approximation of the local gradient of a correlation based similarity measure. Multi-resolution were used to increase the speed.

Evans et al. [8] used anatomical landmarks to drive the deformation, and deformed the solid using a globally elastic model. Miller, Christensen et al. [3, 4] also used a globally elastic model, but derived the driving force from the derivative of a Gaussian sensor model.

These previous approaches to non-rigid registration have all suffered from the use of either *global* transformations or *small deformation* assumptions (as used in linear elasticity).

In [5] Christensen et al. extended their work and described a registration approach in which they use a viscous fluid model to control the deformation. The template image is modelled as a thick fluid that flows out to match the study under the control of the same derivative of a Gaussian sensor model they used in [4]. In [5] Christensen argue that this gaussian sensor model theoretically is better than the correlation based similarity measure used by Bajcsy et al. [1].

Elastic models constrain the possible deformation because the deformation is a compromise between internal and external forces. Elastic displacements do not reach the desired deformation because of internal strain in the elastic continuum. In a viscous fluid model, internal forces disappear over time and the desired deformation can be fully achieved.

Consequently, the fluid registration method satisfies the general requirements of both complex and large deformations and we therefore regard this method as the most advanced registration method available.

Unfortunately, the algorithm proposed by Christensen et al. is rather slow. They originally implemented the algorithm using a massively parallel DECmpp 128x64 MasPar computer on which the algorithm used on the order of 5-10 minutes for 2D and 2-6 hours for 3D registrations. In a recent paper [6] they show estimates of the execution time on a MIPS R4400 processor on the order of 2 hours for 2D and 7 days for 3D. In practice this means that the algorithm is not feasible unless a massively parallel computer is available.

The contribution of this paper is a new fast algorithm based entirely on convolution with filters which gives a *speed-up of at least an order of magnitude.*

2 Theory

In this section we describe our new algorithm for solving the viscous fluid registration problem. Without loss of generality the theory is described in the 2D case, but it is readily extendable to the 3D case.

In the first section we describe the original viscous fluid algorithm by Christensen et al. [5]. We define the template and study images and their relationship. The viscous fluid model is introduced along with the driving force and the numerical solution method. In the second section we discuss the core part of Christensen's numerical solution and introduce the general idea behind the convolution approach that we propose to increase the speed of the method. In the third section, the basic filter for the convolution approach is finally derived.

2.1 Fluid registration

We define the template image as $T(\boldsymbol{x})$ and the study image as $S(\boldsymbol{x})$ where $\boldsymbol{x} \in [0;1]^2$. The purpose of the registration is to determine a warping of $T(\boldsymbol{x})$ onto $S(\boldsymbol{x})$.

Eulerian reference frame In elastic deformation, particles are usually tracked by their initial coordinates, ie. the parametrization of the object. This sort of reference frame is called *Lagrangian*. But in fluid deformation the Lagrangian reference frame is inefficient and an *Eulerian* reference frame is used instead. In the Eulerian reference frame the particles are tracked based on their current/final position. Consequently, a particle at position $\boldsymbol{x} = [x_1, x_2]^T$ in the template image at time t originated at position $\boldsymbol{t}(\boldsymbol{x}, t) = \boldsymbol{x} - \boldsymbol{u}(\boldsymbol{x}, t)$ at time t_0 ($t > t_0$), where $\boldsymbol{u}$ is the displacement. Notice that in the Eulerian reference frame, $\boldsymbol{u}(\boldsymbol{x}, t)$ describes the displacement of the particles as they move through $\boldsymbol{x}$. The Eulerian velocity field is determined by:

$$\boldsymbol{v}(\boldsymbol{x}, t) = \delta \boldsymbol{u}(\boldsymbol{x}, t)/\delta t + \nabla \boldsymbol{u}(\boldsymbol{x}, t)\boldsymbol{v}(\boldsymbol{x}, t) \tag{1}$$

where ∇ is the gradient operator. The term $\nabla \boldsymbol{u}(\boldsymbol{x}, t)\boldsymbol{v}(\boldsymbol{x}, t)$ results from the chain rule of differentiation and accounts for the kinematic non-linearities of the particles.

Viscous fluid model In the Eulerian framework we can write the partial differential equation (PDE) for the viscous fluid deformation of the template as [5]:

$$\mu \Delta \boldsymbol{v}(\boldsymbol{x}) + (\mu + \lambda)\nabla(\nabla \cdot \boldsymbol{v}(\boldsymbol{x})) = \boldsymbol{f}(\boldsymbol{x}, \boldsymbol{u}(\boldsymbol{x})) \tag{2}$$

where $\Delta = \nabla^T \nabla$ is the Laplacian operator and $\nabla(\nabla \cdot \boldsymbol{v})$ is the divergence operator. The force field $\boldsymbol{f}(\boldsymbol{x}, \boldsymbol{u}(\boldsymbol{x}))$ is used to drive the flow. Those familiar with elasticity theory will recognize that for constant force $\boldsymbol{f}$ this is actually the PDE for linear elasticity working on the velocity field $\boldsymbol{v}$. The equation, therefore, works by elasticly smoothing the instantaneous velocity field of the fluid.

The term $\Delta \boldsymbol{v}$ is also called the viscous term because it constrains the velocity field spatially. The $\nabla(\nabla \cdot \boldsymbol{v})$ term allows for contraction or expansion of the fluid.

The force field is defined as the derivative of a cost function C. For MRI images a Gaussian sensor model appears to be an appropriate model of the variation between the template and study image [5, 11]. The cost function and its derivative are:

$$C(T(\boldsymbol{x}), S(\boldsymbol{x}), \boldsymbol{u}) = \frac{1}{2} \int_{\Omega} | T(\boldsymbol{x} - \boldsymbol{u}(\boldsymbol{x}, t)) - S(\boldsymbol{x}) |^2 \, d\boldsymbol{x} \tag{3}$$

$$\boldsymbol{f}(\boldsymbol{x}, \boldsymbol{u}(\boldsymbol{x}, t)) = -[T(\boldsymbol{x} - \boldsymbol{u}(\boldsymbol{x}, t)) - S(\boldsymbol{x})] \nabla T|_{\boldsymbol{x} - \boldsymbol{u}(\boldsymbol{x}, t)} \tag{4}$$

Numerical solution Solution of the viscous fluid registration problem requires solving the PDE [5]:

$$\mu \Delta \boldsymbol{v}(\boldsymbol{x}, t) + (\mu + \lambda)\nabla(\nabla \cdot \boldsymbol{v}(\boldsymbol{x}, t)) = \boldsymbol{f}(\boldsymbol{x}, \boldsymbol{u}(\boldsymbol{x}, t)) \tag{5}$$

$$\frac{\delta \boldsymbol{u}(\boldsymbol{x}, t)}{\delta t} = \boldsymbol{v}(\boldsymbol{x}, t) - \nabla \boldsymbol{u}(\boldsymbol{x}, t)\boldsymbol{v}(\boldsymbol{x}, t) \tag{6}$$

$$\boldsymbol{f}(\boldsymbol{x}, \boldsymbol{u}(\boldsymbol{x}, t)) = -[T(\boldsymbol{x} - \boldsymbol{u}(\boldsymbol{x}, t)) - S(\boldsymbol{x})] \nabla T|_{\boldsymbol{x} - \boldsymbol{u}(\boldsymbol{x}, t)} \tag{7}$$

which includes non-linearities in both the force and the material derivative. To solve this problem we apply Euler integration over time using a forward finite difference estimate of the time derivative in equation 6:

$$\begin{aligned} \boldsymbol{u}(\boldsymbol{x}, t_{i+1}) &= \boldsymbol{u}(\boldsymbol{x}, t_i) + (t_{i+1} - t_i)(\boldsymbol{I} - \nabla \boldsymbol{u}(\boldsymbol{x}, t_i))\boldsymbol{v}(\boldsymbol{x}, t_i) \\ &= \boldsymbol{u}(\boldsymbol{x}, t_i) + (t_{i+1} - t_i)\nabla \boldsymbol{t}(\boldsymbol{x}, t_i)\boldsymbol{v}(\boldsymbol{x}, t_i) \end{aligned} \tag{8}$$

Reliable Euler integration requires a well-conditioned transformation gradient $\nabla \boldsymbol{t}(\boldsymbol{x}, t_i)$. Since the Jacobian $J = \mid \nabla \boldsymbol{t}(\boldsymbol{x}, t_i) \mid$ provides a measure of the condition of $\nabla \boldsymbol{t}(\boldsymbol{x}, t_i)$ we require $J > 0$.

The transformation becomes singular for large curved transformations because of the discretization. To evade this problem we apply the same regridding method as Christensen [5]. Every time the Jacobian J drops below 0.5 we generate a new template by applying the current deformation. In addition the displacement field is set to zero, whereas the current velocities remain constant. The total deformation becomes the concatenation of the displacement fields associated with the sequence of propagated templates.

The complete algorithm for solving the viscous fluid registration problem consequently becomes [5]:

1. Let $i = 0$ and $\boldsymbol{u}(\boldsymbol{x}, 0) = 0$
2. Calculate the body force $\boldsymbol{f}(\boldsymbol{x}, \boldsymbol{u}(\boldsymbol{x}, t_i))$ using equation 7.
3. If $\boldsymbol{f}(\boldsymbol{x}, \boldsymbol{u}(\boldsymbol{x}, t_i))$ is below a threshold for all $\boldsymbol{x}$, then STOP.
4. Solve the linear PDE equation 5 for instantaneous velocity $\boldsymbol{v}(\boldsymbol{x}, t_i)$ and force $\boldsymbol{f}(\boldsymbol{x}, \boldsymbol{u}(\boldsymbol{x}, t_i))$.
5. Choose a timestep $(t_{i+1} - t_i)$ so that $\nabla \boldsymbol{t}(\boldsymbol{x}, t_i)\boldsymbol{v}(\boldsymbol{x}, t_i) < d\boldsymbol{u}_{max}$, where $d\boldsymbol{u}_{max}$ is the maximal flow allowed in one iteration (0.7 in this work).
6. Perform Euler integration using equation 8.
7. If the Jacobian $J = \mid \nabla \boldsymbol{t}(\boldsymbol{x}, t_i) \mid$ is less than 0.5 then regrid the template.
8. $i = i + 1$, goto 1

The only remaining question is how to solve the PDE equation in step 4. We discuss this in the following section.

2.2 Solving the linear PDE

In the algorithm shown above, the core problem is solving the linear PDE:

$$\boldsymbol{\mathcal{L}} \boldsymbol{v} = \mu \Delta \boldsymbol{v} + (\mu + \lambda)\nabla(\nabla \cdot \boldsymbol{v}) = \boldsymbol{f} \tag{9}$$

for constant force and time. In practice, solving this PDE is the time consuming part of the fluid registration. The contribution of the rest of the paper is a fast way of doing this.

As we saw previously, for constant force $\boldsymbol{f}$ and time t this PDE is linear and the linear operator $\boldsymbol{\mathcal{L}}$ is the linear elasticity operator working on $\boldsymbol{v}$. Linear elastic problems are normally solved using implicit finite element or finite difference methods. But in the case of images, we assign nodes in the elastic model to each pixel or voxel. The size of the problem, therefore, is huge and in practice unsolvable with these methods. Instead explicit methods must be used.

Christensen et al. [5] use successive overrelaxation (SOR) with checker board update to solve the linear elastic problem.

We suggest solving the the linear PDE using scale-space convolution. Using the linearity of the PDE and the superposition principle, we create a filter as the impulse response of the linear operator $\boldsymbol{\mathcal{L}}$ and subsequently apply this filter to the force field $\boldsymbol{f}$.

This work has been inspired by the work of Nielsen et al. [12], who show that Tikhonov regularization can be implemented using Gaussian scale-space, and Thirion [13], who propose a 'demon'-based registration algorithm which we will show later, is an approximation to the viscous fluid registration problem.

2.3 Convolution filter for linear elasticity

In this section we develop the convolution filter used to solve the linear PDE. First the displacement field $\boldsymbol{v}$ is decomposed using the eigen-function basis of the linear operator $\boldsymbol{\mathcal{L}}$. Then the impulse response of the linear operator is determined in this basis. We note that the impulse response of a linear operator is a filter that implements the operator. Finally, we discretize the impulse response to get a discrete filter.

Eigen-functions of the linear operator $\boldsymbol{\mathcal{L}}$ The eigen-functions of the linear operator $\boldsymbol{\mathcal{L}}$ using sliding boundary conditions are given by Miller, Christensen et al. in [3, 5] as:

$$\boldsymbol{\phi}_{ij1}(\boldsymbol{x}) = \alpha_1 \begin{bmatrix} ip(\boldsymbol{x}) \\ jq(\boldsymbol{x}) \end{bmatrix} \qquad \boldsymbol{\phi}_{ij2}(\boldsymbol{x}) = \alpha_2 \begin{bmatrix} -jp(\boldsymbol{x}) \\ iq(\boldsymbol{x}) \end{bmatrix} \tag{10}$$

with the eigen-values:

$$\kappa_{ij1} = -\pi^2(2\mu+\lambda)(i^2+j^2) \qquad \kappa_{ij2} = -\pi^2\mu(i^2+j^2) \tag{11}$$

where

$$p(\boldsymbol{x}) = \sin i\pi x_1 \cos j\pi x_2 \qquad q(\boldsymbol{x}) = \cos i\pi x_1 \sin j\pi x_2 \tag{12}$$

$$\alpha_1 = \alpha_2 = \sqrt{\frac{4}{\Gamma_{ij}(i^2+j^2)}} \tag{13}$$

where

$$\Gamma_{ij} = \begin{cases} 1 & \textit{if none of } i,j \textit{ are zero} \\ 2 & \textit{if one of } i,j \textit{ is zero} \end{cases} \tag{14}$$

Using the new orthonormal basis, the velocity $\boldsymbol{v}$ can be decomposed into:

$$\boldsymbol{v}_N(\boldsymbol{x}) = \sum_{ij=0}^{N}\sum_{r=1}^{2} a_{ijr}\boldsymbol{\phi}_{ijr}(\boldsymbol{x}) \tag{15}$$

where a_{ijr} are the coefficients of the decomposition. N determines the number of basis functions included in the decomposition. Note that $\boldsymbol{v}(\boldsymbol{x}) = \lim_{N\to\infty} \boldsymbol{v}_N(\boldsymbol{x})$.

Determining the impulse response of $\mathcal{L}$ We will now determine the impulse response of the linear operator $\mathcal{L}$ for an impulse force $\hat{\boldsymbol{f}}$ in the x_1 direction at $\boldsymbol{c} = [0.5, 0.5]^T$. First the linear operator is applied to the decompostion of $\boldsymbol{v}$:

$$\begin{aligned}\hat{\boldsymbol{f}} = \mathcal{L}\boldsymbol{v}(\boldsymbol{x}) = \mathcal{L}\sum_{ijr} a_{ijr}\boldsymbol{\phi}_{ijr}(\boldsymbol{x}) = \sum_{ijr} a_{ijr}\mathcal{L}\boldsymbol{\phi}_{ijr}(\boldsymbol{x}) \\ = \sum_{ijr} a_{ijr}\kappa_{ijr}\boldsymbol{\phi}_{ijr}(\boldsymbol{x})\end{aligned} \tag{16}$$

We then take the inner product $< \boldsymbol{a}, \boldsymbol{b} >= \int_{\Omega} \boldsymbol{a}^T \boldsymbol{b} d\boldsymbol{x}$ of the equation with $\boldsymbol{\phi}_{lms}(\boldsymbol{x})$:

$$\begin{aligned}&\sum_{ijr} a_{ijr}\kappa_{ijr} < \boldsymbol{\phi}_{ijr}(\boldsymbol{x}), \boldsymbol{\phi}_{lms}(\boldsymbol{x}) > = < \hat{\boldsymbol{f}}, \boldsymbol{\phi}_{lms}(\boldsymbol{x}) > \\ &\Updownarrow \\ &a_{lms}\kappa_{lms} = < \hat{\boldsymbol{f}}, \boldsymbol{\phi}_{lms}(\boldsymbol{x}) > \\ &\Updownarrow \\ &a_{lms} = \frac{1}{\kappa_{lms}} < \hat{\boldsymbol{f}}, \boldsymbol{\phi}_{lms}(\boldsymbol{x}) > = \frac{1}{\kappa_{lms}} \phi_{lms}^{x_1}(\boldsymbol{c})\end{aligned} \tag{17}$$

where $\phi_{lms}^{x_1}$ is the x_1-coordinate of $\boldsymbol{\phi}_{lms}$. In the step from line 1 to line 2, we used the fact that $< \boldsymbol{\phi}_{ijr}, \boldsymbol{\phi}_{lms} >$ is zero for $(i, j, r) \neq (l, m, s)$.

We can now write the decomposition of $\boldsymbol{v}(\boldsymbol{x})$ for the case of an impulse force applied in $\boldsymbol{c}$ as:

$$\begin{aligned}\boldsymbol{v}(\boldsymbol{x}) &= \sum_{ij}^{\infty}\sum_{r=1}^{2} a_{ijr}\boldsymbol{\phi}_{ijr}(\boldsymbol{x}) = \sum_{ij}^{\infty}\sum_{r=1}^{2} \frac{1}{\kappa_{lms}} \phi_{lms}^{x_1}(\boldsymbol{c})\boldsymbol{\phi}_{ijr}(\boldsymbol{x}) \\ &= \frac{4}{\pi^2\mu(2\mu+\lambda)} \sum_{ij}^{\infty} \frac{p(\boldsymbol{c})}{(i^2+j^2)^2\Gamma_{ij}} \begin{bmatrix} -(i^2\mu + (2\mu+\lambda)j^2)p(\boldsymbol{x}) \\ (\mu+\lambda)ijq(\boldsymbol{x}) \end{bmatrix}\end{aligned} \tag{18}$$

This gives us the impulse response of the linear operator $\mathcal{L}$ for an impulse force in the x_1 direction applied in $\boldsymbol{c}$. The impulse response for the x_2 direction is determined by simple rotation of the response for the x_1 direction. In the next section we will see how this impulse response can be used to determine a discrete filter implementing the linear operator $\mathcal{L}$.

Discretizing the impulse response In general the impulse response of the linear operator is the linear filter implementing the operator. But in the continuous case a force applied to a single point yields an infinitely large displacement of this particular point. However, in the discrete case we sample the filter on a discrete grid and apply a lowpass filtering with a cut-off at the Nyquist frequency to eliminate aliasing from higher order frequency components. The force is thereby smoothed over a small area or volume.

We note that the decomposition of the impulse response based on the eigenfunction basis is a frequency based decomposition. Big i and j correspond to high frequencies and small to low frequencies. We can therefore perform an ideal lowpass filtering of the impulse response by truncating the sequence at N instead of summing to infinity.

The sampled filter is defined with dimensions $D \times D$, D odd, in the domain $[0; 1]^2$. The sampling interval is consequently $\theta = 1/(D-1)$ which Shannons sampling theorem relates to the cut-off frequency f by $\theta \leq 1/2f$. From equation 18 the frequencies corresponding to the summation variables are determined:

$$f_i = \frac{1}{2}i \qquad f_j = \frac{1}{2}j \tag{19}$$

and the common truncation point becomes $i = j = N = D - 1$. We can now collect everything in:

Theorem. *Consider a linear filter of size $D \times D$, and let the lattice be addressed by $\boldsymbol{y} = [y_1, y_2]^T$, where $y_r \in [-\frac{D-1}{2}, \frac{D-1}{2}] \cap N$, $r = 1, 2$. The filter implementing the linear elastic operator $\boldsymbol{\mathcal{L}}$ for the x_1 direction is then:*

$$\boldsymbol{v}(\boldsymbol{x}) = \frac{4}{\pi^2 \mu(2\mu + \lambda)} \sum_{ij=0}^{D-1} \frac{p(\boldsymbol{c})}{(i^2 + j^2)^2 \Gamma) ij} \begin{bmatrix} -(i^2\mu + (2\mu + \lambda) j^2) p(\boldsymbol{x}) \\ (\mu + \lambda) ij q(\boldsymbol{x}) \end{bmatrix} \tag{20}$$

where

$$\boldsymbol{x} = \frac{1}{D-1}\boldsymbol{y} + \begin{bmatrix} 1/2 \\ 1/2 \end{bmatrix} \tag{21}$$

□

We leave it to the reader to find the filter component for the x_2 direction.

To show that the filter actually works, we have made some experiments using the filter as the linear elasticity operator and comparing the results with a Finite Element (FEM) implementation of linear elasticity. The results have shown quite similar deformations.

2.4 Summary

In the previous sections we have described the original theory of the viscous fluid registration method and developed a convolution filter for the linear operator used in the core routine of the fluid registration.

Because of the limited span of the filter, we have implemented the viscous fluid registration algorithm using the filter in scale-space. The fluid registration is first performed on a rough scale. The result of this scale is then propagated to a finer scale and the fluid registration restarted here. This process is continued down to the finest scale of the scale-space, yielding the final registration result.

Fig. 1. Circle deforming into a 'C' using viscous fluid registration. From left to right: 1. Template. 2. Study. 3. Deformed template. 4. Grid showing the deformation applied to template.

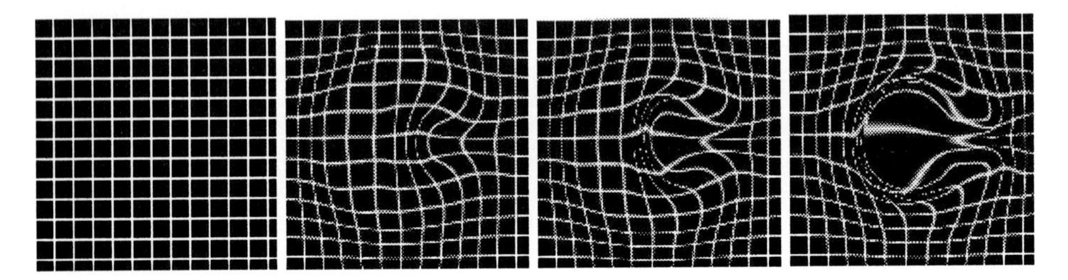

Fig. 2. Development of deformation of grid for viscous fluid registration of circle into 'C'.

3 Results

Figure 1 shows our results of registering a circle to a 'C' using viscous fluid deformation. The grid shows the curved and very large deformations that are applied to the template. Although the deformation is very large, the topology of the template is maintained. This is very important because it ensures that topology is maintained. Figure 2 shows the developing deformation as the fluid circle deforms into the 'C'. These results are very similar to figures 10.20-23 in [5]. Figure 3 show the result for two adjacent CT slices.

In general, our results are very good and similar to those of Christensen et al. But our timings are quite different. We have achieved stable timings on a *single processor* workstation similar to those stated by Christensen et al. for computations on a 128x64 DECmpp 12000 Sx/Model 200 massively parallel computer. When compared to estimates of timings for a MIPS R4400 processor [6] we can conclude that we achieve a speed-up of at least an order of magnitude. We hope to be able to share the data used by Christensen et al. for more elaborate comparison.

3.1 Comparison with 'demon'-based registration

In [13] Thirion proposed a 'demon'-based registration method. This is an iterative algorithm, where forces are determined in the template image based on

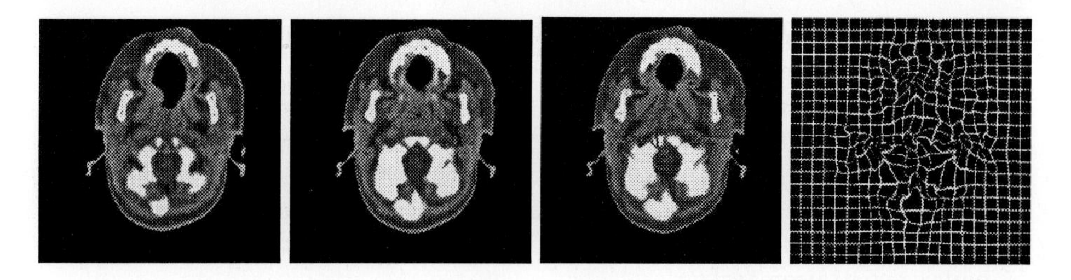

Fig. 3. CT slice registered to another slice using viscous fluid model. From left to right: 1. Template. 2. Study. 3. Deformed template. 4. Grid showing the deformation applied to template.

equations that are very similar to the body force used in this work. His equation 4:

$$\boldsymbol{f} = \frac{-(T-S)\nabla T}{\nabla T^2 + (T-S)^2} \tag{22}$$

is in fact just a normalized version of the body force used here (equation 7).

When Thirion deforms an image to match another, he performs an iterative process where body forces are determined using equation 22, the force field is lowpass filtered using a Gaussian filter and finally integrated over time.

Comparing the 'demon'-based algorithm with the algorithm in this paper, we see that:

- The body forces are almost the same.
- The lowpass filtering using a Gaussian corresponds to our application of the linear elastic filter.
- The time integration of the lowpass filtered force field corresponds to the Euler integration performed using equation 6.

We therefore conclude that the approach proposed in [13] is similar to the viscous fluid registration using convolution which we propose here. It is based on heuristics. Applying the Gaussian filter instead of the real linear elastic filter, is an approximation of the fluid model which could give problems in terms of topology and the stability of the fluid model.

4 Conclusion

In this paper we have shown that it is possible to speed-up the viscous fluid registration algorithm by Christensen et al. [5] by at least an order of magnitude. We achieve this speed-up by implementing the core process of the registration method using convolution with a filter we have developed. Our results are similar to those of Christensen et al.

The speed-up that we show here is based on a software implementation of the convolution. Use of specialized convolution hardware, as found in eg. the

RealityEngine II graphics board from SiliconGraphics, should speed-up the registration even more.

We have also showed that the 'demon'-based registration method by Thirion [13] is similar to the viscous fluid registration method developed here. This insight comes from the implementation of the core routine as a filter. Using this filter, the numerical implementations of the two methods look very similar. Since Thirion's method is a simplification, we suggest that it might have trouble handling more complex registration tasks.

Although we have only showed 2D results, the extension of the filter to the 3D case is straightforward, and we expect to do this in the future.

References

1. R. Bajcsy and S. Kovacic, *Multiresolution elastic matching*, Computer Vision, Graphics and Image Processing, 46:1-21, 1989
2. C. Broit, *Optimal registration of deformed images*, Doctoral dissertation, University of Pennsylvania, August 1981
3. M.I. Miller, G.E. Christensen, Y. Amit and U. Grenander, *Mathematical textbok of deformable neuroanatomies*, Proc. Natl. Acad. Sci. USA, 90:11944-11948, December 1993
4. G.E. Christensen, M.I. Miller and M. Vannier, *A 3D deformable magnetic resonance textbook based on elasticity*, in AAAI Spring Symposium Series: Applications of Computer Vision in Medical Image Processing, pp. 153-156, Stanford University, March 1994
5. G.E. Christensen, *Deformable shape models for anatomy*, Washington University Ph.D. thesis, August 1994
6. G.E. Christensen, M.I. Miller, M. Vannier and U. Grenander, *Individualizing neuroanatomical atlases using a massively parallel computer*, IEEE Computer, 29(1):32-38, January 1996
7. D.L. Collins, T.M. Peters, W. Dai, A.C. Evans, *Model-based segmentation of individual brain structures from MRI data*, Proc. SPIE Visualization in Biomedical Computing (1808), pp. 10-23, 1992
8. A.C. Evans, W. Dai, L. Collins, P. Neelin, S. Marret, *Warping of a computerized 3-D atlas to match brain image volumes for quantitative neuroanatomical and functional analysis*, Proc. SPIE Medical Imaging V (1445), pp. 236-246, 1991
9. K.H. Höhne, M. Bomans, M. Riemer,R. Schubert, U. Tiede and W. Lierse, *A 3D anatomical atlas based on a volume model*, IEEE Computer Graphics Applications, 12(4):72-78, 1992
10. J.C. Gee, M. Reivich and R. Bajcsy, *Elastically deforming atlas to match anatomical brain images*, Journal of Computer Assisted Tomography, 17(2):225-236, March 1993
11. E.R. McVeigh, R.M. Henkelman and M.J. Bronskill, *Noise and filtration in magnetic resonance imaging*, Medical Physics, 12(5):586-591, 1985
12. M. Nielsen, L. Florack and R. Deriche, *Regularization and Scale Space*, INRIA Tech. Rep. RR-2352, September 1994
13. J.-P. Thirion, *Non-rigid matching using demons*, Proc. Int. Conf. Computer Vision and Pattern Recognition (CVPR'96), 1996

A Robust Point Matching Algorithm for Autoradiograph Alignment

Anand Rangarajan[1], Eric Mjolsness[3], Suguna Pappu[1], Lila Davachi[2], Patricia S. Goldman-Rakic[2] and James S. Duncan[1]

[1] Department of Diagnostic Radiology, Yale University School of Medicine
[2] Section of Neurobiology, Yale University School of Medicine
[3] Dept. of Computer Science and Engineering, University of California San Diego

Abstract. Neuroimaging of the human brain has opened the way for a genuine understanding of human cognition; but the circuitry and cellular basis of the extraordinary information processing capacity of humans can be addressed only in experimental animals such as nonhuman primates by using the 2-DG autoradiographic method. This method requires sacrifice of the animal and sectioning of the brain into serial sections followed by production of autoradiographs of individual brain sections which are not in register. We have developed a new automated alignment method to reconstitute the autoradiographs. Our alignment method automatically finds the 2-D spatial mapping and the homologies between the slices and robustly accounts for the natural and artifactual differences by applying the powerful mechanism of outlier rejection adapted from the robust statistics literature.

1 Introduction

Functional studies of cortical circuitry can be performed by observing radiotracer uptake in autoradiographs. Here, local cerebral glucose utilization (LCGU) can be observed in order to study how the brain performs certain cognitive tasks. Current research in nonhuman primates in the Section of Neurobiology, Yale University School of Medicine [4] is concerned with elucidating functional maps of cortical activity using the 2-DG autoradiographic method developed in [9]. This method requires sacrifice of the animal and sectioning of the brain into serial sections followed by production of autoradiographs of extremely fine ($\simeq 10\mu$) slices and observing and quantitatively analyzing radiotracer uptake in each slice. Because of the extremely high spatial resolution of this technique, it is the case that these studies are the likely gold standard to which a number of imaging studies that attempt to measure cognitive function *in vivo* will be compared (e.g. fMRI, PET, SPECT).

One current downside of analyzing these autoradiographs, however, are the very large amounts of data that are produced. Each one of four sections of a monkey brain that are typically analyzed may be used to generate 1000 individual brain slices which are themselves not in register. While the spatial resolution of the individual 2-D slices is very high, the 3-D spatial structure available for example in volumetric MRI is lost. Due to the unavailability of 3-D MRI of the same primate and due to the lack of automated autoradiograph–MRI registration methods, the rich and internally coherent anatomical structure of 3-D MRI remains unutilized in current 3-D primate autoradiograph reconstruction (or "reconstitution" [7]). To date, these difficulties have limited researchers to analyzing only small portions of the brain at a time and thus has not permitted analysis of relationships between spatially disparate (yet possibly cognitively-related) regions. There is a current need for the development of systems that robustly and efficiently portray the globally distributed processing known to exist in the primate brain.

Since we have access to functional (metabolic) 2-DG autoradiographs and 3-D MRI of the same primate, we would like to reconstitute the 3-D autoradiograph volume using the 3-D MRI as an anatomical roadmap. Thereby we overcome the "leaning tower of Pisa" problem inherent in autoradiograph reconstitution in the absence of internally coherent 3-D anatomical information. Reconstitution is broadly divided into two stages: i) Align the autoradiograph slices and ii) Register the slices with MRI.

In this paper, we present a solution to the first problem by designing a new automated autoradiograph alignment method. The method is based on a newly developed *robust point matching* (RPM) algorithm [5, 8] that simultaneously finds homologies (correspondence) and similarity transformation parameters (rotation, translation and scale) between sequential pairwise slices. The inability of previous automated alignment methods to correct for artifacts like cuts and tears in the slices and for systematic variation from slice to slice is hereby accounted for. Our method is robust to this variability in the statistical sense of rejecting artifacts and natural differences as *outliers*—point features in either slice that have no homologies in the other slice [2].

A similarity transformation is sufficient since in stage 2, we plan to register the autoradiographs with 3-D MRI using affine and/or other warping transformations. Other alignment methods which introduce shearing transformations to align the slices face the problem of validation. Without 3-D MRI (or equivalent 3-D volumetric data), there is no 3-D information available to validate shears and other warps of the autoradiograph slices. Since we have 3-D MRI available, our alignment method finds a similarity transformation to approximately register the slices. In Section 3, we present a method to align individual autoradiograph slices. First, edge detection is performed on the slices. Point features (with only location information) are obtained from the edge images. Next, the RPM algorithm is executed on the point sets to obtain a similarity transformation (rotation, translation, scale). At this juncture, we refrain from introducing shear and other warping deformations into the catalog of allowed spatial transformations.

2 Previous Work

Automated Autoradiograph alignment is the primary goal of this study and we will present a novel alignment method in Section 3. If previous automated methods achieve autoradiograph alignment, why do we require another method? The *principal flaw* that other methods share in comparison to our RPM method is that they are not *robust* [2] in the statistical sense of automatically rejecting cuts, tears and natural variations between slices as *outliers*—point features in either set that have no homologies in the other.

The principal autoradiograph alignment methods are i) principal axis/center of mass [6], ii) frequency domain cross correlation [10], iii) spatial domain cross correlation [10] and iv) the disparity analysis [12] methods. (Most of these methods also perform autoradiograph–blockface video alignment. Since our data set does not have blockface video information, we are not including it as a specific aim of this study.) The principal axis method detects the center of mass of the two images, translates one image so that the center of masses coincide, then finds the principal axes using the eigenvector transform and aligns them via a rotation. The method is not robust to tears in the slices and also does not perform well when the images do not have bilateral symmetry. Cross correlation methods (either frequency or spatial domain) maximize the cross correlation with respect to the transformation parameters. This approach assumes densitometic consistency between slices. Consequently, the method is not robust to changes in the intensity patterns and when cuts and tears are present in the slices. A good comparison of these methods can be found in [10]. The disparity analysis method is based on optical flow. An affine transformation is used to regularize the flow field at each point. The method works with either boundary or intensity information. Consequently, the method is dependent either on good boundary detection or on densitometric consistency and is not robust.

General purpose feature matching methods exist in the literature (see [1] and the review in [11]). All these methods are based on minimizing the distances between points (or geometric features) in one set and the closest point (or geometric feature) in the other set with respect to the spatial transformation. Consequently, the distance measures are brittle, sensitive to changes in feature location and to false positives in the detection of closest points. Since good feature extraction is assumed by these methods, they are unable to robustly correct for missing and/or spurious information.

3 Automated Autoradiograph Alignment Approach

Autoradiography: A rhesus monkey was habituated over many months to perform working memory (WM) tasks under restrained conditions. An arterial catheter was introduced into one of the femoral arteries and ^{14}C-deoxy-D-glucose (2DG) of approximately 100μCi/kg was injected through the cannula of the awake rhesus monkey which was kept restrained in a primate chair. Immediately after the injection, the animal performed the task uninterruptedly for an entire 45 minute period. After 45 minutes, the primate was sacrificed, perfused and the brain was frozen at -70°C. 20μm thick frozen sections were cut in a cryostat. The sections along with isotope standards had a seven day exposure after which the film was developed. Autoradiographs of brain sections, togther with each film's set of ^{14}C standards were digitized with a MCID computerized video image processing system. The computer used these standards to quantify radioactivity in each brain image by translating pixel-gray values to ^{14}C levels: these levels were then converted to LCGU rates. (For more details on autoradiograph acquisition, see [4]). While blockface video was not recorded for this dataset, 124 coronal T1-weighted MRI slices were obtained. With the availability of autoradiographs *and* 3-D MRI, our final goal is the simultaneous reconstitution and registration of the autoradiographs with MR.

We now describe the methodology for the alignment of the 2-D autoradiograph slices. In Figure 1, two primate autoradiograph slices (slice 379 and slice 380) are shown. A characteristic feature of the autoradiograph slices that are adjacent (for example slice 379 and 380 shown above) is overall similarity except for local areas of dissimilarity. Cuts, tears, irregular slicing and natural variations create local deformations from slice to slice. Since the slices are handled individually, there is a fairly significant change in orientation between adjacent slices. For all these reasons and others mentioned in Section 2, there is a need for a robust image matching method that finds similarity transformations (rotation, translation, scale) between slices and is stable to local deformations. Such an algorithm is the RPM method and as we shall demonstrate, there is a good payoff in applying this method to our problem. Previous work [5, 8] with the RPM algorithm was confined to data generated synthetically or at best handwritten characters. There is no prior work detailing use of an edge detector to get point features followed by the RPM method. Specific application to autoradiograph alignment is even more novel.

Assume that the edge detection phase is completed and we have edge images corresponding to the autoradiograph slices. (As we shall describe later, a Canny edge detector [3] is sufficient to obtain edges with good localization). High confidence edges alone are chosen after thresholding the edge images. The locations corresponding to the edges are then obtained from the edge images. Denote the edge location information from slice 1 and slice 2, $X_i,\ i = 1, 2, ..., N_1$ and $Y_j,\ j = 1, 2, ..., N_2$ respectively. N_1 and N_2 are the number of edge locations in the sets X and Y respectively. The RPM algorithm minimizes the following objective function (optimization problem):

$$\min_{M,\theta,t,s} E(M, R(\theta), t, s) = \sum_{i=1}^{N_1}\sum_{j=1}^{N_2} M_{ij}||X_i - t - sR(\theta)Y_j||^2 - \alpha \sum_{i=1}^{N_1}\sum_{j=1}^{N_2} M_{ij} \quad (1)$$

$$\text{subject to} \sum_{i=1}^{N_1} M_{ij} \leq 1, \ \sum_{j=1}^{N_2} M_{ij} \leq 1, \text{ and } M_{ij} > 0. \quad (2)$$

Equation (1) describes an optimization problem from which the transformation parameters—rotation matrix $R(\theta)$, translation t, and scale s—can be obtained by minimization. However, (1) also sets up an optimization problem on the point correspondences. The variable M_{ij} is a correspondence variable which indicates when homologies have been found. If M_{ij} is one, feature "i" in slice 1 and feature "j" in slice 2 are homologies. The constraints on M_{ij} enforce one-to-one correspondence between homologies and also robustness [2]: a point feature in one image may have no corresponding homology and should be discarded as a statistical outlier. The degree of robustness is enforced by the parameter α. If α is large, fewer points are discarded and vice-versa.

Automatically finding homologies is important because of the time consuming and labor intensive nature of manual selection. Automatically discarding features that have no homologies is important—otherwise the method would suffer from the same problems that beset intensity correlation.

We use a deterministic annealing [5, 8] method to perform the optimization of (1). The method incorporates a temperature parameter β similar to simulated annealing methods of optimization but in contrast is completely deterministic and more efficient. Poor local minima are avoided by this method. Our presentation will be quite terse. More details can be found in [8, 5].

Algorithm Summary:
Initialize β to β_{min}.
 Solve for the correspondence variables M.
 Solve in closed form for the similarity transformation (rotation, translation and scale parameters.
 Iterate until convergence or until an iteration cap is exceeded.
Increase β and iterate until convergence or until $\beta > \beta_{max}$.

The method solves for the correspondence and the transformation parameters in turn. Given the correspondence (match) matrix M, closed-form solutions are found for the spatial transformation parameters; rotation angle θ, translation vector $[t_x, t_y]$ and scale s. We perform coordinate-descent w.r.t. each spatial transformation parameter until a suitable convergence criterion is achieved.

Given the transformation parameters $(R(\theta), t, s)$, we can solve for the match matrix: M_{ij} is first set to

$$M_{ij} = \exp\left[-\beta\left(\|X_i - t - sR(\theta)Y_j\|^2 - \alpha\right)\right] \tag{3}$$

followed by an *iterated row and column normalization* method to satisfy the row and column constraints in (2):

$$M_{ij} \leftarrow \frac{M_{ij}}{\sum_{a=1}^{N_1} M_{aj}}, \quad M_{ij} \leftarrow \frac{M_{ij}}{\sum_{b=1}^{N_2} M_{ib}} \tag{4}$$

The exponentiation is necessary to obtain non-negative values for the match matrix M. Subsequently, the row and column normalization method performs constraint satisfaction using this non-negative matrix. For more details on the exponentiation used in (3) and on the row and column normalization method of constraint satisfaction, please see [8, 5].

As the parameter β is increased, the match matrix entries "harden"—they approach binary values. Outlier rejection occurs in the limit when $\beta \to \infty$ with the outliers becoming binary valued.

4 Results

Having described the method in general, we now apply it on the two autoradiograph slices (slice 379 and slice 380) shown in Figure 1.

First we run a Canny edge detector [3] on each slice. The edge detector is a single scale (Gaussian filter with width σ) edge detector incorporating hysteresis and non-maximum suppression. Low t_l and high t_h threshold parameters can be specified for edge tracking. The parameters used for both slices were the program defaults—$\sigma = 1$, $t_l = 0$ and $t_h = 255$. The edge images shown in Figure 2 are $\simeq$ 475x350 each which is the same size as the original slices. Some gradations in edge strength are discernible from the edge images. We now threshold the edge images to obtain binary images: the same threshold of 240 was chosen for both slices. The point sets X and Y are obtained from the binary images. The point sets are shown in Figure 3. Every fourth edge was chosen in constituting a point set. There were roughly 800 points in each point set.

Having obtained the point sets, we now execute the RPM algorithm. Exactly as described above, we specify an initial value for the deterministic annealing parameter ($\beta_{initial} = 0.3$). The robustness parameter α was set to 5. An appropriate schedule was prescribed for β; $\beta^{(n+1)} = \beta^{(n)}/0.9$. When β reaches $\beta_{max} = 5$, the algorithm terminates. The maximum number of row and column normalizations in the correspondence step was 30. At each setting of β, we executed the inner loop of the algorithm a maximum of 3 times. Initial conditions are not very relevant. Since $\beta_{min} = 0.3$ is quite low, the algorithm sets the translation parameters to the difference between the centroids of the point sets. The final solution and overlay of the points sets is shown in Figure 4.

The x's and o's are from slice 379 and slice 380 respectively and indicate the coordinate locations of the edge (point) features. Note that a subset of the x's and o's do not match. The method has discarded them as outliers. The matrix M_{ij} has all zero rows (columns) corresponding to the outliers. The recovered registration parameters are shown in Table 1.

Table 1. Registration parameters with and without knowledge of homologies.

	Rotation (in degrees)	Translation (in pixels)
Homologies unknown	5.1	[21,-5]
Homologies known	4.9	[23,0]

The top left corner of the X point set was the center of rotation. The scale factors in both cases were close to unity. The recovered registration parameters with and without assuming knowledge of corresponding homologies are very close to each other.

We have shown an initial study where autoradiograph slices are aligned using a novel edge feature matching method. The method jointly solves for the transformation parameters, rotation, translation and scale as well as the correspondences or homologies. The method also discards features that have no homologies. That the method is robust is demonstrated by the fact that we did not even have to edit the slice index label on the autoradiographs! The label number (380 for example) can be partially seen in Figure 1 and in the edge images as well. The method discarded the label number by employing outlier rejection [2].

We also executed the RPM algorithm on point sets derived from coronal slices 384 and 385 as well as slices 405 and 409. Slices 384 and 385 are displayed in Figure 5. After digitization, the slices were sampled into roughly 475x350 pixels with an approximate pixel size of 100μm. As before Canny edges were obtained with no change in the filter width and the thresholds (low and high). Point sets were extracted from the edge images. There were roughly 3000 points in each point set. In contrast to the previous experiment where every fourth point was taken to yield approximately 800 points, this time every tenth point was chosen from each set for the RPM algorithm. The results are shown in Figure 5. The recovered registration parameters were $\theta = 8.9°$ and $t = [-18, 10]$. The scale factor was again very close to unity. Another difference from the previous experiment is the center of rotation. This time the center of slice 384 was the center of rotation. Since the slices don't match up as well as slices 379 and 380, local differences on the boundary contours can be seen. This experiment serves to demonstrate that the RPM algorithm can also be used with a small subset of the points. A more efficient strategy might be a coarse-to-fine approach with the Canny filter scales chosen to get large scale edges first and then small scale edges.

The final experiment with slices 405 and 409 showcases the robustness property. Examine the slices shown in Figure 6. While there is a much larger rotation difference between the slices, the major difference is the top part of slice 409 which is missing in slice 405. There was no change from the parameters used in the previous experiments. (Slice 405 is 260x160 pixels and the point set has 1000 points. Slice 409 is 275x220 with the point set containing 1500 points.) Every third point was chosen for the RPM algorithm. The results are shown in Figure 6. Despite the large differences between the slices, the RPM algorithm had no difficulty in rejecting the extraneous information as outliers. The recovered transformation parameters were $\theta = 12.8°$ and $t = [-1, 4.6]$. The center of the "o" point set was the center of rotation. The scale parameter was unity.

5 Discussion

The 2-DG method developed in [9] can be used to obtain high resolution functional (metabolic) maps of the primate brain in the form of autoradiographs. While autoradiograph slices are of very high resolution, the large numbers of slices acquired create data processing problems. A block of monkey brain can result in thousands of slices and the slices may have cuts and tears creating further problems for reconstitution. The tedium of data analysis in this area of research is a serious impediment to unique studies that hold promise for bridging human cognition with anatomical and physiological data obtainable from an animal model.

We have developed a new automated alignment method which is based on

Canny edge detection followed by a robust point matching (RPM) algorithm that solves for a similarity transformation between slices. The method automatically finds homologies, and rejects non-homologies as outliers. The alignment method is the first stage in a two stage procedure that i) aligns sequential, pairwise autoradiograph slices and ii) registers and warps the autoradiograph volume onto 3-D MRI. While we have not discussed registration with MRI in this paper, it plays a crucial role in lending 3-D volumetric coherence to the process of autoradiograph reconstitution. Finally, the automated alignment method resulting from this study can be applied to other areas such as registration of MRI to anatomical atlases, and intermodality registration in general.

References

1. Besl, P. J. and McKay, N. D. (1992). A Method for Registration of 3-D Shapes. *IEEE Trans. Patt. Anal. Mach. Intell.*, 14(2):239–256.
2. Black, M. and Rangarajan, A. (1995). On the Unification of Line Processes, Outlier Detection and Robust Statistics with Applications in Early Vision. *Intl. J. Computer Vision.* (in press).
3. Canny, J. (1986). A Computational Approach to Edge Detection. *IEEE Trans. on Pattern Analysis and Machine Intelligence*, 8(6):679–698.
4. Friedman, H. R. and Goldman-Rakic, P. S. (1994). Coactivation of Prefrontal Cortex and Inferior Parietal Cortex in Working Memory Tasks Revealed by 2DG Functional Mapping in the Rhesus Monkey. *J. Neuroscience*, 14(5):2775–2788.
5. Gold, S., Lu, C. P., Rangarajan, A., Pappu, S., and Mjolsness, E. (1995). New Algorithms for 2-D and 3-D Point Matching: Pose Estimation and Correspondence. In Tesauro, G., Touretzky, D., and Alspector, J., editors, *Advances in Neural Information Processing Systems 7*, pages 957–964. MIT Press, Cambridge, MA.
6. Hibbard, L. S. and Hawkins, R. A. (1988). Objective Image Alignment for Three-Dimensional Reconstruction of Digital Autoradiograms. *J. Neurosci. Methods*, 26:55–75.
7. Kim, B., Frey, K. A., Mukhopadhyay, S., Ross, B. D., and Meyer, C. R. (1995). Co-Registration of MRI and Autoradiography of Rat Brain in Three-Dimensions following Automatic Reconstruction of 2-D data set. In Ayache, N., editor, *Computer Vision, Virtual Reality and Robotics in Medicine*, volume 905 of *Lecture Notes in Computer Science*, pages 262–271. Springer-Verlag.
8. Rangarajan, A., Gold, S., and Mjolsness, E. (1996). A Novel Optimizing Network Architecture with Applications. *Neural Computation*, 8(5):1041–1060. (in press).
9. Sokoloff, L., Revich, M., Kennedy, C., DesRosiers, M. H., Patlak, C. S., Pettigrew, K. D., Sakurada, O., and Shinohara, M. (1977). The C14-deoxyglucose method for the measurement of local cerebral glucose utilization: theory, procedure, and normal values in the conscious and anesthetized albino rat. *J. Neurochem.*, 28:897–916.
10. Toga, A. W. and Banerjee, P. K. (1993). Registration revisited. *J. Neurosci. methods*, 48:1–13.
11. Van den Elsen, P. A., Pol, E. and Viergever, M. A. (1993). Medical image matching—A review with classification. *IEEE Engineering in Med. and Biol.*, 12(1):26–39.
12. Zhao, W., Young, T. Y., and Ginsberg, M. D. (1993). Registration and Three-Dimensional Reconstruction of Autoradiographic Images by the Disparity Analysis Method. *IEEE Transactions on Medical Imaging*, 12(4):782–791.

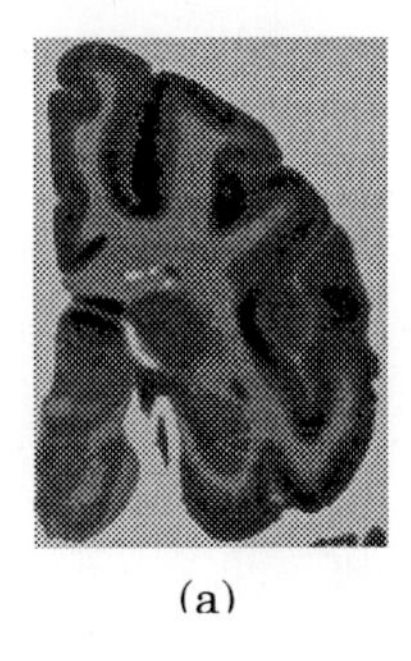

(a)

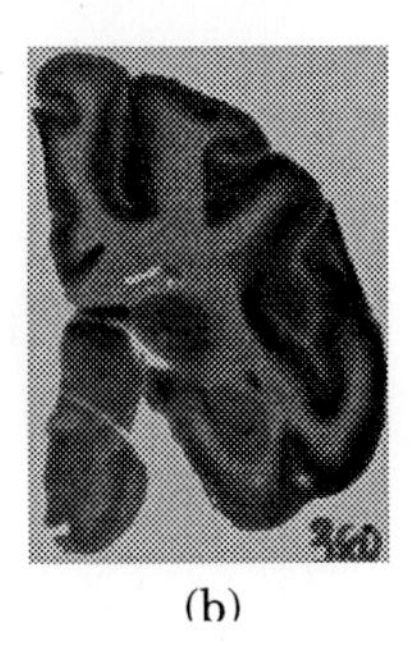

(b)

Fig. 1. (a) Slice 379 and (b) slice 380 of primate autoradiographs of functional (metabolic) activation. The coronal slices correspond to the left parietal and temporal cortex. After digitization, the slices were sampled down into $\simeq$ 475x350 pixels each with a resulting spatial resolution of $\simeq 100\mu$m.

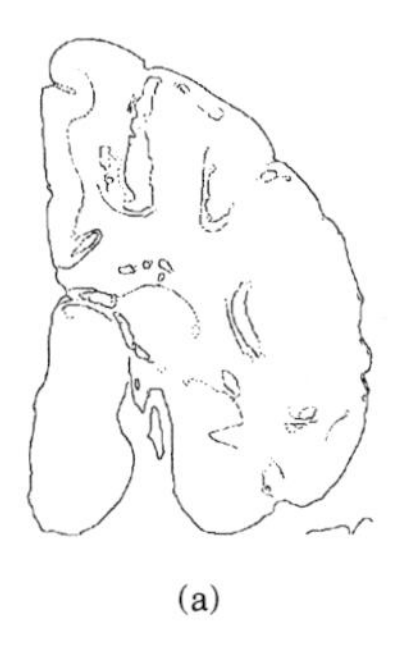

(a)

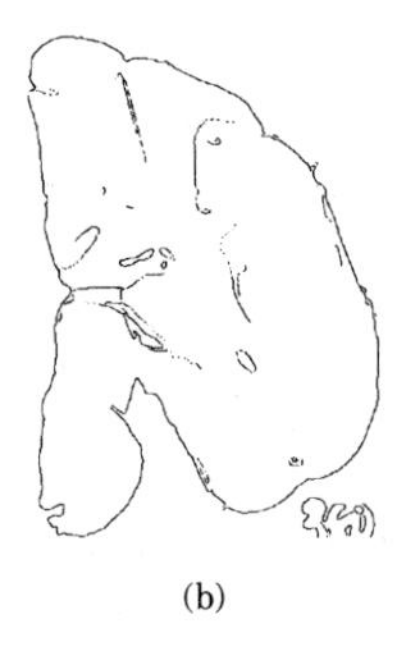

(b)

Fig. 2. Canny edge images corresponding to the autoradiograph slices of Figure 1. Note the number of edges in (a) that do not appear in (b). While we have not dwelled on the instability of edge detection, it is yet another reason for our strong emphasis on robustness.

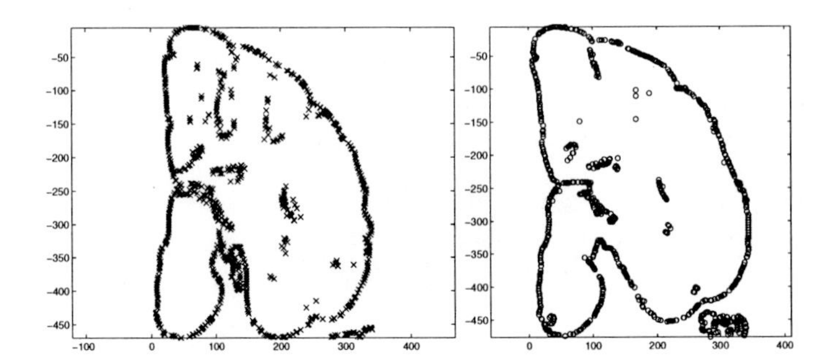

Fig. 3. Point sets corresponding to the Canny edge images. Every fourth edge was taken yielding $\simeq$800 points per point set. Slice 379 and slice 380 are the X and Y point sets respectively.

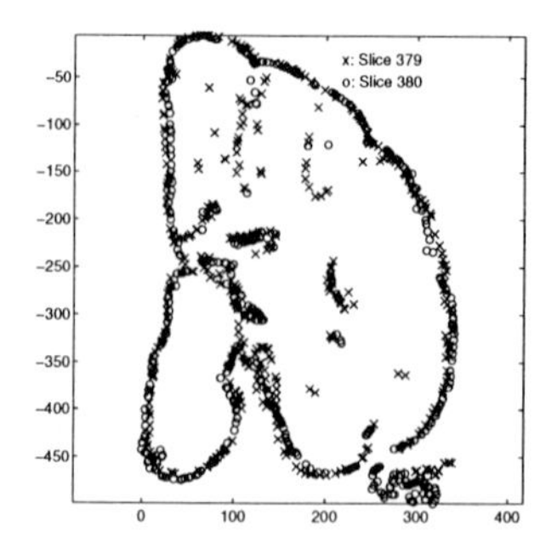

Fig. 4. Final solution found by the method. Note the number of (mostly internal) x's and o's that do not match but remain isolated during overlay. These have been rejected as outliers and play no role towards the final stages of the algorithm.

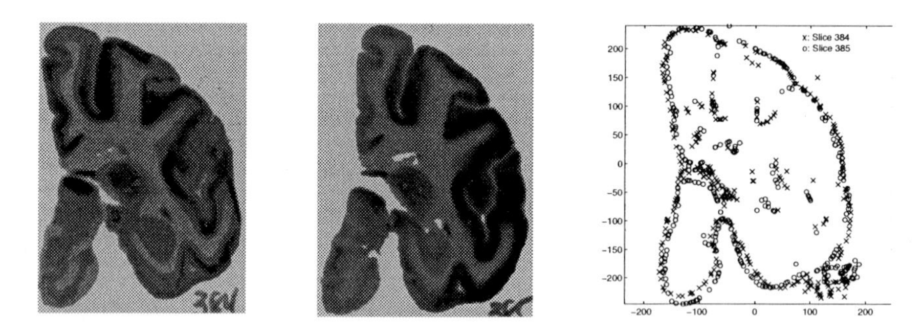

Fig. 5. Slice 384, slice 385 and the final match.

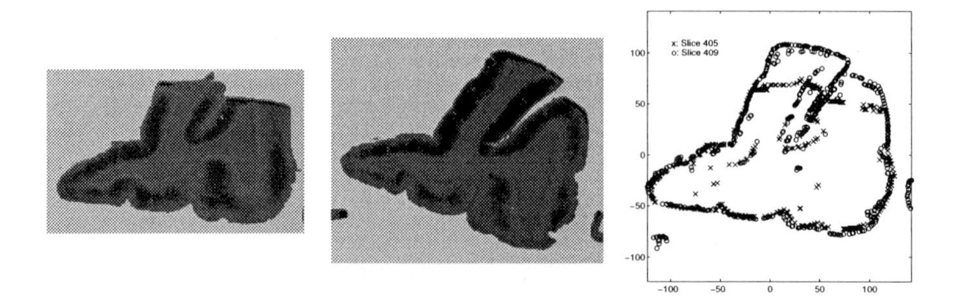

Fig. 6. Slice 405, slice 409 and the final match.

Registration Updating Using Marker Pins in a Video Based Neurosurgical Guidance System (VISLAN)

J Zhao[1], PTE Roberts[2], ACF Colchester[1] and K Holton-Tainter[1]

[1] Department of Neurology, UMDS, Guy's Hospital, London, UK
[2] Roke Manor Research Ltd, Romsey, Hampshire, UK

Abstract. This paper describes registration updating for tracking patient-camera movement using marker pins in a video based neurosurgical guidance system(*VISLAN*). Both the algorithm and the experiment results of accuracy evaluation are given. The results show that the marker pin-based registration updating is a feasible solution for tracking patient-camera movement and has a very reasonable accuracy, particularly at spatial locations close to the markers pins, e.g. it has better than 0.7mm accuracy at distance of 50mm from the centre of the gravity of marker pins.

1. Introduction

The last decade has seen much research and commercial activity in developing various neurosurgical guidance systems[1]. The detailed technology used in systems may vary, but the basic principles are very similar. Firstly, the live patient in theatre is registered with pre-operative image data. This may be achieved by matching the skin surface points obtained by stroking a mechanical arm[2] on the patient or by using a laser range finder[3] or a stereo video system[4] with the surface points segmented from the pre-operative data. An alternative is to use fiducial markers[5], which are implanted before the pre-operative scan and visible to the scanner. The fiducial markers are then located in both the pre-operative images and the intraoperative patient. Registration is obtained by minimising the distance between corresponding objects in the preoperative image and those in the intraoperative patient. Once registration is established, a pointing device (localiser) which is movable within the surgical field is tracked and its tip position is displayed in the same coordinate frame as the pre-operative data. The surgeon uses this information as a guide during the operation. However, many guidance systems require that the patient does not move after the registration, which is limiting because of the risk that the patient could move either accidentally or deliberately during the operation. To overcome this problem, some systems introduce a mechanism of updating the registration once the patient is moved. In the Vanderbilt system[6], an array of LEDs which is attached to a Mayfield clamp is tracked. This is convenient but still requires that there is no movement between the patient's head and the clamp. In the *VISLAN* system, three or more three marker pins are screwed into the skull around the craniotomy window before or immediately after the video skin surface capturing which is used for initial registration. Their positions in intraoperative coordinates are captured. During the operation, the positions of the pins are periodically monitored and used to update the registration when movement is detected.

This paper describes our approach to update registration using intraoperative marker pins and the experiments to test its accuracy. We first briefly describe the complete neurosurgical guidance system (*VISLAN*). We then detail the method of the marker pin location using the stereo video system. Finally the results of experiments to measure accuracy are presented.

2. Description of a video based neurosurgical guidance system (*VISLAN*)

VISLAN consists of two parts: a pre-operative surgical planning and an intra-operative guidance system. The pre-operative system[7] is an integrated software package. It serves two purposes, to process raw images to enhance the visualisation of the data and to prepare objects for intraoperative surgical guidance. The main procedures are segmentation, registration, 3D rendering and cranial window planning. Each function is equipped with a high level graphical user interface. The *VISLAN* intraoperative guidance system is based on stereo video. At the start of the operation, a pair of video cameras mounted about 1m apart above the patient is calibrated using an accurately manufactured tile which has an array of dots. Patterned light is then projected onto the patient's head and the skin surface is reconstructed from the two video images captured. The patterned light serves to enrich the features in the video images and to simplify the surface reconstruction process. The reconstructed skin surface is matched with the skin surface segmented from the patient's pre-operative images to establish the intra-operative to pre-operative registration. Three or more marker pins are then screwed into the skull and their positions are found by the video system using the algorithm described below. After registration, pre-operatively defined objects such as a planned craniotomy window is superimposed on the live video to guide the incision. A passive hand held localiser which has a distinctive pattern is tracked by the video system and its pose is superimposed on the per-operative image to guide the surgery. At the same time, the video system periodically monitors the marker pins and updates the registration when the patient or cameras are moved.

3. Marker pin detection and localisation

3.1 Construction of marker pins

The marker pins (referred to as pins in the following) used for patient position updating are designed to be clinically safe, acceptable to the surgeon, robust, and suitable for optical processing. The pins themselves (shown in figure 2) are aluminium with a 5mm long machine screw thread which screws into the internal thread of a separate bone screw. The bone screws (also shown in figure 1) are of overall length 13mm with a coarse self tapping outer thread length of 9mm which is driven into an appropriate hole drilled into the skull. The visible section of the pin consists of a cylindrical shaft of 2.5mm diameter and 15mm length, on top of which is a machined sphere of 5mm diameter. The shaft is generally black or grey in colour, while the head is white. The spherical head is seen as a circle in the distortion-corrected video image, and the 3D position of the centre of the 2 projected circles, as found by the stereo measurement system, is the centre of the sphere.

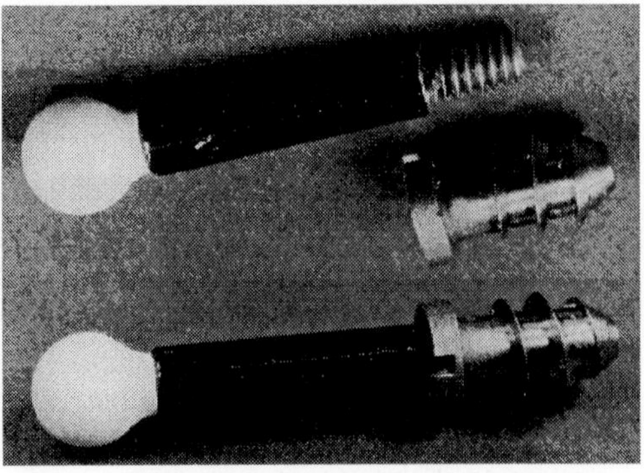

Figure 1 Bone screws and marker pins

3.2 Pin model capture

At the start of surgery the pin positions in the camera coordinate system are measured. The theatre scene is captured by the capture card on the Sparc2 workstation, and the resulting video images from each camera are processed to find the pins. The known camera calibration is then used to find each pin in 3D in the camera coordinate system. The process is as follows.

a) 2D image operations

i) Local mean removal

The local mean removal (LMR) routine applies a convolution filter to the image, whereby, for each pixel, the local average of a square window surrounding the pixel is subtracted from it. This can be expressed as:

$$O(x,y) = I(x,y) - \frac{1}{(2d+1)^2} \sum_{i=-d}^{d} \sum_{j=-d}^{d} I(x+i, y+j),$$

where

O(x,y) is the output intensity of the pixel at position (x,y)
I(x,y) is the initial intensity of the pixel at position (x,y)
2d+1 is the window size in pixels

A class number is assigned to each pixel, by comparing its output level with a pre-set threshold

ii) Connected component finder(CCF):

This finds groups of 4-way connected pixels of the same class (as assigned by the LMR function). Those groups which fall within specified geometric limits are designated to be candidate circles. The only information about the candidate circles which is used in subsequent processing is the ordered list of peripheral pixels defining the boundary.

iii) Circle finder:

This fits a circle to the boundary using a robust least squares minimiser. This is achieved by minimising an objective function which is the sum of squares of the residual radial distances of the boundary points from the circumference of the circle (only inlying points are taken into account in this sum - see below).The minimisation of this objective function is carried out using the Newton technique, with the standard simplifications for a sum squared objective function. The robust least squares minimiser takes account of corrupted portions of the circle by excluding outliers from the objective function. On each cycle of the Newton minimisation boundary points are assessed as being inliers or outliers by examination of the distribution of the residuals on the previous cycle. All boundary points are re-assessed at each iteration. This yields estimates for the pixel co-ordinates of the circle centre, the r.m.s. error in the fit of the object to a circle, and the circle radius. Candidate circles for which the fit is poor are rejected, as are those which lie outside the expected range of radii.

iv) Multi-threshold circle finding :

Applying a single threshold level after LMR results in a process which is sensitive to lighting variations, because the output of LMR varies with illumination. A threshold which is too low relative to the LMR output can result in corruption of candidate circles by adjacent objects in the image; conversely one which is too high can reject desired circles. Lighting across the image can be highly non-uniform so that the threshold required for circle detection can vary across the image. To reduce sensitivity to lighting variations, the LMR, CCF and circle finding processes described above are repeated at several thresholds, yielding a set of circles for each threshold, with estimates of the circle centre position, the r.m.s. error in the fit of the object to a circle, and the circle radius for each one. (Thresholds giving unacceptably high numbers of candidate circles are rejected as being too low).Circle centre position is used to identify multiple detections of the same pin, with the threshold giving the lowest r.m.s error being used for that marker pin.

b) 3D image operations:

i) Epipolar matching.

The centre of each 2D circle in camera 0 is projected onto the camera 1 image, using the known intrinsic and extrinsic camera calibration parameters, to give an epipolar line representing the possible locus of points for that circle in camera 1. For each circle in camera 0, all camera 1 circles lying within a chosen distance (typically 1 pixel) of the epipolar line are considered as possible matches, and that giving the closest unambiguous match kept as the 2D -2D match.

ii) 3D position calculation.

The 3D position of the centre of the circle is then found in 3D using the known intrinsic and extrinsic camera calibration. This creates the initial model of the set of pins in the camera coordinate system which is stored for use in updating the patient position.

It is important to note that this initial model, made as soon as possible after skin surface capture, is the baseline with which later pin positions are compared during registration update.

3.3 Measuring Camera-patient Movement Using Pins

Camera-patient movement can be measured using the marker pins, to verify if registration is still valid. The current pin positions are measured as described above, and their positions compared with those captured at the last registration update. This allows a judgement to be made of whether a registration update is required.

3.4 Registration updating using pins

When the measurement of camera-patient movement shows registration updating to be necessary, update takes place as follows:

1) The new pin model is compared with the initial pin model and all possible pin-pin matches between the two models considered. For each possible match, a quaternian implementation of the Procrustes algorithm is used to find the translation and rotation which best maps the new model onto the old (i.e. which minimises the sum squared distances between the matching points). The match giving the best fit is taken to be correct and the transformation matrix representing the pin movement in the camera 0 coordinate system stored.
2) The initial registration matrix, relating pre-operative and intra-operative (camera) coordinate systems, is modified by the pin movement matrix.

3.5 Interactive pin identification

To allow for the case when the system fails to automatically identify the pins, pins can be identified interactively using images displayed on the user interface. The operator can identify pins which escaped detection or delete incorrectly identified image objects by clicking on the graphically overlay.

4. Experiments and Results

4.1 Experiment 1

This experiment was designed to find the accuracy of pin position measurement using the above algorithm. Six pins were attached to a plastic phantom which has normal human facial features as shown in figure 2. They were placed within the field of view of both cameras in a rough circle with about 25mm radius. Their positions in the camera coordinate system were measured. To test the repeatability of the algorithm under normal ambient lighting condition, we repeated the measurement ten times without deliberate change of any condition. The result shows that the standard deviation was well below *0.1mm*.

Each pin position measurement can be modelled as $\{p_j(i)+e_j(i)\}$, where $p_j(i)$ is the true position of pin i with the phantom at position j and $e_j(i)$ is the measurement error. The difficulty in this experiment, however, is that the true position $p_j(i)$ in camera coordinates is unknown. We therefore obtained another set of pin position measurements by moving the phantom and evaluated a quantity derived from two sets of measurements to indirectly estimate the measurement error. For the experiment, the phantom was deliberately moved (both rotated and translated) by about 53mm. Assume that the amount of phantom movement is known and is represented by a 4x4 matrix M_{21}, the pin position measurements after phantom movement were transformed using matrix M_{21} to estimate their measurements before the movement, and the differences between the estimates and the actual measurements, denoted as CE(i), is

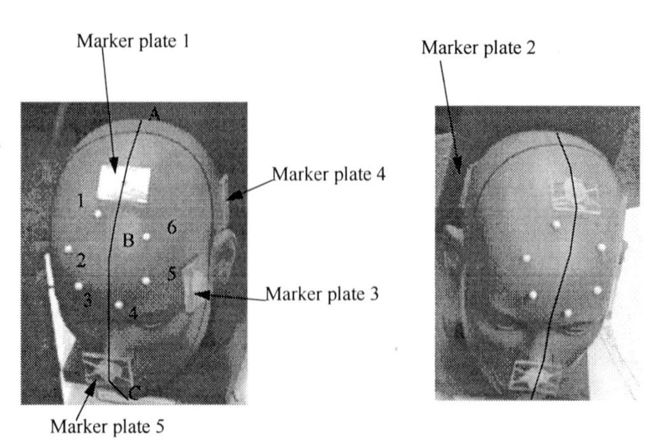

Figure 2. Images captured from two cameras show the locations of marker pins, marker plates and traced line on the phantom

$$\begin{aligned} CE(i) &= (p_1(i)+e_1(i)) - M_{21}(p_2(i)+e_2(i)) = (p_1(i) - M_{21}p_2(i)) + (e_1(i) - M_{21}\, e_2(i)) \\ &= (e_1(i)-M_{21}\, e_2(i)) \end{aligned}$$

CE(i) is thus closely related to measurement errors, reflects the accuracy of pin position measurement and is called as the combined pin position measurement error.

We used the *VISLAN* localiser to provide an independent and accurate estimate of the amount of the phantom movement, and thus M_{21}, since readings from the *VISLAN* localiser have previously been shown to be accurate to within about 0.3mm (0.1-0.5mm) in camera coordinates[8]. Five marker plates (referred to as plates in the following) were attatched in widely spaced positions across the phantom (figure 2). The middle of each plate has a dimple for docking the tip of the *VISLAN* localiser. The position of each plate was measured in camera coordinates by docking the *VISLAN* localiser into the dimple, rotating the localiser and taking the average of about 1000 readings of the tip positions. In addition, each measurement was repeated three times to check consistency in order to avoid any possible mis-operation. It was found that the averages of the readings were very consistent among three trials (difference below *0.1mm*) and the standard deviation in each trial was in the range of *0.14mm* to *0.6mm* depending on the location of the plate in the camera view field. The measurements were obtained both before and after the phantom movement. A matrix, called the plate matrix M_{plate}, was then derived using the Procrustes algorithm and was used as the approximation to M_{21}. A final check of localiser accuracy was provided by the following. The movement measured for each plate separately were collectively consistent with rigid body motion, giving very small residual errors of about 0.1mm, as shown in table 1, after subtracting rigid body motion.

Table 1: Residual Euclidean distance of marker plates

Marker plate	1	2	3	4	5
Residual Eucl. Dist. (mm)	0.1374	0.0610	0.1332	0.0916	0.1484

Table 2 lists the combined pin position measurement errors by using M_{plate} to approximate M_{21}. They are in the range of *0.31* to *0.86mm* and the average is *0.46mm*

Table 2 Combined pin position measurement errors

pin	1	2	3	4	5	6	Average
CE.(mm)	0.3191	0.4447	0.8680	0.3095	0.5129	0.3211	0.4625

4.2 Experiment 2

This second experiment was designed to measure the accuracy of localisation as a function of its spatial location relative to pins after pin-based registration updating. We first used the two sets of pin position measurements in experiment 1, one before and the other after the deliberate phantom movement, to derive a transformation matrix (referred as pin matrix M_{pin}) using the Procrustes algorithm. The residual Euclidean distance errors after substracting rigid body movement are listed in table 3.

Table 3 Residual Euclidean distance of marker pins

Pin	1	2	3	4	5	6
Residual Eucl. Dist(mm)	0.2337	0.2689	0.2957	0.0538	0.3147	0.1654

Assume a point A on the phantom whose true position in the camera coordinate system is $p_1(A)$ before the phantom movement and $p_2(A)$ after the phantom movement, $p_1(A)$ and $p_2(A)$ are related by $p_1(A) = M_{21}\ p_2(A)$. But if $p_2(A)$ is transformed by using M_{pin}, the result $p'1(A) = M_{pin}\ p_2(A)$ is different from $p_1(A)$ and represents a different physical point on the phantom. Their difference, $LE(A)= p_1(A) - p'1(A)$, indicates the localisation error of point A caused by the pin-based registration updating.

As in the experiment 1, the plate matrix M_{plate} was used to approximate M_{21} in this experiment. We obtained data to evaluate localisation errors in two ways. Firstly, the errors were examined along a space curve traced by the *VISLAN* localiser on the surface of the phantom in the mid sagittal plane (curve ABC in figure 2). These errors are plotted as a function of distance from the centre of gravity of the pins (figure 3). The advantage of such data is that their physical location on the phantom is known. Secondly, we computed localisation errors for regularly sampled points (1mm step size in each direction) in a volume which included the whole phantom head. The advantage of this data set is that it gives a more complete picture of the distribution of the error within the space of interest. Three orthogonal slices of the errors through the centre of the gravity of the pins are shown in figure 4.

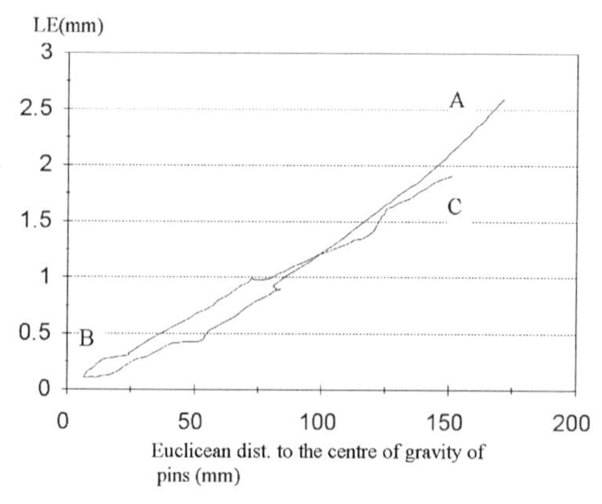

Figure 3. Localisation errors caused by pin-based registration updating as a function of point location relative to the centre of the gravity of the marker pins.

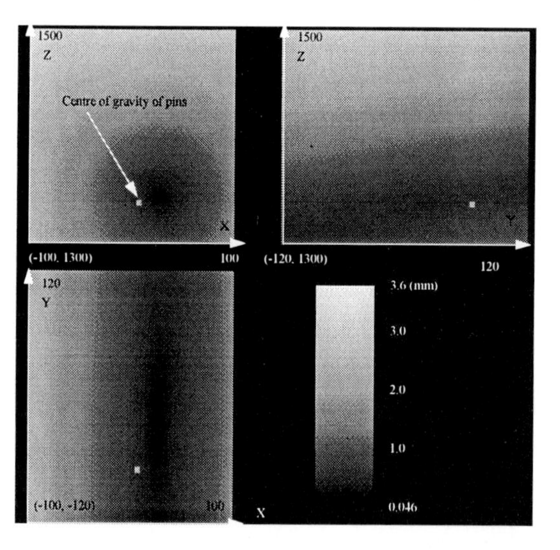

Figure 4. Three orthogonal slices of LE through the centre of gravity of six marker pins. The camera coordinate system has the origin in the centre of projection, the Z axis aligned with the optical axis of the camera, and the X and Y axes parallel with the rows and columns of the CCD array. Top-left: coronal slice. Top-right: sagittal slice. Bottom-left: axial slice. Bottom-right: grey scale of the display. The unit of both the axes and the scale bar is mm.

5. Discussions and Conclusions

Patient movement tracking is important for a clinically acceptable neurosurgical guidance system. It allows the surgeon to move the patient deliberately if he wishes and makes the performance of the system more reliable by detecting and compensating for accidental movement of the patient relative to the cameras. Our marker pin based scheme for patient movement tracking is simple but practical. There is little extra hardware to add for having such a function in the *VISLAN* system. The fixing of pins to the skull is clinically acceptable.

We use two types of measurement to evaluate the accuracy of our automatic marker pin position measurement. One is the residual Euclidean distance assuming rigid body movement and the other is the combined pin position measurement error. Even though both are indirect, they are closely related to the accuracy. The former is independent of the reference while the latter is related to the accuracy of the reference. The 0.2mm residual errors and about 0.5mm combined pin position measurement error indicate that our automatic pin position measurement is reasonably accurate.

Our second experiment measured accuracy of our marker pin-based registration updating at different spatial locations relative to the marker pins themselves. Even though the error is not a single function of the distance as the complex error pattern in figure 4 shows, the distance nevertheless is a major factor. In general, updating is more accurate at points close to the marker pins. The accuracy decreases as it becomes further away from the marker pins. For example, when the distance is below 50mm from the centre of gravity of the marker pins, the accuracy is better than 0.7mm. For a distance of 150mm, the error is about 2mm. The surgical implication of this is that our marker pin-based registration updating is a feasible solution to track the patient movement during surgery when the structures the surgeon wants to localise is close to the marker pins. When the structure of interest is further away from the marker pins and high localisation accuracy is also required, other strategies have to be adopted to track the patient movement, such as to use the localiser to identify some intra-cranial anatomical landmarks or space marker pins more widely across patient's head. Fortunately, the latter case is not common since a surgeon normally takes as a short route as possible to approach the target while our pins are inserted around the craniotomy window.

The *VISLAN* system has been tried in about 15 clinical cases in three surgical units and the pin-based registration updating were included in several recent cases. Figure 5 shows an example of fixation of pins on the patient. The surgeons found that the insertion of pins was easily fitted into their normal surgical procedures. In the theatre environment more user interaction was required due to the complexity of the surgical scene and the lighting conditions. Thus in some cases the user added or, more commonly, deleted pins, by

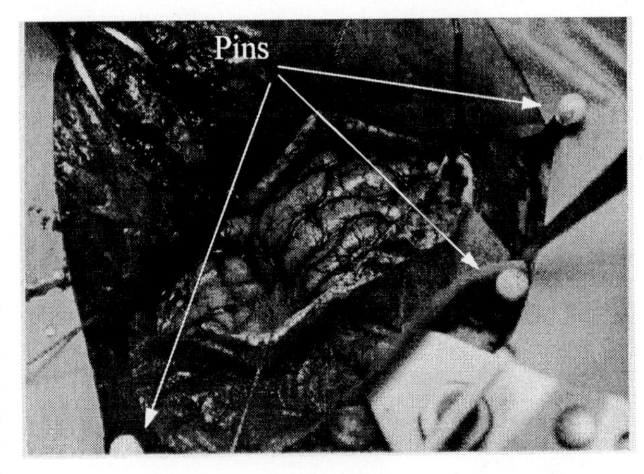

Figure 5 Fixation of pins on the patient. Four pins were used in this case (only three of them are shown in this picture)

interaction with the video images during registration updating. While pin deletion has no implications for the accuracy of the registration updating, manual pin identification is less accurate than automatic pin finding and is used only as a fall-back method. This interaction is carried out at the same time as the user confirmation of the captured pins model, with the whole process typically taking less than one minute.

6. Acknowledgements

We thank our surgical collaborators for their invaluable input: Mr Anthony Strong, King's College Hospital Neurosurgical Unit, and Professor David Thomas, National Hospital for Neurology and Neurosurgery, London. David Carr and Richard Evans at Roke Manor Research Ltd have made important contributions.

VISLAN is a joint academic–industrial collaboration and is partly funded by the UK Department of Trade and Industry.

7. References

1. Maciunas RJ. *Interactive image-guided neurosurgery*. American Association of Neurological Surgeons, 1993:
2. Drake JM, Prudencio J, Holowka S. A comparison of a PUMA robotic system and the ISG viewing wand for neurosurgery. In: Maciunas RJ, ed. *Interactive Image-Guided Neurosurgery*. American Association of Neurological Surgeons, 1993:121-133.
3. Grimson WEL, Ettinger GJ, et al. Evaluating and validating an automated registration system for enhanced reality visualization in surgery. In: Ayache N, ed. *Proceedings of First International Conference on Computer Vision, Virtual Reality and Robotics in Medicine*, Nice France, April 1995, Lecture Notes in Computer Science 905, 3-12.
4. Colchester ACF, Zhao J, et al. Development and preliminary evaluation of VISLAN, a surgical planning and guidance system using intraoperative video imaging, *Medical Image Analysis*, 1996, Volume 1, Number 1..
5. Maciunas RJ, Fitzpatrick JM, Galloway RL, Allen GS. Beyond stereotaxy:Extreme levels of application accuracy are provided by implantable fiducial markers for interactive image-guided neurosurgery. In: Robert J.Maciunas, ed. *Interactive Image-Guided Neurosurgery*. American Association of Neurological Surgeons, 1993:259-270.
6. Galloway, RJ, Jr and etc., Optical localisation for interactive image guided neurosurgery, *SPIE medical imaging*, 1994, Vol. 2164. pp 137-145.
7. Zhao J and Colchester ACF, Preoperative image processing in a computer assisted neurosurgical planning and guidance system (*VISLAN*), *Computer Assisted Radiology 95*, Berlin, H. U. Lemke(ed.).
8. Maitland N, Harris CG. A video based tracker for use in computer aided surgery. In: Hancock ER, ed. *Proceedings British Machine Vision Conference.* Sheffield: BMVA Press, 1994:609-618.

Point-Based Elastic Registration of Medical Image Data Using Approximating Thin-Plate Splines

K. Rohr[1], H.S. Stiehl[1], R. Sprengel[1], W. Beil[1], T.M. Buzug[2], J. Weese[2], and M.H. Kuhn[2]

[1] Universität Hamburg, Fachbereich Informatik, Arbeitsbereich Kognitive Systeme, Vogt-Kölln-Str. 30, D-22527 Hamburg, Germany

[2] Philips Research, Technical Systems Hamburg, Röntgenstr. 24-26, D-22335 Hamburg, Germany

Abstract. We consider elastic registration of medical image data based on thin-plate splines using a set of corresponding anatomical point landmarks. Previous work on this topic has concentrated on using interpolation schemes. Such schemes force the corresponding landmarks to exactly match each other and assume that the landmark positions are known exactly. However, in real applications the localization of landmarks is always prone to some error. Therefore, to take into account these localization errors, we have investigated the application of an approximation scheme which is based on regularization theory. This approach generally leads to a more accurate and robust registration result. In particular, outliers do not disturb the registration result as much as is the case with an interpolation scheme. Also, it is possible to individually weight the landmarks according to their localization uncertainty. In addition to this study, we report on investigations into semi-automatic extraction of anatomical point landmarks.

1 Introduction

In neurosurgery and radiotherapy it is important to either register images from different modalities, e.g. CT (X-ray Computed Tomography) and MR (Magnetic Resonance) images, or to match images to atlas representations. If only *rigid* transformations were applied, then the accuracy of the resulting match often is not satisfactory w.r.t. clinical requirements. In general, *nonrigid* transformations are required to cope with the variations between the data sets. A special class of general nonrigid transformations are *elastic* transformations which allow for local adaptivity and are constrained to some kind of continuity or smoothness.

This contribution is concerned with elastic registration of medical image data based on a set of corresponding anatomical landmarks. Such a feature-based approach comprises three steps: (1) Extraction of landmarks in the different data sets, (2) Establishing the correspondence between the landmarks, and (3) Computing the transformation between the data sets using the information from (1) and (2). Among the different types of landmarks (points, lines, surfaces, and volumes) we here consider point landmarks.

Previous work on point-based elastic registration has concentrated on i) selecting the corresponding landmarks manually and on ii) using an interpolating transformation model (Bookstein [2], Evans et al. [6], and Mardia and Little [10]). The basic approach draws upon thin-plate splines and is computationally efficient, robust, and general w.r.t. different types of images and atlases. Also, the approach is well-suited for user-interaction which is important in clinical scenaria. However, an interpolation scheme forces the corresponding landmarks to exactly match each other. The underlying assumption is that the landmark positions are known exactly. In real applications, however, the localization of landmarks is always prone to some error. This is true for interactive as well as for automatic landmark localization. Therefore, to take into account these localization errors, we have investigated the application of an approximation scheme where the corresponding thin-plate splines result from regularization theory. Generally, such an approach yields a more accurate and robust registration result. In particular, outliers do not disturb the registration result as much as is the case with an interpolation scheme. Also, it is possible to individually weight the landmarks according to their localization uncertainty. We have applied this approach to elastic registration of 2D tomographic images of the human brain. Investigations for 3D images are under way. Additionally, we report on investigations toward the automatic extraction of anatomical point landmarks using differential operators. We will present first experimental results on 2D and 3D tomographic images. Algorithms for this task are important since interactive selection of landmarks is time-consuming and often lacks accuracy.

In the following, we first discuss clinical applications of point-based elastic registration. Then, we briefly review the original thin-plate interpolation scheme and extend this approach to an approximation scheme. Finally, we describe investigations into semi-automatic landmark localization.

2 Clinical Applications for Elastic Registration

Although elastic matching is not routinely used in clinical practice yet, there are several application scenaria in which elastic matching is believed to improve current therapeutical procedures.

2.1 Image-atlas matching

One possible application is trajectory planning for neurosurgical intervention. Pain treatment as well as epilepsy treatment sometimes require to localize a functionally important region not visible in the available image data. There are instructions available in the literature how to construct the position of such a region given landmarks which can be identified in CT or MR images. Hence, it is useful to superimpose an atlas with a medical image as already proposed by Talairach. Due to the individual variability of anatomical structures, rigid registration is generally not sufficient and elastic matching should be applied.

2.2 CT-MR matching

Another application is the registration of CT and MR images for the purpose of radiotherapy planning. Additionally, a template atlas can be superimposed on the MR image to indicate, for example, organs at risk. This superposition result is then overlayed on the CT image prior to dose calculation and isodose visualization on the MR image. It is questionable whether rigid registration is suitable for this purpose since MR images are geometrically distorted. On the one hand, scanner-induced distortions have to be coped with which are caused by, e.g., inhomogeneities of the main magnetic field, imperfect slice or volume selection pulses, nonlinearities of the magnetic field gradients, and eddy currents [11]. These distortions can be reduced by suitable calibration steps: The inhomogeneities of the main magnetic field are minimized by passive and active shimming whereas, e.g., the gradient nonlinearities cannot be completely shimmed. Thus, depending on the scanner protocol, the sum of all remaining distortions leads to a residual error of a few millimeters (for a spherical field of view of 25 cm). On the other hand, there are geometrical distortions in MR images that are induced by the patient and cannot be removed by calibration. Parameters such as susceptibility variations, chemical shift for non-water protons and flow-induced distortions for vessels are very important. While the susceptibility difference of tissue and bone is negligible, the susceptibility difference between tissue and air is approximately 10^{-5}. This can result in a field variation of up to 10 ppm and geometrical distortions of more than 5 mm [9],[4] which is most important for the nasal and aural regions. Consequently, due to the scanner- as well as the patient-induced distortions of the MR image CT and MR images cannot be satisfactorily registered using a rigid transformation.

3 Thin-Plate Spline Interpolation

The use of thin-plate spline interpolation for point-based elastic registration of medical images was first proposed by Bookstein [2]. Here, we briefly describe this method in the general context of d-dimensional images (see also [15]): Given two landmark sets each consisting of n landmarks $\mathbf{p}_i$ and $\mathbf{q}_i$, $i = 1, \ldots, n$ in two images of dimension d, find the transformation $\mathbf{u}$ within a suitable Hilbert-space H^d of admissible functions, which i) minimizes a given functional $J : H^d \rightarrow \mathbb{R}$ and ii) fulfills the interpolation conditions

$$\mathbf{u}(\mathbf{p}_i) = \mathbf{q}_i, \qquad i = 1, \ldots, n. \tag{1}$$

We only consider such functionals $J(\mathbf{u})$ which can be separated into a sum of similar functionals that only depend on one component of the transformation $\mathbf{u}$. Thus, the problem of finding $\mathbf{u}$ can be subdivided into d problems for each component z of $\mathbf{u}$. In the case of interpolation the functional is fully described through the dimension d of the domain and the order m of derivatives used [5] and can be written as

$$J_m^d(z) = \sum_{\alpha_1+\ldots+\alpha_d=m} \frac{m!}{\alpha_1! \cdots \alpha_d!} \int_{\mathbb{R}^d} \left(\frac{\partial^m z}{\partial x_1^{\alpha_1} \cdots \partial x_d^{\alpha_d}} \right)^2 d\mathbf{x}. \tag{2}$$

The functional is invariant w.r.t. translations and rotations.

Let a set of functions ϕ_i span the space $\Pi^{m-1}(\mathbb{R}^d)$ of all polynomials on $\mathbb{R}^d$ up to order $m-1$, which is the nullspace of J_m^d. The dimension of this space is $M = (d+m-1)!/(d!(m-1)!)$ and must be lower than n (this gives the minimum number of landmarks). The solution of the minimization problem can now be written as:

$$z(\mathbf{x}) = \sum_{i=1}^{M} a_i \phi_i(\mathbf{x}) + \sum_{i=1}^{n} w_i U_i(\mathbf{x}), \tag{3}$$

with some basis functions $U_i = U(\cdot, \mathbf{p}_i)$ depending on i) the dimension d of the domain, ii) the order m of the functional J to be minimized and iii) the Hilbert-space H of admissible functions [5, 15]. If we choose the Sobolev space $H = \mathcal{H}^2$, we obtain the kernel

$$U(\mathbf{x}, \mathbf{p}) = \begin{cases} \theta_{m,d} |\mathbf{x} - \mathbf{p}|^{2m-d} \ln |\mathbf{x} - \mathbf{p}| & 2m - d \text{ even positive integer} \\ \theta_{m,d} |\mathbf{x} - \mathbf{p}|^{2m-d} & \text{otherwise} \end{cases}$$

with $\theta_{m,d}$ as defined in [17]. Note, that the basis functions U_i span an n-dimensional space of functions that depends only on the source landmarks.

The coefficient vectors $\mathbf{a} = (a_1, \ldots, a_M)^T$ and $\mathbf{w} = (w_1, \ldots, w_n)^T$ can be computed through the following system of linear equations:

$$\begin{aligned} \mathbf{K}\mathbf{w} + \mathbf{P}\mathbf{a} &= \mathbf{v} \\ \mathbf{P}^T \mathbf{w} &= 0, \end{aligned} \tag{4}$$

where $\mathbf{v}$ is the column vector of one component of the coordinates of the target points $\mathbf{q}_i$, and $K_{ij} = U_i(\mathbf{p}_j), P_{ij} = \phi_j(\mathbf{p}_i)$.

4 Thin-Plate Spline Approximation Based on Regularization Theory

To take into account landmark localization errors one has to weaken the interpolation condition (1). This can be done by combining an approximation criterion with the functional in (2). In the simplest case of a quadratic approximation term, this results in the following functional [17]:

$$J_\lambda(\mathbf{u}) = \sum_{i=1}^{n} |\mathbf{q}_i - \mathbf{u}(\mathbf{p}_i)|^2 + \lambda J_m^d(\mathbf{u}). \tag{5}$$

Such functionals have been used for the reconstruction of surfaces from sparse depth data. Arad et al. [1] recently used a 2D approximation approach of this kind to represent and modify facial expressions. The first term (data term) measures the sum of the quadratic Euclidean distances between the transformed source landmarks and the target landmarks. The second term measures the smoothness of the resulting transformation. Hence, the minimization of (5) yields a transformation $\mathbf{u}$, which i) approximates the distance of the source landmarks

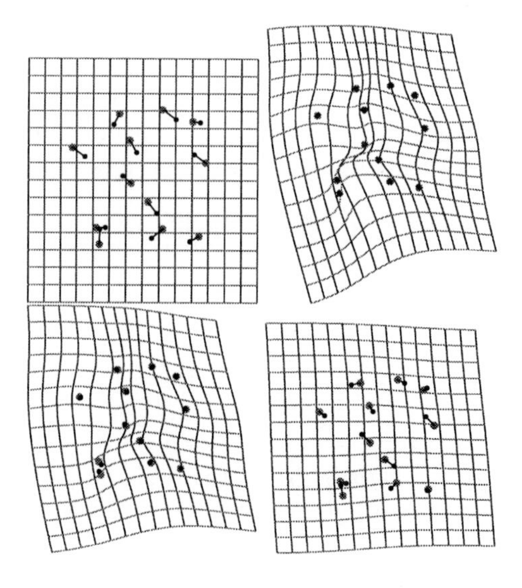

Fig. 1. Thin-plate spline approximation (input data, $\lambda = 0, 0.001$, and 0.1).

to the target landmarks and ii) is sufficiently smooth. The relative weight between the approximation behavior and the smoothness of the transformation is determined by the regularization parameter $\lambda > 0$. If λ is small, we obtain a solution with good approximation behavior (in the limit of $\lambda \to 0$ we have an interpolating transformation). If λ is large, we obtain a very smooth transformation with little adaption to the local structure of the distortions. In the limit of $\lambda \to \infty$ we get a global polynomial of order up to $m - 1$, which has no smoothness energy J_m^d at all.

The interesting fact is that the solutions to the approximation problem have always the same form as in the case of interpolation. Also, the computational scheme is nearly the same. We only have to add λ in the diagonal of the matrix $\mathbf{K}$:

$$
\begin{aligned}
(\mathbf{K} + \lambda \mathbf{I})\mathbf{w} + \mathbf{P}\mathbf{a} &= \mathbf{v} \\
\mathbf{P}^T \mathbf{w} &= 0.
\end{aligned}
\tag{6}
$$

Adding λ in the diagonal also results in a better conditioned system of linear equations than in the case of interpolation.

Fig. 1 shows an example of the thin-plate spline approximation scheme in two dimensions (with $m = 2$) for different values of λ. The small black points and big grey points mark the positions of the source and target landmarks, respectively. The top-left part of Fig. 1 shows the regular grid. The top-right part shows the result for $\lambda = 0$, which is equivalent to the interpolation scheme. At some locations the grid is heavily distorted, especially around the two close landmarks in the bottom-left part of this grid. The two bottom grids visualize results for $\lambda = 0.001$ (bottom-left) and $\lambda = 0.1$ (bottom-right), where the latter one nearly yields a pure affine transformation.

A generalization of the approximation scheme can be made, if we have information about the expected accuracy of the landmarks. Then, we can weight each single data term $|\mathbf{q}_i - \mathbf{u}(\mathbf{p}_i)|^2$ by the inverse variance $1/\sigma_i^2$. If the variance is high, i.e. landmark localization is uncertain, then less penalty is given to the approximation error at this point. The data term now reads

$$\sum_{i=1}^{n} \frac{|\mathbf{q}_i - \mathbf{u}(\mathbf{p}_i)|^2}{\sigma_i^2},$$

and we have to solve the following system of equations:

$$\begin{aligned} (\mathbf{K} + \lambda \mathbf{W}^{-1})\mathbf{w} + \mathbf{P}\mathbf{a} &= \mathbf{v} \\ \mathbf{P}^T \mathbf{w} &= 0, \end{aligned} \tag{7}$$

where $\mathbf{W} = \text{diag}\{1/\sigma_1^2, \ldots, 1/\sigma_n^2\}$. Note, that this approach can be applied to images of arbitrary dimension, i.e. in particular to 2D as well as 3D images.

5 Experimental Results

Within the scenario of CT-MR registration as discussed above we here consider the important application of correcting patient-induced susceptibility distortions of MR images. To this end we have acquired two sagittal MR images of a healthy human volunteer brain with typical susceptibility distortions. In our experiment we used a high-gradient MR image as "ground truth" (instead of clinically common CT images) to avoid exposure of the volunteer to radiation. Both turbo-spin echo images have consecutively been acquired on a modified Philips $1.5T$ MR scanner with a slice thickness of $4mm$ without repositioning. Therefore, we are sure that we actually have identical slicing in space. Using a gradient of $1mT/m$ and $6mT/m$ for the first and second image then leads to a shift of ca. $7.5...10mm$ and ca. $1.3...1.7mm$, respectively (see [14] for details). In our example, we use the second image as "ground truth" to demontrate that the elastic matching approach can cope with these distortions and that the use of approximating thin-plate splines is advantageous.

Within each of the two images we have interactively selected 20 point landmarks. To simulate outliers, one of the landmarks in the first image (No. 3) has been shifted about 15 pixels away from its true position for demonstration purposes (see Fig. 2). Note, however, that interactive localization of landmarks actually can be prone to relatively large errors. Fig. 3 shows the results of the interpolating vs. the approximating ($\lambda = 0.015$) thin-plate spline approach. Each result represents the transformed first image. In the difference image of the two results in Fig. 4 we see that the largest differences occur at the shifted landmark No. 3 which is what we expect. In Fig. 3 on the left it can be seen that the interpolation scheme yields a rather unrealistic deformation since it forces all landmark pairs, including the pair with the simulated outlier, to exactly match each other. Using our approximation scheme instead yields a more accurate registration result (Fig. 3 on the right). The increased accuracy can also be demonstrated by

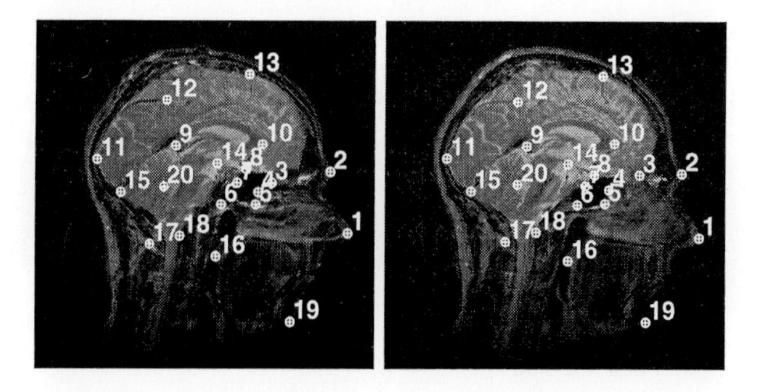

Fig. 2. Original MR images with landmarks: First (left) and second (right) image.

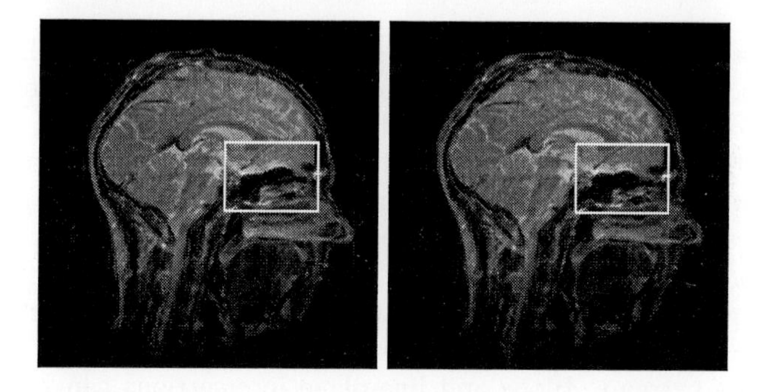

Fig. 3. Registration results: Interpolation (left) and approximation (right).

computing the distance between the grey-value edges of the transformed images and those of the second image. In our case, we applied a distance transformation [12] to Canny edges. The results for the marked rectangular image parts in Fig. 3 can be seen in Fig. 5. Here, the intensities represent the registration error, i.e., the brighter the larger is the error. In particular at the marked circular areas, which indicate the grey-value edges perpendicular to the simulated shift, we see that the registration accuracy has increased significantly. Note, that in this experiment we only used equal weights for all landmarks. However, incorporation of quantitative knowledge about the landmark uncertainties promises a further increase of accuracy.

6 Semi-Automatic Landmark Localization

The problem with point landmarks is their reliable and accurate extraction from 3D images. Therefore, 3D point landmarks have usually been selected manually (e.g., Evans et al. [6], Hill et al. [8]). Only a few automatic approaches have been proposed (e.g., Thirion [16]). In this section, we briefly describe our first investigations into semi-automatic localization of 3D anatomical point landmarks. Semi-automatic means that either a region-of-interest (ROI) or an approximate

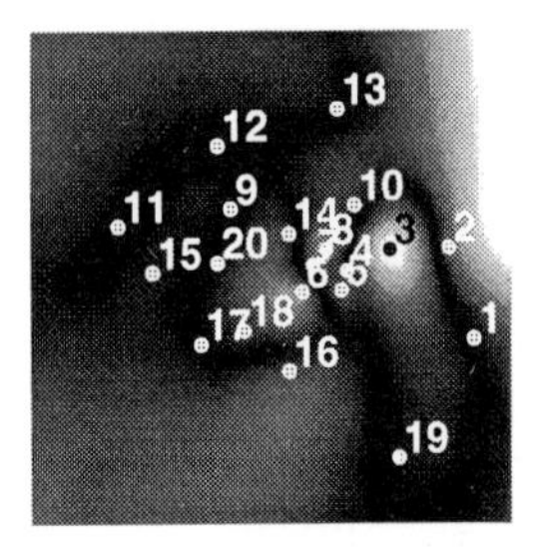

Fig. 4. Difference between the two registration results.

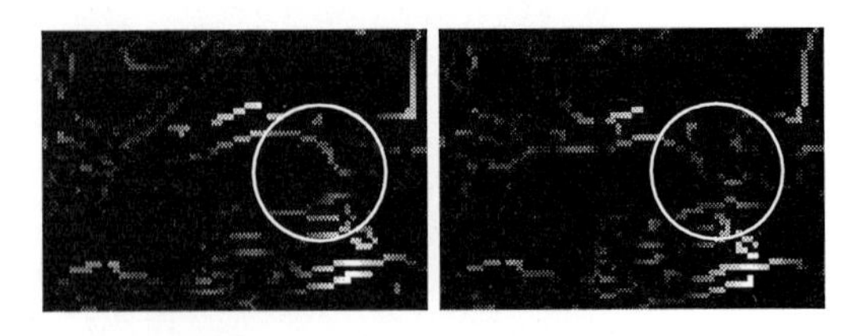

Fig. 5. Registration errors: Interpolation (left) and approximation (right).

position of a specific landmark (or both) is given by the user. Then, an algorithm has to provide a refined position of the landmark. Alternatively, landmark candidates for a large ROI or even for the whole data set may be provided automatically from which the final set of landmarks is selected interactively. Such a semi-automatic approach has the advantage that a user has the possibility to control the results ("keep-the-user-in-the-loop").

Within a ROI we apply specific 3D differential operators such as to exploit the knowledge about a landmark as far as possible, in particular it's geometric structure. Typical landmark types in 3D are blobs, line-plane intersections and curvature extrema. Blobs and line-plane intersections can be determined by exploiting the Eigenvalues of the Hessian matrix $\underline{H}_g$. To localize curvature extrema we exploit the two principal curvatures of the isocontour as is also done in approaches for detecting 3D ridge lines. In our first experiments we used an operator which represents the Gaussian curvature, i.e. the product of the two principal curvatures $K = \lambda_1 \lambda_2$, multiplied with the fourth power of the gradient magnitude $|\nabla g|$. Fig. 6 shows a result of this operator for the right frontal horn in a 3D MR data set. It can be seen that we obtain a strong operator response at the tip of the frontal horn.

We have also investigated 3D differential operators for providing a set of landmark candidates. Since we want to apply these operators on a large ROI or even on the whole data set, it is indispensable to use computationally efficient schemes. To this end we have extended existing 2D differential operators for detecting points of high intensity variations ('corner detectors') to 3D. For a recent analytic study of such 2D operators see Rohr [13]. These operators have the advantage that only low order partial derivatives of the image function are necessary (first or first and second order). Therefore, these operators are computationally efficient and do not suffer from instabilities of computing high order partial derivatives. As an example, in Fig. 7 we show the application of the 2D operator of Förstner [7] vs. a 3D extension of it: $det\underline{C}_g / trace\underline{C}_g \rightarrow max$, where $\underline{C}_g = \overline{\nabla g \, (\nabla g)^T}$ and ∇g denotes the image gradient in 2D and 3D, respectively. Note, that in the 2D case many well detected landmarks agree with the interactively selected landmarks in Bookstein [3]. Note also, that the 3D operator actually takes into account the 3D structure of the landmarks and therefore in a

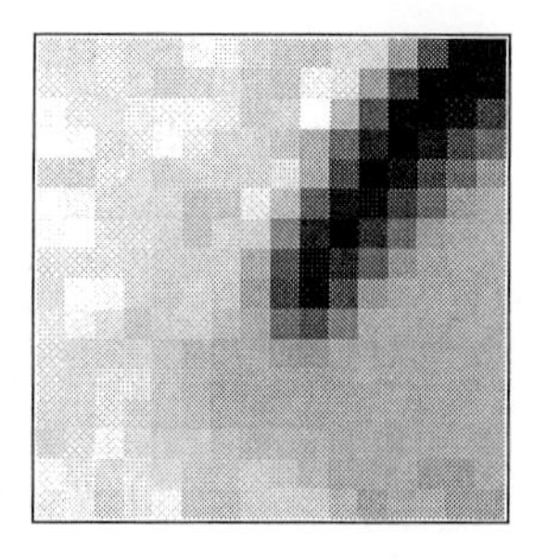

Fig. 6. Right frontal horn in a 3D MR data set (left) and result of computing the 3D Gaussian curvature (right)

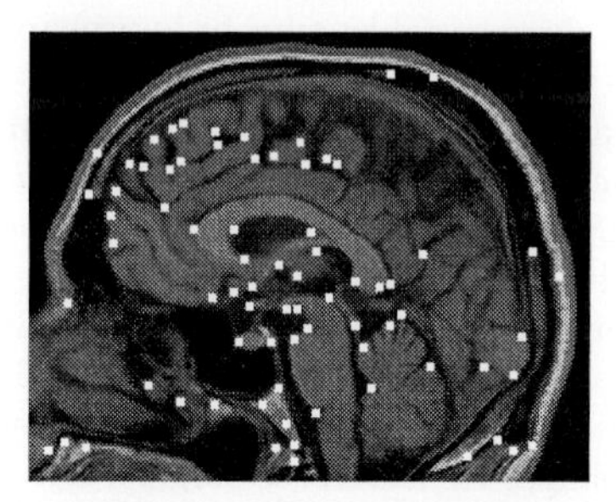

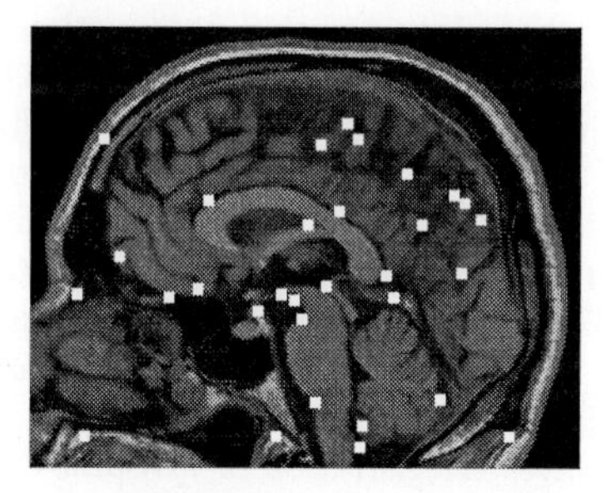

Fig. 7. Landmark candidates: Application of a 2D 'corner' detector (left) vs. a 3D extension (right) on a 2D and 3D MR image, respectively.

single slice of a 3D image only a few of the 3D point landmarks are visible, i.e., other landmarks according to [3] have been detected in different slices.

7 Summary

We have introduced an approximating thin-plate spline approach based on regularization theory to elastic registration of medical image data. In the case of landmark localization errors this scheme generally yields a more accurate and robust registration result. In particular large localization errors, i.e. outliers, do not affect the registration result as much as is the case with an interpolation scheme. In addition, we have reported on investigations into semi-automatic extraction of anatomical point landmarks from 2D as well as 3D tomographic images of the human brain.

Acknowledgement

Support of Philips Research Hamburg, project IMAGINE (IMage- and Atlas-Guided Interventions in NEurosurgery), and of the E.U., project EASI, is gratefully acknowledged. We would like to thank K. Jungnickel (formerly with Philips Research Hamburg) for providing us with the MR images showing the susceptibility distortions.

References

1. N. Arad, N. Dyn, D. Reisfeld, and Y. Yeshurun. Image warping by radial basis functions: Application to facial expressions. *Computer Vision, Graphics, and Image Processing*, 56(2):161–172, 1994.
2. F. Bookstein. Principal warps: Thin-plate splines and the decomposition of deformations. *IEEE Trans. Pattern Anal. and Machine Intell.*, 11(6):567–585, 1989.
3. F. Bookstein. Thin-plate splines and the atlas problem for biomedical images. In A. Colchester and D. Hawkes, editors, *12th Internat. Conf. Information Processing in Medical Imaging*, volume 511 of *Lecture Notes in Computer Science*, pages 326–342, Wye/UK, 1991. Springer-Verlag Berlin Heidelberg.
4. H. Chang and J. Fitzpatrick. A technique for accurate magnetic resonance imaging in the presence of field inhomogeneities. *IEEE Trans. Med. Imaging*, 11:319, 1992.
5. J. Duchon. Interpolation des fonctions de deux variables suivant le principle de la flexion des plaques minces. *R.A.I.R.O. Analyse Numérique*, 10(12), 1976.
6. A. Evans, W. Dai, L. Collins, P. Neelin, and S. Marrett. Warping of a computerized 3-d atlas to match brain image volumes for quantitative neuroanatomical and functional analysis. In M. Loew, editor, *Medical Imaging V: Image Processing*, volume 1445 of *Proc. SPIE*, pages 236–246, San Jose, CA, 1991.
7. W. Förstner. A feature based correspondence algorithm for image matching. *Intern. Arch. of Photogrammetry and Remote Sensing*, 26-3/3:150–166, 1986.
8. D. Hill, D. Hawkes, J. Crossman, M. Gleeson, T. Cox, E. Bracey, A. Strong, and P. Graves. Registration of MR and CT images for skull base surgery using point-like anatomical features. *The British J. of Radiology*, 64(767):1030–1035, 1991.
9. K. Lüdeke, P. Röschmann, and R. Tischler. Susceptibility artifacts in nmr imaging. *MRI*, 3:329, 1985.
10. K. Mardia and J. Little. Image warping using derivative information. In F. Bookstein, J. Duncan, N. Lange, and D. Wilson, editors, *Mathematical Methods in Medical Imaging III*, volume 2299 of *Proc. SPIE*, pages 16–31, San Diego, CA, 25-26 July 1994.
11. J. Michiels, H. Bosmans, P. Pelgrims, D. Vandermeulen, J. Gybels, G. Marchal, and P. Suetens. On the problem of geometric distortion in magnetic resonance images for stereotactic neurosurgery. *Mag. Res. Imag.*, 12:749, 1994.
12. D. W. Paglieroni. A unified transform algorithm and architecture. *Machine Vision and Applications*, 5(1):47–55, 1992.
13. K. Rohr. Localization properties of direct corner detectors. *J. of Mathematical Imaging and Vision*, 4(2):139–150, 1994.
14. K. Rohr, H. S. Stiehl, R. Sprengel, W. Beil, T. M. Buzug, J. Weese, and M. H. Kuhn. Point-based elastic registration of medical image data using approximating thin-plate splines. Techn. Report FBI-HH-M-254/96, FB Informatik, Universität Hamburg, Febr. 1996.
15. R. Sprengel, K. Rohr, and H. S. Stiehl. Thin-plate spline approximation for image registration. 1996. submitted for publication.
16. J.-P. Thirion. Extremal points: definition and application to 3d image registration. In *Proc. IEEE Conf. on Computer Vision and Pattern Recognition*, pages 587–592, Seattle/Washington, USA, 1994.
17. G. Wahba. *Spline Models for Observational Data.* Society for Industrial and Applied Mathematics, Philadelphia, Pennsylvania, 1990.

Cortical Constraints for Non-Linear Cortical Registration

D.L. Collins[1], G. Le Goualher[1], R. Venugopal[2], A. Caramanos[2], A.C. Evans[2], C. Barillot[1*]

[1] Laboratoire SIM, Faculté de Médecine, Université de Rennes I, 35043 RENNES Cedex France

[2] McConnell Brain Imaging Center, Montreal Neurological Institute, McGill University, Montreal, Canada

Abstract. Correspondence of cortical structures is difficult to establish in automatic voxel-based non-linear registration of human brains based solely on gradient magnitude images. Experiments with a set of 9 real MRI data volumes demonstrates that the inclusion of 1) L_{vv}-based features or 2) automatically extracted sulci significantly reduce overlap errors for gyri and sulci on the cortex. Mismatch between gyri can be addressed using manually labelled sulci.

1 Introduction

The interpretation of functional brain images is usually aided by comparison to an atlas or to similar data from other subjects [10]. Unless a matched anatomical (MR or CT) image is available for each subject, anatomical variability must be taken into account to avoid over interpretation of the individuals functional data. While this is the case in many centers (e.g, [8]), in order to fully understand normal cognitive function within a population it is still necessary to compare results among individuals. We have therefore been interested in automatic procedures to capture and quantify anatomical differences between normal brains based on their appearance in 3D MRI. Towards this goal, we have developed a program called ANIMAL (Automatic Nonlinear Image Matching and Anatomical Labelling).

The ANIMAL procedure is based on the assumption that different brains are topologically equivalent and that non-linear deformations can be estimated and applied to one data set in order to bring it into correspondence with another. This registration process is then used 1) to estimate non-linear morphometric variability in a given population [5], 2) to automatically segment MRI data [4],

* We gratefully acknowledge post-doctoral funding from the Human Frontier Science Project Organization (DLC). We also acknowledge Positron Imaging Laboratory of the Montreal Neurological Institute for providing the stereotaxic MRI model and also the MR division of the Department of Radiology for providing the individual MR volumetric data used to test the algorithm. Parts of this project was funded by Medical Research Council (MRC SP-30) and an ICBM grant.

or 3) to remove residual alignment errors remaining after linear registration and thus improve detectability of subtle cognitive activation foci when averaging results among individuals or to establish functional variability.

Validations in [4] have shown that ANIMAL works well for deep structures, but sometimes had difficulty aligning sulci and gyri. In this paper, an extension is described to address this weakness and demonstrate improved results on 9 real MRI examples.

In contrast to existing methods for cortical structure alignment that depend on manual intervention to identify corresponding points [7, 2] or curves [13, 18, 19, 6], the procedure presented here is completely automatic. While other 3D voxel-based non-linear registration procedures exist (e.g., [1, 3, 14, 21, 11], none of these have looked specifically at the question of cortical features for cortical registration. The work presented here has some similarities to that of Subsol et. al., where crest lines are used for registration of the ventricles, however the procedure in [12] has not yet been successfully applied to the cortex.

Since non-linear registration procedures are now being used to align functional images, and given that functional brain imaging is often concerned by cognitive activation at the cortex, it is imperative that cortical structures be properly aligned by these methods.

2 Methods

2.1 Data Acquisition

Ten subjects were scanned on a Philips Gyroscan ACS 1.5 Tesla superconducting magnet system at the Montreal Neurological Institute using a T1-weighted 3-D spoiled gradient-echo acquisition with sagittal volume excitation (TR=18, TE=10, flip angle=30°, 140-180 sagittal slices). Without loss of generality, one of the 10 MRI volumes was chosen to serve as the target and the other 9 were registered to it.

2.2 Registration Algorithm

Registration of one MR data set with another, so that they can be compared on a voxel by voxel basis, is completed in a two step process. First, we begin by automatically calculating the best affine transformation matrix that aligns the two volumes. This matrix is found by optimizing the 9 transformation parameters (3 translations, 3 rotations and 3 scales) that maximize an objective function measuring similarity between Gaussian-blurred features (described in section 2.3 below) extracted from both volumes.

After linear registration, there remains a residual non-linear component of spatial mis-registration among brains. We assume that different brains are topographically equivalent[3], and that the application of non-linear deformations

[3] While this is not strictly true, one brain may have sulcal patterns that another does not, the assumption does not lead to catastrophic deformations of one structure onto another.

to one data set can bring it into exact correspondence with another and thus correct for this mis-registration. The ANIMAL program estimates the required 3D deformation field [5, 4].

The deformation field is built by sequentially stepping through the target volume in a 3D grid pattern. At each node i, the deformation vector required to achieve local registration between the two volumes is found by optimization of 3 translational parameters (tx_i,ty_i,tz_i) that maximize the objective function evaluated only in the neighbourhood region surrounding the node. The algorithm is applied iteratively in a multi-scale hierarchy, so that image blurring and grid size are reduced after each iteration, thus refining the fit. (The algorithm has been previously described in detail in [4].)

2.3 Features

The objective function maximized in the optimization procedure measures the similarity between features extracted from the source and target volumes. The features used in registration are shown in Fig. 1 and described below.

Gaussian Blurring and Gradient Magnitude. Geometrically invariant features are used in the objective function so that the comparison between data sets is independent of the original position, scale and orientation of one data set with respect to the other. Originally [5], we used only blurred image intensity and blurred image gradient magnitude, calculated by convolution of the original data with zeroth and first order 3D isotropic Gaussian derivatives. Convolution with such an operator maintains linearity, shift-invariance and rotational-invariance in the detection of features [20]. Since both linear and non-linear registration procedures are computed in a multi-scale fashion, the original MR data was blurred at two different scales: FWHM =16, and $8mm$, with the FWHM = (2.35σ) of the Gaussian kernel acting as a measure of spatial scale. These features were found to be sufficient to achieve registration of basal ganglia structures [4]. Unfortunately, the gradient magnitude feature did not sufficiently constrain the tangential component of the deformation at the cortex. Other features had to be found to address this weakness.

Differential Geometry-Based Features. Ambiguities in matching may be reduced by using higher order geometrical features. We have used a 3D variant of the L_{vv} operator described by Florack et. al.[9] that we term mean-L_{vv}, or mL_{vv}. In a 2D images, the L_{vv} operator represents the second directional derivative of the image intensity function $L(x, y)$, in the gauge-coordinate direction $\mathbf{v}$, and its value represents the curvature (of the local iso-curve) multiplied by the gradient magnitude (ie, $L_{vv} = -|\mathbf{n}|k$, where $\mathbf{n}$ is the gradient normal to the iso-curve, $|\cdot|$ is the magnitude operator, and k is the curvature).

In 3D, $\mathbf{n}$ is perpendicular to the local iso-surface. Unfortunately, there are an infinite number of directions from which to choose $\mathbf{v}$ in the plane that is tangent to the iso-surface. Differential geometry permits the local definition of the iso-surface by its two principal curvatures $(k1, k2)$ and its two principal directions $(\mathbf{t1}, \mathbf{t2})$. In order to retain a relationship similar to that in 2D, we have chosen to define the value of the mL_{vv} as the product of the gradient magnitude and the mean curvature: $mL_{vv} = -|\mathbf{n}|k_m$; where $k_m = (k1 + k2)/2$ is the mean curvature. As was the case for the gradient magnitude feature, the mL_{vv} was calculated at scales of FWHM =16 and $8mm$.

Explicitly Extracted Sulci. We have previously detailed a method to automatically extract cortical sulci (defined by the 2D surface extending from the sulcal trace at the cortical to the depth of the sulcal fundus) from 3D MRI [15]. In summary, we begin by classifying the MRI volume into grey, white and cerebrospinal fluid (CSF) classes [16]. When restricted to a cortical mask (grey matter and cerebrospinal fluid), negative values of mL_{vv} correspond to gyri and positive values, to sulci. Positive mL_{vv} maxima values correspond to the medial axis of the sulcus.

Each sulci is extracted in a two step process using an active model similar to a 2D Snake that we call an *active ribbon*. First, the superficial cortical trace of the sulcus is modeled by a 1D snake-spline [17]. Second, this snake is then submitted to a set of forces derived from (1) the classification results, (2) the differential characteristics of the iso-surfaces describing the cortex and (3) a distance-from-cortex term to force the snake to converge to the sulcal fundus. The set of successive iterative positions of the 1D snake define the sulcal axis. These loci are then used to drive a 2D active ribbon which forms the final representation of the sulcus.

When applied to an MRI brain volume, the output of this automatic process consists of a set of *Elementary Sulcal Surfaces* (ESS) that represent all cortical sulcal folds. A number of these ESSs may be needed to represent a sulcus as defined by an anatomist. For the experiments described below, six ESS sets were identified manually on 5 of the 9 brains corresponding to six anatomical sulci (the central, lateral (Sylvian) and superior frontal sulci on both the left and right hemispheres) to be used as features in ANIMAL . The use of this type of feature removes possible ambiguity in matching, since homology is defined by the user/expert. In both cases (labelled or unlabelled), the voxel masks representing the sulci were blurred at scales of FWHM =16 and $8mm$.

2.4 Error measures

Four measures of registration quality were used in the experiments described below. First, a simple correlation measure for all brain voxels was computed (with values near 1.0 indicating a perfect fit). The second uses the grey-matter classified data since we are interested in registration at the cortex. An overlap index, Δ, (with a maximum value of 100.0) and a volume difference (δ, in %) for grey-matter voxels were calculated for each experiment:

$$\Delta = \frac{V_s \cap V_t}{V_s \cup V_t}$$

where V_s and V_t represent the set of grey-matter (or sulcal) voxels in the source and target volumes, respectively. These first three statistics give only an global measure of the fit, without regard to specific regions. Indeed, a high intensity correlation value is possible, even when a gyrus is mis-matched to its neighbour.

For the last measure, the cortical trace of 16 sulci[4] were identified manually by a medically-trained expert (RV) on the 10 MRI volumes. From the coordinates of these sulci, we derived an 3D rms minimum distance measure between the labelled sulci of a transformed source and their homologues in the target. This measure is defined on a point-wise basis, computing the root mean square (over all sulcal trace points) of the minimum distance between each source sulcus point and the target sulcus trace.

3 Experiments and Results

Four experiments were completed varying only the features used by ANIMAL. In each experiment, 9 MRI volumes were registered to the chosen target volume. In the first experiment, only the gradient magnitude feature was used in the objective function. In the second, the mL_{vv} was used in addition to the gradient magnitude. In the third, all ESS voxels were in addition to the gradient magnitude. In the last, a set of 6 automatically extracted, manually labelled sulci were used in conjunction with the gradient magnitude feature. The first experiment represents our previously published method, and is used as a baseline. A linear transformation is presented for comparison as well. Figure 3 and 2 summarize the results qualitatively. Table 1 presents the quantitative results for the measures described above.

Baseline - Gradient Magnitude Only. In the first experiment (gradient magnitude only) we can see that the global brain shape is properly corrected and that the main lobes are aligned. Previous experiments in [4] have shown that basal ganglia structures (e.g., thalamus, caudate, putamen, globus pallidus) and ventricular structures are well registered ($\Delta \geq 89\%$). Examination of cross-slices through the volume shows that not all sulci and gyri are properly aligned, particularly in the motor/sensory region (see Fig. 2). Here, the grey-matter overlap was equal to 72% and the rms sulcal distance was $7.6mm$.

With mL_{vv}. In the second experiment (gradient magnitude and mL_{vv}), the global brain shape is maintained. Slightly more sulci appear to be aligned, as indicated by a small increase in the grey-matter overlap to 76.7% and by an equivalent increase in sulcal alignment (rms sulcal distance = 6.7mm). While we can see that some sulci are still misaligned in Fig. 2, overall registration is good.

[4] The sulci identified were: central, pre-central, post-central, superior-frontal, middle-frontal, inferior-frontal, sylvian and superior temporal, on both left and right sides.

With Extracted ESS. Here, all ESS voxels were used as an additional feature in ANIMAL. While there is only a slight increase in the grey matter overlap, when compared to linear (up 74.0% to 75.1%), there is a definite decrease in the volume difference, from 56.6% to 43.3%. However, the sulcal traces in Fig. 2 do not appear to be better grouped than when using mL_{vv}. Figure 3 shows that the internal structures are better registered when using mL_{vv}, probably because the feature is evaluated throughout the volume, and not only at the cortex (as are the ESS).

Labelled ESS. Labelled sulci were used in conjunction with gradient magnitude data for five of the nine source volumes in the third experiment to address problems in establishing correspondence. Once again, the global brain shape is maintained and the alignment of the individual lobes is improved over linear. While the overlap measure does not show improvement, visual inspection of Fig. 2 shows that the previous misalignments in the motor-sensory area have been corrected, and alignment for neighbouring structures has improved. Indeed, when evaluated on the left central sulcus, the rms sulcal distance improves dramatically: 6.9±1.8$mm$for linear; 5.9±2.0$mm$for non-linear with mL_{vv}; 3.5±1.4$mm$for labelled sulci;

Table1: Registration measures

feature	*grey matter* Δ	*grey matter* δ	*intensity correlation*	*rms min sulcal dist*
lin only:	74.0±5.4%	56.6±8.9%	96.4±0.4%	7.2±3.0mm
∇:	72.0±6.4%	38.4±9.4%	95.9±0.6%	7.6±3.3mm
$mL_{vv}+\nabla$:	76.7±6.0%	45.2±8.9%	96.7±0.3%	6.7±3.0mm
all ESS $+\nabla$:	75.1±5.0%	43.3±7.8%	96.6±0.2%	7.2±3.2mm
6 ESS $+\nabla$:	74.2±8.9%	45.1±8.8%	96.2±0.2%	7.2±4.6mm

4 Discussion and Conclusion

Non-linear registration improves cortical structure overlap, when compared to linear registration. Further improvement is possible by incorporating sulcal information from mL_{vv} and the extracted sulcal trace. If one is willing to pay a small price for manual intervention, the use of labelled sulci removes ambiguity when establishing correspondence, and thus reduces errors due to sulcal mis-alignment.

While the images of Figs. 3 and 2 show this qualitative improvement in registration, the quantitative measures of Table 1 do not. There are at least three of reasons for this:

1-The use of explicitly extracted sulcal information should improve registration. Perhaps this data should not be blurred, as is done in the current implementation, but used with a different energy function. We envision using a function evaluated on the sulcal ribbon in association with an rms distance criterion.

2-In our previous experiments [4], the multiresolution scheme ended as a scale of FWHM=$4mm$, instead of the $8mm$used here. Perhaps the difference between the non-linear registration methods will be more evident when the fit is pushed to a finer scale.

3-The first three measures of registration (MRI intensity correlation, grey matter overlap, volume difference) are global measures evaluated over the whole volume. A large improvement in a specific cortical region does not greatly effect these measures. As for the rms minimum sulcal distance measure, it does not take into account end effects. Indeed, two perfectly overlapping sulci, with one longer than the other will have a non-null measure. We are working on better objective measures to quantify the goodness-of-fit for cortical structures.

There are two main directions for future work on this project. First, we plan to apply the technique to a large number of brains for evaluation and validation. Second, we are looking at two ways to remove the need for manual intervention. Preliminary experiments indicate that it may be possible to use sulcal characteristics (e.g., length, depth, orientation) to automatically establish correspondence between (at least) the primary sulci. This would improve the completely automatic non-linear registration procedure presented here.

References

1. R. Bajcsy and S. Kovacic. Multiresolution elastic matching. *Computer Vision, Graphics, and Image Processing*, 46:1–21, 1989.
2. F. Bookstein. Thin-plate splines and the atlas problem for biomedical images. In A. Colchester and D. Hawkes, editors, *Information Processing in Medical Imaging*, volume 511 of *Lecture Notes in Computer Science*, pages 326–342, Wye, UK, July 1991. IPMI, Springer-Verlag.
3. G. Christensen, R. Rabbitt, and M. Miller. 3d brain mapping using a deformable neuroanatomy. *Physics in Med and Biol*, 39:609–618, 1994.
4. D. Collins, C. Holmes, T. Peters, and A. Evans. Automatic 3D model-based neuroanatomical segmentation. *Human Brain Mapping*, 3(3):190–208, 1996.
5. D. L. Collins, T. M. Peters, and A. C. Evans. An automated 3D non-linear image deformation procedure for determination of gross morphometric variability in human brain. In *Proceedings of Conference on Visualization in Biomedical Computing*. SPIE, 1994.
6. D. Dean, P. Buckley, F. Bookstein, J. Kamath, and D. Kwon. Three dimensional mr-based morphometric comparison of schizophrenic and normal cerebral ventricles. In *Proceedings of Conference on Visualization in Biomedical Computing*, Lecture Notes in Computer Science, page this volume. Springer-Verlag, Sept. 1996.
7. A. C. Evans, W. Dai, D. L. Collins, P. Neelin, and T. Marrett. Warping of a computerized 3-D atlas to match brain image volumes for quantitative neuroanatomical and functional analysis. In *Proceedings of the International Society of Optical Engineering: Medical Imaging V*, volume 1445, San Jose, California, 27 February – 1 March 1991. SPIE.
8. A. C. Evans, S. Marrett, J. Torrescorzo, S. Ku, and L. Collins. MRI-PET correlation in three dimensions using a volume-of-interest (VOI) atlas. *Journal of Cerebral Blood Flow and Metabolism*, 11(2):A69–78, Mar 1991.

9. L. M. J. Florack, B. M. ter Haar Romeny, J. J. Koenderink, and M. A. Viergever. Scale and the differential structure of images. *Image and Vison Computing*, 10:376–388, 1992.
10. P. T. Fox, M. A. Mintun, E. M. Reiman, and M. E. Raichle. Enhanced detection of focal brain responses using intersubject averaging and change-distribution analysis of subtracted PET images. *Journal of Cerebral Blood Flow and Metabolism*, 8:642–653, 1988.
11. K. Friston, C. Frith, P. Liddle, and R. Frackowiak. Plastic transformation of PET images. *Journal of Computer Assisted Tomography*, 15(1):634–639, 1991.
12. N. A. G. Subsol, J.-P. Thirion. Application of an automatically built 3d morphometric brain atlas: study of cerebral ventricle shape. In *Proceedings of Conference on Visualization in Biomedical Computing*, Lecture Notes in Computer Science, page this volume. Springer-Verlag, Sept. 1996.
13. Y. Ge, J. Fitzpatrick, R. Kessler, and R. Margolin. Intersubject brain image registration using both cortical and subcortical landmarks. In *Proceedings of SPIE Medical Imaging*, volume 2434, pages 81–95. SPIE, 1995.
14. J. Gee, L. LeBriquer, and C. Barillot. Probabilistic matching of brain images. In Y. Bizais and C. Barillot, editors, *Information Processing in Medical Imaging*, Ile Berder, France, July 1995. IPMI, Kluwer.
15. G. L. Goualher, C. Barillot, Y. Bizais, and J.-M. Scarabin. Three-dimensional segmentation of cortical sulci using active models. In *SPIE Medical Imaging*, page in press, Newport-Beach, Calif., 1996. SPIE.
16. F. Lachmann and C. Barillot. Brain tissue classification from mri by means of texture analysis. In *SPIE Medical Imaging VI*, volume 1652, pages 72–83, Newport-Beach, Calif., 1992. SPIE.
17. F. Leitner, I. Marque, S. Lavalee, and P. Cinquin. Dynamic segmentation: finding the edge with snake splines. In *Int. Conf. on Curves and Surfaces*, pages 279–284. Academic Press, June 1991.
18. S. Luo and A. Evans. Matching sulci in 3d space using force-based deformation. *IEEE Transactions on Medical Imaging*, submitted Nov., 1994.
19. S. Sandor and R. Leahy. Towards automated labelling of the cerebral cortex using a deformable atlas. In Y. Bizais, C. Barillot, and R. DiPaola, editors, *Information Processing in Medical Imaging*, pages 127–138, Brest, France, Aug 1995. IPMI, Kluwer.
20. B. tar Haar Romeny, L. M. Florack, J. J. Koenderink, and M. A. Viergever. Scale space: its natural operators and differential invariants. In A. C. F. Colchester and D. J. Hawkes, editors, *Information Processing in Medical Imaging*, page 239, Wye, UK, July 1991. IPMI.
21. J. Zhengping and P. H. Mowforth. Mapping between MR brain images and a voxel model. *Med Inf (Lond)*, 16(2):183–93, Apr-Jun 1991.

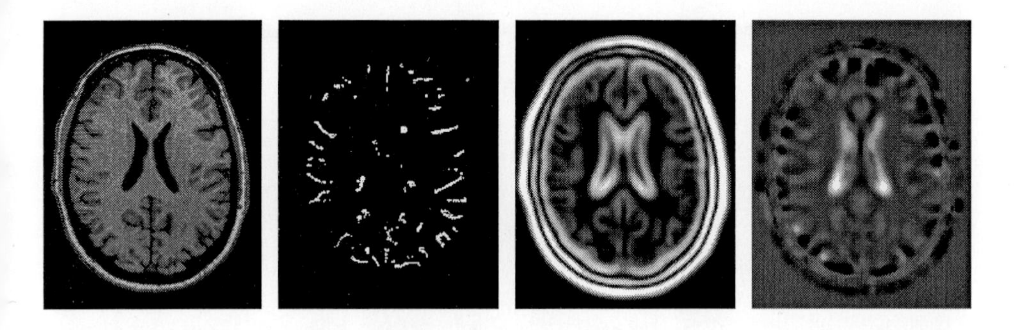

Fig. 1. Features used in registration

This figure shows a transverse slice at the level of the ventricles through the feature volumes used in the experiments. From left to right: The original data; the ESS voxels, the gradient magnitude feature (FWHM = 8mm); the mL_{vv} feature at the same spatial scale. One can see that the gradient magnitude and mL_{vv} feature appear complimentary in the cortical region in the sense that the edges extracted by gradient magnitude are parallel to the cortex while the mL_{vv} constrains the registration tangent to the cortex.

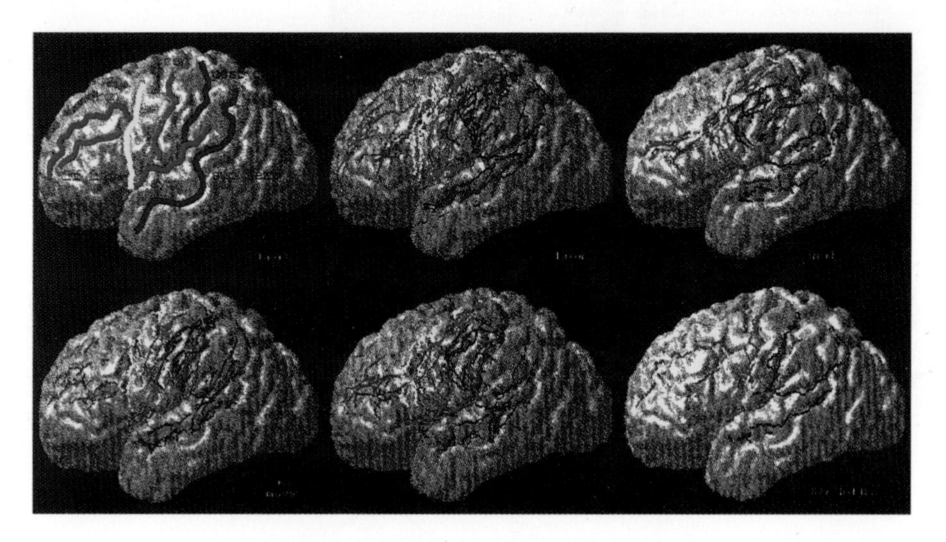

Fig. 2. Results using different features

These images show the sulci on the cortical surface of the target volume. From top to bottom, left to right: target sulci; source sulci mapped through linear tranformation; mapped through non-linear gradient magnitude; mL_{vv}; ESS; and labelled ESS (only 5 of the 9 volumes were labelled). Note that the sulci are better grouped in the non-linear registrations, when compared to linear. In particular, the use of one labelled sulci (central s.) in the region of the motor-sensory area, removes the abiguity and consequently improves the sulcal and gyral matching in the neighbourhood.

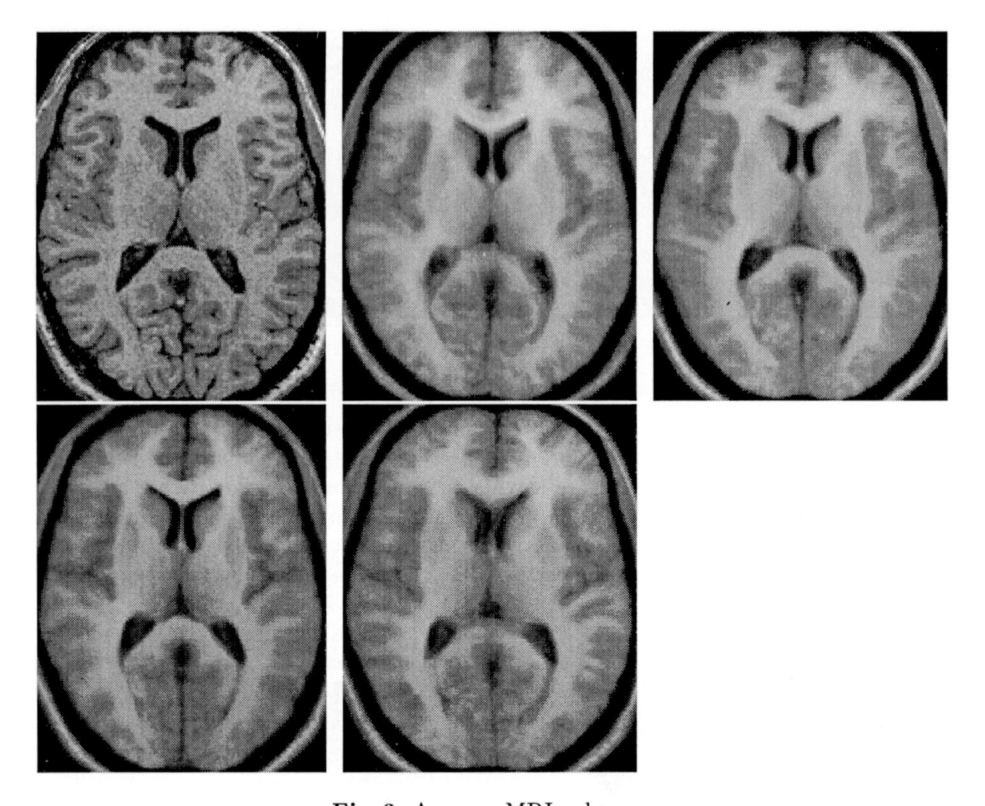

Fig. 3. Average MRI volumes

From left to right, top row: the target volume, the linear average; the non-linear average using the gradient magnitude only. Bottom row: the average using mL_{vv}; using ESS voxels. Not surprisingly, the non-linear registrations appear much better than the linear averages (even though the quantitative measures do not make this clear). The use of mL_{vv} improves the appearance of the cortex near many of the major sulci (e.g., sylvian, superior temporal, cingulate). Extracted sulci also appear to improve registration of some gyri, but only at the surface of the brain.

Image Guided Microscopic Surgery System Using Mutual-Information Based Registration

Nobuhiko Hata[1,2], William M. Wells III [1,3], Michael Halle [1,4]
Shin Nakajima[1], Paul Viola[3], Ron Kikinis[1], Ferenc A. Jolesz[1]

[1] Surgical Planning Laboratory
Department of Radiology
Harvard Medical School and Brigham and Women's Hospital
75 Francis Street, Boston MA 02115
E-mail: noby@bwh.harvard.edu

[2] Faculty of Engineering, The University of Tokyo
[3] Artificial Intelligence Laboratory, Massachusetts Institute of Technology
[4] Media Laboratory, Massachusetts Institute of Technology

Abstract. We have developed an image guided microscopic surgery system for the navigation of surgical procedures that can overlay renderings of anatomical structures that are otherwise invisible. A new histogram-based mutual information maximization technique was applied for alignment of the scope view and three-dimensional computer graphics model. This technique doesn't require any pre-processing nor marker setting but is directly applied to the microscope view and the graphics rendering. Therefore, any special set up in image scanning or preoperative preparation is not necessary.

Graphics technique were implemented to compute three-dimensional scene information by using a graphics accelerator, increasing the algorithm's performance significantly.

Experiments are presented that demonstrate the approach registering a plastic skull to its three-dimensional reconstruction model generated from a CT scan. The tracking performance in the experiments were nearly real-time.

1. Introduction

With the advancement of medical imaging and computer technology, the effective use of diagnostic images has been a focus of discussion. One effective use of medical imaging is image-guided surgery, where information extracted from pre-operative imaging guides the surgical procedures. Previous applications have been presented by transferring medical image data or its three-dimensional (3D) reconstruction to the surgical field and registering it to the patient. Kelly, et al., [4] pioneered computer assisted frame-based and frameless stereotactic surgery, in which region of interest in the operative target can be localized and displayed using the preoperative images. This technique has contributed to a reduction of the invasiveness for the patient.

The surgical microscope is an attractive vehicle for such applications. Roberts, et al., [6] reported the development of a surgical microscope guided by pre-operative CT by using an ultrasonic tracking sensor. The outlines of a tumor were projected into the optics of the microscope. Edwards, et al., [2] presented a stereo operating microscope that projects stereo projections of 3D model into the microscope eyepieces.

All the techniques listed above require alignment of patient and preoperative image to perform intraoperative navigation based on preoperative image Some applications benefit from surgical instruments or pointing devices but these require additional registration. Edward, et al., used an optical tracking sensor to register the microscope and patient in the operating room. The position and orientation of the microscope was tracked by the LED markers attached on the microscope. Patient registration was performed by identifying fiducials or anatomical landmarks on the skin surface. This process is reported to take 1-2 minutes and to extend the duration of the procedure.

Ideally, the method of registration should be non-invasive for the patient and convenient for medical staff.

In this paper, we describe preliminary experiments with an image guided microscopic surgery system that navigates surgical procedures by overlaying an 3D rendering in the microscopic view. A new histogram-based mutual information maximization technique was applied for alignment of scope view and three-dimensional computer graphics model. This technique doesn't require any pre-processing nor marker setting but is directly applied to the microscope view and graphics rendering. Therefore, no special set up in image scanning or preoperative preparation is not necessary.

In addition, the implementation of the graphics techniques used with method is also presented. The general problem is to compute the 3D scene information needed by the registration with the use of a graphics accelerator.

2. Material and Methods

2.1 Registration by Maximization of Mutual Information

"Registration by maximization of mutual information" is applied to a scope view - model registration.

As the method is applied to images directly without any preprocessing, the method is quite general. Therefore, it can be applied for a wide variety of sensors. In previous studies, the method was applied to register medical images of different modalities, 3D model to images and 2D view-based images to images.

Some previous work on image guided surgery has used feature points or markers to register the patient to preoperative medical images and operative tools to preoperative images. The additional preparations and procedures needed for their use are cumbersome not only for patient but also for medical the staff. As the presented method is applied directly to the 3D models and medical images, it is especially suitable for application in video image guided surgery, such as endoscopic surgery, and microscopic surgery.

The detailed description of registration by maximization of mutual information appears in [8][9] and [10].

In this application, x represents the location of a surface patch, T is the transformation placing the object with respect to the camera, u is the surface normal of a patch at the object, and v is the observed image. The general problem of registration is formulated as,

$$\hat{T} = \arg\max_{T} I(u(x), v(T(x))) \tag{1}$$

,where I represents mutual information calculated between u and v. Mutual information is defined in terms of entropy in the following way:

$$I(u(x), v(T(x))) \equiv H(u(x)) + H(v(T(x))) - H(u(x), v(T(x))) \tag{2}$$

,where $H(\cdot)$ is entropy of a random variable, and can be interpreted as a measure of uncertainty, variability or complexity.

The first term on the right is the entropy in the model. The second term is the entropy of the part of the image into which the model projects. It encourages transformations that project u into complex parts of v. The third term, the(negative) joint entropy of u and v, takes on large values if u and v are functionally related. It encourages transformation where v explains u well. Together the last two terms identify transformations that find complexity and explain it well.

2.2 Mutual Information with Histogram

In this section, we describe a histogram-based computation of mutual information.

The entropy of random variable x is described as an expectation of the negative logarithm of the probability density. In [8][9] and [10], the Parzen Window method is used to approximate the underlying probability density $f(z)$ by a superposition of Gaussian densities centered on the elements of a sample B drawn from z.

In this study, we used a histogram for density estimation and smoothed finite differences of the histograms for estimating derivatives of density. Collignon et al., [1] used histograms for computing entropies and mutual information, in a non-derivative optimization using Powell's method. In experiments, we have found significant speed improvements using a simple gradient search model. We can approximate the entropy of random variable z:

$$\hat{H} \equiv -\frac{1}{N}\sum_{z_i \in A} \ln \hat{p}(z_i, B), \tag{3}$$

where A denotes a sample of observations of z and B denotes second sample of observations of z. The derivative of $\hat{H}(z)$ is computed for the usage in finding its local maximal. We compute the derivative of $\hat{H}$ for a gradient search,

$$\frac{d\hat{H}(z)}{dT} \approx -\frac{d}{dT}\frac{1}{N}\sum_{z_i \in A}\ln(\hat{p}(z_i, B))$$

$$= -\frac{1}{N}\sum_{z_i \in A}\frac{\frac{d}{dT}\hat{p}(z_i, B)}{\hat{p}(z_i, B)} \tag{4}$$

$$= -\frac{1}{N}\sum_{z_i \in A}\frac{\frac{d}{dz_i}\hat{p}(z_i, B)\frac{dz_i}{dT} + \frac{d}{dB}\hat{p}(z_i, B)\frac{dB}{dT}}{\hat{p}(z_i, B)}$$

The numerator of the derivative has two components. The first component is the change in entropy that results from changes in z_i. The second one is a measure of the change in the histogram density estimate that results from changes in the sample B. For histograms computed from a finite sample, a differential change in T will move a small number of sample points from one bin to another. As a result, we will assume that the second term is zero.

2.3 Maximization of Mutual Information

In order to maximize the mutual information described in Equation (1), we approximate its derivative of it with respect to transformation matrix T by implementing Equation (4),

$$\frac{d}{dT}I(T) = \sum_{v_i \in A}\frac{\partial \hat{I}}{\partial v_i}\frac{dv_i}{dT} \tag{5}$$

,where A is the intensities of the image pixels that correspond to visible surface patches of the model.

Steps are repeatedly taken that are proportional to the approximation of the derivative of the mutual information with respect to the transformation,

$$T \leftarrow T + \lambda\frac{dI}{dT} \tag{6}$$

where λ is a step size parameter. The steps are repeated a fixed number of times or until convergence is detected.

2.4 Density Estimation with Graphics Techniques

A computer graphics technique is applied to increase the performance of the registration computation.

In order to calculate the histograms of v and u in $I(T)$ of Equation 1 and its derivative in Equation 5, a computer graphics technique is applied. With the help of high performance graphics implemented in acceleration a workstation, computational efficiency is increased.

These equipments were all developed and performed using Ultra 2 with Creator 3D graphics accelerator (SUN Microsystems). In the software development, the Visualization Tool Kit (VTK) [7] is implemented for graphics manipulation and XIL (SUN Microsystems) for video capturing. The VTK stands upper level of the XGL graphics library and controls data loading, rendering and event handling.

The registration algorithm requires the calculation of surface normals for the model at the current position and orientation. These can be performed by manipulating the placement of light sources. Given the view camera for rendering is placed at (0,0,+a[>0]) in the world coordinate system and model at (0,0,0), visible surface patch on the model has all positive z directional normal.

Two of the reflectance component, ambient and diffuse contribution from the light, are activated and the other component, specular contribution, is inactivated. Activated components have ambient reflection coefficient 0.5 and diffuse coefficient 0.5. A directional light on (+a, 0, 0) has diffuse reflection effect according to its x directional normal value. In the same way, a light on the y axis has diffuse reflection effect to according to y directional lighting value.

To achieve the desired reflectance, a positive green light is placed at (+a,0,0), a negative green light at (-a,0,0), a positive blue light at (0,+a,0) and a negative blue at (0,-a,0). The positive light has a RGB color component 1.0 out of its range 0.0 to 1.0, and negative light has -1.0. In addition to setting directional lights, we set the ambient color of the object blue-green using this scheme, the green component of color represents its x directional normal value and blue represents y directional normal value.

This scheme requires placement of negative light sources. In our preliminary experiments, we have verified that XGL and OpenGL can accommodate negative light sources.

Another advantage is that surface normal calculation and user interface can share its graphics property: geometry data storage, position and orientation matrices, pipeline with graphics accelerator, window and event management. By sharing most of this functionality with the already existing rendering platform, development becomes simpler and faster.

Copying the rendered image in the raster memory to conventional memory requires time and that influences the performance seriously. Therefore, rendering the window size for the computation of normal was minimized.

2.5 Display

After the registration is performed, the rendered image of the 3D model is transferred to the microscope though a VGA signal. The microscope has a monochrome display that can be merged to the right eyepiece. Although 3D model is aligned to match the microscope view, additional scaling and translation are required to be displayed on VGA display in the scope.

After the registration with visible anatomical structure is performed, normally invisible internal structure is displayed for surgical navigation. Visible structure may be turned off for better perception of internal objects.

3. Results

3.1 System Configuration

The system consists of a microscope for surgical use (ZEISS) and a workstation (Ultra SPARC 2 with Creator3D, Sun Microsystems, Moutainview, CA) (Fig.1). The Microscope has video (NTSC) output for both right and left microscope view. The video output from the right scope is transferred to the workstation and digitized with video capture board (SUN Video) at the resolution of 600 x 480 [pixels].

After the registration of 3D model with microscope view is performed, the rendering image of the aligned 3D model is transferred to the microscope as a VGA signal. The microscope has monochrome VGA display in the right scope that can be overlaid on the conventional microscope view.

3.2 Tracking Experiment

We performed an experimentation to evaluate the tracking performance of the algorithm. The iteration in Equation (6) is used to perform plastic skull tracking under the configuration described in Section 3.1.

A plastic skull was used as an phantom. 3D model around the right auditory hole of the plastic skull is generated from its CT scan (Siemens Somatom Plus, 512x512[pixels]x274[slices], 0.98x0.98x0.92[mm/voxel]) followed by iso-surface processing[5] and surface smoothing. This experiment also aims the simulation of skull vase surgery that has difficulty in localizing nerves around auditory hole.

The plastic skull was placed under the microscope and lighted up by a light attached to the microscope. The microscopic view had an area of 54.5x 40.9 [mm] at its focal plane. An initial alignment to an images was performed manually. Then the random offset is added to each transitional axis and randomly selected one rotational axis.

The Table 1 shows the result of each experiment with selected range of offset. In each experiment, 50 uniformly randomized offsets lower than the maximal offsets are given as an initial guess. Each trial was terminated either when the number of iteration reached 30 or the convergence was detected.

Additionally, Figure 2 shows an example of tracking result. This convergence required 21 iteration with about 10 seconds.

Table 1. Result of Phantom Experiment

	Maximal Offset		Initial		Final	
	Translation [mm]	Rotation [deg.]	Translatio n [mm]	Rotation [deg.]	Translation [mm]	Rotation [deg.]
Experiment 1	15.0	20.0	5.2	11.6	0.8	7.9
Experiment 2	25.0	20.0	10.0	8.2	2.4	8.2
Experiment 3	15.0	40.0	7.9	18.8	2.3	8.7

The experiment demonstrated that the algorithm performs efficiently and reliably. If the initial guess is placed approximately 15 [mm] and 20 [degrees] away from the result, the model can be aligned correctly within 30 iterations.

4. Discussion

This system has been designed to overcome the difficulties of visibility and visualization that can occur during microsurgery.

A number of factors can cause a surgeon to have difficulties orienting the current microscope view to the anatomical structure of the patient and to previously acquired 3D models. Among these are motions of the microscope, limitations on the field of view, and other obstructions to visibility, such as bleeding.

These difficulties in visual identification and orientation may be directly addressed by the superimposition in the microscope of correctly registered renderings of 3D anatomical structures. This capability is also useful for determining the location of otherwise invisible structures, for example, if a surgeon can detect a nerve inside a bone, it will be very useful while drilling the bone. Similarly, it is also helpful to visualize the position of blood vessels in a tumor during a resection procedure.

Although the CT of today can scan with nearly 1mm in gap and less in pixel size, the microscope provide more precise information. However, the scope used in the experiment has 54.5x40.9[mm] of widest view field, which means the size of the pixel in the digitized and overlaid image (640x480 [pixels]) is 0.08x0.08[mm]. This mismatch of preciseness between microscopic view and 3D model may causes fatal error in registration: the gradient searching cannot escape from a local maxima. The microscope may have additional wider setting for initial registration to restrict the translation and orientation of the model in the following registration procedure.

For a scope view - model registration, "Registration by maximization of mutual information" is applied. In this method, the registration of 3D model and video (microscope) view can be achieved by maximizing the mutual information. One advantage of this method is that it applies directly to the video image and 3D model. In other words, no pre-processing and feature detection is not required. Therefore, surgical procedures should not need to be interrupted for registration processing and surgeons can concentrate on their operative maneuvers. Additionally, the system can be set up by adding workstation with functions of graphics acceleration and video input to the operative microscope. Thus the system is easy to be implemented to a conventional microscope which usually equipped with a video output and sometimes with a video input.

As the registration method only requires a video image and a 3D model, its prospective field of application in medicine potentially broad. Endoscopic and Laparoscopic surgery, where conventional feature based registration is hard to apply, may benefit from the simplicity and efficiency of our technology.

5. Acknowledgment

We are grateful to the authors of "The Visualization Tool Kit", Drs. Will Schroeder, Ken Martin and Bill Lorensen for their technical assistance. We also wish to thank the colleagues in Surgical Planning Laboratory of Brigham and Women's Hospital, and Artificial Intelligence Laboratory of MIT. The assistance and advice of Prof. Dohi of the University of Tokyo is gratefully appreciated.

6. References

1. Collignon A, et al.: Automated multi-modality image registration based on information theory. Proc. Information Processing in Medical Imaging Conf., Kluwer Academic Publishers: 263-274, 1995
2. Edwards PJ, Hawkes DJ, Hill DLG, Jewell D, Spink R, Strong A, Gleeson M: Augumented realty in the stereo operating microscope for optolaryngology and neurosurgical guidance, Proc. CVRMed 95, 1995
3. Friets EM, Strohbehn JW, Hatch JF, Roberts DW: A frameless stereotaxic operating microscope for surgery. IEEE Trans Biomed Eng. **36** (1989) 608-617
4. Kelly PJ: Vlumetric stereotaxis and computer-assisted stereotactic resection of subcortical lesions, in Lunsford LD (ed): Modern Stereotactic Neurosurgery. Boston, Marinus Nijhoff, (1988) 169-184
5. Lorensen WE and Cline HE: Marching cubes: A high resolution 3D surface construction algorithm. Computer Graphics **21** (1987) 163-169
6. Roberts DW, Strohbehn JW, Hatch JF, et al: A frameless stereotaxic integration of computerized tomographic imaging and the operating microscope. J Neurosurg **65** (1986) 545-549
7. Schroeder W, Martin K, Lorensen B: The visualization toolkit : an object-oriented approach to 3D graphics, Prentice-Hall, New Jersy, (1996)
8. Viola PA and Wells WM: Alignment by maximization of mutual information. Proc. 5th Intl. Conf. Computer Vision (1995)
9. Viola PA: Alignment by Maximization of Mutual Information, Ph. D. thesis, Massachusetts Institute of Technology, (1995)
10. Wells WM, Viola PA and Kikinis R: Mutti-Model Volume Registration by Maximization of Mutual Information. Proc 2nd Med. Robotics Comp. Assisted Surg., (1995)

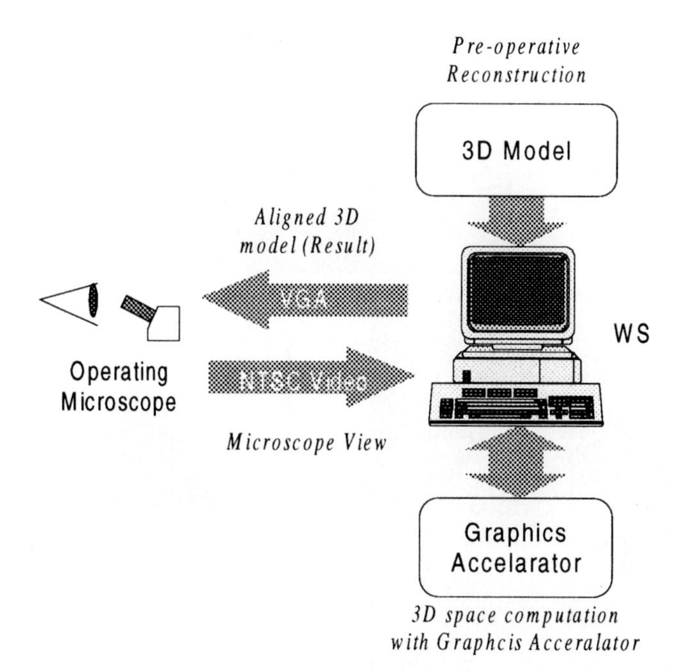

Fig. 1. System Configuration. Microscopic view is digitized and aligned with preoperatively reconstructed 3D model. The use of graphics accelerator in the alignment increases the performance significantly. The aligned 3D model is overlaid in the scope view using VGA monitor implemented in the optical system of the microscope.

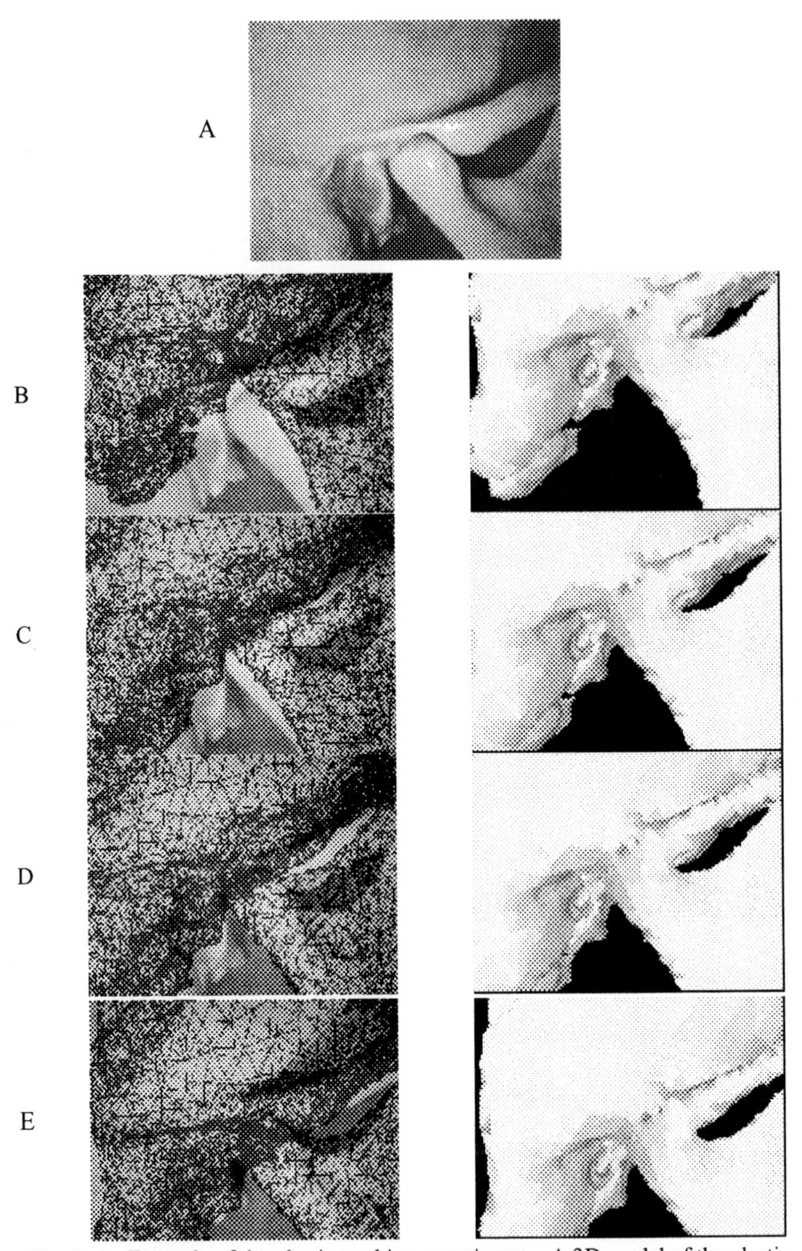

Fig. 2. An Example of the plastic tracking experiment. A 3D model of the plastic skull around right auditory hole is reconstructed and aligned with its original skull in the microscopic view (*A*). (*B*) shows the initial setting of the 3D model by overlaying it in microscope view *(left)* and the same model with surface representation *(right)*. The alignment process is shown in (C), (D) and the final result (E).

Cross Validation of Three Inter-Patients Matching Methods

Jean-Philippe Thirion[1], Gérard Subsol[1] and David Dean[2]

[1]INRIA, Projet Epidaure, 2 004, route des Lucioles, BP 93
06 902 Sophia Antipolis Cedex, France

[2]Department of Anatomy, Case Western Reserve University
10900 Euclid Avenue, Cleveland, Ohio 44106-4930, USA

Abstract. In this paper, we present the cross-validation of three deformable template superimposition techniques a,b and c, used to study 3D CT images of the bony skull. Method (a) relies on the manual identification by anatomists of anthropometric landmarks, method (b) on "crest lines", which have a pure geometric definition, and method (c) is based on 3D non-rigid intensity based matching. We propose to define and compute a distance between methods a, b, c, and also to compute three representations $\bar{I}^a$, $\bar{I}^b$, $\bar{I}^c$ of an "average" skull model based superimposition via these three methods. The overall aim is to determine if the three methods, all developed independently, give mutually coherent image superimposition results.

1 Introduction

The possibility of matching 3D medical images of different patients or of comparing a new patient's image to a diagnostic "normative" image is a challenging issue for the medical community (see [BK89], [CMV94], [CNPE94]). The applications are numerous, ranging from computer-assisted building of 3D electronic atlases, to Computer Aided Surgery (see [CDB+95]) and diagnosis (see [STA95]). More studies such as three dimensional growth of organs (see [Sub95]) or human evolution (see [Dea93]) can be foreseen. But developing useful deformable template-based superimposition techniques is extremely complex, far moreso than for the rigid case: the "ideal" rigid motion can exist, and can be retrieved, up to a residual error transformation (see [PT95]). For inter-patients matching, there can be as many different criteria for image matching as possible applications: the similarities between two patients arise from the sharing of a common phylogeny (i.e., development of the species), and the undergoing of a similar ontogeny (i.e., normal growth and development), both of which are extremely difficult to model. We propose to compare the results of three inter-patients matching techniques, each of them being developed independently and the result of several years of research. First we describe the methods, then we report on cross-validation experiments of both the matching processes and the computation of "average" 3D images.

2 Three superimposition techniques

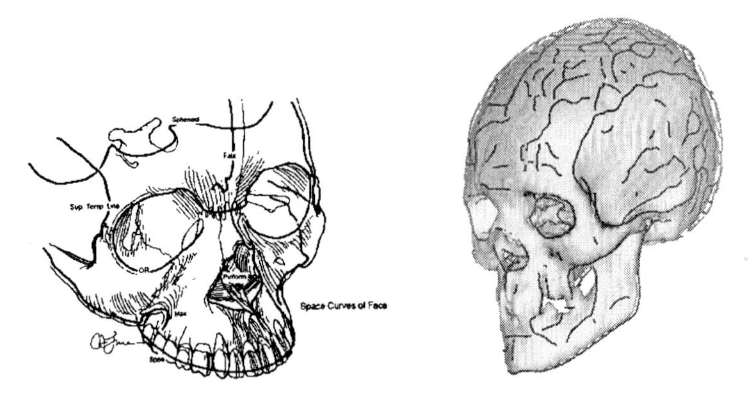

Figure 1. Left, principal anatomical lines (after [BC88]), leading to method *a*. Right, crest lines extracted by Marching Lines (method *b*).

Ridge Curve template matching (method *a*): it has been designed by anatomists, for cranio-facial surgical planning (see [BC88], [Cut91], [Dea93], [CDB$^+$95]). There is a strong need, on one hand, to build an average surface representation of a "normal" patient and on the other hand, to match patients to this average, for diagnosis and simulation. A template of the salient lines of the skull called "ridge curves" (see figure 1, left), has been designed by anatomists. It contains homologous 3D anthropometric landmarks (points), linked by ridge and geodesic curves (see figure 2). The template is matched implicitly by the correspondance of the vertices given by the anatomist. The major advantage is that method *a* relies on homologous features, that is, points or lines which are anatomically invariant, but a new template has to be manually developed for each organ. The selection of landmarks and the final correction are time consuming processes and there can be inter- or intra-observer variabilities.

Crest lines matching (method *b*): "Crest lines" are mathematically defined by differential geometry, and coincide with the intuitive definition of "most salient" lines used in method *a*. The "Marching Lines" algorithm (see [TG95]) automatically extracts crest lines from volumetric images (see figure 1, right). The matching algorithm iteratively computes rigid, affine, and spline transformations, and forms couples of points from the two sets of crest lines, based on their spatial distances after transformation (see [Sub95]). The whole process is fully automatic, leading to a large set of corresponding points within 10 to 20 minutes CPU time (Dec Alpha workstation). The topology of crest lines can vary from patients to patients, the method needs high resolution 3D images and there are no a-priori anatomical information. Their use to build average model representation has been demonstrated in [STA95].

Intensity-based matching (method c): This method has been presented in [Thi95], and [Thi96] and produces a point to point correspondance field between the two 3D images. Its principle is based on an analogy with Maxwell's demons, and relies on polarity (inside/outside) information: the boundaries of the object in the scene image I_s are considered to be hemi-permeable membranes, separating the inside of the object from the outside, and the voxels of the model image I_m are considered to be particles, labeled "inside" or "outside" points. The deformable model I_m is then "diffusing" into the scene image I_s. The approach is entirely automatic and multi-scale. The whole set of voxels is taken into account within 10 to 20 minutes CPU time. No anatomical knowledge is used but the convergence depends on the initial positioning of the two objects, which have to be relatively close.

Each method relies on a privileged set of points. Method a relies on the set A of homologous points (typically several tens of points, or one thousand if the linking curves are taken into account), method b relies on the crest lines (set B, several thousands of points) and method c takes all the voxels with high gradients into account (set C, millions of points). In fact, A,B and C are processes to extract feature points from images and must be distinguished from the matching methods themselves.

3 Cross validation study

Our cross validation relies on two principles: the first is to study the effect of non-rigid deformation on sparse data, and the second is to deny a-priori preference to any of the three methods. The aim is not to determine the "best" of the three methods; that is meaningless. Rather we wish to determine if these three methods produce mutually consistent results. For that purpose, we define a distance between the three methods and compute and compare "average" patients, generated from a reference database of skulls.

3.1 Defining a distance between methods

Let I_1 and I_2 be the images of two specimens. To perform a fair comparison, we compute from the matching result of each technique a B-splines based warp $T^a_{1,2}$, $T^b_{1,2}$, and $T^c_{1,2}$ (see [DSTA95]), with exactly the same parameters (i.e., control points and smoothness constraints). Our distance measurements between methods is close to traditional morphometric studies of skulls (see [ANDB90]) because it is using a sparse set of feature points. It must be contrasted with validation methods based on manually segmented regions (see [GRB93]) or surface distances. We do this to prevent zero distances (between surfaces or regions) while having large local labeling "errors" (i.e. the tip of the nose sliding within the face surface to end up in the middle of the cheek). Our "features of interest" are the points produced by the feature extraction processes A, B or C. We must note that $B_1 = B(I_1)$ is not equivalent to $B_2 = B(I_2)$; in particular,

$T^b_{1,2}(B_1) \neq B_2$. This is true also for c: $(T^c_{1,2}(C_1) \neq C_2)$. However, by definition of a, we have $T^a_{1,2}(A_1) = A_2$ up to the approximation due to warping. A_1 and A_2 are two bijective sets, but correspond to two different inferences of the anatomical features, subject to possible uncertainty and inter-observer variability. In order to compare the different matching methods, one major problem is to reduce the influence of such inferences, for example by considering different types of reference sets A_1, B_1, C_1 in I_1 in our experiments.

Distance definitions For a couple of images (I_1, I_2), we can define a relative distance $d_Z(x, y)$ between two matching methods x and y, relative to set Z of n_Z reference points in image I_1. We choose to compute the average distance between the transformed points $T^x_{1,2}(Z)$ and $T^y_{1,2}(Z)$:

$$d_Z(x,y) = \frac{1}{n_{Z_1}} \sum_{\mathbf{z} \in Z_1} ||T^x_{1,2}(\mathbf{z}) - T^y_{1,2}(\mathbf{z})|| \quad (1)$$

where $T^x_{1,2}$ (resp. $T^y_{1,2}$) is the transformation between I_1 and I_2 obtained with the method x (resp. y), $|| \cdot ||$ is the Euclidean norm, and $Z_1 = Z(I_1)$. It is easy to verify that d_Z corresponds to the mathematical definition of a distance. We can also compute the median distance, that is, the distance $m_Z(x, y)$ for which 50% of the matched points in Z have a distance $||T^x(\mathbf{z}) - T^y(\mathbf{z})||$ less than m_Z, or the maximal distance: $M_Z(x, y) = max\{||T^x(\mathbf{z}) - T^y(\mathbf{z})||, \mathbf{z} \in Z\}$.

To show visually how close the three methods are, we propose to represent them as the vertices of a triangle, the lengths of the edges representing the relative distances between methods (see figure 4). This can be extended to an arbitrary number N of superimposition methods, and leads to a N-simplex representation.

3.2 Computation using a 3D CT skull image database

As we have access to a large number of specimens, we can compute a more robust estimate of the distance $\bar{d}_Z$ between methods a, b, c, by averaging the results over the specimens. We select a particular image I_i, which is considered to be the reference: all the images $I_{j \neq i}$ are matched with I_i, and the average distance $\bar{d}_{Z,i}$ is simply the average of the $n-1$ distances d_Z obtained by comparing the $n-1$ skull images $I_{j \neq i}$ with a reference skull I_i:

$$\bar{d}_{Z,i}(x,y) = \frac{1}{n \times n_Z} \sum_{j \neq i}^{n} \sum_{\mathbf{z} \in Z_i} ||T^x_{ij}(\mathbf{z}) - T^y_{ij}(\mathbf{z})|| \quad (2)$$

where, for example, $T^x_{ij}(\mathbf{z})$ is the transformation between the reference image I_i and image I_j, using method x and applied to a 3D point $\mathbf{z}$ of $Z_i = Z(I_i)$.

4 Validation of average patients computation

Using methods a and b, we have previously computed "average" images from the same 3D CT skull database (see [CDB+95] for method a, [STA95] for method b). The resulting average images $\bar{I}^a$ and $\bar{I}^b$, appeared quite similar on visual inspection. We are now giving quantitative evaluation by measuring the average distances between $\bar{I}_a$, $\bar{I}_b$, $\bar{I}_c$.

4.1 Computing average patients

To measure the average model of n skull's images, using a method x we:

- 1. compute the $n-1$ deformations T^x_{ij} from a reference image I_i toward the other $n-1$ skull images $I_{j \neq i}$.
- 2. subtract the similarity transforms S^x_{ij} (that is, rotation, translation and scaling) from each of these deformations, to get only anatomically meaningful deformations $\tilde{T}^x_{ij}$.
- 3. compute the average transformation: $\bar{T}^x_i = \frac{1}{n}\sum_j \tilde{T}^x_{ij}$ (including $\tilde{T}^x_{ii} = Identity$).
- 4. apply the average transformation $\bar{T}^x_i$ to I_i (or to $Z_i \subset I_i$) to get the average image $\bar{I}^x_i = \bar{T}^x_i(I_i)$ (or the set of average features $\bar{Z}_i = \bar{T}^x_i(Z(I_i))$).

4.2 Comparing the averaging techniques

We can then measure the distance between methods, $\hat{d}_{Z,i}(x,y)$, using the averaged deformations $\bar{T}^x_i$ and $\bar{T}^y_i$. The distance $\hat{d}_{Z,i}(x,y)$ is equivalent to the average of the distances between the averaged features of the different patients:

$$\hat{d}_{Z,i}(x,y) = \frac{1}{n_{Z_i}} \sum_{\mathbf{z} \in Z_i} ||\bar{T}^x_i(\mathbf{z}) - \bar{T}^y_i(\mathbf{z})|| \quad (3)$$

The average deformation $\bar{T}^x_i$ can be computed from the set of deformations T^x_{ij} in several ways. If it is done from point to point correspondences, that is, $\bar{T}^x_i(\mathbf{z}) = \frac{1}{n}\sum_j \hat{T}^x_{ij}(\mathbf{z}), \mathbf{z} \in Z_i$, then it is easy to demonstrate with the triangular inequality that $\hat{d}_{Z,i}(x,y) \leq \bar{d}_{Z,i}(x,y)$, where $\bar{d}$ is the average distance between methods already described. Although each method has a dedicated way to compute average patients, we have verified this inequality experimentally: the distances $\hat{d}(a,b)$, $\hat{d}(b,c)$, $\hat{d}(a,c)$ between the average patients $\bar{I}^a$, $\bar{I}^b$, $\bar{I}^c$ are lesser than the averaged distances $\bar{d}(a,b)$, $\bar{d}(b,c)$, $\bar{d}(a,c)$ between methods. As we can see measures obtained with average patients are more robust, which justifies, at least computationally, the production of average skull images (Many researchers have produced atlases of ideal patients, see [TT88], [ANDB90], [CDB+95])

4.3 Influence of the reference patient

Our averaging methods arbitrarily adopts the frame of a reference image I_i. We can eliminate this partially from the computation of $\bar{d}_Z$, for example by considering the deformations in a circular permutation of the skulls ($I_0 \rightarrow I_1, I_1 \rightarrow I_2$ $\ldots I_n \rightarrow I_0$). However, the result must still be expressed with the same given set of "features of interest", for example $A_0 \subset I_0$. Unfortunately, in the case of "average" patients, we cannot perform this circular permutation. We are obliged to compute, for a method x, the average image $\bar{I}_i^x$ (or the average model $\bar{Z}_i \subset \bar{I}_i^x$). A matching T_{ij}^x is still performed with respect to the same reference specimen I_i. Ideally, we would compute the average of $\bar{I}_i^x$ for each images being the reference $I_i, i \in [1, n]$, but this is too expensive computationally (n^2). Our experiments verify that in method c, the choice of the reference specimen I_i has a negligible influence on the computation of $\bar{I}^c$.

5 Experiments

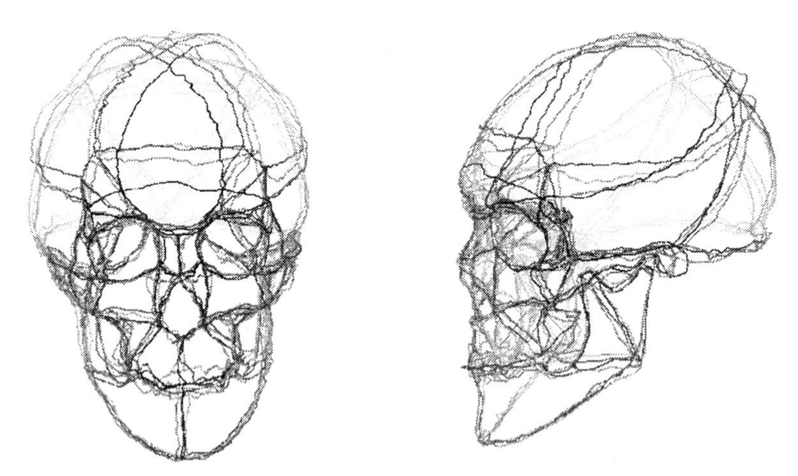

Figure 2. The templates A_2, A_3, A_4 projected on the template A_1 of the reference specimen, using method c (demons). It also illustrates visually $\bar{d}(a, c)$

We have also checked visualy the coherence of methods a, b and c, by projecting in the space of a reference image I_1 all the templates obtained with A: $T_{ji}^c(A_j), j \in [1, 4]$ (see figure 2). The four templates are visually very close one to another. They are for the salient lines of the skull, but can differ significantly in smooth areas, ignored by the ridge curve template.

5.1 Cross-validation

The voxel sizes of our skull images are $1 \times 1 \times 1.5$ mm ($175 \times 200 \times 140$ voxels). To compare the influence of the features of interest (see figure 4, we have measured

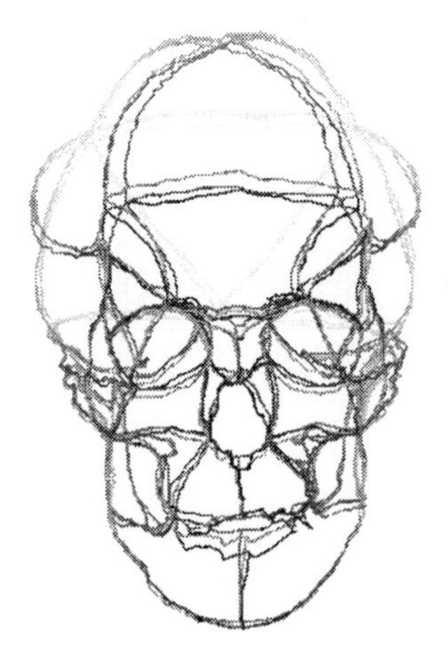

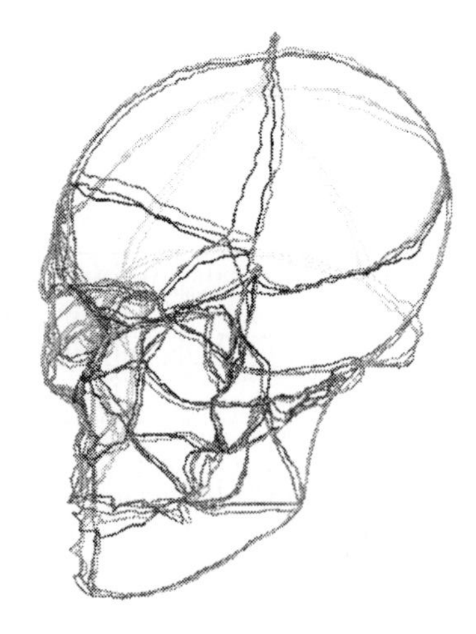

Figure 3. Illustrates visually distances $\hat{d}$: superimposed average templates $\bar{A}^a$, $\bar{A}^b$, $\bar{A}^c$, obtained independently by methods a,b,c.

the average distances between methods using: $B_1 = B(I_1)$, the set of crest lines in image I_1; Bf_1, the set B_1 where only the crest lines with high curvatures are kept; A_1 the template in image I_1 defined by an anatomist. The template or crest lines of the occipital and parietal parts of the skull are much less "well" defined than the face (i.e. no salient lines). This is illustrated by the good performances obtained using the filtered crest lines (Bf), which range from $3mm$ to $6mm$. Figure 4 is a pictorial representation of the average distances, which emphasize this. We can also see that no method is inconsistent with respect to the two other ones.

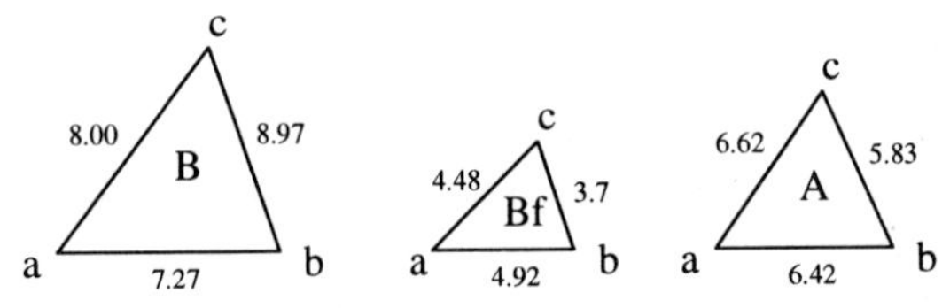

Figure 4. Simplex representation of the average distances between methods a (template), b (crests) and c (demons), in millimeters, for three types of features: B (crest lines), Bf (filtered crest lines), and A (anatomical template).

5.2 Average patients

We have computed the average patients with the three methods (see figure 3), and also the distances between average patients, using different features (see figure 5). The three methods give very close results, and the templates are much smoother than those of each individual skulls. Again, results are more stable around anatomically salient lines. The distances $\hat{d}$ between average patients are much smaller than the average distances $\bar{d}$ between methods. This result support the production of average patients from a computational viewpoint. It should also be noted that the features of interest have only small influence on the results, which suggests that the average is stable in all portions of the skull.

methods	$\hat{d}_B$	$\hat{d}_{Bf}$	$\hat{d}_A$	m_B	m_{Bf}	m_A	M_B	M_{Bf}	M_A
$\hat{d}(a,b)$	2.92	2.59	3.37	2.58	2.45	2.98	11.3	8.58	11.5
$\hat{d}(a,c)$	3.96	3.78	3.61	3.65	3.77	3.12	11.3	8.86	12.7
$\hat{d}(b,c)$	3.33	3.34	3.09	3.18	3.19	2.83	8.87	8.87	8.31

Figure 5. Distances between average features, obtained with method a (templates), b (crests) and c (demons) in millimeters, with three types of features of interest B, Bf and A, and three distances: $\hat{d}$: average distance, m median, M maximal distance.

5.3 Sensitivity to the choice of the reference specimen

We also tested the sensitivity of the choice of the reference specimen with method c (see [Sub95] for similar results with method b). We must be very careful on the choice of the set of features of interest Z. We compute:

- $\bar{A}_1$, using image I_1 as the reference, and A_1 as features, where A_1 is the template, semi-automatically defined by the anatomist in image I_1.
- $\bar{A}_2$, using image I_2 as the reference, and A_2 as features. By definition for method a, $A_2 = T^a_{12}(A_1)$ (this is not true for methods b and c). Hence we can compare directly $\bar{A}_1$ and $\bar{A}_2$
- $\overline{T^c_{12}(A_1)}_2$, using image I_2 as reference, and where $T^c_{12}(A_1)$ is the template A_1, projected onto image I_2 by method c.

We compute the distance $d = \bar{d}_{\mathbf{z}\in A_1}(\bar{A}_1, \bar{A}_2) = \bar{d}_{\mathbf{z}\in A_1}(\bar{A}_1, \overline{T^a_{12}(A_1)}_2)$, which is the average distance between two estimates of average features, using two different reference specimens I_1 and I_2. We also compute $d' = \bar{d}_{\mathbf{z}\in A_1}(\bar{A}_1, \overline{T^c_{12}(A_1)}_2)$ and the results are presented in Figure 6.

	d_A	m_A	M_A
d	4.83	4.09	22.9
d'	1.25	1.13	4.47

Figure 6. Distances between average skulls obtained with two different reference images. d: with two manual template selections A and one application of method c. d': with a single template selection A and two times the application of c.

It should be noticed that d represents the differences in using two distinct reference specimens I_1 and I_2, but also two independent inferences A_1 and A_2

of the segmented anatomical features. This is equivalent to the concatenation of the "errors" of two manual selections, A_1 and A_2, and of one non-rigid matching with c. On the other hand, for d', only method c is used (twice). A_1 is used solely as a list of "features of interest" to produce an average distance. Hence d' is more representative of the perturbation introduced by the choice of the reference specimen alone than distance d. We note that this error ($1.3mm$) is of the same order as the image resolution, which shows that the choice of the reference specimen has very little influence on the result of averaging.

Our personal interpretation of the results are the following: the three methods give mutually coherent results, with an average difference for feature location of $3mm$ to $4mm$ where the skull is highly curved. In smoother places, this average precision is reduced to $6mm$ to $9mm$, but for specific points (outliers) the distance grows to $3cm$ or $4cm$. The computation of "average patients" gives more robust and coherent results, everywhere on the skull surface, with an "average" difference of $3mm$ to $4mm$ and up to $1cm$ for outliers. The computation of the average patient does not depend too much on the choice of the reference specimen.

6 Conclusion

The comparison of three image superimposition methods on a set of four 3D CT-scans of different dry skulls was performed, in order to verify that three methods, all developed independently, give mutually coherent results. To our opinion, the answer is yes. We have established quantitative measurements, such as distances between the individual features produced by these three methods and average features computed by superimposition, which could be used to compare other non-rigid techniques. We intend to use these methods, independently or combined, to classify groups or sub-groups of patients, and possibly to automatically diagnose pathologies.

Acknowledgement: Many thanks to Dr. Bruce Latimer, Curator of Physical Anthropology, Cleveland Museum of Natural History, for the access to the Hamann-Todd collection and also to Dr. Jon Haaga, Chair and Professor, and Elizabeth Russel, Technician, Department of Radiology, Case Western Reserve University, for providing CT-scans of these specimens.

References

ANDB90. A. H. Abbot, D. J. Netherway, D. J. David, and T. Brown. Application and Comparison of Techniques for Three-Dimensional Analysis of Craniofacial Anomalies. *Journal of Craniofacial Surgery*, 1(3):119–134, July 1990.

BC88. F. L. Bookstein and C. B. Cutting. A proposal for the apprehension of curving cranofacial form in three dimensions. In K. Vig and A. Burdi, editors, *Cranofacial Morphogenesis and Dysmorphogenesis*, pages 127–140. 1988.

BK89. R. Bajcsy and S. Kovačič. Multiresolution Elastic Matching. *Computer Vision, Graphics and Image Processing*, (46):1–21, 1989.

CDB+95. C. Cutting, D. Dean, F .L. Bookstein, B. Haddad, D. Khorramabadi, F. Z. Zonneveld, and J.G. Mc Carthy. A Three-dimensional Smooth Surface Analysis of Untreated Crouzon's Disease in the Adult. *Journal of Craniofacial Surgery*, 6:1–10, 1995.

CMV94. G. E. Christensen, M. I. Miller, and M. Vannier. A 3D Deformable Magnetic Resonance Textbook Based on Elasticity. In *Applications of Computer Vision in Medical Image Processing*, pages 153–156, Stanford University (USA), March 1994.

CNPE94. D.L. Collins, P. Neelin, T.M. Peters, and A.C. Evans. Automatic 3d intersubject registration of mr volumetric data in standarized talairach space. *J. of Computer Assisted Tomography*, 18(2):192–205, March 1994.

Cut91. C. B. Cutting. Applications of computer graphics to the evaluation and treatment of major craniofacial malformations. In J. K. Udupa and Herman G. T., editors, *3D Imaging in Medicine*, chapter 6, pages 163–189. CRC Press, 1991.

Dea93. D. Dean. *The Middle Pleistocene Homo erectus/Homo sapiens Transition: New Evidence from Space Curve Statistics.* PhD thesis, The City University of New York, 1993.

DSTA95. J. Declerck, G. Subsol, J.Ph. Thirion, and N. Ayache. Automatic retrieval of anatomical structures in 3D medical images. In N. Ayache, editor, *CVRMed'95*, volume 905 of *Lecture Notes in Computer Science*, pages 153–162, Nice (France), April 1995. Springer Verlag.

GRB93. J. C. Gee, M. Reivich, and R. Bajcsy. Elastically Deforming 3D Atlas to Match Anatomical Brain Images. *Journal of Computer Assisted Tomography*, 17(2):225–236, March 1993.

PT95. X. Pennec and J.Ph. Thirion. Validation of 3-D Registration Methods based on Points and Frames. In *Proceedings of the 5th Int. Conf on Comp. Vision (ICCV95)*, pages 557–562, Cambridge, (USA), June 1995.

STA95. G. Subsol, J.Ph. Thirion, and N. Ayache. A General Scheme for Automatically Building 3D Morphometric Anatomical Atlases: application to a Skull Atlas. In *Medical Robotics and Computer Assisted Surgery*, pages 226–233, Baltimore, Maryland (USA), November 1995.

Sub95. G. Subsol. *Construction automatique d'atlas anatomiques morphométriques à partir d'images médicales tridimensionnelles.* PhD thesis, Ecole Centrale Paris, December 1995.

TG95. J-P Thirion and A Gourdon. Computing the differential characteristics of isointensity surfaces. *Computer Vision and Image Understanding*, 61(2):190–202, March 1995.

Thi95. J-P. Thirion. Fast Non-Rigid Matching of 3D Medical Images. In *edical Robotics and Computer Aided Surgery (MRCAS'95)*, pages 47–54, Baltimore (USA), November 1995.

Thi96. J-P. Thirion. Non-Rigid Matching Using Demons. In *Computer Vision and Pattern Recognition (CVPR'96)*, San Francisco (USA), June 1996. (to appear).

TT88. J. Talairach and P. Tournoux. *Co-Planar Stereotaxic Atlas of the Human Brain.* Georg Thieme Verlag, 1988.

A New Approach to Fast Elastic Alignment with Applications to Human Brains

Thorsten Schormann, Stefan Henn and Karl Zilles

C. and O. Vogt Institute of Brain Research,
Heinrich-Heine University Düsseldorf

Abstract. A technique is presented for elastic alignment applicable to human brains. The transformation which minimizes the distance measure $D(u)$ between template and reference is determined, thereby simultaneously satisfying smoothness constraints derived from an elastic potential known from the theory of kontinuum mechanics. The resulting partial differential equations, with up to $3 \cdot 2^{20}$ unknowns are directly solved for each voxel, that is, without interpolation, by an adapted full multigrid-method (FMG) providing a perfect alignment. For further increases of resolution, the full advantages of the FMG are maintained, that is, parallelization and *linear* effort with $O(N)$, N being the number of grid-points.

1 Introduction

Medical applications require accurate alignment of volumes or images resulting from different imaging modalities, for example, for the integration of histological and MR-images (MR = Magnetic Resonance). Histological information makes it possible to map spatial microstructural information into the corresponding MR-volume which provides only macroscopical resolution. In order to make use of microstructural information, it is necessary to transform with highest accuracy onto the MR-refence volume, that is, with a one-to-one relation of corresponding voxels thereby compensating for the highly distorted histological voxels resulting from preparation procedures. In addition to these tasks, the alignment of individual brains with different morphology is required for the analysis of interindividual variability of brain structures, which is a prerequisite for constructing a human brain atlas. The goal of such an atlas is the fusion of information derived from many single brains into one common coordinate system, the standard human brain.

In a first step, it is necessary to find the transformation $\mathbf{x} \longrightarrow \mathbf{x} - \mathbf{u}(\mathbf{x})$ which maps all points $\mathbf{x} = (x_1, x_2, x_3)$ of the template into the reference by $\mathbf{u} = (u_1, u_2, u_3)$, the local displacements. For this purpose, the objects (brains) are modeled as elastic material, whereby the transformation onto the reference must be smooth so that connected regions remain connected and the global relationships between structures are maintained, that is, tearing or folding are avoided. This can be accomplished by minimizing a distance measure $D(u)$ between template and reference and thus achieving appropriate external forces and simultaneously satisfying smoothness constraints derived from an elastic potential based on the kinematics of continuum mechanics. The correlation of volumes or images with elastic methods has been described in several publications [1][2][3]. Although the elasticity theory has commonly been used in these, as well as in our study, the presented approach extends previous work, in that research was focused on a direct solution and on minimization of computation-time. This was accomplished (i) by direct determination of external forces derived from the variational principle according to Hamilton, that is, without interactive estimation of local diplacements, time consuming application of local similarity functions or stochastic gradient searches, and (ii) with a very fast method for solving the system of equations which results from the application of an elastic potential for each voxel. From these considerations it follows that a volume with a resolution of $64 \times 128 \times 128$ voxels requires the solution of $\sim 3 \cdot 2^{20}$ unknowns which can be carried out by an adapted multigrid method.

2 Theory

The template T is modelled as an elastic medium by applying of an elastic potential or energy function Φ known from the theory according to Navier-Lamé. The potential function is given by [4]:

$$\Phi = -\frac{1}{2}\mu \mathbf{u} \Delta \mathbf{u}^t + \frac{1}{2}(\lambda + \mu) \cdot (\boldsymbol{\nabla} \cdot \mathbf{u})^2$$

whereby $\mathbf{u}$ denotes the local displacements; t the transpose; μ, λ the Lamé constants; Δ and $\boldsymbol{\nabla}\cdot$ are the Laplacian and divergence operators. The goal is to find a transformation such that the template $T(\mathbf{x} - \mathbf{u}(\mathbf{x}))$ and the reference $R(\mathbf{x})$ become as similar as possible. Therefore, the potential function $\Phi = \Phi(\mathbf{u})$ and a squared-error distance measure $D(\mathbf{u})$

$$D(\mathbf{u}) = \int |T(\mathbf{x} - \mathbf{u}(\mathbf{x})) - R(\mathbf{x})|^2 dx$$

is simultaneously minimized by the variational principle according to Hamilton:

$$\delta H = \delta \int L dt$$

whereby the Lagrange function L is given by $L = \Phi - D$. Thus, the external force is the registration distance determined by

$$\mathbf{f} = \mathbf{f}(\mathbf{x} - \mathbf{u}(\mathbf{x})) = 2 \cdot \boldsymbol{\nabla} T \cdot (T - R)$$

The resulting Navier-Lamé partial differential equations (NLE) are solved directly on the original resolution by a multigrid-method, thus avoiding time-consuming calculations of local similarity functions based on Hermite polynomials [1]. Note that the gradient operation for determining the external forces

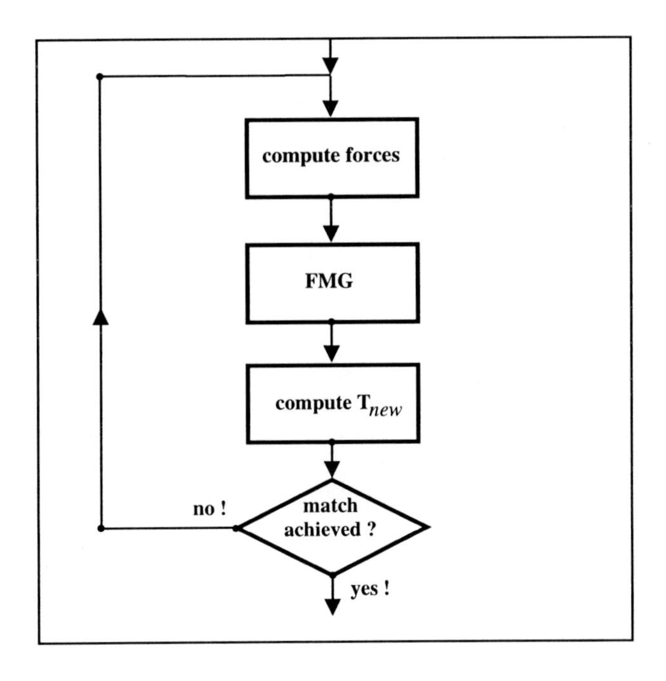

Fig. 1. Schematic for the elastic alignment.

is required only once, on the original resolution, whereby the full advantages of multigrid techniques can be applied.

The next step is the choice of appropriate Lamé constants λ and μ which determines the registration result. The constants λ, μ are chosen, such that the changes in volume are maximal for a selected $\mu > 0$, which corresponds to longitudinal stretch without lateral shrink ($\lambda = 0$, $\mu > 0$). The relation between external forces and the partial differential equation (internal forces) is then determined by one parameter μ with two constraints: (i) for small values of μ the registration result is given predominantly by forces leading to global translations as can be shown theoretically by the Green function of an elastic medium [5] for which the local diplacements decrease with $\frac{1}{r}$ with increasing distance r from a pointlike force; (ii) for large values of μ the object becomes more rigid and thus the model cannot account for local registration distances. It is therefore more convenient to transform the template iteratively by selecting an appropriate μ (Figure 1) which is chosen such that the brain is deformed stepwise. The registration result of the previous iteration is used for the next iteration until a satisfactory match is achieved.

For the numerical solution of the NLE, a full multigrid-like (FMG) process was developed [6], whereby it was necessary to adapt the FMG with respect to optimal parameters. For this reason, the NLE has to be cast in a discrete form, which works on a hierarchy of grids Ω_l^h, where Ω_0^h denotes the finest and $l = 6$ the coarsest grid in our applications and h is a discretization parameter. The discretization of NLE can thus be written as

$$\begin{aligned} \mathrm{NLE}^h \cdot \mathbf{u}^h &= \mathbf{f}^h \quad \Omega_0^h \\ \mathbf{u}^h &= \mathbf{0} \quad \partial\Omega_0^h \end{aligned}$$

where $\mathrm{NLE}^h \in \mathrm{IR}^{3 \cdot 2^{20}} \times \mathrm{IR}^{3 \cdot 2^{20}}$ is a grid operator corresponding to the finest resolution, $\mathbf{u}^h = (u_1^h, u_2^h, u_3^h)$, the deformation field, $\mathbf{f}^h = (f_1^h, f_2^h, f_3^h)$ the forces and $\mathbf{u}^h = \mathbf{0}$ on $\partial\Omega_0^h$ the Dirichlet boundary condition. Due to the efficiency of the FMG, a residual error of only $\pm 10^{-2}$ can be accomplished with two FMG-cycles, thereby achieving sub-voxel accuracy. For further improvement of resolution the advantages of the FMG are maintained: parallelization and linear effort with $O(N)$, N being the number of grid-points.

3 Results

Figure 2 (a,b) shows an example for the alignment of corresponding histological and MR-sections. The contour of the MR reference section is superimposed onto the linear preprocessed [7],[8] histological section before (a) and after alignment (b). Note the misalignment of the right temporal lobe and the upper part of the right hemispheres for which linear algorithms cannot account. Since the sections are preprocessed according to [8] using the Rayleigh-Bessel distribution of local nonlinear deformations, an improved global alignment can be achieved and the computation-time with the present technique is additionally reduced to 90 secs on a SUN SPARC10 workstation. The parameter μ is chosen, such that the maximal displacement calulated by the algorithm is about 2 voxels in the first iteration, a condition which is given by the derivation of forces for which the Taylor expansion up to the first order is taken into account. Therefore, the force calculation is valid only in the local environment of each voxel. The number of iterations are determined by the maximum local displacement, which is in this example given by 5 voxels.

Figure 2 (c,d) shows the alignment of a brain data set with the standard brain of the human brain atlas (different individuals) before (c) and after alignment (d), whereby the contour of the standard brain is superimposed in (c,d). A further example is shown in Figure 2 before (e) and after alignment (f). As can be seen in Figure (c,d), (e,f), a perfect match is achieved with respect to external microscopical structures, e.g., the fissura longitudinalis, primary cortical sulci as the lateral and cingulate sulci as well as for internal structures, e.g., ventricles, whereas the conformation of small structures is predominantly maintained. The latter is an intended effect in order to preserve the individual morphology of

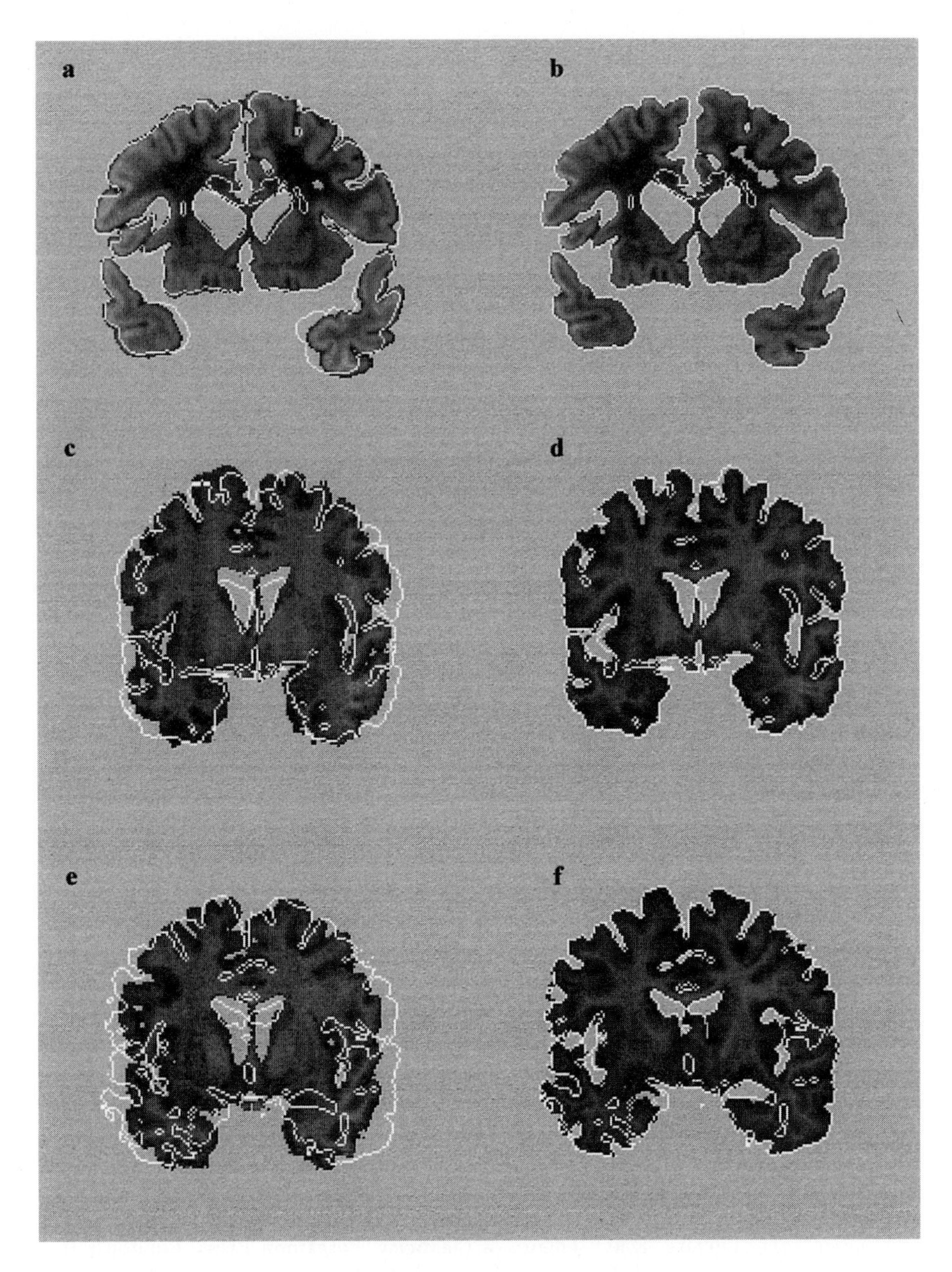

Fig. 2. Matching results of corresponding MR- and histolgical sections (a,b) and two examples (c,d), (e,f) of interindividual nonlinear alignment. Linearly preprocessed histological section with superimposed contour of the MR-reference section before (a) and after nonlinear alignment (b). Section of an individual 3D brain data set (c) with the superimposed contour of the standard brain and after nonlinear alignment (d). A further example is shown in (e,f) with corresponding reference contour.

each brain. It can be attributed to the fact, that these structures are moderately transformed because they are represented by small gray value transitions (low contrast), which in turn determine the strength of external forces.

In order to provide a comparison with another relaxation method (Jacobi's relaxation) used in the literature for determining local displacement fields [1], Table 1 shows the number of iterations required to achieve an accuracy of 10^{-2}. In this context, an FMG iteration is defined according to [6] with n=6 thereby achieving an effort of $O(N)$, whereas Jacobi's relaxation increases with $O(N^2)$.

	number of iterations	
N	Jacobi's relaxation	FMG, F-cycle
$2^{18} = 64^3$	2,946	2
$2^{20} = 64 \cdot 128^2$	6,592	2
$2^{21} = 128^3$	16,541	2

Table 1. Comparison between Jacobi's relaxation and FMG.

For satisfactory results in three dimensions, the algorithm requires 10 iterations each with two 3D FMG-cycles (20 min) including recalculation of forces as shown in Figure 1, so that the total time is given by 200 min for a resolution of $64 \times 128 \times 128$ voxels on a SUN SPARC10 workstation.

ACKNOWLEDGEMENTS
Supported by the DFG (SFB 194 A6). The authors thank K. Witsch for discussion and generous help.

References

1. Bajcsy, R., Kovacic, S.: Multiresolution elastic matching. Comput. Vision. Graph. Im. Proc. **46** (1989) 1–21
2. Miller, M.I., Christensen, G.E., Amit, Y., Grenander, U.: Mathematical textbook of deformable neuroanatomies. Proc. Nat. Acad. Sci. **90** (1993) 11944–11948
3. Christensen, G.E., Rabbitt, R.D., Miller, M.I.: 3D brain mapping using deformable neuroanatomy. Phys. Med. Biol. **39** (1994) 609–618
4. Budó A.: Theoretische Mechanik. Verlag der Wissenschaften, Berlin (1980)
5. Landau, L.D., Lifshitz, E.M.: Theory of Elasticity. Pergamon Press, London (1959)
6. Stüben, K., Trottenberg, U.: Multigrid methods: fundamental algorithms, model problem analysis and applications. In "Multigrid methods" (Hackbusch, W., Trottenberg, U., eds.), Lecture Notes in Mathematics **960**, Springer-Verlag, Berlin (1982).
7. Schormann, T., Matthey, M.v., Dabringhaus, A., Zilles, K.: Alignment of 3-D brain data sets originating from MR and histology. Bioimaging **1** (1993) 119–128
8. Schormann, T., Dabringhaus, A., Zilles, K.: Statistics of deformations in histology and improved alignment with MRI. IEEE Trans. Med. Imaging **14** (1995) 25–35

Individualizing Anatomical Atlases of the Head

Gary E. Christensen[1] and Sarang C. Joshi[2] and Michael I. Miller[2]

[1] Mallinckrodt Institute of Radiology and Department of Surgery
[2] Department of Electrical Engineering
Washington University, St. Louis, MO 63110, USA

Abstract. This paper presents diffeomorphic transformations of 3D anatomical image data in human brains as imaged via magnetic resonance imagery. These transformations are generated in a hierarchical manner accommodating both global and local anatomical detail. The initial low-dimensional registration is accomplished by constraining the transformation to be in a low-dimensional basis. The basis is defined by the Green's function of the elasticity operator placed at predefined locations in the anatomy and the eigenfunctions of the elasticity operator. The high-dimensional, large-deformation registration is accomplished by minimizing the mismatch between the template (atlas) and target image volumes constrained by solving the Navier Stokes fluid model on a vector field.

1 Introduction

For the past several years we have been involved in the development of mathematical and computational software tools for the generation of structural representations of brain anatomy which accommodate normal neuroanatomical variation. To accomplish this, we use the *global shape models* of Grenander [1] to represent the typical global structures in the shape ensemble via the construction of templates (anatomical atlases), and their variabilities by the definition of probabilistic transformations applied to the templates. The transformations form mathematical groups made up of translations, scales, and rotations and are applied locally throughout the continuum of the template coordinate system so that a rich family of shapes may be generated with the global properties of the templates maintained.

2 Coarse-to-fine 3-D Brain Mapping Protocol

Since the full 3-D brains are tremendously complex, we have constructed a basic protocol which proceeds from course-to-fine for the generation of the diffeomorphic maps of the anatomical atlas. It essentially contains three fundamental pieces which proceeds from a low-dimensional, nonrigid, coarse registration to a high-dimensional, fine registration.

The first step is based on a natural extension of Bookstein's landmark work [2] in which the atlas is matched to the target data set by matching operator defined points, lines, surfaces, and subvolumes. As described in Joshi et al. [3], the idea of constraining the mapping via points, lines, surfaces, or volumes can all be viewed as being the solution of a generalized Dirichlet problem which constrains the transformation from one brain to the other. Such a global or coarse registration brings into alignment the major subvolumes and areas of interest, and includes all of the major motions of the affine group. This provides the initial conditions for the second step which refines the coarse alignment using the linear elastic basis model described in [4, 5]. The dimension of the basis flows from low dimension (small number of basis elements) to higher dimensions by incrementing the dimensions in a controlled manner during iteration. Deformation of the template is modeled as an elastic solid which is driven into the shape of the target volume based only on the mismatch between the template and target image volumes. This is fully automatic driven by the volume data itself; the solutions from the lower dimensional landmark solution only work to provide initial conditions. Having completed the coarse step in the transformation, the volumes are roughly aligned so that complete subvolumes can be mapped in great detail. The third and final step is to solve for the transformation at each voxel in the full volume using only the original image data.

Low-Dimensional Transformations The first step in our coarse-to-fine procedure follows the framework described in Joshi et al. [3]. A set of landmarks $\{x_i : x_i \in \Omega, i = 1, 2, \cdots, N\}$ is defined in the atlas which can be easily identified in the target. Define y_i to be the associated points identified in the target associated with each landmark x_i in the template. A Gaussian error of variance σ_i^2 is used to describe the operators certainty of correctly placing each landmark in the target.

The transformation is estimated by solving the Bayesian optimization problem

$$\hat{u} = \arg\min_u \int_\Omega |Lu|^2 + \sum_{i=1}^{N} \frac{|y_i - x_i + u(x_i)|^2}{\sigma_i^2} . \tag{1}$$

The minimizer of this problem has the form

$$\hat{u}(x) = \mathbf{b} + \mathbf{A}x + \sum_{i=1}^{N} \beta_i K(x, x_i)$$

where $\mathbf{A}$ is a 3×3 matrix, $\mathbf{b}$ is a 3×1 vector, $\beta_i = [\beta_i^1, \beta_i^2, \beta_i^3]$ is a weights vector and $K(x, x_i)$ is the Greens function of the operator L^2. The operator L is assumed to be any self-adjoint differential operator such as the Laplacian $L = \nabla^2$, the biharmonic $L = \nabla^4$, or linear elasticity $L = \mu\nabla^2 + (\mu + \lambda)\nabla\nabla\cdot$, $(\nabla\nabla\cdot)u = \nabla(\nabla \cdot u)$.

The landmark transformation is then refined using a linear elastic solid basis transformation [4, 5] which uses the image data of the template (atlas) to match the target data set. Defining the transformation via the vector field according to $h(x) = x - u(x)$, the strain field $u(x)$ for linear elasticity under the small deformation assumption corresponds to energetics of the form $H(u) = \|Lu\|^2$ The variational problem becomes

$$\hat{u} = \arg\min_u \gamma \int_\Omega |T(x - u(x)) - S(x)|^2 dx + \int_\Omega |Lu|^2 \tag{2}$$

where the displacement field u is constrained to have the form

$$u(x) = \sum_{k=0}^{d} \mu_k \phi_k(x) + \sum_{i=1}^{N} \beta_i K(x, x_i) + \mathbf{A}x + \mathbf{b} \tag{3}$$

with the variables $\{\beta_i\}$, $\mathbf{A}$, and $\mathbf{b}$ fixed from the landmark transformation. The basis functions $\{\phi\}$ are the eigenvectors of the operator L.

High-Dimensional Transformation The last step in our coarse to fine procedure is a viscous fluid transformation [6] that accommodates large-distance, nonlinear deformations of the template. For this, an auxiliary random field is introduced, termed the velocity field $v(x,t), x \in \Omega, t \in [0,T]$, which defines the strain vector field $h(x) = x - u(x,T)$ according to

$$v(x,t) = \frac{\mathrm{d}u(x,t)}{\mathrm{d}t} = \frac{\partial u(x,t)}{\partial t} + (\nabla u(x,t))^T v(x,t) \ , \ t \in [0,T] \ . \tag{4}$$

For viscous fluids the stress grows proportionately to the rate of strain $\frac{du}{dt}$, implying that while the viscous fluid kinematics will force the mapping to be 1-1 and onto, large strain distance deformations occur unpenalized as long as the rate of strain during the mapping is smooth as the deformation of the template proceeds. The PDE corresponding to the solution of the variational problem for the fluid formulation is given by

$$\mu \nabla^2 v(x,t) + (\lambda + \mu)\nabla(\nabla \cdot v(x,t)) = b(x - u(x,t)) \tag{5}$$

with the boundary conditions $v(x,t) = 0$, $x \in \partial\Omega$ and $t \in [0,T]$. The coefficients μ and λ are viscosity constants.

3 Results

We have now mapped several 3-D whole brains as imaged via MRI with the intention of studying the shape and volume of deep structures such as the hippocampus. Shown in Figure 1 are the results of mapping an entire MPRAGE volume from a single template to a second individual. This solution involved the computation of $128 \times 128 \times 100 \times 3 \approx 5 \times 10^6$ parameters representing the coordinate system transformation. Notice the phenomenal similarity of the surface rendered template and target (compare middle and right panels).

We emphasize that our solution does not involve mapping the surfaces alone. Rather, we are mapping the entire volume of the heads, which of coarse carries the surfaces along with internal structures. Our express purpose is to study the

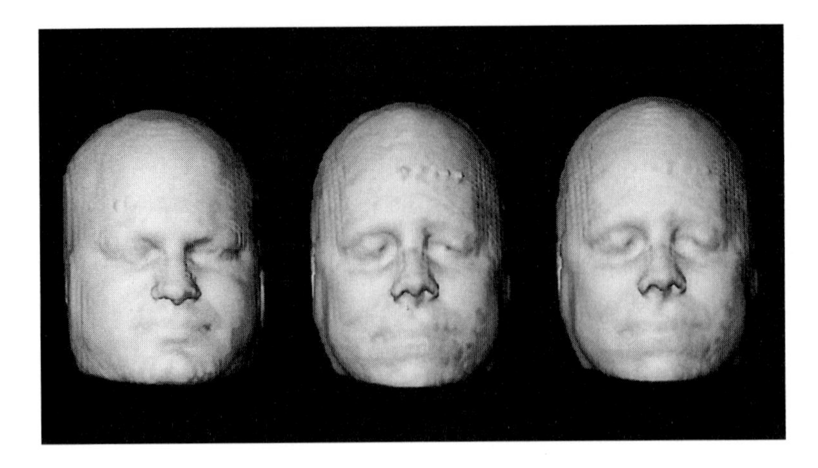

Fig. 1. Surface renderings of the atlas MRI template (left), the MRI of a target individual (middle), and the 3D transformed (individualized) atlas template MRI (right).

shape of deep structures such as the hippocampus. To emphasize this, Fig. 2 shows a slice from the 3-D MRI atlas at its original resolution $1 \times 1 \times 1.25$ mm^3 (left column) and at its interpolated resolution of 0.25^3 mm^3, focusing in on the hippocampus. The segmentation of the template hippocampus was performed by a trained expert at 0.125^3 mm^3 resolution requiring upwards of 20 hours to generate. This segmentation has a smooth surface in 3D so that when the atlas transformation is applied to the segmentation, it generates a smooth 3D segmentation in the target volume.

The atlas was deformed to globally match the target volume by performing a manifold transformation followed by an elastic basis transformation on the entire MRI volume ($256 \times 256 \times 128$ voxel volume, with voxel dimension $1 \times 1 \times 1.25$ mm^3). This initial transformation gave a coarse alignment of the head and a good match in the subregion containing the hippocampus. A $128 \times 128 \times 65$ voxel region at 0.25^3 mm^3 was extracted from the transformed template and target volumes. The template subvolume was then fluidly transformed into the target subvolume (see Fig. 3) Notice that all of the error between the transformed atlas segmentation and the hand segmentation occurs at the edges.

The deformable atlas segmentation procedure was repeated twice on two normal individuals and compared to manual segmentations. Repeated automatic measurements were generated by placing new landmarks in the target image volume. These results demonstrate a good correspondence between the automated and manual measurements; this despite the fact that the hippocampus is a difficult structure to segment because small errors in segmenting the boundary cause large errors in volume. For these two individuals, the average volumes of the transformed atlas hippocampus segmentation were 2846 and 2874 mm^3,

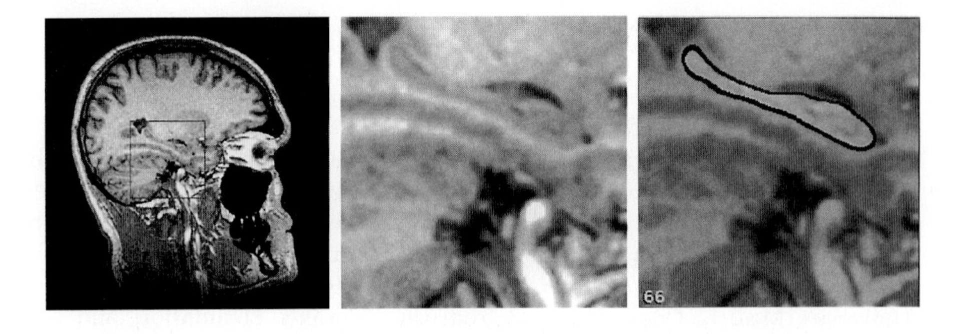

Fig. 2. Sagittal cross section of the MRI atlas template (left); box shows subregion containing the hippocampus. Middle and right panels show the enlargement of the hippocampus subregion with and without hippocampus segmentation superimposed, respectively.

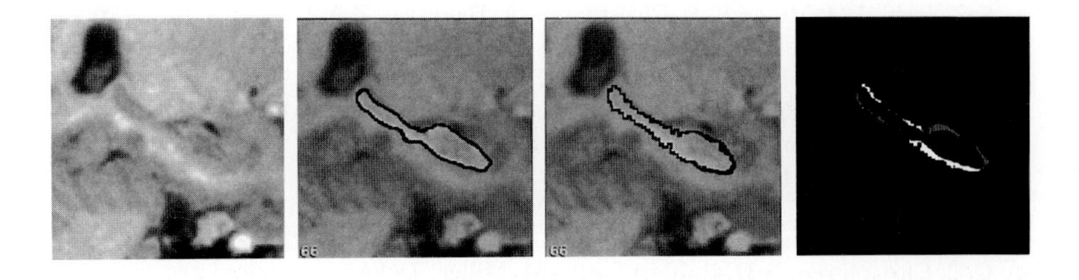

Fig. 3. Sagittal cross section of hippocampus target subregion (left), with transformed atlas segmentation superimposed (middle-left), and with hand segmentation superimposed (middle-right) subvolume of the target image volume. Right panel shows the difference image of the two segmentations.

respectively, while the average of the manual segmentations were 2932 and 2656 mm^3, respectively.

4 Summary

The experiments presented in this paper demonstrate the feasibility of finding diffeomorphic correspondences between anatomies using high-dimensional transformations of a deformable head atlas. A high-dimensional volume transformation was required to accommodate complex neuroanatomical shape. Global transformations provide a coarse correspondence between two anatomies but fail to provide the local correspondence needed.

5 Acknowledgments

We express our appreciation to Drs. John Csernansky, John Haller, and Michael W. Vannier of the Departments of Psychiatry and the Mallinckrodt Institute of Radiology of Washington University School of Medicine for providing the MRI data, and for their efforts on the mapping tools in the hippocampus studies. We also acknowledge Ayananshu Banerjee for his help in the development of the hierarchical algorithm for brain mapping. This research was supported in part by the Craniofacial Imaging Laboratory, St. Louis Children's Hospital, Washington University Medical Center, a grant from the Whitaker Foundation, and NIH grants R01-MH52158-01A1 and NCRR-RR01380.

References

1. U. Grenander. *General Pattern Theory.* Oxford University Press, 1993.
2. F.L. Bookstein. *Morphometric Tools for Landmark Data.* Cambridge University Press, New York, 1991.
3. S.C. Joshi, M.I. Miller, G.E. Christensen, A. Banerjee, T.A. Coogan, and U. Grenander. Hierarchical brain mapping via a generalized dirichlet solution for mapping brain manifolds. *Vision Geometry IV*, SPIE 2573:278–289, 1995.
4. M.I. Miller, G.E. Christensen, Y. Amit, and U. Grenander. Mathematical textbook of deformable neuroanatomies. *Proceedings of the National Academy of Sciences*, 90(24):11944–48, December 1993.
5. G.E. Christensen, R.D. Rabbitt, and M.I. Miller. 3D brain mapping using a deformable neuroanatomy. *Physics in Medicine and Biology*, 39:609–618, 1994.
6. G.E. Christensen, R.D. Rabbitt, and M.I. Miller. Deformable templates using large deformation kinematics. *IEEE Transactions on Image Processing*, To appear in September, 1996.

Mutual Information for Automated Multimodal Image Warping

Boklye Kim[1], Jennifer L. Boes[1], Kirk A. Frey[2], Charles R. Meyer[1]

[1] Department of Radiology, University of Michigan Medical Center, Ann Arbor, Michigan 48109-0553, U.S.A

[2] Department of Internal Medicine, University of Michigan Medical Center

This work supported in part by DHHS PHS NIH 1RO1 CA59412

Abstract. A quantitative assessment of mapping accuracy based on mutual information index (MI), calculated from gray scale 2D histogram, is applied to an automated multimodality (un)warping algorithm. Information rich histological image data, which present non-linear deformations due to the specimen sectioning, is reconstituted into deformation-corrected, 3D volumes for geometric mapping to anatomical data for spatial analyses. Thin-plate-spline (TPS) algorithm has been implemented for automatic unwarping of the distortions using a multivariate optimizer and MI as a global cost function. The MI proves to be a robust objective matching criterion effective for automatic multimodality warping for 2D data sets and can be readily applied to volumetric 3D registrations. The improved performance of TPS warping compared to full affine transformation is quantified by comparison of MI's of both methods.

1 Introduction

Recently, a matching criterion based on gray scale entropy has been applied to multimodal image registrations implementing rigid-body transformations [1-3]. We have investigated development of a multi-modality matching criterion based on the fundamental information theory of mutual information for its application to automatic image warping registration tools. Mutual information index (MI) is calculated based on a two dimensional (2D) gray value histogram of an image pair and provides a user independent and robust assessment of matching accuracy for multimodal registrations involving 2D or 3D, and affine or warping transformations. This study focuses on development of an automated registration algorithm employing an objective matching criterion formulated by MI and its feasibility as a global cost function to be minimized. Accurate registration capabilities of MI based optimization is demonstrated by automated 2D registration of autoradiographic images with video reference images of the block face using affine and thin-plate-spline (TPS) [4] warping transformations.

Histological data sets such as autoradiographs exhibit local deformation artifacts associated with sample slice sectioning. Such data sets require correction of local deformation by image warping. In previous work we have demonstrated the use of TPS warping algorithm based on user defined homologous features for the registration of [^{14}C]-deoxy-D-glucose (2DG) autoradiographic data (AR) with MRI volume, following reconstitution of 2DG volume data set [5]. The registration accuracy of the unwarped volume was then compared with the result of an affine transformation [6]. While markedly better registration was obtained from implementation of TPS warping, identification of homologous features and examining the result of TPS operation were tedious, time consuming processes and critically user-dependent.

Based on the previous finding [5], 2D and 3D registrations of autoradiographic data are presented as a non-affine problems. Automatic registration tools involving the automatic optimization of MI are shown to be effective for registration of autoradiograph data and are truly robust for tasks involving 3D modalities of MRI, PET and 3D reconstituted autoradiograph data.

2 Methods

2.1 Mutual Information

Based on classical theory of entropy [7], mutual information quantifies interdependency of two random variables, i.e. image data. Mutual information index, $MI= -I(x,y)$, where $I(x,y)= -\Sigma p(x,y)log(p(x,y)/p(x)p(y))$, is calculated from the two-dimensional joint density function, $p(x,y)$, of gray values of a geometrically mapped image pair, where $p(x)$ and $p(y)$ are marginal probability distributions. The matching criterion can be obtained by substituting for X and Y the gray values of g_1 and g_2 of the images to be registered and $p(g_1,g_2)$ is obtained by normalizing the 2D histogram of the gray values of the common volume of the two images. MI approaches its lower bound when two images are highly correlated and its upper bound when uncorrelated. The minimum MI for an image pair is achieved when the geometric mapping produces the most correlated, i.e. registered, data sets, independent of modality.

2.2 Multimodal Image Data

Autoradiograph (AR) slice of a rat brain and video images of uncut specimen block face of the same slice (in 512x480 matrix), were acquired as described in literature [6]. For $[^{14}C]$-deoxy-D-glucose (2DG) autoradiography, 20 μm thick horizontal brain sections were cut in an automated cryomicrotome. During the cryomicrotome procedure, video images of the remaining uncut specimen's block face were recorded, prior to every 16th sectioning, with field of view (FOV) of 47 X 44 mm in 512X480 matrix. The autoradiographs were developed after the sections were exposed on film at room temperature for four weeks. The autoradiographs were digitized from film placed on a light box using the same video camera as for the blockface images. Each 2DG AR slice developed into metabolic map exhibited local deformation artifact associated with microtome sectioning. Digitized 2DG AR images were registered to the corresponding reference video images of block face to reconstitute a 2DG volume data for the purpose of 3D registration with MRI volume.

2.3 Registration

Multimodal image data sets were registered employing affine and thin-plate-spline (TPS) warping [4] algorithms implemented using Application Visualization System (AVS) software on DEC 3000/500x OSF1/Alpha.

The automated MI driven registration, as shown in the flow chart, consists of: an *image registration* routine that determines coefficients for coordinate mapping either affine or TPS warping; an image *reconstruction* routine to perform trilinear interpolation for gray values of the mapped pixels; *MI computation* from 2D gray scale histogram of an image pair; an optimizer (MIFMINS) that implements a multivariate minimization algorithm [8].

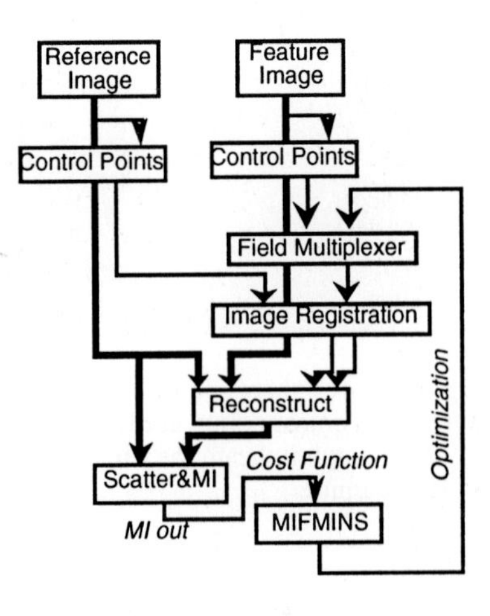

The registration process is initiated by passing the approximate location of feature points in each data set as homologous feature landmarks. These points are used as control points for initialization of vectors that contain the coordinates to define the geometric mapping. MI based automated registration of the two images is achieved by allowing the control points in homologous space (AR image space) to move to positions that minimize the MI. Selecting one point more than the spatial dimension constrains the solution to the affine transformation (e.g. three points for a 2D image) and more points select a general TPS warping solution. The number of variables in the minimization is the product of the number of control points and spatial dimension.

For reconstitution of AR volume, the slice-by-slice 2D registration process, as shown in the flow diagram, is repeated for the number of AR slices constituting the total volume of the brain (60-150 slices). The location of the initial control points are selected arbitrarily by a user once at the initialization step. Since these points are not dependent on features, which vary with geometry, the same set of points is applied to all slices as an automated sequence. Figure 1 displays (a)reference video image (uncut specimen block face) and (b)2DG autoradiograph deformed due to specimen sectioning and control points selected in both images. White dots in both reference and autoradiograph images indicate the location of control points used for initial vectors in TPS registration.

3 Results

Registration of a 2DG autoradiograph slice and its reference video block face image demonstrates the utilization of MI driven automatic mapping developed for a multimodal image pair implementing either affine or warping transformation. In Figure 1, black dots in autoradiograph image indicate the final locations of TPS warping solution achieved by cost function minimization of MI metric using the initial feature points indicated by white dots. Note that some starting points (white) in autoradiograph image were markedly displaced intentionally (indicated by an arrow in Fig. 1b).

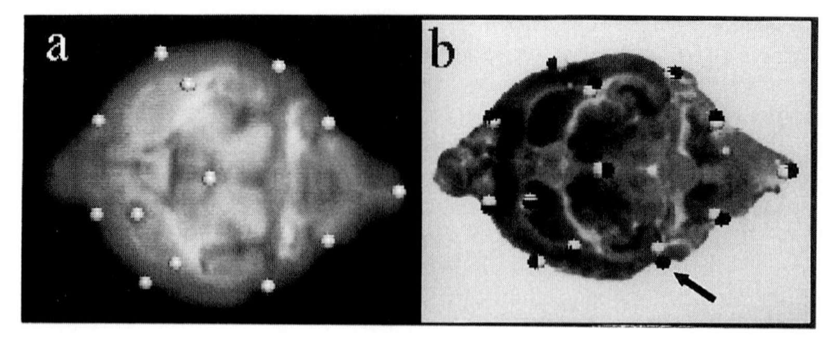

Fig. 1. White spheres on (a) video block face and (b) autoradiograph image indicate locations of feature control points used for the initial vectors in TPS warping and black spheres on (b) autoradiograph indicate the final locations of the control points as a result of MI optimization of TPS warping.

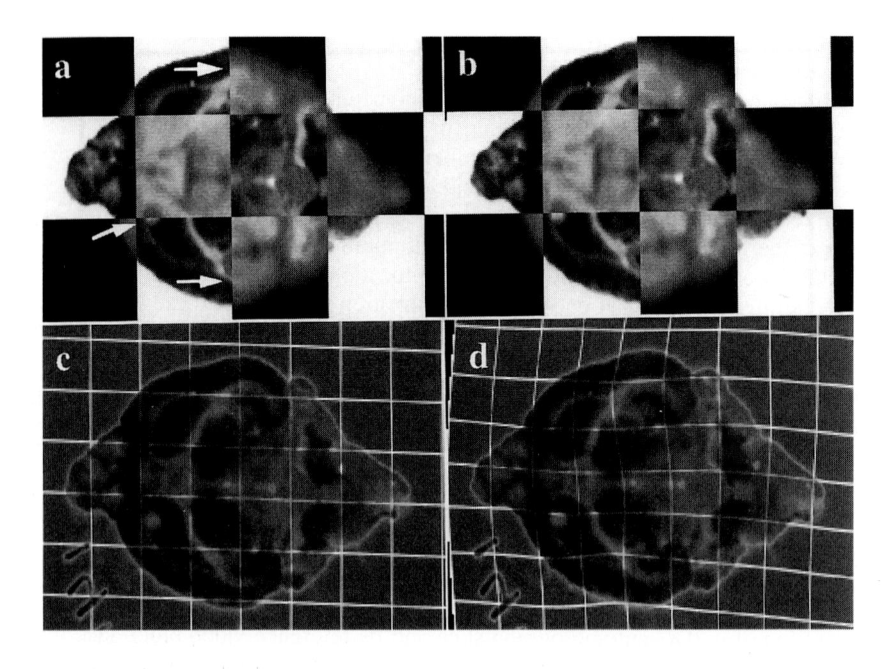

Fig. 2. Results of automatic registration implementing affine and TPS warping transformation of a 2DG autoradiograph slice with the corresponding video reference block face using MI driven optimization: (a) result of affine registration and (b) TPS warping in checker board display of the two images; grid display of (c) the affine and (d) TPS warping solution overlaid with the registered AR image.

Figure 2 displays the result of MI registration using (a)affine and (b)TPS warping transformation algorithms. The mapping quality is visually inspected by images shown in checker board pattern. While the affine transformation (Fig. 2a) results in remarkably good matches for most internal structures of the brain, some features around outer edges denote local deformations that cannot be handled by the affine transformation, e.g. in the mid section of the brain, top and bottom squares, the area around hippocampus and auditory cortex shows internal and outer edge mismatches as well as the lesion in frontal lobe (as indicated by arrows in Fig. 2a). The final TPS warp is illustrated by grid line display overlaid with registered AR in Fig. 2(d). The non-linear warping effect for correction of local deformations due to the slice sectioning is depicted by the deformed grid lines as compared to the global linear transform effect in Fig. 2(c).

The final results are evaluated by comparison of MI values of registered AR images as listed in Table 1. After initialization the MI of the originally deformed AR image with video reference was - 0.462.

	Full Affine	TPS warping	
	automatic	manual	automatic
MI	- 0.691	- 0.709	- 0.718

Table 1. Comparison of MI's from automated full affine and TPS warping, and manually performed TPS warping registrations.

Performance time depends on the number of degrees of freedom. The MI cost function is computed on the total solution including the effect of the warping interpolant as indicated in the flow diagram. On our DEC model 3000/500x OSF/1 Alpha, each iteration for a 236 x 201 image matrix with 13 homologous control points requires 2 seconds and converges in 400 -500 iterations. Each optimization using 26 degrees of freedom takes approximately 15 minutes. The optimization of the TPS warping as shown in Fig. 2 was achieved in two steps. The first used a small number of bins (8x8) in the 2D histogram to find an approximate position close to the global minimum without being trapped by local minima. The second optimization started with the state vector resulting from the first and used a larger number of bins (64x64) to produce the final registration. The larger binning produces a more accurate solution due to the increased sensitivity of the cost function.

3 Conclusions and Discussions

Development of an automatic registration method using a robust matching criterion for matching accuracy proves to be crucial for many image mapping tasks. This work presents the automatic registration using control point based TPS warping algorithm in combination with mutual information index as a matching criterion between two image data sets. Implementation of MI driven optimization loop that converges to the system global minimum to find the accurate registration solution demonstrates the robustness and feasibility of MI based automated multimodal image registrations. The method was utilized for both affine and warping transformation of 2DG metabolic images mapped to video block face references. The result presents improved mapping accuracy by automatic warping algorithm over affine and manually forced TPS mapping as displayed in Table 1. The result demonstrates a multimodal image registration that requires a non-linear correction algorithm consistent with the previous finding [2].

The registration error as determined by final steps of iterations is in the order of 0.01 pixel (92 μ/pixel). Considering the rat brain size (23 mm longitudinal, 15 mm lateral), fine resolution of this order proves to be crucial for analysis of cerebral metabolic functions. Size of the lesion in this particular example is 1.3 mm in diameter, as indicated by a high intensity spot in the forebrain, which requires subvoxel registration accuracy for tumor related functional analysis.

Although it is a common concept that TPS warping algorithm is based on user defined feature landmarks, the current automated procedure needs to be differentiated from the landmark based TPS method. Our algorithm is based on control points, which are defined apart from the feature definition. The control points are allowed to move to find an optimized solution that minimizes the cost metric, MI. Therefore, varying geometry for each slice image falls well within the radial tolerance and once a set of control points is defined the same data set is used through out the whole serial sections with automatically updated slice locations.

References

1. Collignon, A., Maes, F., Delaere, D., Vandermeulen, D., Suetens, P.,Marchal, G.: Automated multimodality image registration using information theory. in Computational Imaging and Vision. 1995. Kluwer Academic Publishers, Ile de Berder, FR.
2. Viola, P.A.,Wells, W.M.: Alignment by maximization of mutual information. in The 5th International Conference on Computer Vision. 1995. MIT.
3. Wells, W.M., Viola, P.,Kikinis, R.: Proceedings of the 2nd annual international symposium on Medical Robotics and Computer Assisted Surgery. in the 2nd annual international symposium on Medical Robotics and Computer Assisted Surgery. 1995. Baltimore.
4. Bookstein, F.,Green , W.D.K.A.: A feature space for edgels in images with landmarks. J of Math. Imag. and Vision, **3** (1993) p. 231-261.
5. Kim, B., Boes, J.L., Frey, K.A.,Meyer, C.R.: Affine versus warping in coregistration of MR imaging and autoradiography. *Radiology*, 197(P) (1995) p. 223.
6. Kim, B., Frey, K.A., Mukhopadhyay, S., Ross, B.,Meyer, C.R.: Co-registration of MRI and autoradiography of rat brain in 3D following automatic reconstruction of 2D data set. in Lecture Notes in Computer Science. 1995. Springer-Verlag, Berlin, Nice, FR.
7. Cover, T.M.,Thomas, J.A.: Elements of Information Theory. Wiley series in Telecomunications, ed. D.L. Schilling. 1991, New York: John Wiley & Sons, Inc.
8. Press, W.H., Flannery, B.P., Teukolsky, S.A.,Vetterling, W.T.: Numerical Recipies in C: The Art of Scientific Computing. 1988, Cambridge: Cambridge Press. 305-309.

Morphological Analysis of Brain Structures Using Spatial Normalization

C. Davatzikos[1], M. Vaillant[1], S. Resnick[2], J.L. Prince[3,1], S. Letovsky[1], and R.N. Bryan[1]
[1] Department of Radiology, Johns Hopkins University
[2] Laboratory for Personality and Cognition, National Institute on Aging
[3] Department of Electrical and Computer Engineering, Johns Hopkins University

Abstract. We present an approach for analyzing the morphology of anatomical structures of the brain, which uses an elastic transformation to normalize brain images into a reference space. The properties of this transformation are used as a quantitative description of the size and shape of brain structures; inter-subject comparisons are made by comparing the transformations themselves.

The utility of this technique is demonstrated on a small group of eigth men and eight women, by comparing the shape and size of their corpus callosum, the main structure connecting the two hemispheres of the brain. Our analysis found the posterior region of the female corpus callosum to be larger than its corresponding region in the males. The average callosal shape of each group was also found, demonstrating visually the callosal shape differences between the two groups.

1 Introduction

The morphological analysis of the brain from tomographic images has been a topic of active research during the past decade [1, 2, 3]. It is important for understanding the normal brain as well as for identifying specific anatomical structures affected by diseases. In this paper we propose a methodology for studying the size and shape of brain structures, based on a spatial normalization procedure [4, 5] which transforms images into a common reference system, bringing them into correspondence with a template image. Associated with this normalization transformation is a deformation function, which is defined at each point of a structure of interest in the template, and which reflects the size difference between an infinitesimal region around a point in the template and its corresponding infinitesimal region in the subject space. Inter-subject comparisons are made by comparing the corresponding deformation functions.

This normalization procedure reduces the effect of the shape variability on size measurements, and provides a common reference system for inter-subject comparisons. More importantly, it does not require the a-priori knowledge of the location of a region of interest within a structure, e.g. a region of abnormality; such regions are readily identified through the deformation function. Finally, as shown in Section 2, our methodology provides a means for obtaining average shapes.

The primary purpose of this paper is to present a morphometric tool. However, we demonstrate its utility in investigating sex differences in the corpus callosum, by applying it to two small groups of right-handed subjects.

2 Materials and Methods

Template/Subject Registration. Consider a structure of interest, $\mathcal{R}$, bounded by the boundary $\mathcal{B}$. Let, also, $\mathcal{R}_a$ be the corresponding structure in a template, which herein is taken from a brain atlas [6], and $\mathcal{B}_a$ be its boundary. We obtain a map between each point in $\mathcal{R}_a$ and a point in $\mathcal{R}$ in two steps. In the first step we find a parameterization, $\mathbf{x}(s), s \in [0,1]$, of $\mathcal{B}$, and a parameterization $\mathbf{x}_a(s)$ of $\mathcal{B}_a$. In conjunction, these two parameterizations define a map from $\mathcal{B}_a$ to $\mathcal{B}$, which corresponds $\mathcal{B}_a \ni \mathbf{x}_a(s), s \in [0,1]$ to $\mathbf{x}(s) \in \mathcal{B}$. Depending on the structure of interest and on the data, we obtain $\mathbf{x}(s)$ and $\mathbf{x}_a(s)$ in the following two ways. 1) If specific landmarks can be identified along $\mathcal{B}$ and $\mathcal{B}_a$, then $\mathbf{x}(s)$ and $\mathbf{x}_a(s)$ are curves parameterized piece-wise by arclength, preserving the landmark correspondences. 2) If specific landmarks cannot be identified along the two boundaries reliably, then $\mathbf{x}(s)$ and $\mathbf{x}_a(s)$ are curves parameterized by arc-length without any intermediate break-points. Since in this case there are no landmarks, the best point-to-point correspondence between $\mathbf{x}(s)$ and $\mathbf{x}_a(s)$ is determined by finding the circular shift of $\mathbf{x}(s)$ which brings its curvature in best agreement with $\mathbf{x}_a(s)$.

In the second step of our registration procedure an elastic warping is applied to the template bringing it into registration with the subject image. This warping is obtained by first deforming $\mathcal{B}_a$ into registration with $\mathcal{B}$; the interior region $\mathcal{R}_a$ in the template is then deformed elastically like a rubber sheet, following the deformation of its boundary (see [4, 5] for details).

Deformation Function. The warping applied to a structure in this registration procedure is reflected in the *deformation function*, denoted by $d(u,v)$, which is defined at each point $(u,v) \in \mathcal{R}_a$. If the point $(u,v) \in \mathcal{R}_a$ is mapped to the point $\mathbf{U}(u,v) \in \mathcal{R}$ in the subject space, then $d(u,v)$ is defined by

$$d(u,v) = \det\left(\nabla \mathbf{U}(u,v)\right) , \tag{1}$$

where ∇ denotes the gradient of a vector function and $\det(\cdot)$ denotes the determinant of a matrix.

The deformation function $d(u,v)$ measures the stretching or shrinking which brings the template into registration with a subject. In particular, from the principles of continuum mechanics [7], if $d(u,v) > 1$ at a point (u,v), then an infinitesimal area around (u,v) expands as a result of the template warping. Similarly, if $d(u,v) < 1$, then a local shrinking around (u,v) occurs.

Consider now a region $\mathcal{P} \subseteq \mathcal{R}_a$ in the template, and its corresponding region $\mathbf{U}(\mathcal{P})$ in the subject image. The area, $\mathcal{A}$, of $\mathbf{U}(\mathcal{P})$ is given by [7] $\mathcal{A} = \iint_{\mathcal{P}} d(u,v) du dv$. Note that, since $dudv$ is the area of an infinitesimal region in the reference space, $d(u,v)$ is a local scaling factor representing area enlargement/shrinkage in the neighborhood of (u,v) with respect to the reference template. Therefore, $d(u,v)$ is a means for quantifying the size of a structure and its subregions by using the template as metric, and is used herein for inter-subject comparisons of shapes.

Population Analyses. Based on the spatial normalization procedure described above, qualitative and quantitative population comparisons can be readily made. Specifically, let $\mathbf{U}_1(u,v), \cdots, \mathbf{U}_N(u,v)$ be the maps from the template to each of the N subjects of a population. Let, also, $\mathbf{U}_p(u,v)$ be the average of the N functions $\mathbf{U}_1, \cdots, \mathbf{U}_N$. Then the average shape of a structure, denoted by $\mathcal{C}_p$, in that population is

$$\mathcal{C}_p = \bigcup_{(u,v)\in\mathcal{C}_a} \mathbf{U}_p(u,v)\,, \tag{2}$$

where $\mathcal{C}_a$ is the collection of points in the template belonging to the structure of interest.

Let $\mu_p(u,v)$ be the point-wise mean of the deformation function of the population. Then the difference between two populations, denoted with subscripts 1 and 2, can be measured as an *effect size* defined as [8]

$$e(u,v) = \frac{\mu_{p_1}(u,v) - \mu_{p_2}(u,v)}{\sigma(u,v)}\,, \tag{3}$$

where $\sigma(u,v)$ is the point-wise standard deviation of the two populations combined. We will use the effect size as a measure of the statistical significance of our experimental results in Section 3.

Impact of the Template. In this section we show that, in principle, any template can be used, without this affecting shape comparisons. Specifically, let $\mathbf{U}(u,v)$ be the map from Template 1 to a subject, and let the corresponding deformation function be $d(u,v)$. Let, also, $\mathbf{T}(u,v)$ be the map from Template 1 to Template 2, and let $s(u,v)$ be the corresponding deformation function. Finally, let $\mathbf{U}'(\mathbf{T}(u,v))$ be the map from Template 2 to the same subject, and $d'(\mathbf{T}(u,v))$ be the corresponding deformation function. Assuming that $\mathbf{U}(\cdot,\cdot)$, $\mathbf{T}(\cdot,\cdot)$, and $\mathbf{U}'(\cdot,\cdot)$ map homologous points to each other, then $\mathbf{U}(u,v) = \mathbf{U}'(\mathbf{T}(u,v))$. From the principles of mathematical analysis it then follows that $\nabla\mathbf{U} = \nabla\mathbf{U}'\nabla\mathbf{T}$, which together with (1) yields

$$d(u,v) = d'(\mathbf{T}(u,v))s(u,v)\,. \tag{4}$$

Now consider two populations having average deformation functions $\mu_{p_1}(u,v)$ and $\mu_{p_2}(u,v)$, respectively, with respect to Template 1. Let, also, $\sigma(u,v)$ be the point-wise standard deviation of the two populations combined. Finally, let the average deformation functions and the combined standard deviation of these populations with respect to Template 2 be $\mu'_{p_1}(\mathbf{T}(u,v))$, $\mu'_{p_2}(\mathbf{T}(u,v))$, and $\sigma'(\mathbf{T}(u,v))$. Using Equation (4) it can be readily shown that

$$\mu'_{p_1}(\mathbf{T}(u,v)) = \frac{1}{s(u,v)}\mu_{p_1}(u,v)\,, \tag{5}$$

with analogous expressions holding for $\mu'_{p_2}(u,v)$ and $\sigma'(u,v)$. Multiplying the numerator and denominator of (3) by $1/s(u,v)$, and using (5), we conclude that $e'(\mathbf{T}(u,v)) = e(u,v)$, which implies the invariance of the effect size from the selection of the template.

A key assumption in the development above is that the warping transformation maps homologous points to each other. In practice there are deviations from this assumption, due to registration errors. These errors, however, are fairly small, assuming that matching the boundaries of two homologous structures yields a good match of the interior of the structures, as well. Moreover, the inherent smoothness of the elastic transformation results in smoothly varying deformation functions, and therefore it reduces the effect of registration errors since neighboring points have very similar deformation functions. This is bolstered by experimental evidence provided in the following section.

3 Results

3.1 Synthetic Images

In order to demonstrate the relationship between the deformation function and shape differences of the underlying structures, in this section we apply our technique to the two synthetic images shown in Fig. 1. In this example we elastically warped the shape in

(a) (b)

Fig. 1. Two synthetic images. The shape in (b) was created from the shape in (a) by expanding the lower left part of the boundary downwards, by squashing the lower right part of the boundary upwards, and by pulling the upper middle part of the boundary downwards.

Fig. 1a to that in Fig. 1b. The resulting deformation function is displayed in Fig. 2a. In Fig. 2a the brighter the deformation function is the more expansion the shape in Fig. 1a underwent to match that in Fig. 1b. Darker regions imply contraction. In Fig. 2b and

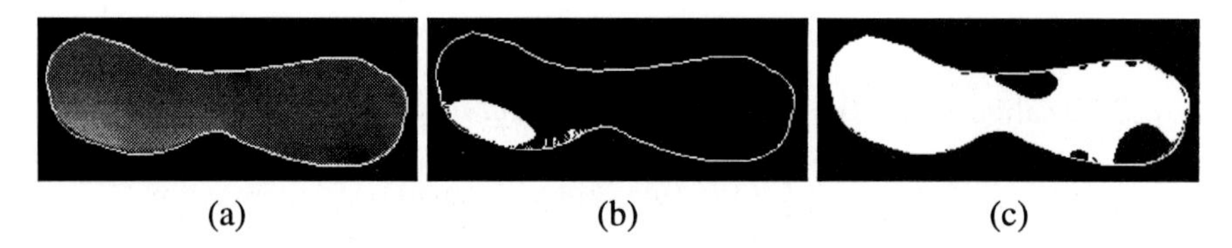

(a) (b) (c)

Fig. 2. (a) The deformation function displayed as a gray scale function superimposed on the boundary of Fig. 1a. The regions in which the deformation function is larger than 1.15 and 0.85 are shown in (b) and (c), respectively, and are consistent with the differences in the two shapes in Fig. 1.

Fig. 2c we show, in white, the regions in which the deformation function was larger than 1.15 and 0.85, respectively; these values correspond to a 15% expansion and shrinkage, respectively. Figs. 2b and Fig. 2c reflect the shape differences between the shapes in Fig. 1a and Fig. 1b. Specifically, the high value of the deformation function at the lower left part of the structure is in agreement with the fact that that part of the boundary was stretched downwards causing the expansion of the structure in that region. Similarly the contraction at the lower right and upper middle is reflected in Fig. 1c in the low values of the deformation function.

3.2 Analysis of the Corpus Callosum

The procedures described in Section 2 were used to compare the shape of the corpus callosum in 8 men and 8 women. The deformation functions were calculated for each of the 16 subjects and averaged. The mean deformation functions were then normalized by the total area, so that the integral of $d(u, v)$ was the same for the two groups. This is equivalent to scaling the original MR images so that the average total callosal area is the same for both groups. In Figs. 3a, 3b, and 3c we show in white the regions having effect size greater than 1, 0.75, and 0.5, respectively.

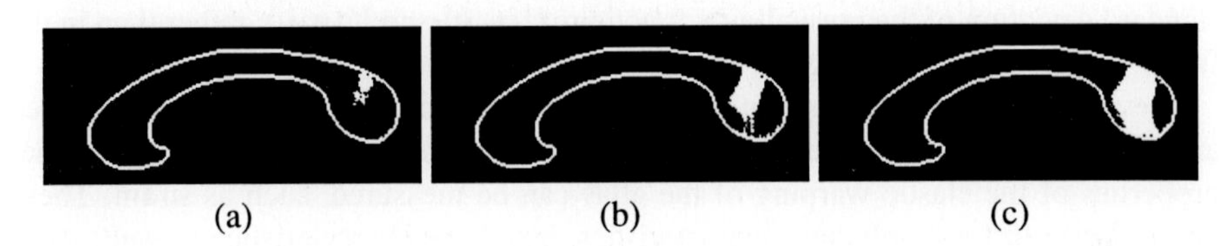

Fig. 3. The regions where the effect size of the difference between males and females was greater than (a) 1, (b) 0.75, and (c) 0.5.

Finally we determined the average corpus callosum for each of the two groups, using Equation (2). The result is shown in Fig. 4, and it qualitatively verifies our quantitative analysis. Specifically, the posterior part of the corpus callosum appears to be more bulbous in females, as opposed to the rest of the corpus callosum which appears to be larger in males.

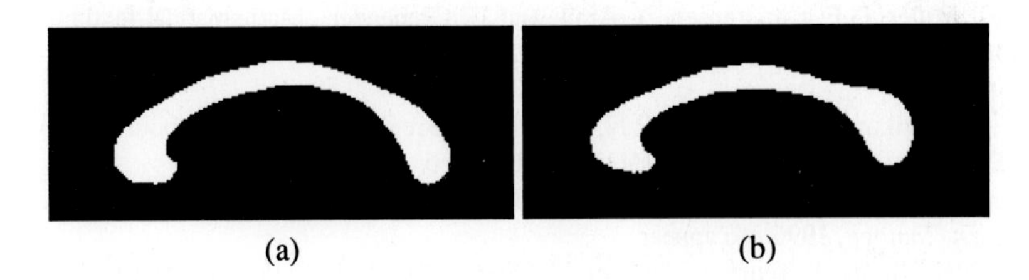

Fig. 4. The average corpus callosum of (a) the male group and (b) the female group.

4 Summary

In this work we developed an approach for quantifying the shape of two-dimensional brain structures, based on a spatial normalization procedure and on the measurement of a deformation function resulting from the registration of an atlas with subject images. This technique was tested by comparing a group of male with a group of female subjects.

Our methodology results in a continuous deformation function which reveals *local* differences between homologous structures in different subjects. Inter-subject comparisons of any sub-region of a structure can be readily obtained by integrating the deformation function in that sub-region. Using an atlas as template in our analysis provides a common reference system for population studies. Within this uniform framework inter-subject comparisons can be readily performed by comparing the deformation functions.

We have also demonstrated the utility of spatial normalization in defining average shapes for populations. The average structure of a group is found by averaging $U(u,v)$ and $V(u,v)$, as opposed to averaging subject images which results in fuzzy average shapes [9, 10]. The shape variability of a structure in a population is reflected in the standard deviation of the normalizing functions $U(u,v)$ and $V(u,v)$, rather than in the fuzziness of the average image.

Several methodological issues of our approach also need to be further investigated and refined in future work. In particular, in addition to the deformation function, other properties of the elastic warping of the atlas can be measured, such as strain. These properties reflect not only size characteristics, but shape characteristics as well, since they reflect angular warping in addition to growth or shrinkage. Moreover, the extension of this methodology to 3D will allow its use in describing 3D shapes. The 3D formulation of our spatial normalization method is reported in [11, 5].

References

1. F.L. Bookstein. Principal warps: Thin-plate splines and the decomposition of deformations. *IEEE Trans. on Pattern Analysis and Machine Intelligence*, 11(6):567–585, 1989.
2. D.L. Collins, C.J. Holmes, T.M. Peters, and A.C. Evans. Automatic 3-D model-based neuroanatomical segmentation. *Human Brain Mapping*, pages 190–208, 1995.
3. M.I. Miller, G.E. Christensen, Y. Amit, and U. Grenander. Mathematical textbook of deformable neuroanatomies. *Proc. of the National Academy of Sciences*, 90:11944–11948, 1993.
4. C. Davatzikos, J.L. Prince, and R.N. Bryan. Image registration based on boundary mapping. *IEEE Trans. on Med. Imaging*, 15(1):112–115, Feb. 1996.
5. C. Davatzikos. Spatial normalization of 3D images using deformable models. *J. Comp. Assist. Tomogr.*, 1996. To appear.
6. J. Talairach and P. Tournoux. *Co-planar Stereotaxic Atlas of the Human Brain*. Thieme, Stuttgart, 1988.
7. M.E. Gurtin. *An Introduction to Continuum Mechanics*. Orlando: Academic Press, 1981.
8. J. Cohen. *Statistical Power Analysis for the Behavioral Sciences*. Lawrence Erlbaum Associates, 1987.
9. A.C. Evans, W. Dai, L. Collins, P. Neeling, and S. Marett. Warping of a computerized 3-D atlas to match brain image volumes for quantitative neuroanatomical and functional analysis. *SPIE Proc., Image Processing*, 1445:236–246, 1991.
10. A.C. Evans, D.L. Collins, S.R. Mills, E.D. Brown, R.L. Kelly, and T.M. Peters. 3D statistical neuroanatomical models from 305 MRI volumes. *Proc. of the IEEE Nucl. Sc. Symposium and Med. Imaging Conf.*, 3:1813–1817, 1993.
11. C. Davatzikos. Nonlinear registration of brain images using deformable models. *Proc. of the Workshop on Math. Meth. in Biom. Image Anal.*, June 1996.

Brain: Description of Shape

Three Dimensional MR-Based Morphometric Comparison of Schizophrenic and Normal Cerebral Ventricles

David Dean[1], Peter Buckley[2], Fred Bookstein[3],
Janardhan Kamath[4], David Kwon[2], Lee Friedman[2], and Christine Lys[2]

[1] Department of Anatomy
[2] Department of Psychiatry
[4] Department of Biomedical Engineering
Case Western Reserve University
10900 Euclid Avenue
Cleveland, OH 44106

[3] Institute of Gerontology, University of Michigan
Ann Arbor, Michigan 48109

Abstract. Enlarged cerebral ventricles are consistently detected in patients suffering from schizophrenia. Several workers report more detailed association of this disease and particular brain areas. In order to assist these efforts at localization we define a tiled-out map of the ventricular surface, a template, based on specialized crest lines that we term: ridge curves. In this analysis we use this template to (1) average 3D MR-based ventricular surface images collected from 12 patients and 12 controls, and (2) locate the position of regularly occurring curvature maxima landmarks along the ridge curves. Statistical analysis of the locations of these curvature maxima landmarks indicate promising localization of a shape difference associated with schizophrenia in the frontal horn of the lateral cerebral ventricle.

1 Ridge Curve-based Multi-image Superimposition

The ventricular surface of the brain presents features of high curvature that can be represented as a wire frame "deformable template". This wire frame can be homologously superimposed on any "normal" and many abnormal cerebral ventricle surfaces. The space curve wires we use in our deformable template are a set of ridge curves (Kent *et al.* [16]; Bookstein and Cutting [5]). A ridge curve, on an idealized smooth surface, is a curve along which the perpendicular principal curvature at every point is a local maximum with respect to the lines of curvature on either side. If we represent ridge curves as space curves, extrema of the lower principal curvature along these edges will often identify recurrent points that can be treated as landmarks. We refer to these recurring points hereafter as curvature maxima landmarks.

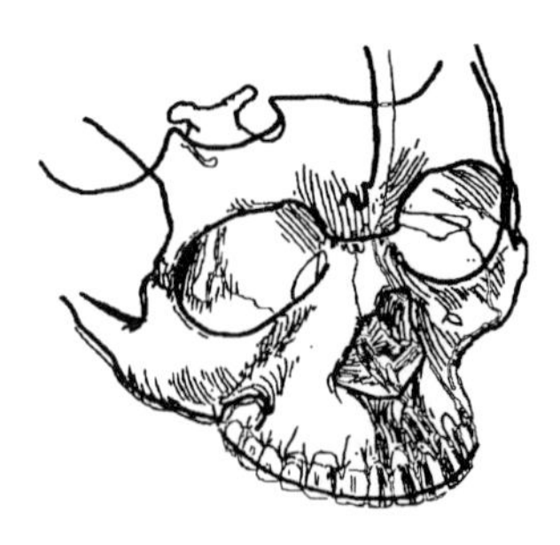

Figure 1. Ridge curves (heavy lines) can be reliably located on different forms, and often correspond to the same morphogenetic processes.

We use the local curvature maxima found along ridge curves as landmarks delimiting geodesic lines that tile out the continuous bony skull or in this case ventricular surface (Figure 2). Software previously developed for work with 3D CT images (Cutting *et al.* [10]) has been used to make a surface template of the cerebral ventricle system as imaged by 3D MR. As far as we know, these ridge curve templates are the first that facilitate the use of biologically significant principal curvature information for image co-registration (Thirion *et al.* [22]) and morphometrics.

It has been argued that ridge curves, for example, Figure 1, found repeatably on biological surfaces are homologous in the same sense as traditional anthropometric landmarks (Bookstein and Cutting [5]; Thirion [21]). Much current research in brain mapping includes image fusion. In many cases this involves either superimposition of images of the same person taken at different times, for example pre- and post-operatively, or matching structural and functional images of the same person, often with the goal of clinical diagnosis. However, our goal is statistical comparison of structural images of two different groups: schizophrenic patients and "normal" controls.

2 Enlarged Cerebral Ventricles in Schizophrenia

Few disorders combine as much distress, economic burden, and loss of human potential as that associated with schizophrenia. There are subtle morphological abnormalities in the brains of some (or most) patients with schizophrenia, as evaluated by magnetic resonance imaging (MRI) (Choh and McKenna [8]). Patients with schizophrenia have been observed to have a smaller cerebrum, less total gray matter, aberrant heterotopic (neurodevelopmental/dysplastic) white matter, and focal abnormalities of the frontal and temporal lobes (Gur and Pearlson [13]; Cannon and Marco [7]; Johnstone *et al.* [14]; Wible *et al.* [25]). However, the most robust finding has been that of enlargement of the lateral ventricular system in schizophrenia (for a review see: Van Horn and McManus [23]).

The search for a more precise diagnostic criterion of schizophrenia has been advanced by MR imaging studies. Several recent MR studies have reported on the increased size or volume of particular portions of the cerebral ventricular system in schizophrenic patients. No compartments appear to demonstrate active enlargement over the course of the illness; rather, all remain relatively static (Waddington and Buckley [24]). The underlying assumption of this work is that regional abnormalities in ventricular volume may reflect regional abnormalities in brain parenchyma. This reasonable hypothesis has stimulated much research and theory. Recent studies indicate specific abnormalities of the frontal (Andreasen *et al.* [1]; Shiraishi *et al.* [20]; Serban *et al.* [19]), and sometimes temporal (DeGreef *et al.* [11]), horns of the lateral ventricles in schizophrenia. Collectively, these findings are consistent with the notion that more sensitive measures of ventricular shape and size may reveal stronger and more consistent correlations with behavior and symptoms.

3 Mapping the Ventricular Surface

Three dimensional isosurface images of ventricles of 12 normals and 12 schizophrenics were extracted from coronal MR volume acquisitions comprising 125 256 x 256 slices. Voxels were approximately 1.0 x 1.0 x 1.5 mm^3. The ventricles were segmented slice by slice using the various arithmetic and morphological tools, for example, geodesic dilation, in ALICE (Hayden Image Processing Systems, Boulder, CO). The process called for substantial user involvement, and at times, required hand editing of the slice in order to prevent the interruption of known connections, for example, the cerebral aqueduct. The surfaces of the 3D images were then semi-automatically digitized to impose a ridge curve-based deformable wireframe.

Our ridge curve wire-frame template for the ventricular surface was created with the help of the "Wrapper", a curvature mapping program (Guéziec and Dean [12]). The user develops hypotheses as to the location of ridge curves that are homologous from specimen to specimen, as opposed to a variant of one individual, by surveying multiple color-coded isosurface images produced by the Wrapper. The surface curvatures are indicated by this color-coding: low curvature areas are purple-to-blue, areas of middling curvature are green-to-yellow, and the highest curvature areas are displayed in red. The color-coded curvature information also facilitates manual tracing of the initial ridge curve wire frame template on the image of a single segmented ventricular surface. (Standard lighting and shading are not adequate as we find the user's eye is always drawn to the edge of the image.) The assembled landmark and ridge curve information is then used to create a new ridge curve deformable template. Eventually an average surface is substituted as a template.

Our approach to the extraction of ridge curves is template-driven (Cutting *et al.* [9, 10]). The processing of an image, skull or brain, begins with the segmentation of the surface of interest from CT or MR tomographic data and its subsequent representation as a topologically connected three dimensional object (Kalvin *et al.* [15]). An operator then locates landmarks on the imaged surface manually. These landmarks drive a thin-plate spline (Bookstein [4]) that warps a previously computed average ridge curve wire frame deformable template so that its landmarks match those of the specimen form. The warping operation carries the ridge curves, and, for that matter, the surface patches, of an "average" wire frame into the coordinate system of the specimen. At the same time that the ridge curves of an average are warped onto the specimen, so are the corresponding normal cross-sections of the averaged surface. Because these cross-sections are in only two dimensions rather than three, fitting can be carried out in closed form by the complex-regression version of Procrustes analysis (Bookstein [4]; Rohlf [17]; Rohlf and Slice [18]). This approach has reduced the number of ambiguities in the ridge curve-finding process (for more details see Cutting *et al.* [10] and Thirion *et al.* [22]). Finally, cubic spline segments are fitted to each ridge segment as a function of linearly estimated arc length.

To complete the wire mesh across which surface patches are lofted, we find geodesic lines connecting landmarks of different ridges across a specimen surface. This scheme of landmarks on adjacent ridges, connected by geodesics, permits the decomposition of the entire ventricular surface into a number of topologically connected surface patches (Figure 2). Surface points in the middle of a surface patch are characterized by two parameters with respect to their border curves. Regardless of whether the patch has 3 or 4 "sides", we use the (u,v) coordinate scheme of rectangular patches; that is, all patches are considered to have four boundary curves, although one may be degenerate (of length zero).

4 Averaging Ridge Curve-Mapped Surfaces

Our averaging algorithm proceeds stepwise from landmarks to curves and then to surfaces (Cutting *et al.* [9]). Our analysis begins with thin plate spline warping of each case's curvature maxima landmarks onto the isotropic Procrustes average of the entire landmark sample. (Bookstein's [2] use of Procrustes analysis superimposes pairs of landmark configurations by minimizing the sum of squared distances over the similarity group acting on one of the forms. The Procrustes average shape is the [unique] shape of minimum such summed squared distances to all the shapes of a sample.) Warping the fully specified average of landmark configurations (Bookstein [4]) drives a partially specified average of curves, here, that assigns a plausible shared coordinate in place of the information along the arc-length that is, properly speaking, absent in each single specimen of the series. The landmark average, in other words, supplies an alignment rule for the correspondence of the coordinate that is deficient along the curve. Similarly, a preliminary average driven by ridge curves and geodesics will assign plausible values to the two "missing" coordinates within surface patches, so as to permit a sensibly aligned averaging of the one coordinate (normal to the mean surface) that we've really got.

Corresponding to this point of view, then, our curve-averaging algorithm proceeds in two steps. First, a rough average based on an estimate of arc-length specimen by specimen is used to construct a consensual coordinate system *for the vicinity of the average curve in the standard Procrustes space*. In the second step, a refined average is computed using differential features of that new localized coordinate system. To the tentative sample mean corresponds a series of *quasinormal planes* bisecting the intersection between successive segments at each of the averaged aliquots. These planes are considered to represent additional information, a "local coordinate system," for the second step of the averaging. Typically each plane intersects each of the splined polygons of the (unwarped) sample just once. Within each of these quasinormal planes, the points of intersection are averaged again. These centroids, all in the space of the averaged landmark configuration, make up the *averaged space curves* (ridge or geodesic) we seek. Note that this averaging process does not use differential properties of the sample curves—for instance, they do not lie at an "averaged" tangent angle to their chord, nor do they represent an "averaged" curvature or torsion. For a preliminary approach to an explicit splining operation preserving this differential structure see Bookstein and Green [6].

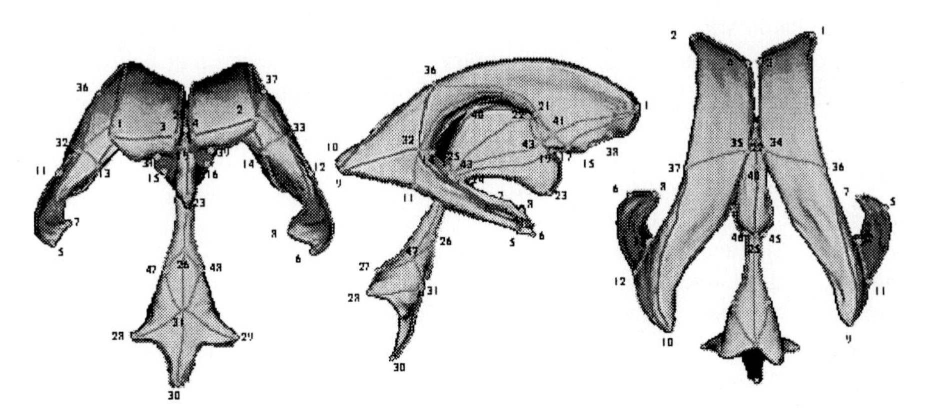

Figure 2. A ridge curve deformable template was used to produce this average schizophrenic ventricle image (N=12). Dark lines are ridge curves; lighter lines are geodesics. From left to right views are: anterior, lateral, and posterodorsal. Numbered landmarks at ridge curve and geodesic intersections referred to in text: 1 r_AntLatFrontPole, 2 l_AntLatFrontPole, 3 r_AntMedFrontPole, 4 AntMedFrontPole, 5 r_LatTempPole, 6 l_LatTempPole, 7 r_MedTempPole, 8 l_MedTempPole, 9 r_OccPole, 10 l_OccPole, 11 r_InfExtNotch, 12 l_InfExtNotch, 13 r_MidInfIntCrest, 14 l_MidInfIntCrest, 15 r_AntInfMedFrontPole, 16 l_AntInfMedFrontPole, 17 r_SupAntForMonroe, 18 l_SupAntForMonroe, 19 InfAntForMonroe, 23 Ant3rdVent, 24 Antinf3rdVent, 25 PosSup3rdVent, 26 AntInf4thVent, 27 PosSup4thVent, 28 r_ForLuschka, 29 l_ForLuschka, 30 ForMagendie, 31 Ant4thVent, 32 r_MedNotchLatCrestIntsect, 33 l_MedNotchLatCrestIntsect, 34 r_MidSupIntCrest, 35 r_MidSupIntCrest, 36 r_LatTorsionMax, 37 l_LatTorsionMax, 38 r_AntmidVentPole, 39 l_AntmidVentPole, 40 Post3rdVent, 41 r_SupMidLatForMonroe, 42 l_SupMidLatForMonroe, 43 r_SupMidLatForMonroe, 44 l_InfMidLatForMonroe, 45 r_Lat3rdVent, 46 l_Lat3rdVent, 47 r_Lat4thVent, 48 l_Lat4thVent.

Finally, the surface patches are parameterized as deviations along normal lines to Coons patches lofted over the average boundary curves. These deviations are averaged to produce a sheet of average deviations along the Coons normal lines for each patch. In this way an average surface can be generated from a knowledge of average border curve segments and average deviations in the interior of the patch. This part of our algorithm, like the others, is described elsewhere in greater detail (Cutting *et al.* [9,10]). This completes the description of how Figure 2 was derived.

5 Results

While both the ventricular surface samples, normal and schizophrenic, were averaged, unlike Cutting *et al.* [10] only the locations of the curvature maxima landmarks were used in this analysis. The goal of this analysis was to demonstrate whether this shape variation showed potentially meaningful differences of mean shape in any regions, large or small, between the two groups. Of the 48 curvature maxima landmarks identified by our current ventricular surface template, 38 landmarks were submitted to a preliminary Procrustes analysis (after Cutting *et al.* [10]) using commercially available statistical software (Splus, StatSoft, Inc., Seattle, WA). There are three basic steps in this procedure. First, a Procrustes average shape is computed; then, each specimen is Procrustes-fitted to that Procrustes average, to produce residual "Procrustes fit coordinates"; finally, these residuals are examined for variance and group differences.

Procrustes analysis includes a preliminary step in which the Centroid Size (root sum of squares around each specimen's centroid) is computed and divided out specimen by specimen. Even though Centroid Size is not itself a volume measure, this 4% difference (which is significant at p~.033 by t-test) might be expected to be associated with about a 12% difference in volume, a finding consistent with many previous studies of ventricular volumes.

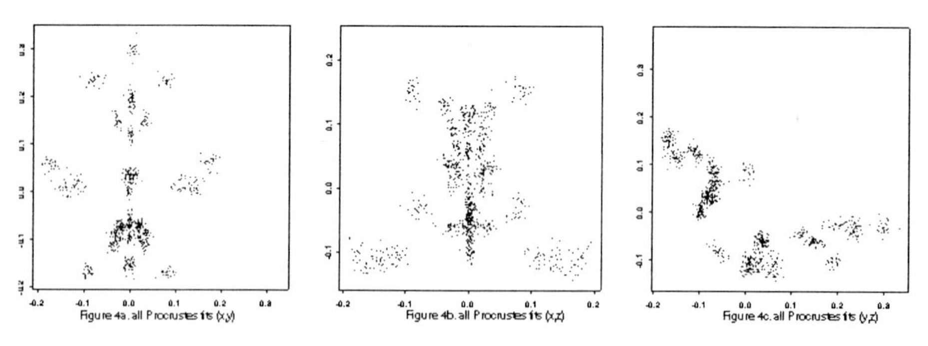

Figure 3. Scatters of all Procrustes fits, 38 landmarks, in xy, xz, and yz projections.

After the Procrustes fitting step (see Figure 3), we survey the variability at every landmark around its Procrustes mean position. It is convenient to summarize these over the whole sphere of directions at every mean landmark position by taking the average of the diagonal entries of the variance-covariance matrix of these residuals (that is, one-third of its trace). Ten of the landmarks have standard deviations that are distinctly larger than those of the other 38. Ten of the landmarks, those numbered 5-10 and 34-37 in figure 2, are distinctly more variable than the others and were dropped from the remainder of the analysis. The rest of this discussion is concerned only with the remaining 38 points. Most of the landmarks have become more reliable once the 10 outliers were discarded.

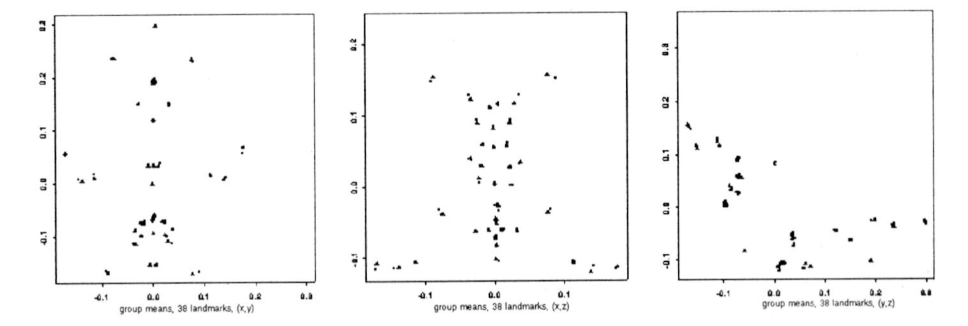

Figure 4. Mean Procrustes fit locations of 38 landmarks for two groups of twelve in, from left to right, xy, xz, yz projections. Note separation between normal and schizophrenic landmark 1 means at lowermost right in x,y and uppermost right in x,z.

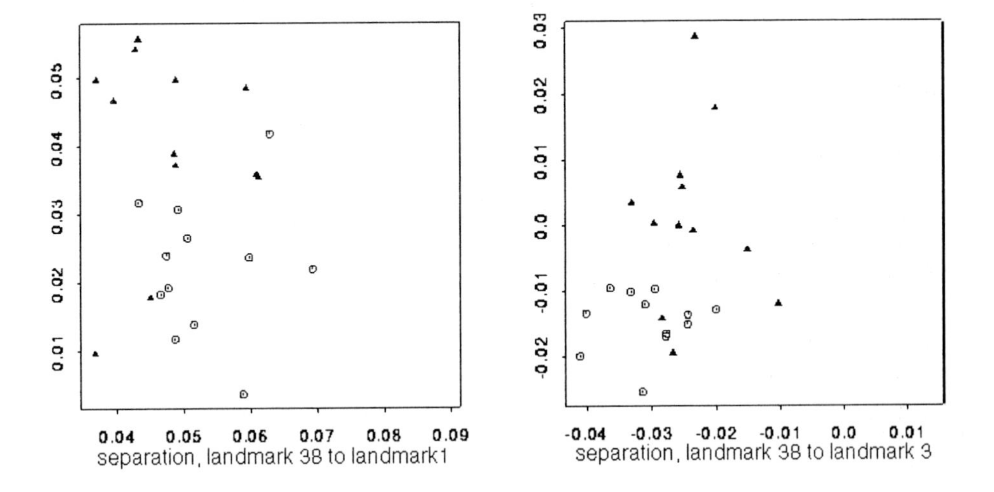

Figure 5. Local shape difference, landmarks 1, 3, and 38, in xz projection. The difference on the right is 2.5 times the standard deviation of the normal subgroup. Circles, ventricles of normals; triangles, ventricles of schizophrenics.

We turn now to group mean differences in these point configurations. Certainly with only 12 cases in each group and a minimum of three dimensions per comparison, we are not likely to have much power for any of these pilot speculations. It is useful, nevertheless, to imagine how such an analysis might go when sample sizes are larger. Figure 3 shows a complete set of two-dimensional projections of the fitted values, which on the whole are gratifyingly precise and reliable over this pilot set of 24 ventricular forms. In the group means, Figure 4, four landmarks show group mean shifts of at least one standard deviation when normalized in this way: landmarks 38 (1.82 s.d.), 1 (1.59 s.d.), 19 (1.34 s.d.), and 11 (1.11 s.d.). The first two of these are adjacent, suggesting a possible regionalization of the change. One can also, separately, compute individual Hotelling's T-squared tests of the shifts of these landmarks between groups using the appropriate within-group covariance structure. The four most significant landmarks arising from this procedure, with the associated tail-probabilities of the T^2, are the same four landmarks: 38 ($p<.007$), 11 (.026), 19 (.027), and 1 (.031). Deviations over 1.0 standard deviations, if maintained in larger samples, are strikingly commensurate with published findings of schizophrenia effects on aspects of the corpus callosum where it abuts this ventricle (Bookstein [2,3]).

Figure 5 shows an enlargement of this region of change in the vicinity of landmarks 1 and 38 with respect to a stable location, landmark 3, nearby. The T^2 using this set of three coordinates is significant at $p\sim.008$. More to the point, the shift here is more than 2.5 times the within-normal standard *deviation*, the largest effect size we have yet seen for a morphometric variable in a neuroanatomical study of schizophrenia. By comparison, a typical "standard" neurometric measure in the vicinity, the (relative) distance between landmarks 1 and 2, differs between groups by only about 135% of its within-group standard deviation, an effect size barely half as large.

6 Conclusions

We can summarize the import of this study as follows. Out of a tentative list of 48 landmarks on the ventricular surface, we are able to find 38 that behave like good point landmarks for typical modern morphometric maneuvers. There were no evident outliers in the pilot sample, and 38 of the 48 landmarks showed standard deviations of Procrustes shape coordinates for these points comparable to the standard deviations of familiar calvarial landmarks (Cutting *et al.* [10]). Selected shape measures, for example, the position of the right anterior lateral frontal pole, show group differences of greater than 1.0 within-group standard deviations. These effect sizes, should they persevere in larger samples, are of magnitude comparable to published findings of group differences in the shape of the schizophrenic brain that are significant when seen in two dimensional data (Bookstein [2]). Such effect sizes would indeed be significant at reasonable power by appropriate fully multivariate omnibus test in the sample sizes to which we now feel this study should be extended. If there is truly such a concentration of clinically useful information in such a small subsampling of the surface as delineated by ridge curves, then the even richer representation that underlies Figure 2 will indeed permit a tremendous increase in the power of three dimensional image co-registration to generate and test important hypotheses about structure and function.

Acknowledgment

This project was supported by NARSAD and Sihler Foundation grants to PB. We thank Krishna Subramanyan, Deljou Khorramabadi, and Betsy Haddad, for technical assistance with the software used for this study. FB's methods of statistical analysis were developed under grant DA09009, part of the NIH's Human Brain Project.

References

1. Andreasen, N., Ehrhardt, J., Swayze, V., Alliger, R., Yuh, T., Cohen, G., Ziebell, S.: Magnetic resonance imaging of the brain in schizophrenia - The pathophysiologic significance of structural abnormalities. Arch. Gen. Psych. **47** (1990) 35-44
2. Bookstein, F.: Biometrics, biomathematics, and the morphometric synthesis. Bull. Math. Biol. **58** (1996) 313-365
3. Bookstein, F.: How to produce a landmark point: the statistical geometry of incompletely registered images. SPIE **2573** (1995) 266-277
4. Bookstein, F.: Morphometric Tools for Landmark Data: Geometry and Biology. Cambridge: Cambridge University Press (1991)
5. Bookstein, F., Cutting, C.: A proposal for the apprehension of curving craniofacial form in three dimensions. In (K. Vig, A. Burdi, eds.): Craniofacial Morphogenesis and Dysmorphogenesis. Ann Arbor, MI: Center for Human Growth and Development. (1988) 127-140
6. Bookstein, F., Green, W.: A feature space for edges in images with landmarks. J. Math. Img. & Vis. **3** (1993) 231-261
7. Cannon, T., Marco, E.: Structural brain abnormalities as indicators of vulnerability to schizophrenia. Schiz. Bull. **20** (1994) 89-102
8. Choh, B., McKenna, P.: Schizophrenia as a brain disease. Br. J. Psych. **167** (1995) 183-94
9. Cutting, C., Bookstein, F., Haddad, B., Dean, D., Kim, D.: A spline-based approach for averaging three-dimensional curves and surfaces. SPIE **2035** (1993) 29-44
10. Cutting, C., Dean, D., Bookstein, F., Haddad, B., Khorramabadi, D., Zonneveld, F.: A three dimensional smooth surface analysis of untreated Crouzon's disease in the adult. J. Craniofac. Surg. **6** (1995) 444-453
11. DeGreef, G., Ashtari, M., Bogerts, B., Bilder, R., Jodi, D., Alvir, J., Lieberman, J.: Volumes of ventricular system subdivisions measured from magnetic resonance imaging in first-episode schizophrenic patients. Arch. Gen. Psych. **49** (1992) 531-537
12. Guéziec, A., Dean, D.: The Wrapper: A surface optimization algorithm that preserves highly curved areas. SPIE **2359** (1994) 631-642
13. Gur, R., Pearlson, G.: Neuroimaging in schizophrenia research [Review]. Schiz. Bull. **19** (1993) 337-353
14. Johnstone, E., Bruton, C., Crow, T., Frith, C., Owens, D.: Clinical correlates of postmortem brain changes in schizophrenia: decreased brain weight and length correlate with indices of early impairment. J. Neurol., Neurosurg., and Psych. **57** (1994) 474-479
15. Kalvin, A., Cutting, C., Haddad, B., Noz, M.: Constructing topologically connected surfaces for the comprehensive analysis of 3D medical structures. SPIE **1445** (1991) 247-258
16. Kent, J., Mardia, K., West, J.: (in press) Ridge curves and shape analysis. Proc. 1996 Brit. Mach. Vis. Conf.

17. Rohlf, F.: Rotational fit (Procrustes) methods. In (F. Rohlf, F. Bookstein, eds.) Proceedings of the Michigan Morphometrics Workshop. Ann Arbor, MI: Spec. Pub. 2 Univ. MI Mus. Zool. (1989) 227-236
18. Rohlf, F., Slice, D.: Extensions of the Procrustes method for optimal superimposition of landmarks. Syst. Zool. **39** (1990) 40-59
19. Serban, G., George, A., Siegel, S., DeLeon, M., Gaffney, M.:Computed tomography scans and negative symptoms in schizophrenia: chronic schizophrenics with negative symptoms and non-enlarged lateral ventricles. Acta Psych. Scand. **81** (1990) 441-447
20. Shiraishi, H., Koizumi, J., Ofuku, K., Suzuki, T., Saito, K., Hori, T., Ikuta, T., Kase, K., Hori, M.: Enlargement of the anterior horn of the lateral ventricle in schizophrenic patients: chronological and morphometric studies. Jap. J. Psych. & Neurol. **44** (1990) 693-702
21. Thirion, J.: The extremal mesh and the understanding of 3D surfaces. Proceedings of the IEEE Workshop on Biomedical Image Analysis. Los Alamitos, CA: Computer Society Press (1994) 3-12
22. Thirion, J., Subsol, G., Dean, D.: Proposal for the Cross Validation of Three Inter-Patients Matching Methods. (this volume)
23. Van Horn, J., McManus, I.: Ventricular enlargement in schizophrenia - A meta-analysis of studies of the ventricular-brain ratio (VBR). Brit. J. Psych. **160** (1992) 687-697
24. Waddington, J., Buckley, P.: The Neurodevelopmental Basis of Schizophrenia. Austin, Texas: Landes Publications (in press)
25. Wible, C., Shenton, M., Hokama, H., Kikinis, R., Jolesz, F., Metcalf, D., McCarley, R.: Prefrontal cortex and schizophrenia. Arch. Gen. Psych. **52** (1995) 279-288

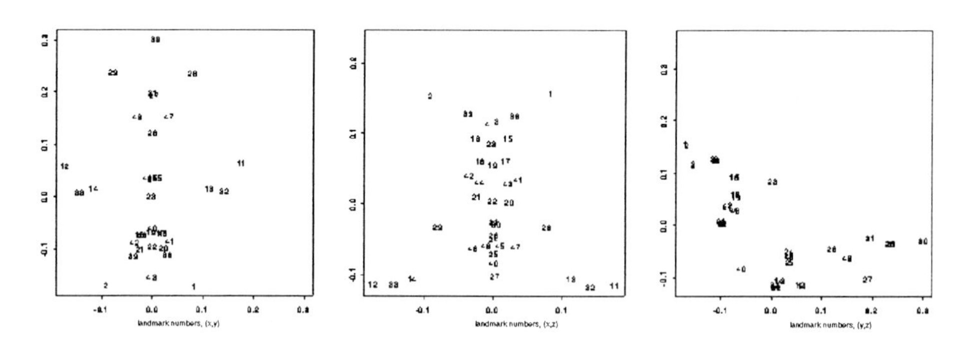

Extra figure. Key to landmark numbers in the panels of Figures 3 and 4.

Application of an Automatically Built 3D Morphometric Brain Atlas: Study of Cerebral Ventricle Shape

Gérard Subsol, Jean-Philippe Thirion and Nicholas Ayache

Institut National de Recherche en Informatique et en Automatique
Projet Epidaure
2 004, route des Lucioles, BP 93
06 902 Sophia Antipolis Cedex, France
E-mail: Gerard.Subsol@epidaure.inria.fr

Abstract. In this paper we present new results on the automatic building of a 3D morphometric brain atlas from volumetric MRI images and its application to the study of the shape of cerebral structures. In particular, we show how it is possible to define "abnormal" deformations of the cerebral ventricles with a small set of parameters.

1 Introduction

In [STA95], we presented a general method for automatically building tridimensional morphometric anatomical atlases from volumetric medical images (CT-Scan, MRI). This method allows automatic computation of a model by synthesizing a database of a given anatomical structure extracted from medical images of different patients. It operates by finding the features that are common to all the patients and computing their average position and a small set of statistical parameters describing their deformations.

What is particularly interesting is that no sophisticated a priori knowledge such as a sulci description or a cortical topography modelization [MRB+95] does need to be introduced. Moreover, no manual process is involved. We hope to obtain a model which takes into account the huge resolution of volumetric medical images (a MRI or CT-Scan image of the head composed of 200 slices of 256 by 256 pixels leads to a resolution of about 1 to 2 millimeters cube). Moreover, as we obtain morphometric parameters, we can envisage their usage in complex medical applications such as assisted diagnosis and therapy.

In previous articles, we have only presented work on a skull atlas and how it could be useful for studying a craniofacial deformity. In this paper, we will focus on a brain atlas. This is a very challenging problem because the cerebral shape is very variable but medical applications are numerous: anatomical or pathological study [RCJSG93] [TTP96], registration with functional data, assisted diagnosis [MPK94] or therapy planning.

In the first part of this paper, we briefly review each step of the atlas building method and we emphasize some specifics about the automatic construction of an atlas of the brain. In the second part, we illustrate the relevance of this atlas by studying data from an individual patient. In particular, we show how it is possible to define "abnormal" deformations of the cerebral ventricles with a small set of parameters.

2 Construction of the brain atlas

The atlas construction method is composed of five steps (see Figure 1). We built the brain atlas from a database of 10 MRI brain images[1].

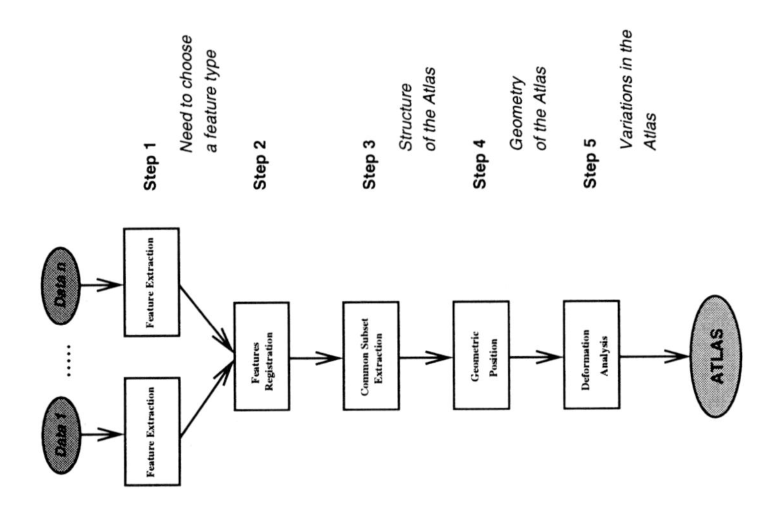

Figure 1. The proposed general scheme for automatically building anatomical atlases.

2.1 Step 1: feature extraction

A brief description. This step consists of automatically extracting 3D image features which will constitute the atlas. It is necessary to find a feature type which is both unambiguous definable (in order to be automatically computed) and anatomical relevant (in order to constitute a meaningful atlas). We choose "crest lines" [MBF92] whose points P are defined by:

$$\overrightarrow{\nabla} k_1 . \overrightarrow{t_1} = 0$$

where k_1 is the principal curvature which has the largest absolute magnitude and $\overrightarrow{t_1}$ the associated principal direction.

Using the "marching lines" algorithm [TG93], one can automatically extract the crest lines lying on a given isosurface directly from the volumetric image.

[1] These MRI data (referenced case1 ... case10) of resolution 123 slices of 256×256 pixels are courtesy of Dr. Ron Kikinis, Brigham & Women's Hospital and Harvard Medical School of Boston (United-States).

A simplified multi-scale scheme. As the mathematical definition of crest lines is based on a continuous description of a surface, it is necessary to convolve the discrete values of the image with a Gaussian function to obtain a continuous representation. The wider is the Gaussian parameter σ, the more regular the intensity is in the continuous representation. This smoothing leads to a modification of the position of the isosurfaces which support crest lines.

So, to obtain regular lines in a significant position, we propose to use a simplified multi-scale scheme:

- Extract a first set $\mathcal{S}_1$ of crest lines with a "low" Gaussian parameter.
- Extract a second set $\mathcal{S}_2$ of crest lines with a "high" Gaussian parameter.
- Register $\mathcal{S}_1$ and $\mathcal{S}_2$ by the line registration algorithm described in the next step "feature registration" and keep the lines of $\mathcal{S}_1$ which are matched with a line of $\mathcal{S}_2$.

This algorithm is quite crude because it uses only two values of the scale parameter and we are hopeful that more sophisticated multiscale methods will considerably improve the quality of crest lines.

In Figure 2, we show the results of the extraction of crest lines on the surface of the lateral cerebral ventricles. With a σ value set to 1.0, we obtain many lines (415) which appear quite noisy; whereas with $\sigma = 3.0$, we extract only 24 longer and smoother lines. However, in the second case, the crest lines do not follow the surface of the ventricles and we lose, in particular, a lot of information along the "horns" of the ventricles. After using the simplified multi-scale scheme, we obtain only 50 lines which seem significant and are precisely localized.

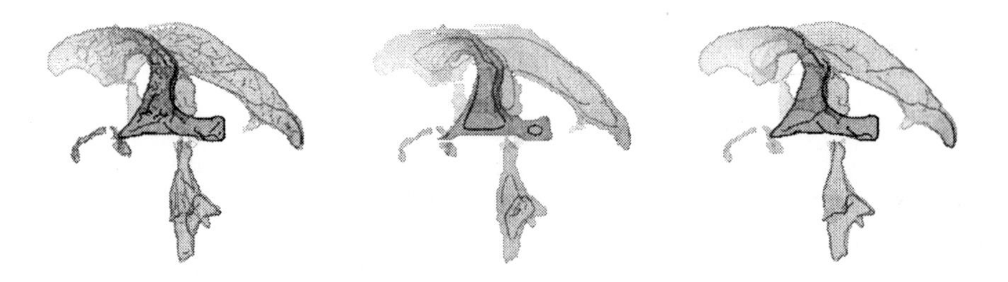

Figure 2. *Crest lines extracted on the surface of the lateral cerebral ventricles. Left, with a low Gaussian parameter, $\sigma = 1.0$. Middle, with a high one, $\sigma = 3.0$. Right, after using a simplified multi-scale scheme.*

We apply this scheme to extract crest lines on the brain (see Figure 3). Notice that we also obtain lines inside along the cerebral ventricles.

An anatomical justification. By their mathematical definition, crest lines are localized along the salient parts of the surface. In particular, on the brain, they follow the convolutions and seem to match the anatomically defined patterns of gyri and sulci such as those referenced in the anatomical atlas [OKA90].

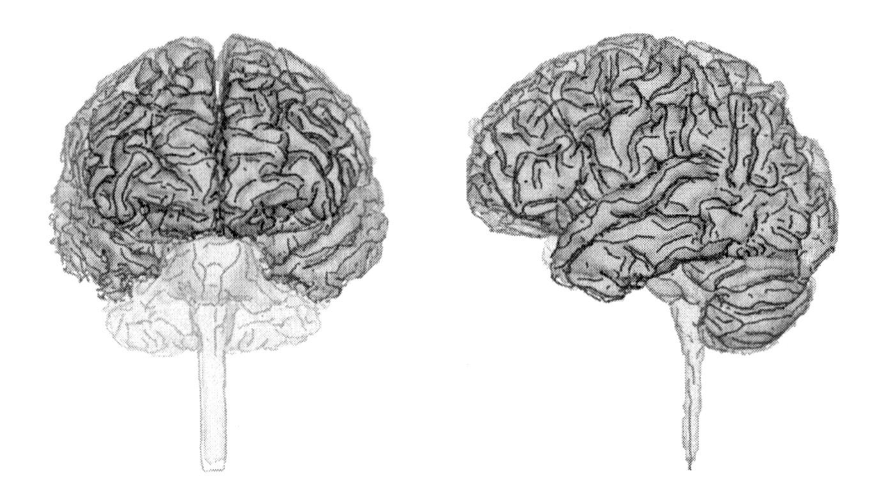

Figure 3. *Crest lines of the brain case1.*

2.2 Step 2: feature registration

A brief description. Given two sets of lines extracted from two different patients, we want a twofold result: line to line correspondences for the step 3 "common feature subset extraction" and point to point correspondences for the step 4 "feature average" and 5 "feature deformation analysis". For that purpose, we have adapted the "Iterative Closest Point" algorithm. We extend the original method by generalizing it to non-rigid deformations (affine and spline transformations) and applying it to lines (a line defines topological constraints over its points). For more details, the reader may refer to [STA95].

Separating sulci and gyri. For the registration process, we separate crest lines into two classes: those with positive "maximal" curvature and negative. Geometrically, this distinguishes crest lines which lie on a convex (the crown of the gyri) and concave surface (the fundi of the sulci).

In the registration algorithm, we separately find correspondences between gyral points and between sulcal points. Then, we use these two lists of matched points to compute a transformation which bring the two sets closer.

2.3 Step 3: common feature subset extraction

A brief description. From the registered lines, we are able to build a graph where each node represents a line of a set and each oriented link represents the relation "is registered with". Then, we can extract the bijective connected parts containing at least one line of each set: they represent the subsets of features which are common to all the data and which will constitute the structure of the atlas.

Some results. From the 10 brains, we found 82 subsets of lines which are present in each brain.

If the lines of one set have been labelled, it becomes possible to automatically label the subset to which they belong and then the corresponding common lines of the other nine sets.

We notice that even if crest lines of the brain are extremely variable, making the registration and graph construction processes difficult, the ventricles and medulla lines are correctly identified [Sub95].

2.4 Step 4: feature average

A brief description. It is necessary to find the mean positions of the features constituting the atlas (i.e., to average the sets of 3D lines defining each common subset).

First, we take a set of crest lines (for example, case1) as a reference set $\mathcal{R}$ and transform all the other sets into the reference frame using the rigid and isotropic scaling transformations computed from point to point correspondences found in step 2. Now, the remaining differences between lines represent meaningful morphological differences.

For each line $\mathcal{L}$ of $\mathcal{R}$, we find the deformations between $\mathcal{L}$ and the corresponding lines of the other sets. Then, we decompose them into a basis of fundamental deformations called modes by "modal analysis" [MPK94]. Once in this basis, the deformations are averaged and the "mean" deformation applied to $\mathcal{L}$ becomes the "mean" line. The set of these mean lines constitutes the atlas (Figure 4).

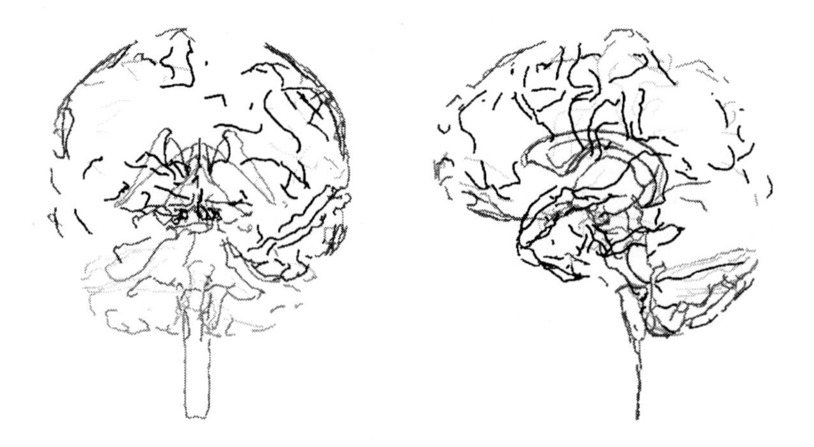

Figure 4. *The "mean" lines constituting the brain atlas. Notice the lines along the lateral ventricles.*

By using point correspondences between $\mathcal{R}$ and the atlas, we can find a volumetric spline transformation [DSTA95] and apply it to the surface of $\mathcal{R}$ in order to obtain a surface representation of the atlas.

2.5 Step 5: feature deformation analysis

A brief description. Let us assume that the the items of the database are "normal". In this step, we want to define the range of "normal" deformations which can be considered as the deformations between the atlas and the items.

We register the crest lines of the atlas with those of the database and we compute the modal decomposition of all these deformations. Then, for each line and for each mode i, we compute its average amplitude d_i and standard deviation σ_i.

If we register the crest lines of the atlas with the crest lines of a set $\mathcal{S}$ extracted from a patient image, we obtain also, for each line, some deformations decomposed in modes with the associated amplitudes $d^{Atlas,\mathcal{S}}[i]$. We can then compare the modal amplitude $d^{Atlas,\mathcal{S}}[i]$ with the "normal" modal amplitude by using an *amplitude distance* defined by (computed for the x, y and z axis);

$$dist(D^{Atlas,\mathcal{S}}, i) = \sqrt{\frac{\left(d^{Atlas,\mathcal{S}}[i] - d_i\right)^2}{\sigma_i^2}}$$

By using a χ^2 test, the larger values of $dist(D^{Atlas,\mathcal{S}}, i)$ characterize the modes which are not in the average range and enable us to find the "abnormal" deformations.

3 Application to the study of cerebral ventricles shape

Introduction. In this section, we use the brain atlas to study an individual brain[2] $\mathcal{CE}$ in the following and, in particular, its cerebral lateral ventricles. Their deformations can be characteristic of Alzheimer's disease or hydrocephalus [MPK94].

First, we perform an automatic segmentation by mathematical morphology and thresholding tools to obtain the cortical surface presented in Figure 5, left.

Automatic labelling. We register the crest lines of the brain atlas with those of $\mathcal{CE}$. Then, we propagate the atlas labels to $\mathcal{CE}$ in order to identify some structures, in particular the lateral cerebral ventricles. In spite of the small number of atlas lines and the important geometrical differences between the atlas and the data, automatic labelling of the medulla, lateral ventricles and the first left lateral convolution appears visually correct in Figure 6.

Moreover, ventricles crest lines underline clearly the asymmetry of the ventricles with the left being much bigger than the right.

[2] MRI data of the head of resolution 254 slices of 256 $\times$ 256 pixels are courtesy of Dr. Neil Roberts, Magnetic Resonance Research Centre, University of Liverpool (Great-Britain).

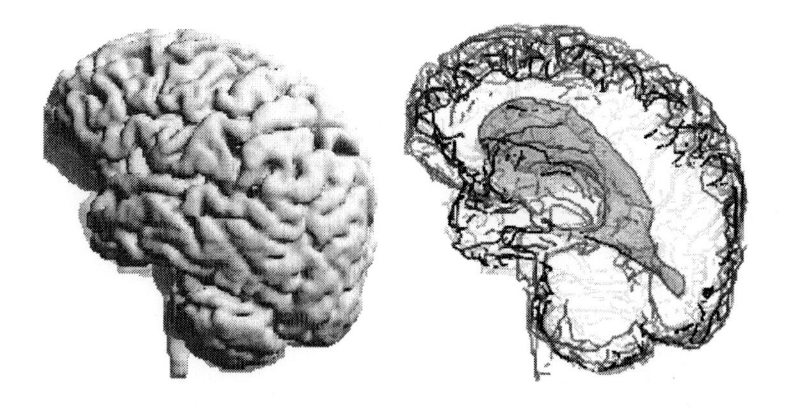

Figure 5. *Left: automatic segmentation of the cortical surface from a MRI of the head. Right: crest lines extracted on the cortical surface and the ventricles.*

Figure 6. *Some lines of $\mathcal{CE}$ have been correctly labelled (medulla, left and right lateral ventricles, first left lateral convolution) thanks to the registration with the atlas.*

Qualitative study. By extracting the points of brain surface which are close to the ventricule crest lines, we can obtain the ventricle surfaces of the atlas and $\mathcal{CE}$. Moreover, from the matching of points found during line registration, we can apply a rigid and scaling transformation to $\mathcal{CE}$ in order to position it in the atlas frame.

In Figure 7, we can notice that not only the patient ventricles are asymmetric but they are also larger that the atlas ones.

Morphometric analysis. We can analyze the deformations of the longest atlas ventricular line. We decompose its deformation towards the correspondent patient ventricular lines and compute the amplitude distances for the first 5 modes.

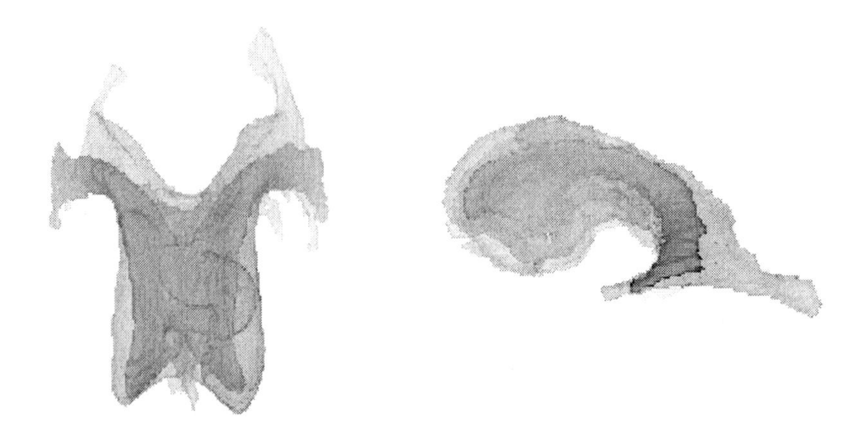

Figure 7. *Qualitative study of $\mathcal{CE}$ (in transparency) ventricles by superimposition with those of the atlas (in solid).*

Significant deformations are those which are further than 3 standard deviations from the average value for the x, y and z axis (see Table 1).

	Mode 0	Mode 1	Mode 2	Mode 3	Mode 4
Left x	-0.447	-5.125	+1.908	-2.140	-1.907
Right x	-1.026	+4.368	-2.254	+1.603	4.723
Left y	-3.687	+8.667	+0.451	+0.736	-5.081
Right y	-1.242	+8.336	-0.999	+0.757	-5.233
Left z	+4.429	-1.287	-3.731	-0.186	-1.442
Right z	+0.309	+1.824	-5.590	+1.986	-0.581

Table 1. *Amplitude distances for the 5 first modes of the deformation of the longest left and right ventricular crest line deformation between $\mathcal{CE}$ and the atlas. The framed modes are detected as "abnormal".*

If we do not consider the mode 0 which corresponds to the translation, we notice how "abnormal" modes tend to apply simultaneously to left and right ventricles (with different amplitudes). It shows that both ventricles are "affected" by the deformations. The y and z "abnormal" modes have the same sign, so they are oriented in the same sense. On the contrary, x "abnormal" modes have opposite sign and the deformations are in inversed directions. If we only consider the first deformation mode which is the more global one, we notice that the x and y amplitudes are larger for the left ventricle. This confirms our visual analysis that the two ventricles are asymmetric and that the left one is the largest.

Towards an assistance to diagnosis Now, let us study more specifically the three "abnormal" modes $1x$, $1y$ and $2z$. For that purpose, we apply successively to the atlas line these three modes and we visualize the results (see Figure 8). We can conclude that:

- the parameter $1x$ models the *horizontal curvature*,
- the parameter $1y$ measures the *vertical enlargement*,
- the parameter $2z$ quantifies the *longitudinal stretching*.

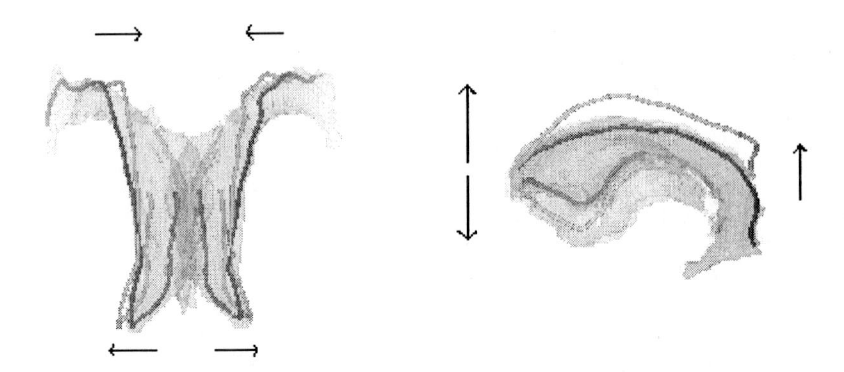

Figure 8. *Quantitative study of the ventricles deformation by analysis of the deformations and decomposition in "abnormal" modes ($1x$ and $1y$ modes). The atlas crest line is in black and the deformed one appears in grey.*

The amplitudes of these 3 modes could characterize the type and severity of deformations of the cerebral ventricles and hence automatically produce a diagnostic hypothesis based on a knowledge database.

4 Conclusion

In this paper, we have automatically built a simple 3D morphometric atlas of crest lines of the brain from 10 MRI images of different patients. We showed how this atlas could be used for some medical applications: automatic labelling, segmentation and assisted diagnosis.

We will concentrate our future work on four objectives:

- studying more precisely statistical tools for shape description: the Finite Element Method [MPK94], Principal Warps [Boo89], Fourier description [SKBG95], Principal Component Analysis [CTCG95].
- using other types of lines, for example, those based on 3D skeletonization [SBK$^+$92].
- working on a larger database (several tens of items) in order better to take into account the variabilities of the human brain.
- validating our first results on a anatomical and medical basis, in particular, within the European Project BIOMORPH about the development and the validation of techniques for brain morphometry.

References

Boo89. F. L. Bookstein. Principal Warps: Thin-Plate Splines and the Decomposition of Deformations. *IEEE Transactions on Pattern Analysis and Machine Intelligence*, 11(6):567–585, June 1989.

CTCG95. T. F. Cootes, C. J. Taylor, D. H. Cooper, and J. Graham. Active Shape Models - Their Training and Application. *Computer Vision and Image Understanding*, 61(1):38–59, January 1995.

DSTA95. J. Declerck, G. Subsol, J.Ph. Thirion, and N. Ayache. Automatic retrieval of anatomical structures in 3D medical images. In N. Ayache, editor, *CVRMed'95*, volume 905 of *Lecture Notes in Computer Science*, pages 153–162, Nice (France), April 1995. Springer Verlag.

MBF92. O. Monga, S. Benayoun, and O. D. Faugeras. Using Partial Derivatives of 3D Images to Extract Typical Surface Features. In *CVPR*, 1992.

MRB+95. J.F. Mangin, J. Regis, I. Bloch, V. Frouin, Y. Samson, and J. Lopez-Krahe. A MRF Based Random Graph Modelling the Human Cortical Topography. In Nicholas Ayache, editor, *CVRMed'95*, volume 905 of *Lecture Notes in Computer Science*, pages 177–183, Nice (France), April 1995. Springer-Verlag.

MPK94. J. Martin, A. Pentland, and R. Kikinis. Shape Analysis of Brain Structures Using Physical and Experimental Modes. In *Computer Vision and Pattern Recognition*, pages 752–755, Seattle, Washington (USA), June 1994.

OKA90. M. Ono, S. Kubik, and Ch. D. Abernathey. *Atlas of the Cerebral Sulci*. Georg Thieme Verlag, 1990.

RCJSG93. J. Rademacher, V. S. Caviness Jr, H. Steinmetz, and A. M. Galaburda. Topographical Variation of the Human Primary Cortices: Implications for Neuroimaging, Brain Mapping, and Neurobiology. *Cerebral Cortex*, 3:313–329, August 1993.

SBK+92. G. Székely, Ch. Brechbühler, O. Kübler, R. Ogniewicz, and T. Budinger. Mapping the human cerebral cortex using 3D medial manifolds. In Richard A. Robb, editor, *Visualization in Biomedical Computing*, pages 130–144, Chapel Hill, North Carolina (USA), October 1992. SPIE.

SKBG95. G. Székely, A. Kelemen, Ch. Brechbühler, and G. Gerig. Segmentation of 3D Objects from MRI Volume Data Using Constrained Elastic Deformations of Flexible Fourier Surface Models. In N. Ayache, editor, *CVRMed'95*, volume 905 of *Lecture Notes in Computer Science*, pages 495–505, Nice (France), April 1995. Springer-Verlag.

STA95. G. Subsol, J.Ph. Thirion, and N. Ayache. A General Scheme for Automatically Building 3D Morphometric Anatomical Atlases: application to a Skull Atlas. In *Medical Robotics and Computer Assisted Surgery*, pages 226–233, Baltimore, Maryland (USA), November 1995.

Sub95. G. Subsol. *Construction automatique d'atlas anatomiques morphométriques à partir d'images médicales tridimensionnelles.* PhD thesis, Ecole Centrale Paris, December 1995. In French.

TG93. J.Ph. Thirion and A. Gourdon. The Marching Lines Algorithm : new results and proofs. Technical Report 1881, INRIA, March 1993. Available by anonymous ftp at ftp.inria.fr /INRIA/tech-reports/RR.

TTP96. A. W. Toga, P. Thompson, and B. A. Payne. *Development Neuroimaging: Mapping the Development of Brain and Behavior*, chapter Modeling Morphometric Changes of the Brain During Development. Academic Press, 1996.

Visualization and Mapping of Anatomic Abnormalities Using a Probabilistic Brain Atlas Based on Random Fluid Transformations

Paul Thompson and Arthur W. Toga

Laboratory of Neuro Imaging, Dept. Neurology, Division of Brain Mapping,
UCLA School of Medicine, Los Angeles, CA 90095-1769.

Abstract This paper describes the design, implementation and preliminary results of a technique for creating a comprehensive probabilistic atlas of the human brain based on high-dimensional fluid transformations. The goal of the atlas is to detect and quantify subtle and distributed patterns of deviation from normal anatomy, in a 3D brain image from any given subject. The algorithm analyzes a reference population of normal scans, and automatically generates color-coded probability maps of the anatomy of new subjects. Given a 3D brain image of a new subject, the algorithm calculates a set of high-dimensional volumetric maps (typically with $384^2 \times 256 \times 3 \approx 0.1$ billion degrees of freedom) fluidly deforming this scan into structural correspondence with other scans, selected one by one from an anatomic image database. The family of volumetric warps so constructed encodes statistical properties and directional biases of local anatomical variation throughout the architecture of the brain. A probability space of random transformations, based on the theory of anisotropic Gaussian random fields, is then developed to reflect the observed variability in stereotaxic space of the points whose correspondences are found by the warping algorithm. A complete system of $384^2 \times 256$ probability density functions is computed, yielding confidence limits in stereotaxic space for the location of every point represented in the 3D image lattice of the new subject's brain. Color-coded probability maps are generated, densely-defined throughout the anatomy of the new subject. These indicate locally the probability of each anatomic point being as unusually situated, given the distributions of corresponding points in the scans of normal subjects. 3D MRI and high-resolution cryosection volumes are analyzed, from subjects with metastatic tumors and Alzheimer's disease. Applications of the random fluid-based probabilistic atlas include the transfer of multi-subject 3D functional, vascular and histologic maps onto a single anatomic template, the mapping of 3D atlases onto the scans of new subjects, and the rapid detection, quantification and mapping of local shape changes in 3D medical images in disease, and during normal or abnormal growth and development.

1 Introduction

Remarkable variations exist, across individuals, in the internal and external geometry of the brain. In the past, comparing data from different subjects or patient subpopulations has been difficult because cortical topography and the internal geometry of the brain vary so greatly. At the same time, much research has been directed towards the development of *standardized 3-dimensional atlases* of the human brain [1,2]. These provide an invariant reference system, and the possibility of template matching, allowing anatomical structures in new scans to be identified and analyzed. Atlases serve as a guide in planning stereotaxic neurosurgical procedures and provide a precise quantitative framework for multi-modality brain mapping.

1.1 Deformable and Probabilistic Brain Atlases

In view of the complex structural variability between individuals, a fixed brain atlas may fail to serve as a faithful representation of every brain. It would, however, be ideal if the atlas could be elastically deformed to fit a new image set from an incoming subject. Transforming individual datasets into the shape of a single reference anatomy, or onto a 3D digital brain atlas, removes subject-specific shape variations, and allows subsequent comparison of brain function between individuals [3]. Conversely, high-dimensional warping algorithms [3-5] can also be used to transfer all the information in a 3D digital brain atlas onto the scan of any given subject, while respecting the intricate patterns of structural variation in their anatomy. Such *deformable atlases* [3,5] can be used to carry 3D maps of functional and vascular territories into the coordinate system of different patients, as well as information on different tissue types and the boundaries of cytoarchitectonic fields and their neurochemical composition. Thirdly, 3D warping algorithms provide a method for calculating local and global shape changes and give valuable information about normal and abnormal growth and development. Deformable atlases not only account for the anatomic variations and idiosyncrasies of each individual subject, but they offer a powerful strategy for exploring and classifying age-related, developmental or pathologic variations in anatomy.

On the other hand, *probabilistic atlasing* [6-8] is a research strategy whose goal is to generate anatomical templates and expert diagnostic systems which retain quantitative information on inter-subject variations in brain architecture. The recent interest in comprehensive brain mapping also stresses that comparisons between subjects, both within and across homogeneous populations, are required to understand normal variability and genuine structural and functional differences. Initial attempts to derive average representations of neuroanatomy have underscored the validity and power of this approach in both clinical and research settings [9,10]. In one such approach, 305 MRI volumes (2-mm thick slices) were mapped by linear transformation into stereotaxic space, intensity normalized, and averaged on a voxel-by-voxel basis [9]. The effect of anatomical variability in different brain areas is illustrated qualitatively by this average-intensity MRI dataset. Nevertheless, the resulting average brain which has regions where individual structures are blurred out due to spatial variability in the population [7,9].

We recently developed and implemented [8] an approach for constructing a probabilistic surface atlas of the brain. This performs a statistical analysis of deep surface structures in the brain (in a reference database of normal scans), and then automatically quantifies and maps distributed patterns of abnormality in the same system of anatomic surfaces in new subjects. Connected systems of parametric meshes model several deep internal fissures, or *sulci*, whose trajectories represent critical functional and lobar boundaries. Many sulci are comparatively stable in their incidence and connectivity in different individuals. They are also sufficiently extended inside the brain to reflect subtle and distributed variations in anatomy between subjects. The parametric form of the system of connected surface elements allowed us to represent the relation between any pair of anatomies as a family of high-resolution displacement maps carrying the surface system of one individual onto another in stereotaxic space. An additional surface analysis algorithm constructed a probability space of random transformations (based on the theory of 3D Gaussian random fields) reflecting the variability in stereotaxic space of the connected system of anatomic surfaces. Automatic parametrization of the surface anatomy of new subjects enabled the detection and mapping of subtle shape and volume abnormalities in the brains of patients with metastatic tumors. These shape changes were visualized in the form of probability maps on a graphical surface model of the subject's anatomy.

1.2 A Hybrid Atlasing Strategy

This paper describes the development of a more comprehensive probabilistic atlas. The technique in [8] quantified anatomic abnormalities in new subjects *only* on a connected system of *surfaces* inside their brains. By contrast, we have now devised a strategy for quantifying, mapping and visualizing the anatomic deviations of any given subject not only on an internal surface system, but throughout the entire volume architecture of their brain. To do so, the strategy invokes:

(1) a spatially accurate, biologically-constrained warping algorithm [11], which computes high-dimensional volume deformations fluidly transforming 3D anatomic images from different subjects and/or modalities into structural correspondence; and

(2) a new 3D probability mapping theory, which generates a visualizable probability measure throughout the 3D volume of the brain. These probability values quantify the severity of structural abnormality at a very local level.

A family of high-dimensional volumetric warps relating the new scan to each normal scan in a brain image database is calculated, and then used by the algorithm to quantify local structural variations. These warps encode the distribution in stereotaxic space of anatomic points which correspond across a normal population, and their dispersion is used to determine the likelihood of local regions of the new subject's anatomy being in their actual configuration. Any discrepancies, which may be of clinical relevance, can therefore be mapped in 3-dimensional space, quantified and evaluated. In deriving the parameters of anatomic variation, the algorithm accounts for the fact that the magnitude of normal anatomic variability, as well as its local directional biases, differ significantly at every single anatomic point in the brain represented in the 3D image lattice of the new subject's scan.

2 Methods

2.1 Anatomic Image Acquisition

A multi-modality, multi-subject reference archive of 3D anatomic images was assembled, by imaging many normal subjects with consistent acquisition criteria. 3D ($384^2 \times 256$ resolution) T_1-weighted fast SPGR (spoiled GRASS) MRI volumes were acquired from ten normal subjects, on a GE Signa 1.5T clinical scanner with TR/TE 14.3/3.2 msec, flip angle 35°, FOV 25cm and contiguous 1mm-thick contiguous slices covering the entire brain. These subjects were age-matched with an identically-imaged patient with clinically determined Alzheimer's disease, whose anatomy was analyzed in our first experiment (*q.v., Preliminary Results*). In addition, six ultra-high resolution (1024^2x1300x24 bit) full-color digital cryosection volumes were acquired, as described previously [12]. Six cadavers (age range 72-91 yrs., 3 males) were obtained, optimally within 5-10 hours *post mortem*, through the Willed Body Program at the UCLA School of Medicine. Standard exclusion procedures were applied [13]. For our second experiment, an additional cryosection volume was identically acquired, from an age-matched subject with two well-defined metastatic tumors (in the high right putamen and left occipital lobe). 3D image data from the 18 heads were corrected for differences in relative position and size by transformation into standardized Talairach stereotaxic space, using the steps specified in the Talairach atlas [1].

2.2. A Generic Model of Brain Geometry using Parametric Surfaces

The external cortical surface was extracted automatically from each scan in parametric form using the 3D active surface algorithm of [14]. Secondly, since so much of the functional territory of the human cortex is buried in the cortical folds or *sulci*, we applied a series of recently-developed parametric strategies for mapping the internal cortex. The construction of extremely complex surface deformation maps on the internal cortex is made easier by building a generic structure to model it (Fig. 1(a)). Connected systems of parametric meshes were used to model major lobar, ventricular and cytoarchitectural boundaries in 3 dimensions. These included complex internal trajectories of the parieto-occipital sulcus, the anterior and posterior calcarine sulcus, the Sylvian fissure, and the cingulate, marginal and supracallosal sulci in both hemispheres. The ventricular system was partitioned into a closed system of 14 connected surface elements whose junctions reflected cytoarchitectonic boundaries of the adjacent tissue.

Interactive outlining of these deep surfaces in sagittally reformatted images resulted in a sampling of approximately 15000 points per structure, capturing the details of each anatomic surface at a very local level. Each surface element was automatically converted into parametric mesh form as described earlier [8,13]. Isolation of corresponding points on these deep surfaces involves the molding of a linearly-elastic lattice-like mesh onto the geometric profile of each surface. The concept is similar to that of a regular net being draped over an object. Points on each surface with the same mesh coordinate occupy similar positions in relation to the geometry of the surface they belong to; under stringent conditions [13] these may also be regarded as homologous. Imposition of an identical regular structure on surfaces from different subjects allows surface statistics to be derived (Fig. 1(i)).

2.3. High-Dimensional Fluid Warping driven by Parametric Surface Systems

Once the surface parametrization and transformation algorithms had been tested and validated on a range of multi-modal image volumes [8,13], the mathematical theory for a surface-based 3D warping algorithm was devised [11]. This algorithm calculates the high-dimensional deformation field (typically with $384^2 \times 256 \times 3 \approx 0.1$ billion degrees of freedom) relating the brain anatomies of an arbitrary pair of subjects. Using complex 3D deformation maps, the connected surface elements in one scan are driven into exact correspondence with their counterparts in a target scan, guiding the transformation of the adjacent brain volume. High spatial acuity of the warp is guaranteed at the surface interfaces used to constrain it, and these include many critical functional interfaces such as the ventricles and cortex, as well as numerous cytoarchitectonic and lobar boundaries in 3 dimensions. The algorithm's spatial accuracy was demonstrated on multi-modality real and simulated data in [11], where full details of the algorithm are presented. Briefly, the algorithm calculates the deformation field required to elastically transform the elements of the surface system in one 3D image to their counterparts in the target scan. Integral distortion functions, which describe the influence of deforming surfaces on points in their vicinity, are then used to extend this surface-based deformation to the whole brain volume.

Let Λ be the image lattice $\{(i,j,k) \mid 0 \le i \le I, 0 \le j \le J, 0 \le k \le K; I, J, K, i, j, k \in \mathbb{N}\}$and let $\Omega = \{(u,v) \mid 0 \le u \le U, 0 \le v \le V; U, V, u, v \in \mathbb{N}\}$ be the lattice of integer-valued grid-points in a rectangular domain of $\mathbb{R}^2$, with fixed size $I \times J$, (for any positive integers I and J). As in [11], let the lth parametric surface in the nth brain be denoted by the mesh $\mathbf{M}_l^n = \{\mathbf{r}_l^n(u,v) \mid 0 \le u \le U, 0 \le v \le V\}$. For each of the L different types of anatomic surface, let $\mathcal{F}_l = \{\mathbf{M}_l^n, \Omega, N\}$ be the family of N parametric meshes $\mathbf{M}_l^n$: $\Omega \subset \mathbb{R}^2 \rightarrow \mathbb{R}^3$ representing that structure in the N scans $\{\mathbf{A}_1,\ldots,\mathbf{A}_N\}$ comprising the reference database. For each pair of scans $\mathbf{A}_p$ and $\mathbf{A}_q$ in the reference database, a 3D displacement field $\mathbf{W}_{pq}(\mathbf{x})$ can be defined, carrying each voxel $\mathbf{x}\in\mathbf{A}_p$ into structural correspondence with its anatomic counterpart in $\mathbf{A}_q$, as follows. Firstly, for each surface mesh $\mathbf{M}_l^p$ in $\mathbf{A}_p$ we define displacement maps on the surfaces which correspond, *i.e.*, $\mathbf{W}_l^p[\mathbf{r}_l^p(u,v)] = \mathbf{r}_l^p(u,v)-\mathbf{r}_l^q(u,v)$. Special algorithms define a dense correspondence vector field between scans on the external cortex. This displacement map matches up major external sulci and is driven by a set of uniformly parametrized external curves, $\mathbf{c}_k(t)$. These curves mark lobar boundaries at which deep sulcal surface meshes interface with the automatically extracted external cortex. The surface warp from $\mathbf{A}_p$ and $\mathbf{A}_q$ is given by $\mathbf{W}_{pq}{}^C:(r_p{}^C,\theta,\phi)\rightarrow(r_q{}^*,\theta+[\Delta\theta(\theta,\phi)],\phi+[\Delta\phi(\theta,\phi)])$, where $r_q{}^*$ is the point on $r_q{}^C(\theta,\phi)$ at parameter location $(\theta+[\Delta\theta(\theta,\phi)],\phi+[\Delta\phi(\theta,\phi)])$. The parameter shifts are given by the solution to a curve-driven warp in the biperiodic parametric space $(\theta,\phi)\in[0,2\pi)\times[0,2\pi)=\Gamma$ of the external cortex [13, *cf.* 15].

These surface maps drive the full 3D volume transformation as follows. For a general voxel $\mathbf{x}\in\mathbf{A}_p$, let $\delta_l{}^p(\mathbf{x})$ be the distance from $\mathbf{x}$ to its nearest point(s) on each mesh $\mathbf{M}_l{}^p$, and let the scalars $\gamma_l{}^p(\mathbf{x})\in[0,1]$ denote the weights $\{1/\delta_l{}^p(\mathbf{x})\}/\Sigma_{l=1 \mathrm{to} L}\{1/\delta_l{}^p(\mathbf{x})\}$. Then $\mathbf{W}_{pq}(\mathbf{x})$ is given by $\mathbf{W}_{pq}(\mathbf{x}) = \Sigma_{l=1 \mathrm{to} L}\gamma_l{}^p(\mathbf{x}).\mathbf{D}_l{}^p(\mathbf{np}_l{}^p(\mathbf{x}))$, for all $\mathbf{x}\in\mathbf{A}_p$. Here the $\mathbf{D}_l{}^p$ are distortion functions due to the deformation of surfaces close to $\mathbf{x}$, given by

$$\mathbf{D}_l{}^p(\mathbf{x}) = \{\int_{r\in B(\mathbf{x};rc)} w_l{}^p(\mathbf{x},\delta_l{}^p(\mathbf{r})).\mathbf{W}_l{}^p[\mathbf{np}_l{}^p(\mathbf{r})]\, \boldsymbol{dr}\}/\{\int_{r\in B(\mathbf{x};rc)} w_l{}^p(\mathbf{x},\delta_l{}^p(\mathbf{r}))\, \boldsymbol{dr}\}.$$

$\mathbf{W}_l{}^p[\mathbf{np}_l{}^p(\mathbf{r})]$ is the (average) displacement vector assigned by the surface displacement maps to the nearest point(s) $\mathbf{np}_l{}^p(\mathbf{r})$ to $\mathbf{r}$ on $\mathbf{M}_l{}^p$. R_c is a constant, and $\mathbf{B}(\mathbf{x};r_c)$ is a sphere of radius $r_c = \mathit{min}\{R_c, \mathit{min}\{\delta_l{}^p(\mathbf{r})\}\}$. The $w_l{}^p$ are additional weight functions defined as $w_l{}^p(\mathbf{x},\delta_l{}^p(\mathbf{r})) = \mathit{exp}(-\{d(\mathbf{np}_l{}^p(\mathbf{r}),\mathbf{x})/\delta_l{}^p(\mathbf{x})\}^2)$, where $d(\mathbf{a},\mathbf{b})$ represents the 3D distance between two points $\mathbf{a}$ and $\mathbf{b}$. The algorithm is greatly accelerated by calculating the fluid deformation fields on a successively refined multi-resolution hierarchy of octree-spline grids [16] in the same space as the target image lattice. Non-singularity of the deformation field is guaranteed by ascribing a non-linear kinematics to the deforming template, and (emulating the method devised by Christensen [3]) discretizing the flow in time. Surface blends $(1-t)\mathbf{r}_l{}^p(u,v)-t\mathbf{r}_l{}^q(u,v)$, $(t\in[0,1])$, were generated for every surface and uniformly reparametrized [8] at times $0\leq..t_m \leq t_{m+1}..\leq 1$. The M warps mapping the full surface system and surrounding volume from one time-point to the next were concatenated to produce the final transformation.

2.4. Probability Mapping Theory

Given a new subject's 3D brain scan, $\mathbf{T}$, our goal was to assign a *probability value* $p(\mathbf{x})$ to each anatomic point $\mathbf{x}$ in $\mathbf{T}$. This probability value indicates how abnormally situated that point is, given where its counterparts are in the normal scans which make up the database. For increasingly extreme deviations of the anatomic structure at $\mathbf{x}$ from its counterparts in the normal scans, the associated probability $p(\mathbf{x})$ of finding the structure so far out will be correspondingly lower.

Probability values are assigned as the result of a 2-stage process, as follows.

(1) For each anatomic point $\mathbf{x}$ in the new subject's scan, $\mathbf{T}$, its counterparts are found in the N normal scans. This is done by calculating the set of warping fields $\{\mathbf{W}_{Tn}(\mathbf{x})\}_{n=1 \mathrm{to} N}$ deforming scan $\mathbf{T}$ into structural correspondence with each of the N scans in the database;

(2) A 3D probability density function is then recovered from the distribution (in stereotaxic space) of the N points corresponding to $\mathbf{x}$. The probability of the anatomic structure at $\mathbf{x}$ is then assessed with reference to the resulting probability distribution.

Firstly, let $\mathbf{W}_{Tn}(\mathbf{x})$ be the 3D displacement vector mapping voxel $\mathbf{x}$ in the new subject's scan, $\mathbf{T}$, onto its counterpart in the nth brain. Its counterparts in stereotaxic space have mean position $\boldsymbol{\mu}(\mathbf{x}) = \mathbf{x}+(1/N)\Sigma_{n=1\mathrm{to}N}\mathbf{W}_{Tn}(\mathbf{x})$, and 3x3 dispersion matrix $\boldsymbol{\Psi}(\mathbf{x})$ whose entries are given by $\sigma^2_{ij}(\mathbf{x})=(1/N)\Sigma_{n=1\mathrm{to}N}\ |\pi_i\mathbf{W}_{T\mu}(\mathbf{x})|.|\pi_j\mathbf{W}_{T\mu}(\mathbf{x})|$, where $1\le i,j\le 3$ and π_1, π_2, π_3 are orthogonal projections onto each of the 3 axes of stereotaxic space. Here $\mathbf{W}_{n\mu}(\mathbf{x})$ is defined as $\boldsymbol{\mu}(\mathbf{x})-(\mathbf{x}+\mathbf{W}_{Tn}(\mathbf{x}))$. As in [8], we let $\boldsymbol{\Psi}(\mathbf{x}) \cong \mathbf{diag}\{\sigma^2_{11}(\mathbf{x}), \sigma^2_{22}(\mathbf{x}), \sigma^2_{33}(\mathbf{x})\}$, where the σ^2_{kk} are the variances of the x, y and z components of the N volumetric warps $\mathbf{W}_{Tn}(\mathbf{x})$ at $\mathbf{x}$, respectively. Then if $\mathbf{W}_{Tn}(\mathbf{x})\sim N_3(\boldsymbol{\mu}(\mathbf{x}),\boldsymbol{\Psi}(\mathbf{x}))$ as $N\to\infty$, the quantity

$$\mathrm{F}(n) = [N(N-3)/3(N^2-1)]\ [\mathbf{W}_{n\mu}(\mathbf{x})]^T[\boldsymbol{\Psi}(\mathbf{x})]^{-1}[\mathbf{W}_{n\mu}(\mathbf{x})]$$

is an F-distributed variable with 3 and N-3 degrees of freedom [17]. For any desired confidence threshold α, $100(1-\alpha)\%$ *confidence regions* in stereotaxic space for possible locations of points corresponding to $\mathbf{x}$ in $\mathbf{T}$ are given by nested ellipsoids $\mathbf{E}_{\lambda(\alpha)}(\mathbf{x})$ in displacement space [8]. Here $\mathbf{E}_\lambda(\mathbf{x})=\{\boldsymbol{\mu}(\mathbf{x})+\lambda[\boldsymbol{\Psi}(\mathbf{x})]^{-1/2}\mathbf{p}|\forall\mathbf{p}\in\mathbf{B}(\mathbf{0};1)\}$, where $\mathbf{B}(\mathbf{0};1)$ is the closed unit sphere in $\mathbb{R}^3$, and $\lambda(\alpha)=[[N(N-3)/3(N^2-1)]^{-1}F_{\alpha,3,N-3}]^{1/2}$, where $F_{\alpha,3,N-3}$ is the critical value of the F distribution such that $\Pr\{F_{3,N-3} \ge F_{\alpha,3,N-3}\}=\alpha$. Finally, if $\mathbf{W}_{T\mu}(\mathbf{x})=\boldsymbol{\mu}(\mathbf{x})-\mathbf{x}$, the appropriate probability measure, defined throughout the new subject's anatomy, is

$$p(\mathbf{x}) = sup\{\alpha\ |\ F_{\alpha,3,N-3} \ge [N(N-3)/3(N^2-1)][\mathbf{W}_{T\mu}(\mathbf{x})]^T[\boldsymbol{\Psi}(\mathbf{x})]^{-1}[\mathbf{W}_{T\mu}(\mathbf{x})]\},\ \forall\mathbf{x}\in\mathbf{T}.$$

This metric quantifies the severity of local discrepancies $\mathbf{W}_{T\mu}(\mathbf{x})$ between an anatomic point in a new subject and its counterparts (found by the warping algorithm) in normal anatomic scans which make up the reference database. This closed form expression, giving the probability measure $p(\mathbf{x})$ for each voxel $\mathbf{x}$ in the image lattice of the new subject's scan, is evaluated and mapped, via a logarithmic look-up table, onto a standard color range. Probability maps are visualized using Data Explorer 2.1 (IBM Visualization Software). All warping and probability mapping algorithms were written in C, and executed on standard 200 MHz DEC Alpha AXP3000 workstations running OSF-1.

3 Preliminary Results

3.1. Visualization of 3D Fluid Deformation Fields A battery of tests was first carried out (Fig. 1(b)-(h); [11]) to evaluate the performance of the warping algorithm on a wide range of real and simulated data. Its capacity to correctly transform images into structural correspondence was investigated, by warping different subjects' anatomic images onto each other, both within and across modalities.

3.2. Probability Maps Derived from a Family of High-Dimensional Warping Fields

A subject with two large, well-defined metastatic tumors, in the high right putamen and left occipital lobe, was cryosectioned and digitally imaged in full color at $1024^2\times1300$ pixel resolution, and a family of deformation fields was recovered relating their anatomy to that of 6 identically-imaged normal subjects in a reference archive. The tumors induced marked distortions in the normal architecture of the brain (Fig. 1(j)). Structures in the immediate vicinity of the lesions exhibit probability values three orders of magnitude lower than normal ($p < 0.0001$; *red colors*), while more distal regions of these structures are normal ($p > 0.05$; *deep blue colors*). Normal results were obtained for all surfaces ($p > 0.05$ [8]), when probability maps were generated for each of the 6 normal subjects which made up the underlying database. The severity of structural herniation, due to the mechanical effects of a lesion, can therefore be highlighted and quantified by probability mapping of structures in each hemisphere.

4 Conclusion

The high-dimensional warping and probabilistic mapping approaches developed here provide a framework for visualizing and mapping complex structural variations throughout the anatomy of new subjects. Preliminary data have illustrated the feasibility of creating probability maps on surface systems, which typically consist of critical functional interfaces and boundaries in 3 dimensions. We have also described a method for calculating probability maps throughout the full volume of a subject's brain, to provide a more comprehensive measure of distributed patterns of structural abnormality. The extension of a probability measure from surfaces to volumes requires the development of an additional algorithm, with almost unlimited degrees of freedom, to transform a scan into structural correspondence with each scan in an image archive. The capacity of this algorithm to estimate the locations of classes of points which correspond across a range of image volumes allows us to characterize the statistical dispersion of these points in stereotaxic space. Algorithms defined on the archive of anatomic data can readily use the family of associated warping fields to produce probability distributions and confidence limits for structure identification. Deviations in the anatomy of new subjects can therefore be analyzed and quantified at an extremely local level. Anisotropic random fields, invoked in the generation of probability maps for new subjects, readily encode local biases in the direction of anatomic variability, and hence quantify the severity of anatomic deviations more effectively than simple distance-based descriptors.

The surface-based modeling, mapping and warping approaches presented in this paper may offer distinct advantages over volume averaging for statistical atlasing applications. Surface representations lend themselves readily to averaging, and subsequent statistical characterization. More particularly though, the averaging procedure itself does not lead to the same type of degradation of structural geometry (and loss of fine anatomic features) as is often apparent in volume averaging approaches. In addition, the retention of an explicit surface topology after averaging is particularly advantageous for subsequent visualization [8,13]. This feature of both the individual and average representations of brain anatomy enables secondary regional information, including local probability maps, to be overlaid and visualized on the underlying surface models [18]. Information about physiology, neurochemistry, and an infinite variety of relevant maps can potentially be layered onto the anatomic atlas and referenced using such a system. In the brain, such surface maps include cytoarchitecture, chemoarchitecture, blood flow distributions and metabolic rates.

In the future, probabilistic mapping is likely to be fundamental to multi-subject atlasing and many other brain mapping projects. Digital probabilistic atlases based on large populations will rectify many current atlasing problems, since they retain quantitative information on the variability inherent in anatomic populations. As the underlying database of subjects increases in size and content, the digital, electronic form of the atlas provides efficiency in statistical and computational comparisons between individuals or groups. The atlas also improves in accuracy over time achieving better statistics as more information is added to the underlying database. In addition, the digital form of the source data enables the population on which probabilistic atlases are based to be stratified into subpopulations by age, gender, by stage of development or to represent different disease types [19].

The ultimate goal of brain mapping is to provide a framework for integrating functional and anatomical data across many subjects and modalities. This task requires precise quantitative knowledge of the variations in geometry and location of intracerebral structures and critical functional interfaces. The high-dimensional warping and probabilistic techniques presented here provide a basis for the generation of anatomical templates and expert diagnostic systems which retain quantitative information on inter-subject variations in brain architecture.

5 Acknowledgments Paul Thompson is supported by the United States Information Agency, under Grant No. G-1-00001, by a Pre-Doctoral Fellowship of the Howard Hughes Medical Institute, and by a Fulbright Scholarship from the US-UK Fulbright Commission, London. Additional support was provided by the National Science Foundation (BIR 93-22434), by the National Library of Medicine (LM/MH05639), by the NCRR (RR05956), and by the Human Brain Project, which is funded jointly by NIMH and NIDA (P20 MH/DA52176).

References

[1]. Talairach J, Tournoux P (1988). *Co-planar Stereotaxic Atlas of the Human Brain*, New York: Thieme Medical Publishers.
[2]. Thurfjell L, Bohm C, Greitz T, Eriksson L (1993). *Transformations and Algorithms in a Computerized Brain Atlas*, IEEE Trans. Nucl. Sci., **40**(4), pt. 1:1167-91.
[3]. Christensen GE, Rabbitt RD, Miller MI (1993). *A Deformable Neuroanatomy Textbook based on Viscous Fluid Mechanics*, 27th Ann. Conf. on Inf. Sciences and Systems, 211-216.
[4]. Thirion J-P (1995). *Fast Non-Rigid Matching of Medical Images*, INRIA Internal Report **2547**, Projet Epidaure, INRIA, France.
[5]. Evans AC, Dai W, Collins L, Neelin P, Marrett S (1991). *Warping of a Computerized 3D Atlas to Match Brain Image Volumes for Quantitative Neuroanatomical and Functional Analysis*, Proc. Int. Soc. Opt. Eng. (SPIE): Med. Imag. III, 264-274.
[6]. Mazziotta JC, Toga AW, Evans AC, Fox P, Lancaster J (1995). *A Probabilistic Atlas of the Human Brain: Theory and Rationale for its Development*, NeuroImage **2**: 89-101.
[7]. Evans AC, Kamber M, Collins DL, MacDonald D (1994). *An MRI-Based Probabilistic Atlas of Neuroanatomy, in: Magnetic Resonance Scanning and Epilepsy,* [Eds: Shorvon SD et al.], Plenum Press, New York, 263-274.
[8]. Thompson PM, Schwartz C, Toga AW (1996). *High-Resolution Random Mesh Algorithms for Creating a Probabilistic 3D Surface Atlas of the Human Brain*, NeuroImage **3**:19-34.
[9]. Evans AC, Collins DL, Milner B (1992). *An MRI-based Stereotactic Brain Atlas from 300 Young Normal Subjects*, in: Proc. 22nd Symp. Society for Neuroscience, Anaheim, 408.
[10]. Andreasen NC, Arndt S, Swayze V, Cizadlo T, Flaum M, O'Leary D, Ehrhardt JC, Yuh WTC (1994). *Thalamic Abnormalities in Schizophrenia Visualized through Magnetic Resonance Image Averaging*, Science, 14 October 1994, **266**:294-298.
[11]. Thompson PM, Toga AW (1996). *A Surface-Based Technique for Warping 3-Dimensional Images of the Brain*, IEEE Transactions on Medical Imaging, Aug. 1996, **15**(4):1-16.
[12]. Toga AW, Ambach KL, Quinn B, Hutchin M, Burton JS (1994). *Post Mortem Anatomy from Cryosectioned Human Brain*, J. Neurosci. Meth., **54**:239-252.
[13]. Thompson PM, Schwartz C, Lin RT, Khan AA, Toga AW (1996). *3D Statistical Analysis of Sulcal Variability in the Human Brain*, Journal of Neuroscience, Jul. 1996.

[14]. MacDonald D, Avis D, Evans AC (1993). *Automatic Parameterization of Human Cortical Surfaces*, Annual Symp. Info. Proc. Med. Imag., (IPMI).
[15]. Davatzikos C (1996). *Non-linear Registration of Brain Images using Deformable Models*, Proc. IEEE Workshop on Math. Methods in Biomedical Image Analysis, June 1996.
[16]. Szeliski R, Lavallée S (1993). *Matching 3D Anatomic Surfaces with Non-Rigid Deformations using Octree-Splines*, SPIE **2031**, Geometric Methods in Computer Vision II, 306-315.
[17]. Anderson TW (1984). *An Introduction to Multivariate Statistical Analysis*. Wiley, New York.
[18]. Sclaroff S (1991). *Deformable Solids and Displacement Maps: A Multi-Scale Technique for Model Recovery and Recognition*. Master's Thesis, MIT Media Laboratory, June 1991.
[19]. Thompson PM, Mega MS, Moussai J, Zohoori S, Xu LQ, Goldkorn A, Khan AA, Coryell J, Small G, Cummings J, Toga AW (1996). *3D Probabilistic Atlas and Average Surface Representation of the Alzheimer's Brain, with Local Variability and Probability Maps of Ventricles and Deep Cortex*, in: Proc. 1996 Symp. of the Society for Neuroscience, Washington DC, [in press].

Fig. 1. *Connected Surface Systems used to Drive the 3D Warp.* The complex internal trajectory of the deep structures controlling the deformation field is illustrated here. Deep sulcal surfaces include: the anterior and posterior calcarine (CALCa/p), cingulate (CING), parieto-occipital (PAOC) and callosal (CALL) sulci and the Sylvian fissure (SYLV). Also shown are the superior and inferior surfaces of the rostral horn (VTSs/i) and inferior horn (VTIs/i) of the right lateral ventricle. Color-coded profiles show the magnitude of the 3D deformation maps warping these surface components (in the right hemisphere of a 3D T_1-weighted SPGR MRI scan of an Alzheimer's patient) onto their counterparts in an identically-acquired scan from an age-matched normal subject.
MRI-to-MRI Experiment. T_1-weighted MR sagittal brain slice images from (b) a randomly-selected normal scan from the reference archive, (c) the target anatomy, from a patient with clinically-determined Alzheimer's disease; and (d) result of warping the reference anatomy into structural correspondence with the target. Due to the high degree of cerebellar atrophy, the cerebellar surface was also used to control the deformation in this case. Note the precise non-linear registration of the cortical boundaries, the desired reconfiguration of the major sulci, and the contraction of the ventricular space and cerebellum. Both global and local differences in anatomy have been accommodated by the transformation. The complexity of the recovered deformation field is shown by applying the two in-slice components of the 3D volumetric transformation to a regular grid in the reference coordinate system. This visualization technique (e) highlights the especially large contraction in the cerebellar region, and the complexity of the warping field in the posterior frontal and cingulate areas, corresponding to subtle local variations in anatomy between the two subjects. To monitor the smooth transition to the surrounding anatomy of the deformation fields initially defined on the surface systems, additional software was developed to visualize the magnitude of the warping field on the surface anatomy of the target brain, as well as on an orthogonal plane slicing through many of these surfaces at the same level as the anatomic sections (f). Note the smooth continuation of the warping field from the complex anatomic surfaces into the surrounding brain architecture, and the highlighting of the severe deformations in the pre-marginal cortex, ventricular and cerebellar areas.
Inter-Modality Warping: Mapping 3D Digital Cryosection Volumes onto 3D MRI Volumes. The result of warping a randomly selected 3D cryosectioned image into the shape of the target MRI anatomy (c) is shown in (h), with cortical and ventricular landmarks of the target anatomy superimposed. Note in particular the reconfiguration of the major sulci (which would only be possible with a high-dimensional warping technique), and note the degree to which the reference *corpus callosum* is deformed into the shape of the target *corpus callosum*.
Distortions in Brain Architecture induced by Tumor Tissue: Probability Maps for Major Sulci in Both Hemispheres. (i) 3D r.m.s. variability maps are shown for major occipital and paralimbic sulci; (j) Color-coded probability maps quantify the impact of two focal metastatic tumors (*illustrated in red*) on the supracallosal, parieto-occipital, and anterior and posterior calcarine sulci in both hemispheres.

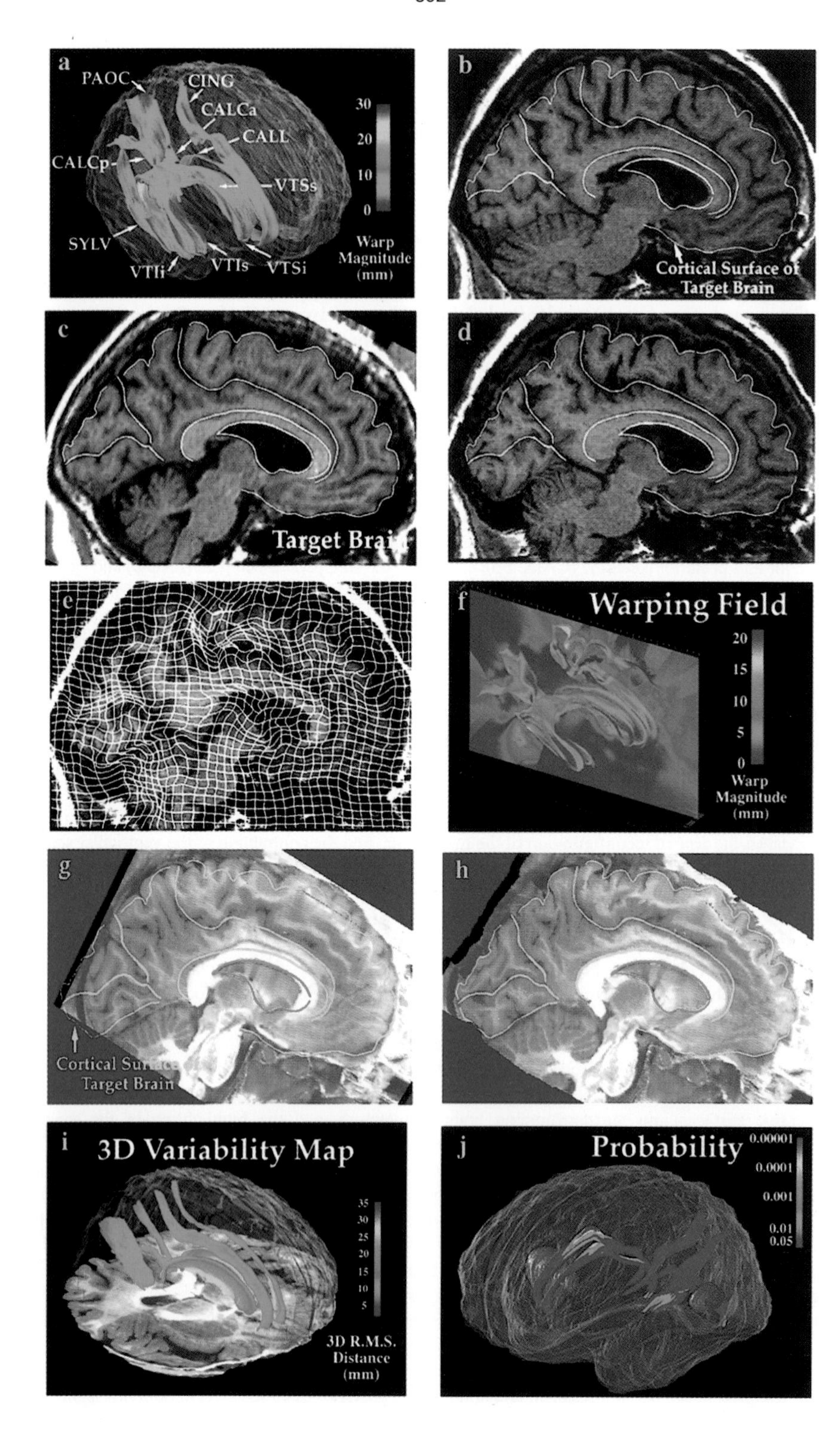
a
PAOC
CING
CALCa
CALL
CALCp
VTSs
SYLV
VTIi
VTIs
VTSi
30
20
10
0
Warp Magnitude (mm)
b
Cortical Surface of Target Brain
c
Target Brain
d
e
f
Warping Field
20
15
10
5
0
Warp Magnitude (mm)
g
Cortical Surface Target Brain
h
i
3D Variability Map
35
30
25
20
15
10
5
3D R.M.S. Distance (mm)
j
Probability
0.00001
0.0001
0.001
0.01
0.05

L-Systems for Three-Dimensional Anatomical Modelling: Towards a Virtual Laboratory in Anatomy

HabibTalhami[1]

[1]Dept. of Electrical and Electronics Engineering
University of Tasmania, G.P.O. Box 252C, Hobart, Tasmania 7001, AUSTRALIA.
H.Talhami@eee.utas.edu.au

Abstract. L-systems are explored as a tool for the modelling of both branching and non-branching anatomical structures. Branching anatomical structures are modelled using parametric L-systems that are normally used to describe tree growth and structure. Non-branching structures are modelled using generalised cylinder L-systems. Both representations produce highly parametric models of the anatomy that can be used for efficient image generation and model matching. Examples are given to illustrate the power of this approach in modelling three-dimensional anatomy such as the ribcage, the lung airway, and the heart.

1. Introduction

L-systems or Lindenmayer systems were first introduced by Lindenmayer [1] for the description of plant structure and development. They can be best described as rewriting systems whereby complex objects can be constructed by successively replacing parts of a simple initial object using a set of rewriting rules or productions. Object generation is governed by two processes: (a) string rewriting, and (b) graphical interpretation of the resulting strings. Frijters and Lindenmayer [2] and Hogweg and Hesper [3] were first to introduce a formally-specified graphical interpretation that allowed the automatic generation of pictures of plants. Realistic-looking plants were later generated by Smith [4]. Prusinkiewicz [5] combined L-system generation with a graphical interpretation based on turtle geometry [6]. This involved the interpretation of symbols as commands controlling the movement of a graphical turtle in the three-dimensional space e.g. "F" means "move forward".

The objective of this study is to investigate the application of L-systems to the modelling of the three-dimensional anatomy. Many concepts can be borrowed from their applications in botany and additional features can be incorporated depending on anatomical requirements. This, we hope, would be a first step towards creating a virtual laboratory in anatomy.

2. A Brief Review of L-Systems

2.1. Formal Definitions

Let V denote an alphabet, V^* the set of all words over V, and V^+ the set of all non-empty words over V.

Definition 2.1.1. A **0L-system** is an ordered triplet $G = < V, w, P >$ where V is the **alphabet** of the system, $w \in V^+$ is a non-empty word called the **axiom** and $P \subset V \times V^*$ is a finite set of **productions**. If a pair (a, c) is a production, we write a $\rightarrow$ c. The letter a and the word c are called the **predecessor** and the **successor** of this production, respectively. It is assumed that for any letter a $\in$ V, there is at least one word c $\in$ V^* such that a $\rightarrow$ c. If no production is explicitly specified for a given predecessor a $\in$ V, the **identity production** a $\rightarrow$ a is assumed. A 0L-system is **deterministic** (noted ***DOL***-system) if and only if for each a $\in$ V, there is exactly one c $\in$ V^* such that a$\rightarrow$c.

Definition 2.1.2. Let G = < V, w, P > be a 0L-system, and suppose that m = $a_1 \ldots a_m$ is an arbitrary word over V. The word n = $c_1 \ldots c_m$ is **directly derived** from m, if and only if $a_i \rightarrow c_i$ for all i = 1,...,m.

Definition 2.1.3. A **T0L-system** is an ordered quadruplet G_T = < V, w, P, T > where V, w, and P are as in the definition of a 0L-system and T is a finite non-empty collection of subsets of P, called **tables**. It is assumed that for each table t $\in$ T, and each letter a $\rightarrow$ V, there is at least one production p $\in$ t with the predecessor a.

Definition 2.1.4. A word n is **directly derived** from the word m in a T0L-system G_T = < V, w, P, T > if n is directly derived from m in 0L-sytem G = < V, w, P > and all production used in this derivation belong to the same table t $\in$ T.

T0L or table-based systems are very useful for the description of complex objects where there is an inherent structuring or a hierarchical representation (e.g. ribcage modelling). Another interesting development of L-systems has been the addition of parametric L-systems whereby numerical parameters can be associated with L-system symbols. This is a most powerful tool since it allows us more flexibility in the modelling of complex objects.

2.2. Turtle Interpretation of Strings

2.2.1. Two-Dimensional Interpretation

The basic idea of the turtle interpretation is illustrated below. A **state** of the turtle is defined as a triplet (x, y, a), where (x, y) represent the Cartesian coordinates of the turtle's *position*, and the angle a, called the *heading*, is the direction the turtle is facing. Given the *step size* d and the angle increment d, the turtle responds to commands represented by the following symbols (Fig. 1a):

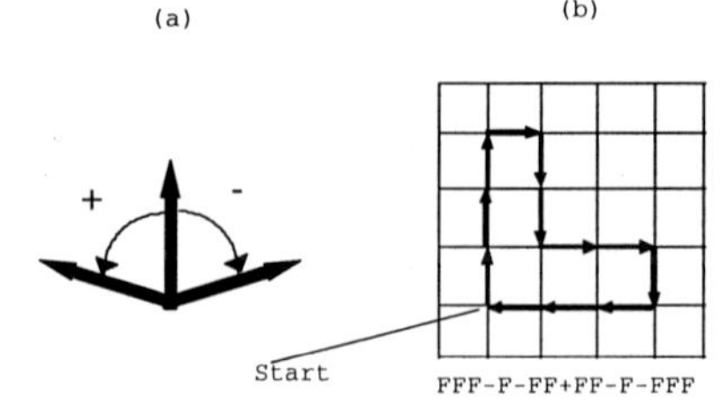

Fig. 1. Turtle Graphics: (a) Turn Left (+) and Turn Right (-) commands. (b) The graphical interpretation of FFF-F-FF+FF-F-FFF

F — Move forward a step. The state of the turtle changes to (x', y', a), where x' = x + d cos a and y' = y + d sin a. A line segment is drawn between points (x, y) and (x', y').

f — Move forward a step d without drawing a line.

\+ — Turn left by angle d. The next state of the turtle is (x, y, a + d). The positive orientation of angles is counter-clockwise.

\- Turn right by angle d. The next state of the turtle is (x, y, a - d).

Given a string n, the initial state of the turtle (x_o, y_o, a_o), and the constants d and d, the **turtle interpretation** of n is the figure drawn by the turtle in response to the turtle commands contained in it (see Fig. 1b).

2.2.2. Three-Dimensional Interpretation

Extending the turtle interpretation to three-dimensions can be achieved following the ideas of Abelson and diSessa [6]. The current **orientation** of the turtle in space is given by three unit vectors $\vec{H}, \vec{L}, \vec{U}$, indicating the turtle's **heading**, the direction to the **left**, and the direction **up**. These vectors are perpendicular to each other, and satisfy the equation $\vec{H} \times \vec{L} = \vec{U}$. The turtle can be rotated according to the following equation:

$$\begin{bmatrix} \vec{H'} & \vec{L'} & \vec{U'} \end{bmatrix} = \begin{bmatrix} \vec{H} & \vec{L} & \vec{U} \end{bmatrix} \mathbf{R} \qquad (1)$$

where R is a 3x3 rotation matrix.

The turtle is oriented in space according to the following symbols:

+ Turn left by angle d, using rotation matrix $\mathbf{R_U}(d)$.
\- Turn right by angle d, using rotation matrix $\mathbf{R_U}(-d)$.
& Pitch down by angle d, using rotation matrix $\mathbf{R_L}(d)$.
^ Pitch up by angle d, using rotation matrix $\mathbf{R_L}(-d)$.
\ Roll left by angle d, using rotation matrix $\mathbf{R_H}(d)$.
/ Roll right by angle d, using rotation matrix $\mathbf{R_H}(-d)$.
| Turn around, using rotation matrix $\mathbf{R_U}(180^o)$.

Three-dimensional graphical interpretation is not normally restricted to these symbols. Other symbols are usually defined to represent say branching, tropism (oriented growth), and collision detection, all defined in the context of plant growth and development.

3. Description of Anatomical Objects using L-Systems

L-systems has been applied successfully for the description of plant structure and growth. Ideally, what is required is a type of L-systems that would capture both the structure and growth (both formation of organs and their later development) of human anatomy. Although this has not been an easy task, the initial part of this work has concentrated on the modelling of structure rather than growth. The idea stemmed from the necessity to model the thorax for the purpose of the model-based processing of chest X-rays (see, for example [7]).

The shape descriptions needed for such a project had to satisfy the following constraints:

(1) They should be natural and should be able to capture variations in anatomy.
(2) They must be expressible in terms of a small number of parameters which can be varied with ease. This extremely useful for matching 3D models to 3D data or even the ill-defined case of 3D models to 2D data (images).
(3) Image generation from models should be efficient for the purpose of both matching and visualization of both normal and abnormal anatomy.

(4) A formal description is necessary since any model-based system would function at a higher level than the "traditional" medical image processing system. A language that can describe using simple words complex object geometries is highly desirable.
(5) Any higher level description has also to be coupled to some kind of graphical representation at the "vertex", "voxel", or "pixel" level. This is especially valuable for representing concepts such as organ adjacency, connectedness and collision.

L-systems satisfy all of these requirements. Their language provides a formal description of the underlying anatomy. The graphical interpretation is efficient and allows complex visualization. Anatomy variations can be captured through either the parametric representation of symbols or through the use of stochastic L-systems where different productions occur with different probabilities.

3.1. Branching Structures

It is not surprising that the human anatomy contains many examples of organs that consist of branching structures such as ribs, lungs, kidneys, and blood vessels. Accurate morphological description of these structures using mathematical models is difficult if not impossible. The L-systems approach adopts a hierarchal description of complex three-dimensional objects. For example, the ribcage can be thought as consisting of layers of ribs. These layers are subdivided into left ribs and right ribs which in term are subdivided into smaller cylindrical elements having diameter W.

The notation used in this study is very similar to the notation of Prusinkiewicz and Lindenmayer[8]. Two-dimensional parametric symbols e.g. layer(j,i) are also used to achieve a compact parametric representation. Also, the step movements of the turtle can be represented using mathematical functions of these parameters. As the L-system tables are iterated a number of times, the turtle draws a complex 3D structure composed of cylindrical elements of the appropriate diameters. Figure 2 shows both (a) the wire-frame graphical interpretation of the ribcage L-system and (b) its rendered image.

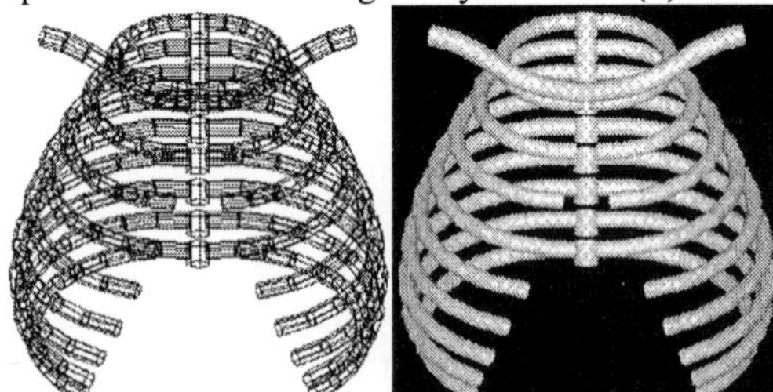

Fig. 2. The wire-frame graphical interpretation and the rendered image of the ribcage.

Another example that can be described nicely using branching L-systems is the internal structure of the lungs. The lung airway is complex in structure with a high degree of recursive division (up to 23 levels). A two-dimensional fractal description has already been attempted by various researchers including Madelbrot [9]. The hierarchy of the airway tree begins at the trachea, leads through bronchi and bronchioles into acinar airways, respiratory bronchioles, alveolar ducts, and ends in the alveolar sacs. Weibel [10] studied the geometry of the lung airway in detail and used two models to describe it: (a) a tree with regular dichotomy (two-branch subdivision), and (b) a tree with irregular dichotomy. We have used his data for the change of branch length and branch diameter with branch level or generation to produce the L-system representation shown in Fig.3.

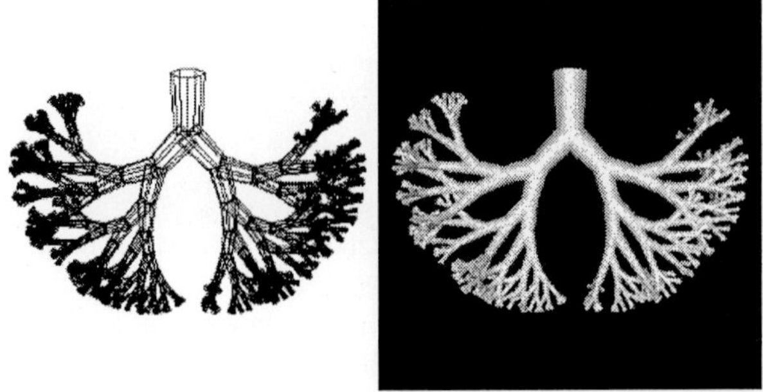

Fig. 3. The wire-frame graphical interpretation and the rendered image of the lung airway.

This study used regular dichotomy and stochastic L-systems but a future study will look more carefully at irregular dichotomy especially for the first few generation. Other important aspects of lung airway geometry that we are now considering is the space-filling growth of the airway tree and collision avoidance (non-intersecting branches).

3.2. Non-Branching Structures

Branching systems such as the ribcage and lung airway lend themselves nicely to an L-system description. The problem that we had to face is how do we represent non-branching organs such as the heart. Most such organs can be defined in terms of a natural medial axis, usually the embryological axis from which organ growth actually took place. Therefore, modelling such organs as generalised cylinders would seem the natural way to proceed. A generalised cylinder (GC) [11] is a 3D object generated by sweeping an arbitrarily shaped contour (usually closed) along an arbitrary 3D axis given by:

$$\vec{a}(u) = (a_x(u), a_y(u), a_z(u)) \tag{2}$$

where $u_1 \leq u \leq u_2$ and the three functions a_x, a_y, and a_z define the shape of the axis. The contour is given by:

$$\vec{c}(u,v) = S(\vec{x}(v), u) \tag{3}$$

where $v_1 \leq v \leq v_2$, S is the sweeping rule that specifies how the cross section, $\vec{x}(v)$, varies along the axis. The orientation of the contour is determined by the Frenet frame which consists of 3 orthogonal unit vectors: tangential, normal and binormal to the contour.

The GC representation has been incorporated in the parametric L-systems by defining a turtle movement which corresponds to a step on the axis in the 3D space. For each step, a contour of a given shape is drawn and the contours are linked together by polygonal faces. Again this scheme lends itself to a highly parametric representation of complex shapes. As an example, consider the model of the heart shown in Fig. 4.

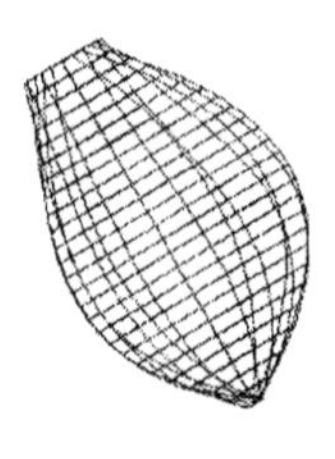
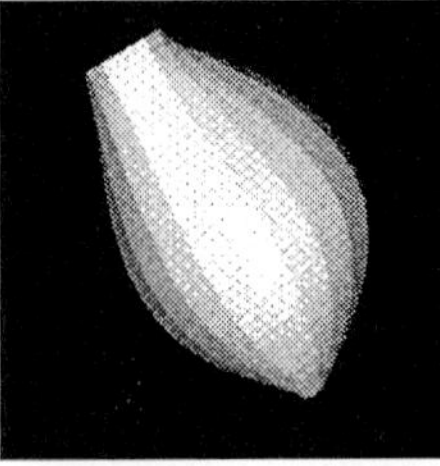

Fig. 4. The wire-frame graphical interpretation and the rendered image of the heart.

Although twenty six contours were used, the GC scheme used Fourier descriptors to parametrize the contours and polynomials to parametrize the axis resulting in a 16 parameter description of the whole volume. This is an example of the considerable reduction in data that can be achieved by using GC L-systems.

4. Conclusion and Future Work

This study has managed to use and extend L-systems to describe complex anatomical objects. Both branching and non-branching objects can be described using a formal language as well as a highly parametric graphical representation. This will be extremely useful for model-based image processing where efficient image generation and fast model matching are needed.

Future work will explore additional aspects of anatomical modelling such as space-filling representations, collision detection, and mutation. Three-dimensional anatomical description through organ growth remains an fascinating challenge that many researchers would reluctantly venture into.

References

1. Lindenmayer, A.: Mathematical models for cellular interaction in development, Parts I and II. Journal of Theoretical Biology **18** (1968) 280-315
2. Frijters, D. and Lindenmayer, A.: A model for growth and flowering of *Aster novaeangliae* on the basis of table (1, 0)L-systems. In G. Rozenberg and A. Salomaa (Eds.) L Systems, Lecture Notes in Computer Science **15 (1974)** 24-52
3. Hogeweg, P. and Hesper, B.: A model study on biomorphological description. Pattern Recognition **6** (1974) 165-179
4. Smith, A.R.: About the cover: "Reconfigurable machines". Computer **11** (1978) 3-4
5. Prusinkiewicz, P.: Graphical applications of L-systems. Proceedings of Graphics Interface **'86** - Vision Interface **'86** (1986) 247-253
6. Abelson, H. and diSessa, A.: Turtle Geometry (1982) M.I.T. Press, Cambridge and London.
7. Brown, M.S., Gill, R.W., Loupas, T., Talhami, H.E., Wilson, L.S., Doust, B.D., Bischof, L.M., Breen, E.J., Jiang, Y., and Sun, C.: Model-Based Interpretation of Chest X-rays, in Computer Applications to Assist Radiology, Boehme, Rowberg and Wolfman, eds, (1994) 344-349
8. Prisinkiewicz, P., and Lindenmayer, A.: The Algorithmic Beauty of Plants (1990) Springer-Verlag, New York
9. Mandelbrot, B.: The Fractal Geometry of Nature (1983), W.H. Freeman Co., New York
10. Weibel, E.R.: Morphometry of the Human Lung (1963) New York, Academic Press
11. IncAgin, G.J., and Binford, T.O.: Computer Description of Curved Objects. IEEE Trans. Computer, Apr. (1976) 439-449

Hierarchical Data Representation of Lung to Model Morphology and Function

Andres Kriete

Institute of Anatomy and Cell Biology
Image Processing Laboratory
Aulweg 123, 35385 Giessen, Germany

Abstract. An initial set of structural properties to model the bronchial tree of the mammalian lung is envisioned, which allows to study morphology and function. Three levels of structural organization are taken into account: 1) the well defined main bronchii of the lung supplying for the lung lobes, 2) the macroscopic, gas conductive segments of the bronchial tree and their random distribution forming the lobes and 3) the respiratory units (acini) at microscopic resolution. The final model combining these structural hierarchies can be used to study gas transport visualized by computer graphical tools.

1 Introduction

Any modeling and functional simulation of organs has to respect the inherent hierarchies of biological structures, their properties and interdependencies from cellular organelles to macroscopical anatomical structures. These basic principles have been reviewed by Hersh [1]. Beside the problems to image and represent these hierarchies, a computer modeling is often complicated by the structural complexity at the various levels. As an example, the mammalian lung requires an investigation of some thousand air conducting bronchial segments and, in addition, the investigation of respiratory units (acini) at microscopic resolution. However, with the progress in 3-D image acquisition, processing, visualization and computer modeling a complete representation of mammalian lungs comes into reach.
As an initial step, a hierarchical model is discussed here which concerns the structure of the rat lung, which has about 4000 bronchial segements. The approach suggested is to first measure the main branches of a lung exactly and to analyse selected parts of the bronchial tree in detail and secondly to use obvious regularities and similarities present in the lung structure to complete a computer model by means of a fractal graphics. The overall workflow is given in Figure 1.

2 Stereoscopic Tracings and Analysis of Casts

Plastic casts of complete lungs are the starting point to investigate the conductive part of the bronchial tree of rat and mouse. The main stem bronchi and the lobes were analysed. One example of such a lung lobe is given in Figure 2. Stereo images

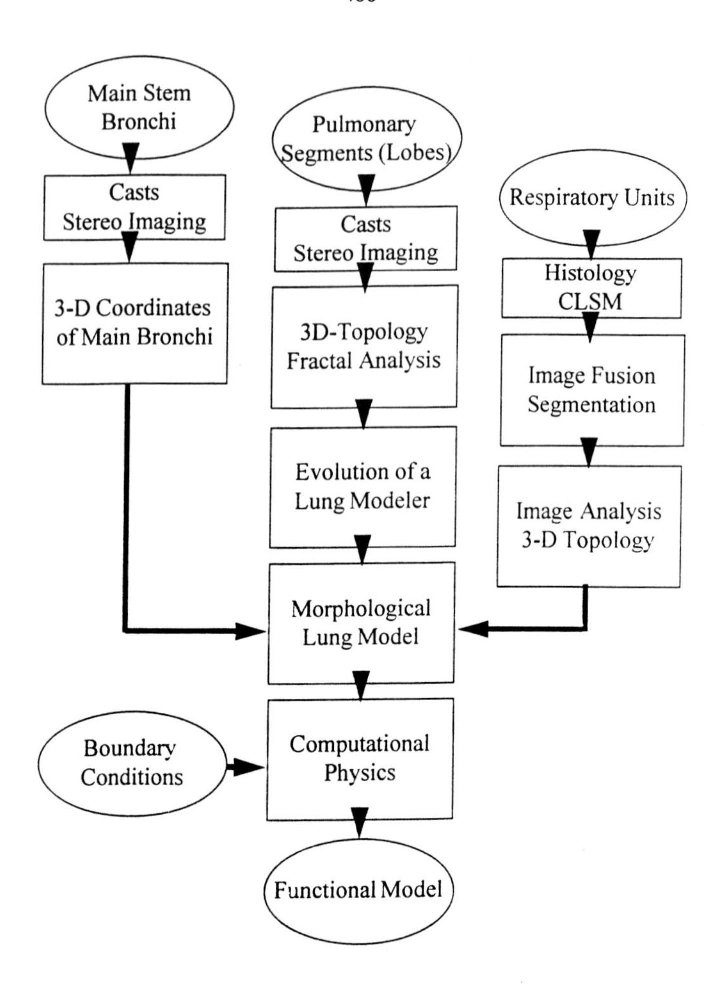

Fig. 1. Workflow required to establish a computer model of lung. Three different levels of structural hierarchies are taken into account. These are imaged and analysed individually and combined into a morphological model. Finally, computational physics is used to add functional properties like gas transport.

of corrosian casts were taken with a macro-lens and analysed with the STERECON system, developed at the NY State Department of Health in Albany [2]. The topology of the tree defined was than objected to a statistical analysis. Classification schemes, such as the so called Strahler order, were also applied. These parameters provide the necessary database before a modeling can commence.

3 Modeling of Lung Lobes

Modeling of the lungs branching pattern has been a subject of the mathematical treatment by fractal geometry. The essential of calculating such forms is to use recursive algorithms. One issue of the fractal modeler presented here is in the application of Lindenmayer (L) - systems. The L-system is used to simulate the growth of structures by reproduction with the help of fractal templates [3,4]. An L-system is adapted to our application relying on L-system grammer which allows a reproducible tuning to match against the available topological databases.
A first and simple way to model the bronchial tree is a regular, dichotomic branching pattern. Such a model has become very popular to assign the structure of lung. However, such models do not represent the casts correctly.
A high degree of self- similarity is of limited value for lung lobes, since such a model would imply a decreasing scaling in the distal segments originating the main stem having the same complexity as branches originating more distal. The overall self-similarity must be limited and the degree of order must be included, in the sense that segments of identical order have identical length and that the order of branches originating distal and asymetric is less. Taking the Strahler ordering scheme, a branch of a certain order has daughter branches of the next higher order, but as well daughter branches with more than just one higher order to assure asymetry. Since there is a relation between stepsize and the branching angle, this fact can be documented in the topological data sets comparing segment length, angle and diameter. Herewith an improved model is given.
Several authors state the fact, that the alveoli of the mammalian lung are more or less of the same size. There is a correlation between the branching angle of bronchi and the proximal volume they supply [5]. One can therefore conclude, that the branching pattern of the conductive part of lung ends in a symetric, more dichotomic form. It is also obvious from inspection of the topology of cast models, that the first branches originating the trachea are monopodialistically arranged, which suggests that there is a transition from a monopodial to a dichotomic branching pattern. Modeling the form of the complete conductive part of lung has to incorporate these facts. Branching models available so far did not consider this transition. Figure 2 gives an example of such a modified model. The form of this lobe very much resembles the original form of the cast. However, all lung lobes show variations from this schema and must be handled individually.

Fig.2. Model of lung generated by using L-system grammar (left). Incorporated are left/right asymetry and transition in bifurcation pattern from a monopodial to a dichotomic branching. As a result, the appearence of this model resembles that of a corresponding cast (right).

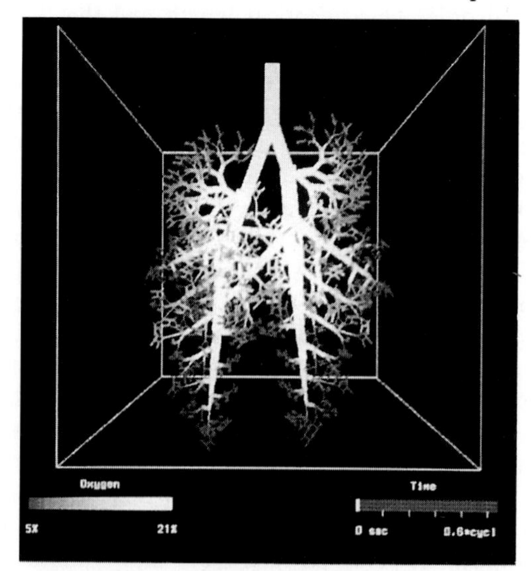

Fig.3. Complete three-dimensional morphological model of lung, consisting of 4090 bronchial segments. Different branching patterns of the various lung lobes are considered. Variation of brightness is due to concentration of oxygen during inhalation, visualized is iteration #1000 out of 10000 iterations for a complete breathing cycle. A color version of this image is available at the following website: http://www.med.uni-giessen.de/~anat/research.html.

4 Three-Dimensional Microscopy of Respiratory Units

To investigate respiratory units, confocal microscopy was applied to histological serial sections. Increase the accessible volume in confocal microscopic imaging, a framework for a computer guided imaging technique was developed. Histological sections of 50-80 micron thickness of a rat lung are the material for our study. First the sections are imaged with a low magnifying lens of 5x (imaged area of 2000x2000 microns). These images are aligned for linear shift and rotation. Transferred back to the confocal microscope, they serve for alignement of the histological sections. Variation of the scanning offset, i.e. shifting the scanning beam electronically away from the center in all 8 possible directions at a lateral distance of 600 microns, gives 9 reference images and at a zoom factor of 4 they construct a frame of 1800 um side length (data holds for a Zeiss Confocal Laser Scanning Microscope CLSM 410). Switching to a 20x magnifying lens, the field of view at zoom factor 1 fits exactly to that of the 9 individual reference images, so that non-overlapping subvolumes can be acquired by optical sectioning. Typically, 819 fused images of 768x768x91 voxels are equivalent to 1800x1800x910 microns. Such a volume encloses one complete acinus, but exhibits as well all structural details like the alveoli. The segmented volumes can be analysed for their structural composition. In particular, the sub-tree of respiratory units was analysed by tracing the acinar ducts [6] and visualizing these lines by embedded geometry. Since the ducts stem from a terminating bronchiolus, the volume these ducts supply can be analysed as a function of distance from the terminating bronchiolus exhibiting a normal (Gaussian) distribution.

5 Completing the Lung Model, Implemetation of Dynamics and Application of Computational Physics

The final step to model the morphology of a complete lung is in the combination of the main stem bronchi (highest level of structural hierarchy) with measured properties and arrangements of lobes and their three-dimensional arrangement (middle level of hierarchy) together with respiratory units (lowest level of hierarchy). Dynamics of the model is due to breathing, which mainly effects the size of the respiratory units. Data concerning the diameter of the acini, together with the surface of the acini, were deduced form microscopical findings and referenced with published data. The amount of gas flowing into the individual units is given by the flow in the corresponding conductive parts.

The complete hierarchical morphological model is extended for the modeling of gas flow and an iterative method to determine gas concentration is implemented. From a physical point of view, the gas concentration can be described as mass transport due to convection, molecular diffusion and uptake [7]. All the bronchial segments can be handled individually as finite elements. Each segment gains mass from the distal segment or losses mass due to transport into the daughter segments or gas uptake. Parameters necessary to perform such calculations include diffusion coefficient, difference of concentration across segment and gradient, way of flow partitioning

and dispersion coefficient.In addition the respiratory units are characterized by a surface-to-volume ratio and a gas uptake coefficient.
For each of the segments, an equation describing conservation of mass can be postulated. A solution of the complete bronchial tree can not be found analytically, instead an iterative method is used, where the breathing cycle is devided up into many time-steps. The number of steps necessary depends on the mathematics used to yield a stable solution, typically 10000 time steps are necessary. For each of the segments at each time step the mass transport equation which is a differential equation is solved by Gauss-Seidel iteration. The calculation proceeds from the first order branches to more distal branches down to the respiratory units, and the resulting concentrations are stored along with the segments. The actual concentration values may be visualized (see Figure 3). Several calculations of the full breathing cycle have to be carried out before a stable solution is achieved.

The resulting concentration values are visualized in a color coded fashion. Calculations can be performed for various amlitudes of the tidal volumes. The amount of gas uptake can be determined. Beside oxygen, other relevant types of gas can be studied as well.

Acknowledgements

I would like to thank M.Marko, NY State Dept. of Health, Albany, for making available the STERECON software for this project and K-P.Valerius, Giessen for loaning corrosion casts and histological sections. Special thanks go to H.-R.Duncker for many elucidating discussions.

References

1. Hersh,J.S.:A survey of modeling representations and their application to biomedical visualization and simulation. in: Proc. of the Conf. on Visualization in Biomedical Computing. Atlanta, IEEE Comp.Soc.Press (1990) 432-440
2. Marko,M.,Leith,A.:Contour-based surface reconstruction using stereoscopic contouring and digitized images. in (Ed.Kriete,A.) Visualization in Biomedical Microscopies. VCH- Publisher, Weinheim (1992) 45-73
3. Rozenberg,G.,Salomaa,A.(Eds):Lindenmayer Systems. Springer, New York (1992)
4. McCormick,B.H.,Mulchandani: L-system modeling of neurons. SPIE Vol. 2359, (Ed.R.Robb) Visualization in Biomedical Computing (1994) 693-705
5. Horsfield,K.:Anatomical factors influencing gas mixing and distribution. in (Eds.)Engel,L.A.,Paiva,M.: Gas mixing and distribution in the lung. M.Dekker (1985) 23-61
6. Kriete,A., Schwebel,T.: 3-D Top - a software package for the topological analysis of image sequences. J. of Structural Biology 116 , Academic Press (1996), 150-154
7. Mercer,R.R.,Anjilvel,S.,Miller,F.J.,Crapo,J.D.:Inhomogenity of ventilatory unit volume and its effects on reactive gas uptake. J.Appl.Physiol. 70,5 (1991) 2193-2205

Visualizing Group Differences in Outline Shape: Methods from Biometrics of Landmark Points

Fred L. Bookstein

University of Michigan, Ann Arbor, Michigan 48109 USA

Abstract. This paper concerns the visualization of group differences in the shapes of outlines without landmarks. Two helpful normalizations, Procrustes registration and the thin-plate spline, are available from the biometrics of landmarks. In this new context, the associated displays have both advantages and disadvantages of legibility or interpretation. I introduce a filter design intended to compromise among the many roles such visualizations are expected to serve. These concerns are illustrated using shapes of the corpus callosum in midsagittal images of the human brain, a data set of some relevance for the neurobiology of schizophrenia.

Over the last decade, advances in the toolkit of *morphometrics*, the quantitative analysis of biological shape and shape change, have been limited principally to the very demanding abstraction of *landmark point data*. The space of shapes of landmark configurations has a natural interpretation as a Riemannian manifold. We now understand very well the structure of null statistical models in this arena and the delicacies of passing from analysis to visualization. The power and elegance of this *morphometric synthesis* (Bookstein, 1996a) derive from symmetries in the group-theoretic underpinning of our vernacular notion of shape that do not extend to the more realistic domain of general informatic structures (edges, surfaces, regions) in routine medical images. Yet there are serious problems in the reduction of image contents to landmark points *sensu stricto*. For instance, such data are much more difficult to locate automatically than more extended structures. Over large stretches of evenly curving parts of a form, landmark points cannot be defined other than arbitrarily, or do not correspond from case to case. More seriously, the discrete structure of landmark data does not particularly suit many of the kinds of explanations biomedical scientists prefer to support via images: for instance, explanations in terms of integral surface area or volume or image contents over extended regions.

The tools introduced in this essay are intended to bridge this gap. They are designed for data sets resembling Figure 1: data from curving forms that, while not featureless, nevertheless need not have any pointlike landmarks anywhere along the arcs. My argument begins with a review of two complementary normalizations borrowed from the landmark synthesis, then explores their interaction for visualization of shape variation and group shape difference.

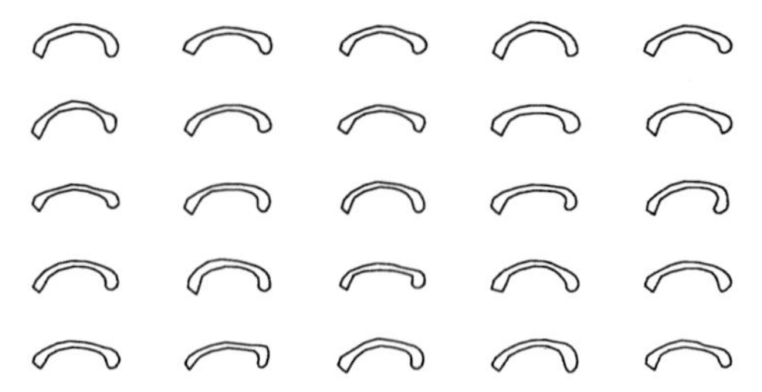

Figure 1. The data set, courtesy of John DeQuardo, MD, and Bill Green: 25 26-point polygons around the callosal border, midsagittal plane, from brain MRI. Medical staff, first twelve forms; schizophrenics, last thirteen. Splenium, rightmost (posterior) end; genu, leftmost (anterior).

1 Two image normalizations

Two normalizations are involved in analyses of outline shape. As with any problem of visualizing a shape phenomenon, one pertains to the Euclidean similarity group, the familiar group of rotations, translations, and changes of scale that leaves shape invariant. For the case of outline data, however (and its counterpart, curving surficial data in three dimensions), there is an additional normalization involving the reparameterization group of each form separately. For this auxiliary normalization, the analyses here rely on a method introduced in Bookstein (1991) and demonstrated for the medical image analysis community in Green (1995).

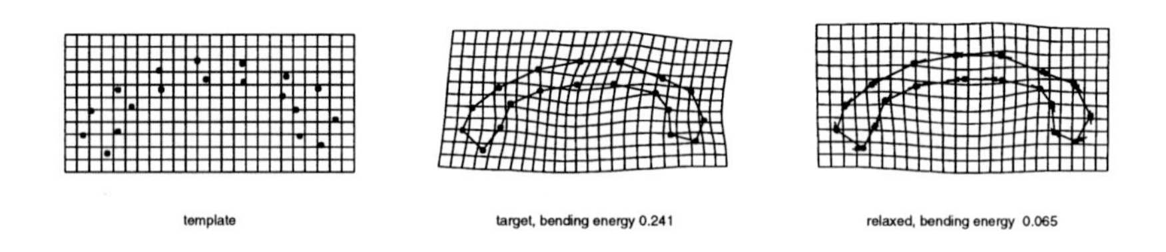

Figure 2. Relaxation by thin-plate spline. Left, callosum-shaped template. Center, ordinary thin-plate spline from the template onto an arbitrary (but similar) polygon. Right: the same after a relaxation along escribed chords (short segments shown through starting positions).

The classical thin-plate spline computes the interpolant of one set of landmarks $X_1 \ldots X_k$ onto another set $Y_1 \ldots Y_k$ that minimizes integral quadratic variation. Collect these coordinatewise as the vector $Y \equiv (Y_x|Y_y) = (Y_{1x}, \ldots, Y_{kx}, Y_{1y}, \ldots, Y_{ky})$. The minimand for the spline turns out to be equal to a quadratic form $Y_x^t L_k^{-1} Y_x + Y_y^t L_k^{-1} Y_y$, where L_k is the upper $k \times k$ submatrix of a matrix L expressing the configuration of the X's in terms of a kernel function $r^2 \log r$. The interpolant is a linear combination of offsets of those kernels (along with an affine part) using coefficients given by $L^{-1}(Y_x|0\ 0\ 0)^t$ and $L^{-1}(Y_y|0\ 0\ 0)^t$.

We modify this setup for the problem of slipping splines as follows. Let there be a "nominal set" of landmarks $Y_1^0, \ldots, Y_k^0$ again collected coordinatewise as the vector $Y^0 = (Y_{1x}^0, \ldots, Y_{kx}^0, Y_{1y}^0, \ldots, Y_{ky}^0)$. We seek the spline of one set of landmarks $X_1 \ldots X_k$ onto another set of landmarks $Y_1 \ldots Y_k$ of which a sublist $Y_{i_1} \ldots Y_{i_m}$ are free to slide away from those nominal positions $Y_{i_j}^0$ along directions $u_j = (u_{jx}, u_{jy})$. To minimize the bending energy $Y_x^t L_k^{-1} Y_x + Y_y^t L_k^{-1} Y_y$ as the landmarks Y_{i_j} of the sublist range over lines $Y_{i_j}^0 + t_j u_j$, collect the parameters $t_1, \ldots, t_m$ in a vector T and the directional constraints $u_1, \ldots, u_m$ in a matrix of $2k$ rows by m columns in which the (i_j, j)-th entry is u_{jx} and the $(k + i_j, j)$-th entry is u_{jy}, otherwise zeroes. The task is now to minimize the form

$$Y^t \begin{pmatrix} L_k^{-1} & 0 \\ 0 & L_k^{-1} \end{pmatrix} Y \equiv Y^t \mathbf{L}_k^{-1} Y$$

over the hyperplane $Y = Y^0 + UT$. The solution to this familiar *generalized* or *weighted least squares* problem is achieved for parameter vector $T = -(U^t \mathbf{L}_k^{-1} U)^{-1} U^t \mathbf{L}_k^{-1} Y^0$. A similar setup applies to relax points upon approximate tangent planes to empirical surfaces and to various special cases (landmarks bound to others in rigid linkages, landmarks restricted to curves in space), and the whole computation can be iterated for greater accuracy. Wherever the curve is adequately sampled, the general effect is to strongly suppress small-scale variations of spacing along the curve in favor of coordinated shifts at larger scale; but displacements perpendicular to the curve remain unaffected, and changes near corners are compromised.

As the energy being minimized by this spline-slip step is very nearly invariant under similarity transformations (indeed, under affine transformations), the Euclidean similarity group has still not been normalized out. Hence there is a second normalization, geometrically orthogonal to the first, of the type that has recently become standard for the morphometrics of landmarks (Bookstein, 1996a, b). Typically, a sample of shapes is normalized by computation of the so-called *Procrustes average* shape followed by superposition of each specimen over that shared average by an appropriate distance-minimizing similarity transformation. First we define a *shape distance*, Procrustes distance, between any two ordered point sets of the same count; the average is then the unique shape of least squared distance from the shapes of a sample. This algorithm can be followed in the panels of Figure 3.

Beginning from an arbitrary but sensibly spaced set of landmarks on one single form, Figure 1 was actually produced by an alternation of the algorithms in Figures 2 and 3—the Procrustes averaging and the spline relaxation—so as to produce each of the 26 vertices as the splined transforms of the same point of their common Procrustes average. The loci that result have no anatomical identifiers but remain "corresponding" points in a sense satisfactory for subsequent morphometric interpretation. Here we will call them **quasilandmarks.**

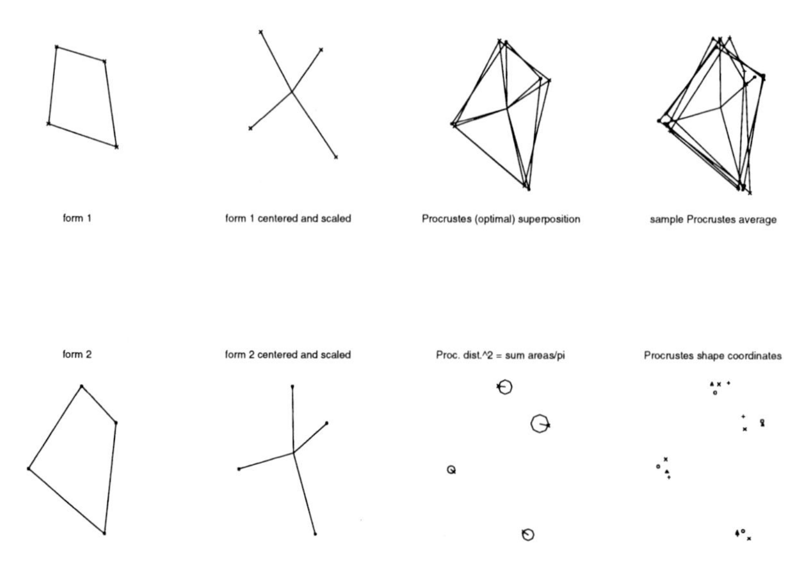

Figure 3. Geometry of Procrustes analysis. Left column: two forms. Second column: Rescale each form so that the sum of squares of the distances to the centroid of its landmarks is 1. Third column: Rotate one form over the other at the centroids (above) to minimize the sum of squared distances between corresponding landmarks (sum of areas of circles below, divided by π). Right column: The Procrustes computation of average shape. (above) Emergence of an average (dots out of center) of four forms from the alternation of two-form superpositions with averages of the resulting superposed locations. This algorithm converges to a global least-squares optimum. (below) Resulting linearized sample scatter of shapes, made up of four separate scatters of *Procrustes shape coordinates.*

2 Visualizations of an entire outline changing

The simplest coordinate system that underlies a valid multivariate statistical method of localizing shape difference is the set of Procrustes-fit coordinates to the Procrustes average shape computed as in Figure 3. The resulting superposition, Figure 4, has now been normalized both in the similarity group and in bending energy. We can show these distributions either by scatters of fitted quasilandmarks individually (left panel) or by explicit rendering of the group mean polygonal outlines (center panel). Note that the mean group difference in landmark position is nearly perpendicular to the outline except at regions of high curvature. Variance, as shown by the "thickness" of these loci, is not far from uniform around the circuit.

The thin-plate spline can sharpen the visual impact of shifts like these. At the right in Figure 4 is the transformation from the normal average to the schizophrenic average (the two curves of the middle panel), magnified threefold. Two features are visually obvious: compression of genu, and upward-rightward translation of the isthmus at right center. The grid suggests that forces near quasilandmark 15 "drag" the form in its vicinity. The enlargement of the bulb of splenium (lower right) is clearly directional, combining vertical extension with horizontal compression; by contrast, the change in genu, far left, is isotropic.

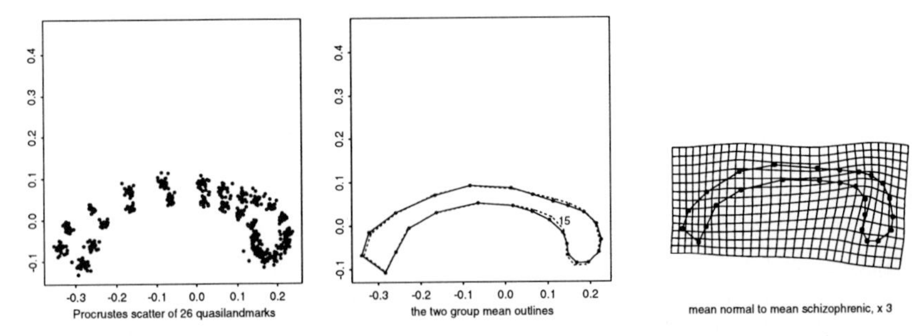

Figure 4. Procrustes fits (least-squares normalization of the Euclidean similarity group) for 26 quasilandmarks after splined slipping. (left) Procrustes scatters for each of the 26 points superimposed over the pooled Procrustes average. (center) The two group means, with quasilandmark 15 indicated. (right) Thin-plate spline for the difference of group mean shapes, magnified threefold.

3 An explicit representation of local displacement

A much more local display of change than either of these was introduced in Bookstein (1996c). The analysis applies a "δ-filter"—a contrast of center against local neighborhood—around the outline at a variety of neighborhood sizes. In this approach, somewhat analogous to contrast-of-Gaussians filters and also related to anisotropic diffusions and other versions of edge-sharpening, the data set is modeled as a common grand mean shape around which specimens vary by large-scale processes, by small-scale noise, and by one or more discrete displacements between groups of small stretches of outline. The small-scale signal is operationalized as the Procrustes residual of each quasilandmark in turn after Procrustes superposition using only the points of some neighborhood around it. A neighborhood consists of the union of the "center" with its j closest neighbors in the mean form, $j = 2, \ldots, 25$ (see Figure 5, left); different similarity normalizations (translation, rotation, rescaling) are computed every time. Because of the original splined slip normalization, the filter's output is usually oriented almost exactly normal to the mean outline.

For the cycle of all 26 quasilandmarks, the signal-to-noise ratio of the implied local displacement is shown as a surface in the second panel of Figure 5. This surface shows one clearly dominant locus of group mean difference, exactly at quasilandmark 15. The center-versus-periphery Procrustes filter output is very robust to variation of neighborhood size but highly sensitive to position along the curve. S/N of the filter atop point 15 is gently maximized at neighborhood size 8. The filter output here now separates the groups quite effectively. In fact, the group separation is 176% of the mean within-group variation in this normal direction, versus only 136% for the Procrustes fit in Figure 3.

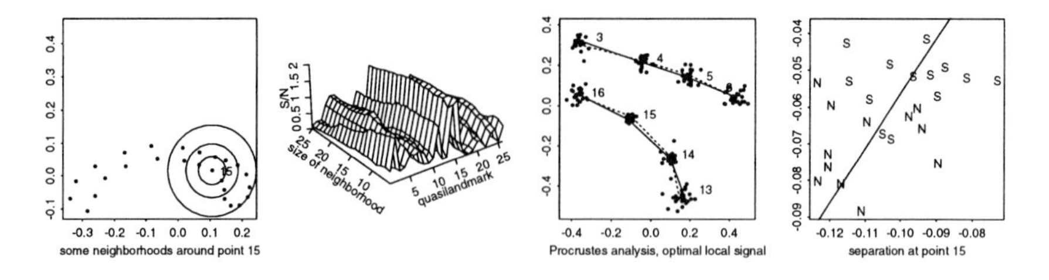

Figure 5. Operation of a δ-filter on these outlines. (left) A few of the 24×26 neighborhoods considered. (left center) The S/N surface by center and radius. (right center) Optimal local analysis of quasilandmark 15 and seven neighbors. The finding appears to be an erosion of the lower (ventricular) border of the isthmus. (right) Output of the δ-filter is read in the direction indicated. N, normals; S, schizophrenics.

The quantitative comparison of biomedical image data from curving form is hardly unprecedented (though I have no room for references on this last page); yet the specific filtering problem posed here, the optimal local description of group differences of outlines, seems novel. We need experience with a variety of data sets to see under what conditions these modes of display suit the clinical and biometric problems of cardiology, neuroanatomy, reconstructive plastic surgery, and other fields in which such data, or their surficial equivalents, arise. This problem of visualizing variability and group differences of outline or surface form, in which both global and local aspects of detection and display are entailed, seems ideal for consideration by the VBC community.

Acknowledgement. Preparation of this contribution was supported by NIH grants DA–09009 and GM–37251 to Fred L. Bookstein.

Minimum possible list of references (sorry—out of space!)

Bookstein, F. L. *Morphometric Tools for Landmark Data: Geometry and Biology*. Cambridge University Press, 1991.

Bookstein, F. L. Biometrics, biomathematics, and the morphometric synthesis. *Bulletin of Mathematical Biology* 58:313–365, 1996a.

Bookstein, F. L. Shape and the information in medical images: A decade of morphometrics. In press in *Proc. Workshop on Mathematical Methods in Biomedical Image Analysis*, I. E. E. E., 1996b.

Bookstein, F. L. Landmark methods for forms without landmarks: Localizing group differences in outline shape. In press in *Proc. Workshop on Mathematical Methods in Biomedical Image Analysis*, I. E. E. E., 1996c.

Green, W. D. K. Spline-based deformable models. Pp. 290–301 in *Vision Geometry IV*. S. P. I. E. *Proceedings*, vol. 2573, 1995.

Visualising Cerebral Asymmetry

P. Marais[1], R. Guillemaud[1,2], M. Sakuma[2], A. Zisserman[1] and M. Brady[1]

[1] Robotics Research Group, Dept. of Engineering Science
[2] Dept. Of Clinical Neurology
University of Oxford, Oxford, UK.

Abstract. We describe techniques for visualising and measuring the asymmetry of the cerebral hemispheres based on MRI scans. These techniques improve on previous approaches in two ways: firstly, measurements are not limited to voxel discretisation scales; secondly, symmetry measurements are inherently 3D. This avoids the errors in slice-based measurements arising from shape distortions introduced by mis-alignment between the head and MRI machine.
We focus on 'sparse' MRI data sets in which the slices are non-contiguous, since these constitute an important source of data in longitudinal neurological studies. Two visualisations of asymmetry are presented. The first is 3D and enables an immediate qualitative appreciation of the disposition of the asymmetry. The second is a 2D rendering of the *symmetry map*, a quantitative measure of asymmetry across the brain.

1 Introduction

The objective of this work is the visualisation and measurement of cerebral asymmetry, i.e. the disparity of the left and right hemispheres at corresponding points about the hemispheric divide. Normal people have a noticeable cerebral asymmetry; this is exhibited (usually) by the left hemisphere being of greater size at the back of the brain than the corresponding region on the right, while the opposite holds at the front of the brain. A number of researchers [1, 3, 4] have suggested that the brains of schizophrenic patients experience a pronounced loss of asymmetry during the progression of the disease. The purpose of this work is to verify and quantify this effect. A large database of *sparse* MRI scans, covering a period of several years, is available for this purpose. In this context, 'sparse' means that the individual images comprising the volumetric data are non-contiguous, with gaps between the slices.

The asymmetry effects are quite subtle in places, around voxel precision. For this reason it is important that accurate, undistorted, geometric measurements are made. Accurate measurements are achieved here at sub-pixel/voxel precision by using snakes to delineate the regions of interest (Section 2). The problem of geometric distortion has often been ignored in previous measurements of asymmetry. If the patient's head is mis-aligned with the principal imaging axes of the machine, then cosine foreshortening severely distorts measurements based on relative areas or lengths. We avoid this problem by computing a symmetry plane in 3 dimensions which is aligned with the head rather than the scans, and referring symmetry measurements to this plane using constructions in 3D. The plane is defined as the mid-saggital (MS) plane and it is obtained by locating it's intersection with each slice (Section 3).

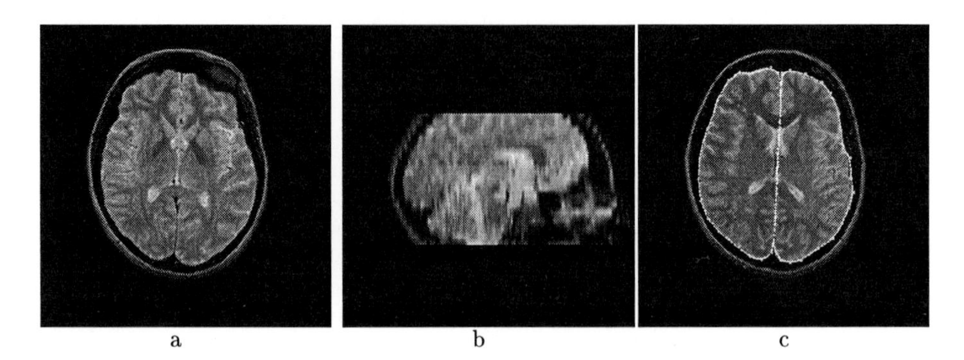

Fig. 1. MRI scans. (a) an individual MRI slice. (b) an orthogonal section through the sparse MRI data set. The affects of poor sampling in the acquisition direction are clear. (c) Snake segmentation. The snakes delimiting the boundary of the left and right hemispheres are shown.

2 Segmentation with Enhanced B-spline Snakes

The available MRI longitudinal study data is acquired as a sequence of (possibly non-contiguous) slices through an object. The slice thickness varies from under one millimetre to about 5mm, while the slice gap, if present may be as large as 2mm. As a precursor to the analysis and visualisation step, we have to segment the objects of interest, in this case the left and right cerebrum. We use approximating B-spline *snakes* [2] to delineate the boundaries of the relevant structures.

The B-spline snakes we employ are different from the 'standard' snake approach of [6]. They utilise 1D edge detectors to search for gradient discontinuities perpendicular to the snake boundary at regular intervals along the snake's length.

This sub-pixel edge detection is restricted to a *search scale*, which limits the distance at which the snake will sense edges. This is important in cluttered data sets, where there are many possible edge associations within slices. Since the problem of finding edges is essentially reduced to one dimension, the computational cost associated with these snakes is low. Different biological structures have different edge characteristics, and the edge search is tuned to these characteristics. Furthermore, the features for which the snake searches are allowed to vary along the curve, permitting *a priori* information about boundary structure to be incorporated into the segmentation process.

After the edge search has been completed, the snake is fitted to the resulting edge data (by spline regression). When the snake has 'locked on' to a boundary in one layer, it proceeds to 'track' the boundary through succeeding slices. A linear prediction scheme is used for tracking, so the the snake is initialised close to its successor on the next slice. The segmentation yields a series of closed curves representing the boundary of the cerebral hemisphere in each cross sectional scan.

These curves are linked to form a surface mesh. A transverse plane is intersected with the curves, and the common intersection point serves as an initial point for the re-sampling of the curve data (the curves may be resampled precisely through the B-spline equations). This method works well provided that the object has cylindrical topology and that the intersecting plane can be chosen to intersect all the curves. In general this has proved satisfactory.

3 Extracting the Mid-Saggital Plane

The definition of the MS plane provides a consistently oriented coordinate frame. To obtain points on the MS plane, a snake is initialised along a mid-saggital line in one slice, and allowed to propagate through the other slices under supervision. Once the snake has propagated through a number of slices, the process is terminated and the mid-saggital line data (control points representing the curve) from these slices are used as the 3D data set for an *orthogonal regression* plane fit. This yields the plane in 3D which minimizes the sum of squared perpendicular distances from the points in the data set to the plane.

The plane-fitting method was evaluated over 25 scans. The standard deviation of the sample points from the plane was between 1 and 1.5 mm (the size of a voxel), while the maximum error of the fit was 4mm, see Figure 2.

The plane constitutes a good approximation to the inter-hemispheric fissure for the study of asymmetry. Previous systems which extracted the MS plane have been limited to voxel precision (for example, the ANALYZE package). For our work the accuracy of the plane must be much higher, since the plane is used to define the reference/basic geometric information for the asymmetry computation.

4 Visualising the symmetry disparity

A direct visualisation of asymmetry is achieved by reflecting the left or right hemisphere through the symmetry plane onto it's partner. This is illustrated in figure 3. The dark surface is the (Gouraud shaded) reflected mesh.

While the reflected surface visualisation is useful for qualitative understanding, we also need a more quantitative description of the asymmetry. One means of doing this is to produce a *symmetry map* — i.e. a surface representing asymmetry over the MS plane. The asymmetry measure used is the signed difference in distances from the MS plane to the left and right cerebral surfaces in a direction perpendicular to the plane. Examples are shown in Figures 4 and 5.

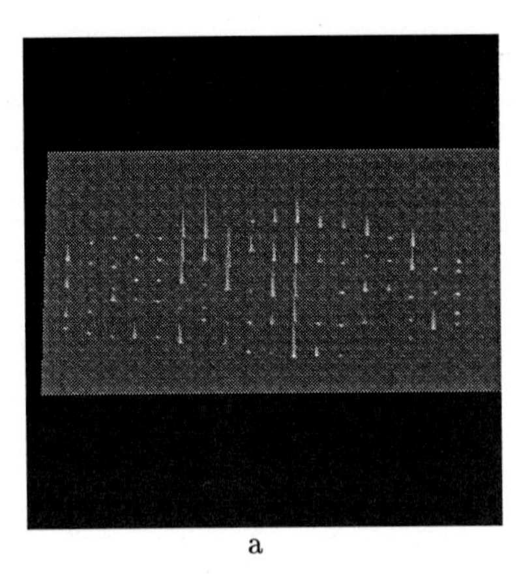

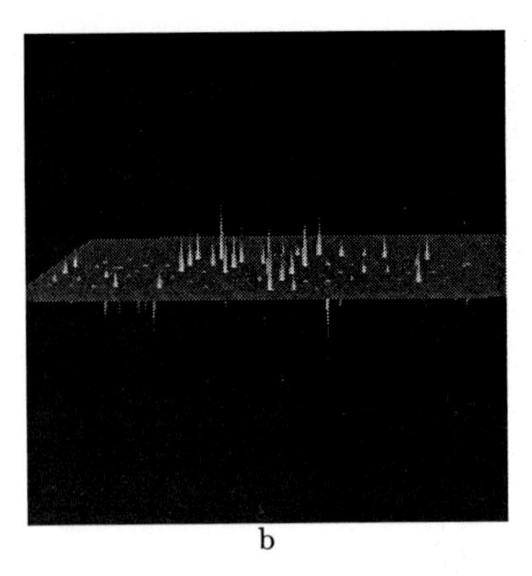

a b

Fig. 2. The error associated with the orthogonal regression plane fit. The peaks are the perpendicular discrepancy between the plane and the individual MS sample points. The max peak size is 2.2mm. The two views illustrate the fact that the error is well distributed on both sides of the plane.

5 Conclusion

The visualisation of cerebral asymmetries has potential as a diagnostic aid for neurological diseases, such as schizophrenia, which alter the morphology of the brain. Because the approach we have developed guards against machine-patient misalignment, we are able to achieve a robust segmentation, and accurate measurements and visualisation of asymmetry. In particular, the techniques are usable for sparse MRI data sets. The generation of appropriate visualisation ensures that the results are displayed in a manner which is easy to assimilate: the reflection provides an indication of the relative sizes of corresponding cerebral regions, while the symmetry map enables measurements of the distribution of the asymmetry.

5.1 Future Work

Improving the robustness of the segmentation The segmentation method we employ already uses *a priori* knowledge to guide its progress. However, we would like to build in model constraints e.g. the notion of expected geometric structure, while providing for deviations from this model caused by inter-patient variability and/or pathology.

Developing more useful symmetry measures The usual L-R symmetry measure, $\frac{M(L)-M(R)}{M(L)+M(R)}$, where $M(.)$ represents some measure applied to the L or R hemisphere, is not able to discriminate really useful asymmetry features. While the symmetry map provides a dense estimate of local asymmetry across the MS plane, we require a deeper measure of brain dissimilarity. It may be that a surface matching technique such as that developed by Feldmar [5], which attempts to find the best (locally) affine fit between the two

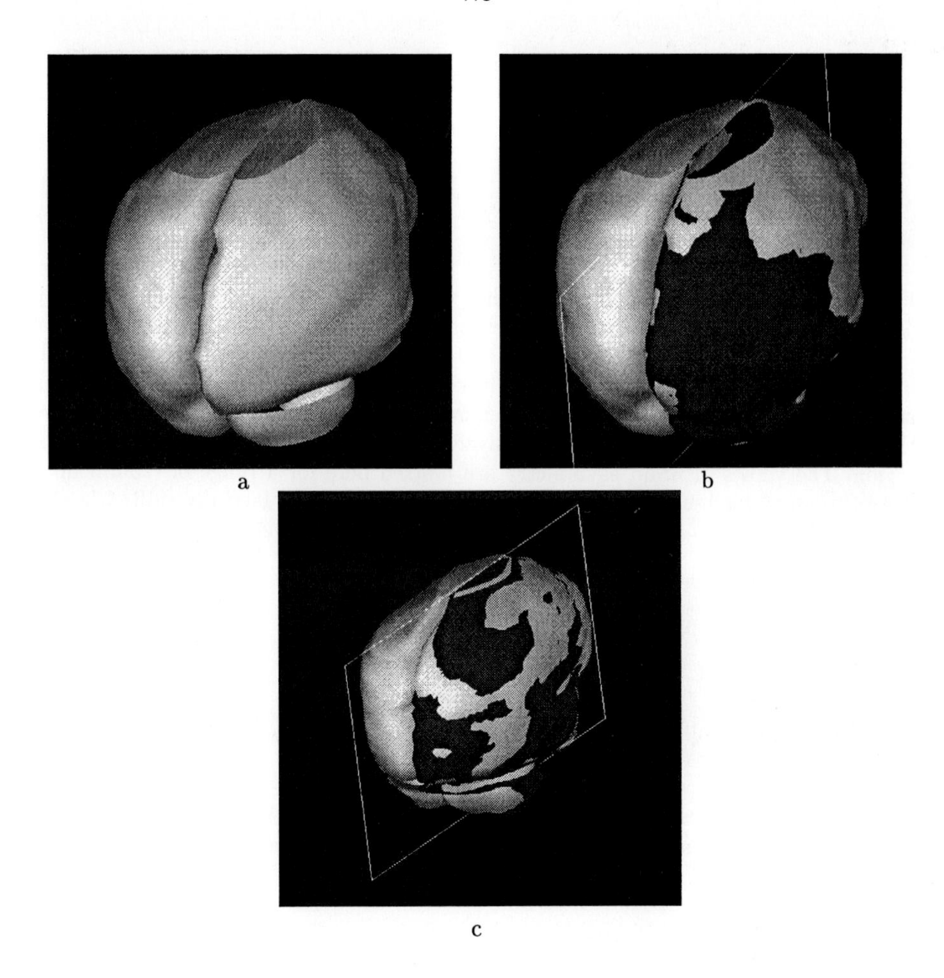

Fig. 3. Visualising asymmetry. (a) the surface meshes representing the left and the right hemisphere obtained via the segmentation technique discussed in Section 2. (b) the left hemisphere is reflected through the MS plane (represented by the rectangle) and superimposed on the right hemisphere. This allows a qualitative appreciation of the L-R asymmetry. Here, the dark surface, which is the reflection of the left hemisphere, is visible in the right posterior region, showing that the left hemisphere is larger than the right in this region. (c) symmetry reflection, applied to a different brain.

surfaces, provides the necessary generalization. One could define a *surface map* on the reference surface which provides data on the normal separation between the two surfaces at any given point.

Acknowledgements

We would like to thank our collaborators Dr. L. De Lisi (SUNY, Stonybrook) and Drs. T. Crow and B. McDonald (Clinical Neurology, Oxford) for invaluable assistance. This work was supported by MRC grant G9220033.

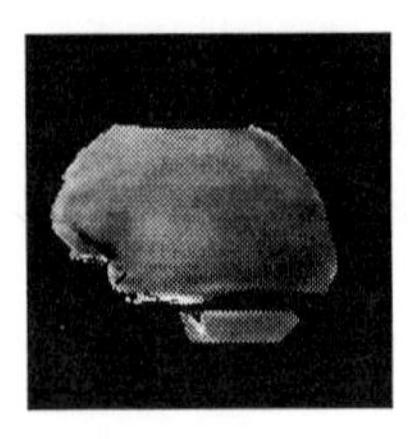

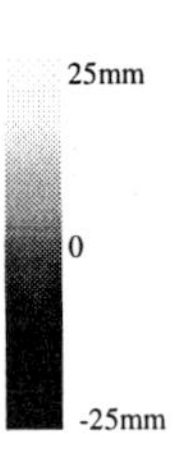

Fig. 4. The symmetry map corresponding to Figure 3b. Dark values correspond to negative asymmetry (the right hemisphere is smaller than the left) whilst lighter values represent positive asymmetry (the right hemisphere is larger than the left). The background has been set to match the darkest value (i.e. the highest negative asymmetry). The map shows that the left hemisphere is larger than the right just above the cerebellum; this asymmetry progressively diminishes as one moves away from this region. The values of brain asymmetry are plotted here for a range of -25mm up to 25mm to allow for border effects (where the hemispheres are laterally shifted, and so misaligned). The key gives the scale of asymmetry to grey-level.

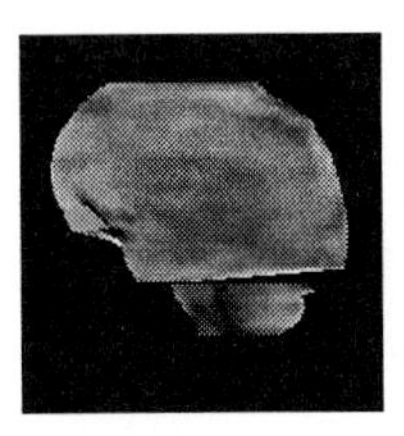

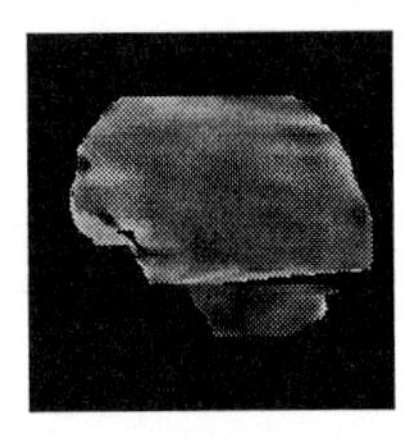

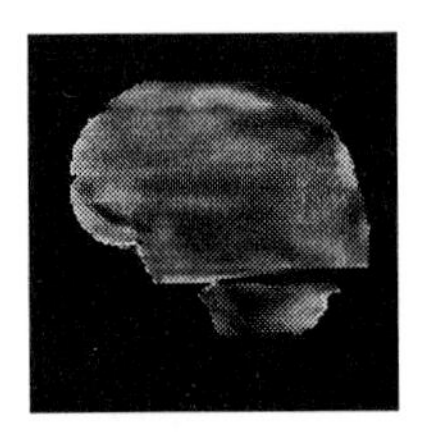

Fig. 5. Evolving symmetry maps. The sequence of symmetry maps shown here are for the same patient at one year intervals. There are small but measurable changes in the asymmetry.

References

1. R. M. Bilder. Absence of regional hemispheric volume asymmetries in first-episode schizophrenia. *American Journal of Psychiatry*, 151(10):1437–1447, Oct 1994.
2. A. Blake, R. Curwen, and A. Zisserman. A framework for spatiotemporal control in the tracking of visual contours. *Intl. Journal of Comp. Vis.*, 11(2):127–145, 1993.
3. T. J. Crow. Temporal lobe asymmetry as the key to the aetiology of schizophrenia. *Schizophrenia Bulletin*, 16:433—443, 1990.
4. L. DeLisi. Loss of normal cerebral asymmetries in first-episode schizophrenia. *To appear: Archives General Psychiatry*, December 1994.
5. J. Feldmar. *Rigid, non-rigid and projective registration of medical images.* PhD thesis, Ecole Polytechnique, 1995.
6. M. Kass, A. Witkin, and Terzopoulus. Snakes: Active contour models. *Proc. First Intl. Conf. on Comput. Vis.*, pages 259–268, 1987.

Brain: Characterization of Pathology

Assessing Patterns and Rates of Brain Atrophy by Serial Magnetic Resonance Imaging: A Segmentation, Registration, Display and Quantification Procedure

P.A. Freeborough and N.C. Fox

Dementia Research Group
The National Hospital, Queen Square, London WC1 3BG, UK and
Imperial College of Science, Technology and Medicine, London W2 1PG

Abstract. We describe a new procedure for assessing rates of brain atrophy by registration of serial 3D MRI incorporating interactive methods for the accurate identification of neuroanatomical structures which are then used in an automated registration procedure to compensate for voxel dimension and intensity inconsistencies. Original methods for visualising and quantifying rates of atrophy are described. In 11 subjects with Alzheimer's disease characteristic patterns of atrophy were seen and rates of atrophy were significantly greater ($p<0.001$) than in 11 age matched controls.

1 Introduction

Brain atrophy (or shrinkage), secondary to neuronal loss, is a characteristic feature of neurodegenerative disorders. These disorders, which include Alzheimer's disease (AD), Pick's disease and frontal lobe degeneration constitute a major and increasing health problem. AD alone accounts for over 400,000 cases in the UK. A definitive diagnosis still requires histology which is usually only possible at post-mortem examination. Significant brain atrophy is an invariant post-mortem finding. The presence of significant atrophy, seen in vivo with CT or MR imaging supports a diagnosis of a degenerative dementia while the pattern of atrophy may point to a specific disease (e.g. frontal and temporal atrophy in Pick's disease). Numerous studies have quantified specific cerebral structures in order to detect atrophy and thereby aid diagnosis. The wide range of biological variability means that cross-sectional studies tend to find overlap between groups of patients and controls. Serial studies overcome some of these problems as individuals form their own controls however measurement of specific structures is time consuming and often lacks reproducibility.

Registration and subtraction of serially acquired 3D Magnetic Resonance (MR) scans of the brain has been used to observe subtle changes in the brain including atrophy [1] and has potential in the measurement of such changes. Practical implementation of this approach in a clinical setting is problematic as voxel dimensions and voxel intensities may be inconsistent between scans acquired at different times. The tracking of atrophy is especially sensitive to these inconsistencies due to the relatively large interval between scans needed to detect change. Furthermore, the atrophic process leads to alterations in brain shape and positional change within the cranium, both of which complicate assessments of atrophy.

2 Interactive Identification of the Brain

The registration procedure requires the brain to be identified on each of the scans to be registered. This permits the exclusion of non-brain voxels from the registration process; non-brain tissue may move with respect to the brain and hence inclusion would compromise the position of best match. Furthermore, the identification of the brain allows the complete set of brain voxels to be used in the normalisation of voxel intensities between scans and in the quantification of rates of global atrophy.

Morphological operators have been shown to be useful in identifying structures from 3D brain scans [2]. The method we present is novel in that it uses conditional (rather than unconditional) morphology in a way that minimises the distortion of structure; this is important when the method is applied to the atrophied brain. In addition, it is truly interactive and does not require excessive computational resources. The method is tailored for scans such as T1-weighted MR on which white matter, grey-matter and cerebral-spinal fluid have high, intermediate and low voxel intensities respectively.

The user is taken through a process in which an approximation to the set of voxels representing the brain is refined in 4 stages; the boundary of this set is overlaid on any selected 2D slice through the data and updated in real time as the user changes the parameters of the step or the viewed slice (position, orientation or magnification). Extensive use is made of the morphological operator **B** which consists of the origin and its 6 immediate neighbours in three dimensions.

The steps involve selecting: 1) an approximate range of voxel intensities for brain tissue, 2) a number of conditional erosions using the operator **B** and an intensity threshold for each, which prevents the removal of voxels with higher intensity (note: after each erosion voxels which are not part of the largest connected set are discarded), 3) a number of dilations using the operator **B** conditional on voxel intensities being within 60%-160% (an arbitrary range which provides consistency) of the mean intensity over the set and 4) a cube size used to define an 'interior' of the brain over which the 60%-160% criteria is reapplied.

The procedure makes use of the knowledge that the brain is the largest structure in a 3D scan of the head and that it is only weakly connected to adjacent structures.

Sufficient erosions at step 2 will disconnect the brain from adjacent structures and subsequent dilations will re-introduce brain voxels removed by the erosions.

Applying the procedure to the atrophied brain without placing conditions on the morphology often results in 'thinned' structures such as cortical gyri being disconnected (fig. 1A). By placing a condition on the erosions such that after the first erosion white matter (i.e. high intensity) is not removed, a 'white matter skeleton' is maintained and thinned structures are not disconnected. In addition placing a condition on the dilations permits the user to apply more dilations than erosions, and so 'fill out' the boundary of the brain; (note the gyri at the top the scan in fig. 1B). Again this uses neuroanatomical knowledge to guide morphological operations. Basing the dilation condition on the mean set intensity leads to a consistent placing of the boundary across different scans. Where an internal brain structure has been removed by the erosions it will be re-introduced at the final interior thresholding step if the structure is sufficiently interior (note the brain stem is included in the centre of fig. 1C).

The whole process typically takes 10 minutes (per scan) with a Sparc 20 workstation (Sun Microsystems Inc., Mount View, CA) and an experienced user.

3 Interactive Identification of the Surface of the Cranium

The registration procedure requires the surface of the cranium to be defined on each scan so as to provide an invariant structure from which to calculate the spatial scaling factors required to match the images. The adult cranium (or skull) is generally unaffected by neurological disorders. The cranial surface is defined simultaneously on the pair of scans to be registered, in a 3 stage process; again, the boundary of the set being defined is overlaid on any selected 2D slice through the data and is updated in real time.

At the first stage the user must select 1) a threshold such that when passing along a row or column from the left, right or top (superior) of the scan the first voxel encountered with intensity above this threshold is at the surface of the scalp, 2) a minimum vertical number of voxels between this point and the surface of the cranium, 3) a minimum horizontal number of voxels and 4) a threshold that is crossed (downwards) when passing from scalp to cranium. Starting with the full set (i.e. all voxels in the scan), the selected values are applied by traversing every row (left to right and right to left) and every column (superior to inferior) and removing all voxels encountered before reaching the edge of the cranium.

In the second stage the user selects the size of a 3D 'closing' operator, used to correct focal errors in the first stage. Such errors occur when the cranial surface is not detected and a whole row or column is removed. At the end of this stage, points not at the surface of the set are removed.

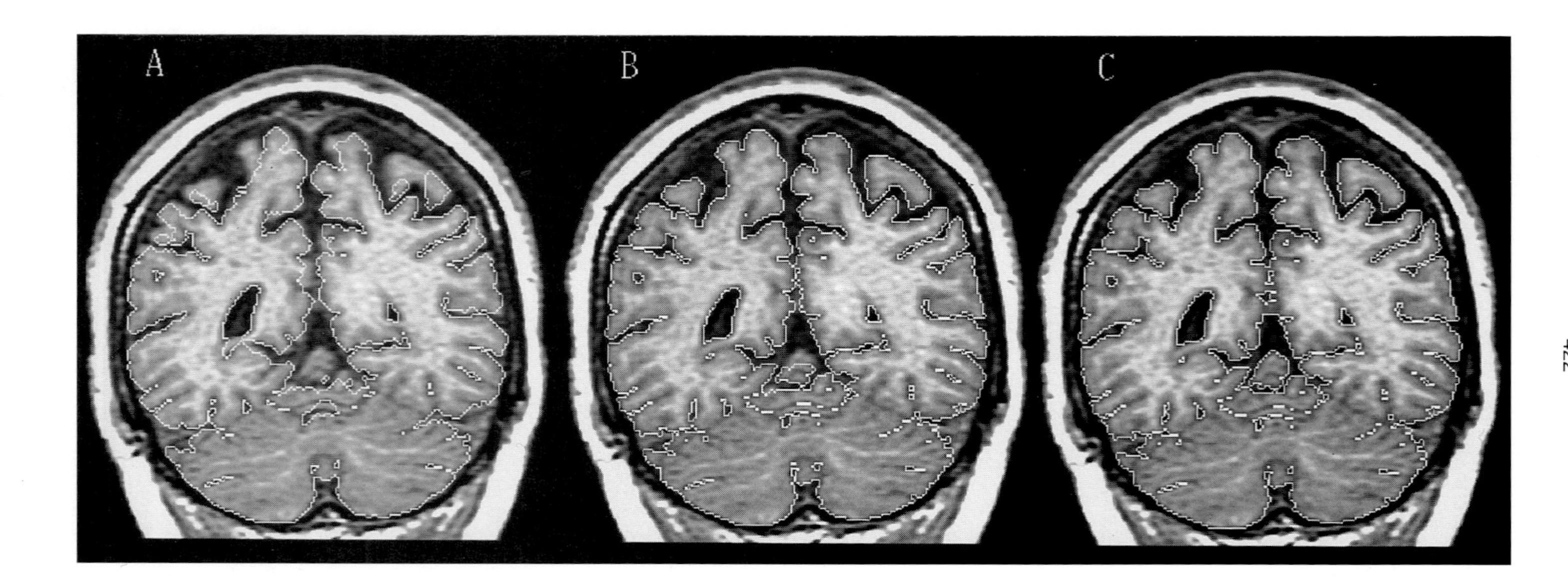

Fig. 1. Interactive segmentation of an atrophied brain (A) using non-conditional morphology, (B) using conditional morphology and (C) using conditional morphology and applying thresholds to the interior

At the third stage, the user defines planes over the scans so as to remove points on one side of the plane; typically these planes are defined so as to remove points in the face where the method fails.

We refer to the set of points identified on each scan using this procedure as a *cap*. The identification of the caps on a pair of scans typically requires 10 minutes, with a Sparc 20 workstation and an experienced user.

4 Registration

The accuracy of registration is critical to the sensitivity of the procedure as it involves subtraction of registered scans. Methods based on optimising a goodness of fit criteria, such as the uniformity of the ratio image [3], have been shown to be more accurate than other approaches [4], and hence we have adopted an optimisation based approach to registration.

Conventional MR image registration methods assume a pair of 3D images to be rigid bodies which can be matched by determining the appropriate set of rotations and translations to apply. We propose that, due to scanner instability, accurate and robust matching also requires small scaling factors to be applied to each dimension; we calculate such factors from the caps (i.e. the surface of the cranium) which should be invariant between successive scans.

Rotations and translations required for matching are determined by iteratively minimising the standard deviation of the ratio image calculated over the whole brain (as identified previously). The rotation and translation determining iterations are interleaved with iterations which calculate scaling factors. The scaling iteration uses least squares regression in each dimension independently to match the closest points on each cap; the constants so determined represent a translation to centre the two caps (compensating for translations of the brain within the cranium) and the gradients represent the scaling required to match the images. In this way the complexity of brain structure is used in the calculation of rotations and translations whereas the invariance (to neurological disease) of the cranium is used in the calculation of scaling.

Image interpolation, necessary for the resampling of one image at each iteration, is performed by warping the image in 3 passes. In each pass the warp is performed using 1D sinc interpolation with a kernel size of R. First and second derivatives are calculated by 1D convolution on the re-sampled image and cross derivatives by 1D convolution on the first derivative images. It has been demonstrated that sinc-based interpolation is necessary for accurate image matching but the computational expense of this approach is large [5]. Posing the interpolation problem as three 1D interpolations dependent on (2R) points rather a single 3D interpolation dependent on

$(2R)^3$ points, permits the use of large interpolation kernels without excessive computational expense.

Registration is performed first at low resolution and then at progressively higher resolutions. At lower resolutions the derivatives of the image, which are used in determining the direction of optimisation, are calculated on a matrix sub-sampled about the point in question; sub-sampling in this way affects the calculation of derivatives but not the number of image points which contribute to the calculation of the ratio image. No scaling iterations are performed and only simple linear interpolation is used in the low resolution iterations, because only an approximate matching is required as initialisation to the next resolution. This multi-resolution approach prevents the procedure converging on local minima in the optimisation function, which are a particular problem where genuine diffuse change has occurred and the global minimum is consequently less deep and narrow; at lower resolutions the gradient of a path on the optimisation function is calculated over a broader range and so local peaks may be skipped, and a solution close to the global minimum found to initialise the highest resolution registration.

5 Display and Quantification

Having registered the brain regions on the serial scans, corresponding brain voxels or slices may be compared, however comparison is complicated by variations in the voxel intensities produced by the scanner on successive acquisitions. We correct for these variations by defining **I** as the intersection of the two brain regions (defined by the previous semi-automated procedure) eroded by the operator **B**, and scaling voxel intensities on each scan such that the mean intensity over the set **I** is unity. Again, the operator **B** consists of the origin and its 6 immediate neighbours in three dimensions.

As the registration procedure matches brain to brain on each scan, and other structures may move relative to brain, meaningful comparisons can only be made between voxels which are within or adjacent to brain. We approximate this set of voxels as **U**, where **U** is the union of the two brain regions dilated by the operator **B**.

We define a normalised voxel pair as having undergone significant atrophy related change if 1) they are members of the set **U**, 2) the difference is greater than 0.2, 3) the minimum is less than 0.95 (the grey matter mode) and 4) the maximum is less than 1.4. Constraint (1) excludes voxels not in or adjacent to brain, (2) reduces noise related change, (3) eliminates change which occurs in the white matter which may not be atrophy related and (4) eliminates change associated with very bright artefacts such as magnetic susceptibility. These values have been empirically determined for the acquisition method used in this study.

The scans are compared using what we refer to as a *difference overlay image*. The early scan is viewed and those voxels which have undergone significant change relative to the later scan are shaded red (loss) or green (gain) using an alternate pixel display. The intensity of colour (red or green) is scaled according to the magnitude of the difference. This approach allows the precise location of change to be visualised and cross-referenced to the original structures in a way which is not possible on conventional difference images. Furthermore, the level at which difference becomes significant is defined objectively and consistently.

Total brain atrophy is quantified by integrating the differences in normalised voxel intensities over the voxels which have 'undergone significant atrophy related change'. This value is converted to a volume of tissue loss by dividing by 0.7, which is the approximate change in intensity corresponding to the loss of a complete voxel of tissue (brain with average intensity 1.0 going to cerebral-spinal fluid with average intensity 0.3); this factor was determined specifically for the acquisition used in this study. Localised measures of atrophy may also be obtained by quantifying over local subsets of **U**.

6 Methods

11 patients with Alzheimer's disease and 11 age matched controls each had two 3D MR scans with matched scan intervals between 6 months and 2 years. Imaging was performed on a GE Signa unit using a spoiled gradient echo technique (256*128*128 matrix, 24*24*19.2cm FOV, acquisition 35/5/1/35 - TR/TE/NEX/FLIP).

The brain and the outer surface of the cranium were identified on each scan using the semi-automated methods described above. Each scan pair was registered using the surface of the cranium for voxel size correction. Change between scans was visualised using difference overlay images. Rates of atrophy were quantified over the whole brain, and were compared using a Mann-Whitney U test.

7 Results

In the normal controls very little change was indicated on the difference overlay images; fig. 2A shows a difference overlay image comparing scans from a 62 year old female control (12 month scan interval). In the AD patients large amounts of red (volume losses) were seen in the neocortex, especially medial and lateral temporal lobe, and around the ventricles (enlargement); fig. 2B shows a difference overlay image comparing serial scans from a 47 year old male with mild to moderate Alzheimer's disease (11 month scan interval).

The median (range) rate of brain atrophy in the AD group of 12.2 (5.6 to 23.5) cc per year was significantly different (p=0.0001) from the rate of 0.3 (-0.9 to 1.6) cc per year in the control group. There was no overlap between the two groups (fig. 3).

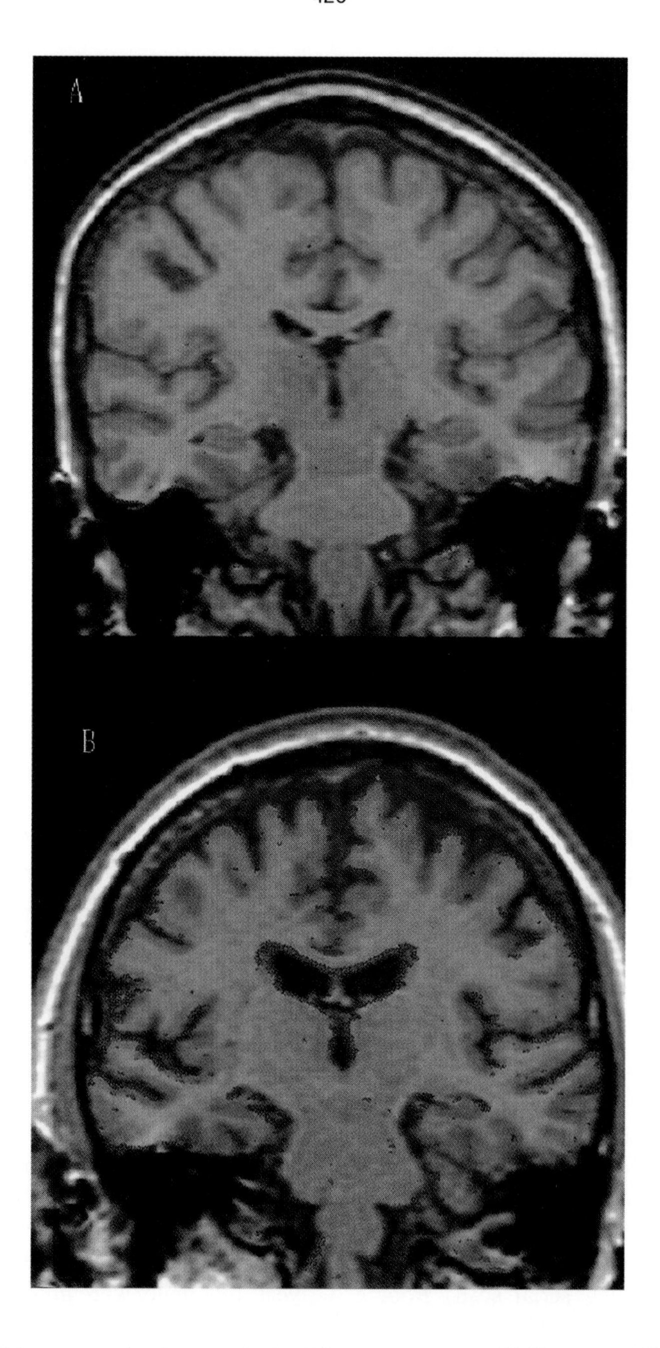

Fig. 2. Difference overlay images obtained from repeat scans of (A) a normal 62 year old female control (scan interval 1 year) and (B) a 48 year old male suffering from mild to moderate Alzheimer's disease (scan interval 11 months)

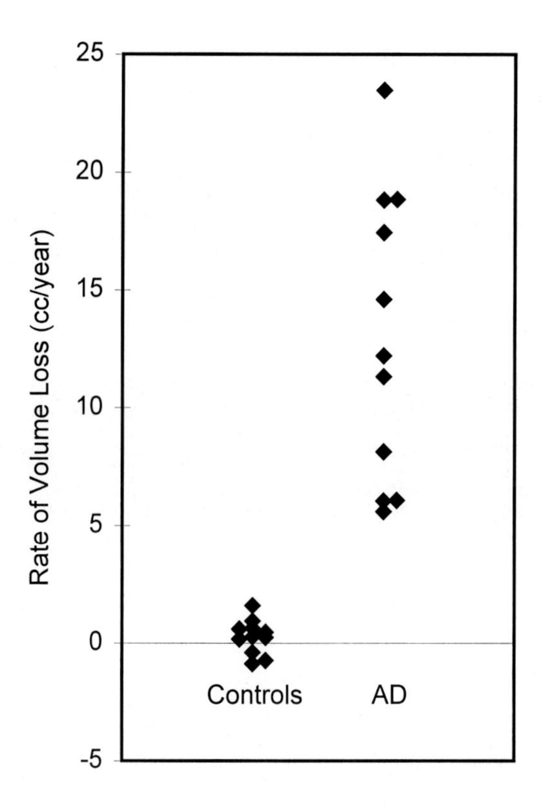

Fig. 3. Rates of brain tissue loss in controls and Alzheimer's disease (AD) measured by the registration procedure

8 Discussion and Conclusion

Registration of serial MRI is potentially a very sensitive method for assessing atrophy. Full exploitation of this sensitivity may be achieved by identifying specific structures on the pair of scans for use in normalisation (voxel dimensions and intensities). We have presented interactive methods for the identification of these structures; the methods incorporate a degree of anatomical knowledge and usefully combine a human's pattern recognition ability with a computer's numerical processing power.

In order to reduce voxel size inconsistencies spatial scaling factors are applied so as to match the surface of the cranium on each scan; this procedure is important as the mismatch which may otherwise occur may mimic or disguise atrophy. Possible sources of voxel size inconsistencies include scanner drift over time and different positioning in the field. The inconsistencies are not necessarily simple scalings, however scaling to fit points near the surface of the brain (i.e. the surface of the cranium) reduces errors where they are most severe.

The facility to view the precise location of loss and to quantify over less easily defined neuroanatomical structures, are particular advantages of the described procedure. Furthermore, the method is not critically sensitive to segmentation errors, such errors only affect sensitivity to change and cannot introduce large biases as occurs when measuring volumes directly.

The procedure is demonstrated to be an effective methodology for assessing and quantifying atrophy and may have applications in the diagnosis of and the evaluation of potential treatments for neurodegenerative disorders such as Alzheimer's disease.

Acknowledgements

P. A. Freeborough is supported by a grant from the Wolfson Foundation and N. C. Fox holds an Alzheimer's Disease Society (UK) Fellowship. We also acknowledge support and advice from Dr Martin Rossor, Dementia Research Group, and Dr Paul Tofts of the Institute of Neurology, London.

Bibliography

1. J.V. Hajnal, N. Saeed, A.. Oatridge et al., Detection of subtle brain changes using subvoxel registration and subtraction of serial MR images, Journal of Computer Assisted Tomography, 19:677-691.
2. K.H. Hohne and W.A. Hanson, Interactive segmentation of MRI and CT volumes using morphological operations, Journal of Computer Assisted Tomography, 16: 285-294.
3. R.P. Woods, J.C. Mazziotta and S.R. Cherry, Rapid automated algorithm for aligning and reslicing PET images, Journal of Computer Assisted Tomography, 16:620-633.
4. S.C. Stropher, J.R. Anderson, X.L. Xu et al., Quantitative comparisons of image registration techniques based on high resolution MRI of the brain, Journal of Computer Assisted Tomography, 18:954-962.
5. J.V. Hajnal, N. Saeed, E.J. Soar et al., A registration and interpolation procedure for subvoxel matching of serially acquired MR images, Journal of Computer Assisted Tomography, 19:289-296.

Characterisation and Classification of Brain Tumours in Three-Dimensional MR Image Sequences

C. Roßmanith[1], H. Handels[1], S.J. Pöppl[1], E. Rinast[2], H.D. Weiss[2]

[1] Institute for Medical Informatics
[2] Institute for Radiology
Medical University of Luebeck, Ratzeburger Allee 160, 23538 Luebeck, FRG
email: rossmani@medinf.mu-luebeck.de

Abstract In this paper a new approach for quantitative description and recognition of brain tumours in three-dimensional MR image sequences is presented. In radiological diagnostics, tumour features like shape, irregularity of tumour borders, contrast-enhancement etc. are used in a qualitative way. The presented image analysis methods extract quantitative descriptions of two- and three-dimensional tumour features. The tumour shape is described by ellipsoid approximations and the analysis of tumour profiles. Fractal features quantify irregularities and self-similarities of tumour contours. The internal tumour structure and especially the homogeneity of tumour tissue is analysed by means of texture analysis. The extracted features reflect 3D information about tumour morphology. The high-dimensional feature information is analyzed by the nearest neighbour classifier. Furthermore, in combination with the leaving-one-out method the nearest neighbour classifier is used to select proper feature subsets discriminating different tumours types. In a clinical study, the developed methods were applied to image sequences with the four most frequent brain tumours: meningiomas, astrocytomas, glioblastomas, and metastases. The clinical accuracy of radiologists in identifying brain tumours is approx. 80% [KWG+89]. The automatic recognition of tumour types achieves a classification rate of 93%.

1 Introduction

Magnetic resonance imaging (MRI) has been of increasing interest in the diagnosis of brain tumours in recent years [HGF90]. During the examination of a patient, several three-dimensional image sequences with varying imaging parameters are usually generated. Of special diagnostic interest are T1 weighted image sequences generated before and after the injection of contrast agents. Tumourous tissue with a defective blood-brain barrier appears bright in contrast-enhanced sequences and thus can be delimited from the surrounding tissue. The clinical accuracy of identifying brain tumours in radiology is approx. 80% [KWG+89]. Image analysis methods have been developed to describe brain tumours in a quantitative way. A clinical study was performed investigating the four most frequent brain tumours: meningiomas, astrocytomas, glioblastomas, and metastases. These tumours contribute 70% to all intracranial tumours. The extracted features reflect 3D information about tumour morphology, the internal textural structure as well as information about the irregularites of tumour boundaries. The following section decribes the first processing step, the segmentation of brain tumours.

2 Segmentation

First of all, the tumour is segmented in each slice of the T1-weighted, contrast enhanced sequence. For that purpose a polygon limited ROI is marked in each slice, containing the entire tumour. Tumours differ with respect to their ability to absorb contrast agents. Well enhanced tumours are segmented with edge oriented algorithms, otherwise threshold based segmentation algorithms are applied. For edge detection a combination of a smoothing exponential filter and a Laplacian operator is applied. In the other case the threshold has to be chosen interactively [RK82, GW87]. Afterwards an edge image is generated, which contains edges belonging to the outer tumour border as well as inner contours of the tumour marked with different indices.

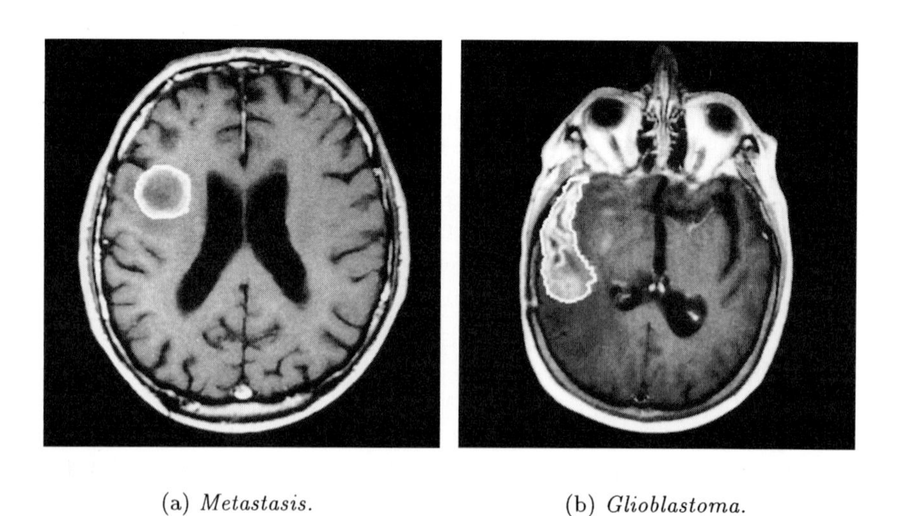

(a) *Metastasis.* (b) *Glioblastoma.*

Fig. 1. *Axial slices of MR image sequences. Additionally the detected external contours of the tumours are marked white.*

If the outer tumour border is not already closed this is done automatically by connecting adjacent contour segments. These edge images contain information about the 3D shape of the tumour and can easily be transformed into masking images $\mathcal{I}_m$:

$$\mathcal{I}_m(x,y) = \begin{cases} 1 & : \quad (x,y) \quad \text{belongs to tumour} \\ 0 & : \quad \text{else} \end{cases} \tag{1}$$

The masking images together with the edge images form the starting point for feature extraction methods described in the following. Fig. 1 examplarily shows two MR images with different types of brain tumours together with the detected outer contours, depicting the result of the segmentation.

3 Fractal analysis

Fractal structures have the particular characteristics to be (statistically) self-similar, i.e. scaled copies appear as substructures of the whole structure. The von Koch snowflake (fig. 2) is a well known fractal which is exactly self-similar by construction [Vos88]. Fractal structures appearing in nature cannot be expected to be exactly self-similar but statistically self-similar. Both kinds of fractals show new details when examined at larger scales. Fractals are of rising interest in image processing and computer graphics [VW92, Müs91, PS88]. In image processing the range of possible scales is limited by the resolution of the image. The relation between the scaling factor and amount of details appearing on that scale is described by fractal dimensions. For exactly self-similar sets, which consist of N copies of themselves, each scaled down by the factor s, the similarity dimension D is defined as follows [Fal90]:

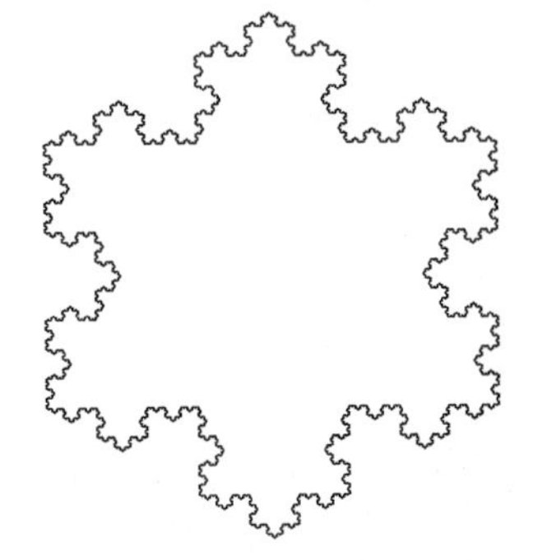

Fig. 2.: *The von Koch snowflake.* $D = \log(4)/\log(3) \approx 1.26$.

$$D = \frac{\log(N)}{-\log(s)}. \tag{2}$$

This dimension is a generalisation of the topological dimension and usually a real number unlike the topological dimension: $D(snowflake) = \log(4)/-\log(\frac{1}{3}) \approx 1.26$. For arbitrary fractal subsets of $\mathbb{R}^n$ the Hausdorff dimension is applicable. The examined set $\mathcal{S}$ is covered with a minimal number of subsets of $\mathbb{R}^n$ with diameter $d \in \mathbb{R}$ equal to or less than $\delta \in \mathbb{R}$. The diameter $d(\mathcal{S})$ of a set $\mathcal{S}$ is given by

$$d(\mathcal{S}) = \sup_{x,y \in \mathcal{S}} \{ |x-y| \}. \tag{3}$$

It denotes the maximal distance between two points in $\mathcal{S}$. The number N of subsets needed to cover $\mathcal{S}$ for $\delta \to 0$ gives information about the amount of details of $\mathcal{S}$. There are several ways to calculate fractal dimensions. In our approach we use the fractal box-dimension D_B [Fal90]. It is obtained by covering a set $\mathcal{S}$ with a minimal number N_δ of spheres with diameter δ, cubes with edge length δ or a grid of mesh size δ and considering δ going to zero:

$$D_B = \lim_{\delta \to 0} \frac{\log(N_\delta)}{-\log(\delta)}. \tag{4}$$

In image analysis, a grid is used to cover $\mathcal{S}$, which can be identified with the set of marked pixels. Unfortunately, we encounter problems as δ goes to zero for the application under consideration. This is due to the discrete and finite structure of digital images, which cannot be examined over an unlimited range of scales. Therefore the number N_{δ_i} of boxes needed to cover $\mathcal{S}$ is determined for different grid sizes δ_i, $i = 1, \dots, n$. Afterwards $\log(N_{\delta_i})$ is plotted versus $-\log(\delta_i)$ and a linear function is determined by means of regression analysis. The resulting regression function optimally fits these points with respect to Gauß' criterion of least squared errors. An estimate $\tilde{D}_B$ of the box-dimension D_B of $\mathcal{S}$ is given by the slope of the linear regression function. Before the regression analysis is performed, a subset of $\{(\log(N_{\delta_i}), -\log(\delta_i)) \mid i = 1, \dots, n\}$ is ascertained by gradually omitting points from the left or the right border in order to achieve a linear correlation coefficient greater than $\rho_0 = 0.99$ for the remaining subset of points. This is performed under the constraint that at least three points remain. Otherwise, if the correlation coefficient is still less than ρ_0, the fractal box-dimension is not estimated. The excluded points are chosen from the border to yield a successive range of scales and besides points at the border have least fractal information: e.g. $N_{\delta_1} = 1$, $N_{\delta_n} = |\mathcal{S}|$ holds for every binary image. The estimate for the fractal dimension of the von Koch snowflake, calculated as described above, is $\tilde{D}_B = 1.25$, differing only slightly from the similarity dimension $D = 1.26$.

In the application in question, the fractal dimension is used to characterise the vital part of a tumour by considering outer and inner tumour contours. The fractal dimensions $\tilde{D}_{B_1,i}$ of each outer tumour contour and $\tilde{D}_{B_2,i}$ of the outer and inner tumour contours are calculated for each slice of a three-dimensional data set and the arithmetic means, $\tilde{D}_{B_1}$ and $\tilde{D}_{B_2}$ are determined to obtain features characterising the entire tumour. Furthermore the difference $\Delta_{D_B} = \tilde{D}_{B_2} - \tilde{D}_{B_1}$ is calculated. Fig. 3(a) shows the contours of a glioblastoma, in fig. 3(b) the accompanying $\log(N)$ vs. $-\log(\delta)$ plot with the determined regression function is examplarily depicted. The resulting fractal dimensions are $\tilde{D}_{B_1} \approx 1.13$ and $\tilde{D}_{B_2} \approx 1.40$.

4 3D shape analysis

Analysis of the 3D shape of the tumour is performed with the intention to differenciate between elongated and spherical shapes using a 3D approximation of the tumour. For that purpose the tumour is approximated by an ellipsoid. The defining parameters of the ellipsoid are calculated by means of principal components analysis, which performs a principal axis transformation of the matrix

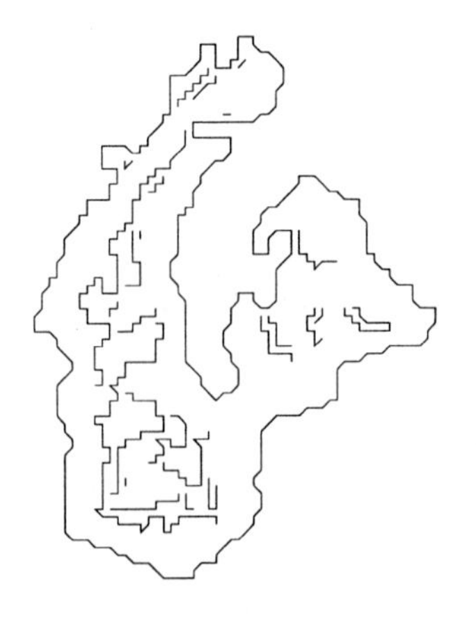

(a) *Outer and inner contours of a glioblastoma.*

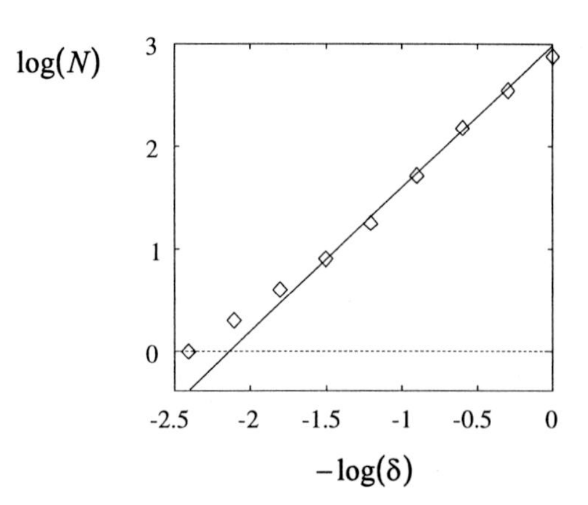

(b) $\log(N)$ *vs.* $-\log(\delta)$ *plot of the tumour contours and the corresponding regression function.* $D_{B_2} \approx 1.40$.

Fig. 3. *Example for the fractal analysis of the vital part of a tumour.*

$\mathcal{M} \in \mathbb{R}^{3\times3}$, the elements of which are the second-order spatial moments of all pixels inside the tumour [HS92]:

$$\mathcal{M} = \begin{pmatrix} \mu_{xx} & \mu_{xy} & \mu_{xz} \\ \mu_{yx} & \mu_{yy} & \mu_{yz} \\ \mu_{zx} & \mu_{zy} & \mu_{zz} \end{pmatrix} \tag{5}$$

where

$$\mu_{ij} = \frac{1}{|\mathcal{V}-1|} \sum_{(v_x,v_y,v_z)\in\mathcal{V}} (v_i - \bar{v}_i)(v_j - \bar{v}_j) \qquad i,j \in \{x,y,z\}, \tag{6}$$

$\mathcal{V}$ denotes the set of all voxels belonging to the tumour, (v_x, v_y, v_z) is the coordinate of a tumour voxel and $\bar{v}_i$ is the mean of v_i[1].

The center of the ellipsoid resides in the center of mass of $\mathcal{V}$, the directions and the lengths of the half-axes of the ellipsoid are calculated from the eigenvectors and the eigenvalues of $\mathcal{M}$, respectively. The normalised lengths $\tilde{a}, \tilde{b}, \tilde{c} \in \mathbb{R}$ of the half-axes are calculated by dividing each length $a, b, c \in \mathbb{R}$ by the maximal length:

$$\tilde{a} = \frac{a}{\max\{a,b,c\}} \qquad \tilde{b} = \frac{b}{\max\{a,b,c\}} \qquad \tilde{c} = \frac{c}{\max\{a,b,c\}}. \tag{7}$$

The normalised lengths are invariant against changes in size and therefore suitable to characterise the 3D shape of the tumour.

[1] In statistics $\mathcal{M}$ is called the covariance matrix of random vectors.

5 Profile analysis

The structure of the tumour border is an important feature for distinguishing different tumour types in medical diagnosis. For shape analysis the idea of comparing the outer tumour contour to a circle [KPF93] is generalised to approximating the outer tumour contour in each slice by an ellipse. Features describing the tumour shape are derived from the profile of the tumour contour relative to the approximated ellipse.
The approximation of the tumour contour by an ellipse $\mathcal{E}$ is performed analogously to the three-dimensional approximation of the tumour by an ellipsoid (see section 4). In this case the elements of the matrix $\mathcal{M} \in \mathbb{R}^{2\times 2}$ are calculated on the basis of all pixels inside the tumour contour. Fig. 4 shows the border of a glioblastoma together with the approximated ellipse.

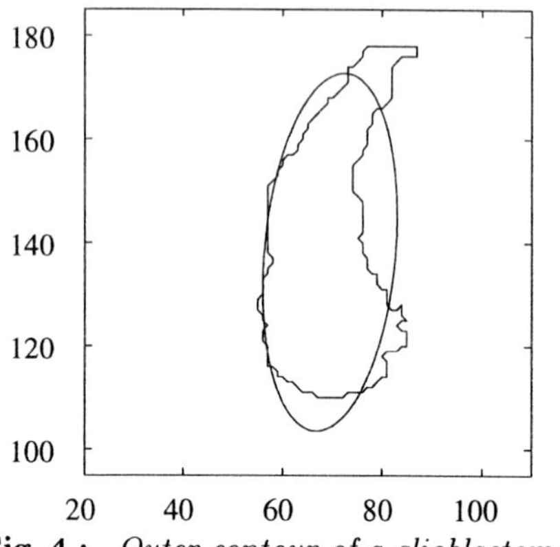

Fig. 4.: *Outer contour of a glioblastoma and the approximated ellipse.*

The profile p of the outer tumour contour is the result of tracing the contour clockwise starting at the leftmost topmost point of the contour and calculating the distance p_i from the ellipse $\mathcal{E}$ to each point P_i of the contour:

$$p = (p_i = d(P_i, \mathcal{E}) \,|\, i = 1, \ldots, N) \,. \tag{8}$$

The absolute value of p_i is given by the length of the shortest line connecting P and $\mathcal{E}$. There is no closed formula for calculating this distance, so that the starting point $Q \in \mathcal{E}$ of the connecting line has to be determined using an iterative numerical algorithm. Q fulfills the condition that $\overline{PQ}$ is perpendicular to the direction of the tangent in Q. Together with the condition $Q \in \mathcal{E}$ it is possible to find a solution for Q using Newton's method. The distance of contour points inside the ellipse is defined to be negative, points outside the ellipse have a positive sign. In fig. 5 the calculated profile for the tumour contour and the ellipse shown in fig. 4 is depicted. Jagged, irregular shapes stand out due to bigger distances whereas the profile of roundly, regularly shapes is of low amplitude.

For shape description of a single tumour contour in slice j the mean μ_j of the absolute values of the distances, the mean sum $\overline{SSQ}_j$ of squared distances as well as the standard deviation σ_j and the entropy H_j of the distribution of distances are determined as follows:

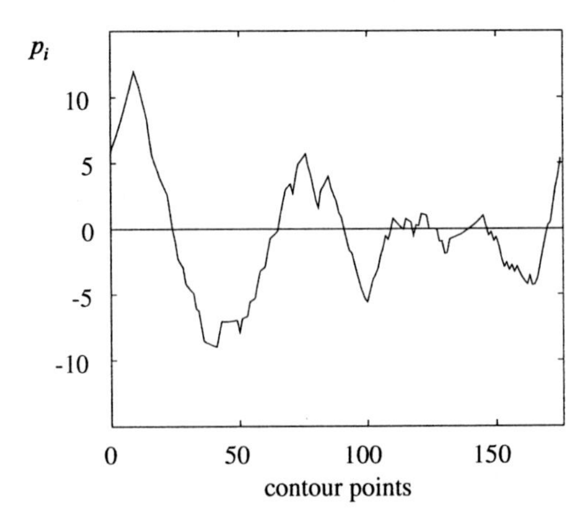

Fig. 5. *Profile of the tumour contour of the glioblastoma (distances from points on the contour to the ellipse). x-axis: index of contour points, y-axis: distances*

$$\mu_j = \frac{1}{N}\sum_{i=1}^{N}|p_i|, \quad \sigma_j = \sqrt{\frac{1}{N-1}\sum_{i=1}^{N}(p_i - \bar{p}_i)^2}, \quad \bar{p}_i = \frac{1}{N}\sum_{i=1}^{N}p_i$$
$$\overline{SSQ}_j = \frac{1}{N}\sum_{i=1}^{N}p_i^2, \quad H_j = -\sum_{i=1}^{N}pr_i\log(pr_i). \tag{9}$$

Here pr_i denotes the relative frequency of distances between $p_{min} + (i-1)\cdot\Delta$ and $p_{min} + i\cdot\Delta$. The value of p_{min} is chosen such that $p_i \geq p_{min}$ holds for $i = 1,\ldots,N$ and $\Delta = 0.1$ is the width of the intervals. Fig. 6 shows the histogram of distances for the profile in fig. 5. These features quantify the quality of the approximation of the tumour contour by an ellipse and hence give information about the smoothness and regularity of the tumour border. The arithmetic means μ, σ, $\overline{SSQ}$, and H of the values calculated in each slice (see eq. 9) are used as tumour describing features. Besides the ratio of the lengths of the half-axes of the approximated ellipses is calculated for each slice and averaged in order obtain the mean ratio η. Circular shapes lead to great values of η whereas elongated structures show smaller values.

6 Texture analysis

Different tumour types vary with respect to their interal structure. Haralick's textural features, which are features of second order statistics, are used to characterise the internal tumour structure and especially to describe the homogeneity of the segmented tumours. These features are derived from the cooccurrence matrix $M^\delta = (m_{ij}^\delta)_{0\leq i,j<g}$ (g denotes the number of grey levels), the definition

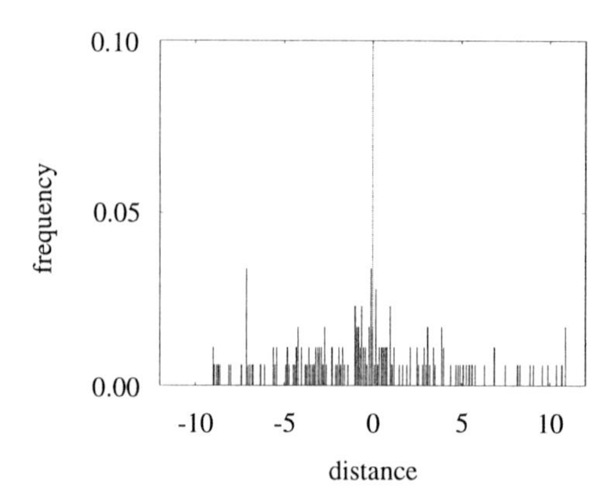

Fig. 6. *Histogram of distances for the glioblastoma.* $\Delta = 0.1$.

of which is based on pixel displacement vectors $\delta = (\delta_x, \delta_y)$. The element m_{ij}^{δ} describes the relative frequency of pixel pairs at a given displacement δ, one pixel having grey level i, the second pixel having grey level j. Haralick published 14 features [HSD73] which characterise the structure of the matrix M^{δ} and hence the distribution of grey level transitions of the underlying image. MR images are stored with a depth of 12 bit per pixel, i.e. $g = 2^{12} = 4096$, which leads to unacceptlable large, very sparse matrices M^{δ}. In order to obtain smaller matrices, which are less sparse and hence are a more stable estimate of grey level transition probabilities, the depth of the images is reduced to 8 bit per pixel using a linear grey level tranformation. In this application one common cooccurence matrix M^{δ_i} is calculated for tumour pixel of all slices for each of the directions 0°, 45°, 90° and 135°, i.e. $\delta_0 = (1,0)$, $\delta_1 = (1,1)$, $\delta_2 = (0,1)$ and $\delta_3 = (-1,1)$. The average of M^{δ_i}, $i = 0, \ldots, 3$, is found in order to achieve invariance against rotation. Afterwards Haralick's 14 textural features are calculated from the resulting matrix.

Besides the inner homogeneity of the tumour tissue the ability of a tumour to absorbe contrast medium is an important feature in the diagnosis of brain tumours. Therefore the amount of contrast medium absorption is quantified by the difference $\Delta_{\bar{g}}$ between the mean signal values in corresponding native and contrast-enhanced T1-weighted images.

7 Results

Up to now 15 cases, including five meningiomas, two astrocytomas, four glioblastomas and four metastases, have been analysed. All diagnoses are verified by histological results. For each case all features described above are calculated. The evaluation of the features shows that glioblastomas typically have great values for both of the fractal features D_{B_1} and Δ_{D_B}. Furthermore the smooth, circular shape of the metastases leads to small values for the shape features μ, σ, $\overline{SSQ}$, H and great values for η. The homogeneity of astrocytomas corresponds to small values for the texture parameters SAM (second angular moment), CON (contrast), and VAR (variance) and to great values for IDM (inverse difference moment), respectively. The lack of absorption of contrast medium of astrocytomas is represented by small values of the feature $\Delta_{\bar{g}}$.

The extracted features were used for tumour type recognition using the nearest neighbour classifier [DH73]: One advantage of the nearest neighbour classifier is that it makes no assumptions about the distribution of features in the considered classes and it can be applied even if only a little number of patient cases has been investigated. Given a labelled set of feature vectors $\mathcal{F} = \{F_1, \ldots, F_m\} \subseteq \mathbb{R}^n$ and a new vector $F_0 \in \mathbb{R}^n$, the nearest neighbour classification labels F_0 with the class of the vector F_i, $1 \leq i \leq m$, which has the minimal distance to F_0[2]. The used distance measure is the Euclidean distance. Because the Euclidean distance is not invariant against changes in scale, the features f_i first are normalised to yield features $\tilde{f}_i$ [JD88]:

$$\tilde{f}_i = \frac{f_i - \bar{f}_i}{\sigma_{f_i}}. \tag{10}$$

Here $\bar{f}_i$ and σ_{f_i} denote the mean and the standard deviation of f_i, respectively. Classification rates have been estimated by means of the 'leaving one out'-method, i.e. each feature vector $\tilde{F}_j$, $1 \leq j \leq 15$ is taken out of the sample in succession and classified with the nearest neighbour classifier [Fuk72]. The classification rate is given by the percentage of correct classified feature vectors. The classification based on all features led to unsatisfying classification rates and motivated incrementally selecting subsets of features. For that purpose all subsets with up to five element were considered. At that number of features the classification rate stabilised with a value of 93%. The result are the following three feature combinations:

$$(VAR, \Delta_{\bar{g}}, \Delta_{D_B}, \eta), \quad (H, \Delta_{\bar{g}}, \Delta_{D_B}, \eta), \quad (VAR, CON, \Delta_{\bar{g}}, \Delta_{D_B}, \eta), \tag{11}$$

In all cases only one meningioma has been misclassified.

References

[DH73] DUDA, R.O. and HART, P.E.: *Pattern Classification and Scene Analysis.* John Wiley & Sons, Inc., 1973.

[Fal90] FALCONER, K.J.: *Fractal Geometry. Mathematical Foundations and Applications.* John Wiley & Sons Ltd., Chichester, 1990.

[Fuk72] FUKUNAGA, K.: *Introduction to Statistical Pattern Recognition.* Academic Press, 2 edition, 1972.

[GW87] GONZALEZ, R.C. and WINTZ, P.: *Digital Image Processing.* Addison Wesley, 2 edition, 1987.

[HGF90] HUK, W.J., GADEMANN, G. and FRIEDMANN, G.: *MRI of Central Nervous System Diseases.* Springer-Verlag, 1990.

[2] F_i is the nearest neighbour of F_0 in $\mathcal{F}$.

[HS92] Haralick, R.M. and Shapiro, L.G.: *Computer and Robot Vision*, volume 1. Addison-Wesley Publishing Company, 1992.

[HSD73] Haralick, R.M., Shanmugam, K. and Dinstein, I.: *Textural Features for Image Classification.* In *IEEE Transaction on Systems Man Cybernetics*, volume SMC-3, pages 610–621, 1973.

[JD88] Jain, A.K. and Dubes, R.C.: *Algorithms for Clustering Data.* Prentice Hall, 1988.

[KPF93] Kilday, J., Palmieri, F. and Fox, M.D.: *Classifying Mammographic Lesions Using Computerized Image Analysis.* IEEE Transactions on medical imaging, 12(4):664–669, 1993.

[KWG+89] Kazner, E., Wende, S., Grumme, T., Stochdorph, O., Felix, R. and Claussen, C.: *Computed Tomography and Magnetic Resonance Tomography of Intracranial Tumors.* Springer Verlag, 2. edition, 1989.

[Müs91] Müssigmann, U.: *Homogeneous fractals and their application in texture analysis.* In *Fractals in the fundamental and applied sciences*, pages 269–283, 1991.

[PS88] Peitgen, H.-O. and Saupe, D. (editors): *The Science Of Fractal Images.* Springer Verlag, 1988.

[RK82] Rosenfeld, A. and Kak, A.C.: *Digital Picture Processing*, volume 2. Academic Press, 2 edition, 1982.

[Vos88] Voss, R.F.: *The Science Of Fractal Images*, pages 21–70. Springer Verlag, 1988.

[VW92] Voss, R.F. and Wyatt, J.C.Y.: *Multifractals and the Local Connected Fractal Dimension: Classification of Early Chinese Landscape Paintings.* In *Application of Chaos and Fractals*, pages 171–184. Springer Verlag, London, 1992.

Automatic Quantification of Multiple Sclerosis Lesion Volume Using Stereotaxic Space

Alex Zijdenbos, Alan Evans, Farhad Riahi, John Sled,
Joe Chui, and Vasken Kollokian

McConnell Brain Imaging Centre, Montréal Neurological Institute, McGill University,
3801 University St., WB-208, Montréal, Canada H3A 2B4.
Email: alex@bic.mni.mcgill.ca

Abstract. The quantitative analysis of MRI data is becoming increasingly important in the evaluation of therapies for the treatment of MS. This paper describes a processing environment for the automatic quantification of lesion load from large ensembles of MR volume data. The main components of this approach are stereotaxic transformation and multispectral classification, supported by pre- and postprocessing techniques to reduce noise and correct for intensity non-uniformities. The results of the automated approach are compared with those obtained by manual lesion delineation, showing a significant lesion volume correlation of 0.94. **Key words:** MRI, multiple sclerosis, tissue quantification, image segmentation, stereotaxic space.

1 Introduction

Although the use of magnetic resonance imaging (MRI) as a qualitative clinical diagnostic tool in the study of multiple sclerosis (MS) has been established for well over a decade, the quantitative analysis of MRI data for this purpose is only now attracting increasing interest. This attention is, among other factors, driven by the increased use of quantitative MRI as a surrogate endpoint in clinical trials aimed at establishing the efficacy of drug therapies. A landmark study in this respect was the interferon beta-1b trial [14], which showed a correlation between MRI measured lesion load and clinical findings. In this study, lesion volume was quantified using manual boundary tracing. The disadvantages of this technique are that it is very labour-intensive and that it suffers from high intra- and interrater variabilities. Various researchers have shown that computer-aided techniques are able to not only reduce operator burden, but also the amount of inter- and intrarater variability associated with the measurement [5, 12, 22].

This paper describes the automatic lesion quantification technique currently under development at the McConnell Brain Imaging Centre (MBIC) for use in a large, multi-center clinical trial. The processing chain employs a conventional multi-spectral (multi-feature) classifier combined with explicit registration of all image volumes to a standardized 3D coordinate space. In this approach, feature values can be obtained from a variety of sources, the most common of which are multiple MRI acquisitions; additional features can be, for instance, prior tissue probability distributions or derived features such as local homogeneity or texture measures. The standard coordinate space, referred to as "stereotaxic space," ensures that the feature values associated with each voxel represent the same physical brain location. The notion of stereotaxic space is crucial to our approach and provides a number of advantages, which will be discussed in the following sections.

2 Methods

This section describes the data processing pipeline used for routine 3D MR data analysis. Shown in Fig. 1 are the main components of this system: preprocessing, registration of the data with a standardized, stereotaxic brain space, postprocessing, and tissue classification. After acquisition and archiving (1), each MRI

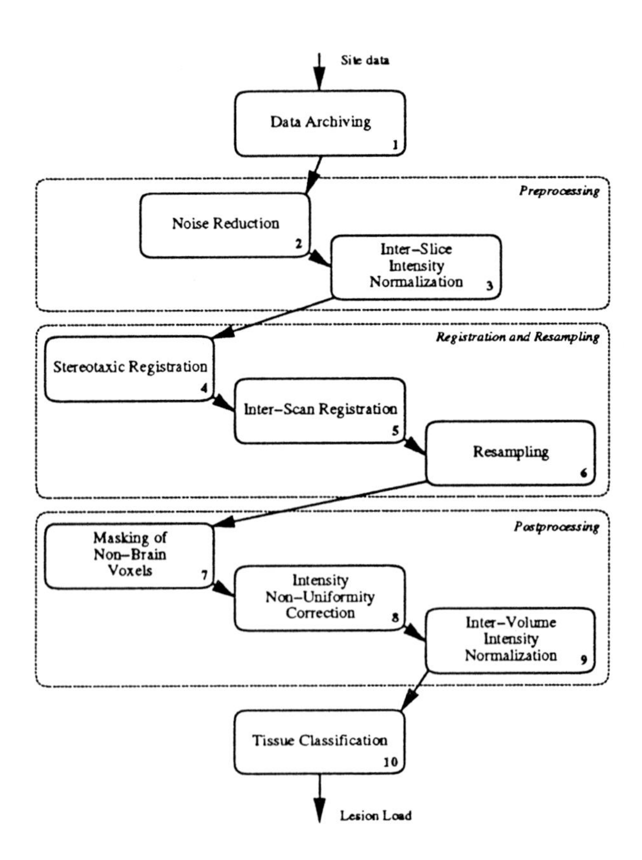

Fig. 1. Diagram of the automatic lesion quantitation algorithm.

volume is first subjected to the preprocessing (prior to registration) stage, which includes noise reduction and correction of inter-slice intensity variations.

Noise Reduction (2). In acquisition sequences for clinical use, the number of excitations (NEX) is usually kept low in a tradeoff with acquisition time, resulting in noisy data. It has been shown that the application of an edge-preserving noise filter can improve the accuracy and reliability of quantitative measurements obtained from MRI [11, 22]. We have selected to use anisotropic diffusion, commonly used for the reduction of noise in MRI [7].

Inter-Slice Intensity Normalization (3). Several researchers have shown variations in intensities between adjacent slices of MRI exams [1, 13, 15, 18, 21], an artifact generally attributed to eddy currents and crosstalk between slices. We are currently using a method described by Zijdenbos *et al.* [21], in which a scaling factor between each pair of adjacent slices is estimated from local (pixel-by-pixel) correction factors.

The key component of the analysis procedure is the registration of each data volume with a standardized stereotaxic brain space (Talairach and Tournoux [16]). This type of standardization

- provides a conceptual framework for the completely automated, 3D analysis across subjects;
- allows longitudinal and cross-site, cross-study, and cross-population analysis;
- allows the use of spatial priors for classification;
- allows the use of spatial masks for postprocessing and anatomically driven hypothesis testing;
- provides a framework for statistical analysis of the results.

Using an automated 3D image registration technique [3], each new source volume is registered with a target volume. Briefly, this method minimizes, in a multi-stage, multi-scale approach, an image similarity function for these two volumes as a function of a 9 parameter (3 translations, 3 rotations, 3 scales) linear geometric transformation. In its current implementation, the image similarity function is the cross-correlation value between two data volumes; other similarity measures, such as the entropy measure proposed by Collignon *et al.* [2], are under investigation.

In many practical applications, including ours, the registration process has two components: stereotaxic registration and inter-scan registration.

Stereotaxic Registration (4). Stereotaxic registration is accomplished by the registration of one of the image modalities to a stereotaxic target. The stereotaxic target used at the MBIC is an average T_1-weighted scan of 305 normal volunteers [3, 4, 6].

Inter-Scan Registration (5). The classification process described here relies on voxel classification in a multi-dimensional feature space. It is vital that the modalities which span up this feature space are in spatial register, i.e., that the feature values obtained from these modalities at (x, y, z) in image space all reflect the same location in physical space. To accomplish this, a similar technique as described for stereotaxic registration is used to register various scans obtained from the same subject.

Following registration, all data volumes are resampled into a 1mm isotropic voxel grid using the established transformations (6). Subsequently, they are subjected to a number of postprocessing steps, including removal of non-brain tissues, correction for intensity non-uniformity, and inter-volume intensity normalization.

Masking of Non-Brain Voxels (7). In order to reduce the computational burden of the following processing steps, non-brain voxels are masked (i.e., labeled as 'background') using a standard brain mask defined in stereotaxic space. The fact that this 'average' brain mask may not exactly fit the individual brain is not a serious concern for the quantification of MS lesions which are typically situated in the white matter, well away from the brain's periphery.

Correction for Intensity Non-Uniformity (8). Low-frequency spatial intensity variations are a major source of errors in the computer-aided quantitative analysis of MRI data. This artifact is caused by a number of machine- and acquisition sequence-specific factors [8, 15], the predominant one being inhomogeneities in the sensitivity of the radiofrequency receiver coil. We are currently using the algorithm proposed by Wells *et al.* [17], an iterative approach which alternates between a classification and a field estimation stage, for the correction of this type of artifact.

Inter-Volume Intensity Normalization (9). Although ideally all data in a trial should be acquired on the same scanner and with the same acquisition sequences, this will rarely be the case in clinical practice. As a result, the intensity distributions of individual volumes may vary considerably over time and between sites. Given that all volumes at this stage are stereotaxically registered (4-6), each volume can be normalized using the technique as described for the correction of inter-slice intensity variations (3). In this case however, a single, global scale factor is estimated from the voxel-by-voxel comparison of two volumes. Clearly, a single scale factor can only be considered a crude, first-order normalization, but it has proven to be useful in practical applications, as will be described in the following sections.

Tissue Classification (10). The central processing step for lesion quantification is the tissue classification, i.e., the labeling of each voxel with a tissue class (in our application background, white matter, gray matter, cerebrospinal fluid, or MS lesion). Many different types of classification algorithms exist, such as statistical pattern recognition techniques, clustering algorithms, decision trees, and artificial neural networks (ANNs). We are currently using a backpropagation ANN [22], but we are also studying the performance of various other supervised and unsupervised classifiers based on MRI simulations [10].

3 Experiments

From a larger pool of MS patient data which is currently being collected in our institute, 30 patients were randomly selected and their MRI scans analyzed, both manually and automatically. Each subject was scanned using a 3D, T_1-weighted gradient-echo sequence (TR=35ms, TE=11ms) and a 2D, T_2/N_H-weighted multi-slice spin-echo (TR=3000ms, TE=30/80ms) sequence. Manual lesion labeling was performed by a combination of threshold-based 3D region growing and voxel-based editing, wherein the user had the freedom to use all three modalities simultaneously. For the automatic analysis, a single backpropagation artificial neural network classifier was trained manually on one of the volumes and subsequently applied to all 30 data sets. In addition to the three MRI acquisitions, spatial tissue probability maps for white matter (WM), gray matter (GM), and cerebrospinal fluid (CSF) were used as additional features in the classification process. These probability maps, shown in Fig. 2 for WM and GM, were obtained by classifying 53 MRI volumes obtained from normal volunteers into WM, GM, and CSF and calculating the frequency with which specific voxel locations in stereotaxic space were labeled as members of each of these three classes (see also [9]).

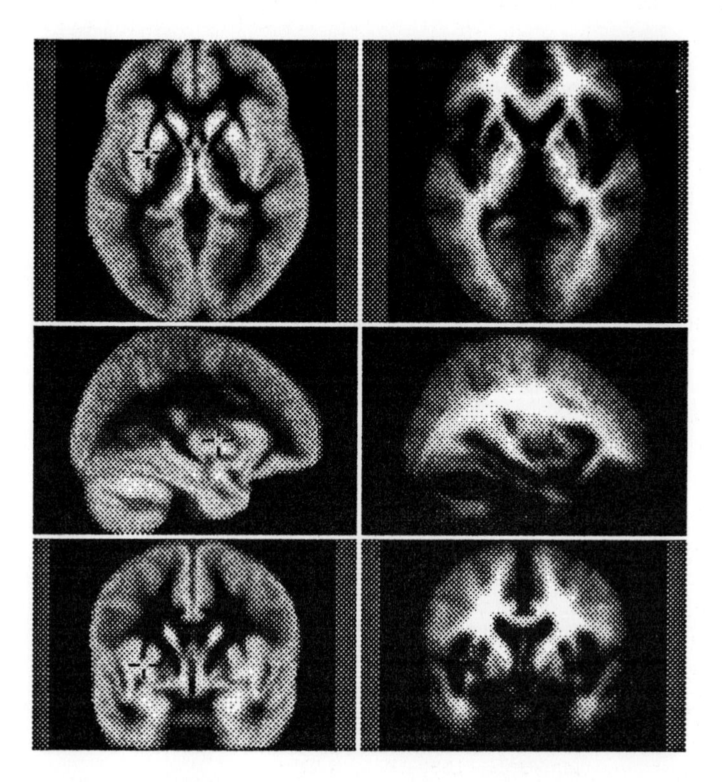

Fig. 2. Stereotaxic probability maps (N=53) for WM (left) and GM (right), shown in transversal, sagittal, and coronal cross sections (top to bottom).

Having obtained lesion labelings both manually and automatically, the total lesion volume in stereotaxic space was calculated for each brain and each method and the results were compared using linear regression. In addition, the correlation and regression parameters between manual and automatic lesion measurements was studied in relation to the setup and parameters of the automatic analysis procedure. This way manual labeling, the only gold standard available for in-vivo studies, can not only be used to validate the technique, but the technique can be 'tuned' to maximize its accuracy with respect to the manual measurements. For instance, the stereotaxic WM probability map can not only be used as an extra feature, but also as a means to reject often occurring false positive lesion voxels by restricting the search volume to voxels for which the white matter probability P_{WM} is above a certain threshold.

4 Results

As an illustration of the similarity between manual and automatic measurements, Fig. 3 shows an example of manually and automatically determined lesion volumes, taken from the set of 30 patients that were studied.

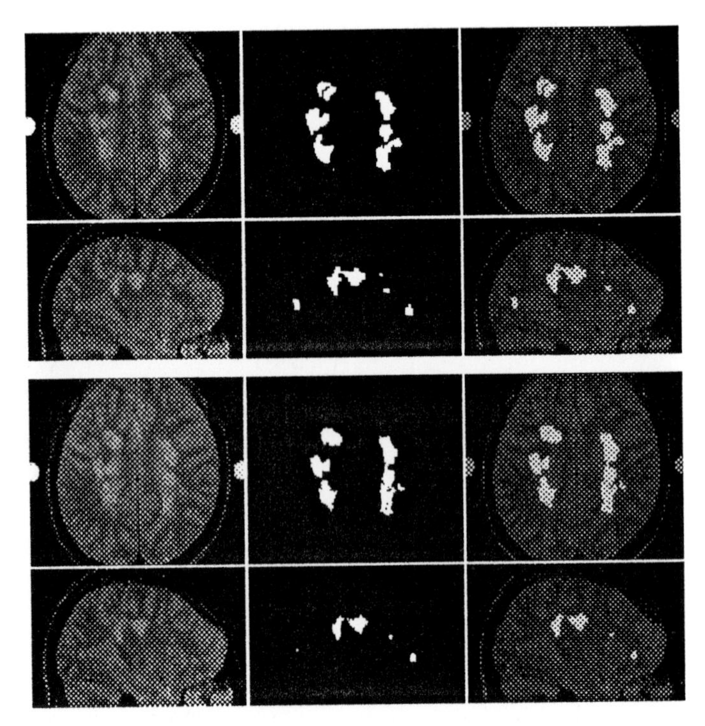

Fig. 3. Manually (top) and automatically (bottom) detected lesions, shown with the N_H-weighted volume in transversal and sagittal cross-sections. The third column shows the lesions overlaid on the image.

Fig. 4 shows the regression of the automatically and manually detected lesion volumes. Also shown in this figure are the regression parameters (slope, intercept, r, and the associated t-value). In this example, only voxels for which $P_{WM} \geq 0.8$ are considered; this value represents the maximum correlation as a function of the minimum value of P_{WM} (see Fig. 5).

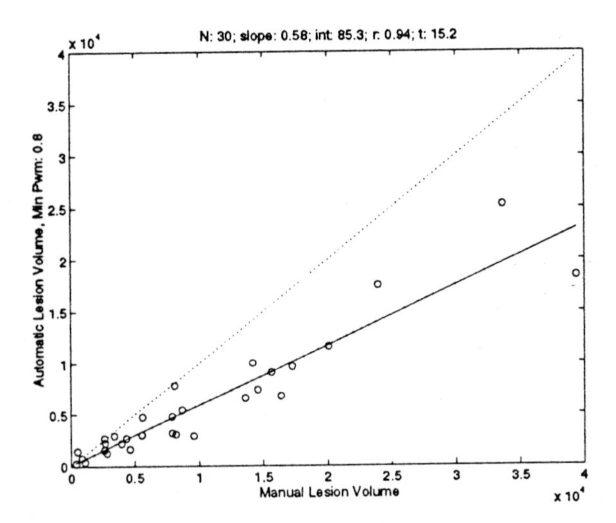

Fig. 4. Automatically detected vs. manually detected lesion volume, with $P_{WM} \geq 0.8$ (see text). The number of data points (N), the slope and intercept of the regression line (solid), the correlation coefficient (r) and the associated t value are also shown. The dotted line illustrates the ideal, one-to-one mapping.

The effect of processing stages and parameter selection is shown in Fig. 5. In this figure, the correlation value r between manually and automatically obtained lesion volumes is shown as a function of minimum white matter likelihood required for a voxel to be included in the measurement. In addition to the curves which reflect the use of the full processing pipeline as described in Section 2, the performance of the automatic analysis is also shown without using the three tissue probability maps as features for the classifier (i.e., a 3-feature classification), and without inter-volume intensity normalization.

5 Discussion

Computer-aided analysis of large amounts of MRI data is becoming increasingly important as the quantitative analysis of MR scans is gaining interest, driven by, among others, clinical trials aimed at the assessment of new drug therapies. With this application in mind, we are currently developing a fully automatic approach to the quantification of lesion load in MR scans of MS patients. The technique relies on multispectral classification techniques and the notion of a standardized, stereotaxic brain space, supported by a number of dedicated pre- and postprocessing routines. The use of multispectral analysis allows local decisions to be made based on the information obtained from multiple data sources, whereas the standard space provides anatomical knowledge and facilitates statistical analysis.

Results obtained with the approach described herein have been presented

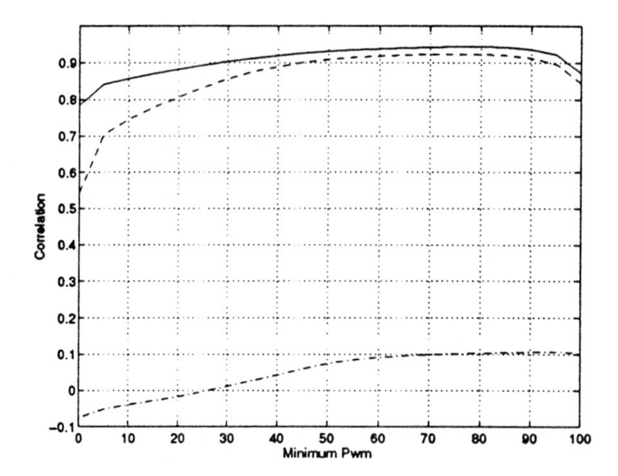

Fig. 5. The correlation between automatically, using 3 intensity features and 3 tissue probability maps, and manually detected lesion volume as a function of the minimum P_{WM} (solid line). The dashed line shows the same results without using the tissue probability maps, while the dash-dotted line shows the results without the application of inter-volume intensity normalization.

and compared against those obtained from manual lesion delineation, showing a significant lesion volume correlation of 0.94. The fact that the slope of the regression (Fig. 4) is not unity shows that the manually obtained lesion volumes were on average larger than those obtained automatically. This is an effect which has been observed before [22] and which appears to be caused by the tendency of many human operators to draw generously around lesions. However, since the effect is systematic, it will not impact on the reliability of the method in, for instance, the context of longitudinal drug treatment studies. In addition, the processing pipeline can be 'tuned' to improve the accuracy of lesion detection, as is illustrated in Fig. 5.

Fig. 5 clearly shows the improvement in the correlation between manual and automatic lesion measurements as a function of the minimum white matter likelihood. The upward trend with increasing threshold value ($0 \leq P_{WM} \leq 0.8$) is caused by the elimination of false positive lesion voxels, which are mainly due to partial volume voxels along the sulci. This effect is even stronger when the stereotaxic tissue probability maps are not used as features for the classifier. It is also obvious from this figure that, even though all data was acquired on the same scanner using the same pulse sequence, the inter-volume intensity normalization is crucial to the use of a single classifier for the classification of multiple data sets.

As another example of the use of stereotaxic space principles, lesion probability maps can be generated from both the manual and the automatic analysis, similar to the WM and GM probability maps shown in Fig. 2. These probability maps can be used in the analysis of image processing or drug treatment effects in ways similar to the established techniques for the analysis of functional image data [19, 20]. Using this type of statistical analysis it is possible to detect not only global differences, but also statistically significant local differences between subgroups.

Although promising, the results presented here should be regarded as preliminary for the following reasons. First, all data sets were classified using a fixed classifier, which was trained by manual supervision on one of the volumes. This approach clearly will not suffice in cases where different scanner types and/or pulse sequences produce contrast variations, resulting in non-linear variations of cluster shape and location in the feature space. The classifier should be adaptive or should be retrained on each data set; the use of stereotaxic tissue proability maps as a means to train supervised classifiers on individual data sets is currently being studied. Second, only one human operator has manually analyzed the data. This validation effort will be extended, including repeated manual measurements by several operators to allow the assessment of intra- and interrater variability associated with manual delineation. As additional validation, the lesion quantification accuracy will be assessed using simulated MRI data [10]. Furthermore, we are currently developing and testing several elements of the processing pipeline, such as various objective functions for the image registration, algorithms for the correction for shading artifact, and different classification techniques. Nevertheless, these corrections are refinements of a process which is now in production usage.

References

1. H. S. Choi, D. R. Haynor, and Y. Kim. Partial volume tissue classification of multichannel magnetic resonance images - a mixel model. *IEEE Transactions on Medical Imaging*, 10(3):395–407, Sept. 1991.
2. A. Collignon, F. Maes, D. Delaere, D. Vandermeulen, P. Suetens, and G. Marchal. Automated multi-modality image registration based on information theory. In Y. Bizais, C. Barillot, and R. D. Paola, editors, *Information Processing in Medical Imaging (IPMI)*, pages 263–274. Kluwer, June 1995.
3. D. L. Collins, P. Neelin, T. M. Peters, and A. C. Evans. Automatic 3D intersubject registration of MR volumetric data in standardized Talairach space. *Journal of Computer Assisted Tomography*, 18(2):192–205, Mar./Apr. 1994.
4. A. Evans, M. Kamber, D. Collins, and D. MacDonald. An MRI-based probabilistic atlas of neuroanatomy. In S. D. Shorvon et al., editors, *Magnetic Resonance Scanning and Epilepsy*, chapter 48, pages 263–274. Plenum Press, 1994.
5. A. C. Evans, J. Frank, and D. H. Miller. Evaluation of multiple sclerosis lesion load: Comparison of image processing techniques: Summary of montreal workshop. *Annals of Neurology*, 1996. In press.
6. A. C. Evans, S. Marrett, P. Neelin, et al. Anatomical mapping of functional activation in stereotactic coordinate space. *NeuroImage*, 1:43–53, 1992.
7. G. Gerig, O. Kübler, R. Kikinis, and F. A. Jolesz. Nonlinear anisotropic filtering of MRI data. *IEEE Transactions on Medical Imaging*, 11(2):221–232, June 1992.
8. R. M. Henkelman and M. J. Bronskill. Artifacts in magnetic resonance imaging. *Reviews of Magnetic Resonance in Medicine*, 2(1):1–126, 1987.

9. M. Kamber, R. Shinghal, D. L. Collins, G. S. Francis, and A. C. Evans. Model-based 3-D segmentation of multiple sclerosis lesions in magnetic resonance brain images. *IEEE Transactions in Medical Imaging*, 14(3):442–453, Sept. 1995.
10. R. K.-S. Kwan, A. C. Evans, and G. B. Pike. An extensible MRI simulator for post-processing evaluation. In *Proceedings of the Fourth International Conference on Visualization in Biomedical Computing (VBC)*, Hamburg, Germany, 1996.
11. J. R. Mitchell, S. J. Karlik, D. H. Lee, M. Eliasziw, G. P. Rice, and A. Fenster. Quantification of multiple sclerosis lesion volumes in 1.5 and 0.5T anisotropically filtered and unfiltered MR exams. *Medical Physics*, 23(1):115–126, Jan. 1996.
12. J. R. Mitchell, S. J. Karlik, D. H. Lee, and A. Fenster. Computer-assisted identification and quantification of multiple sclerosis lesions in MR imaging volumes in the brain. *Journal of Magnetic Resonance Imaging*, pages 197–208, Mar./Apr. 1994.
13. M. Özkan, B. M. Dawant, and R. J. Maciunas. Neural-network-based segmentation of multi-modal medical images: A comparative and prospective study. *IEEE Transactions on Medical Imaging*, 12(3):534–544, Sept. 1993.
14. D. W. Paty, D. K. B. Li, UBC MS/MRI Study Group, and IFNB Multiple Sclerosis Study Group. Interferon beta-1b is effective in relapsing-remitting multiple sclerosis. *Neurology*, 43:662–667, 1993.
15. A. Simmons, P. S. Tofts, G. J. Barker, and S. R. Arridge. Sources of intensity nonuniformity in spin echo images. *Magnetic Resonance in Medicine*, 32:121–128, 1994.
16. J. Talairach and P. Tournoux. *Co-planar Stereotaxic Atlas of the Human Brain: 3-Dimensional Proportional System - an Approach to Cerebral Imaging*. Thieme Medical Publishers, New York, NY, 1988.
17. W. M. Wells III, W. E. L. Grimson, R. Kikinis, and F. A. Jolesz. Statistical intensity correction and segmentation of MRI data. In *Proceedings of the SPIE. Visualization in Biomedical Computing*, volume 2359, pages 13–24, 1994.
18. D. A. G. Wicks, G. J. Barker, and P. S. Tofts. Correction of intensity nonuniformity in MR images of any orientation. *Magnetic Resonance Imaging*, 11(2):183–196, 1993.
19. K. J. Worsley, A. C. Evans, S. Marrett, and P. Neelin. A three-dimensional statistical analysis for CBF activation studies in human brain. *Journal of Cerebral Blood Flow and Metabolism*, 12(6):900–918, 1992.
20. K. J. Worsley, S. Marrett, P. Neelin, A. C. Vandal, K. J. Friston, and A. C. Evans. A unified statistical approach for determining significant signals in images of cerebral activation. *Human Brain Mapping*, 1996. Accepted.
21. A. P. Zijdenbos, B. M. Dawant, and R. A. Margolin. Intensity correction and its effect on measurement variability in the computer-aided analysis of MRI. In *Proceedings of the 9th International Symposium and Exhibition on Computer Assisted Radiology (CAR)*, pages 216–221, Berlin, Germany, June 1995.
22. A. P. Zijdenbos, B. M. Dawant, R. A. Margolin, and A. C. Palmer. Morphometric analysis of white matter lesions in MR images: Method and validation. *IEEE Transactions on Medical Imaging*, 13(4):716–724, Dec. 1994.

3D Skeleton for Virtual Colonoscopy

Yaorong Ge[1,2], David R. Stelts[2], and David J. Vining[2]

[1] Department of Mathematics and Computer Science
Box 7388 Calloway Hall, Wake Forest University, Winston-Salem, NC 27109
[2] Department of Radiology, Bowman Gray School of Medicine
Wake Forest University, Medical Center Blvd, Winston-Salem, NC 27157

Abstract. This paper describes an improved algorithm for generating 3D skeletons from binary objects and its clinical application to virtual colonoscopy. A skeleton provides an ideal central path for an auto-piloted examination of a virtual colon rendered from a spiral computed tomography scan. Keywords: virtual colonoscopy, 3D skeleton, thinning

1 Introduction

Virtual colonoscopy is a new medical technique that allows physicians to examine a patient's colon by literally flying through a computer simulation of his/her colon rendered from a spiral computed tomography (CT) scan. It has been shown in preliminary studies that small protrusions (polyps) on the colon wall, which are early indications of colon cancer, can be detected by the virtual technique as reliably as the conventional colonoscopy [1, 2]. The clinical benefit of this emerging technology is enormous since it will enable millions of people to undergo colon cancer screening with much less discomfort, cost, and risks compared to conventional colonoscopy. However, more development is needed in order to make this technology efficient and available for clinical use.

One challenge involves the time required to analyze a virtual colon. Two approaches exist for examining a simulated colon. In the first approach, a user navigates (flies) through the colon using a mouse or a similar device. This method requires considerable user interaction due to the complex morphology of the colon. As a result, a user may easily take over an hour to analyze a case which is generally not acceptable. The second approach is often referred to as auto-piloting. Internal views of the colon are automatically generated along the view points lying on the central path of the colon. This approach is much more desirable because it eliminates the need for physicians to manually navigate through the colon, thus enabling them to concentrate more on examining the colon's surface for diagnosis. The success of this approach, however, requires the automatic identification of the central path of the colon which remains to be a difficult problem. In this paper we describe an algorithm that generates the central path of a colon based on the concept of a skeleton.

In continuous space, a skeleton is defined as the locus of the centers of maximally inscribed balls (CMBs) [3]. In three dimensions, a skeleton generally consists of both 1D curves and 2D surfaces. For the purpose of this paper, we are not interested in 2D structures. Therefore, 2D surfaces are further reduced to medially placed 1D curves in our skeleton.

In this paper we adopt the commonly used thinning approach for 3D skeletonization in which border points are "peeled off" layer-by-layer while preserving the topology of the original object [4]. Our algorithm improves over existing methods in two aspects. First, we use CMBs as anchor points so that the resulting skeleton is medially positioned. Second, we use a faster algorithm for calculating the number of connected components within a neighborhood surrounding each voxel in the entire object.

In the following section we describe our algorithm for generating skeletons from 3D objects. In Section 3, we present some results from real colon volumes generated from spiral CT scans. In the last section, we state our conclusions and suggest future work.

2 Methods

2.1 Topological Thinning

Topological thinning is a widely used approach for generating skeletons from binary objects. It iteratively converts to background those foreground border points that are not needed to satisfy certain geometrical and topological constraints. Geometrical constraints are used to detect positions where fire fronts meet while topological constraints are needed to maintain topology.

Recent 2D skeletonization techniques have used centers of maximally inscribed disks as geometrical constraints to ensure the medialness of the skeleton [5]. Existing 3D techniques, however, have not employed such constraints. The positions of the skeleton branches in these algorithms are left to the symmetrical fashion of the thinning process. We believe that thinning process itself does not ensure symmetry in all directions. Therefore, we propose to use CMBs as geometrical constraints in our thinning algorithm.

Topological constraints for 3D thinning have been carefully studied in the past decade. It has been shown that topology will be preserved by a thinning process if the border points that are removed (more strictly, converted to background points) during each iteration are simple points but not end points. Recently, Lee and colleagues [4] have shown that a border point is a simple point if and only if the Euler characteristic and the number of connected components in its neighborhood does not change after its removal. Furthermore, they have shown that change of Euler characteristic in a neighborhood can be determined simply by 8 table look-ups, one for each octant, while the number of connected components can be determined by a search method based on the quad-tree data structure.

In this paper, we have implemented a thinning algorithm similar to the one described in [4] except that we use CMBs as anchor points and we employ a simpler and faster algorithm for the determination of the number of connected components in a neighborhood.

2.2 Centers of Maximally Inscribed Balls

In order to detect CMBs we first compute the distance transform of a binary volume. The distance transform for a point in the foreground is the distance from this point to a nearest point in the background. In this paper we use what is called 3-4-5 Chamfer distance transform in which the distance between two direct neighbors that are 1 unit apart is assigned a weight of 3, the distance between two diagonal neighbors that are $\sqrt{2}$ units apart is assigned a weight of 4, and the distance between two diagonal neighbors that are $\sqrt{3}$ units apart is assigned a weight of 5.

With a 3-4-5 distance transform, we can easily determine the CMBs by comparing the distance value of each voxel with its neighboring voxels. It can be shown using the arguments presented in [6] that a voxel v is a CMB if the following conditions are met by all its 26 neighbors q (refer to Figure 1 for the labeling scheme):

$$d(q) < d(v) + t(q, v) \tag{1}$$

where

$$t(q,v) = \begin{cases} 3 \text{ if } q \text{ and } v \text{ are 1 units apart (in positions 0 - 5)} \\ 4 \text{ if } q \text{ and } v \text{ are } \sqrt{2} \text{ units apart (in positions 6 - 17)} \\ 5 \text{ if } q \text{ and } v \text{ are } \sqrt{3} \text{ units apart (in positions 18 - 25)} \end{cases}$$

It can also been shown using similar arguments as in [6] that the only exception to the above condition is when the distance value at v is equal to 3, in which case it should be changed to 1 before using the above formula.

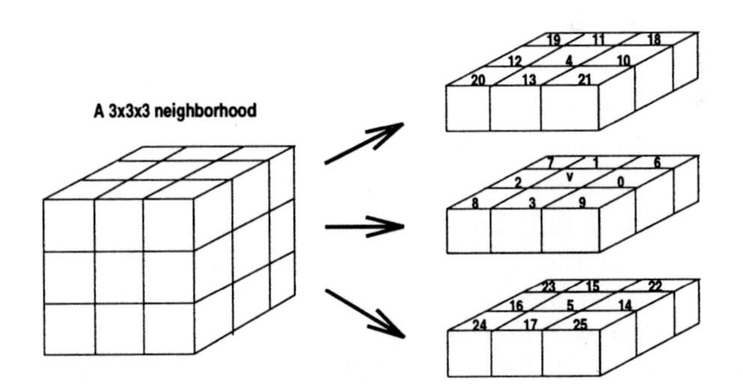

Fig. 1. Labeling the voxels in a neighborhood.

2.3 Connectivity in a Neighborhood

To compute the number of 26-connected components in a neighborhood, we divide the neighborhood into 8 overlapping octants of 2×2 voxels. We observe that if a voxel shared by two octants is in the foreground then all the foreground voxels in these two octants are connected through this shared voxel. Therefore, using the labeling scheme of Figure 1, a foreground voxel at positions 0 - 5 joins four octants while a foreground voxel at positions 6 - 17 joins two octants. If one octant is not joined with any other octant, we need only to check the corner voxel in that octant. If it is in the foreground then that is one more connected component. Otherwise, the number of joined octants is the number of connected components in the neighborhood. Our algorithm can be described as follows:

```
1. For voxels 0 - 5, if it is 1, put the four octants joining at
      that voxel into the same equivalence class.
2. For voxels 6 - 17, if it is 1, put the two octants joining at
      that voxel into the same equivalence class.
3. For each equivalence class of the 8 octants
      If the number of octants is equal to 1 and the corner
         voxel is 1, increase the number of connected
         components by 1
      If the number of octants is greater than 1
         increase the number of connected components by 1
```

This algorithm avoids the use of any extra checking necessary in the existing approaches and it does not require any extra data structure. The number of computations is, therefore, kept at a minimum.

2.4 Selecting a path

Due to the complex shape of a colon, many branches of medially located curves and surfaces will be generated using the above algorithm. To eliminate the unwanted paths, our current system performs three phases of topological thinning. The first phase is fully automatic. The object is thinned using the above algorithm until the remaining points are either CMB points or necessary to retain topology. In the second phase, we ask the user to select a path of interest by manually marking a start point and an end point while viewing the volume slice by slice. With these two points as the only anchor points and the skeleton from the first phase as the foreground object, we apply the thinning algorithm again. The result of this second phase should contain a single open curve that starts and ends at the two designated points. However, extra loops along this path caused by holes in the original object may continue to exist. These holes are caused by problems in image segmentation. To remove these loops we require that the user mark a point on each of the branches that are not wanted by viewing the skeleton together with the rendered colon. Our algorithm then removes those marked points from the existing skeleton and apply another round of thinning. The resulting skeleton is then guaranteed to be a simple path and to be in the center of the colon's lumen.

3 Experimental Results

We have tested our algorithm on 4 abdominal spiral CT scans. Each CT scan consists of up to 500 images (512×512 pixels/image). A region growing algorithm is applied to convert the colon voxels to a value of 1 and all other voxels to a value of 0. Before applying the skeletonization algorithm, the original volume is sub-sampled to 128 × 128 pixels in each image and up to 150 images in a volume.

To create views for auto-piloting, the final skeleton path is sub-sampled every 5 points and then interpolated by a Cat-Rom spline. The resulting spline is sampled 10 points between every two control points to generate smooth flight of the colon lumen using the Marching Cubes algorithm [7].

A typical skeleton with loops is shown in Figure 2 together with a rendering of the original colon. In Figure 3 the loops have been removed by another thinning

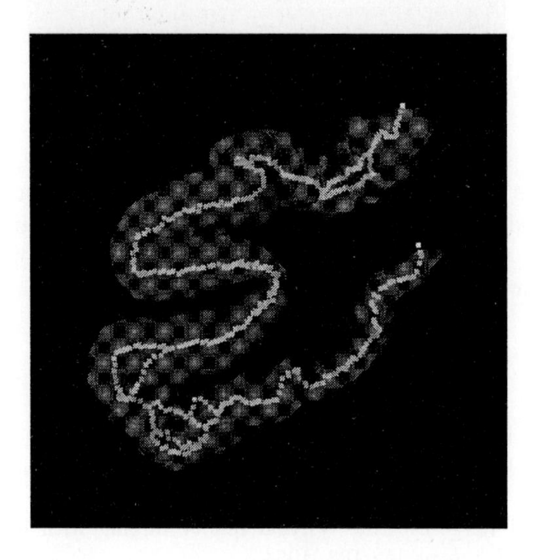

Fig. 2. The skeleton with branches.

process. These loops numbered fewer than 10 in our experiments. Therefore, user interaction for each volume was limited.

The resulting skeletons have been used to guide auto-piloted examination of colons with satisfactory visual effects in all cases.

4 Conclusions

We have implemented an improved algorithm for the generation of 3D skeletons from binary image volumes. We have demonstrated that the resulting skeletons can be successfully used to guide the auto-piloting of virtual colonoscopy.

Our current system may be improved in many aspects. For example, it currently requires some user interaction to remove false loops. We believe that

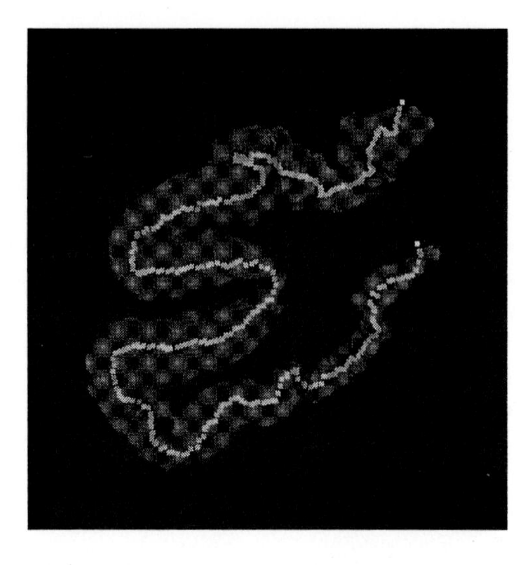

Fig. 3. The skeleton with branches removed.

most of the loops may be eliminated automatically by examining the radius function associated with the skeleton. Furthermore, the thinning process currently requires several scans of the entire volume which can be avoided by using more efficient data structures.

References

1. D. J. Vining, D. W. Gelfand, and R. E. Bechtold et al. Technical feasibility of colon imaging with helical ct and virtual reality. *American Journal of Roentgenology*, 162:104, 1994.
2. D. J. Vining and R. Y. Shifrin. Virtual reality imging with helical ct. *American Journal of Roentgenology*, 162:188, 1994.
3. H. Blum. Biological shape and visual science: Part i. *Journal of Theoretical Biology*, 38:205–287, 1973.
4. Ta-Chin Lee, Rangasami L. Kashyap, and Chong-Nam Chu. Building skeleton models via 3-D medial surface/axis thinning algorithms. *CVGIP: Graphical Models and Image Processing*, 56(6):462–478, 1994.
5. E. R. Davies and A. P. N. Plummer. Thinning algorithms: a critique and a new methodology. *Pattern Recognition*, 14(1):53–63, 1981.
6. Carlo Arcelli and Gabriella Sanniti di Baja. Finding local maxima in a pseudo-Euclidean distance transform. *Computer Vision, Graphics, and Image Processing*, 43:361–367, 1988.
7. W. E. Lorensen and H. E. Cline. Marching cubes: A high resolution 3D surface construction algorithm. *Computer Graphics*, 21(3):163–169, 1987.

Brain: Visualization of Function

Identifying Hypometabolism in PET Images of the Brain: Application to Epilepsy

Tracy L. Faber[1,3], John M. Hoffman[1,2,3], Thomas R. Henry[2],
John R. Votaw[1,3], Marijn E. Brummer[1], and Ernest V. Garcia[1,3]

Departments of Radiology[1] and Neurology[2] and
The Emory Center for PET[3]
Emory University School of Medicine
1364 Clifton Rd N.E., Atlanta, GA. 30322

Abstract. A technique for identifying hypometabolism from Positron Emission Tomography (PET) brain images, which accounts for patient-specific anatomical variations, scanner physical properties, and expected normal variances in metabolism, has been developed and used to identify unilateral temporal lobe seizure foci in epileptics. This method was able to distinguish the epileptogenic focus in three seizure patients with unilateral seizure onset, while demonstrating no hypometabolism in three patients with bilateral seizure onset or in the ten normal volunteers.

1 Introduction

Positron Emission Tomography (PET) images of ^{18}F-fluorodeoxyglucose (FDG) metabolism in patients with intractable complex partial seizures of temporal lobe origin demonstrate hypometabolic regions corresponding with seizure foci in 60-80% of medically intractable epileptics (Figure 1). Distinguishing hypometabolic epileptogenic foci from normal brain tissue requires knowledge of what a "normal" PET FDG image should look like, and how the image might change depending on system physics and patient anatomy. However, the appearance of normal PET FDG images can vary significantly depending on imaging system properties, anatomic differences, and metabolic variabilities. In specific, the point spread function (PSF) of the PET imaging system causes blurring that affects intensity in the final image; this is generally referred to as the detector response effect. Also, apparently identical regions in PET images from different patients may correspond to different anatomical structures because of normal variations in brain anatomy. Finally, normal metabolism varies both globally and locally among individuals. Figure 2 demonstrates these normal effects.

Normal variations like those demonstrated in Figure 2 complicate quantitative as well as qualitative analysis. Most quantitative determinations of normality and/or abnormality in PET and single photon emission computed tomograms (SPECT) of the brain are performed in a highly interactive manner. A neuroanatomy

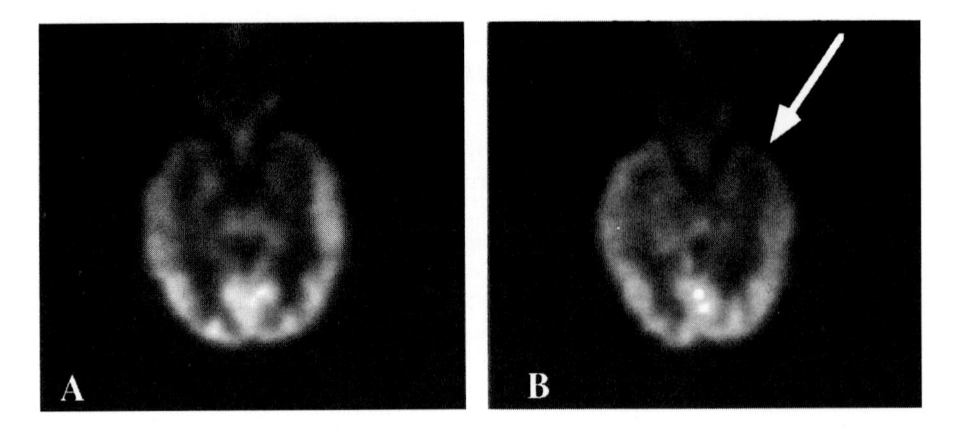

Fig. 1. A. PET FDG image slice taken through temporal lobes of a normal volunteer. Note the symmetric left-to-right temporal lobe intensities. **B.** PET FDG image slice taken through the temporal lobes of an epileptic patient. The decreased intensity in the right temporal lobe (denoted by arrow) corresponds to a hypometabolic epileptogenic focus.

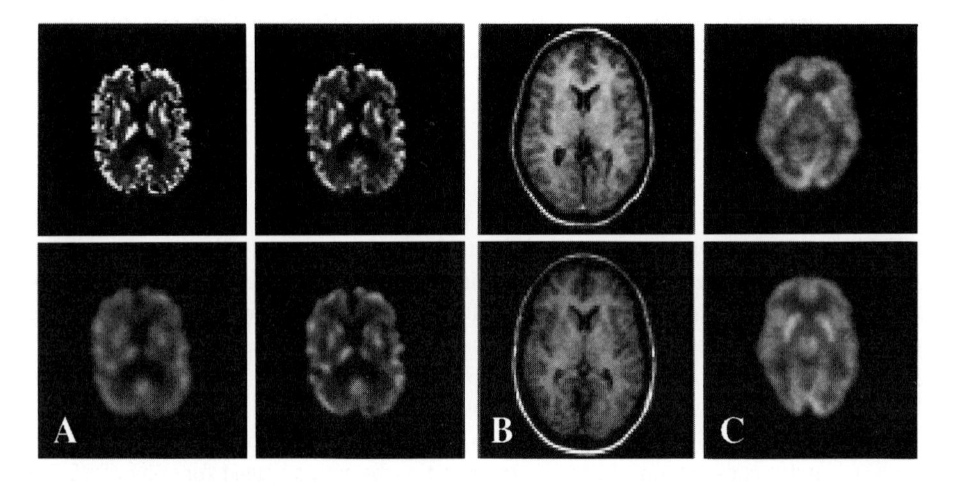

Fig. 2. Numerous factors cause variations in the appearance of normal PET FDG images. **A.** Differences in PET scanner physics cause various degrees of blurring and change the local intensities in the final image. From top left, clockwise: original FDG concentration; simulated PET image obtained from system with a resolution of 5mm; simulated PET image from system with 10mm resolution; simulated PET image from system with 15mm resolution. **B.** Brain anatomy varies among different people. These two MR images were taken from the same region of the brain of two different normal subjects. Note the overall similarities, but the numerous small differences. **C.** Normal metabolism varies from person to person, and within one person's brain. These two PET FDG slices are from the same region of the brain of two different normal subjects. The intensities vary within each image, and similar regions in the two images have different intensities.

expert traces a region-of-interest (ROI) around an area visualized in either the PET image, or in a magnetic resonance (MR) image which has been aligned with the PET. When a region extends into more than one slice, the "region" becomes a volume; thus, the term volume-of-interest (VOI) is often used. Many possible statistical analyses can be applied to the counts contained within the ROI or VOI.

Analysis of metabolism using ROIs or VOIs created from the PET image alone generally can only account for physiological variations, because anatomy is not well-visualized in the low-resolution PET scan. Conversely, creation of the VOIs on an MR or CT image that has been aligned to the PET allows accounting for both anatomical and physiological variabilities. Unfortunately, neither approach accounts for PET scanner physics. In fact, few PET analysis systems account for the detector response effect, even though it is known to affect PET image intensities significantly and unexpectedly [1,2]. The detector response, or partial volume effect, implies that an object must be greater than 2 times the resolution of the imaging system (as measured by the full-width half-maximum (FWHM)) of the point spread function (PSF) in order to appear as its true intensity in the final PET image. The best clinical systems today have a resolution of about 5mm. Many structures in the brain including deep nuclei and even the cortical ribbon of gray matter are smaller than twice this value (10mm); they will therefore appear less intense than their radionuclide concentration would indicate. This effect becomes more important with decreasing system resolution; one study showed that decreasing system resolution increased interobserver variabilities in determining abnormal metabolism in PET brain images [3]. The detector response effect implies that both pathological tissue volume changes and normal variations in structure size will cause localized count variations and increase the variability of any quantitative measurement of metabolism.

The quantitative methods that we have developed account for all of the above "normal" variations. They have been applied to a set of 10 normal subjects and 6 epilepsy patients, 3 of whom have surgically-verified unilateral seizure focus and 3 who were diagnosed with bilateral seizure focus from implanted depth electrodes.

2 Methods

The hypometabolic seizure focus is localized by comparing the subject's PET image to a predicted normal PET metabolism image. The predicted normal PET image is created by integrating patient-specific anatomical information from MR images, PET system-specific physical properties, and a database of normal metabolism obtained from carefully screened normal volunteers in the local population. This predicted normal PET metabolism image estimates how the intensities in a person's PET image should vary given individual anatomy and PET system physics if brain metabolism is normal. Then, the patient's actual PET FDG image is compared statistically to the predicted normal image in order to determine significant differences between the predicted and actual PET images. In this patient population, significantly decreased areas in the actual PET image as compared to the predicted normal images are expected to indicate metabolism abnormalities corresponding to epileptogenic foci.

The creation of a predicted normal metabolism image can be described mathematically:

$$g' = A * h \tag{1}$$

In this equation, g' is the predicted normal PET metabolism image that is to be created. It requires the convolution of an anatomical map of normal metabolism (A), with the known PET system PSF (h). Note that A encapsulates both anatomical and normal metabolism information, as the actual human brain does. Therefore, the creation of A requires knowledge about anatomy and normal metabolism. If the PET PSF, h, is also known, then a normal PET metabolism image, g', can be predicted. Both A and h are computed as described below.

Each patient's anatomy was analyzed using segmented MR images. First, all non-neural tissue was removed from the MR image using an iterative process of interactive thresholding and image morphology [4]. Next, volumes of interest (VOIs) in which metabolism could be considered approximately constant were defined in each subject's MR image by tracing a set of two-dimensional areas in a slice-by-slice manner. The regions which were identified were the lateral temporal, the mesial temporal, and the non-temporal regions. Each of the three sets of slice-based ROIs was stacked to create the three volumes of interest. Finally, because metabolism is normally very different in brain gray matter (nerve cell bodies) and white matter (nerve fibers), these tissues had to be segmented. Slow variations in MR intensity resulting from radio-frequency gradient field inhomogeneities were eliminated by a dividing each image from a strongly filtered version of itself. Then, white and gray matter were segmented using interactive thresholding applied to proton density and T2 weighted MR images.

The physical characteristics of the PET scanning system were determined by a set of phantom studies. Images of axially-oriented line sources in a uniform cylindrical scattering medium (water) were analyzed to determine the width of the gaussian-shaped profiles which represent the PSF. These images were reconstructed using filter and other parameters identical to those utilized with standard clinical FDG brain imaging. From these scans, it was determined that the FWHM of the PET PSF is ~8.mm at the lateral sides of the brain and ~8.5mm near the middle. A homogeneous approximate value of 8mm was used for these preliminary studies.

Normal variations in regional metabolism were computed by studying a set of 10 extensively-screened normal volunteers. Both PET and MR images were acquired so that anatomical variations and PET system physics could be included in the analysis of normal metabolism. Non-neural tissue was removed from the MR images and the result was aligned with the PET image using the methods of Woods, et al. [5]. Temporal VOIs were defined, and white and gray matter were segmented in the MR prior to registration; all were transformed to match the PET image. PET images were normalized to total brain counts.

Local normal metabolism rates in gray and white matter of the major brain VOIs were determined by finding metabolism values to be applied to the segmented MR image (denoted as A in equation 1), such that the difference between the predicted normal and the actual (normal) PET image was minimized for the group of normal volunteers. The function that was minimized was:

$$\sum_{n} \sum_{i} (A*h - g)^2 \tag{2}$$

where A is the high resolution metabolism image created by applying normal metabolism values to the segmented MR image, h is the measured PET PSF, and g is the original PET image. The quantity (A*h) is the predicted normal PET metabolism image. The summation is over all patients (n), and over all non-boundary pixels in each VOI (i). The equation was minimized using least squares techniques; the resulting gray and white matter metabolism values are those needed in each MR VOI to create the intensities observed in the PET image, given the MR anatomy and PET system physics.

The regional normal gray and white metabolism values were used to create predicted normal metabolism images for the 10 normal volunteers as well as for a set of six epilepsy patient studies, as follows. After processing the MR images of each subject to define VOIs and segment gray and white matter, the normal metabolism values were applied to the temporal lobes and surrounding regions. The resulting image was smoothed using a gaussian filter approximating the PET PSF to create a predicted normal PET metabolism image.

Difference images were created by subtracting the predicted normal metabolism values pixel-by-pixel from the actual PET image. Regional percentage differences between actual PET metabolism and predicted normal metabolism were calculated within the previously-defined temporal VOIs as (predicted PET - actual PET)/(actual PET). Differences between an individual's predicted normal and actual PET metabolism image should not be due to anatomy, scanner physics, or normal metabolism variations, since these effects are already incorporated into the predicted normal PET image. Instead, statistically significant differences can be attributed to true metabolism abnormalities.

The difference between the percentage differences in left and right temporal lobes for the normal volunteers, or ipsilateral and contralateral temporal lobes for the epilepsy patients, was calculated. Epilepsy patients were grouped and processed based on whether they had a unilateral seizure focus verified by a seizure-free post-surgical outcome or a bilateral seizure focus verified by depth electrode implantation.

The difference images could be thresholded, based on the pixel-by-pixel differences seen in the normal volunteers. Differences between predicted and actual metabolism greater than that seen in the normal subjects may indicate clinically significant hypometabolism. For each subject, the areas of hypometabolism were overlaid onto his or her MR image, and the result was evaluated subjectively.

3 Results

Mean percentage differences between actual and predicted PET images (Predicted Normal - Actual PET)/(Actual PET) were calculated for both lateral and mesial left and right VOIs, for both white and gray matter. The mesial gray matter ipsilateral to the seizure focus was seen to be the most hypometabolic in epilepsy patients with unilateral seizure onset. The % difference between predicted normal and actual PET images in this region was 3.5% ± 10.78% for the normal subjects, -1.4% ± 18.4 for the bilateral seizure onset patients, and 21.5% ± 9.5% for the unilateral seizure onset patients. This measure is significantly different between epileptic subjects with unilateral seizure focus normal subjects ($p<.05$) and epileptics with bilateral seizure focus ($p<0.10$). In contrast, neither left nor right percentage difference for normal subjects is significantly different from that of epileptics with bilateral seizure foci ($p>.30$)

The difference between left and right percentage differences was also calculated for the three subject groups. These numbers were calculated for both gray and white matter regions, where gray and white matter were defined using the segmented MR images. The results for the ten normal subjects appear in Table 1.

	Lateral Temporal (mean ± st. dev.)	**Mesial Temporal (mean ± st. dev.)**
	Left-Right % Difference	Left-Right % Difference
Gray Matter	4.2 ± 4.4	0.9 ± 3.5
White Matter	4.6 ± 2.6	2.6 ± 5.3

Table 1. Differences between right and left percent differences (Predicted Normal - Actual PET)/(Actual PET) in temporal VOIs of the ten normal volunteers.

The epilepsy patients who were determined to have bilateral seizure onset were grouped together, and the difference between left and right VOIs were computed. These appear in Table 2.

	Lateral Temporal (mean ± st.dev.)	**Mesial Temporal (mean ± st.dev.)**
	Left-Right % Difference	Left-Right % Difference
Gray Matter	3.5 ± 3.3	1.6 ± 2.7
White Matter	-.08 ± 2.3	0.1 ± 4.7

Table 2. Differences between right and left percent differences (Predicted Normal - Actual PET)/(Actual PET) in temporal VOIs of the 3 epilepsy patients with bitemporal seizure foci.

The three epilepsy patients who had verifiable unilateral seizure foci were grouped for analysis. The ipsilateral versus contralateral % differences can be compared in Table 3.

	Lateral Temporal (mean ± st.dev.)	**Mesial Temporal (mean ± st. dev.)**
	Ipsi-Contralateral % Difference	Ipsi-Contralateral % Difference
Gray Matter	13.2 + 2.4	13.9 ± 1.3
White Matter	14.4 ± 7.6	9.1 ± 8.0

Table 3. Differences between ipsilateral and contralateral (to focus) percent differences (Predicted Normal - Actual PET)/(Actual PET) in temporal VOIs of the three epilepsy patients with unilateral seizure foci.

The difference images were thresholded at a level of 1 standard deviation (over all the pixels in the difference image) to allow identification of localized hypometabolism corresponding to the seizure focus. These thresholded images were evaluated subjectively; typical results are shown in Figure 3.

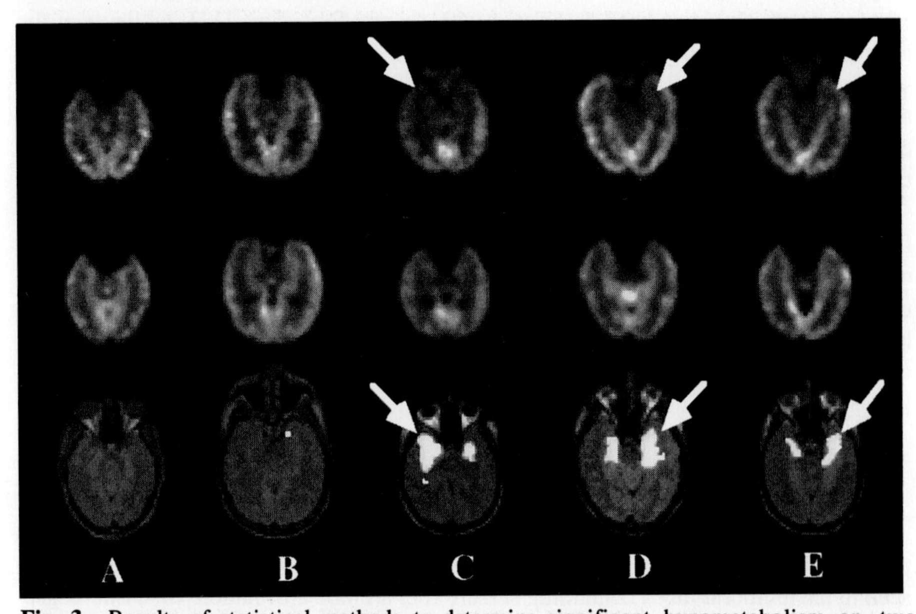

Fig. 3. Results of statistical methods to determine significant hypometabolism on two normal (**A, B**) and three epileptic (**C,D,E**) subjects. The top row shows one representative slice from the original PET metabolism image. The second row shows the corresponding slice of the predicted normal metabolism image. The original PET and predicted normal image can be compared mathematically to determine clinically significant differences. The final row shows significant differences between the original and the predicted normal images overlaid in white on each subject's MR image. The location of the epileptogenic focus for the three epileptic subjects is denoted by an arrow; note that the primary regions of hypometabolism appear in the correct temporal lobe in each case. Also note that non-temporal areas are not being analyzed at this time, and therefore, no conclusions can be drawn about the appearance of the predicted normal metabolism image in those regions.

4 Discussion

Approximately 70-80% of PET studies done on patients being evaluated for temporal lobectomy demonstrate an area of temporal lobe hypometabolism that consistently correlates with subsequent electrophysiological localization of the seizure focus [3,6]. In addition, post-surgical seizure control correlates highly with the degree of hypometabolism in PET studies [7,8]. The value of PET, then, is its ability to predict which individuals will have a good surgical outcome without the risks of depth electrode placement, and to screen patients who are not likely to benefit from the surgery. Automated identification of hypometabolism in general, however, is complicated by anatomical differences. In epilepsy specifically, degeneration of the cortical tissue (anatomy) in the temporal lobe may be partially responsible for the apparently decreased metabolism (physiology) seen in PET images due to the detector response effect. The work of Jack, et al. [9,10] in quantifying hippocampal volumes from MR has shown that this volume is frequently unilaterally decreased in epileptics. Temporal lobe atrophy has also been associated with seizure foci, e.g., [11], while even the normal size of the temporal lobe has been shown to depend on total brain size, age, sex, and handedness [12]. The detector response effect implies that both pathological tissue volume changes and normal variations in structure size will cause localized count variations and increase the variability of any quantitative measurement of metabolism.

Meltzer, et al. [13] have described a method that uses anatomical information from the MR to "correct" PET images for the partial volume effect. They apply an operation similar to inverse filtering of the image, and have shown the results to be useful in diagnosing metabolism abnormalities versus brain volume loss in Alzheimer's disease. Their approach is similar to the one we take; however, our method determines regional normal metabolism values in order to predict the normal PET image, rather than using the problematic inverse filtering operation.

5 Conclusion

The quantitative methods introduced here allow accounting for anatomical and partial volume effect variations. The results indicate that the reduced count values seen in the temporal lobes containing seizure foci are not attributable to tissue degeneration; instead, metabolism is indeed reduced in the temporal lobe ipsilateral to the seizure focus in epilepsy patients. Comparison of left vs. right temporal lobe metabolism in patients with unilateral versus bilateral seizure onset indicates that this method may be appropriate for identifying patients with unilateral temporal lobe foci.

Acknowledgments

This work was supported in part by the Whitaker Foundation.
Thanks to our superior PET and MR technologists, particularly Elizabeth Smith, Delicia Votaw, Bobbie Burrow and Virginia von Allmen.

References

1. Hoffman EJ, Huang SC, Phelps ME: Quantitation in positron computed emission tomography: 1. Effect of object size. J Comput Assist Tomog **3** (1979) 299-308.

2. Mazziotta JC, Phelps ME, Plummer D, Kuhl DE: Quantitation in positron emission computed tomography 5: Physical-anatomical effects. J Comput Assist Tomog **5** (1981) 734-743.

3. Henry TR, Engel JE, Mazziotta JC: Clinical evaluation of interictal fluorine-18-fluorodeoxyglucose PET in partial epilepsy. J Nuc Med **34** (1993) 1892-1898.

4. Serra J. *Image Analysis and Mathematical Morphology*, Academic Press, London, 1982.

5. Woods RP, Mazziotta JC, Cherry SR: MRI-PET registration with automated algorithm. J Comput Assist Tomog **17** (1993) 536-546.

6. Engel J, Henry TR, Risinger MW: Presurgical evaluation of partial epilepsy: Relative contributions of chronic depth electrode recordings vs. FDG PET and scalp-sphenoidal ictal EEG. Neurology **40** (1990) 1670-1677.

7. Swartz BE, Tomiyaso U, Delgado-Escuera AV, et al.: Neuroimaging in temporal lobe epilepsy: Test sensitivity and relationships to pathology and post-operative outcome. Epilepsia **33** (1992) 624-634.

8. Radtke RA, Hanson MW, Hoffman JM, et al: Temporal lobe hypometabolism on PET: Predictor of seizure control after temporal lobectomy. Neurology **43** (1993) 1088-1092.

9. Jack CR, Sharbrough FW, Twomey CK, et al: Temporal lobe seizures: lateralization with MR volume measurements of hippocampal formation. Radiology **175** (1990) 423-429.

10. Jack CR, Sharbrough FW, Cascino GD, et al: Magnetic resonance image-based hippocampal volumetry: Correlation with outcome after temporal lobectomy. Annals of Neurology **31** (1992) 138-146.

11. Kuzniecky R, de la Sayette V, Etheir R: Magnetic resonance imaging in temporal lobe epilepsy: pathological correlations. Ann Neurol. **22** (1987) 341-347.

12. Jack CR, Twomey CK, Zinsmeister AR, et al.: Anterior temporal lobes and hippocampal formations: normative volumetric measurements from MR images in young adults. Radiology **172** (1989) 549-554.

13. Meltzer CC, Leal JP, Mayberg HS, et al.: Correction of PET data for detector response effects in human cerebral cortex by MR imaging. J Comput Assist Tomog **14** (1990) 561-570.

Multi-Array EEG Signals Mapped with Three Dimensional Images for Clinical Epilepsy Studies

C. Rocha, J-L. Dillenseger, J-L. Coatrieux

Laboratoire Traitement du Signal et de l'Image, INSERM
Université de Rennes 1, Campus de Beaulieu, 35042 Rennes Cedex, France

Abstract. The design of methods to simultaneously display signal and image data used in clinical epilepsy research is presented. Electrical sensors, spatially distributed into or over the head of the patient, are basic tools for epilepsy research. The considerable amount of sensors makes the interpretation of the acquired signals difficult when they are displayed as multiple electrical potentials vs. time curves. In addition, the dimension of the representation domain is increased after processing the signals (e.g. time-scale or time-frequency representations). The anatomical reference is always required to understand the mechanisms underlying the brain electrical activity. The solutions reported here make use of two dimensional (2D) or three dimensional (3D) spatio-temporal mappings, surface cartographies projected onto the anatomical structures and compositions of depth/surface/morphology information.

1. Introduction

Epilepsies are neurological disorders that lead to transient and repetitive interruptions in normal electrical brain activity [1]. The medical application domain of this paper is related to the clinical research in epilepsy. We will describe the design of a platform that helps the interpretation of the information collected on the patient or the information a priori known. From a global point of view we can identify four information sources (figure 1):

1 - *Imaging modalities.* Magnetic Resonance Imaging (MRI) and Computed Tomography (CT) are modalities that provide the morphological reference to the brain structures. MRI is preferable to CT because it gives more information concerning brain tissues. Single Photon Emission Computed Tomography (SPECT) and Positron Emission Tomography (PET) give functional information and are used to identify the pathological brain areas.

2 - *Electromagnetic data.* The brain electrical activity is acquired by Electroencephalograms (EEG) (surface recordings), and Stereo-EEG (SEEG) (in-depth recordings). The magnetic components are acquired by Magnetoencephalograms (MEG). These modalities provide a fine temporal description of the underlying process evolution.

3 - *Anatomical and functional models.* After registration with the morphological volumes they help for the understanding and the delimitation of the affected brain structures.

4 - *Clinical observations.* The antecedents of the patients and their medical history are completed by a video monitoring during the seizures in order to objectively record an important part of the epilepsy semiology.

This examination protocol of epileptic patients is very demanding and only specialized medical centers are able to totally achieve it. This protocol is dedicated to highly handicapped patients for whom a surgical intervention must be considered.

The several information sources taking part of the clinical research must be merged and integrated into a multimodal platform that should meet the requirements of the monitoring constraints (signal and real time video observations with the possibility of events marking) and of the 3D visualization of the imaging modalities. The present work is focused on a subset of the overall problem: the simultaneous representation of EEG and SEEG signals on the morphological data. The objective is to facilitate the interpretation of the signals and to have a better understanding of the depth/surface relationships. In Section 2 the problem of matching signals and 3D images is addressed and in Section 3 the different visualization solutions will be described and discussed. The conclusion will set some new perspectives which will complement the platform.

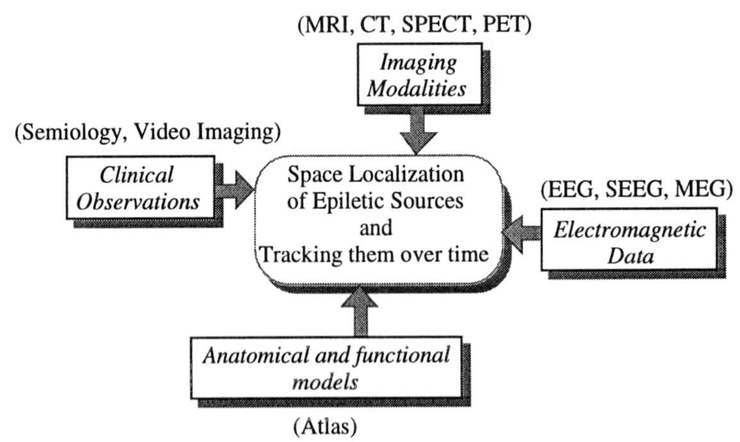

Fig. 1. The four fundamental information sources involved in the monitoring of epileptic patients.

2. Problem Statement: 3D Image and Signal Matching

The EEG and SEEG signals are characterized by:

1 - *The spatial arrangement of the electrodes.* To produce an EEG recording, the electrodes are distributed over the head of the patient according to standard settings (the International 10/20 System is an example). They are sparsely scattered which means that the signals have a low spatial sampling rate. In order to recover the whole electrical mapping over the scalp some interpolation scheme must be applied. The SEEG signals are acquired by means of needle electrodes placed in the brain (figure 2), crossing several brain structures. Because there is a set of recording sensors along each electrode, these sensors form a regular linear array.

2 - *The number and the characteristics of the recording sensors.* The surface signals are usually issued from a small number of electrodes (about twenty). But this number must be increased (up to 60 or more) in order to better identify the localization of the pathological electrical activity. The depth recordings are normally performed by electrodes with fifteen sensors which are geometrically assumed to be points. In order to remove any ambiguity when localizing the epileptic source, five or six electrodes are normally placed in the brain. The characteristics of the sensors are mainly related with their shape and size and with the quality of the contact between the skin and the electrode for the case of surface recordings.

3 - *The sampling frequency.* The sampling frequency of the signals is usually set between 200 and 500 Hz.

After each recording session there are an enormous amount of data to be stored (most of it without significant information). This arises because of the frequency, the total number of sensors and the duration of the monitoring period: during several days and both during epileptical activity (seizures) and interictal periods (between seizures). The graphic representation with multiple electrical potentials vs. time curves (figure 2) does not give information about the spatial relationship between the different electrodes, nor the relationship between the electrodes and the anatomy. It is then advantageous to integrate the representation of the signals into the morphological space of the patient.

The MRI and CT images are suitable for the description of the 3D morphological space and their 3D rendering must be worked out. The methods for the 3D rendering in

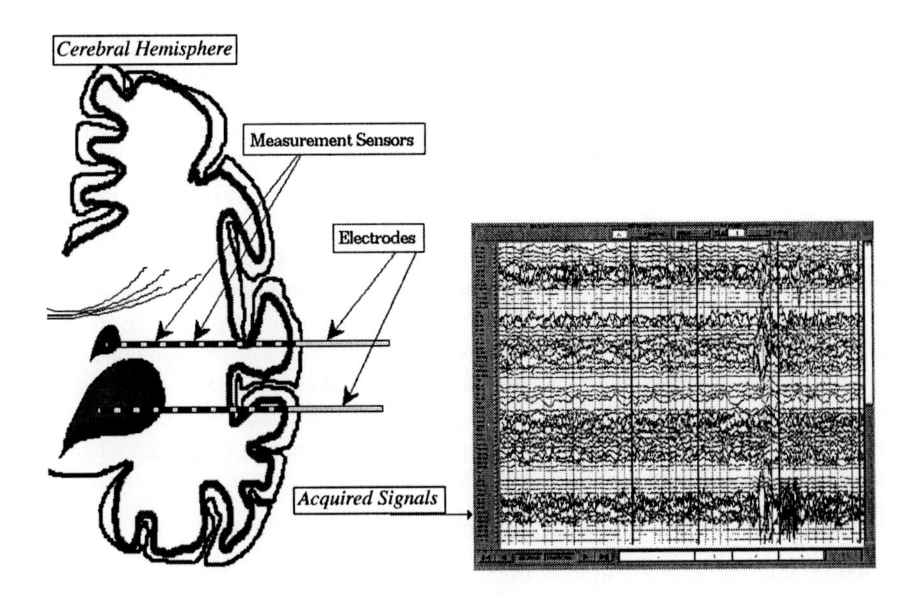

Fig. 2. Schematic drawing showing the brain structures, the localization of the in-depth electrodes, and SEEG signals acquired between seizures.

medical imaging are classified according to their primitive encoding (surface or volume), their projection modes (direct projection or ray casting), and the visual cues that provide the 3D perception (shading, transparency, color, motion, ...). For the state-of-art in this domain, we refer [2] and the special issues [3][4]. An extension of the 3D visualization package X-Image 3D [5] was used for the results reproduced in this paper.

In order to match the electromagnetic data with the morphological structures, the common information of both modalities is used, e.g. the 3D positions of the sensors. Some solutions for the problem of matching the positions of the sensors on the morphology are: measuring the positions of the EEG sensors over the surface of the head by a 3D positioning system (e.g. the Polhemus system), registering the in-depth electrodes from radiological projections or directly from a 3D image modality.

3. Some Representative Visualization Examples

3.1 Visualization of Signals and Its Transforms

To compact and enhance the global perception of the signals, two representations will be presented here. This first type of representation gives no anatomical reference, not even in a schematic form. The spatial relationships are lost and the clinician must mentally reconstruct them.

Multiple Linear Arrays Over Time

To produce a graphic plot [6] that represents the signal amplitudes recorded on each in-depth electrode as a function of time, spatial interpolation is required because of the low spatial sampling of the sensors (15 in the present case). Time interpolation is not necessary since the sampling frequency is high enough. Figure 5 illustrates the obtained space-time map for a single electrode. The sensor positions are placed along the vertical axis. The top of the vertical axis corresponds to the deepest sensor. Spline interpolation was used along this axis. Time is plotted on the horizontal axis. The signal amplitude is encoded through a rainbow color look-up-table: blue for negative values going towards red for positive, values near zero are green.

In terms of visual comprehension, this representation gives: (*i*) a good spatial discrimination between the signals acquired by the different sensors along one electrode, (*ii*) a better perception of the events and their instantaneous delays. A step by step mode or an automatic sequential viewing facilitates the data exploration. The spatial and temporal relations between the different electrodes can be visualized when several windows are displaying the different space-time maps.

Linear Arrays and Transformations of the Raw Data

Another kind of representation of the temporal observations is obtained after applying signal processing techniques in order to transform the raw data domain. These transforms have the particularity of modifying the dimension of the representation domain. Consequently, some new visual clues must be worked out when the space dimension is augmented. A simple illustration example is given by the Fourier

Transform where a time domain observation x(t) is transformed into the frequency domain X(f) and is then described by two variables: modulus and phase (or real and imaginary parts). The most recent research dealing with time-scale and time-frequency representations pointed out the necessity of high-level visualization tools because, most often, only a subjective interpretation of the results is performed. We will illustrate the associated visualization problems with two examples in which the dimension of the original representation is enhanced: a monodimensional graph is converted into a bidimensional diagram.

The first 2D diagram (figure 6) shows the time-scale decomposition of an EEG epoch (top of the figure) which corresponds to an interictal period where some paroxystic transient events (spikes, spike waves) are present. Using a continuous wavelet transform as described in [7], the signal is decomposed into 30 levels and the modulus of each level is plotted (the amplitude is represented by the color look-up-table shown on the right side of the figure). With this representation the transients are enhanced. The purpose of this wavelet decomposition was to design an original detection framework [8]. To complete the representation, the phase information also should be depicted.

The second 2D diagram (figure 3, left side) depicts a time-frequency representation (Smoothed Pseudo Wigner-Ville distribution) that was computed in order to emphasize the non-stationary behavior of the signal [9]. The SEEG epoch (plotted on the top) was recorded during an ictal period. The vertical axis corresponds to the frequency and the horizontal one to the time. Because this representation enhances the time-frequency behavior of the signal components, it is then possible to analyze the different frequency components of the signal and to track them over time.

If we take into account the alignment of the sensors along one in-depth electrode, several 2D diagrams can be stacked and represented as a 3D volume where the coordinates of the axes are the frequency (or the scale level), the time and the sensor positions. This volume is then rendered by the same procedures used to visualize the 3D morphological images. The resulting image (figure 3, right side) shows a 3D visualization of a time-frequency representation of short time segments (5s) of signals acquired by one electrode (12 sensors). The inverse distance weighted interpolation was

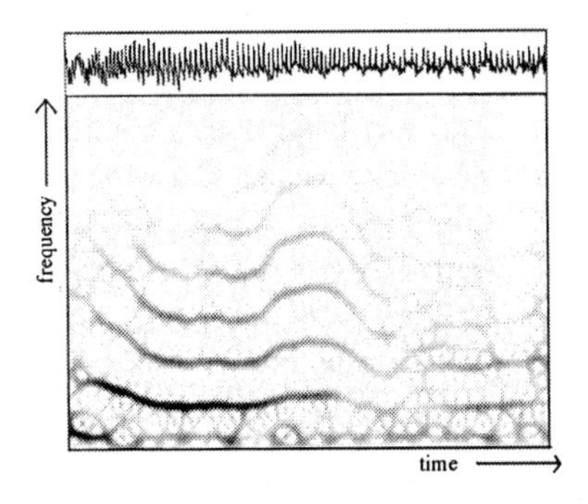

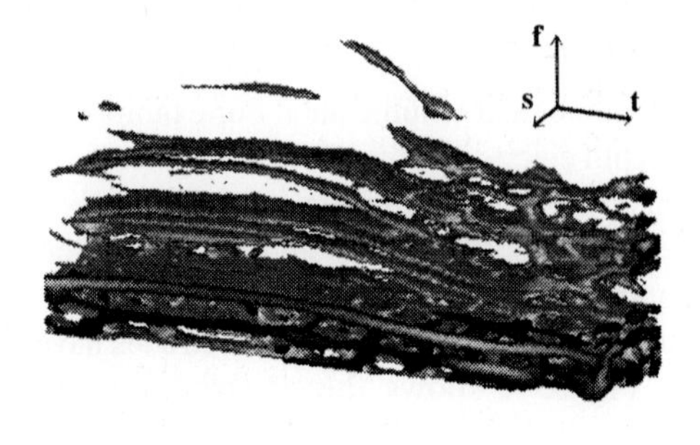

Fig. 3. Time-frequency representation (left) and Iso-surface representation of a volume obtained after stacking the 2D time-frequency representation diagrams (right).

performed along the sensor's positions. The final result is produced by a ray casting technique where the surface threshold was interactively defined by the user. This kind of representation allows the use of several analyzing tools (translation, rotation, transparency), but the visual interpretation of the results is not easy just because of our non-familiarity with the space description.

3.2 Signal and Image Fusion

The knowledge of the spatial relationship between the depth electrodes and the brain structures and also the relationship between the depth electrodes and the surface ones is essential for the interpretation of the epileptic data. The depth electrodes/brain structures relationship is only partially given by the morphological imagery modalities (MRI, CT) because these modalities produce a macroscopic description of the brain structures. A more detailed description of the non-visible structures on the image modalities can be obtained by matching an anatomical atlas on the real volume. As far as we know the surface/in-depth relationship was never analyzed by the help of 3D visualization techniques.

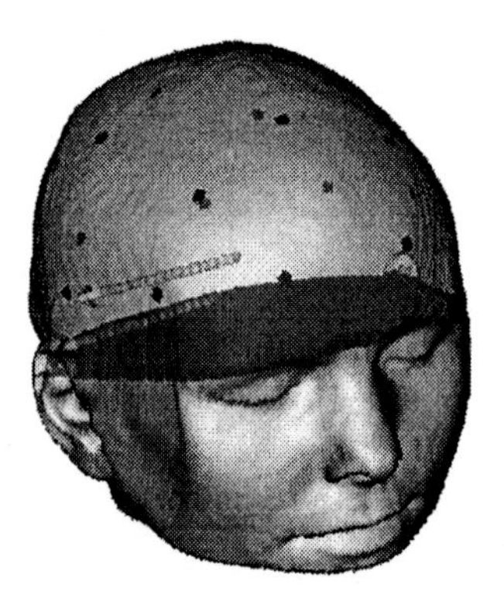

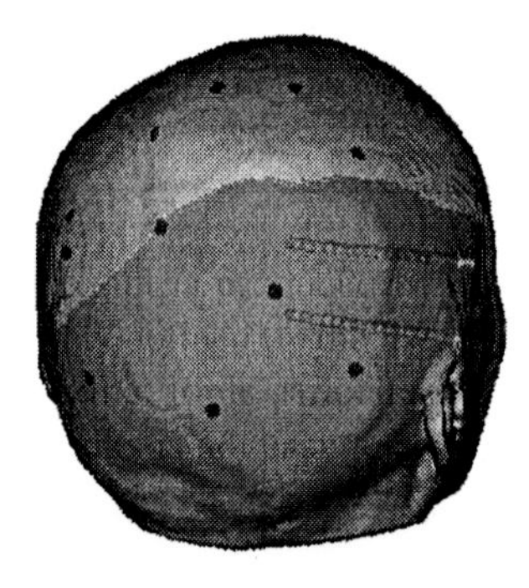

Fig. 4. Visualization of the scalp surface, the surface sensors and the depth electrodes.

Sensor Representation Into/Onto the Image Volume

The electrodes can be modeled by geometrical primitives merged into the morphological volume data. For example, a surface sensor can be modeled as a disc and a depth electrode as a cylinder built from a set of sections perpendicular to the electrode axis, thus simulating the sensors. The sensor positions are estimated in the image reference coordinate system and the electrodes are drawn in the 3D volume by the use of an analytical description or a digitized one. Each surface electrode is orthogonal to the surface normal and the depth electrodes have a voxelized description.

Figure 4 shows the electrodes merged into the morphological volume. The anatomical data is provided by 150 MRI slices, 256x256 pixels. A transparency coefficient is assigned to the region of interest that corresponds to the EEG measurement space (the scalp). The lower part of the head is made opaque, as well as the electrodes. Volume rendering techniques are used to achieve the visualization of the depth and surface electrodes together with the image data. The thickness of the in-depth electrodes has been artificially increased in order to enhance the perception of the representation.

3D Surface Mapping

The 3D morphological surface cartography has been partially described in [10]. The overall problem, reported in [11], includes the interpolation schemes, the transparency effects, the type of the signals to be visualized (original or transformed) and the display modalities protocols. One interpolation procedure is used to estimate the potential values over all the points of the scalp, taking into account the sensors spatial sampling rate. Several procedures can be used [12]: linear interpolation, multiquadratic, spherical splines and B splines. In our case we use the spherical splines. The scalp surface where the potentials representation is projected can be rendered as opaque or transparent (figure 8). The electrical potentials are encoded by a color look-up-table that encodes the signals as blue for negative and red for positive values. The 3D shape is obtained by surface shading.

Various display protocols are available: a temporal sequence is rendered from a given point of view as a sequence of time indexed maps or as a time varying map (i.e. an animation). In this last case the frequency of display must be lower than the signal sampling frequency, in order to let the user analyze the sequence (according to our experiments, a refreshing rate of 2 to 5 images/second is acceptable). Step by step and replay facilities are also included as analyzing tools.

The result of the signal transforms (as described previously) can also be mapped. We refer [7] for an example showing an enhanced localization of a paroxystic event on the frontal area, using a 3D surface mapping of a continuous wavelet transform.

3D Depth and Surface Mapping

We now consider the representation of a function (a linear distribution) along the depth electrodes. The problem concerning the representation of an abstract phenomenon (like the electrical potentials) is related to encoding the visual support. This encoding support can have a geometrical basis and/or one color. With the surface potentials the representation support was the scalp and the information was encoded by the color. On the other hand, the electrical signals acquired within the brain have a pinpoint nature. For this reason, the representation of in-depth signals needs a synthetic support. Several solutions based on simple geometrical primitives and different color look-up-tables have been tested.

The experiments that have been conducted suggest combining two approaches: 1) encoding the amplitude of the signal (the modulus) by a variation of the size of a geometrical support located on each sensor's position, and 2) a color is assigned to each amplitude value. In this case one encoding procedure (the size) enhances the effect of the other (the color). The solution makes use of geometrical supports defined by spheres centered on the sensor positions. The 3D perception is enhanced by the sphere surface shading.

The representation of a short time sequence of the electrical potentials recorded along the depth electrodes is displayed in figure 7, where the scalp is rendered as transparent. Potentials are encoded by a rainbow color scale and the variation of the diameter of the geometrical support (sphere) that is located on each sensor position. The support diameter varies according to the absolute value of the potential. The temporal aspect of the phenomena can then be analyzed. Because the electrical activity is directly represented in the morphological reference, its time and space coherence is immediately perceived. At the beginning of the time sequence, an electrical event occurred on the inner sensors of the electrode (left side of the windows). Some time steps later, a similar event can be observed on the inner sensor of the upper electrode.

The composition of surface and depth measurements is then possible. Figure 9 depicts four different viewing directions at the same time instant where the depth recordings are associated to the semi-transparent surface cartography. This dual representation allows to understand directly the relationship between the internal and surface measurements. This representation has as main characteristics: (*i*) global (surface) and local (depth) views of the phenomena are simultaneously available, and (*ii*) the time consuming examination and correlation of several channels is replaced by a fully integrated image that can be interactively explored.

This work has been oriented toward electrical potentials. However, this representation method can be used for any other variable that is relevant for the interpretation of the phenomena (e.g. phase delay between the sensors, frequency components).

4. Conclusion

The proposed solutions for the display of in-depth or surface signals in clinical epilepsy research present a high interest to facilitate the identification of the source of paroxystic events, as well as their tracking in relation with the cortical structures. The integration in the same 3D morphological space of the signals represented by data arrays distributed along linear or surface structures gives the possibility to preserve the fundamental space-time relations.

It has been shown that the signal-image fusion can handle data of high order multidimensional spaces. The main benefit of these methods is that they can be applied to other medical areas and investigation techniques such as evoked potentials [13], electrocorticography or magnetoencephalography. The visualization of multiple anatomical structures with their respective cartographies and also the representation of volumes generated from time-scale or time-frequency transforms related to their anatomical space are now under study.

Acknowledgments. To Patrick Chauvel, Jean-Michel Badier and Patrick Marquis from the Clinical Epilepsy Research Unit of the University Hospital of Rennes, where the EEG and SEEG signals were acquired. To Lotfi Senhadji and Mohammad Shamsollahi for providing us their results on signal transforms and Fabrice Wendling for the electrical potential vs. time curves, all from the LTSI. To JNICT (Junta Nacional de Investigação Científica e Tecnológica, Portugal), which is supporting the work of the author C. Rocha, scholarship PRAXIS XXI/BD/5312/95.

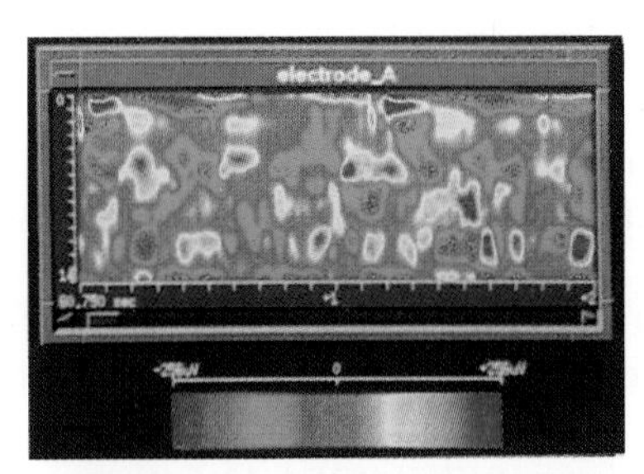

Fig. 5. A space-time map of SEEG signals

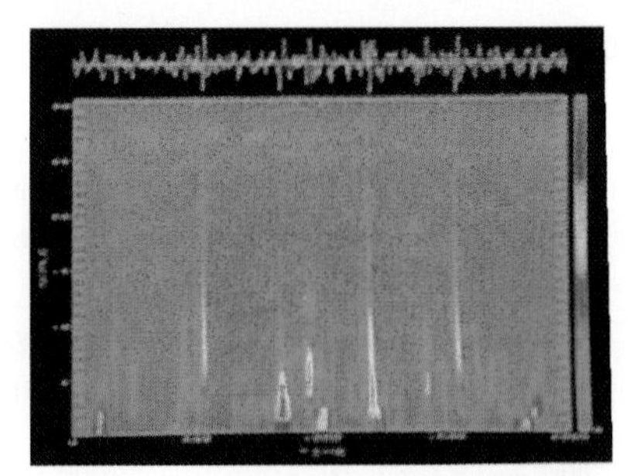

Fig. 6. Time-scale (Wavelet) representation

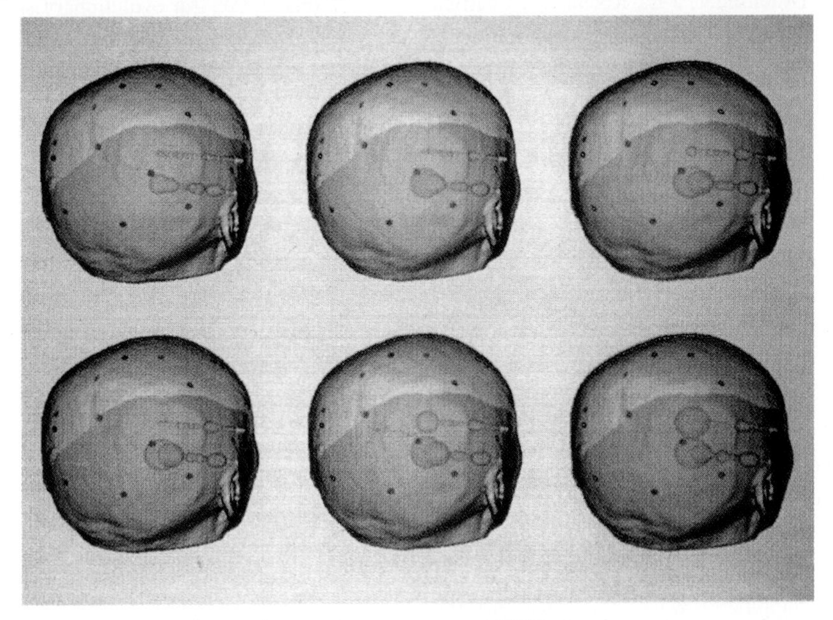

Fig. 7. A short time sequence of SEEG signals

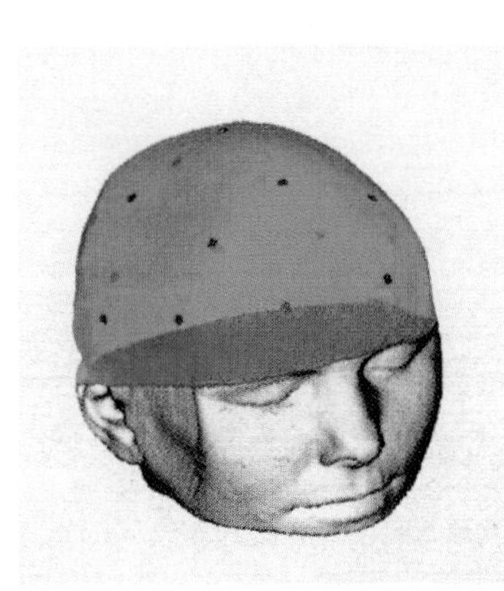

Fig. 8. 3D surface cartography (EEG)

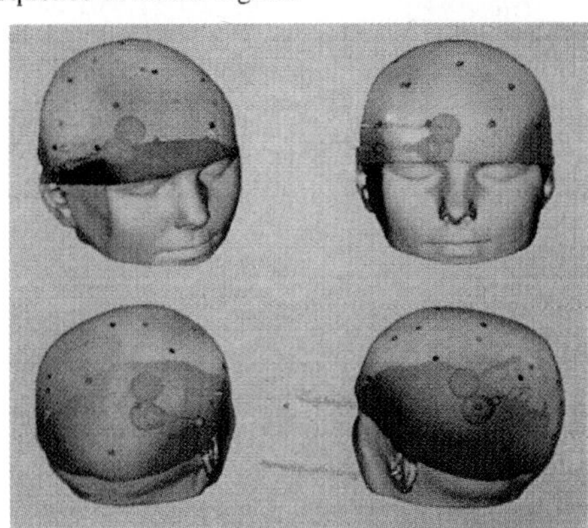

Fig. 9. Composition of EEG/SEEG representations

References

1. Lopes da Silva, F.: Neural mechanisms underlying brain waves: from neural membranes to networks. Electroenceph. and Clinic. Neurophys. **79** (1991) 81-93
2. Kaufman, A. (ed.): Volume visualization. IEEE Computer Society Press, Los Alamitos, California, Order no. 1979 (1990)
3. Coatrieux, J-L., Barillot, C.: A survey of 3D display techniques to render medical data. Höhne, Fuchs, Pizer (eds.): 3-D Imaging in Medicine NATO ASI Series **F60** (1990) 175-195
4. Rhodes, M. L.: Graphics in medicine. Special issue of IEEE CG&A **13**-6 (1993) 49-92
5. Dillenseger, J-L., Rocha, C., Coatrieux, J-L.: 'X-Image 3D': An evolutionary software system for interactive visualization and analysis of multidimensional biomedical data. Proc. International Congress on Medical Informatics MIE **12** (1994) 590-596
6. Badier, J-M.: Étude de la localisation des sources cérébrales d'activité paroxystique par cartographie. Ph.D. Thesis, Université de Technologie de Compiègne, France (1991)
7. Senhadji, L., Dillenseger, J-L., Rocha, C., Wendling, F., Kinie A.: Wavelet analysis of EEG for mapping of epileptic events. Annals of Biomed. Eng. **25**-5 (1995) 543-552
8. Senhadji, L., Bellanger, J-J., Carrault, G.: Détection temps-échelle d'événements paroxystiques intercritiques en électroencéphalographie. Traitement du Signal **12**-4 (1995) 357-371
9. Shamsollahi, M. B., Senhadji, L., Le Bouquin-Jeannes, R.: Time-frequency analysis: a comparison of approaches for complex EEG patterns in epileptic seizures. Proc. IEEE-EMBS **17** (1995)
10. Dillenseger, J-L., Coatrieux, J-L.: Functional and morphological data fusion in electroencephalography. Proc. IEEE-EMBS **14** (1992) 2022-2023
11. Dillenseger, J-L.: Imagerie tridimensionnelle morphologique et fonctionnelle en multimodalité. Ph.D. Thesis, Université François Rabelais de Tours, France (1992)
12. Nielson, G.: Scattered data modeling. IEEE CG&A **13**-1 (1993) 60-70
13. Gevins, A., Brickett, P., Costales, B., Le, J., Reutter, B.: Beyond topographic mapping: towards functional-anatomical imaging with 124-channel EEGs and 3-D MRIs. Brain Topography **3**-1 (1990) 53-64

Visualisation of Pain by Magnetoencephalography in Humans

Burkhart Bromm and E. Scharein

Institute of Physiology, University Hospital Eppendorf, University of Hamburg, Martinistr. 52, D-20246 Hamburg, Germany

Abstract. Magnetoencephalography is a new functional brain imaging technology which allows the non-invasive determination of cortical and subcortical neuronal assemblies involved in the processing of pain. In healthy subjects 4 pain relevant activity centers could be reliably identified which were activated by brief radiant infrared laser heat pulses: the ipsilateral SI area, the bilateral SII areas and a premotoric frontal activity. The localisation these cortical areas was based on the individual head and cortex anatomy.

1 Introduction

With the introduction of magnetometry (MEG) using supra conducting quantum interference devices (SQUIDs) the non-invasive search for brain structures relevant for sensory and mental activities experienced a remarkably strong push. In the following, first results are published which were obtained by a dual cryostat system (Philips; 31 gradiometers). The localisation of pain relevant brain areas was based on the individual head and cortex anatomy obtained by magnetic resonance imaging (MRI) and the application of the boundary element method. The method to induce phasic pain was the same as in a previous study in which electroencephalographic (EEG) recorded activity was evaluated by the brain electrical source analysis (BESA, [2]) to allow comparison of results.

2 Methods

Data were evaluated from 10 healthy male medical students (20-30 years). The pain inducing stimuli were brief radiant heat pulses emitted by an infrared laser (Thulium-YAG crystal), the wavelength (1.8 mm) of which was short enough to be conducted through glass fibre optics [4]. In all sessions the same 4 intensities of 310, 340, 370 and 400 mJ were used, inducing mild to strong pain sensations which were rated by the subjects on a computerised numeric scale. In order to investigate the re-detection accuracy of the localised brain sources, subjects underwent repeated experiments and sessions (for details see following table).

Stimuli:	Heat pulses emitted by infrared Thulium-YAG laser, 0.2 ms, 1.5- and 2-times individual pain threshold, randomly applied to the left and right temple, activating A-delta fibres of upper trigeminal nerve.
Subjects:	10 males (20 - 30 years), repeatedly studied in 3 sessions (1 week apart), each with 3 blocks (30 min apart) of 40 stimuli.
EEG Data:	31 leads, bandpass 0.16-250 Hz, sampling rate 500 Hz; poststimulus EEG segments (500 ms), averaged over 40 stimuli per block (raw data).
MEG Data:	31 gradiometers, 3 cryostat positions (C3, C4, Cz) bandpass 0.16-250 Hz, sampling rate 500 Hz; poststimulus MEG segments (500 ms), averaged over 40 stimuli per block (raw data).

Simultaneously to the measurement of magnetic fields (31 gradiometers) the EEG was recorded by 31 leads. Calculations were based on realistic head models; for this purpose all subjects underwent MR investigations. For an exact assignment of the cryostat system (in which the magnetic fields are measured) to the individual head system (in which brain generators are to be determined), small coils were fixed at defined head sites which broadcast tiny magnetic fields in order to localise the momentary head position in the cryostat system. Afterwards the adjustment coils were replaced by vitamin E capsules to mark the same sites for the subsequent MR scans.

Solutions of the "inverse problem" were performed with the methods of moving dipoles, fixed dipoles and dipole scan procedures (MUSIC and deviation scans), using a sphere or the boundary element method (BEM) as models of volume conduction, and the 3D-reconstruction of the individual cortex as space of solutions (CURRY, Philips software, [3]). With BEM the individual head is modelled with 3 layers: inner and outer scull and scalp surfaces, which were subsequently used for the calculations of the volume currents. After eliminating the noise terms estimated by principal component analysis (PCA), a MUSIC algorithm (multiple signal classification) calculates the contribution of dipoles, free to rotate at each site of the solution space for the explanation of the measured fields. In a first approximation the results can be interpreted as a probability density distribution of independent dipoles within the reconstructed cortex.

3 Results and Discussion

Most cortical activity impressed in the latency range between 100 and 150 ms; later activities at 200 ms and more were considerably less pronounced, in all 3 cryostat positions. Four pain related activity centers could consistently be identified which explained up to 98 % of total variance of the recorded magnetic fields: the first activity center was localised in the primary somatosensory cortex (SI, 80 ms after stimulus onset), two generators with peak maxima around 120 ms were localised in the contra- and ipsi-lateral secondary somatosensory cortex (SII) of the face, a third one with a maximum at 130 ms in the frontal cortex, presumably indicating premotor nocifensive reactions. The late activity between 150 and 230 ms could not by described by MEG measurements, but with the EEG one single central generator in the medial part of the cingulate gyrus was found which may be involved in perceptual activity and cognitive information processing.

Fig. 1 presents the MEG defined cortical activity centers in the 3D-reconstruction of a cortex for one subject, using the scan procedure MUSIC with the first four eigenvalues of the PCA which explained about 98 % of total variance of the measured fields. The yellow arrow (above, left) marks the site of the first superficial activity center in the primary somatosensory cortex at about 80 ms after stimulation.

For the activity centers in the latency range between 100 and 150 ms after stimulation only very tiny activity patches were seen over the cortex surface (above, right). But, with sections in an altitude of 40 and 10 mm above the pre-auricular / nasion level we clearly found 3 marked areas of stimulus induced activity: Two activity maxima were identified in the ipsi- and contra-lateral secondary somatosensory cortex areas of the temple which is on the rostral upper bank of the Sylvian fissure frontally to the insula, corresponding to a somatotopic organisation in SII, where the face is represented more rostrally than the hand or the leg [1]. The third generator was localised in the frontal lobe, very near to the mid-line, the maximum of which was most pronounced in deeper sections. Very similar results were obtained using the deviation scan of the individual cortex, a dipole scan procedure which - unlike MUSIC - includes the noise term.

MUSIC approximations. The 4 activity areas found with MUSIC approximations of the measured magnetic fields were in agreement with our published findings from brain electrical source analysis (BESA) with 31 EEG leads [2]. In that paper we described a serial activation of the two SII areas: Contra-lateral SII was activated approximately 6 ms earlier than the ipsi-lateral one, a very small but significant latency difference. In the present MEG analysis we could not replicate such differences in time. We, therefore, would tend to reject the idea of a transcallosal transmission between both SII cortices, favouring instead the hypothesis of a simultaneous ipsilateral co-activation from uncrossed afferents.

The first results of pain relevant brain generators obtained by MEG and EEG source analysis with individual head anatomy presented here are preliminary. Further investigations are necessary in order to elaborate the re-detection accuracy in repeated measurements within and between individuals and to increase the sharpness of localisation.

Acknowledgement. We wish to thank Dr. Holger Kohlhoff and Katharina Freybott for help with the calculations, Dr. Jens Ellrich, Kriemhild Saha and Gerhard Steinmetz for assistance during the experiments. This investigation was supported by the Deutsche Forschungsgemeinschaft (Br 310/20-2).

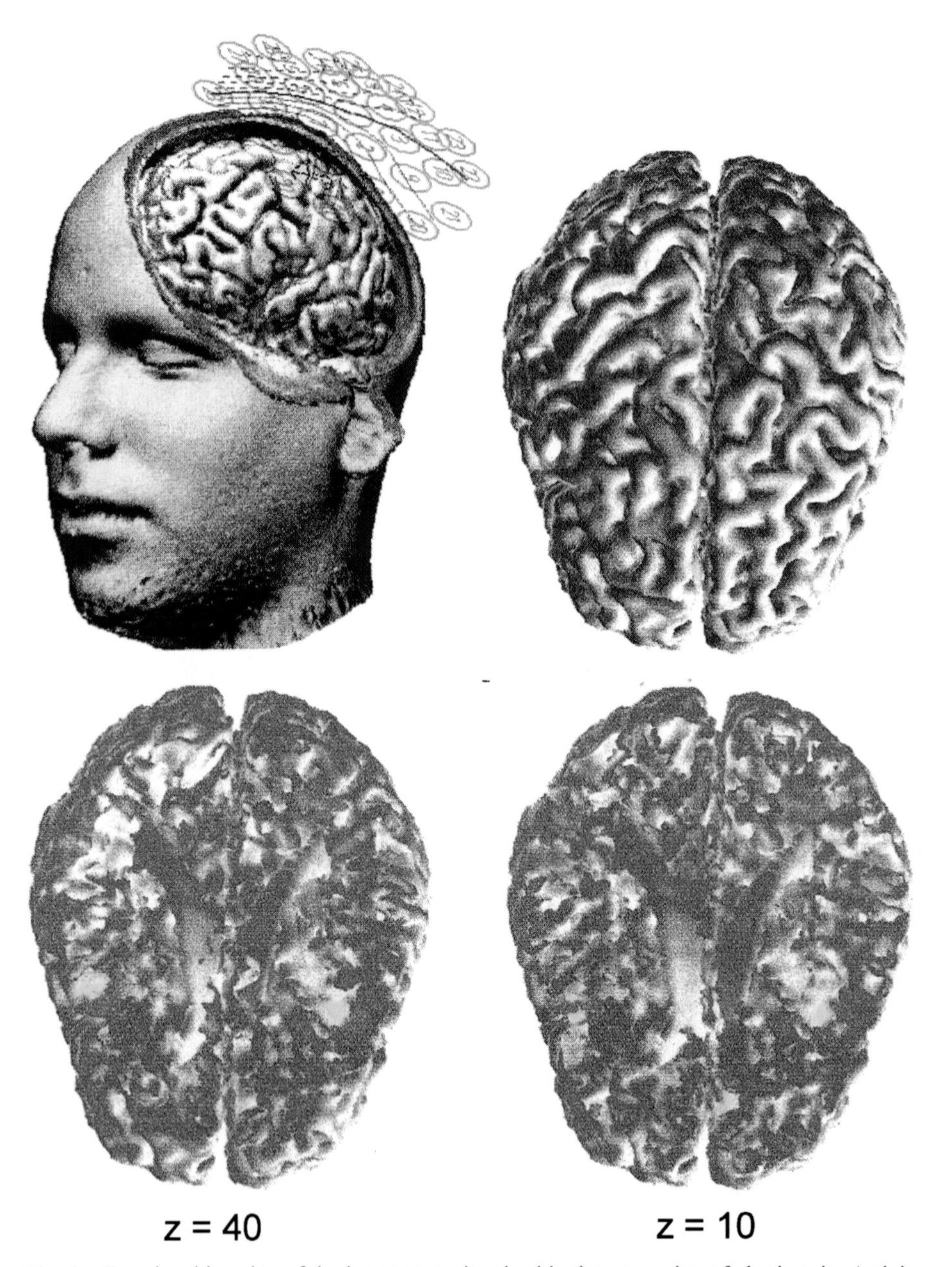

Fig. 1. Functional imaging of brain structures involved in the processing of phasic pain. Activity in the primary (SI, t = 90 ms) and secondary somatosensory cortex (SII, t = 120 ms) was identified by evaluating the magnetic fields in the individual cortex determined by MR scans. Whereas the involved SI area is very superficial (above left, yellow arrow), the SII areas can best be visualised by cutting the cortex in different depths (z = 0 mm: pre-auricular / nasion level). Pain relevant activity impresses by the simultaneous activation of ipsi- and contralateral SII areas, whereas the prefrontal activity describes premotoric activation.

References

1. Brodal, A.: Neurological Anatomy in Relation to Clinical Medicine. Oxford University Press New York (1981)
2. Bromm, B., Chen, A.C.N.: Brain electrical source analysis of laser evoked potentials in response to painful trigeminal nerve stimulation. Electroencephal. Clin. Neurophysiol. **95** (1995) 14-26
3. Fuchs, M., Wischmann, H.-A., Wagner, M., Drenckhahn, R.: Advanced biomagnetic and bioelectric reconstruction algorithms. In: (eds.), Supercomputing in Brain Research: From Tomography to Neural Networks. Word Scientific, London, (1995) 161-173
4. Kazarians, H., Scharein, E. Bromm, B.: Laser evoked brain potentials in response to painful trigeminal nerve activation; reliability and normative data. Int. J. Neurosci. **81** (1995) 111-122

Visualization of Cat Auditory Cortical Functional Organization after Electrical Stimulation with a Multichannel Cochlear Implant by Means of Optical Imaging

H.R. Dinse[1], B. Godde[1], T. Hilger[1], G. Reuter[2], S.M. Cords[2]
T. Lenarz[2], W. von Seelen[1]

[1] Institut für Neuroinformatik, Theoretische Biologie, Ruhr Universität Bochum
D-44780 Bochum, Germany
[2] Medizinische Hochschule Hannover, Exp. HNO, D-30623 Hannover, Germany

Abstract. In order to study the effects of acute electrical cochlear stimulation on the topography of the cat auditory cortex, we measured reflectance changes by means of optical imaging of intrinsic signals. Following single pulse electrical stimulation at selected sites of a multichannel implant device, we found topographically restricted response areas. Systematic variation of the stimulation pairs and thus of the cochlear frequency sites revealed a systematic and corresponding shift of the response areas. Increasingly higher stimulation currents evoked increasingly larger response areas resulting in decreasing spatial, i.e. cochleotopic selectivity. The results indicate that optical imaging intrinsic signals is useful to visualize effects of cochlear stimulation, which results in a profound cochleotopic selectivity. The implications of these findings are discussed in respect to underlying mechanisms of sound sensation mediated by cochlear implants.

1 Introduction

Cochlear prostheses are increasingly used to provide sound perception in patients with profound deafness of sensorineural origin. By electrical stimulation of the acoustic nerve fibers their auditory system can be activated in a systematic way that restores to a considerably degree their capacity of hearing [3,10].
It is a key feature of cortical representations to contain complete and systematic representations of the sensory epithelium. For the auditory system, a cochleotopic organization has been demonstrated for a number of auditory cortical fields [9]. While the representation of single tones within the framework of cortical maps is fairly well understood [12], little is known about the nature and the degree of the topographic aspects of the representation of cochlear stimulation within cortical maps [2,11].
During the recent years, new methods were developed to record optically from the exposed cortex in order to obtain two-dimensional reflectance changes that were shown to correspond to the spatial distributions of the underlying neuronal maps [1,5,7]. These reflectance measurements have the advantage to allow repeated and multiple measurements of functional maps in the same individual animal, with a spatial resolution of up to 50 microns [5].

Here we report about experiments in which the effect of electrical cochlear stimulation on the organization of the cochleotopic organizational maps in the auditory cortex of adult cats were studied by optical recordings of intrinsic signals. We demonstrate that a visualization of associated activity induced by electrical stimulation of selected electrode positions of an acutely implanted multichannel electrode device in AI is possible and that this approach can demonstrate and preserve details about the underlying topography of the frequency representations.

2 Material and Methods

Cats were anesthetized for surgery in accordance with the National Institutes of Health Guide for Care and Use of Laboratory Animals (Revised 1987). For further details see [4,6,11]. The animals were acutely deafened by intracochlear injection of neomycine sulfate. Human multichannel implant electrodes consisting of four pairs of platinum-iridium balls were inserted into the scala tympani (Fig. 1).
Conventional electrophysiological controls were made by recording extracellularly action potentials using glass micro-microelectrodes. Cochlear stimulation was performed using biphasic current pulses of 75 or 100 μs durations/phase. Electrode impedance were in the range of 5 to 50 KΩ. Electrical current levels were expressed in dB relative to 100 μA.

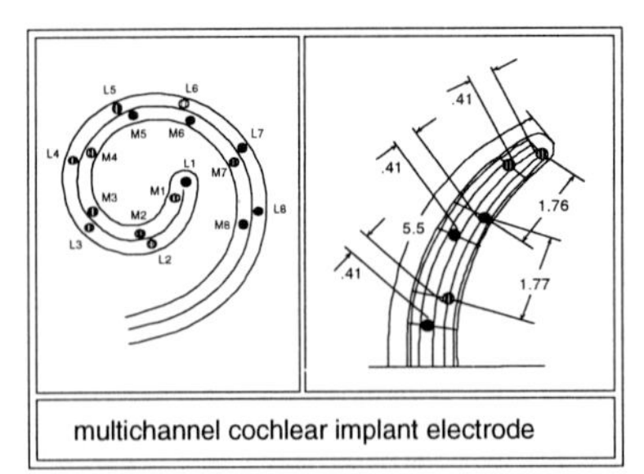

Fig. 1: Schematic illustration of an intracochlear multichannel electrode as used in our cat experiments. Each electrode contained 8 bipolar electrode pairs. Electrode 1 located most distal to the round window. Four of the electrode pairs were inserted via a cochleostomy close to the round window.

I. Topography of activity: In order to measure the possible underlying topography of activity following cochlear implant electrical stimulation, we stimulated different pairs of electrode sites. The different electrode pairs corresponded to different positions in the cochlea and thus to different distance from the round window. **II. Effects of current amplitude:** To measure the effect of increasing stimulation current on the topography of the activity distribution, we systematically varied stimulation currents between threshold and 20 dB above threshold.
For **optical measurements**, we used a Lightstar II imaging and acquisition system (LaVision) with a 2 MHz A/D converter and a Peltier cooled, slow scan 12 bit digital CCD-camera. Images were obtained with acquisition times of 80 ms duration. Averaging was achieved by adding intertrial sequences consisting of 5 images of 80 ms duration, which were averaged to 6 trials (Fig. 2). Each trial was separated by a

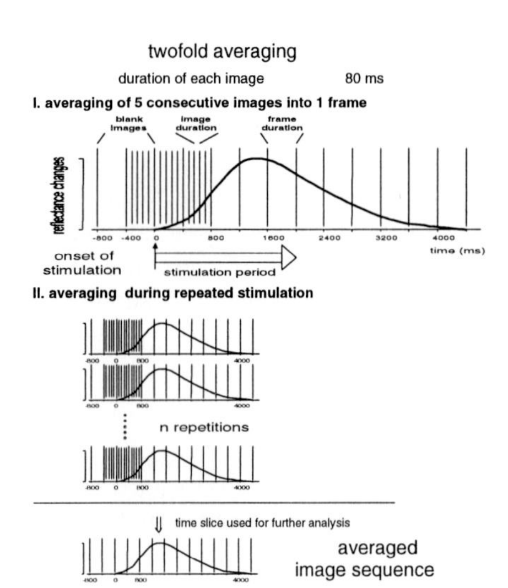

pause of 40 s to avoid intertrial interference. The cortex was illuminated with a 546 nm light source (cf. Fig. 3. for experimental setup). Controls (non stimulus conditions) were taken as blank images prior to each stimulus presentation. Images were computed by subtracting a stimulus from a non-stimulus condition. The spatial distributions of reflectance changes were color-coded and quantitatively computed in terms of cortical area for 25, 50 and 75 % of the maximal reflectance changes.

Fig. 2: Illustration of the twofold averaging scheme used in these experiments.

3 Results

Optical reflectance changes following intracochlear stimulation

We obtained 300 complete optical maps of reflectance changes following different stimulation protocols of the cochlear implant device. Each map consisted of a sequence of 12 images of 400 ms duration providing information about the time course of the signals. Accordingly, a total of 3600 single maps were recorded. Electrical stimulation of radial pairs elicited topographically restricted distributions of optical signals (Figs. 4,5). On average, the amplitude of the intrinsic signals, i.e. the relative reflectance changes measured for the 546 nm light source were in the range of 1.4 % (+/- 1.3 s.d.). The area of the two-dimensional signal distribution increased fairly linearly with amplitude. At 75 % of the maximal reflectance changes (ΔR) we found an average cortical territory of 0.39 mm^2 (+/- 0.32 s.d.), which increased to 2.76 mm^2 (+/- 1.91 s.d.) at 50 % ΔR.

Cochleotopic selectivity

Systematic variation of the stimulation pairs and thus of the cochlear frequency sites revealed a systematic and corresponding shift of the response areas thus producing a profound cochleotopic selectivity. In the example shown, the reflectance measurements were obtained to electrical stimulation of radial electrode pairs 1-2, 3-4 and 5-8 (Fig. 4). There is a systematic shift of the response areas towards more anterior sites that are known to represent successively higher frequencies. A similar pattern of cochleotopic selectivity could be observed when the reference electrode was not one of the implant sites, but the general ground.

Effects of intensity of electrical stimulation

Electrical thresholds for generating optical maps were usually in the range of 150 to 250 μA (3.5 to 8 dB), but could differ between the different electrode pairs. Fig. 5

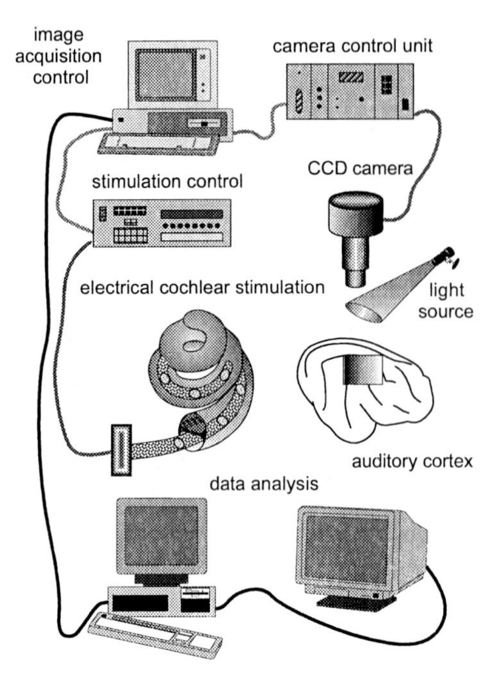

shows the effect of increasing stimulation current on the spatial distribution of the evoked optical maps of reflectance changes. The resulting intensity functions were fairly steep. At about 10 to 15 dB above threshold, a maximal area of excitation is revealed. This enlargement of response areas parallel to the increase of stimulating current reflects a decreasing selectivity of the electrical stimulation.

Fig. 3: Experimental setup used for optical recordings. Images were taken from the exposed cortex illuminated with a 546 nm light source. Rostrocaudal frequency gradient of auditory cortex is schematically illustrated. Digital images were transferred via a camera controller to a PC computer that controlled data acquisition and stimulation. Averaged images were send to a SUN workstation, analyzed and shown on a display monitor.

4 Discussion

The main difference between acoustic and electrical stimulation is that electrically induced activity of the auditory nerve is characterized by high temporal coherence and synchronicity, mainly because no travel times or local mechanical resonance oscillations are involved in their generation. In spite the methodological differences, we were able to demonstrate that by means of optical recording intrinsic signals it is possible to image the topographic representation of acute cochlear stimulation in AI and in surrounding auditory cortical fields, and that stimulation of selected locations of a multichannel implant corresponding to different intracochlear frequency locations results in selective shifts of the evoked response areas, which match the underlying frequency organization and thus preserves the overall cochleotopy.

Advantages of optical imaging of intrinsic signals

There is general agreement about the reliability of optical recording intrinsic signals which was demonstrated either by correspondence of optically and electrophysiologically measured representations or by repeated optical measurements revealing an overall little variance of the maps together with a considerable stability of representational details [1]. It is also well acknowledged that optical imaging of intrinsic signals provides a continuous measurement of two-dimensional activity distributions and allows repeated measurements in one animal, which is not possible when the 2-DG method is used. This is of considerable importance when parametric studies are intended and therefore many variables are to be systematically varied.

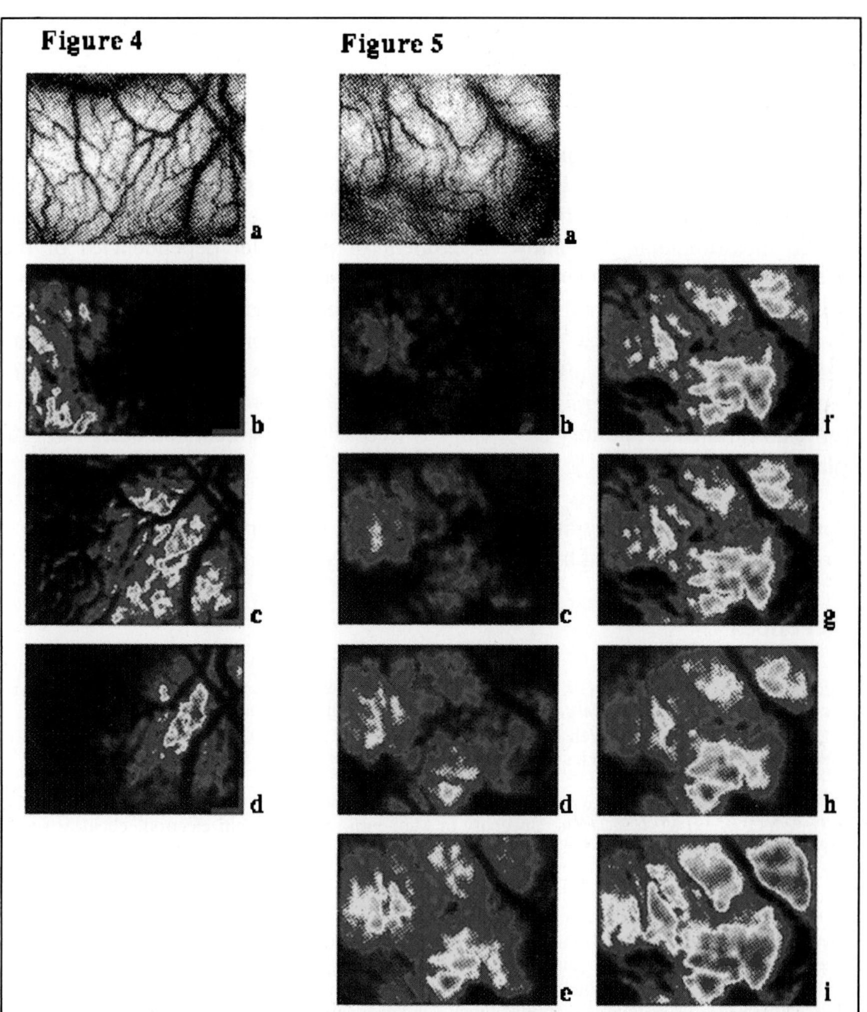

Fig. 4: Cochleotopic selectivity as a result of imaging response maps following three different electrode positions. a. Videoimage of the cortical surface. b. Map of reflectance changes to stimulation of electrode pair 1-2. c. Map of reflectance changes to stimulation of electrode pair 3-4. d. Map of reflectance changes to stimulation of electrode pair 5-8. Stimulation current was 15 dB for 1-2 and 3-4, and 18 dB for 5-8. Scale bar is 0.9 mm.
Fig. 5: Effects of increasing stimulation current on the spatial distribution of reflectance changes to stimulation of electrode pair 1-3. Current is indicated as dB relative 100 μA. a. Videoimage of the cortical surface. b. 3.5 dB. c. 4.2 dB. d. 4.5 dB. e. 4.7 dB. f. 4.9 dB. g. 9.5 dB. h. 14 dB. i. 17 dB. Scale bar is 1 mm.

Electrode placement and intracochlear frequency correspondence

We found the most prominent reflectance changes in the high frequency representation of AI. To provide some absolute values about the frequency ranges stimulated by each electrode, we made use of the average data of the length of the scala tympani and their corresponding frequency sites [8,13]. Assuming an overall length of the inserted electrode portion of about 8 mm, we can estimate the frequency range most probably stimulated to lay within 6 to 50 kHz.

Concluding remarks

By means of optical imaging techniques the effects of acute electrical stimulation with a multichannel cochlear implant device can be visualized. The resulting activity patterns are compatible with known activity maps following acoustic stimulation [4]. This technique provides a fine spatial resolution together with the advantage of repeated measurements that allow assessment of parameter dependencies of these maps. These measurements are the necessary baseline to compare the effects of chronic stimulation in animals with hearing experience and in animals that are deaf from birth. In addition, the way in which cortical systems react to stimulation of cochlear implant devices can shed light on principles of cortical information processing strategies and might therefore be beneficial for our understanding the specific constraints of auditory processing as well as development and design of future implants.

Acknowledgments: We gratefully acknowledge the support of the Institut für Neuroinformatik. Supported by DFG Neurovision Ey 17-3 and BMFT 01VJ 9005.

References

[1] Bonhoeffer T, Grinvald A (1993) The layout of iso orientation domains in area 18 of cat visual cortex: Optical imaging reveals a pinwheel-like organization. J Neurosci 13: 4157-4180

[2] Brown M, Shepherd RK, Webster WR, Martin RL, Clark GM (1992) Cochleotopic selectivity of a multichannel scala tympani electrode array using the 2-deoxyglucose technique. Hearing Res 59: 224-240

[3] Clark GM, Blamey PJ, Brown AM, Gusby PA, Dowell RC, Franz BK, Pyman BC, Sheeperd RK, Tong YC, Webb RL (1987) The Unversity of Melbourne-nucleus multi electrode cochlear implant. Adv Otorhinolaryngol 38: 1-181

[4] Dinse HR, Godde B, Hilger T, Schreiner CE (1996) Optical imaging of topography of frequency and intensity organization of cat auditory cortex using intrinsic signals. Soc Neurosci Abstr 22: in press

[5] Frostig RD, Lieke EE, Ts'o DY, Grinvald A (1990) Cortical functional architecture and local coupling between neuronal activity and the microcirculation revealed by in vivo high resolution optical imaging of intrinsic signals. Proc Natl Acad Sci 87: 6082-6086

[6] Godde B, Hilger T, Berkefeld T, von Seelen W, Dinse HR (1995) Optical imaging of rat somatosensory cortex reveals representational overlap as topographic principle. Neuroreport 7: 24-28

[7] Grinvald A (1992) Optical imaging of architecture and function in the living brain sheds new light on cortical mechanisms underlying visual perception. Brain Topogr 5: 71 5

[8] Hatsushika SI, Shepherd RK, Tong YC, Clark GM, Funasaka S (1990) Dimensions of the scala tympani in the human and cat with reference to cochlear implants. Ann Otol Rhinol Laryngol 99: 871-876

[9] Merzenich MM, Knight PL, Roth GL (1975) Representation of the cochlea within primary auditory cortex in the cat. J Neurophysiol 38: 231-249

[10] Merzenich MM, White M (1977) Cochlear implant: the interface problem. In: Functional electrical stimulation, Hambrecht FT, Reswick JB (eds) New York Decker, pp 321-339

[11] Raggio MW, Schreiner CE (1994) Neuronal responses in cat primary auditory cortex to electrical cochlear stimulation. I Intensity dependence of firing rate and response latency. J Neurophysiol 72: 2334-2359

[12] Schreiner CE (1991) Functional topographies in the primary auditory cortex of the cat. Acta Otolaryngol 491: 7-16

[13] Schuknecht HF (1960) Neuroanatomical correlates of auditory sensitivity and pitch discrimination in the cat. In: Neural mechanisms of the auditory and vestibular systems. Rasmussen GL, Windle WF (eds) Thomas, Springfield pp 76-90

Simulation of Surgery and Endoscopy

The Brain Bench: Virtual Stereotaxis for Rapid Neurosurgery Planning and Training

Tim Poston Wieslaw L. Nowinski Luis Serra Chua Beng Choon Ng Hern
CIEMED*, Institute for Systems Science
Kent Ridge, National University of Singapore, Singapore 119597
Prem Kumar Pillay
The Brain Centre, Singapore General Hospital
College Road, Singapore 169608

Abstract. This surgical planning and training system integrates an electronic atlas of brain structure with a virtual stereotactic frame and patient data, for intuitive, reach-in manipulation. The objective is plans prepared faster; better, more accurate choice of target points; improved avoidance of sensitive structures; fewer sub-optimal frame attachments; and speedier, more effective training. If validated by clinical study now under way, this will improve medical efficacy and reduce costs.

Keywords Brain atlas, image-guided surgery, neurosurgery planning, frame-based stereotaxis, virtual reality, virtual workbench

mpeg video: http://ciemed.iss.nus.sg/research/3dinterfaces/mpg/brainBench.mpg

1 Introduction

3D views have become common in medical software, though many scans are still displayed as a spread of 2D slices. Stereo display adds to the sense of depth, but—in most software—manipulation is still done as through a glass, clumsily; an object that does not move and turn with the hand is awkward to control. 3D display demands a 3D interface, but is not in itself such an interface.

Merging the display space and the user's personal space constitutes Virtual Reality, but VR alone is not enough for a clinical tool fit for routine use.

Dextrous tasks require a high resolution display, a well calibrated relation between hand and displayed object, and minimal conflict between depth cues. Our collaboration, between a neurosurgeon and the creators of a hand-eye co-ordinated interface and of an electronic successor to clinically standard brain atlases, aims to achieve this in a life-critical area of high-precision surgery.

We first describe the standard approach to planning brain probe insertion, and some of its problems. We outline the 'Virtual Workbench' general purpose 3D interface scheme and the electronic Brain Atlas, then the Brain Bench tool that incorporates them. An account of some planning procedures for particular types of intervention is followed by summaries of the advantages we see in the new system and of our approach to validating these advantages, and future plans.

* The Centre for Information-enhanced Medicine, jointly created by the Institute of Systems Science of the Nat[l.] University of Singapore and Johns Hopkins University.

2 Background

Stereotactic neurosurgery dates from 1906. Using X-rays, surgeons hand-computed the (x, y, z) coordinates of target lesions and adjusted a stereotactic frame to guide the delivery of a penetrating instrument (biopsy needle, electrode,...; we will merely say 'probe') to the target location chosen. The guiding frame is movably—and precisely controllably—attached to a fixation device, rigidly screwed to the skull, in a position established with millimetric precision. Only a narrow hole is opened in the skull, and no sight path opened in the brain.

Volume scans such as CAT and MR have made target localization far more accurate, increasing the range of cases treatable, and reducing morbidity and cost. Software support has still been based on fundamentally 2D display paradigms, displaying volume data as a set of slices (a surface reconstruction of the lesion itself may appear in one or more monoscopic 3D views, with shading techniques as the principal depth cue). In a typical interface the surgeon chooses a target in one slice, an entry point on another, and studies slice by slice the surroundings of the point where the line joining them passes through. No planning software has included a 3D brain atlas to assist the surgeon in locating brain structures unrevealed by the scan. When 2D (slice by slice) brain atlases have been included, they have used a purely linear fit based on the two inter-commissure landmarks of the Talairach-Tournoux (TT) scheme[11], rather than the whole piecewise-linear TT proportional grid. No software has assisted in the initial placement of the fixation device for frame-based surgery which during scan holds fiducial markers, and during surgery the guiding frame. Some disadvantages have been:

- Spatial data have the six degrees of freedom of Euclidean motion. Control by a two-degree-of-freedom mouse is necessarily indirect, making any interface laborious to use, and gives it a slow learning curve. Long training is costly, as are errors by a surgeon who is trained but not yet long in experience.
- When a target point should be deep in an asymmetric lesion, where the deepest point does not appear central in any one slice, selection can be hard.
- If the selected path is oblique to the displayed slices, it is not easy to judge by how far (obliquely to the slice planes) the probe will miss a sensitive structure, such as a nerve, blood vessel or eloquent brain area; particularly for unseen structures, whose position is anatomically inferred. This makes it hard to estimate dangers due to errors in MR localization or registration.
- Structures such as blood vessels are hard to perceive as 3D networks from the pointwise way they meet separately displayed slices, with gaps easily crossed in 3D.
- The initial placement of the fixation device on the skull involves directly all six Euclidean degrees of freedom, so that nobody has yet designed a software planner for it that surgeons find better to use than placement 'by eye'. Mouse-based planning tools all start with data from a scan with the fixation device (and fiducials) already in place. For a less experienced surgeon, in a centre new to thse techniques, a poor fixation placement can force a sub-optimal probe path, or even rescheduling the entire procedure.

Goble *et al.*[2] describe a 3D interface for image exploration, using hand-held props, though these are not used in their probe path selection, which does not include an atlas. Davey *et al.*[1] display insertion of a probe in a stereo view of patient data, clarifying its relation with DSA-revealed arteries in 3D, but manipulation is by 2D mouse and atlas work is not yet integrated with their planning system. Höhne *et al.*[3] gives a mouse-based interface to a probe entering a 3D brain atlas, not registered with patient data or frame position. we describe the Brain Bench, created at ISS and now starting further development and clinical testing at the Brain Centre of Singapore General Hospital. It is based on the Virtual Workbench[7] general environment for careful work in 3D, and the Electronic Brain Atlas[5], a combined electronic version of brain atlas books standard in stereotaxis, reorganized as a system of 3D structures. We describe these briefly below, with an account of their combination in the Brain Bench, and validation plans.

3 The Virtual Workbench

Conventional VR immerses the user in a computer-constructed environment via a head-mounted display. We have found it more productive for the user to 'reach in' to a 3D display where hand-eye coordination allows careful, dextrous work in examination of data and manipulation of computer constructs. The user reaches behind a mirror in which a 3D display shows both the patient data and a visual echo of the physical tool handle that the user's hand controls, in the same apparent place. This 'virtual' tool handle interacts with the data as a grasper, cutter, marker... as arranged by software. The 'see it where you feel it' coordination contrasts with the remote props in [2].

A more technical account is in [7]. Informally, the Virtual Workbench is a box one reaches into, with invisible hands holding visible tools. The left tool typically grasps objects and moves them (using the motor habits learned in childhood), while the dominant hand performs delicate work. A curve or other shape may be created, cut, bent,..., or one can make measurements with a virtual ruler; just place it between points of interest, as with a real one. The ruler typifies the contrast between the Virtual Workbench and a slice-by-slice 2D display; for the distance between a point P and an extended object E, a stereo view makes it easy to select the point in E nearest to P. For the mutually nearest points in E and another object E'—usually best estimated as between data slices—with a reach-in tool one can measure their distance directly. Finding these nearest points can be hard in a 2D system.

This contrast increases when real-time volume display of data becomes possible, for instance by the use of '3D textures' (see, *e.g.*, [8]) to display a structure. Instead of contour-outlining it slice by slice, one directly displays it as a 3D solid (not an extracted surface) by making the material around it transparent. Picking a point on the (easily perceived) surface is hard with a 2D mouse, easy by reaching in.

4 The Electronic Brain Atlas

The brain atlas database[4] developed at CI$_E$MED contains electronic versions of several paper brain atlases used in day-to-day clinical practise:

- *Atlas of Stereotaxy of the Human Brain*; Schaltenbrand and Wahren[9]
- *Co-Planar Stereotactic Atlas of the Human Brain* (TT); Talairach and Tournoux[10]
- *Referentially Oriented Cerebral MRI*; and *Anatomy. Atlas of Stereotaxic Anatomical Correlations for Gray and White Matter*; Talairach and Tournoux[11]
- *Atlas of Cerebral Sulci*; Ono, Kubik and Abernathey[6]

The atlas data from the paper atlases are digitized, enhanced, color coded, labelled, and organized into volumes. Various 3D extensions of them are in effect, and more are in progress. The workstation system [5] supports registration, 3D display and real-time manipulation, object extraction/editing, quantification, image processing and analysis, reformatting, anatomical index operations, and file handling. Its primary goal is automatic labelling and quantification of human cerebral structures. Its main applications are (3D) neuroeducation, (quantitative) neuroradiology, (stereotactic) neurosurgery, and neuroscience.

Each source atlas has its own strengths. The TT holds anatomical and functional information, with axial, coronal, and sagittal sections taken from a single brain specimen. Its proportional grid system lets one localize cortical and subcortical structures in 3D space, and to compare different brains. The TT and its grid are a clinical standard, mainly used for functional MRI and PET studies.

The Schaltenbrand-Wahren atlas[9], based on 111 brains, contains macroscopic and microscopic frontal, sagittal and horizontal sections through the hemispheres and the brain stem. The macroscopic sections give the extent of brain structure variation. The microscopic ones detail cerebral deep structures rarely displayed well by MRI or CT, such as the thalamic nuclei.

The *Atlas of Cerebral Sulci*[6] studies 25 brain specimen for variation and consistencies in location, shape, size, dimensions, and relationships to the internal structures. It contains two types of information: drawings of the sulcal patterns (showing variability of side branches, *etc.*), and variation incidence rates.

The mutual pre-registration of these atlases merges their information (such as different parcellations of the thalamus), and makes them all applicable to a scan when the user brings any one of them into register with it.

5 The Brain Bench

We have constructed an initial version of a system in which a Virtual Workbench user can manipulate brain scan data, with a model of the frame by which probes are controlled (Fig. 1), and brain structures from the electronic Brain Atlas (coloured surfaces), registered by the standard surgical landmarks with the scan data. This 'Brain Bench' currently contains the TT deep brain structures, where patient variability is relatively small once adjustment to its standard coordinates is made. We are working on support for the other datasets of the multi-source atlas above, with its tools for analysis and managing variation.

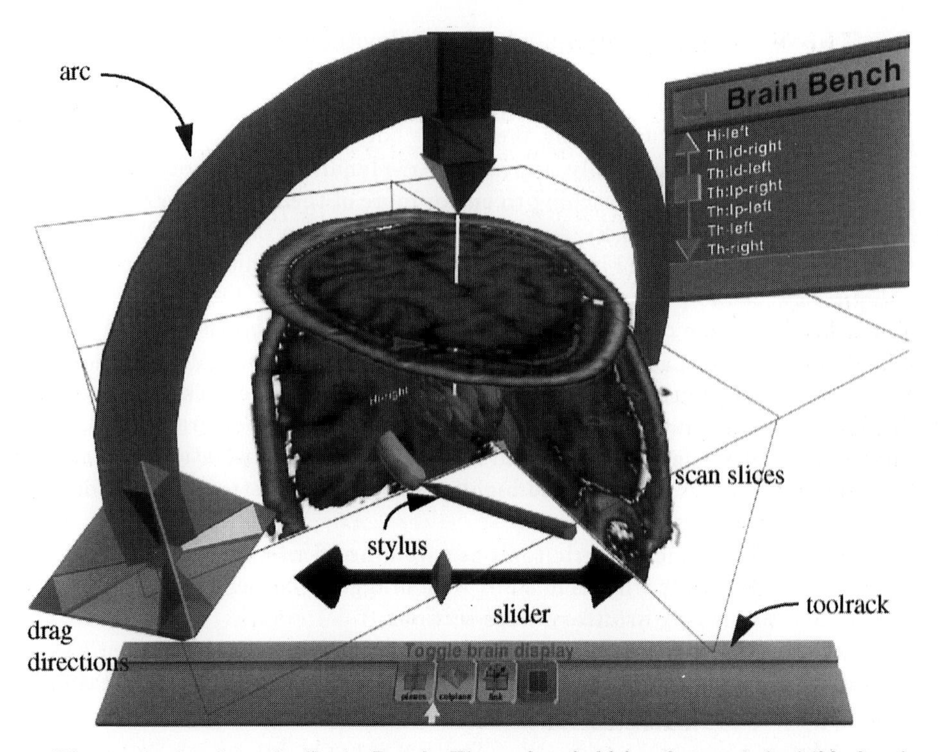

Fig. 1. A view into the Brain Bench. The stylus, held by the user's invisible hand, can move brain structures, data plane slices, the zoom slider, and the arc (a preliminary model of a stereotactic frame). The button rack controls its mode of action.

We have aimed to make this environment straightforward, intuitive and quick to use and learn. The user's grasping hand controls the entire displayed complex, turning and placing it as one does an object held in real space. The other does detailed manipulation, using 3D widgets (interaction gadgets) to interact with the surgical planning tools. In all interactions, 'reach in for the object of interest and press the switch on the stylus to interact with it'. When the stylus tip enters the volume of influence around an object, the object signals 'Reactive!' by a colour change or other highlighting. At any one time, the effect of the stylus is signalled by its appearance; a slice-selector, a point-mover, a volume-of-interest selector, *etc.* When it enters a widget's volume of influence, it becomes the manipulator associated with that widget, that knows how to interact with it. For example, while it is in the volume of the slider, the point-mover transforms to the slider-dragger; it will not point-move the slider widget but drag its bead.

The deep brain structures from the atlas behave like widgets. When the stylus moves near them, the structure nearest it remains opaque and is labelled with its abbreviated TT name; the others turn translucent, revealing its shape in context. The user can then press the stylus switch to grip it, and examine it as if on a skewer held by the stylus hand.

Two sets of widgets control the stereotactic frame: one controls the target and the other the entry point. The arc's hinge direction is fixed by the base frame, whose virtual form can only be dragged without turning; selecting one or more of the arms at the end of the arc in Figs. 1 and 3 allows dragging it with those degrees of freedom. The entry path is adjusted by moving the arc on its hinge, and the probe guide along the arc, or both at once.

The patient's data mainly appear as tri-planar cross-sections. The cross-sections are controlled by reaching to one or more of them, and dragging them to other, parallel planes (Fig. 2). Another cross-section can be enabled, orthogonal to the path of the probe (Fig. 3). Sliding it up and down this path shows a non-oblique view of the path's passage through the scan data. One can also enable a cross-section plane containing the path, and turn it with this path as axis.

A 'toolrack' menu at the bottom switches between stylus modes. As the stylus moves over it, the button nearest the tip is highlighted. The rack announces the button's function, enabled by clicking. A slider scales the whole frame/data/atlas ensemble, allowing detailed or overall inspection and manipulation. A scroll list widget lets one select, identify and highlight chosen brain structures (often hard to find among the rest).

We also provide such generic tools as a volume-of-interest widget (which lets one volume render only what is in a box with draggable corners), and a hand-held slicing tool for viewing arbitrary cross-sections.

Modelling the initial attachment of the fixation device to the patient's skull, a natural extension of this toolkit, will allow the surgeon to test the feasibility of a proposed combination of placement and adjustment.

6 Surgery Planning Using the Brain Bench

A stereotactic procedure is usually performed in four stages: (*i*) frame fixation, (*ii*) diagnostic imaging, (*iii*) computerized neurosurgery planning, and (*iv*) neurosurgery. The Brain Bench addresses (*i*) and (*iii*).

The frame model shown, a simplified version of the Leksell system, will be replaced by exact models of several frames in clinical use. A semi-circular arc can rotate about to an axis, which slides in three dimensions relative to the fixation device, controlled by the widget at left in Fig. 1. The probe guide moves along the arc, and constrains the probe to move along a straight line; in several systems, along a radius through the arc centre (often placed at the chosen target).

A procedure involves two steps, placement of the fixation device and of the probe path. Where the scan data include a system of fiducials attached to an already fixed device (the current normal procedure), locating it can be automated. When planning support begins earlier, as we propose here, a virtual version of the device can be placed freehand over any 3D model of the patient's head (as volume or surface) in which the target can be moderately well localized. If an approximate plan for the second step gives a path securely far from obstruction by screws or other frame constraints, the real fixation device can be similarly attached and the fiducial-equipped scan taken, with a much reduced likelihood of obstruction later. After this scan, the normal precise registration of the device position with the scan data is done.

Next, the probe path is selected (Fig. 4). The chosen target becomes the arc centre, then the probe path changes with the angle of the arc and, on it, of the probe holder. The tools already described help in judging candidate targets and paths, as one varies the choice by dragging around the target (select the robe tip) or the approach angle (select the guide tip). The entry path is chosen to minimize damage to the structures along the penetrated path (eloquent brain, blood vessels, critical deep structures, etc). Beside visual observation of the data with the tools above, and of the path's relation to brain atlas information, the system lists (and optionally highlights) the atlas structures intersected. A cortical surface labelled with the Brodmann (functional) areas is under development.

Once a path is selected, the system prints out the mechanical adjustment values to which the frame must be set.

Two major Brain Bench applications are tumour stereotaxis and functional neurosurgery. In tumour stereotaxis, the target is selected using the three orthogonal views; if the tumour shape has been extracted from the scan as a set of voxels, a set of contours, or a surface, this can be displayed also. These cues guide choice of target(s) within the tumour. In functional neurosurgery, *e.g.*, thalamotomy, pallidotomy, or hippocampoerectomy, the target is an anatomically defined point in the structure to be treated, chosen on functional grounds. The atlas acts as a guide to the target structure in the patient data (as well as segmenting and labelling the structures surrounding it), for a more informed initial guess in placing an electrode to locate the target by functional response (along a good line for sampling at several depths), hopefully leading to a good placement in fewer steps.

7 Projected Advantages

- The reach-in interface allows natural manipulation, easier to learn and faster to use than moving 3D objects by pushing a 2D mouse around a pad.
- The surgeon adjusts the whole probe trajectory as a visual object, among atlas structures and volume displayed scan data.
- Good visual selection of a deep point within a structure becomes easier.
- An easily-used placement planner will reduce the effort, time and patient discomfort now caused by misplaced fixation devices.
- The 3D system will provide an improved learning tool for the mental 3D anatomy model that trainees must still acquire, creating a better one faster than 2D images can provide.

8 Validation plan

The first test of the system will be of its use in the education and training of neurosurgeons. We will compare

- surgeons trained with the traditional method, involving hands-on experience with the stereotactic system, observation of patient treatment, and finally performing an actual case under supervision
- surgeons trained by combining these with the Brain Bench for planning.

Each will be tested both on the ability to carry out a procedure efficaciously and safely, and on the time taken. We will also test the young neurosurgeons on the accuracy with which they identify anatomical structures, and their localization of target centres, with traditional methods and with the Brain Bench.

We will compare target and trajectory selection with that of trainees using traditional methods, against a 'gold standard' best choice provided by the senior neurosurgeon in charge, with experience of more than 1,000 cases.

Accuracy testing will be performed with sharply defined landmarks. Choosing a target point that can be anatomically located with submillimetric precision, we will test the frame settings found against those found with traditional software, for access at the same angle. Predictions of the point at which the probe will meet different data slices will be compared for the two systems.

Finally, using an established neurosurgical team with a good track record in stereotactic neurosurgery, we will monitor the extent to which use of the Brain Bench can improve their surgical results. This will be carried out in two stages.

In the first, the team will perform their surgery using the traditional planning methods, while in parallel making a plan using the Brain Bench Target and trajectory plans will be compared for accuracy and appropriateness with those made by traditional methods; in the case of any difference (for instance, in the choice of a central point), the Brain Bench plan must be considered "at least as good" by reviewers who examine it only with the established software. This will be carried out for a total of 15 patients.

Once accuracy has been verified and the choices made with it found satisfactory, increased reliance will be placed on the Brain Bench for surgical plans actually executed. Measurements to compare it with traditional methods will require a study of 50 patients. Specifically, we will measure

1. time taken to identify the anatomical target.
2. number of electrode insertions before a target is functionally confirmed
3. time taken to identify an appropriate trajectory.
4. time taken to identify alternate targets and trajectories.
5. the accuracy with which a target volume centre is selected.
6. a surgeon's evaluation from 0 to 10, for each case, of the improvement in the surgeon's ability to carry out the operation.
7. a surgeon's evaluation from 0 to 10, for each case, of user-friendliness.
8. a surgeon's evaluation from 0 to 10, for each case, of the convenience in teaching other members to use the system, and to supply relevant information during the procedure.

9 Further Work

The Brain Bench promises to be an important tool in neurosurgery time and cost reduction, in both training and clinical practise. Further developments will depend strongly on clinician input, but we anticipate adding features such as

- **Local adaptive registration**: such fitting is now too slow for clinical use on a whole brain, but can be much faster if applied to a subregion of interest

and supplied with a good initial linear position adjustment; such selection and adjustment tools become natural to design in the Virtual Workbench.
- **Atlas extension tools**, by which structures not easily extracted from scan data can be added and edited by expert neuroanatomists. (In this environment it is easy [7] to hand map and edit 3D curves within a volume data set; work continues on intuitive hand shaping of surfaces and solids.)
- With such tools we will **extend the basic electronic brain atlas** to major blood vessels and cranial nerve pathways, and improve the existing account of the white matter 'cabling' within and from the 'processor' grey matter of the cortex; now sketchily indicated by dotted lines within the planes of the atlas pages, suppressing the aspect of between-plane connectivity.
- Brain structures in the atlas, will be **coded for sensitivity** to particular probe types (a fine probe being in general less damaging). The system can then not merely report contact, but issue proximity warnings.
- **Frameless stereotaxis** presents closely related planning problems, and the Brain Bench can straightforwardly extend to providing numerical guidance to such systems as well as settings for frame positions.
- Planning **stereotactic radiotherapy** (targeted X or γ radiation with sub-millimetric precision) can similarly benefit from combining sensitivity-coded brain atlas guidance with dextrous, intuitive control.

References

1. B. Davey, R. Comeau, C. Gabe, A. Olivier & T. Peters, Interactive Stereoscopic Image-Guided Neurosurgery, *SPIE* vol. 2164, 1994, pp.167–176.
2. J. C. Goble, K. Hinkley, R. Pausch, J. W. Snell & N. F. Kassell, Two-handed Spatial Interface Tools for Neurosurgical Planning, *IEEE Computer*, July 1995, pp. 20-26.
3. K. H. Höhne, M. Bomans, M. Riemer, R. Schubert, U. Tiede, W. Lierse, A 3D anatomical atlas based on a volume model, *IEEE Comput Graphics Appl.* **12**, 4 (1992), pp. 72–78.
4. W. L. Nowinski, R. N. Bryan & R. Raghavan, eds., *The Electronic Clinical Brain Atlas. Three-Dimensional Navigation of the Human Brain*. Thieme, NY, 1996.
5. W. L. Nowinski, A. Fang, B. T. Nguyen, J. K. Raphel, L. Jagannathan, R. Raghavan, R. N. Bryan & G. Miller, Multiple brain atlas database and atlas-based neuroimaging system, *Journal of Image Guided Surgery* (in press).
6. M. Ono, S. Kubic & C. D. Abernathey, *Atlas of the Cerebral Sulci*. Georg Thieme Verlag, Stuttgart 1990.
7. T. Poston & L. Serra, Dextrous Virtual Work, *Communications of the ACM*, May 1996, **29**:5, pp. 37-45.
8. T. Poston, L. Serra, M. Solaiyappan & P. A. Heng, The Graphics Demands of Virtual Medicine, *Computers & Graphics*, vol. **20** (1996), pp. 61–68.
9. G. Schaltenbrand & W. Wahren, *Atlas of Stereotaxy of the Human Brain*. Georg Thieme Verlag, Stuttgart 1977.
10. J. Talairach & P. Tournoux, *Co-Planar Stereotactic Atlas of the Human Brain*. Georg Thieme Verlag, Stuttgart 1988.
11. J. Talairach & P. Tournoux, *Referentially Oriented Cerebral MRI Anatomy. Atlas of Stereotaxic Anatomical Correlations for Gray and White Matter*. Georg Thieme Verlag, Stuttgart 1993.

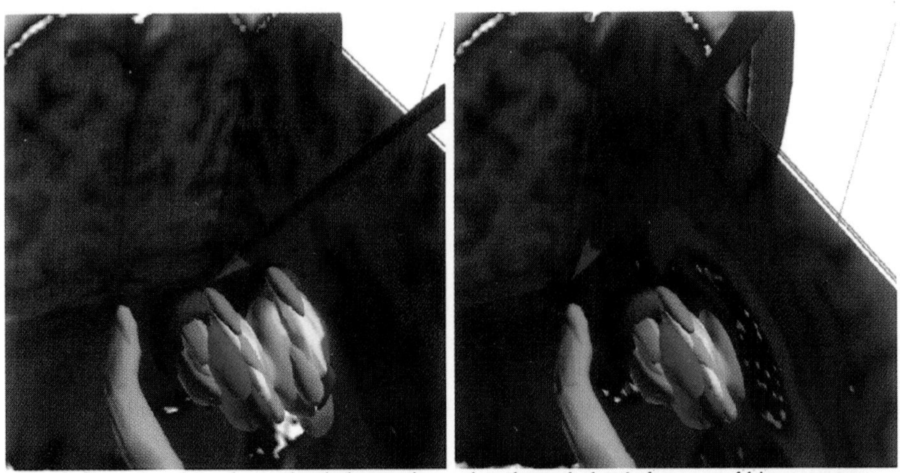

Fig. 2. Dragging data planes, by their meeting point, through the thalamus and hippocampus.

Fig. 3. A data slice orthogonal to the probe, dragged along it to study impact

Fig. 4. Swinging the arc and dragging the probe around it. Impacted structures are highlighted.

A Flexible Environment for Image Guided Virtual Surgery Planning

Johan Van Cleynenbreugel*, Kris Verstreken, Guy Marchal, Paul Suetens

Laboratory for Medical Imaging Research, K.U.Leuven, Belgium

Abstract. We present an object-oriented environment built on top of OpenGL and intended for virtual planning of image-guided surgery. We enumerate typical requirements from this field and show how they are coped with by our developments. A number of already derived surgery planning applications demonstrates the flexibility of our environment.

1 Introduction

Although the capabilities of 3D visualization and modelling techniques have expanded vastly over the past decade, their actual use in diagnostic radiology and presurgical planning is still limited. A main reason for this poor level of acceptance is certainly the negative cost-benefit balance of the solutions proposed. For long time, the investments in powerful hardware, needed to obtain an acceptable throughput, have been too prohibitive to incorporate 3D visualization in daily routine. Moreover the real need for true 3D representations still is a topic of discussion. So, given the ever decreasing costs of general purpose and graphics specific hardware, a fruitful research option in this area is to explore the possibilities of flexible and easily adaptable software environments.

Image guided virtual surgery planning is a very promising field to gain the utmost from this approach (e.g. [5]). Indeed in many surgical disciplines the wealth of information extractable from medical imaging data remains rather underutilized. The static format under which these data are delivered (often only printed on film) accounts for their relative disuse. In the operating theatre, a surgeon does not work along preformatted 2D slices but dynamically observes 3D tangible structures. Hence a first major requirement for better presurgical exploitation of medical volume data (acquired by modalities as CT and MR) is to provide the surgeon with suitably adapted access paths through the data volume. Furthermore during surgery additional elements (implants, instruments) are introduced in the OR scene. Therefore a second major requirement is that CAD-like descriptions of geometric objects can be introduced, visualized and manipulated in the space spanned by the data volume.

This paper describes a flexible environment designed to cope with both prerequisites. First we elaborate on the requirements for virtual planning of

* Corresponding author: Johan Van Cleynenbreugel, ESAT + Radiologie, University Hospital Gasthuisberg, Herestraat 49, B-3000 Leuven, Belgium. e-mail: Johan.VanCleynenbreugel@uz.kuleuven.ac.be

image-guided surgery and survey a number of comparable approaches. Afterwards generic implementation options are detailed. Their power is illustrated by a number of applications. A discussion about lessons learnt and future extensions concludes the paper.

2 Background

In our opinion, an environment devoted to image guided virtual surgery planning must satisfy the following set of requirements.

- The 3D space covered by the medical imaging data volume(s) must be represented as a 3D scene.
- The environment must support ways to derive (properly reformatted) 2D slices by easy, interactive manipulation. This requirement bridges the gap between many image-based surgery planning (still performed on 2D slices in daily routine) and current 3D environments. Support is also needed to display such slices on appropriate geometrical locations in the scene.
- Many surgical actions (e.g. drill paths) and accessories (e.g. implant screws) can be modelled by axial-symmetric or membrane-like 3D geometrical structures. Hence such representations must be introduced easily, manipulated flexibly and visualised correctly on appropriate views of the 3D scene.
- Image derived representations such as surfaces and volume visualizations must be co-presented with the geometrical abstractions and the reformatted 2D slices.
- Constraints related to planning and execution of surgery must be representable. Examples are constraints derivable from the image data (such as the course of a nerve or a blood vessel) or constraints typical to surgical actions and surgical instruments (such as the stiffness of an osteotomy saw).
- It must be possible to view different locations in the same scene from independent viewing angles.
- Finally, the environment must be flexible to program, allowing for easy adaptation of its concepts to different applications.

Before going into details on our developments, we survey how current biomedical visualization systems meet this set of requirements.

In [1] a surgery rehearsal system is discussed. It offers virtual tools (such as knives, saws and rasps) to operate on the VOXEL-MAN [4]) representation of an image data set (which results from a prior segmentation step). Although this system meets many of the requirements, it is intended for free-form cutting; hence no constraint handling is taken into account. Similar free-form cutting principles (without constraints) to sculp surface or volume data are available in commercial systems (Analyze, 3DViewnix, VoxelView). However the inclusion of geometrical structures, resembling surgical actions or surgical materials (drill paths, membranes or implants) can hardly be found in general. For specific applications such as neurosurgery planning, e.g. [6], or craniofacial surgery simulation, e.g. [3], specific solutions involving joint visualization and 3D manipulation of image derived and CAD-like data have been reported.

In [9] a fruitful approach towards flexible programming is described, by providing a front-end scripting language to a powerful object-oriented medical visualization system. However its target application is diagnostic radiology rather than surgery planning. The same flexibility can be found in current (CAD-like) 3D graphics programming environments. Furthermore an environment such as OpenInventor, [8], does provide support for constraint handling between graphic objects. Usually however, the incorporation of medical images is not straightforward in such systems, so they lack to meet many of the requirements stated above. Nevertheless surgical simulators, combining object manipulation to restricted visualization (relatively small textures mapped over geometric surfaces) are currently appearing (e.g. [2] describes a simulator authoring environment based on OpenInventor). The apriori segmentation step required to generate models is still too prohibitive to make their principles useful in a clinical setting.

3 Generic Options

Conceptually our environment is designed to bridge the gap between CAD-like object manipulation (reflecting the point of view of the surgeon) and huge data set visualization (ditto for the radiologist). It is implementted in C++ on top of OpenGL. A class hierarchy incorporating common principles for both CAD objects and image data sets is provided. In principle every "object" must know how to behave, whether it is a geometrical model, satisfying surgical specifications, or an image data set. In the object-oriented design thus obtained, several classes can be distinguished.

The root class provides identification and housekeeping. Object-specific parameters (color, transparency, position in space and rotation) and generic methods (such as drawing and interaction) are defined at this level. Shape and data set subclasses are derived from this class. Examples are spheres, cones, (the classic CAD things) ..., adaptable curves and surfaces, surface models, and MPR slices. These subclasses have different specializations for their methods. For example, drawing of CAD-like objects needs recalculation whereas drawing of raster-like structures requires storage and data transfer.

One of our fundamental design options is to build surgery planning systems that behave in an intuitive way. The analogy of a movie studio scene is used: a collection of objects in a 3D space is first lighted and then inspected by one or more cameras. Each camera produces a different view of the scene. Lights and cameras (as well as cutting planes) are subclasses of the root class too. The principle of a *camera-generated view* has three important consequences. First, several projection transformations are possible. For example, stereoscopic viewing is readily implemented by a pair of adjacent cameras. Second, a view is composed in the sense that objects lying in front of the camera can deliberately be made (in)visible. Third, if the same object is presented in many views, the result of a manipulation on it in one view is immediately updated in the others.

Based on this basic machinery for scene description, provisions have been made to relate individual scene objects through the concepts of link and constraint. By a *link* we mean that attributes of one object are functional dependent of attributes of another object: changes in the former immediately adjust the latter. For example, the location of a sphere can be coupled to the position of a camera (e.g. by a translation); moving the sphere will reposition the camera. The concept of *constraint* is meant to restrict the degrees of freedom a manipulable object normally has in a 3D scene. For example, an axis can be related to an object, in such a way that the latter is only allowed to rotate around (or slide along) this axis. Constraints do not only express ways a surgeon utilizes his instruments but also how anatomical parts move with respect to each other. For example, an osteotomy is the movement of a saw along a well defined path; the insertion of an implant into bone is done with a fixed direction that may not change during drilling. The incorporation of constraints in preoperative surgery planning can prove to be a major advancement, when it is supported by appropriate visualisation.

4 Applications

Four applications are illustrating some of the possibilities derivable from our approach. First we show how membrane-like structures can be modeled to bone surfaces. A true 3D oral implant planning system is a second (and currently most important) instantiation of our generic principles. The latter examples relate to orthopaedics surgery.

4.1 Virtual Sculpting of Personalised Membranes

A number of surgical interventions is based on the availability of patient-specific implants. For example in oral surgery, personalised membrane-like titanium implants are employed for purposes of osseoregeneration. As explained in figures 1 and 2, our environment can easily support the virtual planning of such personalised membranes.

4.2 Oral Implant Planning

Presurgical planning is a critical step towards the success of oral implants that have to carry later on a fixed denture. Currently this planning is most often based on images on film resulting from so-called Dental CT programs (although 2D based commercial planning systems do exist[2]). Dental CT allows to reslice maxillar or mandibular data according to specific views of the jaws (axial, panoramic, cross-sectional). Based on these representations and on geometric measurements made on the films, a surgery plan is established. In this way the surgeon decides on quality of bone and location and orientation of implants in it. During

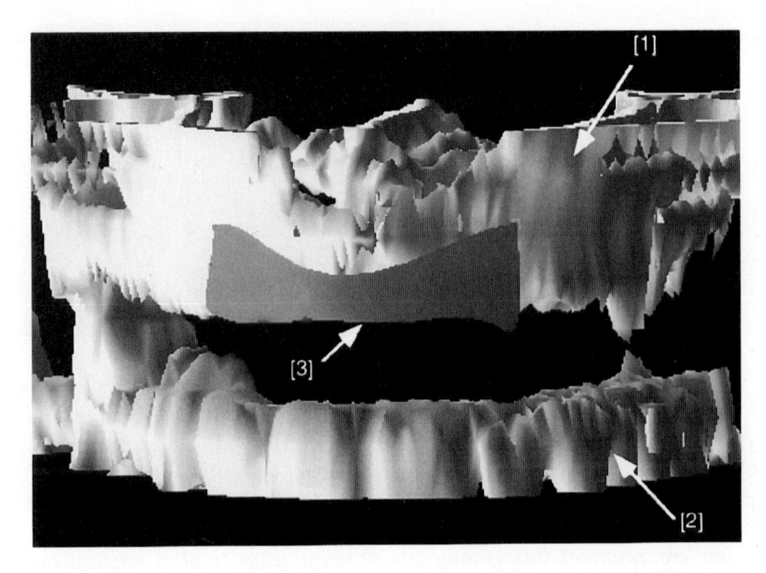

Fig. 1. *This figure shows surfaces from a maxilla region obtained from CT imaging. The region pointed out by 1 corresponds to the maxillar bone, the region indicated by 2 is obtained from an existing denture (painted by a radio-opaque contrast medium) worn by the patient during image acquisition. The region pointed out by 3 results from a virtual membrane sculpted around the surface of the jawbone. In principle this virtual membrane can be converted to an actual implant.*

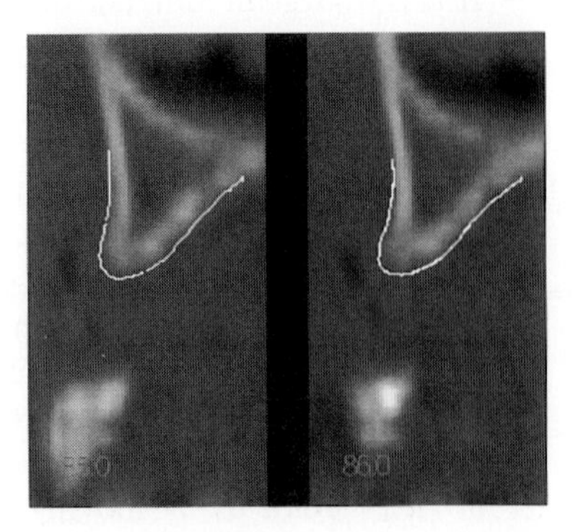

Fig. 2. *On the left the virtual membrane from figure 1 is shown. Being a* NURBS *surface, it can be folded and twisted by using a set of control points (depicted as small spheres), until the appropriate shape fitting around the jawbone is obtained. In our environment, the instantaneous intersection of this membrane with a properly MPR from the CT dataset can be shown at will. The intersections on the right are with cross-sectional slices through the maxilla, perpendicular to the scan planes, see also figures 3 and 4.*

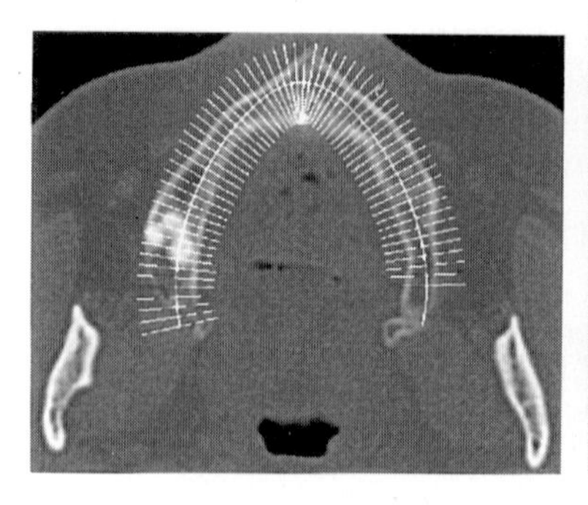

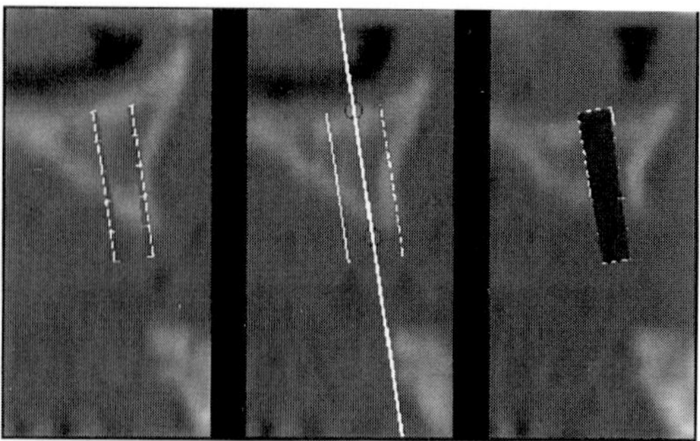

Fig. 3. *The figure on the left shows one axial slice from a maxilla Spiral CT data set. The centerline through the jawbone can be adjusted freely. The shorter lines, locally perpendicular to the centerline, indicate the intersections of the cross-sectional slices. On the right the planning of one cylindric implant is shown on three succesive cross-sectional slices along the centerline. Introducing the implants via these cross-sectional reslicings closely ressembles the current planning practice of oral surgeons; in our environment this is only one of many ways to plan implant locations in a 3D scene.*

actual surgery this plan must be mentally reconstructed. Due to the expanding application of the technique, there is an increasing demand for better planning.

Starting from the options defined above an oral implant planning system has been derived (see also [7]). Its striking features are:

- It is the surgeon him/herself who decides on the geometry of the three dental views. These views are then generated (and can be altered) on the fly by the surgical user, see figure 3.
- The 3D space spanned by the CT volume can be viewed from different camera positions, leading to clinical opportunities unconceived before, e.g. as on figure 4, left.
- Oral implants are modeled by cylinders and can be placed by indicating a pair of target and entry point on any appropriate representation of the 3D space, e.g. as on figure 3, right. The radius and the length of the cylinder can be chosen interactively according to the industry standards. The position of the oral implant model can be changed either by repositioning the target or entry point or by moving the cylinder axis. This symmetry axis extends from both ends of the cylinder enabling functional and aesthetic assements of multiple implant placements, see figure 4, right.
- A bone surface (at the moment precalculated from the CT data) describing the jaw bone(s) (and possibly the radio-opaque teeth) can be copresented with the implant model and the CT data, see figure 4, right.
- The viewpoint of the entire scene (CT data, surface model, implant models) can be interactively adjusted to optimally match that of the actual operation.

[2] e.g. SIM/Plant, Columbia Scientific Inc

Furthermore the option of stereoscopic visualization is available.
- Finally the software can naturally be extended to support per-operative guiding.

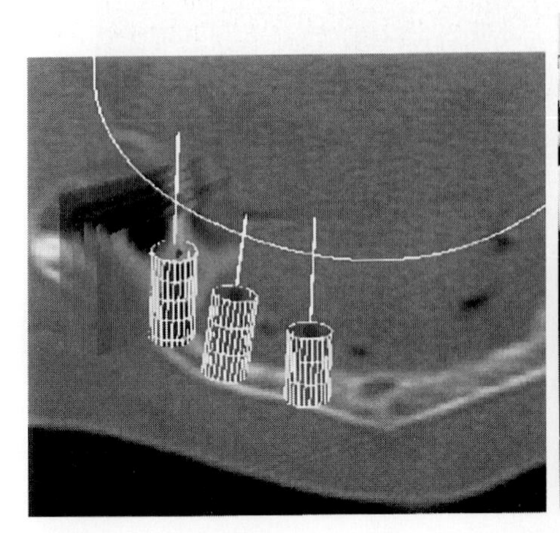
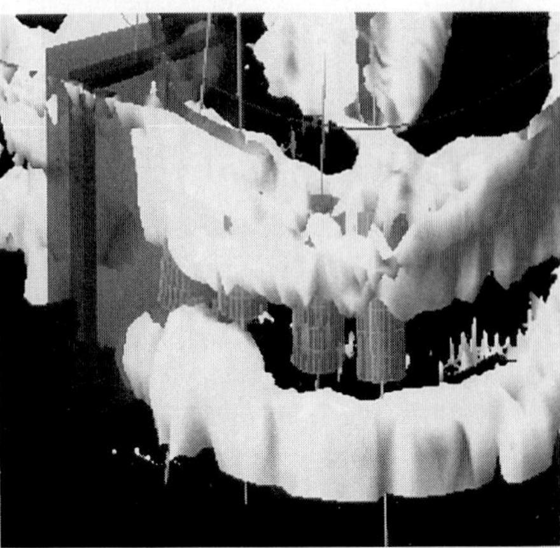

Fig. 4. *On the left a perspective view over the 3D scene of the Spiral CT data set (one axial and a number of cross-sectional slices) is shown . It will be clear that the implants are modeled as true 3D objects. Moving an implant on this view would immediately update its representation on other views, such as the one in figure 3, right. On the right the surface of maxillar bone and radio-opaque denture is shown together with a number of implants and a number of cross-sectional slices.*

4.3 Transpedicular Trajectory Planning

In spine surgery transpedicular screw insertion planning involves the determination of a feasible trajectory inside the pedicle of a vertebra. Critical adjacent structures such as the spinal chord or blood vessels may not be touched. In figures 5 and 6, we show how our environment can be employed to generate optimal trajectories.

4.4 Relating 2D to 3D Locations

In practice, simple solutions are required to relate the well-known 2D slice representation of tomographic acquisitions to the less-known derived 3D visualizations. In particular in situations where 3D is reported to give superior information, the link to the golden standard of 2D is often difficult to make. In figure 7 we illustrate the advantages of our environment in the context of acetabular fractures.

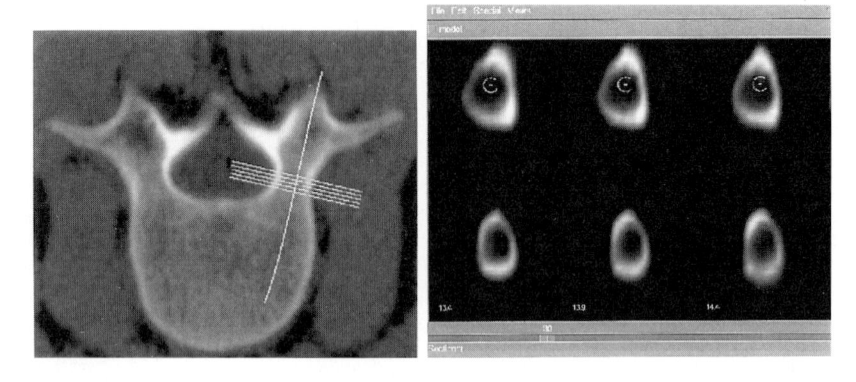

Fig. 5. *Comparable to the oral implant application, a cross-sectional reslice path (left) is defined over an axial slice of a spine CT set (containing two vertebrae in this case). Some consecutive reslices are shown on the right. In the upper pedicle the intersection of a planned transpedicular drilling path (modeled by a long cylinder) is visible.*

5 Discussion

Although the oral implant planning problem has been the main impetus, the current state of the environment described here enables the generation of multiple applications. The movie studio analogy copes with many of the data representation requirements put forward in this text. We believe our environment to be easily extendible towards other types of surgery.

In order to make improvements with respect to constraint handling, we are currently working on the problem of "collision detection". E.g., insertion of a long rod in a "hole" without touching the walls can be very difficult if the rod can move straight through this same wall. To prevent this, collision detection must be introduced and the corresponding constraints must be calculated and added to the object in question.

Finally, we remark that pre-operative planning is only a first step towards computer assisted surgery. Therefore, the goal of easy intra-operative employment of the planning data has been one of our design options from the start.

Acknowledgements

The work on oral implant planning is in cooperation with Prof. D. van Steenberghe, Dept. Periodontology, K.U. Leuven, holder of the Nobelpharma-sponsored Brånemark Chair in Osseointegration. The orthopaedics related and virtual sculpting work is part of PHIDIAS, (laser PHotopolymerisation models based on medical Imaging, a Development Improving the Accuracy of Surgery), project nr. BE/5930 of the BRITE-EURAM programme of the European Commission. Finally IBM, Belgium is acknowledged for support through a joint academic study contract.

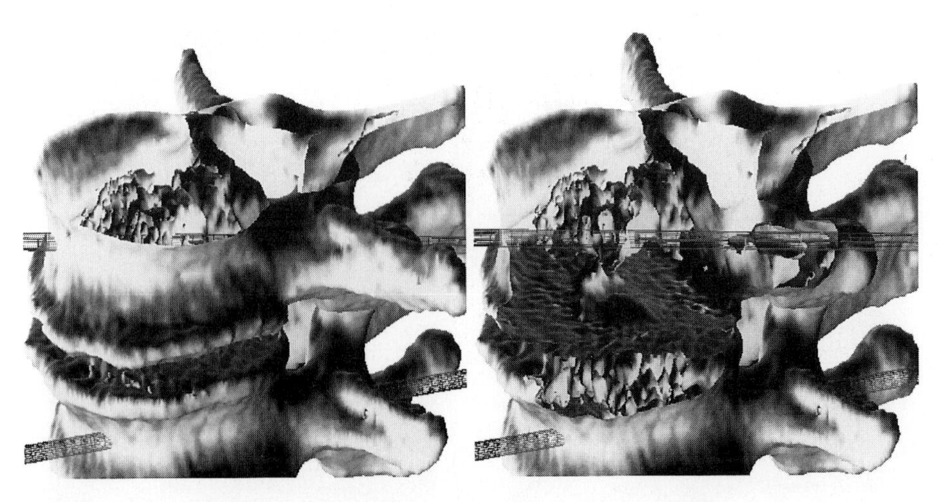

Fig. 6. *On the left a view of two planned transpedicular trajectories in relation to the vertebrae bone surfaces are shown for the example of figure 5. On the right, a more elucidating view of how the upper trajectory passes through the upper pedicle is generated by using the cutting plane facility of a view. Indeed this type of composed view removes the part of the vertebra cortex that is obstructing the observation of inner structures.*

References

1. B. Pflesser, U. Tiede and K.H. Höne. Towards Realistic Visualization for Surgery Rehearsal. In N. Ayache, editor, *Computer Vision, Virtual Reality and Robotics in Medicine, Lecture Notes in Computer Science 905*, pages 487–491. Springer, April 1995.
2. D.A. Meglan, R. Raju, G.L. Merill, et al. The Teleos Virtual Environment Toolkit for Simulation-based Surgical Education. In K.S. Morgan S.J. Weghorst, H.B. Sieburg, editor, *Medicine Meets Virtual Reality 4*, pages 346–351. Health Technology and Informatics 29, IOS press, January 1996.
3. H. Delingette, G. Subsol, S. Cotin, and J. Pignon. A Craniofacial Surgery Simulation Testbed. In R.A. Robb, editor, *Third Intern. Conf. on Visualization in Biomedical Computing*, pages 607–618. SPIE (The International Society for Optical Engineering), Vol. 2359, October 1994.
4. K.H. Höne, M. Bomans, A. Pommert, et al. 3D Visualization of tomographic volume data using the generalized voxel-model. *Visual Computing*, 6(1):28–36, 1990.
5. R. Kikinis, H. Cline, D. Altobelli, et. al. Interactive Visualization and manipulation of 3D reconstructions for the planning of surgical procedures. In R.A. Robb, editor, *Second Intern. Conf. on Visualization in Biomedical Computing*, pages 559–563. SPIE (The International Society for Optical Engineering), Vol. 1808, October 1992.
6. R. Verbeeck, D. Vandermeulen, J. Michiels, P. Suetens, G. Marchal, J. Gybels, and B. Nuttin. Computer Assisted Stereotactic Neurosurgery. *Image and Vision Computing*, 11(8):468–485, October 1993.

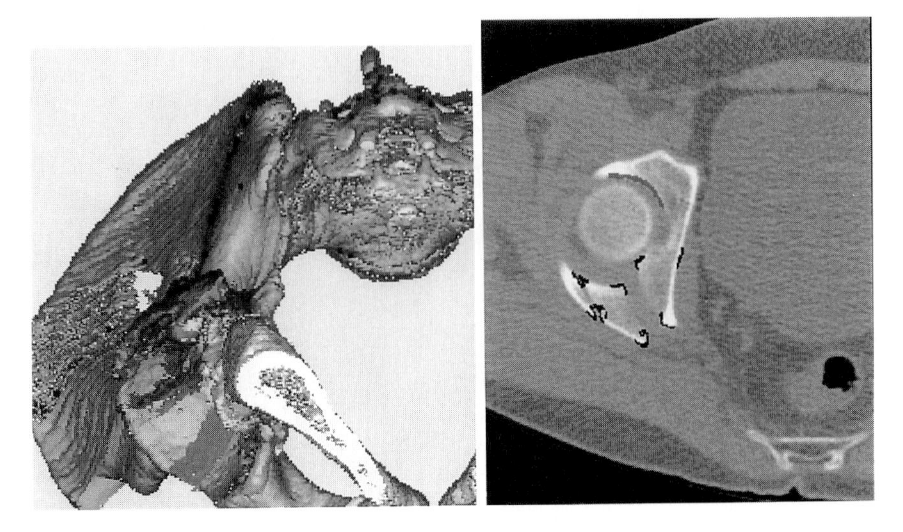

Fig. 7. *On the left a surface derived from a pelvis CT set is shown. By using a free-hand 3D paintbrush, the transtectal course of severe acetabular fractures are outlined. A view through this scene via an axial CT slices is shown on the left. Fractures are labeled by color.*

7. K. Verstreken, J. Van Cleynenbreugel, G. Marchal, D. van Steenberghe, and P. Suetens. Computer Assisted Planning of Oral Implant Surgery. In K.S. Morgan S.J. Weghorst, H.B. Sieburg, editor, *Medicine Meets Virtual Reality 4*, pages 423–434. Health Technology and Informatics 29, IOS press, January 1996.
8. J. Wernecke. *The Inventor Mentor.* Addison-Wesley, 1994. ISBN 0-201-62495-8.
9. K. Zuiderveld. *Visualization of Multimodality Medical Volume Data using Object-Oriented Methods.* Ph.D. Thesis, Utrecht, The Netherlands, 1995. ISBN 90-393-0687-7.

An Advanced System for the Simulation and Planning of Orthodontic Treatments

Alcañiz, M.[1] ; Chinesta, F.[2] ; Monserrat, C.[1] ; Grau, V.[1] ; Ramón, A.[2]

[1] Departamento de Expresión Gráfica en la Ingenieria
[2] Departamento de Mecánica de los Medios Contínuos y Teoria de Estructuras,
Universidad Politécnica de Valencia, Camino de Vera, s/n, 46071, SPAIN.

Abstract. This paper describes a new method for 3D orthodontic treatment simulation developed for an orthodontic planning system (MAGALLANES). We develop an original system for three-dimensional reconstruction of dental anatomy. Data are acquired directly from the patient with low cost 3D digitizers avoiding use of dental casts in orthodontic treatments. We apply these 3D dental models to simulate three-dimensional movement of teeth, including rotations, during orthodontic treatment. We develop an original simplified model of arch-wire behavior and a viscoplastic behavior law for the alveolar bone, to simulate teeth displacements during orthodontic treatments. The proposed algorithm enables to quantify the effect of orthodontic appliances on teeth movement. Preliminary results are very promising.

1 Introduction

Orthodontics is the branch of dentistry that is concerned with the study of the craniofacial complex growth. Detection and correction of malocclusions and other dental abnormalities is one of the most important and critical phases of orthodontic diagnosis. In orthodontics, information concerning the shape and positional change of a tooth is a fundamental reference for diagnostic and treatment evaluation. When an orthodontic force is applied to a tooth over a long period of time, it moves owing to resorption of the alveolar bone on the pressurized side and apposition of new bone on the opposite side. In daily practice, orthodontists carry out patient diagnosis and treatment evaluation analyzing tooth shape and movement during treatment. To do this, they base their diagnosis on the craniofacial information provided by X-ray sagital images (cephalographies) and plaster dental casts. Cast analysis is carried out using simple instruments such as a caliper and a ruler, which do not give accurate measures. Furthermore, plaster model obtention is a complex operation for the orthodontist, and a very unpleasant experience for the patient. Plaster cast storage is also a huge problem in orthodontics consultancies. Once plaster casts have been obtained, the clinician plans the orthodontic treatment to be applied to the patient (dental extraction, teeth movements and appliances). Actually treatment planning is based exclusively on clinician experience, as there is no simulation tool that allows him to visualize the time evolution of the treatment on an arcade model of the patient. To modelize orthodontics treatments, the rehological behavior of both the arch-wire and the alveolar bone will be considered. The bending model of the arch-wire shows as main peculiarities the capability to operate with geometrical shapes represented by curves in space; an elastic behavior that can be linear (steel) or non-linear (NiTi, β-Titanium); and the necessity to solve an inverse problem, as we know both the unstressed and the strained configurations of the arch-wire in contact with teeth, but we don't know the value of the loads applied on each tooth by the arch-wire. The quantification of these loads has a great importance to evaluate the possibility of teeth displacement in the bone, and to simulate numerically the teeth trajectory during the orthodontic treatment. At this point we will obtain a one-dimensional simplified model from the minimum complementary energy theorem, resulting in the linear case the Castigliano theorem, which is generalizable to the non-linear case. On the other hand, to modelize the bone rehological behavior, a viscoplastic law is used. It is assumed that the strain takes place in the direction of the load transmitted by the tooth. It is also assumed that teeth interact only through the coupling induced by the arch-wire. The coupled solution will be obtained solving uncoupled problems with the following iterative scheme:

1. Being known the unstressed configuration and being established the strained configuration.
2. Strain values will be obtained in the points where the arch-wire is fixed to teeth (u_i) (fig. 1).
3. The arch-wire model will be solved, obtaining the loads (P_i) transmitted to the teeth by the arch-wire. These loads will be finally transmitted to the bone by the teeth (fig. 2).
4. Displacement for each tooth will be obtained in a time step by solving the rehological model of the bone (fig. 3).
5. Teeth positions will be actualized by obtaining the new strained configuration for the arch-wire (fig. 4).
6. The process continues iteratively until the acting stress in each tooth is lower than critical stress required to start (or maintain) the motion.

In this paper we propose an accurate and optimized 3D dental anatomy reconstruction system, starting from data captured directly from the patient's mouth with a low cost 3D digitizer, as well as a treatment editor. Furthermore, we propose a simplified model of the biomechanical behavior and a computational algorithm to simulate numerically teeth displacements in orthodontic treatments.

2 Dental anatomy 3D reconstruction

In order to reconstruct three-dimensional dental anatomy, many plaster cast measurement techniques have been attempted. An indirect technique such as stereophotogrammetry has been applied for the morphological study of the tooth [4]. Although this technique has been used successfully to measure three-dimensional shapes, it requires a sophisticated procedure to digitize shape. Direct measurement techniques have also been applied. Van der Linden et al [6] have described the Optocom, which can measure x and y coordinates using a microscope, and z coordinate by mechanical contact with a dental cast. The Duret system [2] makes use of a hand-held optical probe to produce images of the patient's mouth, that are processed using Moire fringe techniques. There are two main problems with this system: the need of commercial solutions for the 3D reconstruction and the high cost of the capturing peripheral. The CEREC system [3] also uses an optical topographic scanning procedure. This system has demonstrated a great accuracy, but it can only capture and reconstruct oclusal teeth faces, and not the global 3D tooth anatomy. Yamamoto [7] has described a system for measuring three-dimensional profiles of dental casts and three-dimensional tooth movement during orthodontic treatment. The profile measurement is based on a triangulation method which detects a laser spot on a cast using an image sensor. This system does not capture directly the patient teeth anatomy, thus not avoiding the use of plaster casts, and has the additional problem of the high cost of the laser capturing system. The proposed algorithm performs an accurate 3D dental anatomy reconstruction from a few points directly captured from the patient by using an anatomical database of dental pieces. The process is divided in three phases:

2.1 Anatomy capturing

Two curves are directly captured for each tooth, as shown in fig. 6: the mesio-distal oclusal curve and the vestibular-lingual external curve. These two curves define the position and orientation of each tooth and, according to clinician suggestions and several realized tests, give a sufficiently accurate dental shape description for the orthodontists. The curves can be captured on the plaster casts or directly on the patient's mouth by means of a 3D digitizer. We have tested an ultrasonic device (Freedom from SAC, CT, USA) and a mechanical arm (Space-Arm from FARO Technologies, FL, USA), both with a customized dental probe attached. The accuracy is 0.5 mm for the sonic device and 0.3 mm for the mechanical arm. Measures taken with the mechanical arm have proved more satisfactory due to the environmental and air speed sensibility of the sonic device.

2.2 Model preparation

Once dental curves have been captured, the system adjusts automatically a standard dental model to the patient anatomy. We make use of a standard dental anatomy database in which each tooth is defined by several contours (level curves). The number of contours depends on the tooth type. Spacing between contours is lower in the region where the tooth geometry is highly variable (fosae and cuspid regions). Taking the two captured curves as preliminary information, the algorithm retrieves from the dental database the contours corresponding to the tooth type. These two curves define the mesio-distal and vestibulo-lingual ratios of the patient dental anatomy at different heights. The system adjusts and scales each contour in order to fit these ratios to obtain an accurate anatomical description of each patient tooth (fig. 7).

2.3 3D reconstruction

For the 3D reconstruction from dental contours, we have implemented a version of the Delaunay 3D reconstruction algorithm [1]. This algorithm has been chosen for its important data reduction, polyhedral representation and low data processing time.

3 Biomechanic modelling

3.1 Rehological behaviour of the arch-wire

For each material we know the local stress-strain relationship:

$$\sigma = g(\epsilon) \tag{1}$$

The bending moment - curvature relationship is obtained from the equation 1, Bernoulli Kinematic hypothesis and the cross-section geometry (see figure 5):

$$\frac{-y}{\Delta dx} = \frac{\rho}{dx} \rightarrow \epsilon = -\frac{y}{\rho}$$

From this, the stress distribution results

$$\sigma = g(-\frac{y}{\rho})$$

and integrating in the cross-section it results

$$M_z = -\int_A y\, g(-\frac{y}{\rho})\, dA \tag{2}$$

We can write the bending moment - curvature relationship as:

$$\left.\begin{array}{l} M_z = f(\frac{1}{\rho}) \\ \frac{1}{\rho} = f^{-1}(M_z) \end{array}\right\}$$

Simplified model. We start from the complementary work variational formulation of the elastic problem

$$\int_\Omega \mathrm{Tr}(\underline{\underline{\epsilon}}\, \delta\underline{\underline{\sigma}})\, d\Omega = \int_{\partial\Omega} \delta\underline{F}^T\, \underline{u}\, dS$$

being $\underline{\underline{\sigma}}$ and $\underline{\underline{\epsilon}}$ the stress and strain tensors.

If we denote the complementary work by R, and a system of N concentrated loads $\underline{P}_1, \cdots, \underline{P}_N$ is applied on the elastic body, from the above equation results

$$\delta\tilde{R} = \int_{\Omega} \delta R \, d\Omega = \sum_{i=1}^{i=N} \delta \underline{P}_i \, \underline{u}_i = \sum_{i=1}^{i=N} \left(\frac{\partial \tilde{R}}{\partial \underline{P}_i} \right)^T \delta \underline{P}_i$$

From this results Castigliano's theorem

$$\frac{\partial \tilde{R}}{\partial \underline{P}_i} = \underline{u}_i \qquad \forall i \in [1, \cdots, N]$$

Since $\underline{P}_i$ and $\underline{u}_i$ have the same direction, the relationship above may be written in a scalar form, although with the direction fixed by the vector load $\underline{P}_i$.

In the bending of the arch-wire, stresses are in the axes direction, thus $\delta R = \epsilon_x \, \delta\sigma_x = \epsilon \, \delta\sigma$ and from the integration in the elastic body it results

$$\delta\tilde{R} = \int_L \int_S \delta R \, dS \, ds = \int_L \delta R_S \, ds$$

being L the arch-length of the arch-wire and S the cross-section. For δR_S we have the following relationship

$$\delta R_S = \int_S \delta R \, dS = \int_S \epsilon \, \delta\sigma dS = \int_S \frac{-y}{\rho} \, \delta\sigma \, dS = \frac{1}{\rho} \int_S -y \, \delta\sigma \, dS = \frac{\delta M_z}{\rho}$$

Thus, it results

$$\tilde{R} = \int_L \{ \int_0^{M_z} \delta R_S \} \, ds = \int_L \{ \int_0^{M_z} \frac{\delta M_z}{\rho} \} \, ds$$

Applying Castigliano's theorem, we obtain

$$u_k = \frac{\partial \tilde{R}}{\partial P_k} = \int_L \{ \frac{\partial}{\partial P_k} \int_0^{M_z} \frac{\delta M_z}{\rho} \} \, ds$$

taking into account $M_z(s) = P_1 \, \overline{M}_1(s) + \cdots + P_N \, \overline{M}_N(s)$ being $\overline{M}_i(s)$ the unitary bending diagram associated with a unitary load applied in the point i in the direction of the real load $\underline{P}_i$, results

$$\frac{\partial M_z}{\partial P_k} = \overline{M}_k$$

and using the Leibnitz integration rule

$$u_k = \int_L \frac{\partial M_z}{\partial P_k} \frac{1}{\rho} (M_z(s)) \, ds$$

in conclusion

$$\left. \begin{array}{rcl} u_k & = & \int_L \frac{1}{\rho} \, \overline{M}_k \, ds \\ \frac{1}{\rho} & = & f^{-1}(M_z) \end{array} \right\} \tag{3}$$

Linearization. Equation 3 depends on the loads P_i, $i \in [1, \cdots, N]$, and this relationship is non-linear. In this case a linearization of equation 3 is required. Using a Newton-Raphson method result

$$R_k^{(n)} = u_k - \int_L \left(\frac{1}{\rho}\right)^{(n)} \overline{M}_k \, ds \quad \forall \, k \in [1, \cdots, N]$$

where (n) es the iteration number and R_k is the residual vector.
We obtain the tangent matrix from

$$R_k^{(n+1)} = 0 \approx R_k^{(n)} + \sum_{i=1}^{i=N} \left(\frac{\partial R_k}{\partial P_i}\right)^{(n)} \Delta P_i^{(n)} \quad \forall \, k \in [1, \cdots, N]$$

where

$$\left(\frac{\partial R_k}{\partial P_i}\right)^{(n)} = - \int_L \left(\frac{\partial f^{-1}(M_z)}{\partial M_z}\right)^{(n)} \overline{M}_i \, \overline{M}_k \, ds$$

Solving $\forall \, k$ we obtain $\Delta P_i^{(n)}$, and $P_i^{(n+1)} = P_i^{(n)} + \Delta P_i^{(n)}$

3.2 Bone rehological behaviour

The alveolar bone, where the teeth are located, has a behaviour characterized by an elastic component and other viscous component, with irreversible deformation. Due to this, we can consider its behaviour to be equal to an elastoviscoplastic solid. We evaluate in an uncoupled form the displacement of each tooth, with respect to the rest of teeth because they only interact in a direct form through the contact zone or through the arch wire. In this way, the analysis is reduced to an onedimensional simplified model. Asuming the initial hypothesis, we can say that the global deformation is given by the expresion:

$$\epsilon = \epsilon^e + \epsilon^v$$

The elastic component is:

$$\epsilon^e = \frac{\sigma}{E}$$

and the viscous component is, according to law:

$$\frac{d\epsilon^v}{dt} = \left(\frac{\sigma}{\lambda}\right)^N$$

where N is the viscous law exponent and λ is the viscous law coeficient.
Then:

$$\frac{d\epsilon}{dt} = \frac{1}{E}\frac{d\sigma}{dt} + \left(\frac{\sigma}{\lambda}\right)^N$$

This differential equation is solved with Finite Differences :

$$\frac{\epsilon_{i+1} - \epsilon_i}{\Delta t} = \frac{1}{E}\frac{\sigma_{i+1} - \sigma_i}{\Delta t} + \left(\frac{\sigma_{i+1}}{\lambda}\right)^N \tag{4}$$

On the other hand, the stress that tooth makes on the bone is determined by the arch wire deformation; and that is in relation to teeth position, so we obtain a coupled sistem.

For each time step we need to solve the bending moment of the arch wire, that is a consequence of teeth determinated position; and the deformation that we can obtain in a following time step, is a consequence of this deformation.

The expression 4 is an implicit model that results to be a non linear and coupled one. We can write this equation as an explicit model:

$$\frac{\epsilon_i - \epsilon_{i-1}}{\Delta t} = \frac{1}{E}\frac{\sigma_i - \sigma_{i-1}}{\Delta t} + \left(\frac{\sigma_i}{\lambda}\right)^N$$

expression that we can solve in an uncoupled form.

4 The 3D treatment simulation editor

In the 3D treatment simulation editor, the user can simulate different orthodontics treatments and visualize the effect on teeth movements. For this task, the system presents a 3D realistic dental arcade model of the patient and the orthodontic appliances used. Based on these models, the editor facilitates treatment simulation by providing realistic visualization of dental arcades, as shown in figure 8, and 3D navigation tools (zoom, rotations, translations, etc.). The editor allows to obtain morphometric measurements through several implemented algorithms. Using the digital arcade models, the system has information about relevant morphometric points of dental pieces, making the clinician labor in data capture and arcade analysis less time consuming. Furthermore, the system can compare automatically the morphometric measurements against ergonomic established models, enabling an automated selection of the best suited devices. In addition, the user can make direct measurement of distances on the arcade model. The editor implements several teeth movement functions. With these functions, the user can apply to each teeth several standard operations performed in orthodontics, like teeth extraction, mesio-distal translations, rotations, etc. For this task, the user can select orthodontics appliances from system libraries and place them on teeth trough placement functions (Fig 8). Then the system automatically simulates teeth movements. This feature allows to visualize on the screen the time evolution of the dental pieces due to the treatment, the movements of the dental pieces due to the application of orthodontic devices, and a space-time digital animation of a period of the treatment or the whole of it. Stereoscopic visualization has been implemented, trough stereoscopic glasses directly connected to the computer. Proposed system has been implemented in C++ language using Open-GL graphic library and can be run in PC compatible platform under Windows 95 operating system using low cost graphic accelerator cards.

5 Discussion and conclusions

Preliminary results are very promising showing a standard deviation of the data of 0,2 mm, and 92% of the data distributed with an error of less than 0.4 mm for 3D reconstruction method.

We have presented a new method for low cost digitizing of 3D dental anatomy andtooth movement prediction, thus enabling a low cost orthodontics global simulation system. Obtained results are very promising. The proposed techniques are being integrated in an advanced system for simulation and planning of orthodontics treatments (Magallanes). This system, in addition to the 3D treatment simulation module described here, includes an algorithm for automatically locating certain characteristic anatomical points called landmarks on cephalographies (sagital skull X-ray images). The algorithm uses digital image processing and feature recognition techniques to locate the landmarks. These features increase considerably the diagnostic and assistential quality in the orthodontic field, avoiding routine patient visits for arcade measurements and/or orthodontic devices checking.

6 Acknowledgments

MAGALLANES is a joint academic-industrial project supported by IBM Corp (Healthcare Industry Europe, Advanced Health Applications Division, Proj. contract n ° 20067). The authors express our gratitude to our clinical collaborators, the team lead by Dr. Jose Antonio Canut from the Dentistry Faculty of Valencia.

References

1. **Boissonat.** "Shape reconstruction from planar cross-sections". Comp Vis Graph and Image Processing. Vol. 44. Pp 1-29. 1988
2. **Duret, J.L. Blouin and B. Duret.** "CAD/CAM in dentistry". Journal of American Dental Association.. vol. 117. Pp. 715-720. 1988.
3. **Moermann, H. Jans, M. Brandestini, A Ferru and F. Lutz.** "Computer machined adhesive porcelain inlays: margin adaptation after fatigue stress" [Abstract]. J. Dent. Res. Vol. 65. Pp 762. 1986
4. **Rekow.** "CAD/CAM for dental restorations-some of the curious challenges". IEEE Trans. Biomedical Engineering. Vol. 38. Pp. 314-318.1991
5. **Ricketts, R. W. Bench, J. J. Hilgers, and R. Schulhof.** "An overview of computerized cephalometrics". Amer. J. Orthod. Vol. 61. Pp 1-28. 1972.
6. **Van der Linden, H. Boersma, T. Zelders, K.A. Peters, and J.H. Raaben.** "Three-dimensional analysis of dental casts by means of the Optocom". J. Dent. Res., vol. 51, p 1100. 1972
7. **Yamamoto, S. Hayashi, H. Nishikawa, S. Nakamura and T. Mikami.** "Measurement of dental cast profile and three-dimensional tooth movement during orthodontic treatment". IEEE Trans. Biomed. Eng. Vol. 38-4 pp 360-365. 1991.
8. **Nikolai, Robert J.** "Bioengineering Analysis of orthodontic Mechanics".

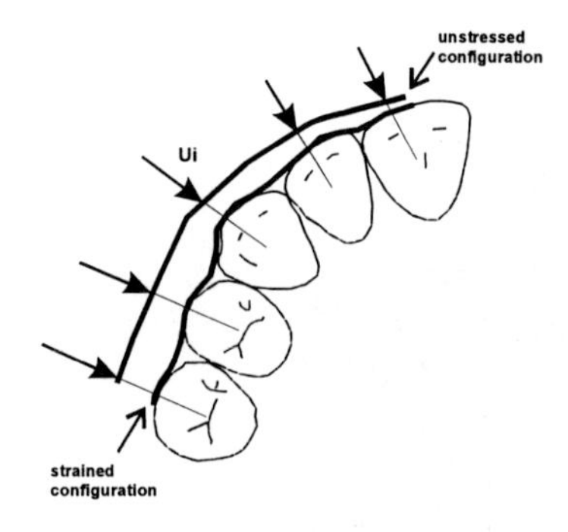

Fig. 1. Arch-wire fixed to teeth

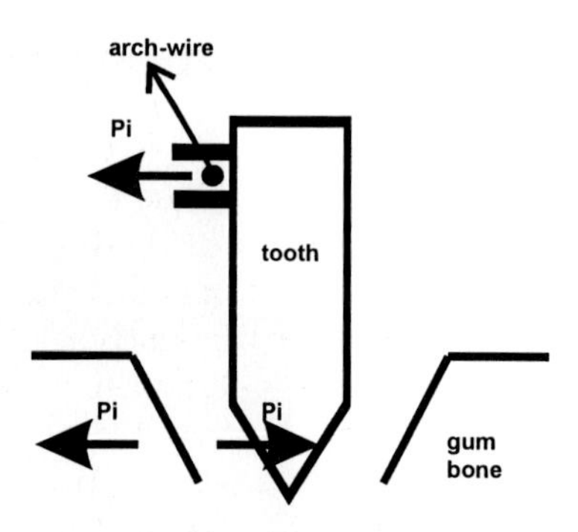

Fig. 2. Loads transmitted to teeth by arch-wire

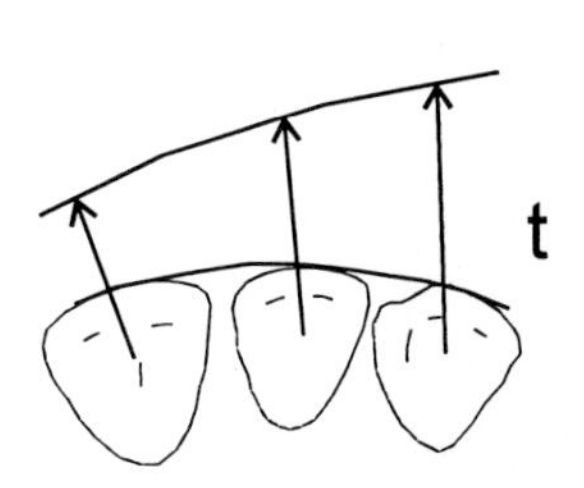

Fig. 3. Tooth position update

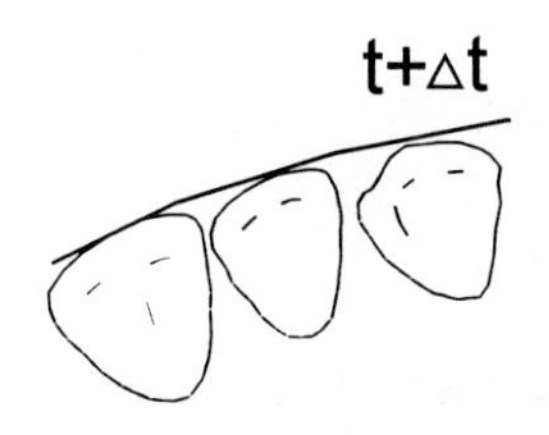

Fig. 4. Strained arch-wire update

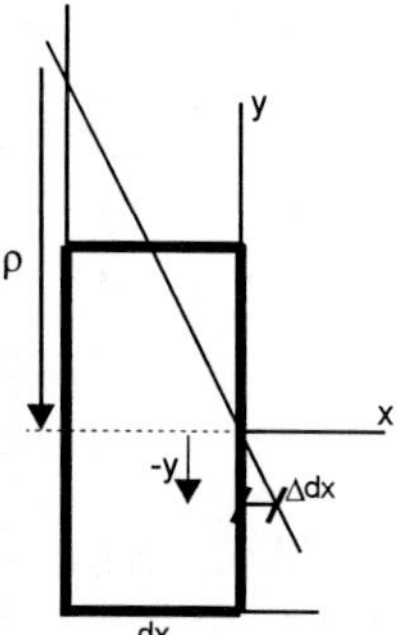

Fig. 5. Bending moment-curvature relationship

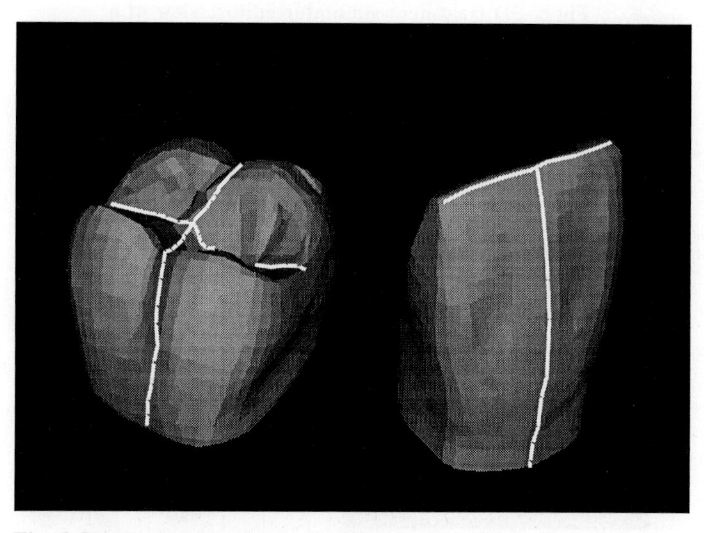

Fig. 6. Image showing the mesio-distal and vestibulo-lingual curves captured on patient mouth for first molar and central teeth

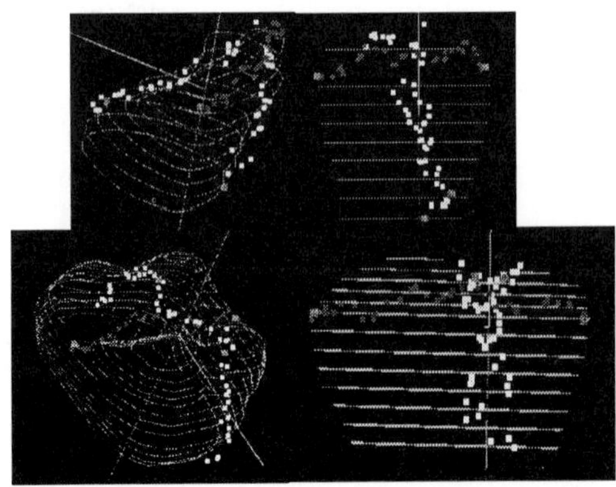

Fig. 7. Fitting process showing captured curves and tooth contours for a first molar and a central teeth

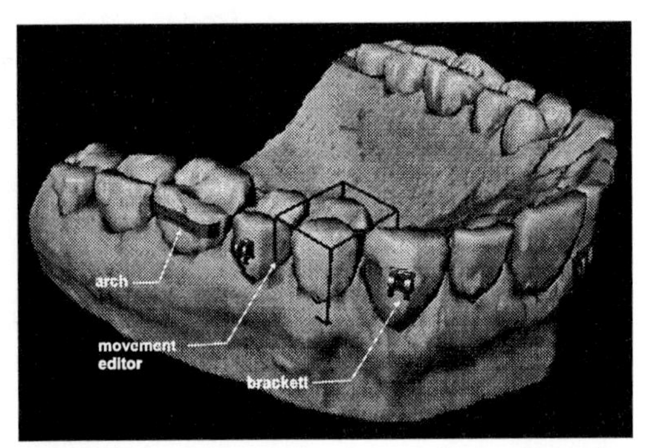

Fig 8. 3D treatment simulator editor: view of a reconstructed arcade with orthodontic appliances and movement editor tool

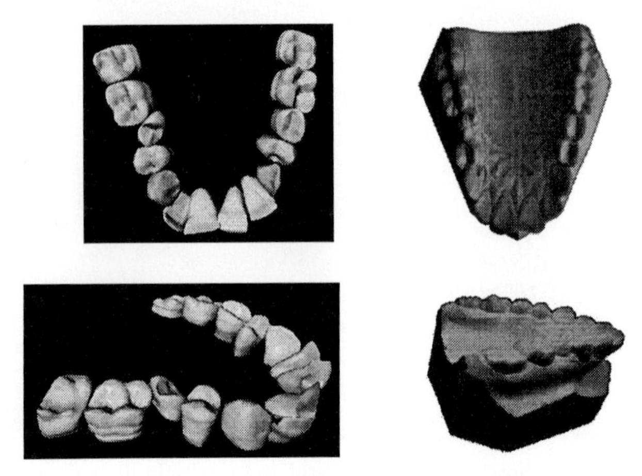

Fig. 9. Computer graphic display, using a shading method, of 3D reconstructed arcade and plaster cast of the same arcade.

Assessment of Several Virtual Endoscopy Techniques Using Computed Tomography and Perspective Volume Rendering

R Shahidi[1], V Argiro[1], S Napel[2], L Gray[3], H P McAdams[3], G D Rubin[2], C F Beaulieu[2], R B Jeffrey[2], A Johnson[3]

[1] Vital Images, Inc., Fairfield, IA

[2] Department of Radiology, School of Medicine, Stanford Univ., Stanford, CA

[3] Department of Radiology, Duke Univ. Medical Center, Durham, NC

Abstract

Advanced medical imaging techniques available today, such as CT, MRI, and PET, are capable of producing large quantities of near isotropically-sampled data. These datasets coupled with advanced techniques in Computer graphics can create new ways to explore cross-sectional medical images. We hypothesize that 3-D renderings of the imaging data may provide enhanced diagnostic interpretation and superior visualization for treatment planning purposes. We have investigated the application of Perspective Volume Rendering (PVR) as a method to facilitate interactive exploration of various tubular structures. In this paper we will explore the results from four noninvasive applications of PVR: 1) detection and studies of aneurysms in the circle of Willis, 2) studies of thoracic aorta aneurysms, 3) assessment of bronchial anastamoses and 4) detection of polyps in colonic lumen.

Introduction

Recent developments in computation technology and 3-D imaging are fundamentally improving the role of medical imaging in patient care systems. Previously restricted to slice-based diagnostics, the use of various imaging modalities such as MRI, CT, PET have matured to systems that permit the physicians to visualize and quantify the extent of disease in the volumetric form to plan therapeutic interventions.[1,2,3,4,5] This manuscript describes a system for accomplishing this using Perspective Volume Rendering (PVR). Using PVR, the user "flies" through and around the data and, in effect, generates virtual endoscopic views. PVR additionally allows viewing of tissues inaccessible to conventional endoscopy, such as in soft tissues and behind bones. The main advantage of PVR, in comparison with surface rendering, is the accessibility of the whole volume at any stage of the fly through. This way, there is no need to edit out structures, as the user can simply lower the opacity of obscuring objects to observe the anatomical structures behind them. In addition, when using PVR, there is no need for preprocessing in order to extract surfaces of interest prior to visualization. In this paper we describe this technique and our initial experience with virtual fly through of the tracheobronchial tree, the colon and the vascular system of the brain and thoracic cage.

Correspondence should be addressed to: Ramin Shahidi, Stanford Medical Center, Image Navigation Labs, Dept. of Neurosurgery, 300 Pasteur Dr., Room 155, Stanford, CA 94305 USA.

Methods

We have developed a technique that enables radiologists and surgeons to begin with a conventional cross sections, proceed to selectively emphasized structures within the image volume and then to visualize the context and optimal path to abnormalities with cinematic tour inside the image volume. We have applied this technique to forty eight CT data sets (High Speed Advantage GE Medical Spiral and Helical CT Systems, Milwaukee, Wisconsin and Somatom Plus 4, Siemens Medical Systems, Iselin, New Jersey). Thirty were of the aneurysm studies in circle of Willis, four of the thoracic aorta, ten of the tracheobronchial tree and four of the colon. The following sections describe the CT data acquisition, post-processing protocols and virtual endoscopy techniques. Post-processing was performed with VoxelView software (Vital Images, Inc. Fairfield, IA) on a Silicon Graphics (Mountain View, CA) Onyx-RE2 and Indigo2-Impact workstations.

Sources of Patient Data

CT Angiograms of the base of the brain were obtained with 1.0 mm thick sections, 60 seconds, pitch of one and with 0.5mm reconstructions on a 512x512 matrix. Spiral CT scanning began after a 100cc contrast (Isovue 300) intravenous injection at the rate of 3 cc/sec. The images were transferred to an Onyx RE2 workstation and converted to 8 bits covering a range of approximately 1000 Hu in 256 gray-scale values. The reconstruction of the images into the 3D volume was performed with 0.5 mm intervals. The 0.5 mm intervals were obtained using interpolation of every other image. The initial images of the vasculature were obtained by optimizing opacification and transparency of the selected tissues (values higher than 120). The areas of interest (circle of Willis) were sub-volumed, followed by seed-fill for creating contrast against the skull. In addition, color and lighting were applied to the volumetric image for better visualization. Flying through the volume of interest was then accomplished using perspective rendering with the camera's field of view (FOV) at 40 degrees.

CT Angiograms of the thoracic aorta were obtained during breath holding with 3mm collimation, 6 mm/s table speed at 120 KV and between 260 and 280 mA. Exposures lasted between 30 and 40 seconds as required to cover the anatomical structures of interest. Non-ionic contrast media (100-160 ml) was injected intravenously at 4 ml/sec. Spiral CT scanning began after a delay, determined for each patient by using a test injection. Images were reconstructed on a 512 x 512 matrix using a 25 cm field of view and a soft reconstruction kernel. All images were transferred to an Onyx-RE2 workstation and converted to 8 bits covering the range of gray-scale values from 0 to 256. We created a subvolume enclosing the aorta and its major branches. Initial opacity curves were chosen to render the transition regions from tissue to iodine (roughly 50-100 Hu) and iodine to bone (350 Hu and above) fairly opaque and all other structures transparent. Flying through the aorta with the virtual camera (FOV at 25 degrees) was accomplished using perspective rendering.

Tracheobronchial CT scans were obtained during a single breath-hold. All patients had recent bronchoscopic correlation. All scans were performed on a Helical CT using 3 mm collimation, pitch 1.0, and overlapping 1.0 mm reconstructions. Depending upon the length of breath hold, 7 to 10 slabs were obtained through the region of interest and then combined into a single volume. Later redundant transverse sections were removed by a radiologist. The combined images were transferred to an Indigo2-Impact workstation and converted to 8 bits. The resolution of the reconstructed images (512 x 512) was then increased using one interpolation between each axial gap. The opacity curves were chosen to clear out the airways and the soft tissue covering the bronchial branches. No subvolume was used for this technique. Customized color tables and lighting parameters were created for better visualization of the interior of the tracheobronchial tree. We used a wider FOV (50-60) for flying through the bronchial tree for better visualization of narrow passages and strictures.

Colorectal CT scans were obtained from patients already prepared to undergo colonoscopy. Prior to CT scanning, the colon was inflated using a squeeze bulb and a small rectal catheter. Spiral scanning was performed with 3 mm collimation, 6 mm/sec table feed, and an exposure lasting from 40-60 seconds (sufficient for complete coverage of the colon) at 120 KV and 260 mA. Images were reconstructed every 1 mm with a 30cm field of view over 512 x 512 pixels. All images were transferred to an Onyx-RE2 workstation and converted to 8 bits covering the range of approximately 1000 Hu with 256 gray-scale values. We created a subvolume enclosing the complete colon and excluding the other structures as much as possible. Initial opacity curves were chosen to render air (-1000 Hu) completely transparent, and soft tissue (-850 Hu and above) fairly opaque, with a smooth transition in between. Optimized camera FOV (40 degrees), opacity curves, coloring tables and lighting parameters were all manually selected for better visualization of colonic lumen.

Perspective Volume Rendering

Perspective volume rendering (PVR) was being conducted on all patient data. We used the VoxelView volume rendering package for visualization of the inner and outer walls of the colon, as well as the soft tissues of the lumen (figures A-D). Volume rendering represents 3-D objects as collections of cube-like building blocks called voxels, or volume elements. Each voxel is a sample of the original volume, a 3-D pixel on a regular 3-D grid or raster. Each voxel has associated with it one or more values, quantifying some measured or calculated property of the original object, such as luminosity, density, flow velocity or metabolic activity. Using this technique, one can assign a range of transparency and color values to each voxel representing a tissue type in the volume. This allows one to selectively view embedded pathologies and organs of interest. The main advantage of this method of rendering is its ability to preserve the integrity of the original data during the rendering process{6}.

Perspective Surface Rendering

The surface visualization approach to virtual endoscopy involves three steps: 1) determining which voxels comprise the anatomical structures of interest 2) generating a set of polygons that represent the anatomical surface, and 3) displaying the three-dimensional model representation{7}. Identifying regions of interest, referred to as segmentation, is generally a very difficult problem for medical image analysis. However, virtual endoscopic applications focus on the air/tissue interface seen on computed tomography images. Visualizing hollow organs simplifies the segmentation process since the voxels corresponding to the air column can be detected using a threshold value. Voxels with an intensity value less than the threshold (air) can then be grouped into a single connected object using a three-dimensional region growing technique. Polygons representing the outer surface of this object can then be calculated using a variant of a marching cubes algorithm{8}. The surfaces generated with this approach are displayed from either an external or internal viewpoint. Figure E shows a view of the outer surface and figure F shows a view of the inner surface of a colon generated with this technique.

Path Planning

In the case of PVR, rendering time was not short enough to obtain a manual exploration within the data set. Therefore, a "key frame" technique was used to create the animations. Key framing is a technique for specifying viewing parameters at intermediate positions along a flight path. One uses the flight path to define a smooth path through the volume. Once the key frames are chosen, the computer generates additional images (steps) in between to create a smooth animation. The information for additional images are defined using spline interpolation between any of the two adjacent key frames. This creates the intermediate variables such as changes in camera path, color, opacity, threshold, lighting and other rendering parameters. The resulting set of images can then be replayed using computerized video programs (typically at 30 frames/sec).

For all of the above, we manually selected up to 200 "key frames" and optimized the tissue classifications at each location. Up to 1800 frames/minute were rendered, along a smooth path connecting the key frame parameters. All frames were compressed using JPEG with approximately 20x compression and recorded on video tape at 30 frames/sec.

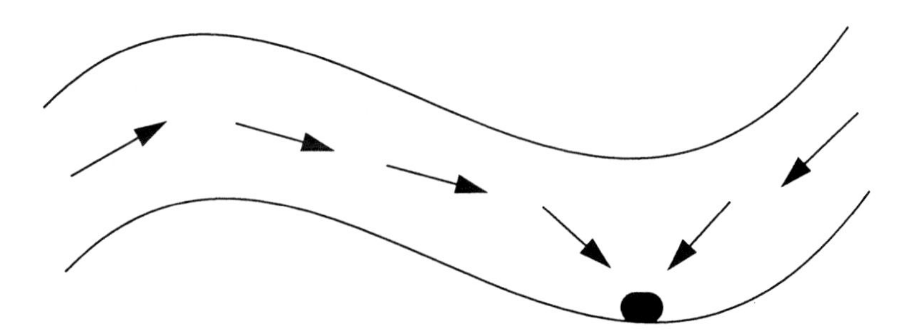

Fig. 1. Key framing is used to define specific nodes for the fly path. Once the nodes are defined spline curves are generated to create a smooth passage through the nodes.

Results

In the circle of Willis studies, twenty five aneurysms in thirty patients were evaluated. Using PVR, we were able to identify all aneurysms, the smallest of which measured 1.5-2.0 mm in size. Twenty two aneurysms involved the anterior cerebral circulation and three aneurysms involved vertebral basilar arteries. The ability to detect aneurysms and their features (e.g., aneurysm neck, overlying and underlying vessels) was enhanced by opacifying vessels, making the skull transparent, generating a subvolume and finally performing a fly through which allowed for an unlimited number of perspectives (Figure 2).

In the thoracic aorta studies, four patients were evaluated. PVR provided detailed exploration of the interior of the aorta and its major branches. Aortic aneurysms and calcifications were clearly depicted. Also, the superior mesuteric and brachiocephalic branches were clearly seen. Figure 3 illustrates one case where a thoracic aortic aneurysm was repaired with an endoluminal stent graft. The chosen color scale resulted in clear delineation of the stent graft supports from the aortic wall. This enabled a clear appreciation of the adequacy of the stent graft deployment and demonstrated regions where the metallic struts of the stent graft projected into the aortic lumen.

In the tracheobronchial studies, ten patients who had received single or bilateral lung transplantation were evaluated for bronchial anastamotic strictures. All patients had recent bronchoscopic correlation. Using PVR, we were able to identify all bronchial anastamoses. In two cases, endobronchial stents had been placed. In all cases, 3D imaging of the tracheobronchial tree better localized the site of the anastamosis, better judged the degree of narrowing, and better assessed the placement of the stent than did conventional axial images. Figure 4 illustrates a lung transplantation in the left bronchial branch.

In the colorectal studies, four patients were evaluated. PVR of the colorectal CT data resulted in clear depiction of several polyps, approximately 5 mm in diameter that were confirmed with conventional colonoscopy (Figure 5). In addition, in one patient a large constrictive mass was well depicted and compared favorably with findings at colonoscopy. Initial settings of the opacity function were adequate to visualize most of the mucosal surface of the colon; however, in some regions thin lumen walls were semi-transparent. Adjustment of the opacity function was easily accomplished to verify intact surfaces.

Conclusions

Interactive analysis of spiral CT Angiography in the evaluation of intracranial aneurysms will enhance our ability to detect aneurysms and will provide additional information for surgical planning. Ultimately, it may be possible to replace conventional angiography in many cases and thereby substantially shorten the imaging time and cost prior to sending the patient to surgery.

Angiographic studies with spiral CT and PVR is a promising technique for evaluation of the aorta and its major branches. Though the spatial resolution is limited compared to conventional angiography, the relative safety and economics may justify its use in some cases. such as follow-up of surgical and endoluminal therapies.

Volumetric studies of helical CT data shows promise as a tool for virtual bronchoscopy. PVR is superior to axial CT for evaluating the status of the bronchial anastamosis in patients who have undergone lung transplantation. In certain instances, it can also obviate invasive, bronchoscopic assessment. Before that time comes, however, PVR will likely need to become a real-time technique and be well integrated with other tools such as rapid image reformation.

PVR has also shown promise for visualization of the colonic lumen. In our experience, volume rendered images of the colon are less susceptible to artifacts that may be caused by threshold selection with surface rendering techniques, such as pseudo-holes and tears in the mucosal surface. However, extensive comparisons between PVR, review of tomographic sections, and pathological results must be made to optimize and validate this technique.

References

1. Rubin GD, Beaulieu CF, Argiro V, Ringl H, Norbash A, Feller J, Dake M, Jeffrey RB, Napel S: Perspective volume rendering of CT and MR images: Applications for endoscopic imaging, Radiology, 1996; 199:321-330.
2. Naidich DP, Lee JJ, Garay SM, et al. Comparison of CT and fiber-optic bronchoscopy in the evaluation of bronchial disease. AJR 1987; 148:1 - 7.
3. Quint LE, Whyte RI, Kazerooni EA, et al. Stenosis of the central airways: evaluation by using helical CT with multi-planar reconstructions. Radiology 1985; 194:871-877.
4. Vining DJ, Gelfand D, Betchtold R, Scharling E, Grishaw E, and Shifrin E, Technical Feasibility of Colon Imaging with Helical CT and Virtual Reality". Presented at the annual Meeting of the American Roentgen Ray Society, New Orleans, April 1994.
5. Napel S, Rubin GD, Beaulieu CF, Jeffrey RB, Argiro V., Perspective volume rendering of cross-sectional images for simulated endoscopy and intra-parenchymal viewing, Proc. of Medical Image. SPIE, Newport Beach Cal. 1996
6. Argiro V, Van Zandt W. Voxels: Data in 3-D, BYTE, May 1992.
7. Lorensen WE, Jolesz, and Kikinis R., " The Exploration of Cross-Sectional Data with a Virtual Endoscope". Editors Satava R, and Morgan K, Interactive Technology and the New Medical Paradigm for Health Care, 221-230. IOS Press, Washington D.C., 1995.
8. Lorensen WE, Cline HE. Marching Cubes: a high resolution 3-D surface construction algorithm,. Comp. Graph., 21(4):163-169.1987.

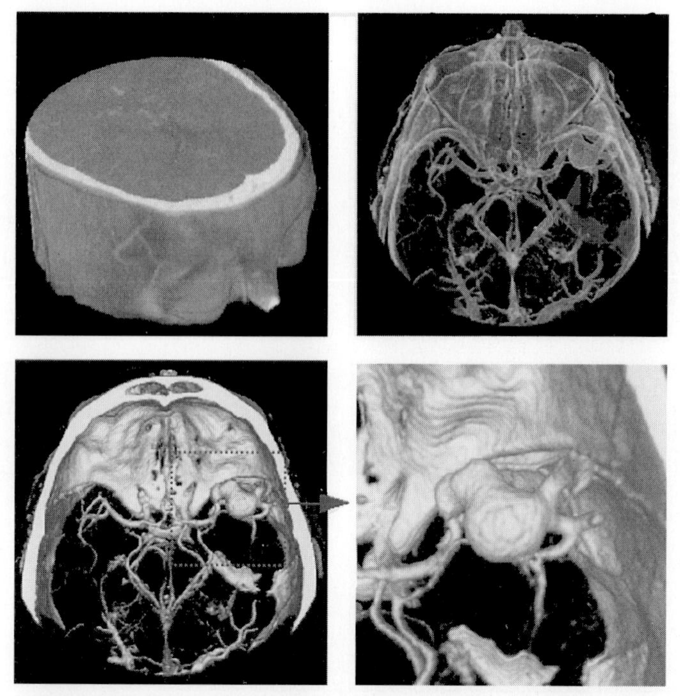

Fig. 2. Flying through and around an aneurysm in the middle cerebral artery.

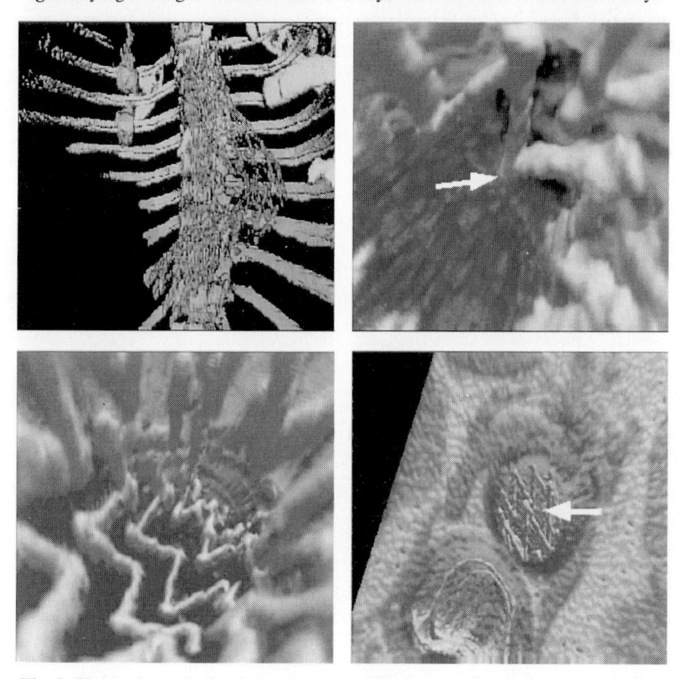

Fig. 3. Flying through the thoracic aorta with the metallic stent–graft struds and their patial blockage of the lumen (arrows).

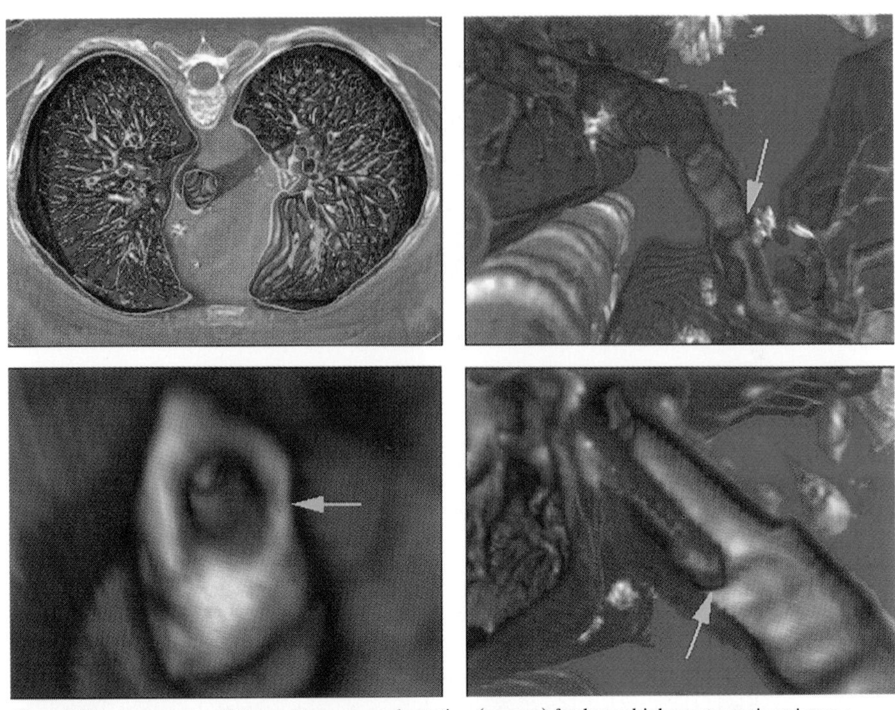

Fig. 4. PVR assessment of bilateral lung transplantation (arrows) for bronchial anastamotic strictures.

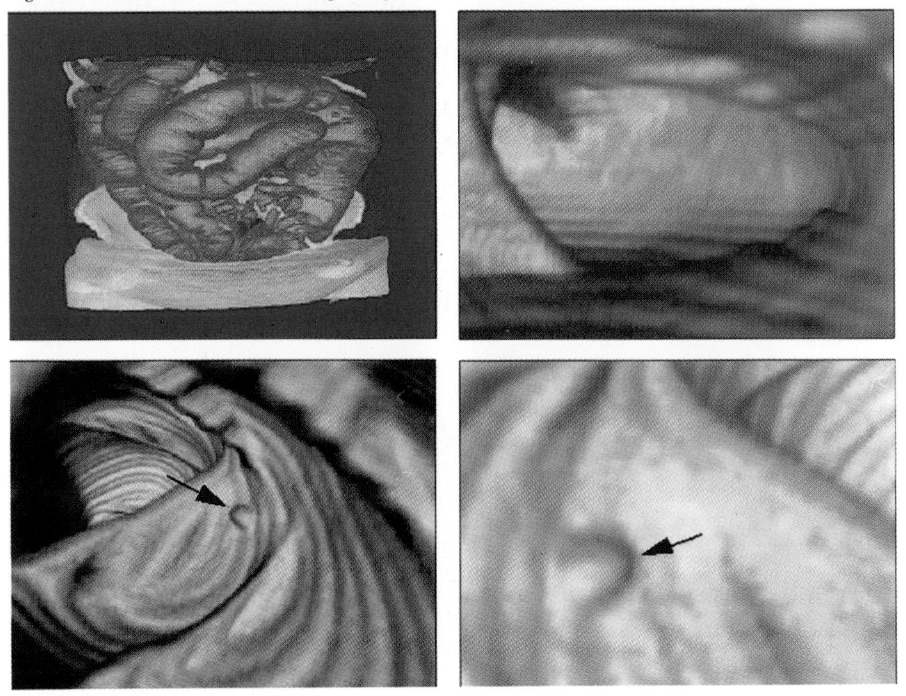

Fig. 5. PVR images viewed from the cecum towards the ascending colon showing a 5mm polyp (arrows).

Surgery Simulation Using Fast Finite Elements

Morten Bro-Nielsen[1,2]

[1] Dept. of Mathematical Modelling
Technical University of Denmark, Bldg. 321
DK-2800 Lyngby, Denmark
[2] 3D-Lab, School of Dentistry, Univ. of Copenhagen,
Nørre Alle 20, DK-2200 Copenhagen N, Denmark
e-mail: bro@imm.dtu.dk WWW http://www.imm.dtu.dk/~bro

Abstract. This paper describes our recent work on real-time Surgery Simulation using Fast Finite Element models of linear elasticity [1]. In addition we discuss various improvements in terms of speed and realism.

1 Introduction

Real-time surgery simulation using general *volumetric* models has just recently become possible [2, 3, 5]. With the development of Fast Finite Element (FFE) models it is now possible to simulate the elastic behaviour of a solid with video frame rates, ie. 20 frames/second.

Other attempts have used *surface* models as the basic modeling method [6]. The problem with surface models, besides the obvious non-solid behaviour, is the lack of an defined interior when surgical procedures are modeled. The surgeon cannot cut a virtual organ modeled using a surface model since there is nothing inside the surface. Some simple cuts can be modeled, such as cutting an artery or other thin structures, but general surgical incisions are impossible.

This paper discusses real-time simulation of deformable objects using 3D solid volumetric Fast Finite Element which result in linear matrix systems. In addition we discuss parallelization of the FFE models using domain decomposition and empirical derivation of non-linear forces to account for material non-linearities.

2 Theory

In this section we describe the model which we use to simulate elastic deformation of a volumetric solid in real-time [3]. To develop the model we formulate a number of requirements that the model should fulfil:

1. Speed is everything. Deformation should be calculated in the smallest amount of time possible.
2. We do not care about the time taken for one-time pre-calculation such as setting up equations, inverting matrices etc. If something takes 24 hours extra in the pre-calculation stage, but will save 0.01 second in the simulation stage, we should do it.

3. The elastic model should be *visually* convincing. The model may be physically incorrect if it looks right.
4. In the long run we want to be able to make cuts in the model to accommodate surgical procedures. This involves changing the topology of the model and most importantly requires models that have defined interiors, ie. volumetric models.

In particular the last requirement lead us to select mesh-based 3D Finite Element (FE) models. The alternative would be parametric models such as [6]. But these models do not provide the needed freedom to perform topology changes to allow cutting. Although some of the models can handle simple cuts we aim towards being able to make completely general cuts in the models. We are convinced that only mesh-based models will allow this.

To meet the first two requirements we choose the linear elastic deformation model which is also known as Hooke's law. Using linear elasticity as the basic model involves a number of assumptions regarding the physical material that is modeled. Most importantly linear elastic models are only valid for very small deformations and strains. They are typically correct for such rigid structures as metal beams, buildings etc. Although they are used extensively in modeling, the visual result of large deformation modeling using linear elasticity is seldom satisfactory

But when used with FE these models lead to linear matrix systems $\boldsymbol{K}\underline{\boldsymbol{u}} = \underline{\boldsymbol{f}}$ which are easy to solve and fast. There is, therefore, a trade-of between the speed of the system and the visual deformation result.

Linear elastic models are used here because modeling general elastic volumetric deformation using FE is only *just* possible with todays computers. With faster computers in the future we expect more realistic models, such as incompressible Mooney-Rivlin material models [4], to be used.

2.1 Condensation

The linear matrix system $\boldsymbol{K}\underline{\boldsymbol{u}} = \underline{\boldsymbol{f}}$ models the behaviour of the *solid* object. This includes both surface nodes as well as the internal nodes of the model. But for simulation purposes we are usually only interested in the behaviour of the surface nodes since these are the only *visible* nodes. We, therefore, use condensation [9] to remove the internal nodes from the matrix equation.

The matrix equation for the condensed problem has the same size as would result from a FE *surface* model. But, it is important to understand that it shows *exactly* the same behaviour for the surface nodes as the original *solid volumetric* system.

Without loss of generality, let us assume that the nodes of the FE model have been ordered with the surface nodes first, followed by the internal nodes. Using this ordering we can rewrite the linear system as a block matrix system (*s*urface / *i*nternal):

$$\begin{bmatrix} \boldsymbol{K}_{ss} & \boldsymbol{K}_{si} \\ \boldsymbol{K}_{is} & \boldsymbol{K}_{ii} \end{bmatrix} \begin{bmatrix} \underline{\boldsymbol{u}}_s \\ \underline{\boldsymbol{u}}_i \end{bmatrix} = \begin{bmatrix} \underline{\boldsymbol{f}}_s \\ \underline{\boldsymbol{f}}_i \end{bmatrix} \tag{1}$$

From this block matrix system we can create a new linear matrix system $\boldsymbol{K}^*_{ss}\underline{\boldsymbol{u}}_s = \underline{\boldsymbol{f}}^*_s$ which only involves the variables of the *surface* nodes:

$$\boldsymbol{K}^*_{ss} = \boldsymbol{K}_{ss} - \boldsymbol{K}_{si}\boldsymbol{K}_{ii}^{-1}\boldsymbol{K}_{is} \qquad \underline{\boldsymbol{f}}^*_s = \underline{\boldsymbol{f}}_s - \boldsymbol{K}_{si}\boldsymbol{K}_{ii}^{-1}\underline{\boldsymbol{f}}_i \tag{2}$$

The displacement of the internal nodes can still be calculated using $\underline{\boldsymbol{u}}_i = \boldsymbol{K}_{ii}^{-1}(\underline{\boldsymbol{f}}_i - \boldsymbol{K}_{is}\underline{\boldsymbol{u}}_s)$ Notice, that if no forces are applied to internal nodes, $\underline{\boldsymbol{f}}^*_s = \underline{\boldsymbol{f}}_s$.

Generally the new stiffness matrix will be dense compared to the sparse structure of the original system. But, since we intend to solve the system by inverting the stiffness matrix in the pre-calculation stage, this is not important. Without loss of generality, we will understand that both the original system and the condensed system $\boldsymbol{K}^*_{ss}\underline{\boldsymbol{u}}_s = \underline{\boldsymbol{f}}^*_s$ can be used when the following text refers to the original system $\boldsymbol{K}\underline{\boldsymbol{u}} = \underline{\boldsymbol{f}}$.

2.2 Solving $\boldsymbol{K}\underline{\boldsymbol{u}} = \underline{\boldsymbol{f}}$ using Selective Matrix Vector Multiplication

Formally, solving the linear matrix system using the inverted stiffness matrix is performed using $\underline{\boldsymbol{u}} = \boldsymbol{K}^{-1}\underline{\boldsymbol{f}}$. If only a few positions of the force vector are non-zero, clearly standard matrix vector multiplication would involve a large number of superfluous multiplications. We note that

$$\underline{\boldsymbol{u}} = \boldsymbol{K}^{-1}\underline{\boldsymbol{f}} = \sum_i \boldsymbol{K}_{*i}^{-1}\underline{\boldsymbol{f}}_i \tag{3}$$

where $\boldsymbol{K}_{*i}^{-1}$ is the i'th column vector of $\boldsymbol{K}^{-1}$ and $\underline{\boldsymbol{f}}_i$ the i'th element of $\underline{\boldsymbol{f}}$. Since the majority of the $\underline{\boldsymbol{f}}_i$ are zero, we restrict i to run through only the positions of $\underline{\boldsymbol{f}}$ for which $\underline{\boldsymbol{f}}_i \neq 0$'[3]. If n of the N positions in $\underline{\boldsymbol{f}}$ are non-zero this will reduce the complexity to $o(n/N)$ times the time of a normal matrix vector multiplication. We call this approach *Selective Matrix Vector Multiplication* (SMVM).

2.3 Parallelization $\boldsymbol{K}\underline{\boldsymbol{u}} = \underline{\boldsymbol{f}}$ using domain decomposition

Let us assume that the domain of the solid Ω has been decomposed into a number of non-overlapping sub-domains Ω_i with a common boundary Γ.

If we order the nodes of the global stiffness matrix $\boldsymbol{K}$ with the nodes of the boundary Γ first followed by sections of nodes corresponding to the sub-domains Ω_i, we can use the condensation technique described in the previous section to separate computation of the individual sub-domains. Each sub-domain can now handled by one processor and the result assembled on a final processor. This way we are able to parallelize solution of the linear matrix system.

3 Simulation system

In this section we describe how we generate the FE mesh model of the physical organ, limb etc. and show the simulation system that we have implemented.

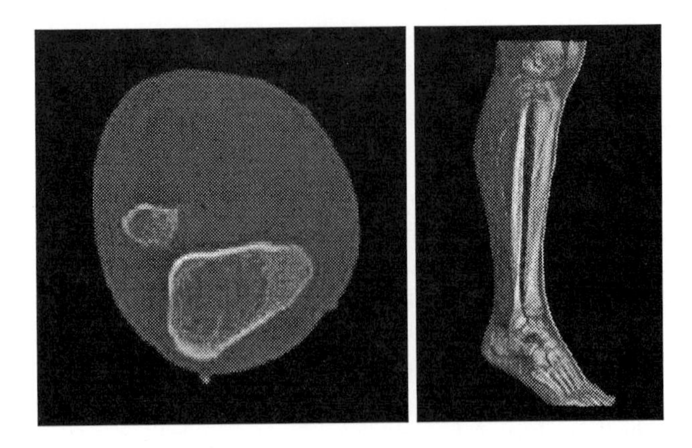

Fig. 1. Voxel data from the visible human data set.

In addition to a range of simple box-like structures we have used data from the Visible Human project [10] to make a model of the lower leg.

Since the Visible Human data set is voxel-based (see figure 1) it was necessary to generate a mesh model of it. To do this, we first used the Mvox software [1] to manually draw contours on the boundary of the skin and bone in the voxel data. We then applied the Nuages software [7] to create a 3D tetrahedral mesh model of the leg. The result was the FE mesh model shown in figure 3.

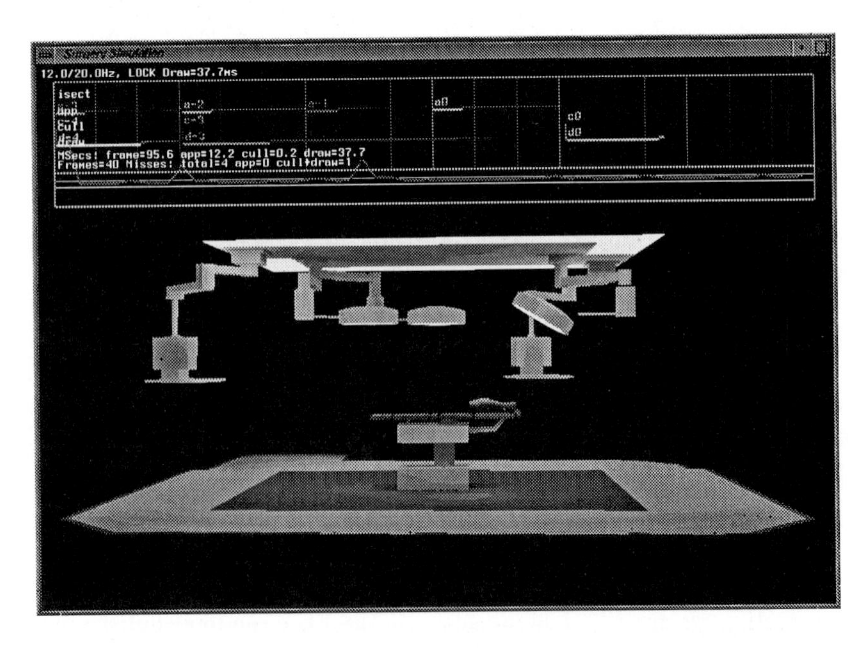

Fig. 2. Simulation system implemented using SGI Performer.

The simulation system has been implemented on an Silicon Graphics ONYX with four Mips R4400 processors using the SGI Performer graphics library. SGI Performer helps the programmer create parallel pipe-lining software by providing the basic tools for communication, shared memory etc.

Figure 2 shows a screen dump with the Virtual Operating room environment and the leg lying on the operating table. Figure 3 shows the surface of the FE mesh shown in the simulator.

4 Conclusion

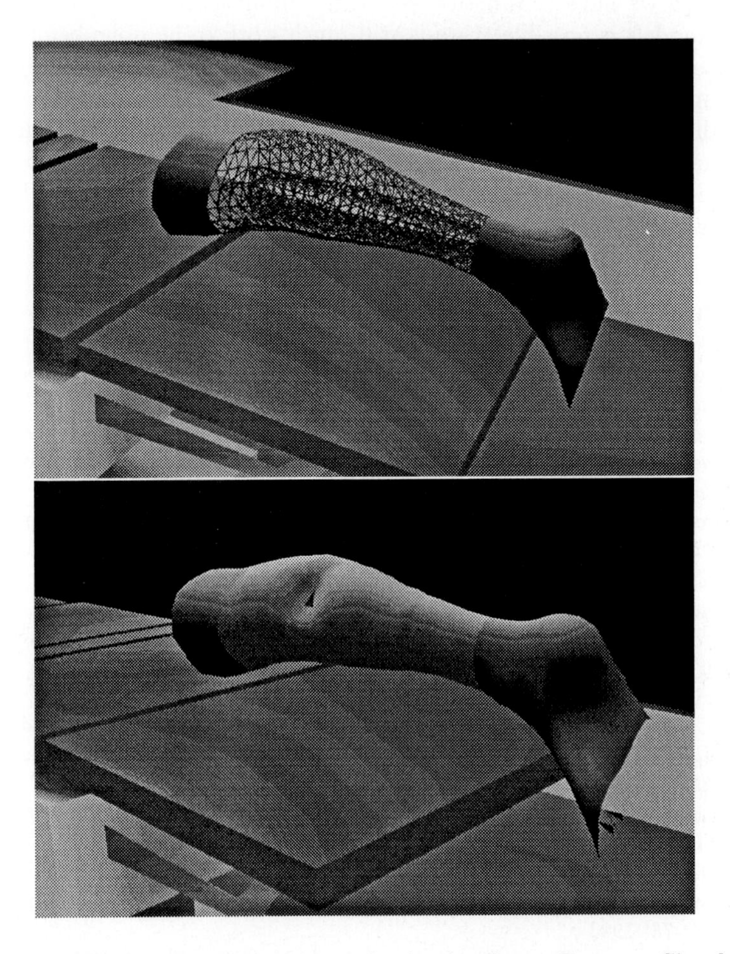

Fig. 3. Top: Wireframe model of lower leg in simulator. Bottom: Simulation of pushing on a the lower leg.

In this paper we have described a method for real-time simulation of elastic deformation of a volumetric solid based on linear elastic Fast Finite Flements

(FFE). We have discussed how solution of the linear matrix system $\boldsymbol{K}\underline{u} = \underline{f}$ can be implemented in parallel using domain decomposition. The simulation system we have developed for surgery simulation was finally described.

The example using a leg from the Visible Human data set with 700 system nodes (condensed system with only surface nodes) ran comfortably with a single processor using only 1/3 of a frame (20 frames/second) when forces were applied to 3 nodes. This included calculation of the deformation and also basic processing. So although both more nodes and more surface nodes with forces applied would increase the time requirement, we believe bigger models could be accommodated using the SMVM method.

Although we have shown that real-time simulation of solid volumetric deformable models is possible there is still much work to be done before realistic surgery simulation can be performed. Most importantly we are currently working on the implementation of cutting in a FE mesh.

In addition to more realistic tissue models we also need detailed segmentation of the organs, limbs etc. to allow different material properties and models to be used. The current parallel research in digital atlasses such as the VoxelMan atlas [8] is a significant step in this direction.

References

1. M. Bro-Nielsen, *Mvox: Interactive 2-4D medical image and graphics visualization software*, submitted to CAR'96, 1996
2. M. Bro-Nielsen, *Modelling elasticity in solids using Active Cubes - Application to simulated operations*, Proc. Computer Vision, Virtual Reality and Robotics in Medicine (CVRMed'95), pp. 535-541, 1995
3. M. Bro-Nielsen and S. Cotin, *Real-time Volumetric Deformable Models for Surgery Simulation using Finite Elements and Condensation*, Proc. Eurographics'96, 1996
4. P.G. Ciarlet, *Mathematical elasticity, vol. I: Three-dimensional elasticity*, North-Holland, ISBM 0-444-70529-8, 1988
5. S. Cotin, H. Delingette, M. Bro-Nielsen, N. Ayache, J.M. Clément, V. Tassetti and J. Marescaux, *Geometric and Physical representations for a simulator of hepatic surgery*, Proc. Medicine Meets Virtual Reality, 1996
6. S.A. Cover, N.F. Ezquerra and J.F. O'Brien, R. Rowe, T. Gadacz and E. Palm, *Interactively deformable models for surgery simulation*, IEEE Computer Graphics & Applications, pp. 68-75, Nov. 1993
7. B. Geiger: *Three-dimensional modeling of human organs and its application to diagnosis and surgical planning*, INRIA Tech. Rep. 2105, Dec. 1993
8. K.H. Höhne, M. Bomans, M. Riemer,R. Schubert, U. Tiede and W. Lierse, *A 3D anatomical atlas based on a volume model*, IEEE Computer Graphics Applications, 12(4):72-78, 1992
9. H. Kardestuncer, *Finite element handbook*, McGraw-Hill, ISBN 0-07-033305-X, 1987
10. *The Visible Human Project*, WWW http://www.nlm.nih.gov/extramural_research.dir/visible_human.html

Real Time Volumetric Deformable Models for Surgery Simulation

Stéphane Cotin, Hervé Delingette and Nicholas Ayache

INRIA, EPIDAURE Group
2004, route des Lucioles
BP 93, 06 902 Sophia Antipolis Cedex, France
e-mail: Stephane.Cotin@sophia.inria.fr

Abstract. Surgical simulation increasingly appears to be a mandatory aspect of tomorrow's surgery. From today's gesture training, surgical simulators will evolve in order to perform surgical planning with the ability to rehearse the various steps of the operation, thus reducing operating time and risks. In this paper, we describe a virtual environment for surgical training and more specifically a model based on elasticity theory which conveniently links the shape of deformable bodies and the forces associated with the deformation while achieving real time performance.

1 Introduction

A recent advance in surgery is minimally-invasive techniques. In laparoscopic surgery, the basic technique is to make small incisions through which different surgical tools and a micro video camera can be inserted, avoiding the large incisions needed for open surgery. However, although this method has several advantages, the working conditions are substantially modified for the surgeon in comparison with a classical operation. The major difficulty lies in the hand-eye coordination. Therefore, training is a prerequisite and a simulator would assist surgeons by providing a new way to practice. Much recent work focuses on surgery simulation [4] [9] [2], etc. but none of them are able to simultaneously achieve the goals of real time computation and complex modeling of the biomechanical properties. Our simulator project attempts to develop a system for *gesture training* in hepatic surgery. The advantage of our method lies in the capacity to deform a model of an organ in real-time, using realistic volumetric models of non-homogeneous elastic tissues and including force feedback. This is achieved by using a volumetric mesh, which in addition is well suited for the simulation of tissue cutting.

2 Force Feedback

Recent work has shown that the sense of presence in virtual environments is highly correlated with the degree of perception of that environment [1]. In particular, the sensation of forces and the sensation of textures are very important in medical applications. Since in minimally invasive surgery the surgeon's hands remain outside of the patient's body, we do not need tactile sensing, but only force sensing. The realism of the sensation of forces is highly dependent on the physical realism of the model. Our system, uses a volumetric mesh with non-homogeneous elasticity. We believe this to be necessary for physical realism. The flow of information in the simulator should form a closed loop [6]: the model deforms according to the surgeon induced motion of the force feedback device, this deformation allows us to compute the force of contact and finally the loop is closed by generating this force through a mechanical transmission (see Fig. 3). The main difficulty is related to the real time constraint imposed by such a system: to remain realistic, the forces must be computed at a very high frequency, at least equal to 500 Hz and the deformations at about 50 Hz.

3 Linear elasticity

Most papers dealing with deformable models [11] [12] [8] use physically-based models or realistic dynamics [2] [3]. Linear elasticity is often used [10] [9] [3] as a way to obtain a good approximation of the behavior of a deformable body. The stress-strain relation provided by the theory of elasticity allows us to model various physical behaviors. These relations are importants in our situation since they will give a physical link between the deformation of the body and the force induced in the force feedback system. However, the use of a force feedback system for the interaction with the virtual organ implies several constraints. First, the boundary conditions are different than in a classical approach (a set of external forces applied to the model) since we cannot measure the forces exerted on the virtual organ. Second, the dynamic solving of the time-dependent differential equations do not allow for fast deformations with complex models.

3.1 Theory

We consider a volumetric deformable object and its configuration (shape) Ω before deformation. Under the action of a field of forces, this object will deform and will take a new configuration Ω^*. Then the problem consists of determining the displacement field $\vec{u}$ which associates to the position P of any point of the object before deformation, its position P^* in the final configuration. If we assume that the Euclidean space $\mathbb{R}^3$ is referred to an orthogonal system $(O, \vec{e_1}, \vec{e_2}, \vec{e_3})$, the displacement $\vec{u}(P)$, that describes the motion of any point P in the model, can be written as:

$$\vec{u}(P) = \overrightarrow{PP^*} = \sum_{i=1}^{3} u_i(x_1, x_2, x_3)\vec{e_i}. \tag{3.1}$$

From the knowledge of the displacement field $\vec{u}$, we deduce the components $\varepsilon_{jk}(\vec{u})$ of the *strain* tensor ε in linear theory to be:

$$\varepsilon_{jk} = \frac{1}{2}\left(\frac{\partial u_j}{\partial x_k} + \frac{\partial u_k}{\partial x_j}\right) \qquad j = 1,2,3 \quad k = 1,2,3. \tag{3.2}$$

The Hooke's law gives the components $\sigma_{jk}(u)$ of the *stress* tensor σ as:

$$\sigma_{jk} = \sum_{l=1}^{3} \sum_{m=1}^{3} E_{jklm} \varepsilon_{lm}(u). \tag{3.3}$$

For an isotropic material, the coefficients E_{jklm} are given by:

$$E_{jklm} = \lambda \delta_{jk} \delta_{lm} + \mu(\delta_{jl}\delta_{km} + \delta_{jm}\delta_{kl}) \quad \forall\, \varepsilon_{jk} = \varepsilon_{kj} \tag{3.4}$$

where the scalars λ and μ are the *Lamé coefficients* and δ the Kronecker symbol.

3.2 The finite element method

More and more researchers tend to use finite element methods to solve the equations governing deformable models [5] [8]. In our approach, we have used a classical finite elements scheme, i.e. with Lagrange tetrahedral elements of type P_1 (linear elements with zero order continuity, only involving the values of the function at the vertices of the element). The use of this class of elements implies the decomposition of the domain Ω into a set of tetrahedral elements. Through variational principles, the resolution of the problem of elasticity theory becomes equivalent to the resolution of a linear system $\boldsymbol{K}\,\boldsymbol{U} = \boldsymbol{F}$ where $\boldsymbol{K}$ is the stiffness matrix, $\boldsymbol{U}$ is the unknown displacement field and $\boldsymbol{F}$ the external forces. The size of the matrix $\boldsymbol{K}$ is $3\,N \times 3\,N$, where N is the number of vertices (or nodes) of the mesh, each having 3 degrees of freedom. Clearly, the size of the mesh (in terms of number of nodes) is an important parameter which influences the computation time. Since we want to model organs with anatomical precision, it appears that the resolution of such a linear system cannot be computed in real time without the implementation of a speed up algorithm (see Section 4).

3.3 Boundary conditions

Let u be the displacement field and suppose we know the value of the displacement of the node i on the boundary: $\overrightarrow{U} = \{U_1, U_2, U_3\}$. Practically, $\overrightarrow{U}$ corresponds to the displacement of the node i of the model in contact with the virtual tool. Then, to take into account this boundary condition in the resolution step of the linear system, we modify the matrix $\boldsymbol{K}$ in the following way: $K_{3i+j,k} = 0$, for $j \in \{1,2,3\}$, $k \in \{1,..,n\}$ and $k \neq 3i+j$. $K_{3i+j,3i+j} = 1$ for $j \in \{1,2,3\}$. In matrix $\boldsymbol{F}$, the following substitution is done: $F_{3i} = U_1$, $F_{3i+1} = U_2$ and $F_{3i+2} = U_3$. However, although the previous boundary conditions are well suited for mechanical systems, they are usually too much restrictive for anatomical modeling. For this reason, we have added another boundary condition that can be viewed as attaching a spring of infinite length to the node i (see Fig. 2). According to the stiffness k of the spring, we can simulate *weak* or *strong* constraints, i.e. the node i can move off freely or not from its equilibrum position. Many of the ligaments attached to an organ can be modelled this way. This constraint, expressed as an external force $\overrightarrow{f}$, can be written: $\overrightarrow{f} = k(\overrightarrow{v} \cdot \overrightarrow{n})\overrightarrow{n}$, where k is the spring stiffness, $\overrightarrow{n}$ is a normalized vector in the direction of the spring and $\overrightarrow{v}$ the displacement of the node i. This relation is valid under the assumption $\|\overrightarrow{l_0}\| \gg \|\overrightarrow{u}\|$ ($\|\overrightarrow{l_0}\|$ is the spring rest length) and governed by the necessary linearity of the previous expression in order to achieve real time deformations (see Section 4).

3.4 Non homogeneous material

Although we have poor knowledge at the biomechanical properties of most organs, we are aware that linear elasticity is just an approximation. However, to take into account some anatomical characteristics in the deformation process, we use non-homogeneous elasticity. At the present time, we use *a priori* knowledge to set the elastic coefficient of any node in the volumetric model, according to the grey levels in the 3D medical image. Nevertheless, some works [7] suggest it will be possible to compute these parameters from medical image modalities.

4 Real time deformations

The degree of realism required in surgical simulation necessitates a complex model of the organ. The complexity (number of vertices) of the mesh has a direct impact on the size of the matrices involved in the linear system $\boldsymbol{KU} = \boldsymbol{F}$. Then the computation time required for resolution of the system becomes too high for real-time deformation of the mesh. To speed up the interactivity rate, we take advantage of the linearity of the equations combined with a pre-processing algorithm. This pre-processing algorithm can be described as follow:

- set all the constrained nodes of the model
- for each free node n on the surface of the mesh, apply an elementary displacement (dx, dy, dz) and compute:
  - the volumetric displacement field corresponding to this deformation.
  - store the displacement of every free node in the mesh as a set of 3×3 tensors T_{nk} expressing the relation between the displacement of node k ($k \neq n$) in the mesh and the elementary displacement of the node n.
  - compute the components of the elementary force f_n.
  - store this result as a 3×3 tensor F_n.

To decrease the memory required for storage of the tensors, we keep only the significant displacements i.e. tensors such that: $\lambda >$ threshold, with $\lambda = max\{|\lambda_1|, |\lambda_2|, |\lambda_3|\}$ where $\lambda_1, \lambda_3, \lambda_3$ are the eigenvalues of the tensor T_{nk}. In the case of a liver model, composed of 940 nodes, a reduction of 60% was obtained this way. Finally, the computation of the displacement field and the reaction forces are reduced to a linear combination, according to the displacement $(\delta x, \delta y, \delta z)$ imposed at a set of nodes on the surface of the mesh. After that, it becomes possible to deform, *interactively*, the model of any object, whatever its complexity (see Fig. 1).

5 Experiments

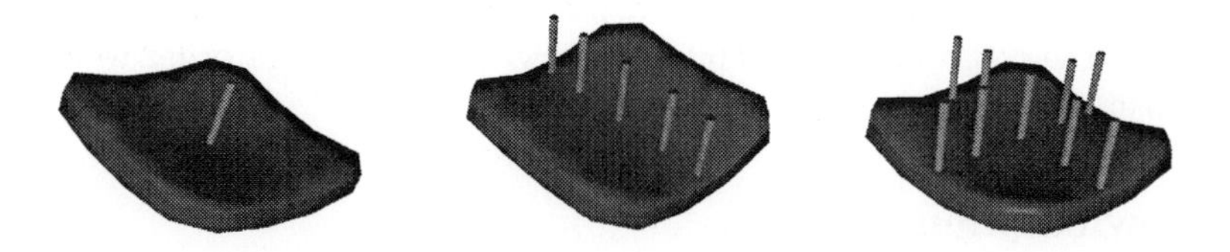

Figure 1. Deformation of a plate. The nodes on corners are rigidly fixed. (*Left*) the plate is deformed under a simple contact. (*Middle*) and (*right*) several constraints are combined to give the final deformation, in real-time.

a) high spring stiffness b) low spring stiffness

Figure 2. The left and right sides of the plate are linked to springs of various stiffness. In *a)* the deformation is close to what occurs with fixed nodes whereas in *b)* the deformation is smaller but coupled with a translation in the direction of the tool.

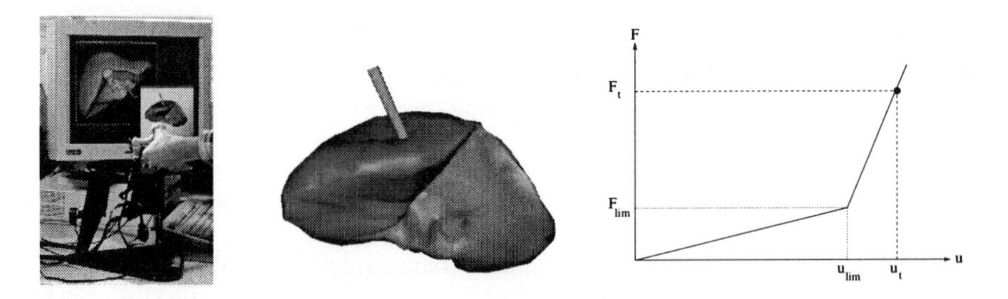

Figure 3. (*Left*) the surgeon moves a virtual tool via the force feedback device (*middle*) after collision detection the model of the organ deforms in real-time and (*right*) a non-linear reaction force is computed and sent back to the force feedback system.

6 Conclusion - Perspectives

In this paper we have proposed a method for deforming in real time a volumetric model of an object. Because our models are physically-based, they respond in a natural way to the interactions provided by the user. These interactions are guided via a force feedback device, and elasticity theory allows computation of the contact forces which are input in the mechanical system.

We are now working on dynamical tissue cutting simulation, using spring-masses models in combination with the method presented here. We also plan to add a dynamical behavior to our deformations in order to incorporate the velocity of the tool when it gets into touch with the organ in order to increase the realism of deformations.

Acknowledgment

The authors are grateful to J. Marescaux, J.M. Clément, V. Tassetti and R. Wolfram from *IRCAD* (Institut de Recherche sur les Cancers de l'Appareil Digestif, Strasbourg, France) for their collaboration in building the liver model. We are also grateful to Lewis Griffin for proof reading this paper.

References

1. W. Barfield and C. Hendrix. *Interactive Technology and the New Paradigm for Healthcare*, chapter 4: Factors Affecting Presence and Performance in Virtual Environments, pages 21–28. IOS Press, 1995.
2. M. Bro-Nielsen. Modelling elasticity in solids using active cubes - application to simulated operations. In *Computer Vision, Virtual Reality and Robotics in Medecine*, volume 905 of *Lecture Notes in Computer Science*, pages 535–541. Springer, 1995.
3. D. Chen and D. Zeltzer. Computer Animation of a Biomechanically Based Model of Muscle using the Finite Element Method. In *Computer Graphics*, volume 26, No 2. ACM SIGGRAPH'92, 1992.
4. S. A. Cover, N. F. Ezquerra, and J. F. O'Brien. Interactively Deformable Models for Surgery Simulation. *IEEE Computer Graphics and App.*, pages 68–75, 1993.
5. J. P. Gourret, N. Magnenat-Thalmann, and D. Thalmann. Modeling of Contact Deformations between a Synthetic Human and its Environment. In *Computer Aided Design*, volume 23 No 7, pages 514–520, September 1991.
6. B.G. Jackson and L.B. Rosenberg. *Interactive Technology and the New Paradigm for Healthcare*, chapter 24: Force Feedback and Medical Simulation, pages 147–151. IOS Press, 1995.
7. R. T. Mull. Mass estimates by computed tomography: physical density from CT numbers. *American Journal of the Roentgen Ray Society*, 143:1101–1104, November 1984.
8. J. C. Platt and A. H. Barr. Constraint Methods for Flexible Models. In *Computer Graphics (SIGGRAPH '88)*, volume 22 No 4, pages 279–288, 1988.
9. G. J. Song and N. P. Reddy. *Interactive Technology and the New Paradigm for Healthcare*, chapter 54: Tissue Cutting in Virtual Environment, pages 359–364. IOS Press, 1995.
10. Thomas H. Speeter. Three Dimensional Finite Element Analysis of Elastic Continua for Tactile Sensing. *The International Journal of Robotics Research*, 11 No 1:1–19, February 1992.
11. D. Terzopoulos, J. Platt, A. Barr, and K. Fleisher. Elastically Deformable Models. In *Computer Graphics (SIGGRAPH '87)*, volume 21 No 4, pages 205–214, July 1987.
12. K. Waters and D. Terzopoulos. A Physical Model of Facial Tissue and Muscle Articulation. *IEEE*, 1990.

Craniofacial Surgery Simulation

Erwin Keeve [1], Sabine Girod [2], Bernd Girod [1]

[1] Telecommunications Institute, University of Erlangen-Nuremberg
Cauerstr. 7, 91058 Erlangen, Germany, {keeve, girod}@nt.e-technik.uni-erlangen.de
[2] Department of Oral and Maxillofacial Surgery, University Erlangen-Nuremberg
Glückstr. 11, 91054 Erlangen, Germany, sgirod@nt.e-technik.uni-erlangen.de

Abstract: Craniofacial surgery requires careful preoperative planning in order to restore functionality and to improve the patient's aesthetics. In this article we present two different tissue models, integrated in an interactive surgical simulation system, which allow the preoperative visualization of the patient's postoperative appearance. We combine a reconstruction of the patient's skull from computer tomography with a laser scan of the skin, and bring them together into the system. The surgical procedure is simulated and its impact on the skin is calculated with either a *MASS SPRING* or a *FINITE ELEMENT* tissue model.

1. Introduction

Computer-based three-dimensional visualization techniques have made a great impact on the field of medicine in the last decade. Physical models of the skull can be created through computer-generated reconstruction using stereolithography on which planned surgery can be simulated, or individual implants produced. Such modern visualization methods also allow the use of surgery robots that can help with the exact positioning of surgical instruments, especially in neurosurgery. Because of their wide-ranging surgical impact, craniofacial operations require careful preoperative planning. The goal is not only to improve the functionality, but also to restore an aesthetically pleasing face. Currently, only limited prediction of the tissue appearance after a jaw realignment operation is possible. The relationship between bone and tissue movements can be found for forward and backward shifts of the lower jaw. However, in operations involving the upper jaw or other displacements of the lower jaw, there is presently no satisfactory method of predicting tissue changes *[Ste94]*.

In this paper we present two different physical-based tissue models, a *MASS SPRING* and a *FINITE ELEMENT* approach, which are applied to individual anatomical structures. Both models are integrated into our interactive surgical simulation system, described in *[Kee96a]*. These models allow a precise preoperative three-dimensional visualization of the patient's appearance after craniofacial surgery. We present simulation results of the *MASS SPRING* approach and discuss them with respect to first results of the *FINITE ELEMENT* model. Section 2 and 3 describe the data acquisition and the surgical simulation. Section 4, the *MASS SPRING* tissue model, which transfers the bone realignments to the skin surface. Section 5 describes the *FINITE ELEMENT* approach. Both approaches are discussed and compared with respect to the presented results in Section 6.

2. Data Acquisition

The *MARCHING CUBES* method is used to reconstruct the skull from computer tomography *[Lor87]*. This skull reconstruction provides the basis for the interactive surgical simulation, as described in Section 3. A Cyberware scanner obtains the geometry of the patient's skin surface. By providing texture information as well, this 3D laser scanner makes an additional contribution to the photorealistic appearance of the face. After data acquisition, an adaptive reduction technique is used to reduce the size of both datasets up to 80% without sacrificing the visible detail of the objects *[Kee96b]*.

3. Surgical Simulation

In contrast to some systems that allow only predetermined surgical procedures, our system provides the flexibility to apply any craniofacial operation on the bone structure and manipulate it at will. These changes are made with a *CUTTING PLANE* with which the bone is split in half. Afterwards, each bone segment is defined as a unique object that can be further manipulated. The *CUTTING PLANE* can be changed in shape and orientation to enable the simulation of different procedures. As an example of a surgical procedure, Figure 1 shows the simulation of a Dal-Pont Osteotomy, which splits the ascending branches of the mandible. The interactive simulation is analogous to a real operation and can be accomplished on a common graphics workstation in approximately 10 minutes.

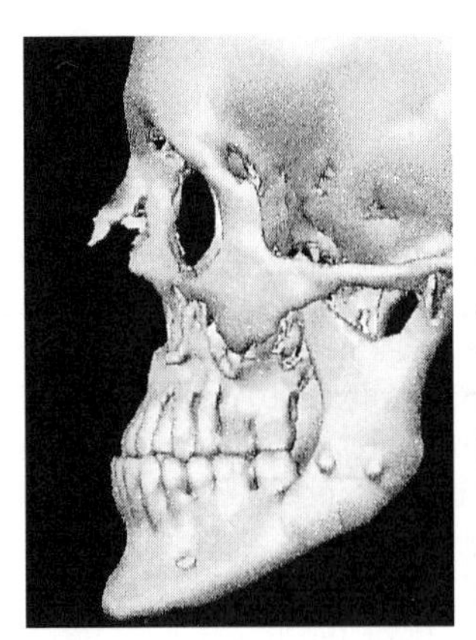

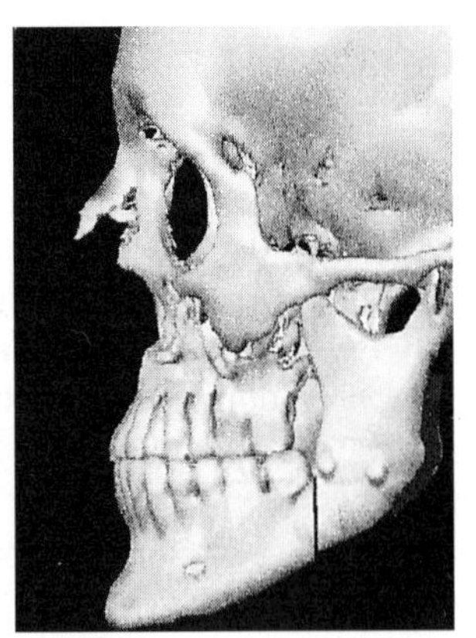

Preoperative bone structure

Simulated bone realignment

Figure 1: Simulated bone realignment

4. *MASS SPRING* Tissue Model

A layered *MASS SPRING* tissue model for facial animation was first introduced by Waters in 1992 *[Wat92]*. In order to use this model for surgical simulation, it was modified so that individual patient skin and bone structures are taken into account. The basic element of this model is shown in Figure 2. The different mechanical properties of the homogeneous layers and surfaces are represented by a *MASS SPRING* model. To approximate the non-linear stress-strain relationship of human tissue each spring constant is defined biphasic. To consider the incompressibility of human tissue, a *VOLUME PRESERVATION FORCE* was added to the model. Also an additional force is added to each basic element to consider the natural prestress of human tissue. The mathematical formulation of such a model is described by Newton's motion equation. For each node, a differential equation of second degree can be given that depends on mass, damping and the sum of all spring forces affecting it. To simulate the dynamics of the spring network, the motion equations are numerically integrated over time. The simulation is finished when all nodes meet minimal velocity and minimal acceleration criteria.

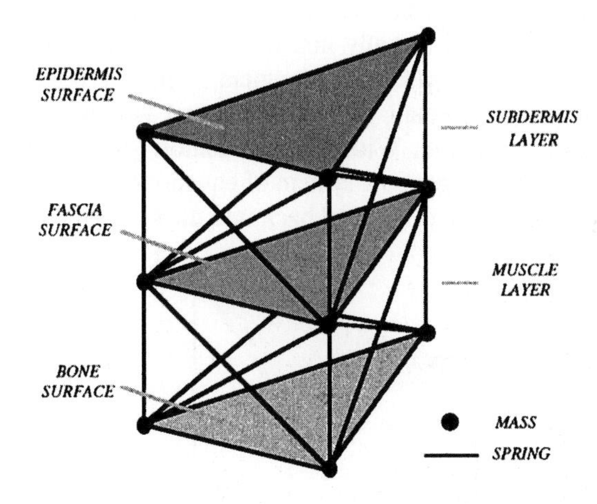

Figure 2: Basic tissue element of *MASS SPRING* model

Figure 4a and b show the preoperative scan and the simulated postoperative appearance of the patient. The actual postoperative scan is given for purposes of comparison in Figure 4d. Considering 3,080 basic tissue elements, this simulation of the tissue changes takes approximately one minute using a SGI High Impact workstation.

5. *FINITE ELEMENT* Tissue Model

Anatomy-based models are being developed in different fields of medicine that simulate the physical properties of human soft tissue and skeleton. The goal of these efforts is the computer-based realistic simulation of the deformation of soft tissue or the skeleton, such as in plastic surgery or in minimally invasive surgery. The basis of such simulations is often the finite element method which is increasingly being applied in the field of computer graphics and animation *[Bat82]*. The basic idea of the finite element method is that a continuum is approximated by dividing it into a mesh of discrete elements. Different element types are appropriate for different applications; in our system we use a grid of *SIX NODE PRISMS*, as shown in Figure 3, to model a layer of human tissue. Such an element is defined by its corner nodes. In our displacement-based

FINITE ELEMENT approach, given displacements are specified for certain nodes and the resulting positions of the other nodes are calculated. The given displacements cause strain, which is related to stress through Hooke's law, which in turn creates internal forces. In order to bring these forces into equilibrium, a system of differential equations is solved delivering the displacements of the unconstrained nodes.

We have developed an interface between our surgical simulation system and the finite element package *DIFFPACK [Lan95]*. The interface provides *DIFFPACK* with a tissue model, which includes element coordinates, their boundary conditions and their given displacements due to bone realignment. Our *DIFFPACK* application uses a linearly elastic, time-independent, isotropic formulation to calculate all resulting displacements. In this first application, we assume the stress-strain relationship to be linearly elastic and isotropic. Human tissue is actually non-linearly elastic and anisotropic, and these improvements are currently under development. The formulation is also time-independent because we have small displacements and we are interested in the static result, not in dynamic animation. With the information given by the interface and the above formulation, *DIFFPACK* sets up a system of equations and solves it, providing the final displacements. These results are read back into the surgical simulation system.

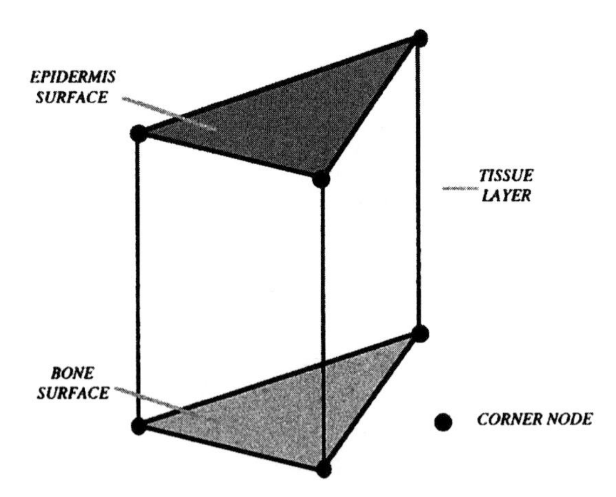

Figure 3: Basic tissue element of *FINITE ELEMENT* model

As an example, we used the same patient data as in the *MASS SPRING* approach. This provided 3,080 *SIX NODE PRISM* elements to the linear *FINITE ELEMENT* model. We again simulated an operation of cutting the frontal mandible and pushing it backwards. Figure 4a and c show the patient's skin before surgery and the *FINITE ELEMENT* simulation. It took one minute to set up the initial configuration using a SGI High Impact workstation. The finite element calculations were done in about 10 minutes.

6. Comparison

The *MASS SPRING* approach can not model the exact physical properties of human tissue, instead the physical behaviour has to be expressed with tools of masses and springs. However, the presented results show how realistic this model simulates tissue changes according to bone realignments. On the other hand, the *FINITE ELEMENT* approach can model the physical properties of human tissue more precisely. But the *FINITE ELEMENT* simulation is time consuming and cannot be done interactively using common workstations. Therefore we plan to combine both methods: The *MASS SPRING* approach gives the surgeon the ability to realize quick interactive simulations of tissue changes to improve his planning process. If one simulation does not yield his expectations, he is able to undo it and try another operation. On the other hand, we provide a more precise *FINITE ELEMENT* model which considers the exact physical properties of human tissue. This model can be used off-line, to verify the chosen surgical procedure.

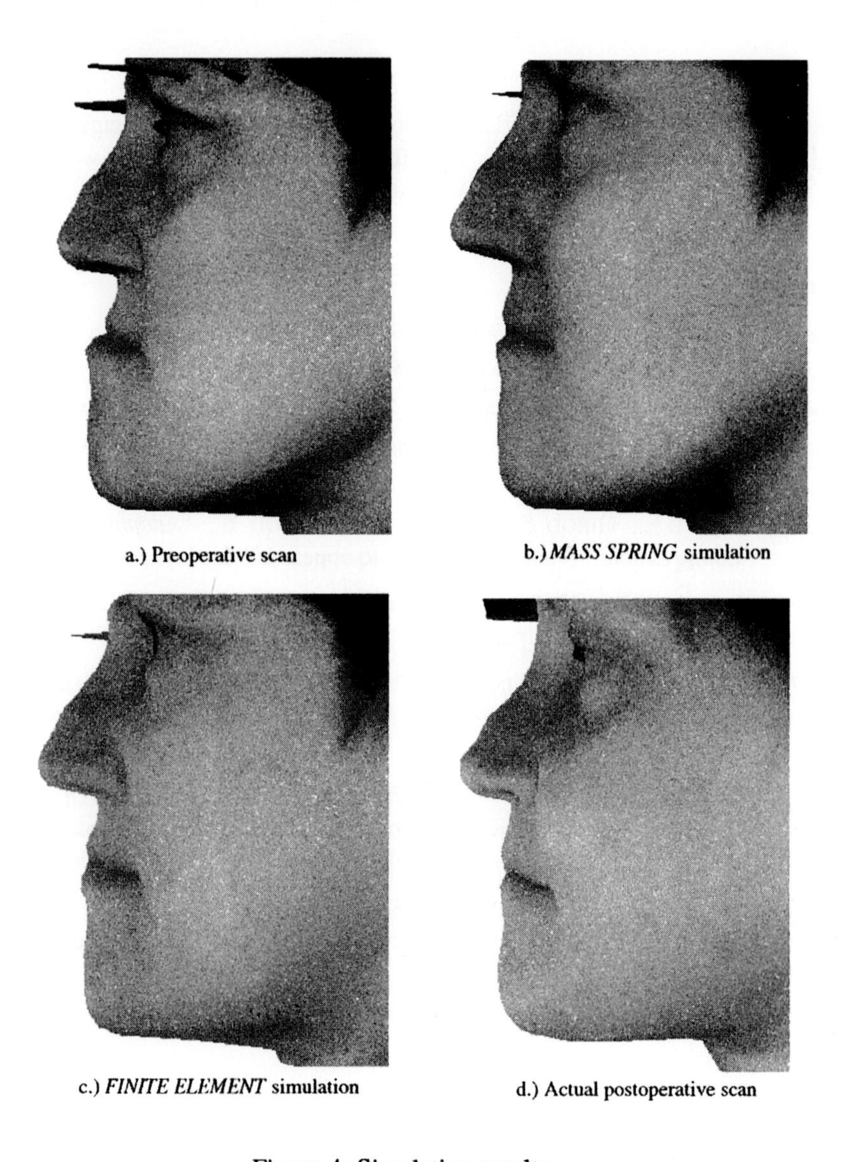

a.) Preoperative scan

b.) *MASS SPRING* simulation

c.) *FINITE ELEMENT* simulation

d.) Actual postoperative scan

Figure 4: Simulation results

7. Conclusions

The presented tissue models, which are both integrated into our surgical simulation system, allow the prediction of the postoperative appearance of a patient after craniofacial surgery. The *MASS SPRING* tissue model has already been used for several case studies at the Department of Oral and Maxillofacial Surgery at our university and has aided the preoperative planning procedure. To improve our system we also developed a *FINITE ELEMENT* model which describes the exact physical behaviour of the human tissue. The presented first results of the *FINITE ELEMENT* model show the potential of this new approach which is currently under further improvement.

Acknowledgements: This work is supported by Deutsche Forschungsgemeinschaft (DFG) research grant Gi 198/2-1. The finite element calculations were done by *DIFFPACK*, which was developed by *SINTEF* and the University of Oslo.

References:

[Bat82] BATHE K.-J., *"Finite Element Procedures in Engineering Analysis,"* Prentice Hall, Englewood Cliffs, New Jersey, 1982.

[Kee96a] KEEVE E., GIROD S., PFEIFLE P., GIROD B., *"Anatomy Based Facial Tissue Modeling Using the Finite Element Method,"* Proc. Visualization '96, San Francisco, CA, Oct. 27 - Nov. 1, 1996.

[Kee96b] KEEVE E., GIROD S., SCHALLER S., GIROD B., *"Adaptive Surface Data Compression,"* Signal Processing, to appear 1996.

[Lan95] LANGTANGEN H.-P., *"A Solver for the Equations of Linear Thermo-Elasticity,"* The Diffpack Report Series, SINTEF, June 24, 1995.

[Lor87] LORENSEN W. E., CLINE H. E., *"Marching Cubes: A High Resolution 3D Surface Construction Algorithm,"* ACM Computer Graphics, Vol 21, No. 4, pp. 38-44, July 1987.

[Ste94] STEINHÄUSER E. W., *"Weichteilvorhersage bei bimaxillären Operationen,"* Fortschritte in der Kiefer- und Gesichts-Chirurgie, Sonderband: Die bimaxilläre Chirurgie bei skelettalen Dysgnathien, Thieme-Verlag, 1994.

[Wat92] WATERS K., *"A Physical Model of Facial Tissue and Muscle Articulation Derived from Computer Tomography Data,"* SPIE Visualization in Biomedical Computing, Volume 1808, pp. 574-583. 1992.

Creation and Validation of Patient Specific Anatomical Models for Prostate Surgery Planning Using Virtual Reality

Paul A. Kay, Richard A. Robb PhD, Robert P. Myers MD, Bernie F. King MD

Mayo Clinic and Foundation, Rochester, MN 55901, USA

Abstract. Prostate surgery to remove cancer can be associated with morbidity due to complex and variable individual anatomy. We have developed and demonstrated the ability to accurately extract patient-specific anatomic regions from MRI pelvic scan data of pre-prostatectomy patients and to effectively convert these volume images to faithful polygonal surface representations of anatomic models for use in interactive virtual surgical planning. Models of the prostate gland and surrounding structures from several patients were created then evaluated by a radiologist and a surgeon who found them to be faithful and consistent with real anatomy. The surgeon provided confirming post-operative validation of the patient-specific models in planning radical prostatectomies.

1. Introduction

Serious morbidity may follow radical prostatectomy. Attempts to spare the external urinary sphincter and neurovascular bundles are confounded by the complexity and variability of individual pelvic anatomy. Surgical simulation in an immersive environment using virtual reality (VR) systems offers promise for effective pre-operative understanding of individual patient anatomy and realistic assessment of candidate surgical scenarios. We have developed methods of extracting patient-specific anatomical models from MRI volume images of the pelvic region in a format compatible with currently available VR modeling systems [1]. These patient-specific models are used in a 3-D interactive surgical planning environment by surgeons. We have created patient-specific models for several radical prostatectomy patients and a urologic surgeon has validated the faithfulness and usefulness of the patient specific models for presurgical planning.

2. Background

The anatomical variability and the presence of adjacent vital structures, including the external urinary sphincter and neurovascular bundles, renders surgery on the prostate gland a delicate and challenging procedure. There is significant risk of morbidity in the form of urinary incontinence and impotence [2]. The crux of the practice dilemma [3] is the need for improved non-invasive pre-operative techniques that can more precisely assess the exact anatomical relationship of the prostate gland to the bladder, membranous urethra and neurovascular bundles, and thereby more clearly indicate the optimal approach to the surgery. Pre-operative 3-D visualization of the *in situ* prostate gland and its anatomic environment will give surgeons a detailed understanding of the individual patient's specific anatomy and thereby potentially decrease the rate of morbidity and the time required in the operating room. Such visualizations are feasible using high-end VR graphics workstations. These systems can achieve real-time interactive speeds by using relatively sparse geometric surface descriptions of objects to be displayed. There is a need to develop techniques to accurately segment the anatomical data from the MRI scan and create patient specific models of this data. These models need to be validated in comparison to what a surgeon actually sees during an operation, and for their usefulness to the surgeon in planning an effective surgical approach.

Volumetric MRI scan images contain much of the anatomic information the surgeon is likely to see during surgery. MRI scan protocols need to be developed which optimize the information obtained from the scans for use in 3-D model building. Once a scan protocol is developed the images must first be converted from sets of sequential 2-D sections traversing several structures in thin planes to specific 3-D objects with unambiguous contiguous surfaces. In order to use an interactive VR system to display these objects in real-time, the 3-D regions representing organs in the volumetric MRI scan must be transformed into geometric representations of their surfaces. Furthermore, these models must not only faithfully represent the actual anatomic structures in terms of size, shape and location, but the must also be sufficiently sparse to permit real-time display update rates for interactive manipulation. We have developed techniques to accurately extract anatomical surfaces from volumetric MRI scans and have developed methods to use two different algorithms [4,5] which produce faithful geometric models of pelvic anatomy suitable for virtual reality display and pre-operative prostate surgery planning.

3.1 Methods and Results

3.1.1 MRI Scan Optimization

Certain anatomical characteristics of the prostate gland and surrounding anatomy are important to visualize in developing a surgical plan for prostatectomies [6]. MRI is currently the best way to view soft tissue anatomy of the prostatic region, and it is the most accurate method to view and stage primary prostate tumors [7]. A Mayo surgeon has specified the structures and measures that he wants to view and apply preoperatively. A Mayo radiologist has optimized the scanning protocol (i.e., pulse sequences) to capture the relevant anatomy and pathology, but also to provide a 3-D image data set that can be accurately and reliably segmented in order to produce realistic models for surgical planning. The radiologist has evaluated the accuracy of the scan segmentations to ensure the validity of the resultant models. MRI scan optimization is perhaps the most critical step in the overall process. In some cases patient motion limited the quality of the images and hence the models.

3.1.2 Volume Segmentation and “Smoothing”

ANALYZE is a comprehensive software package developed in the Biomedical Imaging Resource at Mayo for interactive comprehensive display, manipulation and analysis of biomedical images [8] and which has been used in a variety of surgical simulation applications. Image editing techniques have been optimized for efficiency and accuracy for specific anatomical structures. Automated (3-D math morphology), semi-automated (guided 3-D region growing), and manual segmentation procedures may be used, depending on the structures and scan quality. Once the segmented volumes are derived from the volumetric MRI scans, several image processing techniques were utilized to "smooth" the surface to prepare for surface model generation with the minimum number of artifacts. The techniques used were different for each anatomical structures and were developed to maximize the faithfulness of each anatomic structure. Methods used included mathematic morphology, median filtering and lowpass filtering in various combinations and order. Care was taken to compare the faithfulness of the processed anatomic structures to the segmented structures. On average, an entire set of anatomical objects takes approximately two hours for a trained user to segment and prepare for modeling.

3.1.3 Images-To-Polygons Algorithms

The use of virtual reality for surgery planning requires real-time displays of realistic objects representing a patient's internal organs. However, the use of medical image data as the basis for patient-specific models within a virtual environment poses unique problems. While currently available rendering algorithms can generate photo-realistic images [8] from the rather dense volumetric data generated by medical imaging systems, ray tracing algorithms cannot sustain the visual update rate required for real time display, and most surface display algorithms generate polygonal surface representations with extremely high numbers of polygons [9].

To be successful, a medical VR system needs to accurately segment a desired object from volumetric data, detect its surface and generate a good polygonal representation of this surface from a limited polygonal "budget" (defined as the constraint on the number of polygons which can be rendered and displayed in real time). Currently available hardware is able to render and manipulate 30,000 - 50,000 complex polygons (a complex polygon includes shading, texture mapping and anti-aliasing) per frame at real time display rates. Many polygonization algorithms produce high resolution surfaces that may contain several million polygons. Current techniques for pruning these polygons (called "decimation") to a more manageable number are cumbersome and do not always yield satisfactory results.

We have used two different algorithms to create accurate surface models. The first algorithm was developed in Biomedical Imaging Resource at Mayo [4] and features the novel and useful ability to *pre-specify the* number of polygons to be used for the model. The algorithm has four main steps described in Figure 1. A surface representation is obtained from the segmented objects and is used to calculate a set of curvature weights. These curvature weights are used as the input to a 2-D Kohonen network for tiling. The adaptation of the network to the input vectors results in a display surface that preserves useful detail relative to the number of polygons used. The algorithm tiles smaller and long stringy objects well (See Figure 2a). The other algorithm [5] used to tile anatomical structures was developed at the University of Washington for CAD/CAM systems and was designed to use laser range data surface points as input. It is straightforward to convert volumetric image data to a set of surface points to be used in this algorithm. It is useful in tiling large objects and complex branching structures such as seen in Figure 2b. The user may specify the distance at which neighboring points are sampled which gives some control over the number of polygons produced, however, the number of polygons created with this algorithm may be to numerous to be used for real-time virtual reality environments, and requires a decimation step.

3.1.4 Model Validation

A urologic surgeon and a radiologist at Mayo have used and validated the faithfulness and usefulness of the patient specific models we generated by these algorithms. Several patients were scanned and individual anatomic models were produced for each of them. After surgery, the surgeon evaluated the models and their characteristics through visual inspection of images created from the models at several different orientations. The surgeon and the radiologist compared the information in the MRI scans and the models and found the models to convey accurate anatomical relationships. The anatomical structures that were useful to the surgeon were noted and more fully developed. The surgeon determined that the models were most true to what he encountered in the OR, and noted that "These [images of the models] are more useful than the MRI scans." Several different models from various patients in the study are compared in Figure 3. It can be observed from these images the variation in individual anatomy - the difference in the bladder relationship to the prostate, the location and extent of the tumor when it is localizable from the MRI scans, the shape and size of the prostate itself and the length of the membranous urethra (that part of the urethra that extends from the prostate to the penis).

4. Discussion

In order to perform virtual pre-operative surgical planning or surgery simulation using anatomic models, the models must first be validated by the surgeon. In an attempt to validate the accuracy and usefulness of patient specific pelvic anatomy models, we have worked closely with a Mayo urologic surgeon and Mayo radiologist in an attempt to optimize the input and output for the model building techniques. The ultimate utilization of the pre-operative planning and surgery simulation system will be determined by the confidence a surgeon has in the representations that we create of their patients. The surgeons have determined that the models we create are faithful representation of the information they require in a pre-surgical planning/simulation environment, and are exploring their use both pre-operatively and post-operatively.

References

1. Kay PA, Robb RA, King BF, Myers RP, Camp JJ: Surgical planning for prostatectomies using three-dimensional visualization and a virtual reality display system. Proceedings of Medical Imaging 1995, San Diego, California, February 26 - March 2, 1995.
2. Starney TA, McNeal JE: Adenocarcinoma of the prostate, Campbell's Urology, 6th ed., vol. 2, pp. 1159-1221, 1992.
3. Garnick MB: The dilemmas of prostate cancer. Scientific American, pp 72-81, April 1994.

4. Cameron BM, Manduca A, Robb RA: Surface generation for virtual reality displays with a limited polygonal budget. Proceedings International Conferences on Image Processing, October 1995.
5. H. Hoppe: Surface Reconstruction from unorganized points. Doctoral Dissertation, University Washington, 1994.
6. Myers RP: Practical pelvic anatomy pertinent to radical retropubic prostatectomy. AUA Update Series, vol. 13, np. 4, pp. 26-31, 1994.
7. Rifkin MD, Zerhouni EA, Gatsonis CA, Quint LE, Paaushter DM, Epstein JI, Hamper U, Walsh PC, McNeil BJ: Comparison of magnetic resonance imaging and ultrasonography in staging early prostate cancer. New Engl J Med, vol. 323, pp. 621-626, 1990.
8. Robb RA: A software system for interactive and quantitative analysis of biomedical images, In: Höhne KH, Fuchs H, and Pizer SM: 3D Imaging in Medicine, NATO ASI Series, vol. F60, pp. 333-361, 1990.
9. Lorensen WE, Cline HE: Marching Cubes: a high resolution 3-D surface construction algorithm. Computer Graphics 21(4), pp 163-169, 1987.

Figures

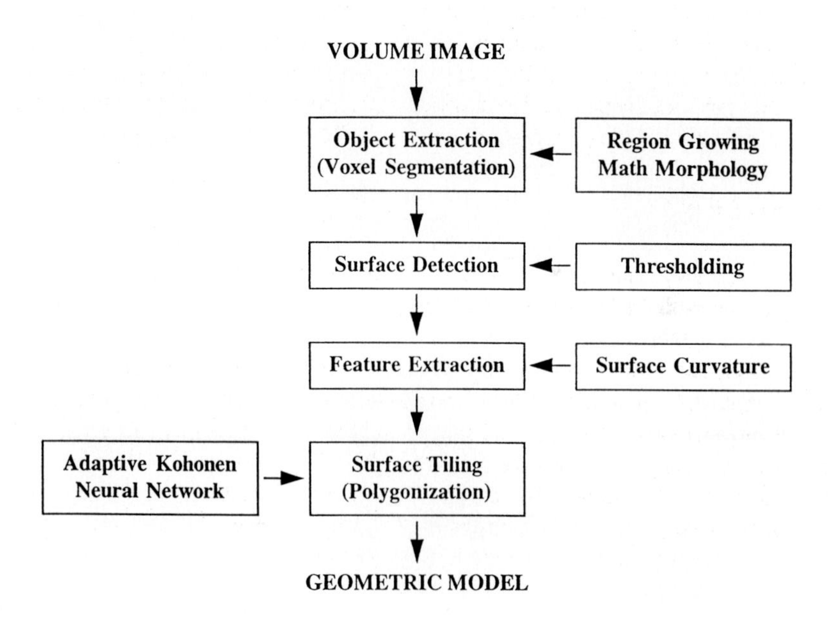

Fig. 1. Procedure for producing 3-D geometric models from volume images

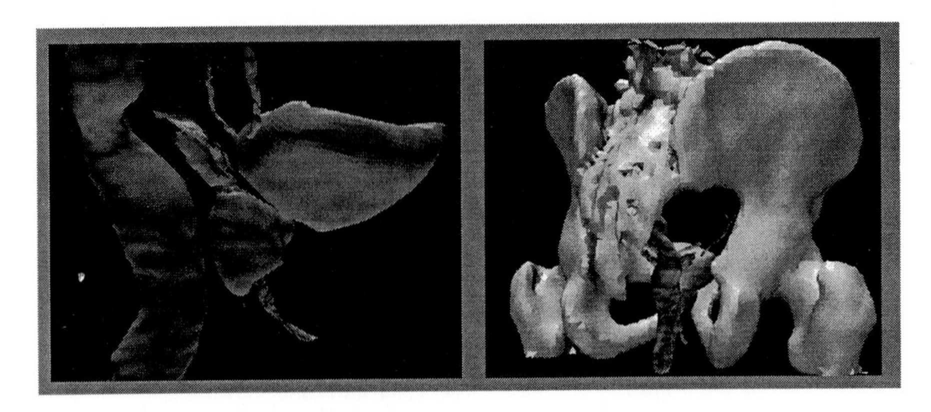

Fig. 2. A) Bladder, prostate, rectum and associated anatomical structures tiled by the Kohonen algorithm. B) Model of pelvic girdle produced by the Hoppe Algorithm. The ureters, urethra and other tubular structures were tiled by the Kohonen tiler.

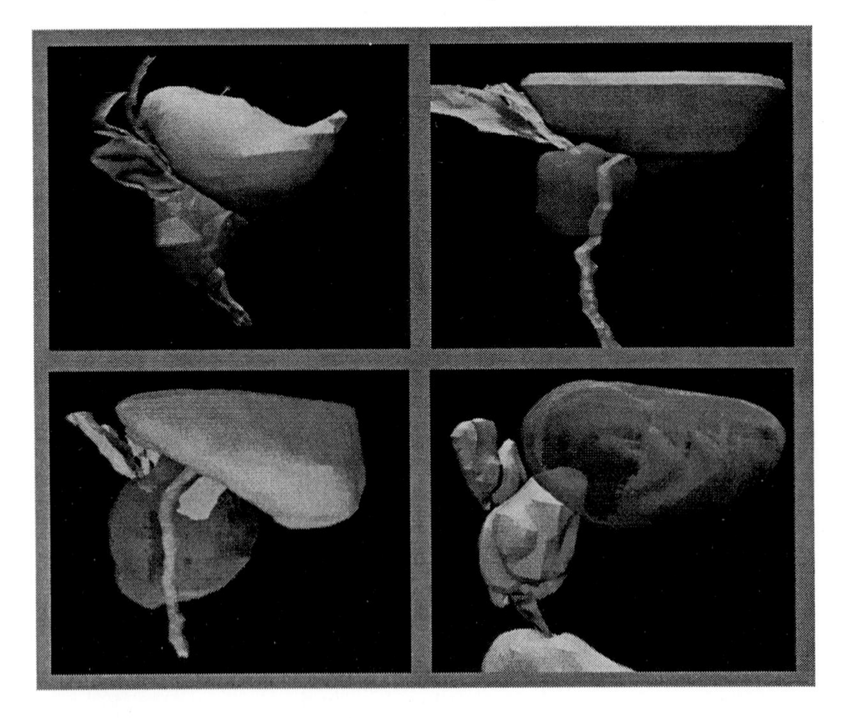

Fig. 3. Four different patients and individual anatomic models of the prostate gland and associated structures. A radiologist determines the existence and location of the tumor.

Use of Shape for Automated, Optimized 3D Radiosurgical Treatment Planning

J. D. Bourland[1, 2] and Q. R. Wu[2]

[1]Radiation Oncology, Wake Forest University, Winston-Salem, NC 27157, USA
[2]Radiation Oncology, Mayo Clinic, Rochester, MN 55905, USA

Abstract. Conformal treatment planning for radiosurgical dose delivery is challenging for irregularly shaped target volumes. Spheres of radiation dose, called "shots", must be positioned within a target image to yield an acceptable cumulative dose distribution. An automated solution has been developed that uses target shape, as represented by the 3D medial axis transform, to determine optimal shot positions, sizes, and number. These parameters are determined from successive 3D skeletons of unplanned target volumes. Essentially, target shape is replicated by the arrangement of shots comprising the 3D skeleton, forcing a conformal solution. Optimization is determined by the ratio of uncovered to total target volume and the smallest number of shots. Use of the medial axis transform reduces a 3D optimization problem to 1D with corresponding savings in computation time. Clinical cases in 3D demonstrate this approach as a replacement for manual planning.

1 Introduction

Multiple isocenters, or "shots", are used for gamma unit treatment to deliver a conformal dose to an irregular radiosurgical target. Each shot can be considered as a spherical distribution of dose centered at the convergence of 201 radiation beams at a particular coordinate in stereotactic space (Fig. 1). The planning problem is to cover a target volume with a collection of shots having the appropriate number, positions, and sizes, such that the prescription dose surface conforms to the 3D surface of the target with minimal margin. Manually planning an optimal configuration of shots is an iterative and sometimes difficult process. Often, as a shot is added to the plan, the once conformal isodose surface deforms due to renormalization. Also, accounting for changes in target shape in the z direction (from axial slice to slice) can be difficult. Automated optimization approaches for radiosurgery planning have included simulated annealing and other random methods. These methods can fail because the number of shots must be optimized simultaneously with position and size. Instead of random methods, a method based on target shape has been developed. The new method mimics and replaces the manual process by considering the shape of the target volume as the information required to determine a shot's irradiation parameters.

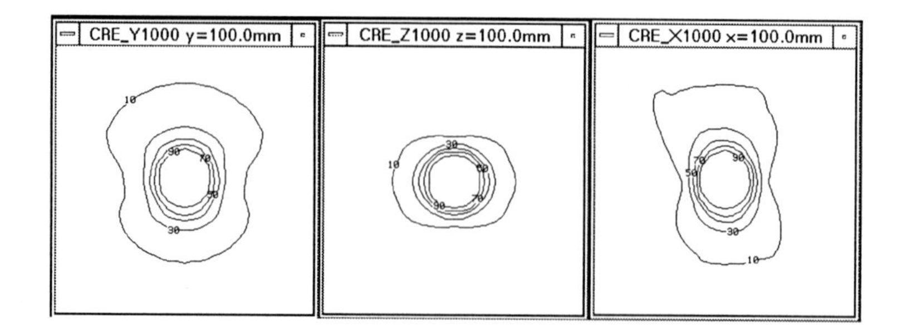

Fig. 1. The "spherical" dose distribution from 201 radiation beams through three anatomical planes. Lines represent regions of equal dose, normalized to the "shot" center. Diameter at the 50% isodose level is nominally 18 mm.

A region's medial axis (Blum), or skeleton, uniquely characterizes the region by its symmetric axis (Fig. 2). Each point has an associated maximum internal disk (2D) or sphere (3D). The maximum internal disk is tangential to at least two boundary points and cannot be fully covered by any other internal disk. A region's shape and medial axis are direct transformations: given the medial axis, a region's shape can be recovered, and *vice versa*.

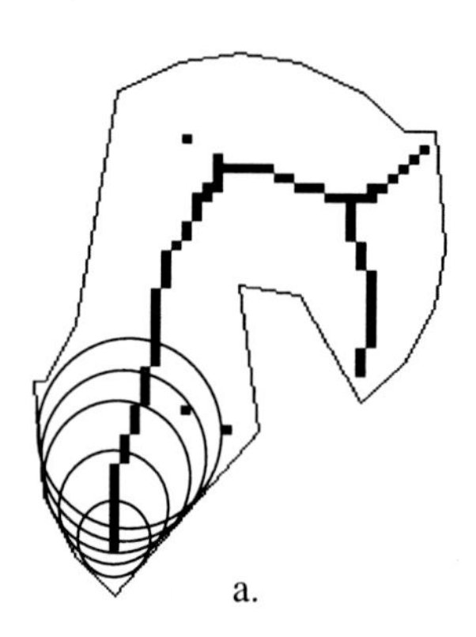

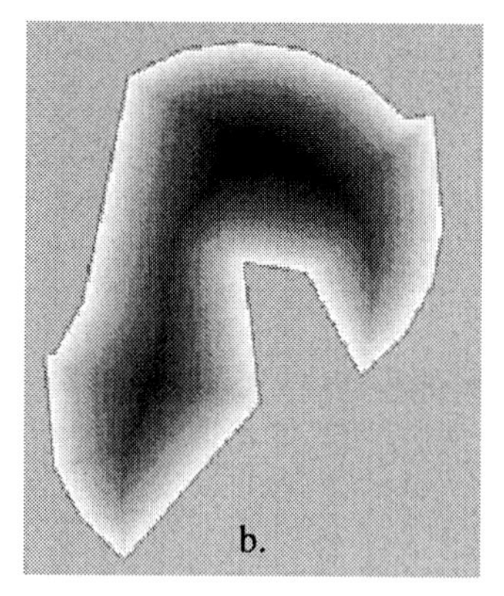

Fig. 2. The skeleton of an irregular 2D region. a. position of internal disks on medial axis. b. Euclidean distance transform, with distance represented by intensity. Ridge detection of the distance transform yields the medial axis in (a).

In conformal treatment, the prescription dose surface (in this work, the cumulative 50% isodose line or surface) closely corresponds to the target volume. To obtain this dose coverage a target volume must be "filled" with an arrangement of shots (Fig. 3). With a nominal shot diameter equal to the average width of the dose profile at 50% height, an ideal shot has a position and size such that the shot diameter matches as best as possible the target boundary. Shot diameters can be varied using physical collimators and relative dosimetric weights.

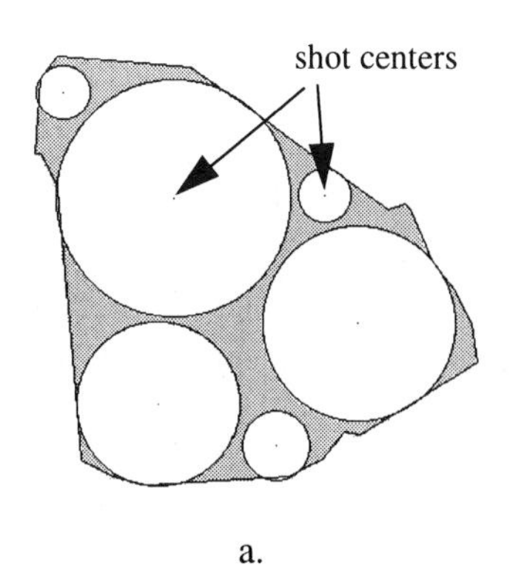

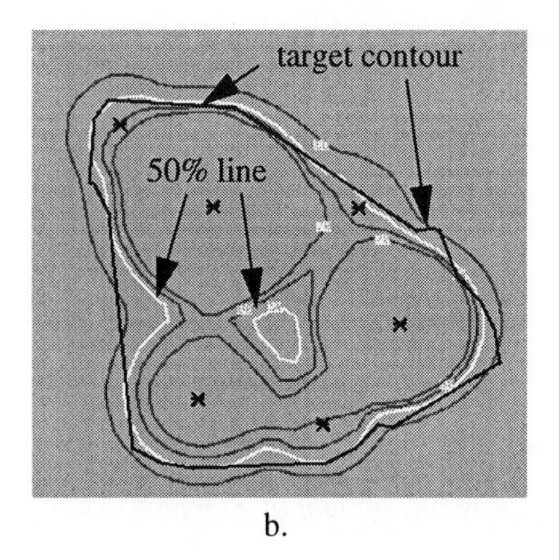

Fig. 3. An irregular target shape is covered by filling it with multiple shots. Large and small shots are used to produce a conformal dose distribution. a. Relative shot positions and sizes. b. The resulting dose distribution, shown with isodose curves. Note that gaps between shot boundaries in (a) are covered in most cases with adequate dose in (b).

Given conformal treatment by filling a tumor volume with a number of approximately spherical shots, the hypothesis for this work follows: the arrangement of shots (number, positions, and sizes) can be determined from the target skeleton and its sub-volumes and their sub-skeletons (Wu). Thus, candidate shot positions reside on target skeleton loci with shot diameters equal to those of the maximum internal spheres. The analogy is straightforward - the internal disks or spheres are shots of dose. One deviation in the analogy is that internal disks may overlap while dose shots are allowed to overlap only a small amount, due to the creation of "hot" spots and dose renormalization.

2 Method

In this work, ridges in the 3D Euclidean Distance Transformation (EDT) of the target volume are detected and characterized as medial axes or surfaces. Ridge detection is performed by determining the curvature, K_p at each point p in the EDT according to:

$$K_p = (\hat{n} \bullet \nabla)(\hat{n} \bullet \nabla) f = -\hat{t}^T Hess(f)\hat{t}$$

where $\hat{n}$ is the unit vector normal to the iso-EDT surface, f is the EDT map, $\hat{t}$ is the unit vector tangent to the iso-EDT surface, and $Hess(f)$ is the Hessian matrix of f. 2D EDT ridge detection is shown in Fig. 2. K_p is a vector at three and higher dimensions, since multiple orthogonal curvatures are then possible.

Given the 3D skeleton, for optimal target coverage the center of each shot is located on a skeleton or subsequent sub-skeletons. The objective of optimization is to find the best shot configuration with the minimum number of shots, such that the set covers the target volume within a fractional uncovered volume of ε. Denote a plan for target region R as $P_N(R)$ and the optimal plan as $P_{N_{opt}}(R)$, with N number of shots. The optimization is formulated as:

$$P_{N_{opt}}(R) = P_N(R) \,\Big|\, N = min(n),\ \forall n,\ \frac{V(R) - V\left(\bigcup_{i=1}^{N} S^{d_i} x_i\right)}{V(R)} \leq \varepsilon$$

where $V(R)$ is the region volume, $S^{d_i}(x_i)$ is the ith shot with diameter d_i and position x_i, $V\left(\bigcup_{i=1}^{N} S^{d_i}(x_i)\right)$ is the volume covered by the set of N shots, and ε is the fractional amount of R which is not covered by the set of shots.

The optimization is cast as a dynamic process according to:

$$P_{N_{opt}}(R) = S^{d_i}(x_i) + P_{N_{opt}-1}\left(R - S^{d_i}(x_i)\right)$$

where $x_i \in x(R_i)$, $R_i = R - \bigcup_{j=1\,(j \neq i)}^{N} S^{d_j}(x_j)$, and $x(R_i)$ is the medial axis of R.

Implementation of this optimization consists of the following steps: 1) Calculate the medial axis of a target volume. 2) Find all skeleton end points. Shots are placed at these points, initiating potential plans P1, P2, ... 3) For each potential plan, record the shot, delete the shot-covered region from the original target volume, and calculate the skeleton of the remaining target volume. 4) Detect cross points of the (sub) skeleton. Inherently, a shot based on a cross point will match multiple target boundaries and, thus, be the best shot for that portion of the region. If a single cross point is found, the shot is placed there. If several cross points are found, a combinatorial search is performed. If no cross point is found, return to Step 2). 5) Iterate Steps 3) and 4) until target coverage is within tolerance. 6) The plan with the least number of shots is chosen as the optimal plan. Skeleton-based optimization is shown in Figure 4 for an irregularly shaped region. The caption describes the steps being illustrated.

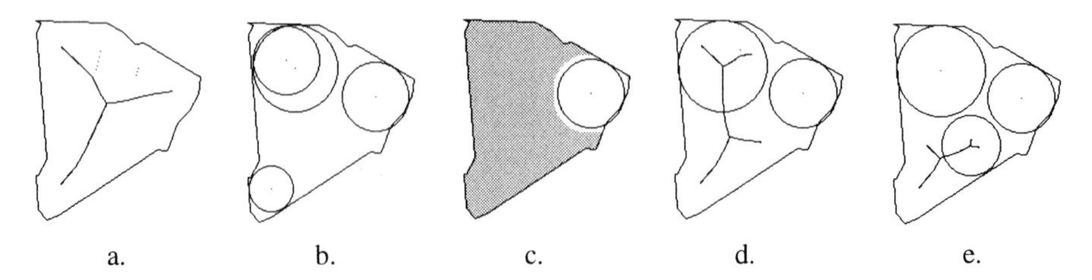

Fig. 4. Steps of the optimization process. a) Initial target and skeleton. b) Potential shots at each end point. c) Potential plan for the first shot - shading indicates remaining target. d) The skeleton for the uncovered region in (c) is calculated and an optimal shot is placed on one of two cross-points. e) A third shot is added and the process continues.

3 Results

A treatment plan determined by the method is shown in Figure 5 for a complex sinus tumor. The boundary of the enhancing target region has been contoured in white. The automated skeleton approach yields shot parameters that result in a clinically acceptable cumulative dose distribution (Fig. 5b). This result is comparable in dose coverage to the plan actually used for treatment (Fig. 5a).

4 Discussion and Conclusion

This work depends on the calculation of the 3D medial axis transform. While 2D skeletons can be found by several means, determining 3D skeletons is more difficult. A unique approach of ridge detection of the 3D Euclidean Distance Transform was developed, instead of using morphological thinning or other techniques.

Given the target's skeleton, the position, diameter, and number of shots each are determined optimally and simultaneously, using dynamic programming and a progressively cropped search space. These automated treatment plans rival those determined by expert manual planners. Mathematical optimization can be proved based on the two constraints of: 1) 50% dose conformation and 2) the minimum number of shots. Clinical optimization remains to be investigated, since competing plans may yield similar clinical outcomes. Additional constraints or measures to be modeled include those for dose inhomogeneity within the target, avoidance and morbidity of adjacent critical structures, and target biological response.

Shape is an important target characteristic. In manual planning the planner is placing shots based on target shape and knowledge of dosimetry. Shape-based planning works because the skeleton reduces the search space from 3D to 1D and also automates the process. Instead of sphere packing or random techniques, target shape determines the parameters for each shot as its region is being planned. This approach forces the dose distribution to be conformal locally, at each shot center, resulting in a conformal cumulative distribution.

References

Blum, H., Nagel, R., "Shape description using weighted symmetric axis features," Pattern Recognition **10**:167-180 (1978).

Wu, Q. R., "Treatment planning optimization for gamma unit radiosurgery," Ph.D. Dissertation, Mayo Foundation/Graduate School, Rochester, Minnesota (1996).

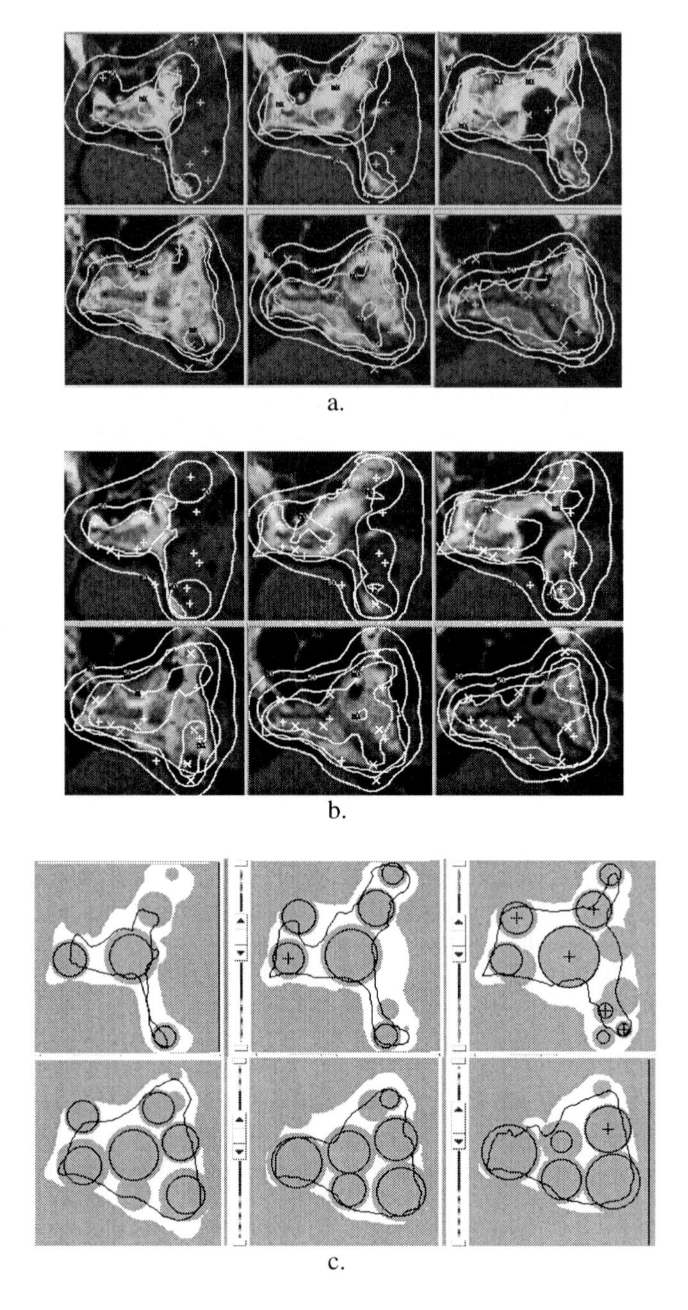

Fig. 5. Treatment plans for complex sinus tumor. Enhanced region and white contour delineates the target volume on transverse slices with corresponding z positions in (a), (b), and (c). Shot centers are marked by crosses. a) Manually performed dose plan used in treatment. b) Automated dose plan. c) Shot configuration for automated plan in (b).

Image Guided Surgery and Endoscopy

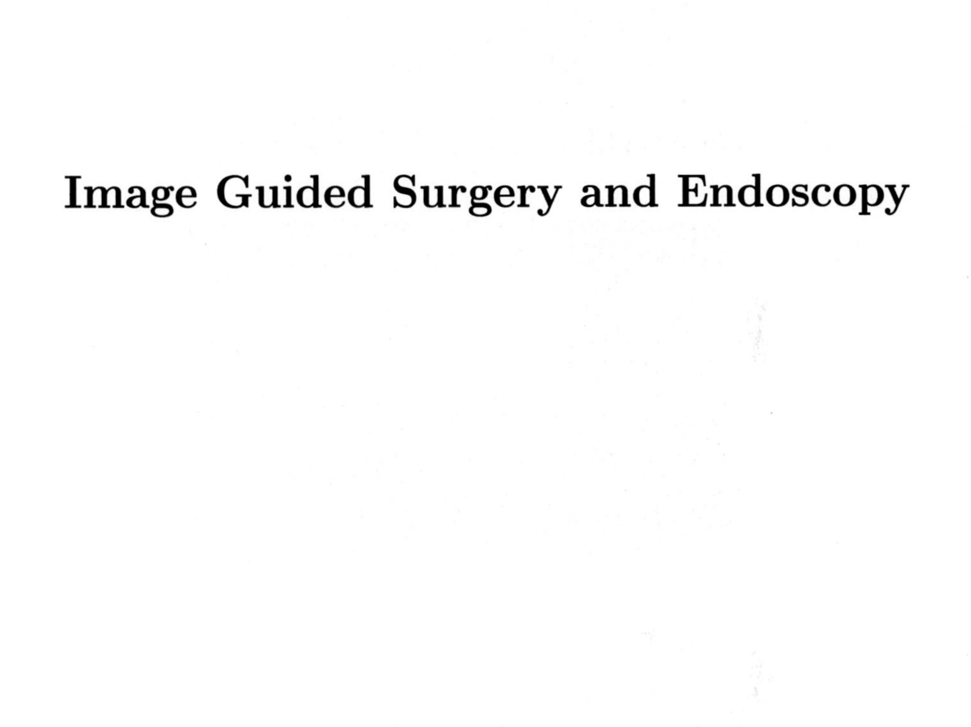

MRI Guided Intranasal Flexible Micro-Endoscopy

Derek LG Hill, Paul N Poynter Smith, Claire L Emery, Paul E Summers, Stephen F Keevil, J Paul M Pracy, Rory Walsh, David J Hawkes, Michael J Gleeson

UMDS, Guy's & St Thomas' Hospitals, London, UK
D.Hill@umds.ac.uk

Abstract. Interventional, fast MR images were registered to pre-operative CT and MR images of a primate cadaver. Localisation of an MR compatible, visible, flexible endoscope was achieved to within 2mm in the nasal cavity. Orthogonal views of the endoscope's position can be displayed alongside oriented video endoscope images to give 3D visualisation.

1 Introduction

The ability of MRI to provide both anatomical and functional information without exposure of patients and staff to ionising radiation makes interventional MR an attractive goal. The vastly superior soft tissue resolution of MRI, along with its multi planar imaging capability, make it an interesting alternative to CT, X-ray fluoroscopy and ultrasonography. Recent advances in MR hardware, and alternative k-space sampling algorithms [2,17,18], have enabled much faster scanning times. The speed of scanning has now reached the stage where almost real time imaging is a possibility.

Proposed uses for interventional MR include percutaneous breast biopsies [14], aspiration cytology, laser ablation of liver metastases [8,9], MR guided intravascular procedures [6,19], interstitial focused ultrasound surgery [16], needle placement [4,10,15], as well as endoscopic imaging [11]. Many of these have only been demonstrated *ex vivo*.

Conventional endoscopic sinus surgery uses rigid endoscopes to view the sinuses and osteomeatal complex. There is a small but significant risk of damage to structures which are in close relation to the sinuses. At particular risk are the orbital contents (including the optic nerve), the cribiform plate, skull base and the internal carotid arteries which indent the roof of the sphenoid sinus. Access to the sinuses can be limited by the use of rigid endoscopes in such a confined space, and it is sometimes necessary to use 30° and 70° angled lenses to look into the maxillary and frontal sinuses. The use of angled lenses requires considerable experience and expertise as the view afforded is disorientating. Access and safety could be improved by the use of a flexible endoscope, but in this case tracking would be helpful to provide orientation within the sinuses. A conventional localiser (eg: mechanical arm) cannot be used to track a flexible tool, and CT, fluoroscopy and ultrasound are all unsatisfactory for tracking a tool that can move in 3D through air and tissue.

The objective of the work described here is to demonstrate the feasibility of tracking a flexible micro-endoscope using an MR scanner, and use image registration techniques to relate video images acquired with the endoscope to high quality CT images acquired before the intervention.

1.1 MR compatibility

Instruments inserted into the patient must not degrade the image quality. Ferromagnetic materials should be kept well away from MR systems as they are attracted to the bore of the magnet and can be dangerous to both patient and clinician. Other materials have magnetic properties characterised by their magnetic susceptibility. Any material that has different magnetic susceptibility to the tissues being imaged will create a local geometric and intensity distortion within the image (especially using gradient echo imaging). Surgical instruments introduced into the patient should, therefore, have a susceptibility as close to that of tissue as possible. The image degradation caused by susceptibility differences is dependent on the imaging sequence being used. This degradation is minimised using spin-echo sequences, in which spin dephasing caused by susceptibility differences is rephased using a 180° RF pulse.

1.2 MR visibility

The tip of the instruments must be visible in the MR images. Solid objects (eg: surgical instruments) normally appear black in MR images, and are usually surrounded by a wider area of signal loss due to susceptibility differences. When introduced into a tissue that appears bright on an MR image, they can be clearly seen as signal voids. However, the size and shape of signal voids caused by magnetic susceptibility can change with orientation with respect to the magnetic field. In addition, not all of the structures within a patient are bright on MRI, for example the air sinuses, bone, and some fluids are black in certain MR sequences. A signal void would be indistinguishable from its surroundings. Adding a bright marker (eg: one containing an appropriate concentration of the paramagnetic contrast material Gadolinium DTPA) to the end of the instruments can make it visible when surrounded by low intensity tissue or air [3], but the instrument can cease to be visible if moved into a region where the surrounding tissue produces a similar MR signal to the marker. These problems can be overcome using an active marker which incorporates a small RF receive or send/receive coil surrounded by MR visible material [7,13,16,18]. Very simple imaging sequences, made up of a small number of projections, can be used to identify the 3D position of these active markers. Active markers are, however, as yet untested *in vivo* and the metal RF coils can cause local heating which may, in turn, cause additional as yet unquantified trauma to the patient. Maier et al. [5] have assessed the increase in tissue temperature that coils of this sort generate. Heating is caused primarily by the RF field, with a doubling of the RF field amplitude leading to a 4 fold increase in tissue temperature. It is also dependent on magnetic field strength and the sequence used, with the greatest effects seen with turbo spin echo sequences, in high field magnets and with the catheter orientated in the plane perpendicular to the direction of the static magnetic field.

In the work described here, we have chosen to use a passive, high intensity marker, though we acknowledge that, once safety concerns have been overcome, active markers will be much more straightforward to track automatically.

2 Method

2.1 Design of passive endoscope marker

In order to track a bright passive marker, it is necessary to select the optimum concentration of Gadolinium DTPA (Gd-DTPA) for use with the fast MR sequence to be used during the intervention. Concentrations of between 0 and 30 mmol l^{-1} of Gd-DTPA were placed in 20ml plastic vials, and imaged using a turbo spin-echo (TSE) pulse sequence (TR=106ms, TE=11ms, TF=8). The concentration that provided the highest intensity was selected. The degradation in signal intensity caused by partial volume effect and susceptibility artefacts was then assessed by placing the optimum concentration of marker fluid into plastic tubing and re-imaging. The intensity of the marker relative to surrounding tissue, for the selected pulse sequence, was then assessed using a fresh porcine cadaver.

2.2 Modification of microendoscope for MR compatibility

We constructed a modified microendoscope, using the 6000 pixel imaging fibres and the illumination fibres of an Omegascope OM2/070 flexible microendoscope. This was made MR compatible by encasing it in a bio-compatible polythene catheter with an outer diameter of 2.3mm. The final 2cm of the catheter had the guidewire and working channels filled with Gd-DTPA contrast agent, of the concentration determined using the experiments described in section 2.1, and the ends were sealed with wax [1]. A single chip colour CCD camera was attached to the endoscope eye-piece, and could operate satisfactorily in the bore of a 1.5T scanner. The video output was connected, via a long video lead, to a SunVideo frame grabber in a workstation in the scanner control room.

2.3 Construction of MR compatible light source

An MR compatible endoscope light source was constructed. The light source was housed in an aluminium alloy box, with a DC rather than AC power input, avoiding the need for the light source to contain a transformer. Long power cables were used to keep the power supply transformer outside the RF shield. A 150W 15V reflective xenon bulb was used to focus light onto the end of the illumination fibres. A small 6V fan was attached to the side of the box to cool the bulb. Although the fan contained a small amount of ferromagnetic material, the light source could be attached to the scanner couch, approximately 50cm away from the magnet bore without being attracted into the scanner, or degrading the images. The endoscope was connected to the light source using a machined PVC socket, and was protected from the heat of the bulb by heat absorbing glass.

2.4 Display of images in the scanner room

A 1024 x 768 pixel active colour LCD projector panel was used to project video from the endoscope and processed images displayed on a Sun workstation, into the scanner room, where it could be seen by the clinician. The projector could be placed in a field strength of less than 1mT, with the images projected onto a screen placed in an arbitrary field strength.

2.5 Cadaver experiments

The expected clinical protocol for the use of the system is described below, and was carried out using a primate cadaver head, from a separate research project for which the temporal bones had been removed.

"Pre-operative" imaging. CT scanning is the modality of choice for visualising the paranasal sinuses and osteomeatal complex. "Preoperative" imaging was carried out using a Philips SR7000 spiral CT scanner. 61 slices, 1.5mm thick, with a field of view of 200mm and a 512 X 512 matrix were acquired. A high resolution pre-operative T_1 weighted spin echo (SE) MR scan was also acquired using the 1.5T Philips Gyroscan ACS II. This was registered to the CT scan by optimising the mutual information of the joint probability distribution [20].

"Interventional" imaging. Prior to the interventional scan some dissection of the primate cadaver was performed by an ENT surgeon in order to allow access to the maxillary sinus. This is the same procedure as carried out in conventional sinus surgery, and in clinical use, we envisage that this would be done while the patient is lying on the scanner couch. The cadaver head was then placed in the scanner, and the endoscope inserted into the nasal sinus. The orientation of the endoscope was ascertained before insertion so the operator could orient himself while entering the sinus system. The marker attached to the flexible microendoscope was visualised within the sinus system using the fast imaging sequence for which the marker concentration had been optimised. These "interventional" images were registered to the preoperative MR images, and hence to the preoperative CT images, by optimising mutual information of the joint probability distribution.

Although the image quality of the fast sequence was much reduced, the endoscope tip could be clearly seen. The contrast to noise ratio was not always sufficient to locate the marker by thresholding alone, so a small amount of user interaction was required to identify the endoscope tip and orientation.

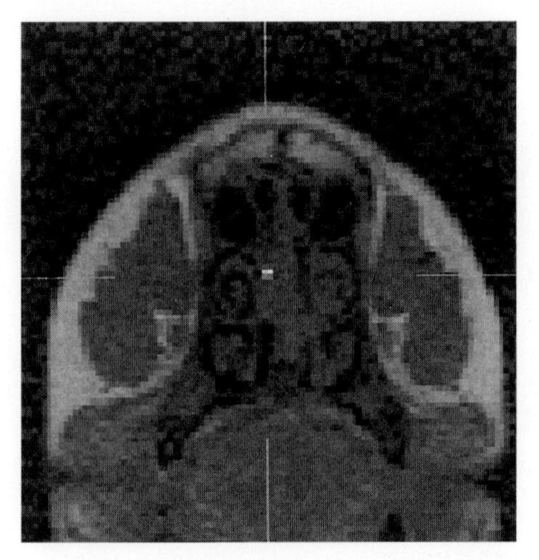

Fig. 1. Turbo spin echo image, showing endoscope marker visible with high intensity entering the nasal cavity of a porcine cadaver.

Having calculated the registration transform relating the interventional MR images to the preoperative CT and MRI images, the position of the endoscope tip could be displayed in orthogonal slices or rendered views generated from the preoperative images. Video images obtained during the intervention from the CCD camera could thus be displayed alongside rendered CT images produced from the same position.

3 Results

3.1 Design of endoscope

The concentration of Gd-DTPA that produced the highest intensity signal when imaged with the TSE sequence described was 7.5mmol l^{-1}. The intensity was reduced dramatically if the marker diameter was less than 1mm. Figure 1 shows a TSE image acquired with the endoscope incorporating a marker inserted in a porcine cadaver. The MR compatible light source is shown in figure 2.

3.2 Cadaver Experiment

The orthogonal views from the "preoperative" CT of the primate head are seen in Figure 3. The nasal sinuses are very clearly visualised. Figure 4 shows the high resolution MRI, with the thresholded bone boundary from CT overlaid. Visual inspection suggested that the accuracy of registration was within 2mm. This is in agreement with accuracy measurements for the registration algorithm made using human subjects [21].

Figure 5 shows a composite view comprising orthogonal views, and a rendered image showing the relationship between the endoscope localised in the high

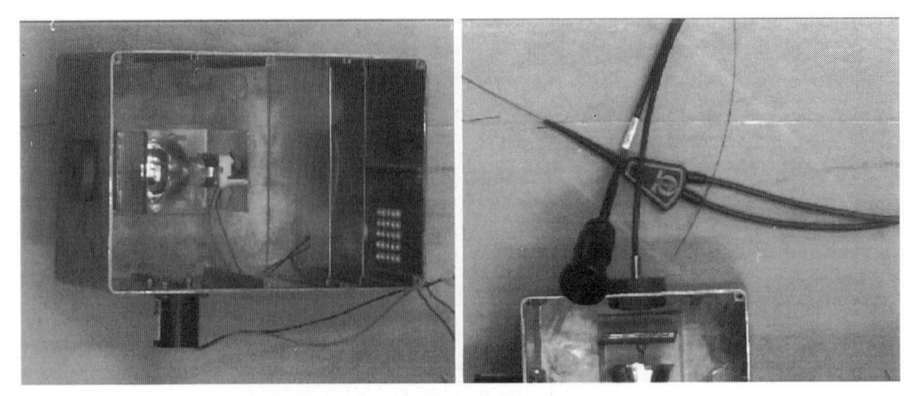

Fig. 2. The endoscope light source with lid removed (left), and attached to endoscope (right).

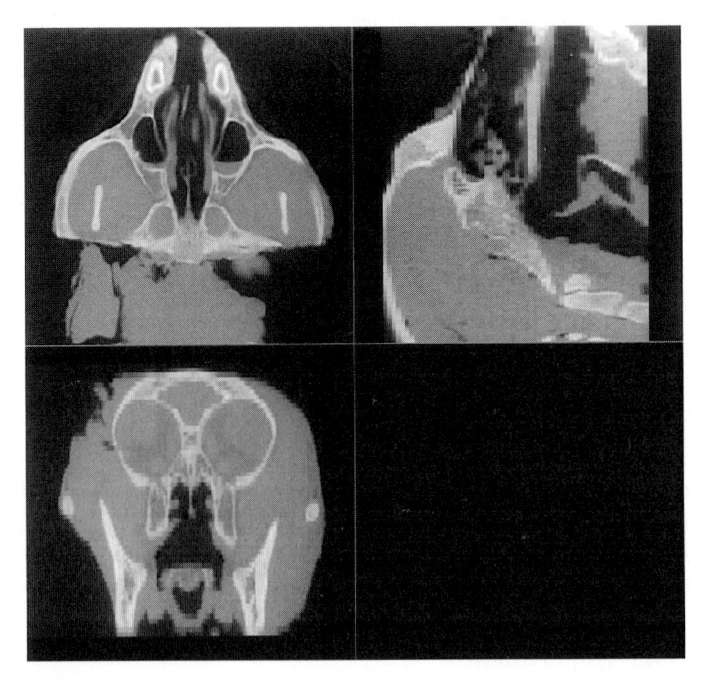

Fig. 3. Orthogonal views from the CT scan of a monkey cadaver.

speed TSE imaging sequence, and the pre-operative CT scan. In the orthogonal views the endoscope marker is shown as a thresholded boundary overlaid on the CT image. In the perspective rendered scene the endoscope is shown as a dark cylinder protruding from the nasal cavity.

Knowing the position of the tip of the endoscope relative to the preoperative CT, it is possible to produce a rendered scene along the axis of the endoscope. A frame of video endoscope image is shown alongside a CT image obtained from the same position in figure 6.

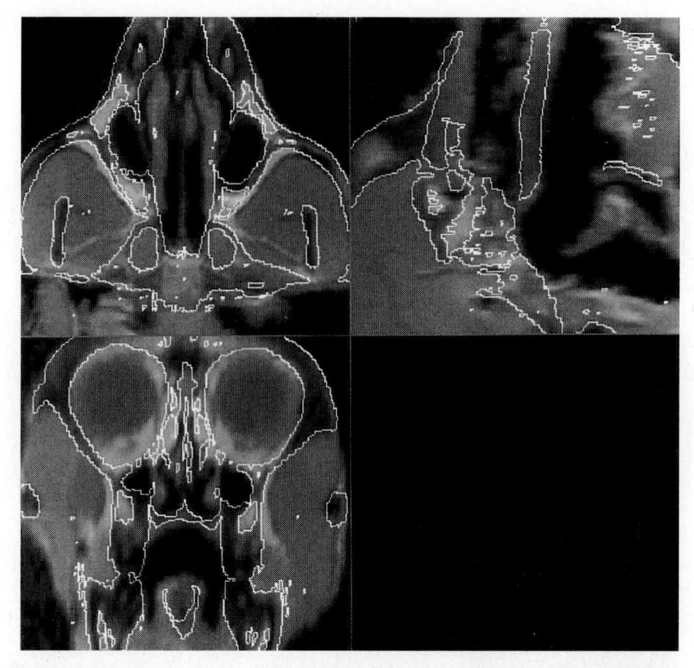

Fig. 4. Orthogonal views from the MR, with the thresholded bone boundary from CT overlaid.

4 Discussion

The results show that it is possible to locate the tip of a flexible endoscope accurately in a fast SE sequence and to obtain video pictures from the same location. Current surgical guidance systems, using mechanical arms, ultrasound, or optical tracking are capable of localising the position of a rigid pointer or rigid surgical instruments, with respect to preoperative images. These guidance systems are not appropriate for use with flexible instruments, or where significant tissue deformation is expected during the course of the procedure.

Interventional radiology is also unsatisfactory for tracking flexible instruments. Conventional x-rays provide poor contrast, and, even with a bi-plane system, only limited three dimensional information. Interventional CT or conventional ultrasound provides only a single plane, which is insufficient for tracking flexible tools. Three dimensional ultrasound may have applications in some parts of the body, but the mixture of air, bone and soft tissue in the paranasal sinuses makes ultrasound unsuitable for tracking flexible tools in this application.

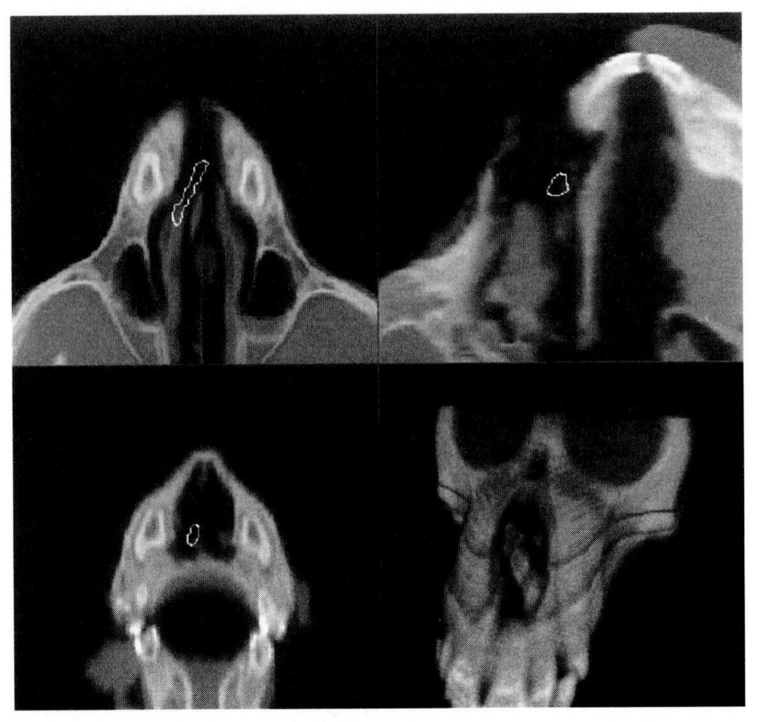

Fig. 5. Orthogonal views and rendered scene from the CT scan. The thresholded boundary of the endoscope marker from the registered MR is overlaid in the orthogonal CT views, and the endoscope is displayed as a dark cylinder prodtruding from the nose in the perspective rendered view.

We have shown that interventional MR can be used to track flexible instruments, in a manner that is analogous to the use of a conventional guidance system, and by means of voxel-based registration algorithms, to display the position of the instrument in the context of the preoperative images. A great deal of work is required before these techniques can enter routine use. The cost of conventional MR imaging systems, and the relatively poor patient access are major obstacles. More specifically, in our application, it is necessary to automate the tracking of the endoscope. This might be done by tracking the passive marker in a series of fast images, or by means of safe active markers. Appropriate active markers are being developed by the scanner manufacturers.

Improvements in image processing are also necessary to accurately relate the video endoscope images to the preoperative radiological images. To achieve this, it is necessary to calibrate the endoscope optics, and produce rendered views using this same distorted perspective geometry.

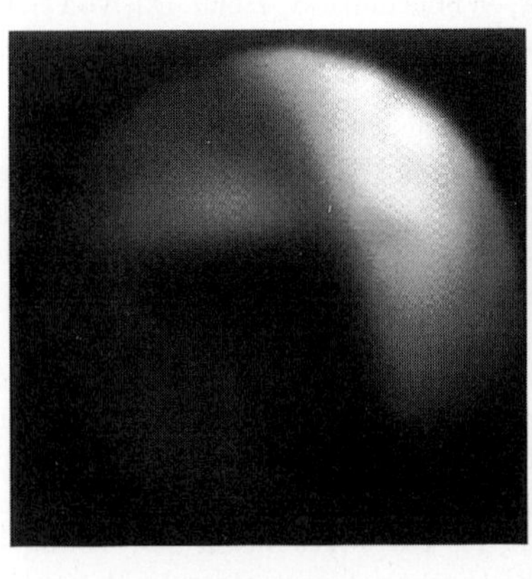
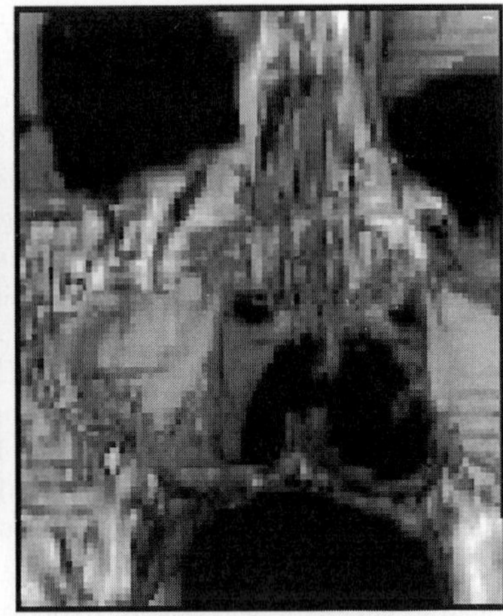

Fig. 6. Frame-grabbed video-endoscopy image alongside a rendered CT view along the same axis as the endoscope marker. The rendered view is produced using a parallel projection, and has a much larger field of view than the video image. The aim is to provide context for the endoscope image, rather than generate a corresponding rendered view.

5 Acknowledgements

We are grateful for the assistance of Dr Daryl DeCuhna, Mr Julian McGlashan, Dr Eddie Boyd, and John Walliker for their help in experimental design and equipment configuration. We are also grateful to the Medical Physics workshop for assisting in construction of equipment.

References

1. Emery CL, Hill DLG, Keevil SF, Summers PE, McGlashan JA, Walsh R, Diamantopoulos C. Liepens PJ, Hawkes DJ, Gleeson MJ: Preliminary work on MR Guided Flexible Micro-endoscopic Sinus Surgery. Proc Soc. Magn. Reson. (1995) p 1158.
2. Friebe MH, Brusch M, Hellwig S, Wentz KU, Groenmeyer DHW.: Keyhole sequence with rotational K-space updating as a means of fast high quality low-field interventional MR imaging. Proc Soc. Magn. Reson. (1994) p 161
3. Kochli VD, McKinnon GC, Hofmann E, von Schulthess GK: Vascular interventions guided by ultrafast MR imaging: Evaluation of different materials Magn. Reson. Med. **31** (1994) 309-314
4. Leung DA, Debatin JF, Wildermuth S, Heske N, Davis CP, Hauser M, Dumoulin CL, Darrow RD, Alder N, von Schulthess GK: Active biplanar MR-Guidance of needle placements in real time. Proc. Soc. Magn. Reson. (1995) p 495.
5. Maier SE, Wildermuth S, Darrow RD, Watkins RD, Debatin JF, Dumoulin CL. Safety of MR Tracking Catheters. Proc. Soc. Magn. Reson. (1995) pg 497
6. Leung DA, Wildermuth S, Debatin JF, Holtz D, Dumoulin CL, Darrow RD, Hofmann E, Schulthess GK: Active visualisation of intravascular catheters with MRI: In vitro and In vivo evaluation. Proc. Soc. Magn. Reson. (1995) p 428.

7. McKinnon GC, Debatin JF, Leung DA, Wildermuth S, Holtz DJ, von Schulthess GK: Towards visible guidewire antennas for interventional MR. Proc. Soc. Magn. Reson. (1993) p 429.
8. Roberts HRS, Paley M, Lees WR, Hall-Craggs MA, Brown SG: MR Control of laser destruction of hepatic metastases. Proc. Soc. Magn. Reson. (1995) pg 1590.
9. Vogl TJ, Weinhold N, Muller P, Roggan A, Balzer JO, Bechstein W, Lobeck H, Felix R: Evaluation of the Therapeutic Potential of MR guided Laser Induced Thermotherapy (LITT) in the treatment of liver metastases before surgery. Proc. Soc. Magn. Reson. (1995) p 1589.
10. Wildermuth MD, Debatin JF, Leung DA, Hofmann E, Dumoulin CL, Darrow RD, Schopke WD, Uhlschmid MD, McKinnon GC, von Schulthess GK: MR guided percutaneous intravascular interventions: In vivo assessment of potential applications. Proc. Soc. Magn. Reson. (1995) p 1161.
11. Desouza NM Hall AS, Coutts GA, Puni R, Taylor-Robinson SD, Young IR: Endoscopic magnetic resonance imaging of the upper gastrointestinal tract using a dedicated surface receiver coil. Proc. Soc. Magn. Reson. (1995) p 1157.
12. Lewin JS Duerk JL, Haaga JR: Needle localisation in MR guided therapy: Effect of field strength, sequence design, and magnetic field orientation. Proc. Soc. Magn. Reson. (1995) p 1155.
13. McKinnon G: Tracking and monitoring the effects of interventional MR instruments Proc. Soc. Magn. Reson. (1995) p 490.
14. DeSouza NM: MRI directs localisation, biopsy of breast lesions. Diagnostic Imaging Europe. (Feb 1995) 40-43.
15. Leung DA, Debatin JF, Wildermuth S, McKinnon GC, Holtz D, Dumoulin Cl, Darrow RD, Hofmann E, von Schulthess GK: Intravascular MR tracking catheter: Preliminary experimental evaluation. American Journal of Roentgenol. **164** (1995) 1265-1270.
16. Dumoulin CL, Souza SP, Darrow RD: Tracking of an invasive device within an MR imaging system. Proc. Soc. Magn. Reson. Med. (1992) pg 104.
17. Van Vaals JJ. Van Vyperen GH, Hoogenboom TLM, Duijvestijn MJ: Real-Time MR imaging using LoLo (Local Look) method for interactive and interventional MR at 0.5T and 1.5T. Journ. Magn. Reson. Imag. **4(P)** (1994) 38.
18. Dumoulin CL, Souza SP, Darrow RD: Real time position monitoring of invasive devices using Magnetic Resonance. Magn. Reson. Med. **29** (1993) 411-415
19. Martin AJ, Henkelman RM: Intravascular MR Imaging in a porcine animal model. Magn. Reson. Med. **32** (1994) 224 - 229.
20. Studholme C, Hill DLG, Hawkes DJ: Automated 3D registration of truncated MR and CT images of the head. Proc. British Machine Vision Conference ed. Pycock, BMVA press Sept 1995 vol 1. p27-36
21. West J, Fitzpatrick JM, Wang MY, Dawant BM, Maurer CR, Kessler RM, Maciunas RJ, Barillot C, Lemoine D, Collignon AM, Maes F, Suetens P, Vandermeulen D, van den Elsen PA, Hemmler PF, Napel S, Sumanaweera TS, Harkness BA, Hill DLG, Studholme C, Malandain G, Pennec X, Noz ME, Maguire CQ, Pollack M, Pelizzari CA, Robb RA, Hanson DP, Woods RP.: Comparison and evaluation of retrospective intermodality image registration techniques Medical Imaging 1996: Image Processing. Proc. Soc. Photo Opt. Instrum. Eng. **2710** (1996) 332-347

Computer-Assisted Insertion of Pedicle Screws

Qing Hang Li,* Hans J. Holdener,** Lucia Zamorano,* Paul King,* Zhaowei Jiang,* Federico C. Vinas,* L. Nolte,*** H. Visarius,*** and Fernando Diaz*

* Dept. of Neurological surgery, Wayne State University, Detroit, MI 48201, USA
** Dept of Orthopaedic Surgery, Wayne State University, Detroit, MI 48201, USA
*** Muller Institute for Biomechanics, Bern, Switzerland.

Abstract. The possible complications of a pedicle screw fixation system include injury to neurologic and vascular structures resulting from inaccurate placement of the instrumentation. In a review of 617 surgical cases in which pedicle screw implants were used, Esses and co-authors reported a overall complication rate of 27.4%.[1] The most common intraoperative problem was unrecognized screw misplacement (5.2%). Fracturing the pedicle during screw insertion and iatrogenic cerebrospinal fluid leak occurred in 4.2% of cases. Such a complication rate is not acceptable in clinical practice. In this paper, we discuss a computer-assisted spine surgery system designed for real-time intraoperative localization of surgical instruments on precaptured images used during surgery. Localization was achieved by combining image-guided stereotaxis with advanced opto-electronic position sensing techniques. The insertion of pedicle screws can be directly monitored by interactive navigation using specially equipped surgical tools. Our preliminary results showed no misplacement of pedicle screw, which have further confirmed the clinical potential of this system.

1. Introduction

The accuracy and precision of transpedicle screw insertion and placement is critically important to clinical results. Consequently, correct orientation to the unexposed spinal anatomy has become an even greater concern. In particular, the various screw fixation techniques that require placing bone screws into pedicles of the thoracic, lumbar, and sacral spine, into the lateral masses of the cervical spine, and across joint spaces in the upper cervical spine require "visualization" of the unexposed spinal anatomy. Although intraoperative fluoroscopy, serial radiography, and ultrasonography have proven useful, they are all limited in that they provide only two-dimensional (2-D) imaging of a complex three-dimensional (3-D) structure. Consequently, special experience is required for the surgeon to extrapolate the images and a knowledge of the complex anatomy of the spine. Inaccuracy during insertion of pedicle screws can be a direct result of this inference. Several studies have shown the unreliability of routine radiography in assessing insertion of pedicle screws in the lumbosacral spine. The rate of penetration of the pedicle cortex by an inserted screw ranges from 21% to 31% in these studies.[3,9]

Intraoperative interactive frameless stereotaxy using surface reference marks in place of a stereotactic frame has been successfully applied to intracranial surgery.[12,13] The method provides an accurate and reliable means of localizing lesions within the brain to reduce operative time, costs, and morbidity. However, the application of this technology to other areas of surgery, particularly spinal surgery, has not been widely accepted.

The purpose of this study was to establish a computer-assisted spine surgery system (CAS) for pre-operative planning and for enhanced intraoperative guidance of transpedicle screw insertion. The system uses a space digitizing system to locate instruments and tomographic image reconstruction to image the area of surgical interest and project the trajectory and 3-D placement of the screws. The system is conceptually based on stereotaxis, a method for localization of objects within a body without direct access to its interior. All elements, the spine surgery tools and pointer, the surgical object (e.g. vertebrae), and the associated virtual object (computed tomographic (CT) scan or other medical image) are treated as rigid bodies. In this report, we describe the application of this technology to the insertion of spinal pedicle screw.

2 Materials and Methods

2.1 Registration and Image Transformation

The most critical problem in computer-assisted image-guided spinal surgery is the stereotactic transformation and coordinate matching between the patient's anatomical landmark and medical images such as CT. In this section, we describe in detail the algorithm for the functions involved in the CAS system.

Image Registration. The two major types of image data registration are fiducial matching and surface matching. The goals are to correlate, data from the medical image to the 'real world', which refers to the coordinate space of the surgical instruments. A tracking device is attached to the instruments to continually relay information regarding position to the CAS system. Coordinate matching ensures that any point seen in a medical image corresponds to an actual point in the real world (the patient's spinal anatomy).

The next step is to correlate different images with each other. Each imaging modality displays anatomical structures and lesions in a unique way. This benefits the surgeon by providing several different ways of viewing the same anatomical structure, but requires the development of an interactive relationship between the images and the real world. Registration is used to build the relationship and enable the surgeon to use each imaging modality to its greatest advantage for localizing the anatomical structure.

Registration is a two-step process involving the extraction of common features between imaging modality and matching these features using an algorithm. In the CAS system, matching is based on anatomical landmarks (fiducials). The anatomical landmarks usually used for matching are spinal and transverse processes.

Matching Anatomical Landmarks. Anatomical landmark matching involves constructing a transformation matrix based on a list of anatomical landmarks. For each fiducial there are two sets of three coordinate values *x*, *y*, and *z*; one set forms a coordinate space. Basically fiducial matching defines a matrix, uses it to transfer one set of coordinate values from the medical image, and calculates the difference between this set and the second of coordinate values. Each fiducial F1, F2, ... F*n* has two sets of coordinate values, C1[*i*](*x*,*y*,*z*) and C2[*i*](*x*,*y*,*z*), from coordinate space 1 and 2 respectively. Ideally, the two sets should be match as Equation 1:

$$\begin{matrix} C1(i)x \\ C1(i)y \\ C1(i)z \\ 1 \end{matrix} = M(2 \rightarrow 1) * \begin{matrix} C2(i)x \\ C2(i)y \\ C2(i)z \\ 1 \end{matrix} \qquad \text{.......(1)}$$

For $i = 1 \cdots n$.

where $M\ (2 \rightarrow 1)$ is the transformation matrix from coordinate space 2 to coordinate space 1. In clinical practice, it is impossible to match exactly one coordinate space of a landmark to another one for two reasons: one, the instrumentation used for digitization has a small margin of inaccuracy, and two, because the fiducial itself is not actually a point but a volume, no matter how small it is. Therefore digitization of anatomical landmarks proceeds from the center of the tiny volume of the point, which because of the irregular shape of the landmark is not a perfect center. For these reasons, eq. 1 is not completely suitable; there is no transformation matrix, $M\ (2 \rightarrow 1)$, that can perfectly fit all landmarks in one coordinate space to those in another.

Therefore the concept of approximation is used to describe a not-perfect fit but one with some slight tolerance that exists for every case. Our system uses several methods to approximate $M\ (2 \rightarrow 1)$, such as least square approximation, non-linear programming, and principle axes fitting. Principle axes fitting reorients and repositions the origin and X, Y, Z axes of one coordinate space so that it fits the origin and axes of another space. With multiple anatomical landmarks, this is achieved in two steps: three anatomical landmarks are used to make an initial rough match, then downhill iteration refines the match. With three landmarks, we derive mean and orthogonal axis construction as follows:

1) The mean positions of all landmarks are calculated in both coordinate spaces, source coordinate space, and destination coordinate space.

$$\mathrm{MEAN1} = (x, y, z)$$
$$\mathrm{MEAN2} = (x, y, z) \qquad (2)$$

2) Two specific landmarks (Lm, Ln) are selected and satisfy:

$$\{(Lm\_1 - \mathrm{MEAN1})*(Ln\_1 - \mathrm{MEAN1})\}+\{(Lm\_2 - \mathrm{MEAN2})*(Ln\_2 - \mathrm{MEAN2})\}$$
$$= \max\{[(Li\_1 - \mathrm{MEAN1}) * (Lj\_1 - \mathrm{MEAN1})] + [(Li\_2 - \mathrm{MEAN2})*(Lj\_2 - \mathrm{MEAN2})]\}$$
$$(\text{For } i, j = 1 \ldots n) \qquad (3)$$

where * is the cross product of vectors. Equation 3 shows that Lm, Ln are two landmarks that can generate maximum cross product with mean position. These specific landmarks reduce possible error by selecting the long arm of vectors.

3) From landmarks Lm, Ln and the mean positions, we can define three orthogonal axes in each coordinate system, as follows:

$$\mathrm{X1} = \mathrm{normalize}\ (Lm\_1 - \mathrm{MEAN1})$$
$$\mathrm{Z1} = \mathrm{normalize}\ (\mathrm{X1} * (Ln\_1 - \mathrm{MEAN1}))$$
$$\mathrm{Y1} = \mathrm{Z1} * \mathrm{X1}$$
$$\mathrm{X2} = \mathrm{normalize}\ (\mathrm{Lm\_2} - \mathrm{MEAN2})$$
$$\mathrm{Z2} = \mathrm{normalize}\ (\mathrm{X2} * (Ln\_2 - \mathrm{MEAN2}))$$
$$\mathrm{Y2} = \mathrm{Z2} * \mathrm{X2} \qquad (4)$$

where * is cross product, and "normalize" is the function that normalizes the input vector.

If we shift the coordinate system of coordinate space 1 to the new origin MEAN1, and rotate it so that x, y, and z axes are superimposed on X1, Y1, Z1, the new coordinate system is identical to the coordinate system generated by shifting the coordinate system of coordinate space 2 to MEAN2, and rotating it so that x, y, and z are superimposed on X2, Y2, and Z2 respectively. With this understanding, the transformation matrix from space 2 to space 1 can derived in the following way:

$$M_{\mathrm{initial}}\ (2 \rightarrow 1) = M\ (2\_\mathrm{shift}) * M\ (2\_\mathrm{rotate}) \times M^{-1}(1\_\mathrm{rotate}) * M^{-1}\ (1\_\mathrm{shift}) \qquad (5)$$

In order to go down a hill the fastest possible route, one must choose the steepest direction on each step. This is the concept behind downhill iteration. Obviously M_{initial} (2

$\rightarrow$ 1) is not accurate enough and is sometimes skewed. This occurs because defining an orthogonal axis is based upon the mean and two specific points, so these two points have a greater effect on the rotation matrix than other fiducials. We use downhill iteration to solve this problem. To migrate from one iteration to another requires a measurement of matching. With a calculated error measurement as the criteria of accuracy of a certain transformation matrix, we can refine the transformation matrix step by step, and finally come to the global minimum. The series of modifications to the transformation matrix alone is the direction of greatest decrement of error. The modification of the transformation matrix in each step is as follows:

$$\begin{aligned} &\text{MEAN}\ \{(X_{1\_0}, {}_0Y_{1\_0}, Z_{1\_0}),\ (X_{2\_0}, Y_{2\_0}, Z_{2\_0})\} \\ &Lm\ \{(X_{1\_1}, Y_{1\_1}, Z_{1\_1}), (X_{2\_1}, Y_{2\_1}, Z_{2\_1})\} \\ &Ln\ \{(X_{1\_3}, Y_{1\_3}, Z_{1\_3}), (X_{2\_3}, Y_{2\_3}, Z_{2\_3})\} \end{aligned} \quad (6)$$

In each step we just move one fiducial's position in one coordinate space in a single direction. If we choose to move fiducials in space 1, it is the equivalent of modifying one variable of X_{1_0}, Y_{1_0}, Z_{1_0}, X_{1_1}, Y_{1_1}, Z_{1_1}, X_{1_2}, Y_{1_2}, and Z_{1_2} by a fixed amount D, either $+D$ or $-D$ (generally $D = 0.1$ mm). After the modification, the new transformation matrix can be calculated in the same method mentioned above. The new transformation matrix is generated according to the modification, and three new fiducials are used as initial values for the next step of iteration.

2.2 Intraoperative Matching

The use of an intraoperative digitizer optimizes the movement and pathway of the surgical instrument by providing on-line image-guided position and orientation feedback. This feedback is very helpful in determining the pathway of a pedicle screw in the pedicle canal in order to avoid damaging the spinal cord, root, and major blood vessels surrounding the spinal vertebra.

For intraoperative localization various surgical tools are defined as rigid bodies with coordinate reference assigned at the tip of these tools. At least four light emitting diodes (LED), mounted on a specially designed plate, are attached to these tools. During surgery, at least three of these markers must remain visible to the camera. The process of changing the instrument simply involves choosing the needed instrument from the menu.

The points to be matched during the surgery are bone anatomical landmarks, most commonly the spinal processes, transverse processes, and spinal facet joints. These points are absolutely identified on the CT images as well as on the patient's spine. The local rigid dynamic reference is attached on the spinal process and keeps it stable during the whole procedure of inserting pedicle screws. For different vertebrae, there are two possibilities. We have to check the landmarks with the surgical tool, and if these points are correctly showing on the image that means the relative position between two vertebrae has not changed after CT scan. The process of inserting a pedicle screw can continue. If the relative position changed, a new matching procedure is needed, which normally takes 3 - 5 minutes. The total transformation calculated by the system is presented by the following equation:

$$T(I\text{-}CT)_i = T^{-1}(C - I)_i * T(C - P)_i * T(P - L) * T(L - CT) \quad (7)$$

where $T^{-1}(C - I)_i$ is the transformation matrix between instrument and camera. $T(C - P)_i$ is the transformation matrix between camera and rigid body dynamic reference. $T(P - L)$ is the transformation matrix between dynamic reference and landmarks. $T(L - CT)$ is created by touching the landmarks on the patient's spine and matching them with their projection as seen through the CT image.(Figure 1)

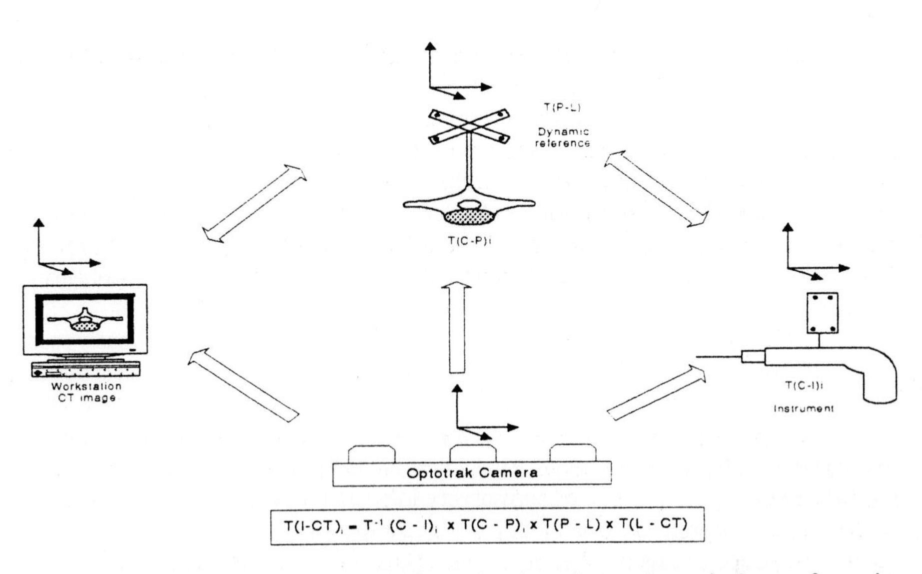

Figure 1. Landmarks matching, showing the mathematical transformation calculated by the system to find the position with respect to the CT images data. 'T', transformation matrix; '$T()_i$', transformation being calculated in real-time.

3 The CAS System

The following hardware and software components are used:

Space Digitization. A contact-less, pre-calibrated motion analysis system chosen to locate the instruments. It can track up to 256 pulsed infrared LED markers with a maximum sampling rate of about 3500 markers per second (OPTOTRAK 3010, Northern Digital, Waterloo, CAN). Within a field of view of 1.0 m x 1.2 m at a distance of 2.0 m the camera can locate each LED with an accuracy of 0.15 mm. By attaching at least three markers onto each rigid body its location and orientation in space can be determined. The system is precalibrated, reducing set-up time and allowing better tolerance to environmental changes. The limitation of the system is the need of a direct viewing line between the camera and the LED's. To overcome this limitation we have designed a special holder to allow for different heights and angulation.

The OPTOTRAK system has the ability to define several rigid bodies. Each rigid body is assigned a separate dynamic reference. It provides relative coordinates from the camera to any one of these rigid body's dynamic reference on the fly, by attaching to different vertebrae or other rigid body. By calibrating the relative coordinates the transformation matrix is established. This eliminates the need for new matching during spine surgery. The motion analysis system is controlled by DOS-based software running on a personal computer networked with a Unix workstation (SUN Microsystems, Inc. Mountain View, CA) that is used for image data acquisition/reconstruction and real-time instrument visualization. A single monitor provides all the computer-based information.

Tool Set. The tools necessary for pedicle hole preparation, i.e. pedicle probe, pedicle awl, and pneumatic drill (based on the AO Spine Tool Set, Synthes, Waldenburg,

Switzerland) are equipped with CAS tools, holding three or more LEDs. All connectors are color-coded to avoid errors.

Calibration. In order to use the tools with the CAS system, a calibration or geometric registration of each tool is required. The calibration remains valid as long as the marker set is rigidly attached to the tool and no deformation occurs.

Dynamic Reference Base (DRB). During surgery, the system must compensate for camera as well as patient motion. Thus, a dynamic reference with LED markers is attached to the spinal process of the vertebra undergoing surgery. All mathematical transformations are then performed relative to the DRB.

Software. The basic module is based on the NSPS, a software previously used to aid neurosurgery.[12,13] It allows for image acquisition, pre-operative planning and simulation, and skeletal registration as well as tracking and real-time visualization of the surgical tools.

Trajectory Module. The trajectory, i.e. the desired screw axis, can be defined by selection of an entry and target point, using a best fit algorithm, or using the intraoperative real-time trajectory module of the neurosurgery software program. The module allows the surgeon to intraoperatively select a point of entry and trajectory axis with any tool and immediately displays cuts perpendicular to the chosen trajectory . This enables a fast and convenient method to simulate the screw insertion and 3-D placement.

4 Surgical Preplanning

Axial tomographic scans (CT) in 2 mm slides of all vertebrae through the proposed surgical field are obtained. The basic module of the software program reconstructs and displays 3-D vertebral images or multiple 2-D views, i.e. frontal, sagittal, and transverse sections. Using the mouse cursor it is possible to reach any point in correlated with different sections directly . Surgical preplanning proceeds as follows:

Recording the Matching Points. The bone anatomical landmarks for intraoperative matching are recorded before the surgery. The principle for choosing these points is ease of identification both on the images and during surgery. The spinal processes and transverse processes are normally used for this purpose.

Determining the Screw Length and the Width of Pedicle . In this part of preplanning, the main purpose is to provide more accurate information for the insertion of the pedicle screw. Even the angle of insertion can be measured during the preplanning.

Creating a Ideal Trajectory for the Pedicle Screw. In order to create an approximate axis for pedicle screw insertion an entry and a target point are defined by the surgeon preoperatively. This is performed using the software trajectory module by capturing coordinates from any of the sections with the mouse cursor. A serial sections orthogonal to the chosen trajectory (surgeon's perspective) can be displayed to check the insertion axis along the entire pedicle. The optimized chosen trajectory is saved in a file and used as guidance to insert the pedicle screws intraoperatively.

5 Intraoperative Interactive Image Guidance

The images are presented intraoperatively to the surgeon in a multiplanar reconstructed fashion. In addition to the standard coronal, sagittal, and axial planes, a set of surgeon's perspective views showing the axial view of the screw trajectory can be displayed, starting from the entry point to the end point of insertion. (Fig 2)

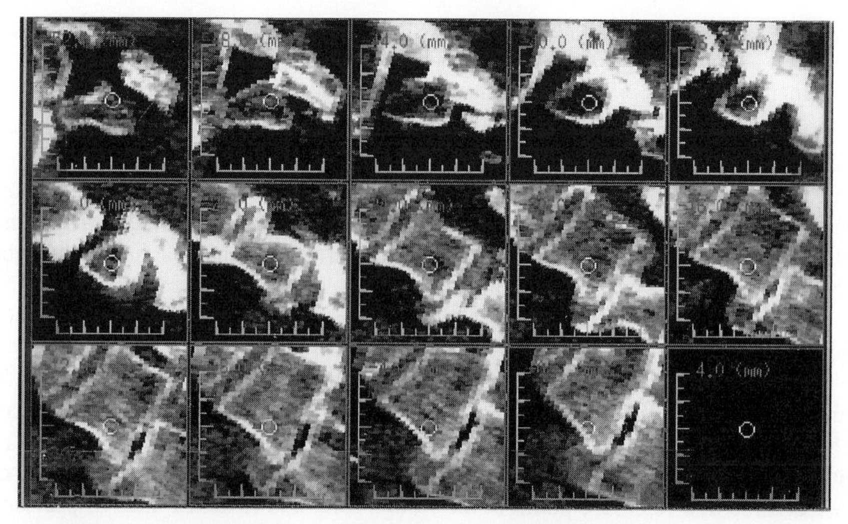

Figure 2. The surgeon's perspective view of the pedicle.

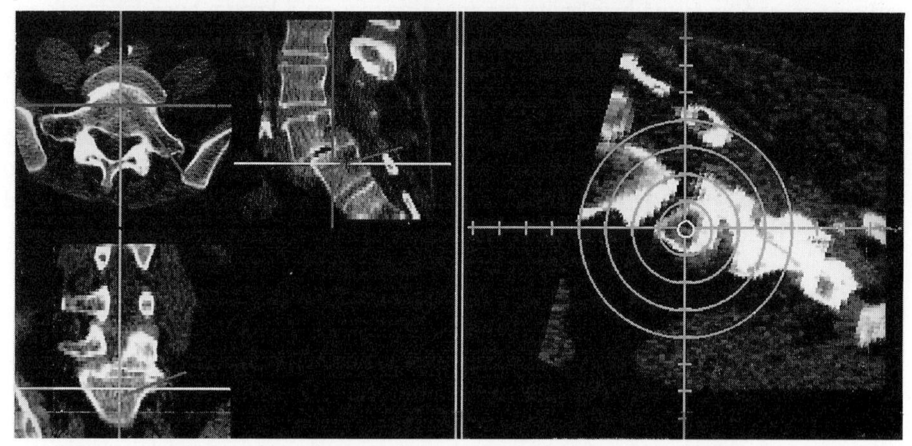

Figure 3. The real-time trajectory and surgeon's perspective views.

This view is very helpful to observe to whole procedure of the pedicle screw inserts through the pedicle. In the operating room, the patient is placed in the prone or kneeling position. A standard exposure of the spinal levels to be instrumented is performed. During the exposure of the dorsal vertebral elements the assistant prepares the CAS instruments. The colored-coded plugs are dropped and connected to the strober unit . The instrumented tools are calibrated (hand-held) without any delay in the surgical procedure. A lateral radiograph is obtained to confirm the appropriate level. The DRB is fixed on the chosen spinal process and keeps it stable during the whole procedure of inserting the pedicle screws. At least three reference points previously selected from the CT image set are identified in the operative field and digitized with the space pointer. After the coordinate of these reference points are obtained, the relationship between the image data and the intraoperative anatomy is defined, allowing rapid mathematical conversion between the two using a 3-D coordinate transformation algorithm described previously. This procedure is performed in less 5 minutes. The accuracy of this conversion can be assessed by localizing other bone landmarks in the field and comparing their location to the corresponding point identified on the CT image. Once the matching process is complete the surgeon can precisely locate and mark the entry point with the pedicle probe using the pre-operatively defined or the real-time trajectory module within the software program. The orientation of each pedicle can be showed quickly and easily. (Fig. 3)

Any errors in trajectory or entry point can be determined and corrected by minor adjustments of the drill guide position. After the drill guide position is satisfactorily aligned with the optimum trajectory, a pilot hole is drilled through the pedicle into the vertebral body. Then the appropriate length and diameter screw is inserted. The process is repeated for each pedicle to be instrumented; C-arm fluoroscopy or serial radiographs are not required.

There are two options for using the system at a second spinal level. The first method is to check the landmarks by using the space pointer on that level. If these landmarks are accurately matched with the CT image, the procedure of inserting a pedicle screw on that level can be continued. If the accurate matching cannot be reached, the second method is to select three new reference points at the second level and to match them as previously described. At any point during the localization and orientation procedure, the accuracy of the system can be tested by placing the space pointer on a known bone landmark and confirming its location on the CT image showing on the computer screen. If there is any discrepancy between the space pointer and cursor position, the matching process can easily be repeated. Accuracy is dependent on how closely the space pointer is placed to the exact site selected as a reference point as well as on the proper immobilization of the patient and the DRB during the matching, localization, and orientation procedures. The ability to localize reference points serially during surgery is not as easily performed with intracranial procedures but confirms and enhances the accuracy of the system during spinal surgery.

6 Results

Since May 1995 to April 1996, the CAS system has been used to assist insertion of 36 pedicles into the spine of ten patients requiring spinal fixation following the procedure we have just described. All ten spine patients underwent fusion of the level T9-S1 as a primary surgical intervention for severely degenerative changes, spondylolisthesis, or a spinal vertebral fracture. Intraoperatively, the system provided rapid and relatively precise identification of the pedicle and its orientation to the exposed surface anatomy of the spinal column. This reduced the operative time for the screw insertion by approximately 30 minutes. It also reduced the need for intraoperative fluoroscopy and serial lateral radiographs. It was very helpful for identifying pedicles when traditional bone landmarks were absent (due to previous surgery). Following transfer of the digitized CT image and establishing a file of specific points or "landmarks" on the area of the spine to be exposed during surgery, the portable equipment was arranged in the operating room. All components carrying markers underwent gas sterilization prior to surgery. Matching was done of the previously selected virtual landmarks with their actual analogues on the patients. The mean time of matching was 6.4 minutes. Guided placement of transpedicle screws was then performed as described above. In one case, only unilateral placement was possible because the position of the DRB was changed and tracking failed. Postoperative CT showed all pedicle screws were in position and in good agreement with the intraoperative simulation. Clinically, all 10 patients reported improvement in their preoperative back pain or other discomfort. Of the 36 screws placed, all were determined to be in the ideal position. There was no screw perforation of lateral cortex, and no occurrence of spinal cord or nerve root injury secondary to screw insertion. A-P and lateral plain film radiographs or CT scans of the instrumented levels were obtained at each subsequent follow-up visit in all patients. To date, there has been no incidence of pseudoarthrosis or fixation failure secondary to pedicle screw breakage or displacement.

7 Discussion

In the past decade, pedicle screw fixation of the spinal column has gain greater acceptance as improved instrumentation and clinical effectiveness has become apparent.[10] However, because of the complexity of the spinal anatomy involved as well as the need of individualize the construction to each patient, insertion of a pedicle screw has a significant learning curve. Traditionally, the approximate location of each pedicle was determined by identifying specific bone landmarks in the surgical field. Before inserting the screws, the surgeon could only approximate the pedicles' dimensions and estimate angulation in the sagittal and axial plane. This resulted in varying incidences of neural injury and fixation failure secondary to screw misplacement or fracture. These complications can be reduced if the surgeon is properly oriented to each pedicle to be instrumented. Failure in pedicle screw placement involving pedicle cortex perforations have been reported at rates ranging from 5 - 6%, with an overall complication rate of 27.4%.[1,3] One of the reasons for these failure rates may be the significant morphometric variability in pedicle dimensions, particularly the transverse width and angle.[1]

Various techniques have been proposed to more safely insert pedicle screws.[7] Identification of the entry point is commonly based on anatomic landmarks. With the help of CT and X-rays the proper angulation of the screw axis may be gained.[2] Mechanical drill guides have been reported to reduce the risk of pedicle and vertebral cortex perforation.[11] However, these procedures are limited due to their lack of ability to visualize the precise 3-D position of the screw within the pedicle. Vaccaro and Garfin reviewed 54 articles and concluded that the application of pedicle screw fixation requires a substantial "practice effect," because of the difficulty of placing a screw "blindly" down a small hollow tube, that is, the pedicle surrounded by neural elements.[8] Only through appropriate training, a thorough knowledge of lumbar pedicle structure, and an appreciation of spinal biomechanics, can pedicle screw instrumentation be a safe, useful, and important adjunctive means of rigidly fixing the lumbar spine.[8] Stereotaxy for intracranial localization has been widely accepted as an effective and accurate method.[12,13] However, the application of this technology to spinal surgery has not been widely accepted.[4] Nolte and colleagues reported preliminary results from 3 cases using the Optotrak system for pedicle screw placement.[5,6]

For application of stereotactic technology to spinal surgery, the spine itself provides the reference points. The CAS system enables planning and simulation of orthopedic surgeries as well as intraoperative real-time control. The matching process "links" the reconstructed image set with the intraoperative anatomical structures. The appropriate screw entry point, angle of trajectory in the axial and sagittal planes, and depth of insertion can be determined by simultaneous analysis of the reconstructed images. The system functions as a confirmation tool to assist the surgeon in identifying the pedicle and relating its position and orientation to the exposed spinal anatomy. Preliminary use of this system has shown that it provides useful and accurate information for identifying pedicles and their orientation. Maintaining the dynamic reference base fixed on the spinal process during the procedure is critical. If the DRB changes position, the whole transformation matrix will be wrong. Rematching is required under this circumstance.

8 Conclusion

Our initial experience with the CAS system appears to confirm its clinical potential. We are in the process of assisting additional transpedicular screw insertions in order to statistically analyze our results, and we will attempt to extend the system to other orthopedic interventions such as periacetabular fractures, cervical spinal fractures, high level tibial osteotomy, and impacted femoral neck fractures.

9 References

1. Esses SI, Sachs BL, Dreyzin V: Complications associated with the technique of pedicle screw fixation - A selected survey of ABS members. Spine **18** (1993) 2231-2239
2. Farbar GL, Place HM, Mazur RA, Casey Jones DE, Damiano TR: Accuracy of pedicle screw placement in lumbar fusion by plain radiographs and computer tomography. Spine **20** (1995) 1494-1499
3. Gertzbein, SD, Robbins SE: Accuracy of pedicle screw placement in vivo. Spine **15** (1990) 11-14
4. Kalfas IH, Kormos DW, Murphy MA, et al: Application of frameless stereotaxy to pedicle screw fixation of the spine. J Neurosurg **83** (1995) 641-647
5. Nolte LP, Zamorano L, Jiang Z, Wang Q, Langlots F, Berlemann U: Image-guided insertion of transpedicular screws: a laboratory set-up. Spine **20** (1995) 497-500
6. Nolte LP, Visarius H, Arm E, Langlots F, Schwarzenbach O: Computer-aided fixation of spinal implants. J Image Guided Surgery **1** (1995) 88-93
7. Steinmann JC, Herkowitz HN, El-Kommos H, Wesolowski DP: Spinal pedicle fixation -Confirmation of an image-based technique for screw placement. Spine **18** (1993)1856-61
8. Vaccaro AR, Garfin SR: Internal fixation (pedicle screw fixation) for fusions of the lumbar spine. Spine **20** (1995) 157S-167S
9. Weinstein JN, Spratt KF, Spengler D, et al: Spinal pedicle fixation: Reliability and validity of roentgenogram-based assessment and surgical factors on successful screw placement. Spine **13** (1988) 1012-1018
10. Wiltse LL: History of pedicle screw fixation of the spine. Spine State Art Rev **6** (1992) 1-10
11. Wu SS, Liang PL, Pai WM, Au MK, Lin LC: Spinal transpedicular drill guide: Design and application. J Spinal Disorders **4** (1991) 96-103
12. Zamorano L, Nolte LP, Kadi M, Jiang Z: Interactive intraoperative localization using an infrared-based system. Neuro Res **15** (1993) 290-298
13. Zamorano L, Kadi M, Dong A: Computer-assisted neurosurgery: Simulation and automation. Stereotact Funct Neurosurg **59** (1992) 115-122

ACKNOWLEDGMENT: This research was supported by the Detroit Receiving Hospital and Wayne State University, Bioengineering Center.

PROBOT - A Computer Integrated Prostatectomy System

Q. Mei*, S.J. Harris*, F. Arambula-Cosio*, MS Nathan†, R.D.Hibberd*, JEA Wickham†, B.L.Davies*

* MIM (Mechatronics in Medicine) Laboratory, Imperial College, London SW7 2BX, England
† Department of Surgery, Guys Hospital, London SE1, England

Abstract: A computer integrated prostatectomy system named **PROBOT** has been produced to aid in the resection of prostatic tissue. The system is image guided, model based, with simulation and online video monitoring. The development and trial of the system have not only demonstrated the successful robotic imaging and resection of the prostate, but have also shown that soft tissue robotic surgery in general, can be successful.

1 Introduction

Computer integrated surgery is an emerging active field. It integrates computer assisted surgery and robotics with imaging & modelling to offer the possibility both to significantly improve the efficacy, safety, and cost-effectiveness of existing clinical procedures and to develop new procedures that cannot otherwise be performed.

Fig. 1. PROBOT is in operation

Most computer integrated surgery systems that are under development are hard tissue or bone related[7,8,9]. Because of their inherent rigidity, bones can reliably be immobilised and their location repeatably and consistently correlated to the computer model.

Research into the computer integrated prostatectomy system, which incorporates a purpose-built robot to create a cavity in the prostate (i.e., a soft tissue), has been undertaken at the **Mechatronics in Medicine (MIM)** lab, Imperial College, London for a number of years[**1-6**]. The system has evolved into the current version named **PROBOT** (**PRO**STATECTOMY **ROBOT**). The task of the robot is to position and control specific devices. Currently, the system is undergoing a series of clinical trials at Guy's Hospital, London [**Fig. 1**].

The current system is designed to have the following features:
- Automatic transurethral US(ultrasound) scan of the prostate at 5mm intervals between bladder neck and verumontanum(a visible landmark) using the robot.
- Generation of an optimised cutting sequence based on a surgeon defined cutting volume and cutter data.
- Dynamic visualisation of final realised cutting volume.
- Real time simulation of what has been cut and what is proposed to be cut.
- Real time video display showing the actual cutting progress inside the prostate.
- Re-generate the cutting sequence, allowing the cavity to be redefined on recorded image slices during the operation. The cutting sequence thus generated excludes the region which has already been cut.
- Run time hardware check with control data test and rectification.
- Light pen or mouse operation for all data and command inputs via a virtual keyboard on screen to suit the requirements of the operating theatre.
- Hardware constraints to prevent gross movements outside the prostate region[2].
- Use of VaporTrode roller cutter (a diathermy cutter developed by **CIRCON ACMI**) for non-bleeding cutting and clear video image monitoring.
- Highly interactive graphical user interface using VESA standard under DOS.

This paper mainly focuses on the interactive procedure, the graphical user interface and the general working principles of the **PROBOT** system.

2 The PROBOT System

Fig. 2 shows a simplified block diagram of the system. Data from US scanner and video camera have been integrated into the computer by means of a frame-store card. As the robot has 6 degrees of freedom, and all the end effectors (US probe, cutter, resectoscope, video camera, etc.) can be easily mounted on it, different functions can be performed successively or concurrently under the control of the computer.

The system works in an active way. The process of imaging, planning and cutting all use the same coordinate system of the robot. The US probe is inserted along the resectoscope sheath and applied with the patient in the same anatomical position as that subsequently used for cutting, thus making registration between imaging and cutting a simple procedure. Although some distortion of the prostate may occur, when the US probe is passed through it, this will be of largely the same type as that experienced by the passage of the resectoscope itself. Thus, any distortion in the image is the same as the distortion introduced by the cutter.

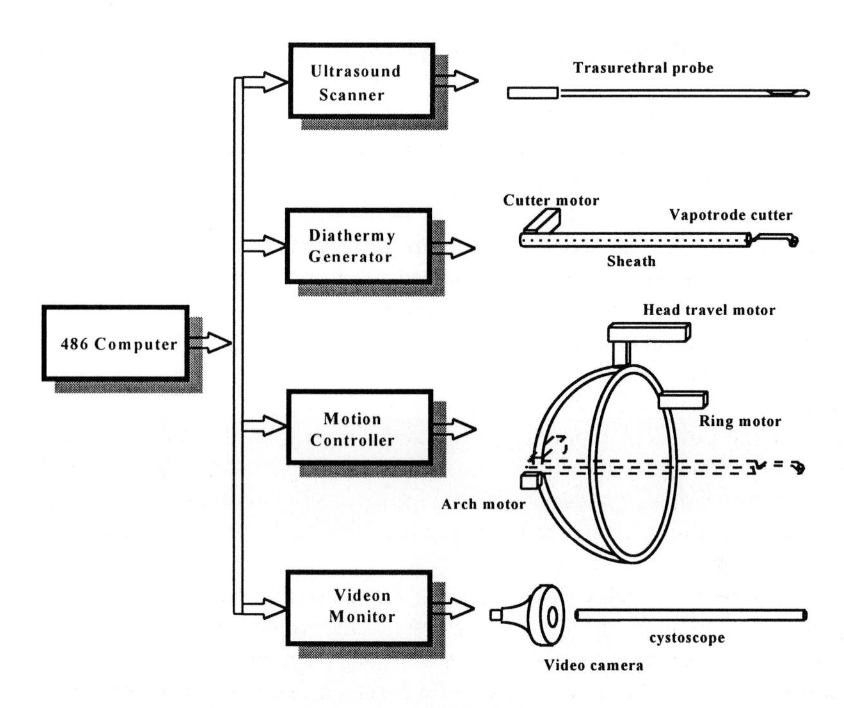

Fig. 2. A simplified block diagram of PROBOT

3 Interactive Procedure of the System

When the patient is anaesthetised, the surgeon enters the cystoscope, with the sheath, transurethrally into the patient's prostate. Then, while maintaining the sheath tip at the bladder neck and in an appropriate orientation, the surgeon will bring the robot over and lock it to the sheath base. The video camera is then mounted on the cystoscope so that the situation inside the prostate is displayed on the screen. After that, the surgeon visually inspects the prostate by controlling the robot through an interactive menu and registers the two key positions (the bladder neck and verumontanum) to the coordinate system of the robot. The verumontanum is an identifiable landmark and must not be cut.

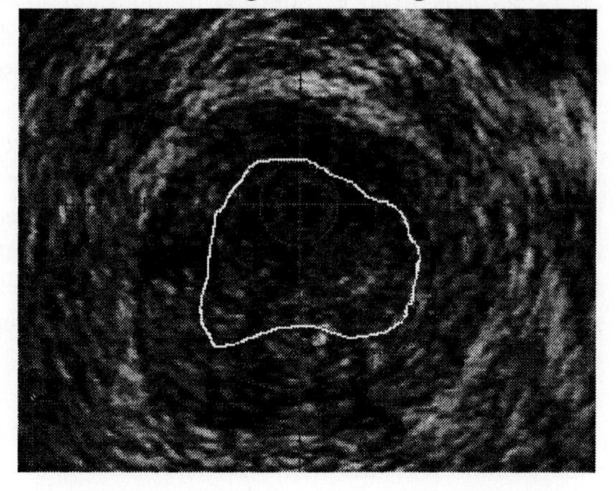

Fig.3. Drawing the cutting area on image slice

The cystoscope is then replaced by a transurethral US probe (**B&K**, 5.5Mhz), and an automatic imaging process carried out by the robot to take prostate images at 5mm intervals. After this is done, the surgeon draws and edits, using a light pen, an appropriate cutting area on each recorded image slice (**Fig. 3**).

During cutting, the sheath of the cutter is positioned at different eccentric angles inside the prostate. It compresses and deforms the prostate tissue to some extent. Thus a parameter called *press_depth* is introduced to characterise this phenomenon. By changing *press_depth*, the surgeon can, to some extent, control the size of the opening of the cutting volume (**Fig. 4**).

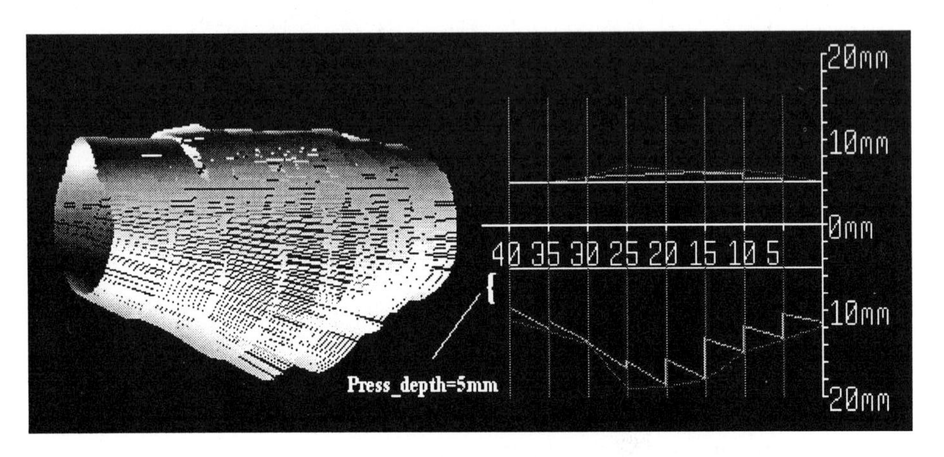

Fig. 4. Display of mechanically feasible cutting volumes

Next, the software will generate the cutting sequence. The total number of cuts or the total time spent on cutting (as time for a single cut is almost a constant and known to be 14 seconds) can be adjusted by both *cut_depth* and *cut_width*. The smaller the value for *cut_depth* or *cut_width* is, the smaller the amount of tissue each cut will remove, and therefore the larger the number of cuts(**Fig. 5**).

The whole interactive procedure allows the surgeon to adjust or refine the cutting details of the operation to his satisfaction. After the cutting is started, the cutting procedure can be interrupted and the cutting speed adjusted. The cutting equipment can be changed if necessary as the coordinate system of **PROBOT** is maintained until the operation is finished. The cutting area on each slice can be modified and those regions already cut are remembered by the program and are excluded in the regeneration of the cutting sequence.

The PROBOT supports two cutting schemes: one is *basic multi-cone cutting* and the other is *multi-extended-cone optimised cutting*.

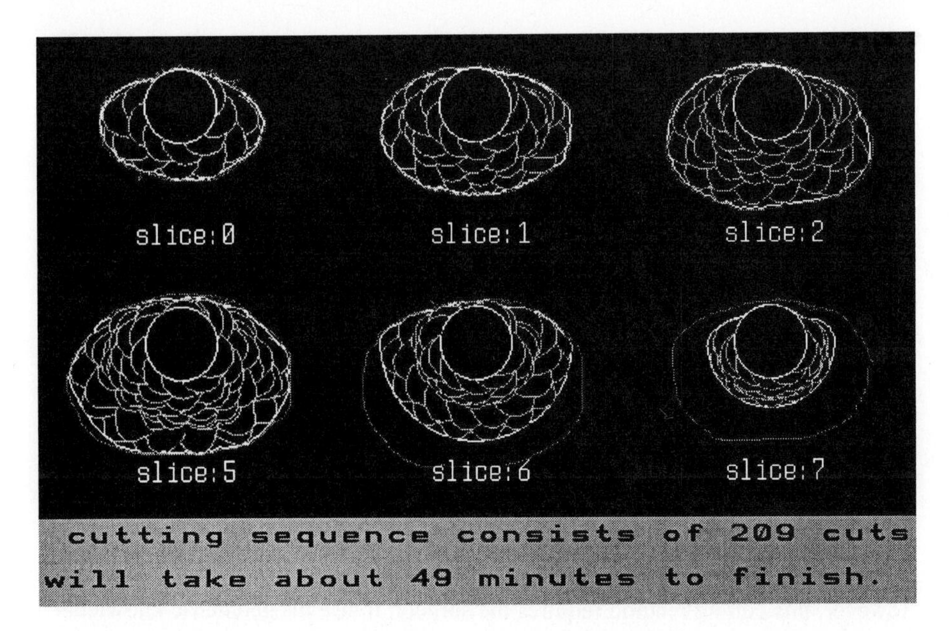

Fig. 5. Generation of optimised cutting sequence before cutting (part of the display)

4 Basic Multi-Cone Cutting

The principle of basic cone cutting is shown in **Fig. 6.** The surgeon is required to choose how many cones and where (an image slice) each cone starts. Each single cone obeys the following rules:

1. the maximum radius of the surgeon defined cutting area on the start slice will be the radius of the cone base;
2. the maximum distance between sheath tip and cutter tip (24mm) will be the edge of the cone where possible.
3. the pivot position of each cone to be fixed and aligned with the central axis.
4. the arch angle to be within 30°.
5. the last cone to be positioned at verumontanum.

An example is shown in **Fig.** 7. For a single cone, the cutting sequence is generated from inside to outside, and in a first clockwise then counterclockwise way (to avoid entangling the cables around the robot frame).

When using basic multi-cone cutting, the cutting details of the subsequent cones can be adjusted after the current one is finished. This increases the system's flexibility. Furthermore, as the cone only runs from bladder neck to verumontanum, this means that the sheath will never go towards the bladder neck. This is believed to be a safer procedure considering the prostate's anatomical structure.

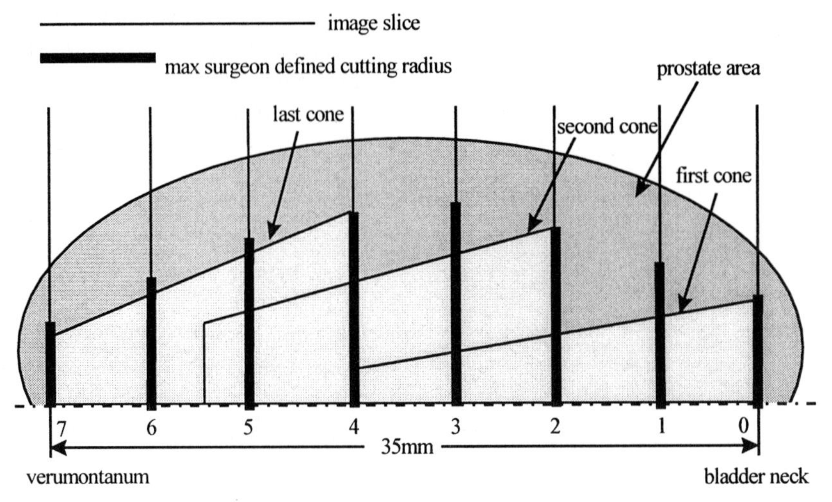

Fig. 7. Example of basic multi-cone cutting

However, this cutting method requires the surgeon to use his geometric knowledge of the cutter and the motorised frame to select the appropriate number of cones and where each cone should start. Besides, there may be some region shared by different cutting cones, so that repetitive and unnecessary cuts may be generated resulting in extra cutting time. Also, situations could exist where a slice with a large defined cutting area is neither marked as the start slice nor is fully covered by any cutting cone, resulting in under-cutting.

5 Multi-extended Cone Optimised Cutting

The time taken by the cutting sequence is mainly determined by the total number of cuts. To minimise the total number of cuts, it is necessary to make a minimum number of cutting cones and remove a maximum amount for each cut (proportional to the cutting stroke, *cut_width* and *cut_depth*). It is also necessary to avoid repetitive cuts, which requires remembering where all the previous cuts have reached so that the region can be avoided and the following cuts can be generated more effectively.

Compared with the basic multi-cone cutting, this scheme has three different aspects. First, each cone is extended to the maximum mechanical reach, which uses where possible, the maximum mechanically feasible base radius and cutting stroke, and one more covering slice next to the cone's base[**Fig. 8**]. Secondly, the software will decide how many necessary overlapping cones are to be used to cover the surgeon defined cutting volume. Thirdly, these extended cones are taken as a whole to generate an optimised cutting sequence before the actual cutting is started.

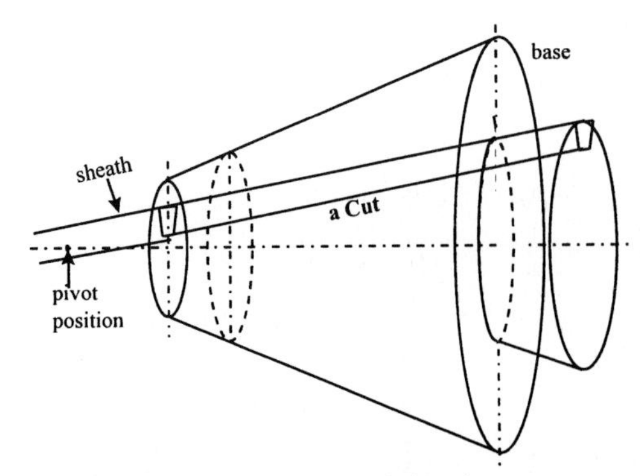

Fig. 8. Principle of the extended cutting cone

Fig. 9 shows an example of using the scheme where three cones are generated. Principally these cones consist of a *set* of cuts covering the surgeon defined cutting volume. The optimisation of cutting sequence generates, from the *set,* the least number of cuts which cover the same cutting volume so as to minimise the cutting time. **Fig. 10** is an example showing the principle of this. The three overlapping cones are changed into three non-overlapping cones through a set operation, thus the repetitive cuts can be avoided.

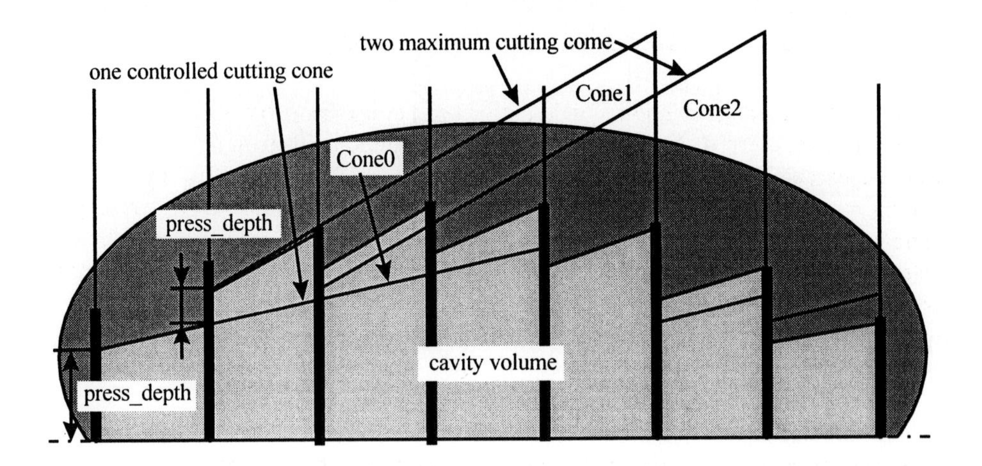

Figure 9- Example of using multi-extended cutting cone method

The algorithm for optimised cutting sequence generation is based on the above principle. The method of *max-stroke cutting by cone* simply refers to generating cuts cone by cone, while the method of *max-stroke cutting by ring* refers to generating cuts by rings in the full length from bladder_neck to verumontanum. The former only requires the sheath to go towards bladder-neck once, whereas the latter requires the sheath to go back and forth frequently. Although *max-stroke cutting by ring* also produces fast cutting, it is not considered to be safe as prostate may be distorted so that the registration can be ruined. The compensation for possible distortion of the prostate by *max-stroke cutting by cone* when going towards the bladder-neck is to let the sheath tip reach the bladder-neck first and then retreat to the desired cone position rather than going there directly.

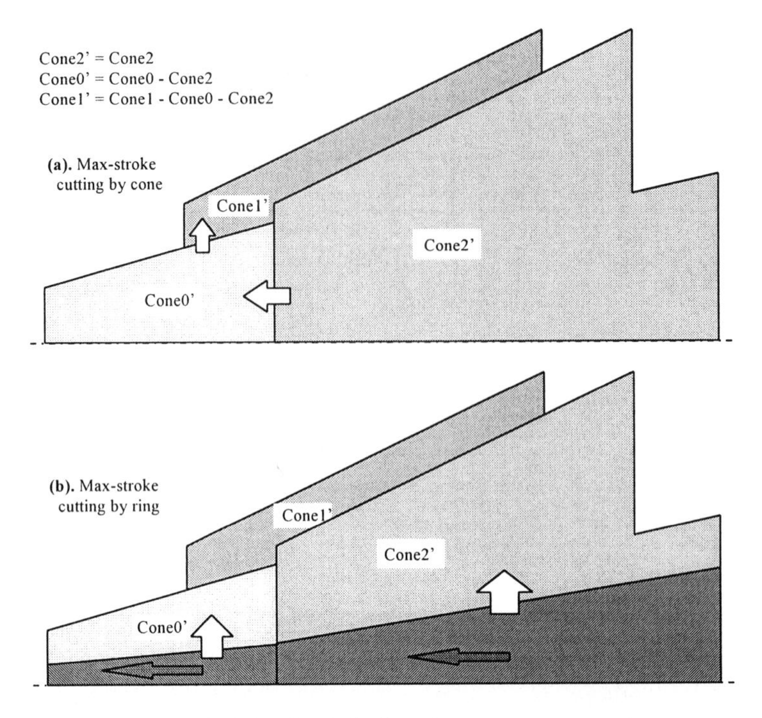

Fig. 10. Principle of optimised cutting sequence generation

By optimising the cutting sequence generation, the system has the following advantages:

- The surgeon is only required to tell the computer where the verumontanum and where the bladder neck are and subsequently draw the cutting area on each recorded image. All the other work is done by the computer, thus relieving the surgeon from knowing the engineering details of the system.
- All the cutting cones are arranged in the most efficient way and the cutting sequences are generated optimally. The maximum amount of tissue cutting and the least amount of time taken can thus be ensured.
- The whole cutting sequence is generated before the actual cutting starts. Simulation of the actual cutting process can be fully realised. The total cutting time left can be shown, since the average time for each cut is known.

During cutting, the cutting sequence is simulated on the screen showing what has been done, what has not been done, and where and how much the next cut is going to remove and how much time is left for the rest of cuts. Cuts are clearly marked on the recorded US image where the current cone base stays. The live video from the cystoscope shows clearly the cutting process inside the prostate. The VaporTrode roller(from **Circon ACMI**) cutting is virtually blood free, since it coagulates very well while cutting. This is essential for the clear video. An example is shown in **Fig.11.**

The system also allows the surgeon to stop at any time if necessary, to either change the speed of cutting or redefine the cutting area on each slice while maintaining the original registration. The optimised cutting sequence can be regenerated excluding the area which has already been cut.

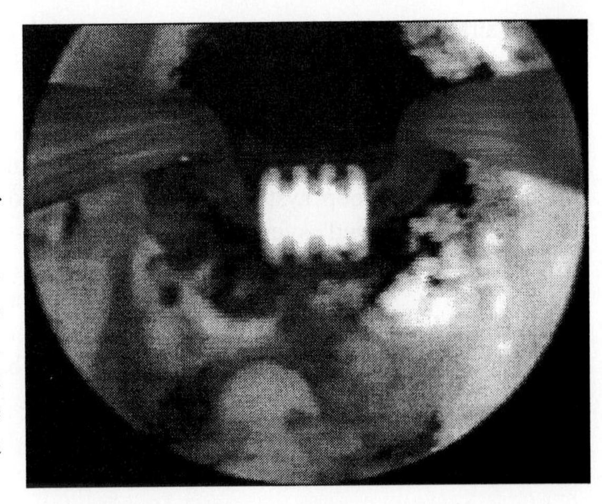

Fig. 11. Live video image from cystoscope

6 Conclusion, Discussion and Future Research

The graphical user interface has been designed to allow all functions of the software to be easily accessible to the surgeon, and information is presented to the surgeon in an easy to read graphical format. Careful design of this interface removes the possibility of the surgeon making any numerical input errors. The display of models of the cut region, both as a prior simulation and as a display of current cuts, enables him to be sure that the cuts made are correct.

During cutting, the clear video display shows the actual situation of the cutting progress. The US image slice display will show the tissue, while the overall simulation of cutting progress shows the running state of the cutting sequence. By monitoring these, the surgeon can know exactly what the surrounding area of the cutter is; how much has been cut and where the cutter will go. If there is any cutting action which looks too close to the outer limit of the prostate or looks dangerous to the patient, the surgeon can pause, interrupt or cancel the process. It is always essential for the surgeon to be able to smoothly interrupt and modify the cutting sequence, to ensure he is always in control.

So far, the **PROBOT** using the basic multi-cone cutting method, has been successfully used on 16 patients. Among them 6 were purely for setup and imaging trials, 9 were for partial robotised operation, and one was fully finished by **PROBOT**. More trials are being arranged and the results will be reported later.

By nature the prostate can be regarded as a fairly hard object which is fixed in position inside the human body anatomically. A rigid model can thus be generated and processed which makes the robotised operation possible. This model may differ slightly from the actual prostate during operation, as the US probe or cutter sheath tends to straighten the prostate. On the other hand, the prostate tends to restore to its original shape when tissue is being removed or the sheath is withdrawn and the tension is released. Thus the surgeon needs to keep in mind this dynamic procedure, and try to predict and compensate when he draws the cutting areas and be ready to intervene in case an emergency situation occurs. In terms of control, the integrated system is an open loop system although the video and simulation monitoring and checking can be viewed as feedback.

Future work is being focused on the segmentation of the US images so that the defined cavity region of the prostate can also be generated automatically and a more consistent operation result achieved. In addition a barrel shape, instead of a stepped one of the cutting volume, is also being pursued. It is believed that the computer integrated surgery system can help the surgeon to perform operations faster, better, and safer.

References

1. Davies, B. L., Hibberd, R.D., Ng, W.S., Timoney, A., Wickham, J.E.A., **"Mechanical Constraints-The Answer to Safe Robotic Surgery?"** J.de Innovation et Technoogie en Biologie et Medecine, France, Vol. 13 No. 4. pp425-436, Sept. 1992.
2. Gomes, M.P.S., Arambula-Cosio, F., Mei, Q., Harris, S., Hibberd, R.D., Davies,B.L., "**Computer Assisted Ultrasound Segmentation for Robotic Prostatectomy**", 7th National Conference on Pattern Recognition, Aveiro, Portugal, 23-24 March 95.
3 Harris, S., Mei, Q., Arambula-Cosio, F., Hibberd, R.D., Nathan, MS, Wickham, J.E.A., Davies, B. L., **"Transurethral Resection of the Prostate: The evolution of a robotic system to meet users' needs**", Proc. of MEDIMEC'95, pp7-14, Sept. 1995.
4. Harris, S., Mei, Q., Arambula-Cosio, F., Hibberd, R.D., Nathan, MS, Wickham, J.E.A., Davies, B. L., "**A Robotic Procedure for Transurethral Resection of the Prostate**". Second Computer-assisted Medical Robotics International Conference, John Wiley Press, Baltimore, USA, November 1995.
5. Ng, W.S., **"Development of a Surgeon Assistant Robot for prostatectomy"**, *Ph.D. Thesis*, University of London, 1992.
6. Davies, B.L., **"Medical Robotics"**, *Ph.D. Thesis*, University of London, 1995.
7. LAVALLÉE, S, et al., **Image-Guided Operating Robot: A Clinical Application in Stereotactic Neurosurgery,** p343-352, COMPUTER INTEGRATED SURGERY Technology and Clinical Applications, The MIT Press, 1996.
8. Kall, B. A. **Computer-Assisted Surgical Planning and Robotics in Stereotactic Neurosurgery,** p353-362, COMPUTER INTEGRATED SURGERY Technology and Clinical Applications, The MIT Press, 1996.
9. Taylor, R. H., et al., **An Image-Directed Robotic System for Precise Orthopaedic Surgery**, p379-398, COMPUTER INTEGRATED SURGERY Technology and Clinical Applications, The MIT Press, 1996.

Towards Performing Ultrasound-Guided Needle Biopsies from within a Head-Mounted Display

Henry Fuchs, Andrei State, Etta D. Pisano MD*,
William F. Garrett, Gentaro Hirota, Mark Livingston, Mary C. Whitton,
Stephen M. Pizer

Department of Computer Science
*Department of Radiology, UNC School of Medicine
and Member, UNC-Lineberger Comprehensive Cancer Center

University of North Carolina at Chapel Hill
Chapel Hill, North Carolina, 27599

Abstract. Augmented reality is applied to ultrasound-guided needle biopsy of the human breast. In a tracked stereoscopic head-mounted display, a physician sees the ultrasound imagery "emanating" from the transducer, properly registered with the patient and the biopsy needle. A physician has successfully used the system to guide a needle into a synthetic tumor within a breast phantom and examine a human patient in preparation for a cyst aspiration.

1 Introduction

In recent years, ultrasound-guided needle biopsy of breast lesions has been used for diagnosis and has replaced open surgical intervention. Ultrasound guidance is also often used for needle localization of some lesions prior to biopsy, as well as for cyst aspiration. However, ultrasound guidance for such interventions is difficult to learn and perform. A physician needs good hand-eye coordination and three-dimensional visualization skills to guide the biopsy needle to the target tissue area with the aid of ultrasound imagery; typically, the ultrasound data is viewed separately from the patient, displayed on a conventional video monitor. We believe that the use of augmented reality (AR) technology can significantly simplify both learning and performing ultrasound-guided interventions.

AR combines computer-synthesized images with the observer's view of her "real world" surroundings [Bajura 92]. In our application, the synthetic imagery consists of computer-processed echography data, acquired through ultrasound imaging, and geometric representations of the ultrasound probe and of the patient's skin surface. The real-world surroundings are the physician's view of the patient, acquired by miniature video cameras mounted on the physician's head. Tracking systems acquire position and geometry information for patient and physician. A high-performance graphics computer generates the combined imagery in real time. The composite images are presented to the physician user via a video-see-through head-mounted display (abbreviated as HMD in the following). The physician sees the ultrasound imagery "in place," registered with the patient.

With conventional methods, the physician has only a two-dimensional ultrasound image to aid her in the inherently three-dimensional task of guiding a needle to a biopsy target. We hope that our system, by presenting echography data registered within the patient, will make these widely-practiced procedures easier to perform, both in the breast and eventually also in other, less accessible parts of the body.

In the following sections we discuss medical issues and motivation and give a brief description of our AR system. We conclude with preliminary experimental results and a discussion of near-term as well as longer-term future work.

2 Clinical Background

While screening mammography in conjunction with breast physical examination has been demonstrated to reduce breast cancer mortality, this test generates a tremendous number of breast biopsies [Feig 88, Shapiro 77, Shapiro 82], of which relatively few cases are actually malignant. In fact, among non-palpable lesions that are submitted to needle localization and open surgical biopsy, only 10-30% prove to be malignant [Schwartz 88, Skinner 88, Landerscaper 82, Wright 86, USPSTF 89]. In recent years, ultrasound-guided percutaneous biopsy (both fine needle aspiration cytology and large bore core biopsy) of non-palpable breast lesions has been used as a supplemental diagnostic test, in large part replacing open surgical intervention for lesions [Fornage 87, Harper 88, Hogg 88]. In addition, ultrasound guidance is frequently used for needle localization of occult lesions prior to surgical biopsy and to drain probable cysts [Kopans 84, Muller 85, Kopans 82, Tabar 81].

Unfortunately, ultrasound guidance for these interventions is not an easy method to learn or perform; it requires extensive hand-eye coordination and a good sense for three-dimensional relationships. During the procedure, the biopsy needle is guided to the target tissue area with the aid of echography images. The data, in the form of real-time, single-slice ultrasound images, is displayed in two dimensions on the sonography unit screen (a conventional video monitor). The operator must learn to alter the position and orientation of the needle with respect to the lesion location in real time. Part of the difficulty in guiding the needle is keeping both the needle and the lesion simultaneously in view, that is, within the ultrasound image slice, as the needle approaches the lesion. Also, both position and orientation of the needle (5 degrees of freedom) must be ascertained, in part, from the two-dimensional ultrasound image. As mentioned before, the ultrasound slice is not registered with the patient; it is viewed separately from the patient. The slice's position and orientation within the patient must in turn be ascertained by the physician using visual and tactile feedback. Establishing correspondence between the biopsy needle and structures within the patient's body is thus an indirect process as far as echography image guidance is concerned. On the one hand, the physician must assess the geometric relationship between needle and ultrasound image, a difficult task since the ultrasound image has non-negligible “thickness.” On the other hand, she must assess the geometric relationship between the (“mostly” cross-sectional) ultrasound image and the body of the patient. Precision in both parts of this process is essential for the accuracy of the biopsy procedure but difficult to achieve. With the chest wall, pleura and pericardium in close proximity to the breast, and with lesion size ranging down to only a few millimeters, it is quite important for the needle to be accurately positioned, both in order to avoid complications and to obtain the correct diagnosis [Fornage 87, Gordon 93].

Ultrasound is preferable to stereotactic (mammographic) guidance in some circumstances because it does not use ionizing radiation, the guidance is utilized in "real time" which can significantly shorten the procedure, and there are some parts of the breast for which stereotactic guidance is impracticable (namely, in lesions directly against the chest wall and in the superficial subaureolar region) [Parker 93].

3 Motivation

We believe that an AR system displaying live ultrasound data in real-time and properly registered to the part of the patient upon which the intervention is performed could be a powerful and intuitive tool. In contrast with the traditional display method for echography imagery, this tool presents a stereoscopic, three-dimensional view of the patient fused with the ultrasound data, which is displayed not only correctly positioned and aligned within the patient, but also at its true scale (life size). (In contrast, in the traditional technique the ultrasound image occupies an entire video display and is viewed at varying degrees of magnification, often larger than life size.) The AR paradigm replaces the indirect two-part process described in Section 2 with a single correlation step, in which the computer-enhanced visual feedback matches the physician's view of patient, needle, and needle entry point, as well as the physician's tactile feedback. We believe the physician performing ultrasound-guided needle biopsy will benefit from such an integrated, unified sensory environment, which we postulate will require significantly reduced "intuition" on the physician's part. We anticipate the following specific benefits for such biopsy procedures:

- Reduced average time for the procedure (benefits both physician and patient)
- Reduced training time for physicians learning to perform needle biopsies
- Greater accuracy in sampling of and distinguishing between multiple targets
- Reduced trauma to the patient through shorter and more accurate procedures
- Wider availability of the procedure due to ease of performing it.

We selected breast biopsy as the first application of this new image display technology for several reasons. First, the breast is relatively accessible since it is located on the outside of the body. This simplifies the logistics of the procedure itself for this pilot developmental phase. Second, the patients are generally healthy and able to cooperate with the use of the new equipment. Third, biopsies can be performed outside a hospital setting (i.e., in the computer science department) without undue risk. The breast contains no major blood vessels or vital organs that might become accidentally damaged if the equipment were to unexpectedly malfunction.

We also chose breast biopsy because it presents realistic clinical challenges. Patient motion has a more pronounced effect on the location of an abnormality than it would with deeper organs. In addition, the lesions within the breast vary significantly in size and location but tend to be smaller than those sampled in deeper locations.

In summary, breast biopsy is procedurally convenient but poses realistic technical challenges. Consequently, if AR technology can meet these challenges, we can expect to eventually apply it to other organs in less accessible locations.

4 Augmented Reality System

We have designed and implemented a prototype AR system—described in detail in [State 96]—to aid physicians performing ultrasound-guided needle biopsies. It has been used for phantom experiments as well as human subject experiments and consists entirely of commercial components:

- A stereoscopic video-see-through HMD equipped with miniature video cameras, assembled from a Virtual Research VR4 HMD and two Panasonic GP-KS102 color CCD video cameras with Cosmicar 12.5 mm lenses.

• A PIE Medical Model 200 ultrasound machine with a 7.5 MHz linear transducer.

• An Ascension Flock of Birds™ with Extended Range Transmitter tracking system for tracking the observer's head.

• A FARO Technologies Metrecom IND-01 mechanical arm for precise (tethered) tracking of the ultrasound transducer.

• A Silicon Graphics Onyx Reality Engine2™ workstation with 4 high-capacity raster managers, a Sirius Video™ real-time video capture device and a Multi-Channel Output unit. The Sirius is used for both HMD video acquisition and for ultrasound video acquisition.

The prototype system makes heavy use of the high-speed image-based texturing capability available in the graphics workstation. The video capture subsystem acquires HMD camera video and ultrasound video. The display presented to the user resembles the display offered by an earlier on-line volume reconstruction system [State 94, 95], but the images obtained are superior to the older system's. The new system can sustain a frame rate of 10 Hz for both stereo display update and ultrasound image capture. It provides high-resolution ultrasound slice display and rendering for up to 255 ultrasound slice images at 256-by-256-pixel resolution. Registration errors (due to distortion and lack of precision in the magnetic head-tracker) are corrected dynamically, by videometric tracking of landmarks in the video image. The live video and computer graphics elements are composited digitally, which prevents artifacts introduced by the external chroma key device used in previous systems.

Each ultrasound slice is presented in its correct location and orientation at the moment of acquisition and displayed properly intersecting with other slices. Old slices dim and fade away in time, controlled by a user-definable decay parameter. This "3D radar display," expected to be useful in scanning moving structures such as fetuses or biopsy needles, reflects decreasing knowledge about imaging targets which have not been recently "visited" by the scanner. A modified Binary-Space-Partition (BSP—[Fuchs 80]) tree algorithm handles intersecting slices properly and manages expiration of old slices efficiently [Garrett 96]. A large number of directly rendered ultrasound slices can give the appearance of a volume data set.

The ultrasound probe is tracked by a high-precision, rigidly tethered mechanical arm (visible in Figs. 1, 4). The arm provides sub-millimeter positioning accuracy and thus guarantees registration between individual ultrasound slices.

The system is also capable of acquiring patient geometry for a specific area of the skin surface. This is currently done via a manually assisted "sweep" before the actual scanning. During the subsequent scanning the acquired geometry is used to render a synthetic opening—a virtual "pit"—embedded within the patient. The ultrasound data (consisting of slices) is displayed within this opening (Figs. 2, 3, 4, 5).

5 Results

In April 1995, during an ultrasound-guided needle biopsy procedure on a live human subject, the physician (Pisano) did not wear the HMD as the system did not yet have stereo capability. The HMD was worn by a separate observer. The physician used conventional ultrasound display for guidance. During this experiment's preliminary exploration phase, however, the physician did wear the HMD (Fig. 1, left) and was able to observe correct "registration" between tactile feedback and a volumetric display of a cyst presented within the HMD (Fig. 1, right).

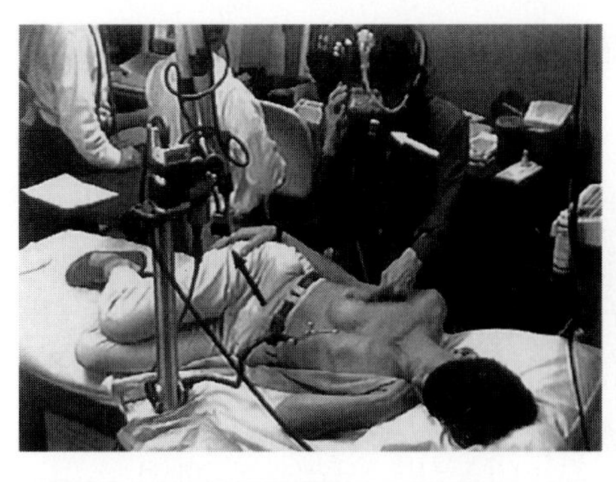

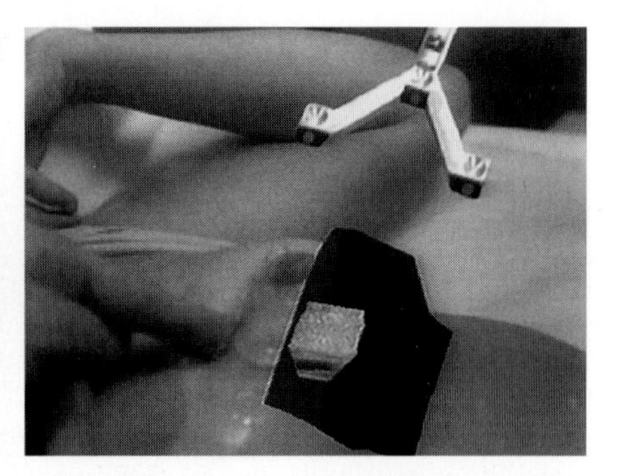

Fig. 1. Left: human subject experiment on April 7, 1995. The physician, wearing a HMD (marked by white arrow) examines the subject's right breast. The black arrow points to the ultrasound transducer attached to the mechanical arm for precise 6 degree-of-freedom tracking. Right: view inside HMD during the experiment in Fig. 1. The synthetic pit contains several ultrasound slices; the frontmost slice is a cross-section through a cyst (dark spot). The physician's finger points to the cyst as she perceives it via tactile feedback. The Y-shaped arm holds landmarks used to maintain correct registration between synthetic imagery and the patient.

In January 1996 the physician, wearing the HMD, successfully guided a biopsy needle into a synthetic tumor within a life-sized breast phantom (Fig. 2). Fig. 3 shows the volume data set acquired by the scanner. It contains images of the cyst and of the needle. The entire procedure was performed using the AR system. In a subsequent human subject experiment, the physician wore the HMD during the exploratory and planning phases preceding the actual biopsy on a human subject, all the way through partial insertion of a cyst aspiration needle. Fig. 4 shows the physician examining the patient and the image she sees inside the HMD.

During these experiments it became clear that the new AR paradigm (or rather, its imperfect implementation in our current system) also has a number of drawbacks:

• The tracked probe is sometimes cumbersome. The accuracy of the mechanical tracker attached to the ultrasound probe (visible in Figs. 1, left, and 4, left) is superior to that of all tracking systems we have experimented with in the past. This accuracy is necessary for correct registration, which is a key element of the AR paradigm. The arm is jointed arm and provides 6 degrees of freedom. Each joint, however, does have a range of motion which is limited to approximately 330°, with hard stops at either end. The physician occasionally was unable to quickly position the probe exactly as she desired due to running into a stop on one of the joints of the arm.

• The physician user must attempt to keep the head tracking landmarks in view at all times. Accurate tracking of the user's head is provided by a hybrid system using a conventional magnetic tracker in combination with vision-based landmark tracking. The techniques require that the landmarks be stationary. The landmarks are mounted on a fixture which is positioned near the patient, but not in the sterile field. (The fixtures are visible all HMD images in this paper) To be effective, the physician has to keep the landmarks in view of the head-mounted cameras, which represents a significant burden. In fact, keeping the landmarks in view proved difficult unless they were positioned so close to the working area as to often physically encumber the physician.

• The pixel resolution of ultrasound slice as it appears within the HMD is poor. Due to the pixel resolution of the HMD, the overall resolution of the images presented to the user is only about 240 x 200. In addition, the ultrasound data, as

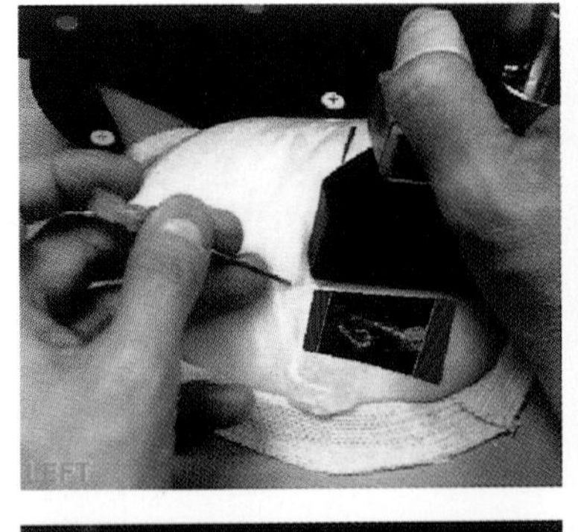

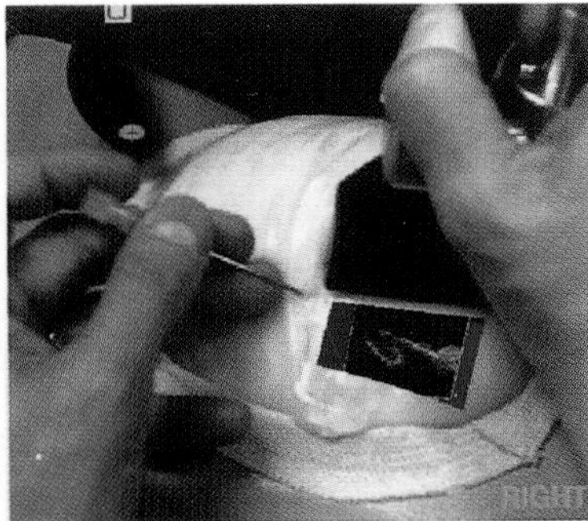

Fig. 2. The physician guides a needle into a synthetic lesion within a breast phantom. Note correct registration between the needle and its image in the ultrasound slice.

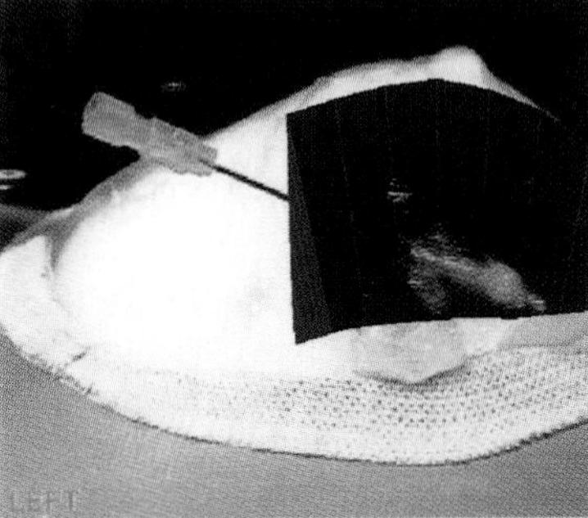

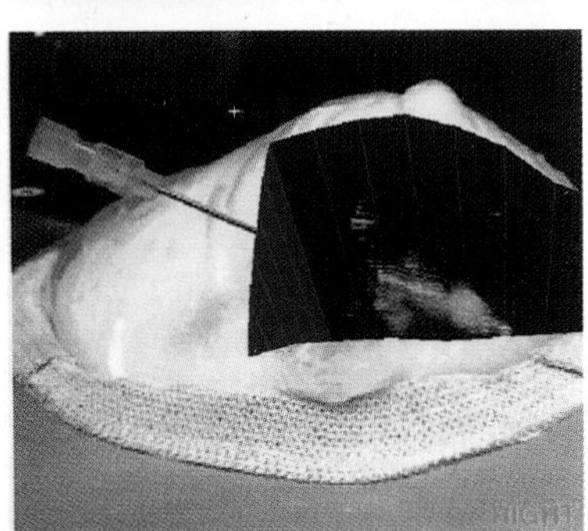

Fig. 3. Multiple ultrasound slices show a needle inserted into a synthetic lesion within the phantom. Note correct alignment of the needle and its volumetric ultrasound image.

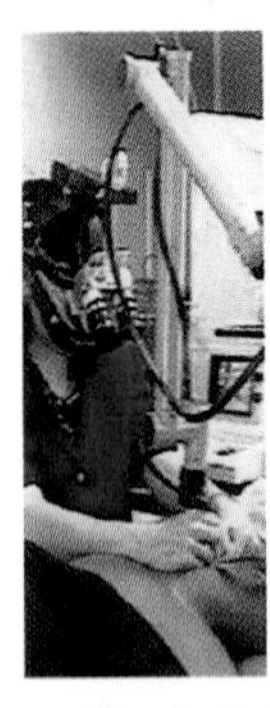

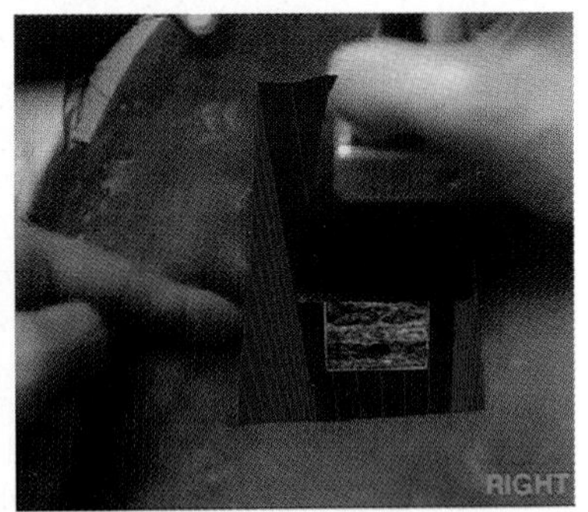

Fig. 4. Left: human subject experiment on January 5, 1996. During preliminary exploration, the physician uses the AR system. The mechanical tracker for the transducer is visible in the foreground. Center and right: stereo view within HMD. The system was not used for the actual intervention.

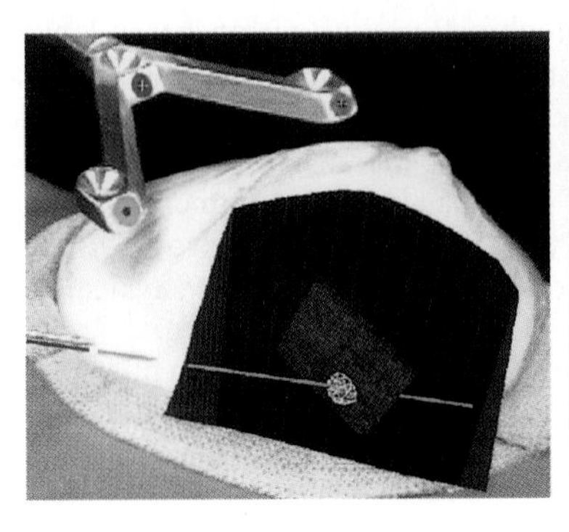
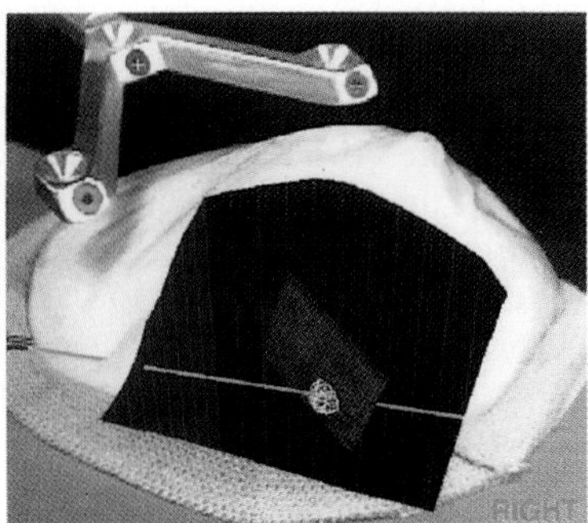

Fig. 5. The tracked needle is shown in red, the needle's projected path towards the target in green. Note slight misalignment between the red line and the actual needle.

noted above, is only a fraction of the entire viewing area. Together, these characteristics of the system mean that the pixel resolution of the slice is on the order of 40 x 40 HMD pixels (Fig. 4), compared to the typical video resolution of 500×500 on the display of a conventional ultrasound machine. Small features within the echography image are not resolved by our system as they are less than one pixel in size. For example, while a 14-gauge biopsy needle was visible within the HMD, a 22-gauge aspiration needle could not be seen in the HMD.

• From the physician user's point of view, the ultrasound slices are sometimes displayed edge-on. At other times, they are occluded by the probe (if the probe is between the physician's head and the ultrasound image). Also, the ultrasound slice occupies only a small portion of the display. These observations are in fact inherent characteristics of the AR paradigm. We believe a simple change of technique is necessary in these situations: the physician must move her head and/or body in order to be able to "see around" visual obstacles or move closer, as one would when a mechanical tool obscured an element that the tool is operating on or when one wanted to have a closer look at a small part. In our current system, the HMD is relatively cumbersome (see above) and other factors such as landmark tracking (see above) also restrict user movement. We therefore expect these visibility problems to become less significant once the usability issues are addressed and the physician will be able to move about unencumbered.

• The video-see-through HMD weighs nearly six pounds, including counterweights for balance, when in use. The unit is heavy, bulky, and somewhat cumbersome. In addition, there is a 4 inch offset between the lenses of the cameras and the user's eyes. While the user can adapt to this offset, it requires a level of training which would better be eliminated from the system.

• Good visual registration of the computer-generated imagery with the patient requires that the patient be still during the whole procedure once we acquire patient geometry data (the patient "sweep"). This data is gathered before the sterile field is created and cannot be re-acquired without compromising sterility. Holding still generally proves uncomfortable for the patients.

• The AR system requires a fairly lengthy set-up process before each patient experiment, on the order of nearly an hour including a series of per-patient calibrations and measurements. While marginally acceptable in this experimental system, clinical use should require easier start-up.

In addition to addressing all the above issues (see next section), it is obvious that our AR paradigm would greatly benefit from the incorporation of a needle tracker into the system. This would allow us to further enhance the display by showing the needle inside the pit and by displaying a projected trajectory of the needle which could be used for "aiming." We therefore conducted a (preliminary) experiment with a tracked needle in a breast phantom and with simple target enhancement techniques (Fig. 5). Since the mechanical tracker is our only high-precision tracker, we had to switch tracking from the ultrasound probe to the needle during the procedure. First, while tracking the probe, we positioned a single ultrasound slice imaging a cross-section through a cyst into the breast phantom. Then we disconnected the mechanical tracker from the ultrasound probe and attached it to the needle. Tracking the needle allows us to display the projected path of the needle during insertion, which we hope will significantly improve guidance. The reason for this is that the physician can aim the needle with the (green) projected trajectory displayed in stereo within the HMD. The (yellow) wireframe sphere also visible in Fig. 5 is an interactively positioned target enhancement marker; the needle is aimed at this marker. Since the marker is fixed in space, it can also be used to re-acquire the target after exploration of neighboring tissue with the ultrasound scanner. The marker was initially positioned by aiming at the ultrasound slice with crosshairs displayed within the HMD. This experiment was only marginally successful: despite visibly aiming directly at the target, upon verification with the ultrasound probe after the experiment we determined that we had missed the target in one instance and that we had only tangentially hit the boundary of the target in the second instance. We are currently investigating the sources of these errors, which we speculate may be due to compounded errors from the mechanical tracker (which is used twice, once for the ultrasound slice and once for the needle).

6 Conclusions and Future Work

We have implemented an AR system sufficiently robust and accurate that a physician can successfully perform needle insertion into a target inside a standard breast phantom and report that the procedure was easy. We anticipate our physician colleague performing the entire procedure on human subjects using the AR system—i.e., while wearing the HMD—as the system's accuracy and usability are improved in the near future.

Our primary near-term goal is to improve tracking and registration accuracy and incorporate needle tracking. We are currently investigating commercial high-accuracy optical systems for this purpose. We are also on the brink of integrating a new lightweight HMD into the system [Colucci 95]. This unit is characterized by video-resolution displays and a folded optical path for the video cameras. The latter eliminates the eye offset problem. The increased resolution addresses the slice display problem mentioned above.

In order to achieve proper depth relationships between synthetic imagery and the patient's body, the system must acquire and maintain a geometric model of the relevant regions of the skin surface. This is currently done before the actual surgical procedure, in the manually assisted "sweep" phase, during which the skin surface is swept with the mechanical tracker. The drawback of this technique is that it doesn't track skin surface deformations or patient motion. If these occur, the sweep must be repeated. We plan to enhance this component of our system by collaborating with UNC's telepresence research group.

Another goal will be to improve the visualization of intervention targets. The visualization of simple targets such as nodules and cysts is relatively trivial with our current ultrasound slice(s) rendering technique and the simple target enhancement method shown in Figure @@. More complex image analysis and rendering techniques for the ultrasound data will have to be used in order to visualize complex structures or in order to visualize motion of (or detail within) simple structures. We have begun to investigate the applicability of methods such as Whitaker's active blobs technique [Whitaker 94], core-based methods [Pizer 96], and Kalman-filter-based methods. We eventually hope to be able to automatically construct surface models of targets such as cysts from a set of ultrasound slices and then to track their motion and/or deformation from a set of *n* most recent slices.

In the more distant future, we hope to be able to expand our work to include a new visualization for sampling the abdominal viscera, which are frequently biopsied percutaneously with sonographic guidance using current technology. For example, the application of an advanced version of our AR technology could allow relatively easier sampling of lesions within the liver and kidneys. As our methods mature and become considerably more robust, we hope to expand the use of these visualization methods for less accessible organs such as the pancreas and adrenals. Other likely future targets for AR systems are small targets such as foreign body fragments resulting from accident and trauma emergencies. Eventually head-mounted displays and other AR devices may replace conventional displays altogether for some applications.

Acknowledgments

It takes many people to realize a complex system like the one described here. We wish to express our gratitude to John Airey, Ronald T. Azuma, Michael Bajura, Andrew Brandt, Gary Bishop, David T. Chen, Nancy Chescheir (MD), D'nardo Colucci, Mark Deutchman (MD), Darlene Freedman, Jai Glasgow, Arthur Gregory, Stefan Gottschalk, Ricardo Hahn (MD), David Harrison, Linda A. Houseman, Marco Jacobs, Fred Jordan, Vern Katz (MD), Kurtis Keller, Amy Kreiling, Shankar Krishnan, Dinesh Manocha, Michael North, Ryutarou Ohbuchi, Scott Pritchett, Russell M. Taylor II, Chris Tector, Kathy Tesh, John Thomas, Greg Turk, Peggy Wetzel, Steve Work, the patients, the Geometry Center at the University of Minnesota, PIE Medical Equipment B.V., Silicon Graphics, Inc., and the UNC Medical Image Program Project (NCI P01 CA47982).

We thank the anonymous reviewers for their comments and criticism.

This work was supported in part by ARPA DABT63-93-C-0048 ("Enabling Technologies and Application Demonstrations for Synthetic Environments"). Approved by ARPA for Public Release—Distribution Unlimited. Additional support was provided by the National Science Foundation Science and Technology Center for Computer Graphics and Scientific Visualization (NSF prime contract 8920219).

References

Bajura M, Fuchs H, Ohbuchi R (1992). "Merging Virtual Objects with the Real World." Proceedings of SIGGRAPH '92 (Chicago, Illinois, July 26-31, 1992). In Computer Graphics, 26, 2 (July 1992), ACM SIGGRAPH, New York, 1992; 203-210.

Colucci D, Chi V (1995). "Computer Glasses: A Compact, Lightweight, and Cost-effective Display for Monocular and Tiled Wide Field-of-View Systems." Proceedings of SPIE; **2537**: 61-70.

Feig SA (1988). "Decreased Breast Cancer Mortality through Mammographic Screening: Results of Clinical Trials." *Radiology* 1988;**167**: 659-665.

Fornage BD, Faroux MJ, Sumatos A (1987). "Breast masses: ultrasound guided fine-needle aspiration biopsy." *Radiology* 1987; **162**: 409-411.

Fuchs H, Kedem ZM, Naylor BF (1980). "On Visible Surface Generation by A Priori Tree Structures." *Proceedings of SIGGRAPH '80*; 124-133.

Garrett WF, Fuchs H, State A, Whitton MC (1996). "Real-Time Incremental Visualization of Dynamic Ultrasound Volumes Using Parallel BSP Trees." To appear in Proceedings of IEEE Visualization 1996.

Gordon PB, Goldenberg SL, Chan NHL (1993). "Solid Breast Lesions: Diagnosis with US-Guided Fine-Needle Aspiration Biopsy." *Radiology* 1993; **189**: 573-580.

Harper AP (1988). "Fine needle aspiration biopsy of the breast using ultrasound techniques - superficial localization and direct visualization." *Ultrasound Med Biol,* suppl. 1, 1988; **14**:5-12.

Hogg JP, Harris KM, Skonick ML (1988). "The role of ultrasound-guided needle aspiration of breast masses." *Ultrasound Med Biol,* suppl. 1, 1988; **14**:13.

Kopans DB, Meyer JE (1982). "Versatile springhook-wire breast lesion localizer." *AJR* 1982; **138**: 586.

Kopans DB, Meyer JE, Lindfors KK, Buchianieri SS (1984). "Breast sonography to guide cyst aspiration and wire localization of occult solid lesions." *AJR* 1984; **143**: 489.

Lalouche, R.C., Bickmore, D., Tessler, F., Mankovich, H. K., and Kangaraloo, H (1989). "Three-dimensional reconstruction of ultrasound images." *SPIE'89, Medical Imaging.* SPIE, 1989. 59-66.

Landerscaper J, Gunderson S, Gunderson A, et al. (1982). "Needle localization and Biopsy of Nonpalpable Lesions of the Breast." *Surg Gynecol Obstet* 1982; **164**: 399-403.

Muller JWTh (1985). "Diagnosis of breast cysts with mammography, ultrasound and puncture. A review." *Diagn Imag Clin Med* 1985; **54**:170.

Parker SH, Jobe W, Dennis MA, Stavros AT, Johnson KK, et al. (1993). "US-Guided Automated Large-Core Breast Biopsy." *Radiology* 1993; **187**: 507-511.

Pizer SM, Eberly D, Morse BS, Fritsch DS (1996). "Zoom-Invariant Vision of Figural Shape: The Mathematics of Cores." Submitted to *Computer Vision and Image Understanding.* Also University of North Carolina at Chapel Hill, Department of Computer Science, Technical Report TR96-004.

Sakas G, Walter S (1995). "Extracting Surfaces from Fuzzy 3D-Ultrasound Data." *Computer Graphics: Proceedings of SIGGRAPH '95*, (Los Angeles, CA, August 6-11, 1995), Annual Conference Series, 1995, ACM SIGGRAPH; 465-474.

Schwartz G, Feig S, Patchefsky A (1988). "Significance and staging on Nonpalpable Carcinomas of the Breast." *Surg Gynecol Obstet* 1988; **166**: 6-10.

Shapiro S (1977). "Evidence of Screening for Breast Cancer from a Randomized Trial." *Cancer* 1977; **39**: 2772-2782.

Shapiro S, Venet W, Strax P, Venet L, Roeser R (1982). "Ten-to-Fourteen-Year Effect of screening on Breast Cancer Mortality." *JNCI* 1982; **69**: 349-355.

Skinner M, Swain M, Simmon R, et al. (1988). "Nonpalpable Breast lesion at Biopsy." *Ann Surg* 1988; **208**: 203-208.

State A, McAllister J, Neumann U, Chen H, Cullip T, Chen DT, Fuchs H (1995). "Interactive Volume Visualization on a Heterogeneous Message-Passing Multicomputer." *Proceedings of the 1995 Symposium on Interactive 3D Graphics.* Monterey, CA, April 9-12, 1995; 69-74, 208.

State A, Livingston M, Garrett WF, Hirota G, Whitton MC, Pisano ED, Fuchs H (1996). "Technologies for Augmented-Reality Systems: Realizing Ultrasound-Guided Needle Biopsies." To appear in Proceedings of SIGGRAPH 96 (New Orleans, Louisiana, August 5-9, 1996). In Computer Graphics Proceedings, Annual Conference Series, 1996, ACM SIGGRAPH.

Tabar L, Pentek Z, Dean PB (1981). "The diagnostic and therapeutic value of breast cyst puncture and pneumocystography." *Radiology* 1981; **141**: 659.

US Preventive Services Task Force (1989). "Guide to Clinical Preventive Services: An Assessment of the Effectiveness of 169 Interventions." Williams and Wilkins, Baltimore, MD, 1989. Chapter 6, 39-62.

Whitaker, RT (1994). "Volumetric deformable models: active blobs." *Proceedings of Visualization in Biomedical Computing 1994,* Rochester, Minnesota, October 1994; 122-134.

Wright C (1986). "Breast Cancer Screening." *Surgery* 1986; **100**(4): 594-598.

Automated Multimodality Registration Using the Full Affine Transformation: Application to MR and CT Guided Skull Base Surgery

Colin Studholme, John A. Little,Graeme P. Penny,
Derek L.G. Hill, David J. Hawkes

Division of Radiological Sciences
United Medical and Dental Schools of Guy's and St. Thomas' Hospitals Guy's Hospital, London Bridge, London, SE1 9RT, UK

Abstract. We have used a twelve degrees of freedom global affine registration technique incorporating a multiresolution optimisation of mutual information to register MR and CT images with uncertain voxel dimensions and CT gantry tilt. Visual assessment indicates improved accuracy, without loss of robustness compared to rigid body registration.

1 Introduction

This papers presents an extension of voxel similarity based registration techniques to estimating global affine parameters between MR and CT images of the head. Current surgical planning and guidance work at our site has made use of accurate rigid registration of MR and CT images. Although estimates of machine scaling parameters are confirmed by phantom based image measurements, inspection of some clinically registered images has indicated visually detectable scaling differences in some cases.

Manufacturers typically claim a tolerance of around 3% in calibration accuracy when setting up medical imaging equipment. A discrepancy of $6mm$ may therefore be experienced over a $200mm$ field of view, which will be unacceptable in many applications. In addition we sometimes obtain patient images from other sites where such checks are not regularly applied and parameters such as CT gantry tilt are unknown. In order to reduce additional radiation dose to the patient, as well as costs, it is important that the need to re-scan locally is kept to a minimum.

Image distortion in MR cannot be corrected by using scaling and skew factors alone. In our experience, however, scaling and skew errors are often larger than errors arising from other components of MR distortion for many clinical datasets. The ability to automatically detect and estimate discrepancies in scaling and skew parameters between images would therefore be a useful check to ensure spatial integrity of the data as part of normal clinical registration protocols.

Voxel similarity based approaches [3, 5], particularly those making use of mutual information measures between images [6, 2, 7, 4] have been shown to provide a robust and fully automatic method of estimating rigid registration parameters. We demonstrate that the estimate of an affine rather than rigid transformation can improve accuracy, without compromising robustness, and without a prohibitive increase in computation time.

2 Method

2.1 Mutual Information and Image Registration

In this work we have used the measure of mutual information $I(M;C)$ as proposed by Viola and Wells [6, 7] and Collignon [2]. This was evaluated from a discrete estimate of the joint, $p\{m,c\}$, and separate, $p\{m\}$ and $p\{c\}$ probability distributions, of the MR and CT intensities M and C occurring in the volume of overlap of the two images:

$$I(M;C) = \sum_{m \in M} \sum_{n \in C} p\{m,c\} \log \frac{p\{m,c\}}{p\{m\}p\{c\}} \tag{1}$$

2.2 Registration Parameters and Optimisation

In order to efficiently optimise affine parameters we need to be able to equate changes in translations, rotations, scalings and angles of skew. Global voxel similarity measures provide a bulk measure of voxel displacement. A reasonable choice of relationship between equivalent small changes in the parameters is one which results in the same mean displacement of voxels in the overlap of the images. For simplicity we assume a cuboidal volume of overlap $\mathcal{F}_x \times \mathcal{F}_y \times \mathcal{F}_z$ between MR volume V_m and CT volume V_c. For translations take the step sizes in each direction $\{\delta t_x, \delta t_y, \delta t_z\}$ to be equal to δt. For changes in rotation, scale and skew, we can integrate the displacements of points within the volume of overlap to find the mean displacement which we equate to δt. This gives us,

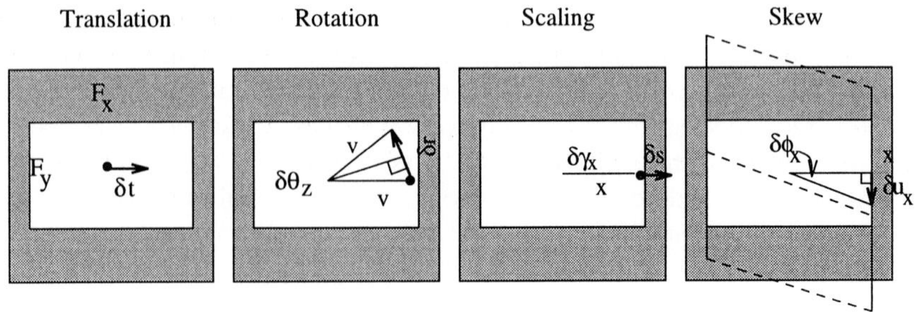

Fig. 1. Equivalent displacements for small changes in different affine parameters

$$\delta\theta_z = 2\sin^{-1}\frac{6\delta t \mathcal{F}_x \mathcal{F}_y}{\mathcal{K}}, \ \delta\gamma_x = 4\delta t/\mathcal{F}_x, \ \delta\phi_x = \tan^{-1}(4\delta t/\mathcal{F}_x). \tag{2}$$

and equivalent formula for parameters in the other planes and where,

$$\mathcal{K} = \mathcal{F}_x^3 \left\{ \frac{\sin\alpha_{xy}}{\cos^2\alpha_{xy}} + \ln|\tan(\frac{\pi}{4} + \frac{\alpha_{xy}}{2})| \right\} + \mathcal{F}_y^3 \left\{ \frac{\sin\beta_{xy}}{\cos^2\beta_{xy}} + \ln|\tan(\frac{\pi}{2} - \frac{\alpha_{xy}}{2})| \right\}$$

and $\alpha_{xy} = \tan^{-1}(\mathcal{F}_y/\mathcal{F}_x)$ and $\beta_{xy} = \pi/2 - \alpha_{xy}$.

To optimise the parameters we evaluate $I(M;C)$ over the the set of 13 (for rigid) or 25 (for affine) transformations $\mathcal{T}(T_0)$. These are the current starting estimate and the starting estimate with increments and decrements of each of the 3 translations $\{\delta t_x, \delta t_y, \delta t_z\}$, 3 rotations $\{\delta\theta_x, \delta\theta_y, \delta\theta_z\}$, 3 scaling factors $\{\delta\gamma_x, \delta\gamma_y, \delta\gamma_z\}$ and 3 skew angles $\{\delta\phi_x, \delta\phi_y, \delta\phi_z\}$. We choose the best estimate of the registration transformation T_1 for the value of I such that:

$$T_1 = \min_{T \in \mathcal{T}(T_0)} \{I(M(V_m \cap TV_c), C(V_m \cap TV_c))\} \tag{3}$$

If $T_{n+1} \neq T_n$ then we can repeat the search with $\mathcal{T}(T_{n+1})$ until $T_{n+1} = T_n$. The step sizes can then be reduced and the search continued, the minimum values of step size determining how close we get to the optimum transformation. Starting with low resolution versions of the images and large step sizes, we apply this optimisation, reducing the step size and increasing the image resolution. Experimentally we have found this provides both computational efficiency and robustness to starting estimate.

2.3 Test Data

Three clinically acquired image pairs were used for the tests. Firstly an MR and CT volume covering a significant portion of the skull with slice thicknesses of 1.2mm and 2.0mm respectively. Secondly a more typical truncated CT volume limited to an axial extent of 31mm in the skull base with corresponding full brain volume MR. Both these images originated from scanners at another site and contained errors in slice thickness and skew angle estimates. Thirdly a locally acquired pair of images including a relatively large volume CT for which scanner quality control had been performed.

Following optimisation of mutual information the results were visually inspected, and for patient A compared to phantom measurements. In order to assess the precision of the estimate for patient A, randomised rigid starting estimates were created by adding random translations (of magnitude 10 mm) and random rotations (of magnitude of 10 degrees) to the six rigid estimates. Optimisation was then initiated from 50 of these orientations and the results recorded.

3 Results

Table 1 shows transformation estimates following optimisation of rigid and affine parameters. The right of figure 2 shows example slices through the MR volume for patient A illustrating the improvement provided by registration including scale estimates. It is interesting to note that when scaling is not included the rigid registration is biased toward better alignment around the skull base. This is presumably because of the larger amount of registration information available in this region. The left of figure 2 shows (a) the poor rigid estimate for patient B

Registration Parameter Estimates

Parameter	Patient A Rigid	Patient A Affine	Patient B Rigid	Patient B Affine	Patient C Rigid	Patient C Affine
$t_x$mm	-2.83 (0.13)	-2.83 (0.26)	-1.70	-1.32	8.71	8.53
$t_y$mm	-2.50 (0.59)	-2.71 (0.09)	9.30	9.82	32.83	32.65
$t_z$mm	-26.34 (0.20)	-26.34(0.10)	-45.8	-45.0	-0.24	0.08
θ_x°	1.09 (0.32)	2.09 (0.31)	38.6	38.24	24.41	24.08
θ_y°	-2.38 (0.33)	-1.59 (0.31)	-0.88	-1.0	-6.29	-6.15
θ_z°	0.00 (0.18)	0.00 (0.24)	5.05	5.17	-0.65	-1.23
γ_x	1.00	1.00 (0.0014)	1.00	1.00	1.00	0.99
γ_y	1.00	0.98 (0.0026)	1.00	0.99	1.00	1.00
γ_z	1.00	1.06 (0.0069)	1.00	1.65	1.00	0.96
ϕ_x°	0.00	0.00 (0.11)	0.00	0.23	0.00	0.51
ϕ_y°	0.00	0.71 (0.43)	0.00	-0.04	0.00	-0.11
ϕ_z°	0.00	1.71 (0.27)	0.00	-8.0	0.00	-0.74

Table 1. Comparison of registration estimates for Patient A, B and C following optimisation of rigid parameters, and all affine parameters.

using the supplied scaling and gantry tilt angles, and (b) the significant improvement provided by a full affine parameter optimisation. In order to investigate the origin of the scaling error for patient A, an MR and CT visible point landmark phantom was imaged in the scanners using the same protocols. Distances between points in the images could then be measured by interactive location of the centres of markers. Equivalent distances on the phantom were measured and indicated a seven percent reduction between the phantom and its CT image in the axial direction. This figure was later confirmed as a calibration error in the bed speed for spiral acquisitions.

The figures in brackets in table 1 show the standard deviations of final parameter estimates from 50 randomised starts around the first estimate. The standard deviation of the registration parameters is small for both the rigid and affine solutions and no significant increase is noted when the number of parameters is increased, with in one case (t_y) an appreciable decrease.

4 Discussion

We have shown experimentally that when extending the transformation estimate to 12 parameters, the robustness to starting estimate, robustness to limited field of view, and the precision of the final results are maintained. Typical execution times are increased from 10 to 24 minutes on a Sparc Station 20/61 for dataset C covering a large volume of the head. Where computing time needs to be minimised and some parameters are known accurately, the dimensionality of the search space can be reduced appropriately.

Given two images we can only estimate the discrepancy in scaling and skew estimates between the images, we do not know which has the correct dimensions, so we cannot rely on this approach to replace routine quality assurance procedures. This is particularly important when the images are to be registered to the

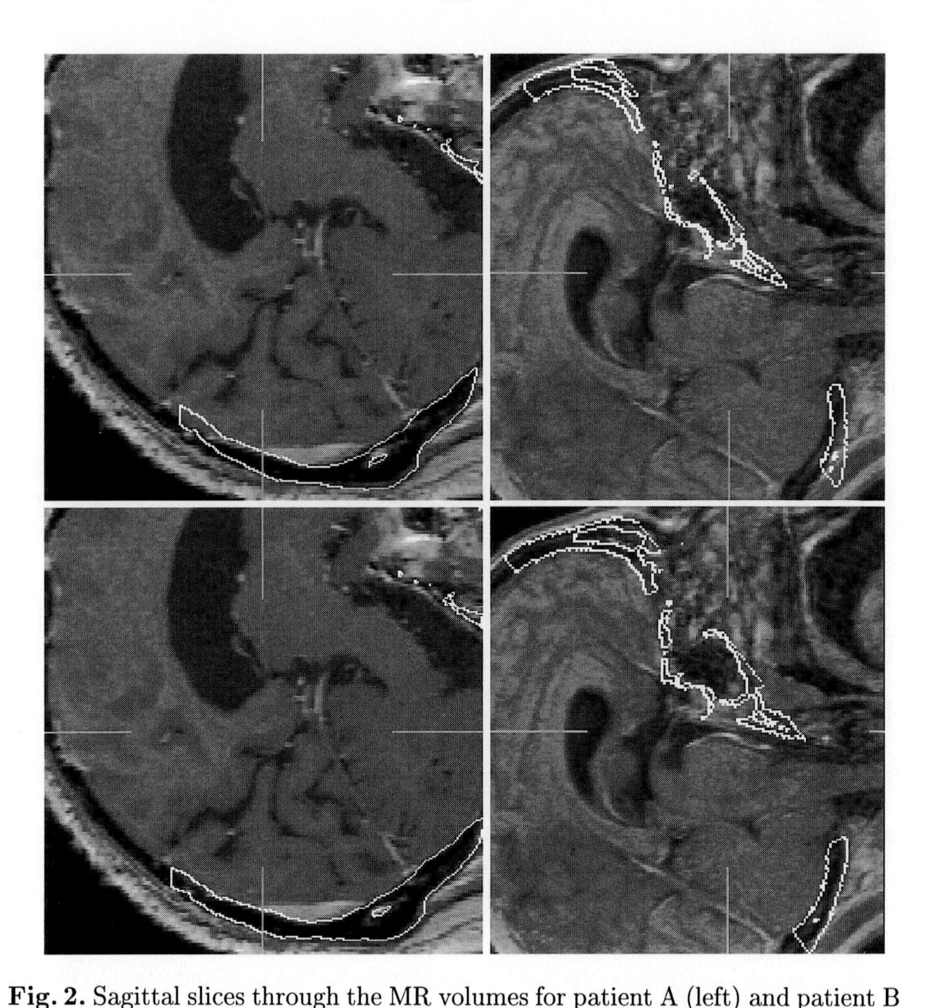

Fig. 2. Sagittal slices through the MR volumes for patient A (left) and patient B (right) with the CT bone boundary, obtained by intensity threshold, overlaid in white showing (top) rigid transformation estimate, and (bottom) affine estimate

patient in surgery. However we have found that this approach is very useful in highlighting discrepancies in clinical data that merit further investigation. This is particularly true when combining MR with CT acquired in the skull base, where limited axial extent leads to difficulties in visually distinguishing small errors in skew or scale factors from rigid alignment errors.

Most concern has been expressed about MR image integrity, because of the process by which scaling is determined in MR and because patient induced distortion can affect image quality [1]. In two of the three examples given in this paper subsequent investigation demonstrated that the major source of the error was in the CT acquisition. In one case a hardware maladjustment in bed movement led to a 7 percent error in axial scaling during a helical scan sequence, while in the other an incorrect slice interval was provided in the image header information. We have not found any significant errors in transaxial scaling in CT data.

One of the risks of using fully automated image registration techniques, is that the operator will fail to spot deficiencies in the data, and that the resulting incorrectly registered images will be used inappropriately for diagnosis, treatment planning or image guided surgery. MR distortion, and artifacts caused by patient motion may still lead to significant errors in some cases. Further work is underway using phantoms to provide a better assessment of the accuracy of skew and scaling estimates provided by the approach. Additional degrees of freedom could be added to the registration algorithm to make it possible to automatically detect, and perhaps even correct, significant MR distortion, or the slice alignment errors that arise when a patient moves between slices in a CT, or 2D multi-slice MR acquisition.

Acknowledgements

We are grateful to EPSRC, Guy's and St Thomas's Hospital NHS Trust and Philips Medical Systems (ICS/AD) for their financial support of this work. We are grateful for the support and encouragement of Prof. Michael Maisey, Dr. Tim Cox, Dr. Alan Colchester, Prof. Mike Gleeson, Mr Tony Strong and Mr Charles Polkey in this work. We thank Dr Jason Zhao for assisting in making the data for patient B available to us and the radiological staff at Guy's, St Thomas's and the Maudsley Hospitals for their assistance in acquiring the data for patient A and the phantom.

References

1. H. Chang and J.M. Fitzpatrick. A technique for accurate MR imaging in the presence of field inhomogeneities. *IEEE Trans. Med. Imaging*, 11:319–329, 1992.
2. A. Collignon, F. Maes, D. Delaere, D. Vandermeulen, P. Suetens, and G. Marchal. Automated multimodality image registration using information theory. In Bizais Y., Barillot C, and Di Paola R., editors, *Proceedings of Information Processing in Medical Imaging*, pages 263–274, 1995. Brest, France.
3. D.L.G. Hill, C. Studholme, and D.J. Hawkes. Voxel similarity measures for automated image registration. In *Proceedings of Visualisation in Biomedical Computing*, pages 205–216, 1994. Rochester Mn.,U.S.A.
4. C. Studholme, D.L.G. Hill, and D.J. Hawkes. Automated 3D registration of truncated MR and CT images of the head. In D. Pycock, editor, *Proceedings of the British Machine Vision Conference*, pages 27–36. BMVA, 1995. Birmingham.
5. P.A. Van den Elsen, E.J.D. Pol, T.S. Sumanawaeera, P.F. Hemler, S. Napel, and J.R. Adler. Grey value correlation techniques used for automatic matching of CT and MR brain and spine images. In *Proceedings of Visualisation in Biomedical Computing*, pages 227–237, 1994. Rochester Mn.,U.S.A.
6. P.A. Viola and W.M. Wells. Alignment by maximisation of mutual information. In *Proceedings of the 5th International Conference on Computer Vision*, pages 15–23, 1995.
7. W.M. Wells, P. Viola, and R. Kikinis. Multimodal volume registration by maximization of mutual information. In *Proceedings of the 2nd annual international symposium on Medical Robotics and Computer Assisted Surgery*, pages 55–62, 1995. Baltimore, U.S.A.

Author Index

Lecture Notes in Computer Science

For information about Vols. 1–1075

please contact your bookseller or Springer-Verlag

Vol. 1111: J.J. Alferes, L. Moniz Pereira, Reasoning with Logic Programming. XXI, 326 pages. 1996. (Subseries LNAI).

Vol. 1112: C. von der Malsburg, W. von Seelen, J.C. Vorbrüggen, B. Sendhoff (Eds.), Artificial Neural Networks – ICANN 96. Proceedings, 1996. XXV, 922 pages. 1996.

Vol. 1113: W. Penczek, A. Szałas (Eds.), Mathematical Foundations of Computer Science 1996. Proceedings, 1996. X, 592 pages. 1996.

Vol. 1114: N. Foo, R. Goebel (Eds.), PRICAI'96: Topics in Artificial Intelligence. Proceedings, 1996. XXI, 658 pages. 1996. (Subseries LNAI).

Vol. 1115: P.W. Eklund, G. Ellis, G. Mann (Eds.), Conceptual Structures: Knowledge Representation as Interlingua. Proceedings, 1996. XIII, 321 pages. 1996. (Subseries LNAI).

Vol. 1116: J. Hall (Ed.), Management of Telecommunication Systems and Services. XXI, 229 pages. 1996.

Vol. 1117: A. Ferreira, J. Rolim, Y. Saad, T. Yang (Eds.), Parallel Algorithms for Irregularly Structured Problems. Proceedings, 1996. IX, 358 pages. 1996.

Vol. 1118: E.C. Freuder (Ed.), Principles and Practice of Constraint Programming — CP 96. Proceedings, 1996. XIX, 574 pages. 1996.

Vol. 1119: U. Montanari, V. Sassone (Eds.), CONCUR '96: Concurrency Theory. Proceedings, 1996. XII, 751 pages. 1996.

Vol. 1120: M. Deza. R. Euler, I. Manoussakis (Eds.), Combinatorics and Computer Science. Proceedings, 1995. IX, 415 pages. 1996.

Vol. 1121: P. Perner, P. Wang, A. Rosenfeld (Eds.), Advances in Structural and Syntactical Pattern Recognition. Proceedings, 1996. X, 393 pages. 1996.

Vol. 1122: H. Cohen (Ed.), Algorithmic Number Theory. Proceedings, 1996. IX, 405 pages. 1996.

Vol. 1123: L. Bougé, P. Fraigniaud, A. Mignotte, Y. Robert (Eds.), Euro-Par'96. Parallel Processing. Proceedings, 1996, Vol. I. XXXIII, 842 pages. 1996.

Vol. 1124: L. Bougé, P. Fraigniaud, A. Mignotte, Y. Robert (Eds.), Euro-Par'96. Parallel Processing. Proceedings, 1996, Vol. II. XXXIII, 926 pages. 1996.

Vol. 1125: J. von Wright, J. Grundy, J. Harrison (Eds.), Theorem Proving in Higher Order Logics. Proceedings, 1996. VIII, 447 pages. 1996.

Vol. 1126: J.J. Alferes, L. Moniz Pereira, E. Orlowska (Eds.), Logics in Artificial Intelligence. Proceedings, 1996. IX, 417 pages. 1996. (Subseries LNAI).

Vol. 1127: L. Böszörményi (Ed.), Parallel Computation. Proceedings, 1996. XI, 235 pages. 1996.

Vol. 1128: J. Calmet, C. Limongelli (Eds.), Design and Implementation of Symbolic Computation Systems. Proceedings, 1996. IX, 356 pages. 1996.

Vol. 1129: J. Launchbury, E. Meijer, T. Sheard (Eds.), Advanced Functional Programming. Proceedings, 1996. VII, 238 pages. 1996.

Vol. 1130: M. Haveraaen, O. Owe, O.-J. Dahl (Eds.), Recent Trends in Data Type Specification. Proceedings, 1995. VIII, 551 pages. 1996.

Vol. 1131: K.H. Höhne, R. Kikinis (Eds.), Visualization in Biomedical Computing. Proceedings, 1996. XII, 610 pages. 1996.

Vol. 1132: G.-R. Perrin, A. Darte (Eds.), The Data Parallel Programming Model. XV, 284 pages. 1996.

Vol. 1133: J.-Y. Chouinard, P. Fortier, T.A. Gulliver (Eds.), Information Theory and Applications. Proceedings, 1995. XII, 309 pages. 1996.

Vol. 1134: R. Wagner, H. Thoma (Eds.), Database and Expert Systems Applications. Proceedings, 1996. XV, 921 pages. 1996.

Vol. 1135: B. Jonsson, J. Parrow (Eds.), Formal Techniques in Real-Time and Fault-Tolerant Systems. Proceedings, 1996. X, 479 pages. 1996.

Vol. 1136: J. Diaz, M. Serna (Eds.), Algorithms - ESA '96. Proceedings, 1996. XII, 566 pages. 1996.

Vol. 1137: G. Görz, S. Hölldobler (Eds.), KI-96: Advances in Artificial Intelligence. Proceedings, 1996. XI, 387 pages. 1996. (Subseries LNAI).

Vol. 1138: J. Calmet, J.A. Campbell, J. Pfalzgraf (Eds.), Artificial Intelligence and Symbolic Mathematical Computation. Proceedings, 1996. VIII, 381 pages. 1996.

Vol. 1139: M. Hanus, M. Rogriguez-Artalejo (Eds.), Algebraic and Logic Programming. Proceedings, 1996. VIII, 345 pages. 1996.

Vol. 1140: H. Kuchen, S. Doaitse Swierstra (Eds.), Programming Languages: Implementations, Logics, and Programs. Proceedings, 1996. XI, 479 pages. 1996.

Vol. 1141: H.-M. Voigt, W. Ebeling, I. Rechenberg, H.-P. Schwefel (Eds.), Parallel Problem Solving from Nature – PPSN IV. Proceedings, 1996. XVII, 1.050 pages. 1996.

Vol. 1142: R.W. Hartenstein, M. Glesner (Eds.), Field-Programmable Logic. Proceedings, 1996. X, 432 pages. 1996.

Vol. 1143: T.C. Fogarty (Ed.), Evolutionary Computing. Proceedings, 1996. VIII, 305 pages. 1996.

Vol. 1144: J. Ponce, A. Zisserman, M. Hebert (Eds.), Object Representation in Computer Vision. Proceedings, 1996. VIII, 403 pages. 1996.

Vol. 1145: R. Cousot, D.A. Schmidt (Eds.), Static Analysis. Proceedings, 1996. IX, 389 pages. 1996.

Vol. 1146: E. Bertino, H. Kurth, G. Martella, E. Montolivo (Eds.), Computer Security – ESORICS 96. Proceedings, 1996. X, 365 pages. 1996.

Vol. 1147: L. Miclet, C. de la Higuera (Eds.), Grammatical Inference: Learning Syntax from Sentences. Proceedings, 1996. VIII, 327 pages. 1996. (Subseries LNAI).

Vol. 1148: M.C. Lin, D. Manocha (Eds.), Applied Computational Geometry. Proceedings, 1996. VIII, 223 pages. 1996.

Vol. 1149: C. Montangero (Ed.), Software Process Technology. Proceedings, 1996. IX, 291 pages. 1996.

Vol. 1150: A. Hlawiczka, J.G. Silva, L. Simoncini (Eds.), Dependable Computing – EDCC-2. Proceedings, 1996. XVI, 440 pages. 1996.

Vol. 1151: Ö. Babaoğlu, K. Marzullo (Eds.), Distributed Algorithms. Proceedings, 1996. VIII, 381 pages. 1996.